Student's Solutions Manual

to accompany

Calculus

Third Edition

Robert T. Smith
Millersville University

Roland B. Minton
Roanoke College

Prepared by

Rebecca Torrey   **James Seibert**   **Blake Thornton**
Brandeis University   *Regis University*   *Washington University in St. Louis*

with contributions by
Ji Li - *Brandeis University*
and
C Trimble & Associates

McGraw-Hill Higher Education

Boston   Burr Ridge, IL   Dubuque, IA   New York   San Francisco   St. Louis
Bangkok   Bogotá   Caracas   Kuala Lumpur   Lisbon   London   Madrid   Mexico City
Milan   Montreal   New Delhi   Santiago   Seoul   Singapore   Sydney   Taipei   Toronto

McGraw-Hill
Higher Education

Student's Solutions Manual to accompany
CALCULUS, THIRD EDITION
ROBERT T. SMITH AND ROLAND B. MINTON

Published by McGraw-Hill Higher Education, an imprint of The McGraw-Hill Companies, Inc., 1221 Avenue of the Americas, New York, NY 10020. Copyright © 2008 by The McGraw-Hill Companies, Inc. All rights reserved.

2 3 4 5 6 7 8 9 0 QPD/QPD 0 9 8 7

ISBN: 978-0-07-326845-3
MHID: 0-07-326845-3

www.mhhe.com

Contents

16 Second-Order Differential Equations   571

Preliminaries

0.1 THE REAL NUMBERS AND THE CARTESIAN PLANE

1. $3x + 2 < 11$

$3x < 9$

$x < 3$

$(-\infty, 3)$

3. $2x - 3 < -7$

$2x < -4$

$x < -2$

$(-\infty, -2)$

5. $4 - 3x < 6$

$-3x < 2$

$x > -\dfrac{2}{3}$

$\left(-\dfrac{2}{3}, \infty\right)$

7. $4 \le x + 1 < 7$

$3 \le x < 6$

$[3, 6)$

9. $-2 < 2 - 2x < 3$

$-4 < -2x < 1$

$2 > x > -\dfrac{1}{2}$

$-\dfrac{1}{2} < x < 2$

$\left(-\dfrac{1}{2}, 2\right)$

11. $x^2 + 3x - 4 > 0$

$(x + 4)(x - 1) > 0$

$x < -4 \text{ or } x > 1$

$(-\infty, -4) \cup (1, \infty)$

13. $x^2 - x - 6 < 0$

$(x - 3)(x + 2) < 0$

$(-2, 3)$

15. $3x^2 + 4 > 0$

$3x^2 > -4$

$x^2 > -\dfrac{4}{3}$

This is always true. $(-\infty, \infty)$

17. $|x - 3| < 4$
$-4 < x - 3 < 4$
$-1 < x < 7$
$(-1, 7)$

19. $|3 - x| < 1$

$-1 < 3 - x < 1$
$-4 < -x < -2$
$4 > x > 2$
$2 < x < 4$

$(2, 4)$

21. $|2x + 1| > 2$

$2x + 1 < 2 \quad$ or $\quad 2x + 1 > 2$
$2x < -3 \qquad\qquad 2x > 1$
$x < -\dfrac{3}{2} \quad$ or $\quad x > \dfrac{1}{2}$

$\left(-\infty, -\dfrac{3}{2}\right) \cup \left(\dfrac{1}{2}, \infty\right)$

23. $\dfrac{x + 2}{x - 2} > 0$

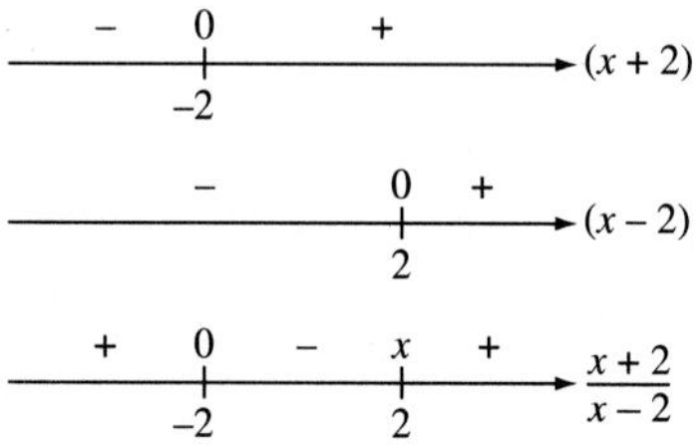

$x < -2$ or $x > 2$
$(-\infty, -2) \cup (2, \infty)$

25. $\dfrac{x^2 - x - 2}{(x + 4)^2} > 0$

$\dfrac{(x - 2)(x + 1)}{(x + 4)^2} > 0$

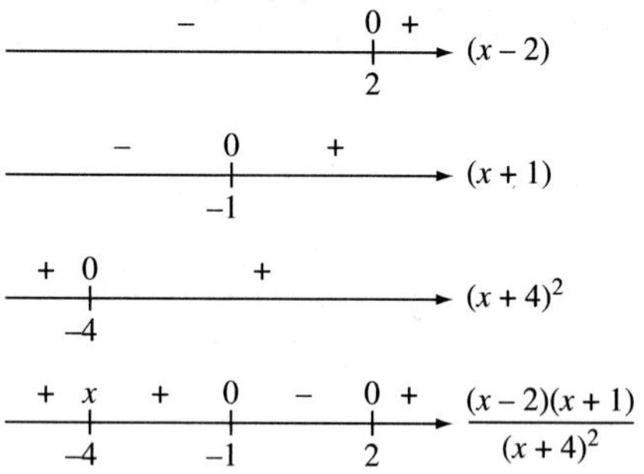

$x < -4$ or $-4 < x < -1$ or $x > 2$
$(-\infty, -4) \cup (-4, -1) \cup (2, \infty)$

27. $\dfrac{-8x}{(x + 1)^3} < 0$

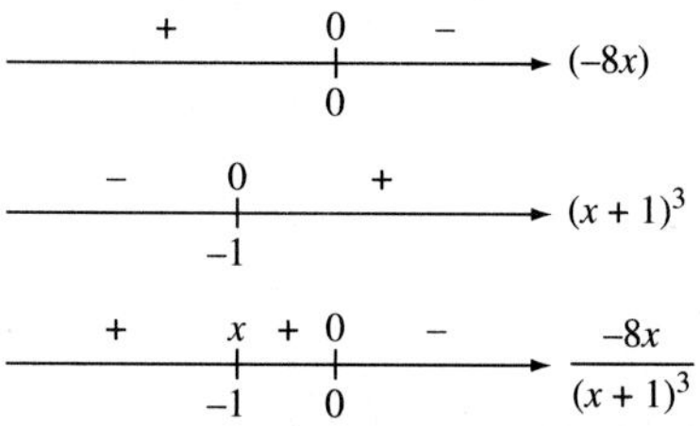

$x < -1$ or $x > 0$
$(-\infty, -1) \cup (0, \infty)$

29. $d\{(2, 1), (4, 4)\} = \sqrt{(4 - 2)^2 + (4 - 1)^2}$
$= \sqrt{4 + 9}$
$= \sqrt{13}$

31. $d\{(-1, -2), (3, -2)\}$

$= \sqrt{[3 - (-1)]^2 + [-2 - (-2)]^2}$
$= \sqrt{16 + 0}$
$= 4$

33. $d\{(0, 2), (-2, 6)\} = \sqrt{(-2 - 0)^2 + (6 - 2)^2}$
$= \sqrt{4 + 16}$
$= \sqrt{20}$

35.
$$d\{(1, 1), (3, 4)\} = \sqrt{(3-1)^2 + (4-1)^2}$$
$$= \sqrt{4+9} = \sqrt{13}$$
$$d\{(1, 1), (0, 6)\} = \sqrt{(0-1)^2 + (6-1)^2}$$
$$= \sqrt{1+25} = \sqrt{26}$$
$$d\{(3, 4), (0, 6)\} = \sqrt{(0-3)^2 + (6-4)^2}$$
$$= \sqrt{9+4} = \sqrt{13}$$
$$(\sqrt{13})^2 + (\sqrt{13})^2 = (\sqrt{26})^2$$

This statement is true, so yes, this is a right triangle.

37.
$$d\{(-2, 3), (2, 9)\} = \sqrt{[2-(-2)]^2 + (9-3)^2}$$
$$= \sqrt{16+36} = \sqrt{52}$$
$$d\{(-2, 3), (-4, 13)\} = \sqrt{[-4-(-2)]^2 + (13-3)^2}$$
$$= \sqrt{4+100} = \sqrt{104}$$
$$d\{(2, 9), (-4, 13)\} = \sqrt{(-4-2)^2 + (13-9)^2}$$
$$= \sqrt{36+16} = \sqrt{52}$$
$$(\sqrt{52})^2 + (\sqrt{52})^2 = (\sqrt{104})^2$$

This statement is true, so yes, this is a right triangle.

39.

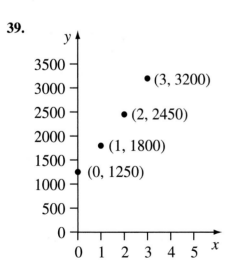

The y-values are increasing by 550, 650, 750, . . .

The next population is $3200 + 850 = 4050$.

41.

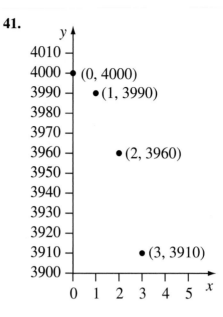

The y-values are decreasing by 10, 30, 50, . . .

The next population is $3910 - 70 = 3840$.

43. When a calculation is done using a finite number of digits, the result can only contain a finite number of digits. Therefore, the result can be written as a quotient of two integers (where the denominator is a power of 10), so the result must be rational.

45.
$$\frac{\left(\frac{1}{2}\right)^7 - \left(\frac{2}{3}\right)^{12}}{\left(\frac{1}{2}\right)^7} \approx 0.013 \text{ or } 1.3\%$$

47.

P	win %
0.551	0.568
0.587	0.593
0.404	0.414
0.538	0.556
0.605	0.615

0.2 LINES AND FUNCTIONS

1. Yes. The slope of the line joining the points $(2, 1)$ and $(0, 2)$ is $-\frac{1}{2}$, which is also the slope of the line joining the points $(0, 2)$ and $(4, 0)$.

3. No. The slope of the line joining the points (4, 1) and (3, 2) is -1, while the slope of the line joining the points (3, 2) and (1, 3) is $-\frac{1}{2}$.

5. Slope is $\dfrac{6-2}{3-1} = \dfrac{4}{2} = 2.$

7. Slope is $\dfrac{-1-(-6)}{1-3} = \dfrac{5}{-2} = -\dfrac{5}{2}.$

9. Slope is

$$\dfrac{-0.4-(-1.4)}{-1.1-0.3} = \dfrac{1.0}{-1.4} = -\dfrac{5}{7}.$$

11. $y = 2(x-1) + 3 = 2x + 1$
For $x = 2$, $y = 2(2) + 1 = 5$. (2, 5) is a second point on the line.

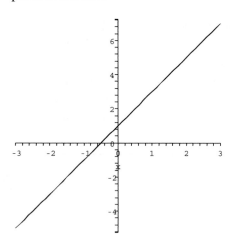

13. $y = 1$
(0, 1) is a second point on the line.

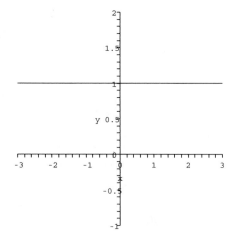

15. $y = 1.2(x - 2.3) + 1.1 = 1.2x - 1.66$
For $x = 3.3$, $y = 1.2(3.3 - 2.3) + 1.1 = 2.3$.
(3.3, 2.3) is a second point on the line.

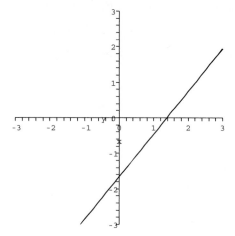

17. Parallel. Both have slope 3.

19. Perpendicular. Slopes are -2 and $\frac{1}{2}$; their product is -1.

21. Perpendicular. Slopes are 3 and $-\frac{1}{3}$; their product is -1.

23. (a) $y = 2(x - 2) + 1$
(b) $y = -\frac{1}{2}(x - 2) + 1$

25. (a) $y = 2(x - 3) + 1$
(b) $y = -\frac{1}{2}(x - 3) + 1$

27. Slope $\dfrac{3-1}{2-1} = \dfrac{2}{1} = 2$ through the given points.
One possibility:

$$y = 2(x - 1) + 1 = 2x - 1.$$

When $x = 4$, $y = 7$.

29. Slope $\dfrac{3-4}{1-0.5} = \dfrac{-1}{0.5} = -2$ through the given points. One possibility:

$$y = -2(x - 1) + 3 = -2x + 5.$$

When $x = 4$, $y = -3$.

31. Yes, passes vertical line test.

33. No. The vertical line $x = 0$ meets the curve twice; nearby vertical lines meet it three times.

35. Both: This is clearly a cubic polynomial and also a rational function because it can be written as

$$f(x) = \frac{x^3 - 4x + 1}{1}.$$

(This shows that all polynomials are rational.)

37. Rational; both the numerator and the denominator are polynomials.

39. Neither: Contains square root.

41. We need the function under the square root to be nonnegative.
$x + 2 \geq 0$ when $x \geq -2$. The domain is $\{x \in \mathbb{R} | x \geq -2\} = [-2, \infty)$.

43. Negatives are permitted inside the cube root. There are no restrictions, so the domain is $(-\infty, \infty)$ or all real numbers.

45. The denominator cannot be zero. $x^2 - 1 = 0$ when $x = \pm 1$. The domain is
$\{x \in \mathbb{R} | x \neq \pm 1\}$
$= (-\infty, -1) \cup (-1, 1) \cup (1, \infty)$.

47. $f(0) = 0^2 - 0 - 1 = -1$
$f(2) = 2^2 - 2 - 1 = 1$
$f(-3) = (-3)^2 - (-3) - 1 = 11$
$f\left(\frac{1}{2}\right) = \left(\frac{1}{2}\right)^2 - \frac{1}{2} - 1 = -\frac{5}{4}$

49. $f(0) = \sqrt{0 + 1} = 1$
$f(3) = \sqrt{3 + 1} = 2$
$f(-1) = \sqrt{-1 + 1} = 0$
$f\left(\frac{1}{2}\right) = \sqrt{\frac{1}{2} + 1} = \sqrt{\frac{3}{2}} = \frac{\sqrt{6}}{2}$

51. The only constraint we know is that the width should not be negative, so a reasonable domain would be $\{x | x > 0\}$.

53. We know that x should not be negative, that x must be an integer and that x cannot exceed the number of candy bars made, i.e., a reasonable domain would be $\{x | 0 \leq x \leq \text{number made}, x \text{ an integer}\}$.

55. Answers vary. There may well be a positive correlation (more study hours = better grade), but not necessarily a functional relation. You may study the same amount of time for two different exams and get different grades.

57. Answers vary. While not denying a negative correlation (more exercise = less weight), there are too many other factors (metabolic rate, diet) to be able to quantify a person's weight as a function just of the amount of exercise. Two people may exercise the same amount of time but have different weights.

59. A flat interval corresponds to an interval of constant speed; going up means that the speed is increasing while the graph going down means that the speed is decreasing. It is likely that the bicyclist is going uphill when the graph is going down and going downhill when the graph is going up.

61. The x-intercept occurs where
$0 = x^2 - 2x - 8 = (x - 4)(x + 2)$,
so $x = 4$ or $x = -2$; y-intercept at
$y = 0^2 - 2(0) - 8 = -8$.

63. The x-intercept occurs where
$0 = x^3 - 8 = (x - 2)(x^2 + 2x + 4)$,
so $x = 2$ (using the quadratic formula on the quadratic factor gives the solutions
$x = -1 \pm \sqrt{-3}$, neither of which is real so neither contributes a solution); y-intercept at
$y = 0^3 - 8 = -8$.

65. The x-intercepts occur where the numerator is zero, at $0 = x^2 - 4 = (x-2)(x+2)$, so $x = \pm 2$; y-intercept at $y = \dfrac{0^2 - 4}{0 + 1} = -4$.

67. $x^2 - 5x + 6 = (x - 3)(x - 2)$, so the zeros are $x = 2$ and $x = 3$.

69. Quadratic formula gives
$$x = \frac{4 \pm \sqrt{16 - 8}}{2} = 2 \pm \sqrt{2}.$$

71.
$$\begin{aligned} x^3 - 3x^2 + 2x &= x(x^2 - 3x + 2) \\ &= x(x - 2)(x - 1) \end{aligned}$$
so the zeros are $x = 0$, 1, and 2.

73. With $t = x^3$, $x^6 + x^3 - 2$ becomes $t^2 + t - 2$ and factors as $(t + 2)(t - 1)$. The expression is zero only if one of the factors is zero, i.e., if $t = 1$ or $t = -2$. With $x = t^{1/3}$, the first occurs only if $x = (1)^{1/3} = 1$. The latter occurs only if $x = (-2)^{1/3}$, about -1.2599.

75. If $B(h) = -1.8h + 212$, then we can solve $B(h) = 98.6$ for h as follows:
$$\begin{aligned} 98.6 &= -1.8h + 212 \\ 1.8h &= 113.4 \\ h &= \frac{113.4}{1.8} = 63 \end{aligned}$$

This altitude (63,000 feet above sea level, more than double the height of Mt. Everest) would be the elevation at which we humans boil alive in our skins. Of course the cold of space and the near-total lack of external pressure create additional complications, which we shall not try to analyze.

77. This is a two-point line-fitting problem. If a point is interpreted as $(R, T) = $ (chirp rate, temperature), then the two given points are $(160, 79)$ and $(100, 64)$. The slope being $\dfrac{64 - 79}{100 - 160} = \dfrac{-15}{-60} = \dfrac{1}{4}$, we could write $T = \frac{1}{4}(R - 100) + 64$ or $T = \frac{1}{4}R + 39$.

79. Her winning percentage is calculated by the formula $P = \dfrac{100w}{t}$, where P is the winning percentage, w is the number of games won, and t is the total number of games. Plugging in $w = 415$ and $t = 415 + 120 = 535$, we find her winning percentage is approximately $P \approx 77.57$, so we see that the percentage displayed is rounded up from the actual percentage. Let x be the number of games won in a row. If she doesn't lose any games, her new winning percentage will be given by the formula $P = \dfrac{100(415 + x)}{535 + x}$. In order to have her winning percentage displayed as 80%, she only needs a winning percentage of 79.5 or greater. Thus, we must solve the inequality $79.5 \le \dfrac{100(415 + x)}{535 + x}$:

$$\begin{aligned} 79.5 &\le \frac{100(415 + x)}{535 + x} \\ 79.5(535 + x) &\le 41{,}500 + 100x \\ 42{,}532.5 + 79.5x &\le 41{,}500 + 100x \\ 1032.5 &\le 20.5x \\ 50.4 &\le x \end{aligned}$$

(In the above, we are allowed to multiply both sides of the inequality by $535 + x$ because we assume x (the number of wins in a row) is positive.) Thus she must win at least 50.4 times in a row to get her winning percentage to display as 80%. Since she can't win a fraction of a game, she must win at least 51 games in a row.

0.3 GRAPHING CALCULATORS AND COMPUTER ALGEBRA SYSTEMS

1. Intercepts $x = \pm 1$, $y = -1$. Minimum at $(0, -1)$. No vertical asymptotes.

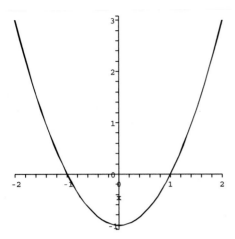

3. Intercepts $y = 8$ (no x-intercepts). Minimum at $(-1, 7)$. No vertical asymptotes.

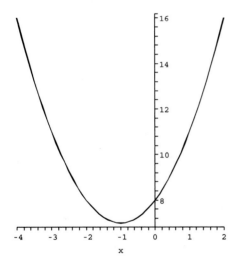

5. Intercepts $x = -1$, $y = 1$. No extrema or vertical asymptotes.

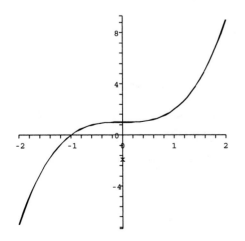

7. Intercepts $x \approx 0.453$, $y = -1$. No extrema or vertical asymptotes.

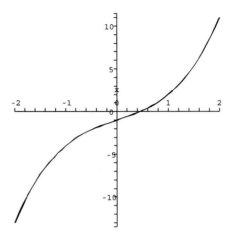

9. Intercepts $x = \pm 1$, $y = -1$. Minimum at $(0, -1)$. No vertical asymptotes.

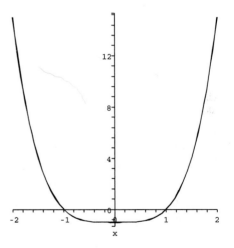

11. Intercepts $x \approx 0.475$, $x \approx -1.395$, $y = -1$. Minimum at (approximately) $(-0.794, -2.191)$. No vertical asymptotes.

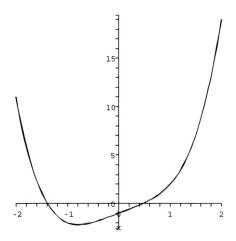

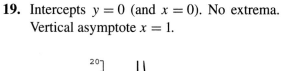

17. Intercepts $y = -3$ (no x-intercepts). No extrema. Vertical asymptote $x = 1$.

13. Intercepts $x \approx -1.149$, $y = 2$. No extrema or vertical asymptotes.

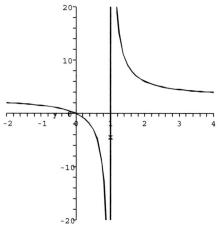

19. Intercepts $y = 0$ (and $x = 0$). No extrema. Vertical asymptote $x = 1$.

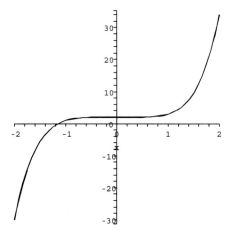

15. Intercepts $x \approx 0.050$, $y = -1$. The two local maxima occur at approximately $(-1.930, -8.866)$ and $(1.036, 12.018)$, while the two local minima occur at approximately $(-1.036, -14.018)$ and $(1.930, 6.866)$. No vertical asymptotes.

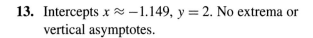

21. Intercepts $y = 0$ (and $x = 0$). Local maximum at $(0, 0)$. Local minimum at $(2, 12)$. Vertical asymptote $x = 1$.

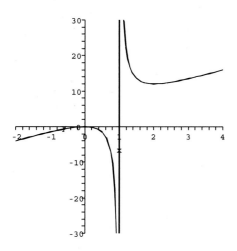

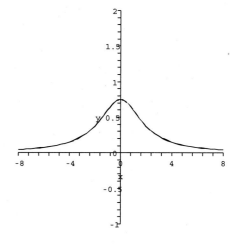

23. Intercepts $y = -1/2$ (no x-intercepts). Local maximum at $(0, -1/2)$. Vertical asymptotes $x = \pm 2$.

27. Intercepts $x = -2$, $y = -1/3$. No extrema. Vertical asymptotes at $x = -3$ and $x = 2$.

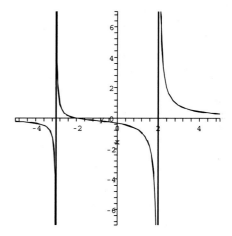

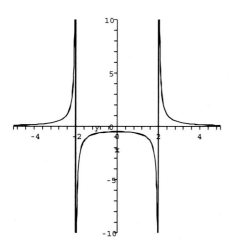

29. Intercepts $y = 0$ (and $x = 0$). No extrema. No vertical asymptotes.

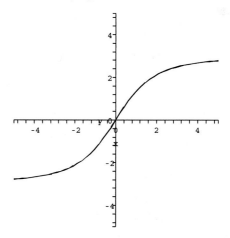

25. Intercepts $y = 3/4$ (no x-intercepts). Local maximum at $(0, 3/4)$. No vertical asymptotes.

31. Vertical asymptotes where $x^2 - 4 = 0 \Rightarrow x = \pm 2$.

33. Vertical asymptotes where
$$x^2 + 3x - 10 = 0$$
$$\Rightarrow (x + 5)(x - 2) = 0$$
$$\Rightarrow x = -5 \text{ or } x = 2.$$

35. A vertical asymptote may occur when the denominator is zero. This denominator, however, is never zero, so there are no vertical asymptotes.

37. Vertical asymptotes where
$$x^3 + 3x^2 + 2x = 0$$
$$\Rightarrow x(x^2 + 3x + 2) = 0$$
$$\Rightarrow x(x + 2)(x + 1) = 0$$
$$\Rightarrow x = 0, x = -2, \text{ or } x = -1.$$
Since none of these x-values make the numerator zero, they are all vertical asymptotes.

39. A window with $-0.1 \le x \le 0.1$ and $-0.0001 \le y \le 0.0001$ shows all details.

41. A window with $-14 \le x \le 14$ and $-80 \le y \le 80$ shows all details.

43. From graph $x = 1$.

45. From graph $x = \pm 1$.

47. From graph $x = 0$.

49. The graph makes it appear that there are two intersection points. Solving algebraically, $\sqrt{x - 1} = x^2 - 1$ (for $x \ge 1$) when

$$x - 1 = (x^2 - 1)^2 = ((x - 1)(x + 1))^2$$
$$= (x - 1)^2(x + 1)^2.$$

We see that $x = 1$ is one solution (obvious from the start), while for any other, we can cancel one factor of $x - 1$ and find

$$1 = (x - 1)(x + 1)^2 = (x^2 - 1)(x + 1)$$
$$= x^3 + x^2 - x - 1.$$

Hence $x^3 + x^2 - x - 2 = 0$.

By solver, this equation has only the one solution $x \approx 1.206$.

51. The graph does not clearly show the number of intersection points. Solving algebraically,

$$x^3 - 3x^2 = 1 - 3x$$
$$\Rightarrow x^3 - 3x^2 + 3x - 1 = 0$$
$$\Rightarrow (x - 1)^3 = 0 \Rightarrow x = 1.$$

So there is only one solution: $x = 1$.

53. After zooming out, the graph shows that there are two solutions: one near zero, and one around ten. Algebraically,
$$(x^2 - 1)^{2/3} = 2x + 1$$
$$\Rightarrow (x^2 - 1)^2 = (2x + 1)^3$$
$$\Rightarrow x^4 - 2x^2 + 1 = 8x^3 + 12x^2 + 6x + 1$$
$$\Rightarrow x^4 - 8x^3 - 14x^2 - 6x = 0$$
$$\Rightarrow x(x^3 - 8x^2 - 14x - 6) = 0.$$

We thus confirm the obvious solution $x = 0$, and by solver or spreadsheet, find the second solution $x \approx 9.534$.

55. The graph shows that there are two solutions: $x \approx \pm 1.177$ by calculator or spreadsheet.

57. Calculator shows zeros at approximately -1.879, 0.347, and 1.532.

59. Calculator shows zeros at approximately 0.5637 and 3.0715.

61. Calculator shows zeros at approximately -5.248 and 10.006.

63. The graph of $y = x^2$ on the window $-10 \le x \le 10$, $-10 \le y \le 10$ appears identical (except for labels) to the graph of $y = 2(x - 1)^2 + 3$ if the latter is drawn on a graphing window centered at the point $(1, 3)$ with
$$-9 \le x \le 11$$
$$-17 \le y \le 23.$$

65. $\sqrt{y^2}$ is the distance from (x, y) to the x-axis. $\sqrt{x^2 + (y - 2)^2}$ is the distance from (x, y) to the point $(0, 2)$. If we require that these be the same, and we square both quantities, we have
$$y^2 = x^2 + (y - 2)^2$$
$$y^2 = x^2 + y^2 - 4y + 4$$
$$4y = x^2 + 4$$
$$y = \frac{1}{4}x^2 + 1$$

In this relation, we see that y is a quadratic function of x. The graph is commonly known as a parabola.

0.4 TRIGONOMETRIC FUNCTIONS

1. **(a)** $\left(\dfrac{\pi}{4}\right)\left(\dfrac{180°}{\pi}\right) = 45°$

   (b) $\left(\dfrac{\pi}{3}\right)\left(\dfrac{180°}{\pi}\right) = 60°$

   (c) $\left(\dfrac{\pi}{6}\right)\left(\dfrac{180°}{\pi}\right) = 30°$

   (d) $\left(\dfrac{4\pi}{3}\right)\left(\dfrac{180°}{\pi}\right) = 240°$

3. **(a)** $(180°)\left(\dfrac{\pi}{180°}\right) = \pi$

   (b) $(270°)\left(\dfrac{\pi}{180°}\right) = \dfrac{3\pi}{2}$

   (c) $(120°)\left(\dfrac{\pi}{180°}\right) = \dfrac{2\pi}{3}$

   (d) $(30°)\left(\dfrac{\pi}{180°}\right) = \dfrac{\pi}{6}$

5. $2\cos x - 1 = 0$ when $\cos x = 1/2$. This occurs whenever $x = \frac{\pi}{3} + 2k\pi$ or $x = -\frac{\pi}{3} + 2k\pi$ for any integer k.

7. $\sqrt{2}\cos x - 1 = 0$ when $\cos x = 1/\sqrt{2}$. This occurs whenever $x = \frac{\pi}{4} + 2k\pi$ or $x = -\frac{\pi}{4} + 2k\pi$ for any integer k.

9. $\sin^2 x - 4\sin x + 3 = (\sin x - 1)(\sin x - 3) = 0$ when $\sin x = 1$ ($\sin x \neq 3$ for any x). This occurs whenever $x = \frac{\pi}{2} + 2k\pi$ for any integer k.

11. $\sin^2 x + \cos x - 1 = (1 - \cos^2 x) + \cos x - 1 = (\cos x)(\cos x - 1) = 0$ when $\cos x = 0$ or $\cos x = 1$. This occurs whenever $x = \frac{\pi}{2} + k\pi$ or $x = 2k\pi$ for any integer k.

13. $\cos^2 x + \cos x = (\cos x)(\cos x + 1) = 0$ when $\cos x = 0$ or $\cos x = -1$. This occurs whenever $x = \frac{\pi}{2} + k\pi$ or $x = \pi + 2k\pi$ for any integer k.

15. The graph of $f(x) = \sin 3x$.

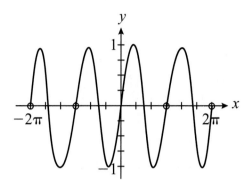

17. The graph of $f(x) = \tan 2x$.

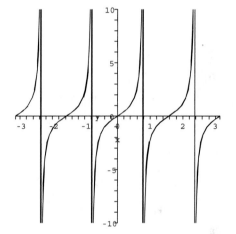

19. The graph of $f(x) = 3\cos(x - \pi/2)$.

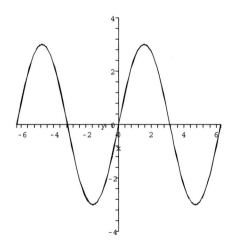

21. The graph of $f(x) = \sin 2x - 2\cos 2x$.

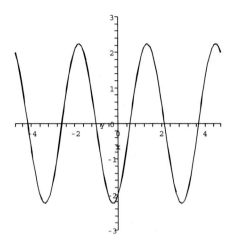

23. The graph of $f(x) = \sin x \sin 12x$.

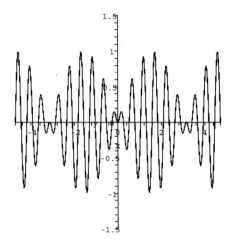

25. Amplitude is 3, period is $\frac{2\pi}{2} = \pi$, frequency is $\frac{1}{\pi}$.

27. Amplitude is 5, period is $\frac{2\pi}{3}$, frequency is $\frac{3}{2\pi}$.

29. Amplitude is 3, period is $\frac{2\pi}{2} = \pi$, frequency is $\frac{1}{\pi}$.

31. Amplitude is 4 (the graph oscillates between -4 and 4, so we may ignore the minus sign), period is 2π, frequency is $\frac{1}{2\pi}$.

33. $\sin(\alpha - \beta) = \sin(\alpha + (-\beta))$
$= \sin\alpha\cos(-\beta) + \sin(-\beta)\cos\alpha$
$= \sin\alpha\cos\beta - \sin\beta\cos\alpha$

35. (a) $\cos(2\theta) = \cos(\theta + \theta) = \cos(\theta)\cos(\theta) - \sin(\theta)\sin(\theta)\cos^2\theta - \sin^2\theta$
$= \cos^2\theta - (1 - \cos^2\theta) = 2\cos^2\theta - 1$

(b) Just continue on, writing $\cos(2\theta) = 2\cos^2\theta - 1 = 2(1 - \sin^2\theta) - 1 = 1 - 2\sin^2\theta$

37. Use the formula
$\cos(x + \beta) = \cos x\cos\beta - \sin\beta\sin x$.
Now we see that $\cos\beta$ must equal $4/5$ and $\sin\beta$ must equal $3/5$. Since $(4/5)^2 + (3/5)^2 = 1$, this is possible. We see that $\beta \approx 0.6435$ radians, or $-36.87°$.

39. $\cos(2x)$ has period $\frac{2\pi}{2} = \pi$ and $\sin(\pi x)$ has period $\frac{2\pi}{\pi} = 2$. There are no common integer multiples of the periods, so the function $f(x) = \cos(2x) + 3\sin(\pi x)$ is not periodic.

41. $\sin(2x)$ has period $\frac{2\pi}{2} = \pi$ and $\cos(5x)$ has period $\frac{2\pi}{5}$. The smallest integer multiple of both of these is the fundamental period, and it is 2π.

43. $\cos^2\theta = 1 - \sin^2\theta$
$= 1 - \left(\frac{1}{3}\right)^2 = 1 - \frac{1}{9} = \frac{8}{9}$.

Because θ is in the first quadrant, its cosine is nonnegative. Hence
$\cos\theta = \sqrt{\frac{8}{9}} = \frac{2\sqrt{2}}{3} = 0.9428$.

45. $\cos^2\theta = 1 - \sin^2\theta = 1 - \left(\frac{1}{2}\right)^2 = \frac{3}{4}$. Because θ is in the second quadrant, its cosine is negative, $\cos\theta = -\sqrt{\frac{3}{4}} = -\frac{\sqrt{3}}{2}$.

47. From graph the three solutions are -1.868, -0.538, and 2.585.

49. From graph the two solutions are ± 1.455.

51. Let h be the height of the rocket. Then $\frac{h}{2} = \tan 20°$.
$h = 2\tan 20° \approx 0.73$ (miles)

53. Let h be the distance from the ground to the top of the steeple. Then $\dfrac{h}{80+20} = \tan 50°$
$h = 100 \tan 50° \approx 119.2$ (feet).

55. The period is $\pi/30$ (seconds). So the frequency is $f = 30/\pi$ (cycles per second) and the meter voltage is $\dfrac{170}{\sqrt{2}} \approx 120.2$.

57. There seems to be a certain slowly increasing base for sales $(110 + 2t)$, and given that the sine function has period $\dfrac{2\pi}{\pi/6} = 12$ months, the sine term apparently represents some sort of seasonally cyclic pattern. If we assume that travel peaks at Thanksgiving, the effect is that time zero would correspond to a time one-quarter period (3 months) prior to Thanksgiving, or very late August.

The annual increase for the year beginning at time t is given by
$s(t + 12) - s(t)$ and automatically ignores both the seasonal factor and the basic 110, and indeed it is the constant $2 \times 12 = 24$ (in thousands of dollars per year and independent of the reference point t).

59.

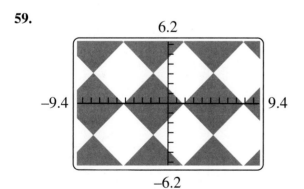

0.5 TRANSFORMATIONS OF FUNCTIONS

1. $(f \circ g)(x) = f(g(x))$
$$= f(\sqrt{x-3}) = \sqrt{x-3} + 1$$
with domain $\{x \mid x \geq 3\}$.

$(g \circ f)(x) = g(f(x))$
$$= g(x+1)$$
$$= \sqrt{(x+1) - 3} = \sqrt{x - 2}$$
with domain $\{x \mid x \geq 2\}$.

3. $(f \circ g)(x) = f(g(x)) = f(x^3 + 4) = \dfrac{1}{x^3 + 4}$
with domain $\{x \mid x \neq -\sqrt[3]{4}\}$
$(g \circ f)(x) = g(f(x)) = g\left(\dfrac{1}{x}\right) = \dfrac{1}{x^3} + 4$
with domain $\{x \mid x \neq 0\}$.

5. $(f \circ g)(x) = f(g(x)) = f(\sin x)$
$= \sin^2 x + 1$ with domain $(-\infty, \infty)$ or all real numbers.
$(g \circ f)(x) = g(f(x)) = g(x^2 + 1)$
$= \sin(x^2 + 1)$ with domain $(-\infty, \infty)$ or all real numbers.

7. $\sqrt{x^4 + 1} = f(g(x))$ when $f(x) = \sqrt{x}$ and $g(x) = x^4 + 1$, for example.

9. $\dfrac{1}{x^2 + 1} = f(g(x))$ when $f(x) = 1/x$ and $g(x) = x^2 + 1$, for example.

11. $(4x + 1)^2 + 3 = f(g(x))$ when $f(x) = x^2 + 3$ and $g(x) = 4x + 1$, for example.

13. $\sin^3 x = f(g(x))$ when $f(x) = x^3$ and $g(x) = \sin x$, for example.

15. $\dfrac{3}{\sqrt{\sin x + 2}} = f(g(h(x)))$ when $f(x) = 3/x$, $g(x) = \sqrt{x}$ and $h(x) = \sin x + 2$, for example.

17. $\cos^3(4x - 2) = f(g(h(x)))$ when $f(x) = x^3$, $g(x) = \cos x$ and $h(x) = 4x - 2$, for example.

19. $4 \cos(x^2) - 5 = f(g(h(x)))$ when $f(x) = 4x - 5$, $g(x) = \cos x$ and $h(x) = x^2$, for example.

21. Graph of $f(x) - 3$: The graph of f is translated down by 3 units.

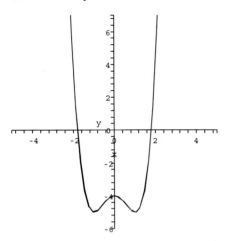

23. Graph of $f(x - 3)$: The graph of f is translated to the right by 3 units.

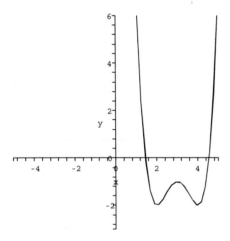

25. Graph of $f(2x)$: The horizontal scale is divided by 2.

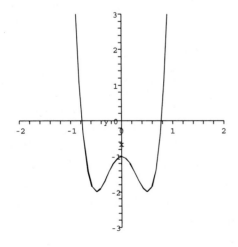

27. Graph of $4f(x) - 1$: The vertical scale is multiplied by 4 and then the graph is shifted down by 1 unit.

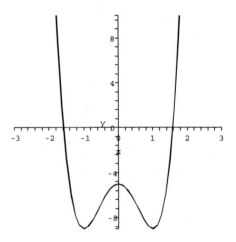

29. Graph of $f(x - 4)$: The graph of f is shifted to the right by 4 units.

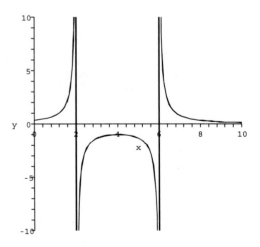

31. Graph of $f(2x)$: The horizontal scale is divided by 2.

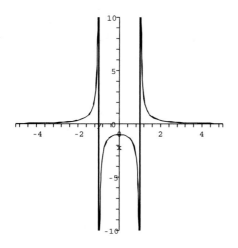

33. Graph of $f(3x + 3)$: The graph of f is shifted 3 units to the left and then the horizontal axis is divided by 3.

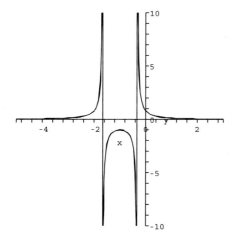

35. Graph of $2f(x) - 4$: The vertical scale is multiplied by 2 and then the graph is shifted down by 4 units.

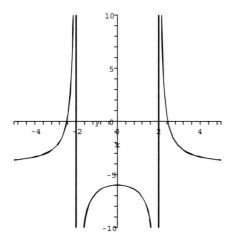

37. $f(x) = x^2 + 2x + 1 = (x + 1)^2$. Shift $y = x^2$ to the left 1 unit.

39. $f(x) = x^2 + 2x + 4 = (x^2 + 2x + 1) + 4 - 1$
$\qquad = (x + 1)^2 + 3.$

Shift $y = x^2$ to the left 1 unit and up 3 units.

41. $f(x) = 2x^2 + 4x + 4$
$\qquad = 2(x^2 + 2x + 2)$
$\qquad = 2(x^2 + 2x + 1 - 1 + 2)$
$\qquad = 2[(x + 1)^2 + 1]$

Shift $y = x^2$ to the left 1 unit, then up 1 unit, then multiply the scale on the y-axis by 2.

43. Graph is reflected across the x-axis and the scale on the y-axis is multiplied by 2.

45. Graph is reflected across the x-axis, the scale on the y-axis is multiplied by 3, and the graph is shifted up 2 units.

47. Graph is reflected across the y-axis.

49. Graph is reflected across the y-axis and shifted up 1 unit.

51. The graph is reflected across the x-axis and the scale on the y-axis is multiplied by $|c|$.

53. The graph of $y = |x|^3$ is identical to that of $y = x^3$ to the right of the y-axis because for $x > 0$ we have $|x|^3 = x^3$. For $y = |x|^3$ the graph to the left of the y-axis is the reflection through the y-axis of the graph to the right of the y-axis. In general to graph $y = f(|x|)$ based on the graph of $y = f(x)$, the procedure is to discard the part of the graph to the left of the y-axis, and replace it by a reflection in the y-axis of the part to the right of the y-axis.

55. The rest of the first 10 iterates of $f(x) = \cos x$ with $x_0 = 1$ are:

$x_4 = \cos 0.65 \approx 0.796$

$x_5 = \cos 0.796 \approx 0.700$

$x_6 = \cos 0.700 \approx 0.765$

$x_7 = \cos 0.765 \approx 0.721$

$x_8 = \cos 0.721 \approx 0.751$

$x_9 = \cos 0.751 \approx 0.731$

$x_{10} = \cos .731 \approx 0.744$

Continuing in this fashion and retaining more decimal places, one finds that x_{36} through x_{40} are all 0.739085. The same process is used with a different x_0.

57. They converge to 0. For $x_0 = 1$, $x_1 = \sin x_0 \approx$ 0.841, $x_2 = \sin x_1 \approx 0.745$, $x_3 = \sin x_2 \approx$ 0.678 and so on.

59. If the iterates of a function f (starting from some point x_0) are going to go toward (and remain arbitrarily close to) a certain number L, this number L must be a solution of the equation $f(x) = x$. For the list of iterates $x_0, x_1, x_2, x_3, \ldots$ is, apart from the first term, the same list as the list of numbers $f(x_0), f(x_1), f(x_2), f(x_3), \ldots$. (Remember that x_{n+1} is $f(x_n)$.) If any of the numbers in the first list are close to L, then the f-values (in the second list) are close to $f(L)$. But since the lists are *identical* (apart from the first term x_0 which is not in the second list), it must be true that L and $f(L)$ are the same number.

If conditions are right (and they are in the two cases $f(x) = \cos(x)$ (#55) and $f(x) = \sin(x)$ (#57)), this "convergence" will indeed occur, and since there is in these cases only *one* solution (x about 0.739085 in #55 and $x = 0$ in #57) it won't matter where you started.

Chapter 0 Review Exercises

1. $m = \dfrac{7 - 3}{0 - 2} = \dfrac{4}{-2} = -2$

3. These lines both have slope 3. They are parallel unless they are coincident. But the first line

includes the point (0, 1), which does not satisfy the equation of the second line. The lines are not coincident.

5. Let $P = (1, 2)$, $Q = (2, 4)$, $R = (0, 6)$.

Then PQ has slope $\dfrac{4 - 2}{2 - 1} = 2$

QR has slope $\dfrac{6 - 4}{0 - 2} = -1$

RP has slope $\dfrac{2 - 6}{1 - 0} = -4$

Since no two of these slopes are negative reciprocals, none of the angles are right angles. The triangle is not a right triangle.

7. The line apparently goes through (1, 1) and (3, 2). So the slope would be $m = \frac{2-1}{3-1} = \frac{1}{2}$. The equation would be $y = \frac{1}{2}(x - 1) + 1$ or $y = \frac{1}{2}x + \frac{1}{2}$. Using the equation with $x = 4$, we find $y = \frac{1}{2}(4) + \frac{1}{2} = \frac{5}{2}$.

9. Using the point-slope method, we find $y = -\frac{1}{3}(x + 1) - 1$.

11. The graph passes the vertical line test, so it is a function.

13. The radicand cannot be negative, hence we require $4 - x^2 \geq 0 \Rightarrow 4 \geq x^2$. Therefore, the natural domain is $\{x \mid -2 \leq x \leq 2\}$ or $[-2, 2]$.

15. Intercepts at $x = -4$ and 2, and $y = -8$. Local minimum at $x = -1$. No vertical asymptotes.

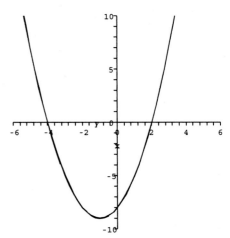

17. Intercepts at $x = -1$ and 1, and $y = 1$. Local minimum at $x = 1$ and at $x = -1$. Local maximum at $x = 0$. No vertical asymptotes.

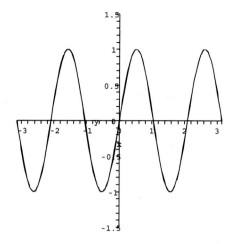

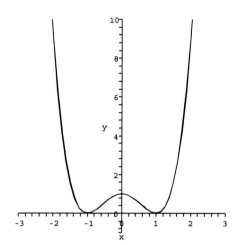

19. Intercept at $y = 0$ and at $x = 0$. No extrema. Vertical asymptote $x = -2$.

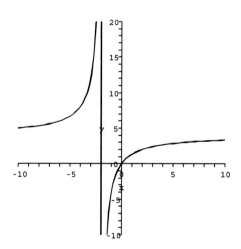

23. Intercept at $y = 2$. x-intercepts, local maxima and local minima occur regularly. No vertical asymptotes.

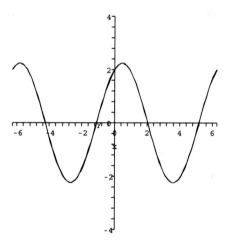

21. Intercept at $y = 0$ and $x = \frac{k\pi}{3}$ for integers k. Extrema: y takes maximum 1 and minimum -1 with great predictability and regularity. No vertical asymptotes.

25. $f(0) = 1$, so y-intercept at $(0, 1)$. The equation $\sec 2x = 0$ has no solutions, so there are no x-intercepts. Local minima of 1 occur at $x = k\pi$ where k is an integer, and local maxima of -1 occur at $x = \frac{\pi}{2} + k\pi$ where k is an integer.

There are vertical asymptotes at $x = \frac{\pi}{4} + \frac{k\pi}{2}$ where k is an integer.

17. $\lim\limits_{x\to 1}\left(\dfrac{1}{x-1}-\dfrac{2}{x^2-1}\right)$

$=\lim\limits_{x\to 1}\left(\dfrac{1}{x-1}-\dfrac{2}{(x-1)(x+1)}\right)$

$=\lim\limits_{x\to 1}\left(\dfrac{x+1}{(x-1)(x+1)}-\dfrac{2}{(x-1)(x+1)}\right)$

$=\lim\limits_{x\to 1}\left(\dfrac{x-1}{(x-1)(x+1)}\right)$

$=\lim\limits_{x\to 1}\left(\dfrac{1}{x+1}\right)=\dfrac{1}{2}$

19. $\lim\limits_{x\to 0}\dfrac{1-\cos^2 x}{1-\cos x}$

$=\lim\limits_{x\to 0}\dfrac{(1-\cos x)\,(1+\cos x)}{1-\cos x}$

$=\lim\limits_{x\to 0}(1+\cos x)=2$

21. $\lim\limits_{x\to 0^+}\dfrac{\sin(|x|)}{x}=\lim\limits_{x\to 0^+}\dfrac{\sin(x)}{x}=1$

$\lim\limits_{x\to 0^-}\dfrac{\sin(|x|)}{x}$

$=\lim\limits_{x\to 0^-}\dfrac{\sin(-x)}{x}$

$=\lim\limits_{x\to 0^-}\dfrac{-\sin(x)}{x}=-1$

Since the limit from the left does not equal the limit from the right, we see that $\lim\limits_{x\to 0}\dfrac{\sin(|x|)}{x}$ does not exist.

23. $\lim\limits_{x\to 2^-}f(x)=\lim\limits_{x\to 2^-}2x=2(2)=4$

$\lim\limits_{x\to 2^+}f(x)=\lim\limits_{x\to 2^+}x^2=2^2=4$

$\lim\limits_{x\to 2}f(x)=4$

25. $\lim\limits_{x\to 0}f(x)=\lim\limits_{x\to 0}(3x+1)=3(0)+1=1$

27. $\lim\limits_{x\to -1^-}f(x)=\lim\limits_{x\to -1^-}(2x+1)$

$=2(-1)+1=-1$

$\lim\limits_{x\to -1^+}f(x)=\lim\limits_{x\to -1^+}3=3$

Therefore $\lim\limits_{x\to -1}f(x)$ does not exist.

29. $\lim\limits_{h\to 0}\dfrac{(2+h)^2-4}{h}$

$=\lim\limits_{h\to 0}\dfrac{(4+4h+h^2)-4}{h}$

$=\lim\limits_{h\to 0}\dfrac{4h+h^2}{h}=\lim\limits_{h\to 0}(4+h)=4$

31. $\lim\limits_{h\to 0}\dfrac{h^2}{\sqrt{h^2+h+3}-\sqrt{h+3}}$

$=\lim\limits_{h\to 0}\dfrac{h^2(\sqrt{h^2+h+3}+\sqrt{h+3})}{(h^2+h+3)-(h+3)}$

$=\lim\limits_{h\to 0}\dfrac{h^2(\sqrt{h^2+h+3}+\sqrt{h+3})}{h^2}$

$=\lim\limits_{h\to 0}\left[\sqrt{h^2+h+3}+\sqrt{h+3}\right]=2\sqrt{3}$

To get from the first line to the second, we have multiplied by

$$\dfrac{\sqrt{h^2+h+3}+\sqrt{h+3}}{\sqrt{h^2+h+3}+\sqrt{h+3}}.$$

33. $\lim\limits_{t\to -2}\dfrac{\frac{1}{2}+\frac{1}{t}}{2+t}$

$=\lim\limits_{t\to -2}\dfrac{\frac{t+2}{2t}}{2+t}$

$=\lim\limits_{t\to -2}\dfrac{1}{2t}=-\dfrac{1}{4}$

35.

x	$x^2\sin(1/x)$
-0.1	0.0054
-0.01	5×10^{-5}
-0.001	-8×10^{-7}
0.1	-0.0054
0.01	-5×10^{-5}
0.001	8×10^{-7}

Conjecture: $\lim\limits_{x\to 0}x^2\sin(1/x)=0$.

Let $f(x)=-x^2$, $h(x)=x^2$. Then

$$f(x)\le x^2\sin\left(\dfrac{1}{x}\right)\le h(x)$$

$\lim\limits_{x\to 0}(-x^2)=0$, $\lim\limits_{x\to 0}(x^2)=0$

Therefore, by the Squeeze Theorem,

$$\lim_{x \to 0} x^2 \sin\left(\frac{1}{x}\right) = 0.$$

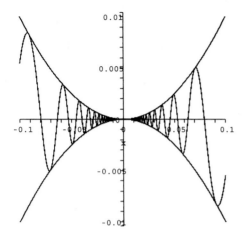

37. Let $f(x) = 0$, $h(x) = \sqrt{x}$. We see that

$$f(x) \le \sqrt{x} \cos^2(1/x) \le h(x),$$

$$\lim_{x \to 0^+} 0 = 0, \quad \lim_{x \to 0^+} \sqrt{x} = 0$$

Therefore, by the Squeeze Theorem,

$$\lim_{x \to 0^+} \sqrt{x} \cos^2\left(\frac{1}{x}\right) = 0.$$

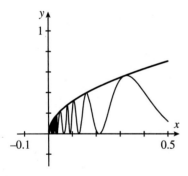

39. $\lim_{x \to 4^+} \sqrt{16 - x^2}$ does not exist because the domain of the function is $[-4, 4]$.

41. $\lim_{x \to -2^-} \sqrt{x^2 + 3x + 2} = 0$

43. $\lim_{x \to 0^+} \frac{\sqrt{1 - \cos x}}{x} = \sqrt{\frac{1}{2}} = \frac{\sqrt{2}}{2}$

45. $\lim_{x \to a^-} f(x) = \lim_{x \to a^-} g(x) = g(a)$ because $g(x)$ is a polynomial. Similarly,

$$\lim_{x \to a^+} f(x) = \lim_{x \to a^+} h(x) = h(a).$$

47. (a) $\lim_{x \to 2} (x^2 - 3x + 1)$
$$= 2^2 - 3(2) + 1 \quad \text{(Theorem 3.2)}$$
$$= -1$$

(b) $\lim_{x \to 0} \dfrac{x - 2}{x^2 + 1}$
$$= \frac{\lim_{x \to 0} (x - 2)}{\lim_{x \to 0} (x^2 + 1)} \quad \text{(Theorem 3.1(iv))}$$
$$= \frac{\lim_{x \to 0} x - \lim_{x \to 0} 2}{\lim_{x \to 0} x^2 + \lim_{x \to 0} 1} \quad \text{(Theorem 3.1(ii))}$$
$$= \frac{0 - 2}{0 + 1}$$
(Equations 3.1, 3.2, and 3.4)
$$= -2$$

49. Velocity is given by the limit
$$\lim_{h \to 0} \frac{f(2 + h) - f(2)}{h}$$
$$= \lim_{h \to 0} \frac{(2 + h)^2 + 2 - (2^2 + 2)}{h}$$
$$= \lim_{h \to 0} \frac{4h + h^2}{h}$$
$$= \lim_{h \to 0} (4 + h) = 4.$$

51. Velocity is given by the limit
$$\lim_{h \to 0} \frac{f(0 + h) - f(0)}{h}$$
$$= \lim_{h \to 0} \frac{(0 + h)^3 - (0)^3}{h}$$
$$= \lim_{h \to 0} \frac{h^3}{h}$$
$$= \lim_{h \to 0} h^2 = 0.$$

53. $m = \lim_{h \to 0} \dfrac{\sqrt{1 + h} - 1}{h} \dfrac{\sqrt{1 + h} + 1}{\sqrt{1 + h} + 1}$
$$= \lim_{h \to 0} \frac{1 + h - 1}{h(\sqrt{1 + h} + 1)}$$
$$= \lim_{h \to 0} \frac{h}{h(\sqrt{1 + h} + 1)}$$

$$= \lim_{h \to 0} \frac{1}{\sqrt{1+h}+1}$$

$$= \frac{1}{\sqrt{1+0}+1} = \frac{1}{2}$$

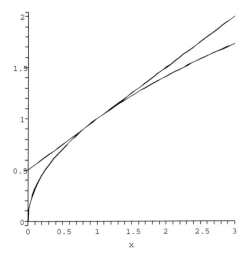

55. $\lim_{x \to 0^+} \cot x$ does not exist. As x approaches 0 from the right, the values of $\cot x$ increase without bound.

57. $\lim_{x \to a} \left[3f(x)g(x) \right]$
$= 3(\lim_{x \to a} f(x))(\lim_{x \to a} g(x))$
$= 3(2)(-3) = -18$

59. $\lim_{x \to a} \left[\dfrac{3f(x)+2g(x)}{h(x)} \right]$ cannot be determined from the given information since

$$\lim_{x \to a} \left[3f(x)+2g(x) \right]$$
$$= 3 \lim_{x \to a} f(x) + 2 \lim_{x \to a} g(x)$$
$$= 3(2) + 2(-3) = 0$$

and $\lim_{x \to a} h(x) = 0$.

61. Since we have a starting place, and we have shown that we can always get from one step to the next, the theorem must be true for any positive integer.

Given that $\lim_{x \to a} f(x) = L$.

Assume that $\lim_{x \to a} [f(x)]^k = L^k$.
Now $\lim_{x \to a} [f(x)]^{k+1} = \lim_{x \to a} [f(x)]^k f(x)$
$= \lim_{x \to a} [f(x)]^k \lim_{x \to a} f(x) = L^k L = L^{k+1}$.

Therefore $\lim_{x \to a} [f(x)]^n = L^n$ for any positive integer n.

63. The limit of a quotient is not the quotient of the limits if the denominator is 0. The fraction $\frac{0}{0}$ is undefined, not equal to 1.

65. $f(x) = \frac{1}{x}$, $g(x) = x$. $\lim_{x \to 0} f(x)g(x) = 1$, but $\lim_{x \to 0} f(x)$ does not exist.

67. False. For example, let $f(x) = 1/x$. $\lim_{x \to 0} f(x)$ does not exist, but

$$\lim_{x \to 0} \frac{1}{f(x)} = \lim_{x \to 0} x = 0.$$

69. If $\lim_{x \to 0^+} T(x) = 0$, then $a = 0$. If $\lim_{x \to 20,000}$ exists, then b must be 2400. These limits should exist so that \$0 income corresponds to \$0 tax, and so that the tax function doesn't have sudden jumps.

71. (a) $\lim_{x \to 1} [x]$ does not exist. Approaches 0 from left, 1 from right.

(b) $\lim_{x \to 1.5} [x] = 1$.

(c) $\lim_{x \to 1.5} [2x]$ does not exist. Approaches 2 from left, 3 from right.

(d) $\lim_{x \to 1} (x - [x])$ does not exist. Approaches 1 from left, 0 from right.

1.4 CONTINUITY AND ITS CONSEQUENCES

1. Discontinuous at $x = -2$ (limit does not exist), and at $x = 2$ (function undefined).

3. Discontinuous at $x = -2$ (function undefined), at $x = 1$ (function undefined), and at $x = 4$ (limit does not exist).

5. Discontinuous at $x = -2$ (limit does not exist), at $x = 2$ (function undefined), and at $x = 4$ (limit does not exist).

7. $f(1)$ is not defined and $\lim\limits_{x \to 1} f(x)$ does not exist.

9. $f(0)$ is not defined and $\lim\limits_{x \to 0} f(x)$ does not exist.

11. $\lim\limits_{x \to 2^-} f(x) = \lim\limits_{x \to 2^-} (x^2) = 4 \ \lim\limits_{x \to 2^+} f(x) =$
$\lim\limits_{x \to 2^+} (3x - 2) = 4 \ \lim\limits_{x \to 2} f(x) = 4; f(2) = 3$
$\lim\limits_{x \to 2} f(x) \neq f(2)$

13. $f(x) = \dfrac{x - 1}{(x + 1)(x - 1)}$ has a removable discontinuity at $x = 1$ and a nonremovable discontinuity at $x = -1$; the removable discontinuity is removed by $g(x) = \dfrac{1}{x + 1}$.

15. No discontinuities.

17. $f(x) = \dfrac{x^2 \sin x}{\cos x}$ has nonremovable discontinuities at $x = \frac{\pi}{2} + k\pi$ for any integer k.

19. $f(x)$ is undefined where $x^3 - x^2 \leq 0$, which corresponds to $x \leq 1$. Thus, $f(x)$ is discontinuous for $x = 1$. (Not removable.)

21. $f(x)$ has a nonremovable discontinuity at $x = 1$.

23. $f(x)$ has a nonremovable discontinuity at $x = 1$:
$\lim\limits_{x \to -1^-} f(x) = \lim\limits_{x \to -1^-} (3x - 1) = -4$
$\lim\limits_{x \to -1^+} f(x) = \lim\limits_{x \to -1^+} (x^2 + 5x) = -4$
$\lim\limits_{x \to 1^-} f(x) = \lim\limits_{x \to 1^-} (x^2 + 5x) = 6$
$\lim\limits_{x \to 1^+} f(x) = \lim\limits_{x \to 1^+} (3x^3) = 3$

25. Continuous where $x + 3 \geq 0$, i.e., on $[-3, \infty)$.

27. Continuous where $x + 1 > 0$ i.e., on $(-1, \infty)$.

29. Continuous everywhere, i.e., on $(-\infty, \infty)$.

31.
$\lim\limits_{x \to 0^-} f(x) = \lim\limits_{x \to 0^-} 2\dfrac{\sin x}{x}$
$= 2 \lim\limits_{x \to 0^-} \dfrac{\sin x}{x} = 2$

Hence a must equal 2 if f is continuous.
$\lim\limits_{x \to 0^+} f(x) = \lim\limits_{x \to 0^+} b \cos x$
$= b \lim\limits_{x \to 0^+} \cos x = b,$

so b and a must equal 2 if f is continuous.

33. $\lim\limits_{x \to 0^-} a\sqrt{9 - x^2} = 3a$
$\lim\limits_{x \to 0^+} (\sin bx + 1) = 0 + 1 = 1$
If f is continuous at $x = 0$, then $3a = 1$ or $a = \dfrac{1}{3}$.
$\lim\limits_{x \to 3^-} (\sin bx + 1) = \sin 3b + 1$
$\lim\limits_{x \to 3^+} \sqrt{x - 2} = 1$
If f is continuous at $x = 3$, then $\sin 3b + 1 = 1$ or $\sin 3b = 0$. Thus, $3b = k\pi$ or $b = \frac{k\pi}{3}$ where k is any integer.

35. $\lim\limits_{x \to 10,000^-} T(x) = \lim\limits_{x \to 10,000^-} 0.14x$
$= 0.14(10,000) = 1400$
$\lim\limits_{x \to 10,000^+} T(x) = \lim\limits_{x \to 10,000^+} (c + 0.21x)$
$= c + 0.21(10,000)$
$= c + 2100$
$c + 2100 = 1400$
$c = -700$

A small change in income should not result in a big change in tax, so the tax function should be continuous.

37. For $T(x)$ to be continuous at $x = 141,250$, we must have
$\lim\limits_{x \to 141,250^-} T(x) = \lim\limits_{x \to 141,250^+} T(x)$
$= T(141,250).$

Now
$\lim\limits_{x \to 141,250^-} T(x) = \lim\limits_{x \to 141,250^-} (0.30x - a)$
$= (0.30)(141,250) - 5685$
$= 36,690.$

On the other hand,

$$\lim_{x \to 141,250^+} T(x) = \lim_{x \to 141,250^+} (0.35)x - b$$
$$= (0.35)(141,250) - b$$
$$= 49,437.50 - b.$$

Hence
$b = 49,437.50 - 36,690 = 12,747.50$.
For $T(x)$ to be continuous at $x = 307,050$, we must have

$$\lim_{x \to 307,050^-} T(x) = \lim_{x \to 307,050^+} T(x)$$
$$= T(307,050).$$

Now $\lim_{x \to 307,050^-} T(x)$

$$= \lim_{x \to 307,050^-} (0.35x - b)$$
$$= (0.35)(307,050) - 12,747.5$$
$$= 94,720.$$

On the other hand, $\lim_{x \to 307,050^+} T(x)$

$$= \lim_{x \to 307,050^+} (0.386x - c)$$
$$= (0.386)(307,050) - c$$
$$= 118,521.3 - c.$$

Hence $c = 118,521.3 - 94,720 = 23,801.3$.

39. The first two rows of the following table (together with the Intermediate Value Theorem) show that $f(x)$ has a root in [2, 3]. In the following rows, we use the midpoint of the previous interval as our new x. When $f(x)$ is positive, we use the left half, and when $f(x)$ is negative, we use the right half of the interval. (Because the function goes from negative to positive. If the function went from positive to negative, the intervals would be reversed.)

x	$f(x)$
2	-3
3	2
2.5	-0.75
2.75	0.5625
2.625	-0.109375
2.6875	0.223
2.65625	0.557

The zero is in the interval [2.625, 2.65625].

41. The first two rows of the following table (together with the Intermediate Value Theorem) show that $f(x)$ has a root in $[-1, 0]$. In the following rows, we use the midpoint of the previous interval as our new x. When $f(x)$ is positive, we use the right half, and when $f(x)$ is negative, we use the left half of the interval.

x	$f(x)$
-1	1
0	-2
-0.5	-0.125
-0.75	0.578
-0.625	0.256
-0.5625	0.072
-0.53125	-0.025

The zero is in the interval $[-0.5625, -0.53125]$.

43. The first two rows of the following table (together with the Intermediate Value Theorem) show that $f(x)$ has a root in [0, 1]. In the following rows, we use the midpoint of the previous interval as our new x. When $f(x)$ is positive, we use the right half, and when $f(x)$ is negative, we use the left half of the interval.

x	$f(x)$
0	1
1	-0.46
0.5	0.378
0.75	-0.018
0.625	0.186
0.6875	0.085
0.71875	0.034

The zero is in the interval [0.71875, 0.75].

45. $\lim_{x \to 2^+} f(x) = \lim_{x \to 2^+} (3x - 1) = 5$
$f(2) = 3(2) - 1 = 5$

Thus $f(x)$ is continuous from the right at $x = 2$.

47. $\lim\limits_{x \to 2^+} f(x) = \lim\limits_{x \to 2^+} (3x - 3) = 3$

$f(2) = 2^2 = 4$

Thus $f(x)$ is not continuous from the right at $x = 2$.

49. A function is continuous from the left at $x = a$ if $\lim\limits_{x \to a^-} f(x) = f(a)$.

    (a) $\lim\limits_{x \to 2^-} f(x) = \lim\limits_{x \to 2^-} x^2 = 4$ $f(2) = 5$ Thus $f(x)$ is not continuous from the left at $x = 2$.

    (b) $\lim\limits_{x \to 2^-} f(x) = \lim\limits_{x \to 2^-} x^2 = 4$ $f(2) = 3$ Thus $f(x)$ is not continuous from the left at $x = 2$.

    (c) $\lim\limits_{x \to 2^-} f(x) = \lim\limits_{x \to 2^-} x^2 = 4$ $f(2) = 4$ Thus $f(x)$ is continuous from the left at $x = 2$.

    (d) $f(x)$ is not continuous from the left at $x = 2$ because $f(2)$ is undefined.

51. Need $g(30) = 100$ and $g(34) = 0$. We may take $g(T)$ to be linear. $m = \dfrac{0 - 100}{34 - 30} = -25$.

$y = -25(x - 30) + 100$ $g(T)$
$= 100 - 25(T - 30)$.

53.

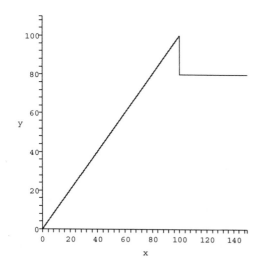

The graph is discontinuous at $x = 100$. This is when the box starts moving.

55. Let $M(t)$ be her distance from home as a function of time on Monday. Let $T(t)$ be her distance from home as a function of time on Tuesday. M (1:59 P.M.) $- T$ (1:59 P.M.) $= M$ (1:59 P.M.) > 0
M (7:13 A.M.) $- T$ (7:13 A.M.) $= T$ (7:13 A.M.) < 0. By the Intermediate Value Theorem, there is a time between 7:13 A.M. and 1:59 P.M. for which $M(t) - T(t) = 0$. At this time, her distance from home was the same on both days.

57.

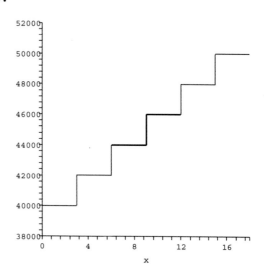

The function $s(t)$ has jump discontinuities every three months when the salary suddenly increases by $2000. In the function $f(t)$, the $2000 increase occurs gradually over the 3 month period, so $f(t)$ is continuous. It might be easier to do calculations with $f(t)$ because it is continuous and because it is given by a simpler formula.

59. We already know $f(x) \neq 0$ for $a < x < b$. Suppose $f(d) < 0$ for some d, $a < d < b$. Then by the Intermediate Value Theorem, there is an e in the interval $[c, d]$ such that $f(e) = 0$. But this e would also be between a and b, which is impossible. Thus, $f(x) > 0$ for all $a < x < b$.

61. $\lim\limits_{x \to 0} x f(x) = \lim\limits_{x \to 0} x \lim\limits_{x \to 0} f(x)$

                   $= 0 f(0) = 0$

63.
$$\lim_{x\to a} g(x) = \lim_{x\to a} |f(x)| = \left|\lim_{x\to a} f(x)\right|$$
$$= |f(a)| = g(a).$$

65. Let $b \geq a$. Then

$$\lim_{x\to b} h(x) = \lim_{x\to b} \left(\max_{a\leq t\leq b} f(t)\right)$$
$$= \max_{a\leq t\leq b} \left(\lim_{t\to b} f(t)\right)$$
$$= h(b)$$

since f is continuous. Thus, h is continuous for $x \geq a$.

No, the property would not be true if f were not assumed to be continuous. A counterexample is

$$f(x) = \begin{cases} 1 & \text{if } a \leq x < b \\ 2 & \text{if } b \leq x \end{cases}.$$

Then $h(x) = 1$ for $a \leq x < b$, and $h(x) = 2$ for $x \geq b$. Thus, h is not continuous at $x = b$.

1.5 LIMITS INVOLVING INFINITY

1. **(a)** $\displaystyle\lim_{x\to 1^-} \frac{1-2x}{x^2-1} = \infty$

(b) $\displaystyle\lim_{x\to 1^+} \frac{1-2x}{x^2-1} = -\infty$

(c) Does not exist.

3. **(a)** $\displaystyle\lim_{x\to 2^-} \frac{x-4}{x^2-4x+4} = -\infty$

(b) $\displaystyle\lim_{x\to 2^+} \frac{x-4}{x^2-4x+4} = -\infty$

(c) $\displaystyle\lim_{x\to 2} \frac{x-4}{x^2-4x+4} = -\infty$

5. $\displaystyle\lim_{x\to 2^-} \frac{-x}{\sqrt{4-x^2}} = -\infty$

As x approaches 2 from below, the numerator is near -2 and the denominator is small and positive, so the fraction goes to $-\infty$.

7. $\displaystyle\lim_{x\to -\infty} \frac{-x}{\sqrt{4+x^2}} = \lim_{x\to -\infty} \frac{-x}{-x\sqrt{\frac{4}{x^2}+1}} =$

$$\lim_{x\to -\infty} \frac{1}{\sqrt{\frac{4}{x^2}+1}} = \frac{1}{\sqrt{1}} = 1$$

9. $\displaystyle\lim_{x\to \infty} \frac{x^3-2}{3x^2+4x-1} = \lim_{x\to \infty} \frac{x^2\left(x-\frac{2}{x^2}\right)}{x^2\left(3+\frac{4}{x}-\frac{1}{x^2}\right)}$

$$= \lim_{x\to \infty} \frac{\left(x-\frac{2}{x^2}\right)}{3+\frac{4}{x}-\frac{1}{x^2}} = \infty$$

11. $\displaystyle\lim_{x\to \infty} \frac{2x^2-x+1}{4x^2-3x-1} = \lim_{x\to \infty} \left[\frac{2x^2-x+1}{4x^2-3x-1} \cdot \frac{(1/x^2)}{(1/x^2)}\right]$

$$= \lim_{x\to \infty} \frac{2-\frac{1}{x}+\frac{1}{x^2}}{4-\frac{3}{x}-\frac{1}{x^2}}$$

$$= \frac{2}{4} = \frac{1}{2} = 0.5$$

13. $\displaystyle\lim_{x\to \infty} \sin 2x$ does not exist because the values of $y = \sin 2x$ continue to oscillate between -1 and 1 as $x \to \infty$.

15. $\displaystyle\lim_{x\to 0^+} \frac{3-\frac{2}{x}}{2+\frac{1}{x}} = \lim_{x\to 0^+} \left[\frac{3-\frac{2}{x}}{2+\frac{1}{x}} \cdot \frac{x}{x}\right]$

$$\lim_{\to 0^+} \frac{3x-2}{2x+1}$$

$$= \frac{3\cdot 0-2}{2\cdot 0+1} = -2$$

17. $\displaystyle\lim_{x\to \infty} \frac{3x+\sin x}{4x-\cos 2x} = \frac{3}{4} = 0.75$

Note that $\displaystyle\frac{3x-1}{4x+1} \leq \frac{3x+\sin x}{4x-\cos 2x} \leq \frac{3x+1}{4x-1}$

for all x and that $\displaystyle\lim_{x\to \infty} \frac{3x-1}{4x+1} = \lim_{x\to \infty} \frac{3x+1}{4x-1}$

$= \dfrac{3}{4}$. The result follows by the Squeeze Theorem.

19.
$$\lim_{x \to \pi/2^+} \frac{\tan x - x}{\tan^2 x + 3} = \lim_{x \to \pi/2^+} \left[\frac{\tan x - x}{\tan^2 x + 3} \cdot \frac{\cos^2 x}{\cos^2 x} \right]$$

$$= \lim_{x \to \pi/2^+} \frac{\sin x \cos x - x \cos^2 x}{\sin^2 x + 3 \cos^2 x}$$

$$= \frac{1 \cdot 0 - \frac{\pi}{2} \cdot 0^2}{1^2 + 3 \cdot 0^2} = 0$$

21. Since $4 + x^2$ is never 0, there are no vertical asymptotes. We have $\lim\limits_{x \to \infty} \dfrac{x}{\sqrt{4 + x^2}}$

$$= \lim_{x \to \infty} \frac{x}{x \sqrt{\frac{4}{x^2} + 1}}$$

$$= \lim_{x \to \infty} \frac{1}{\sqrt{\frac{4}{x^2} + 1}}$$

$$= \frac{1}{\sqrt{1}} = 1$$

and

$$\lim_{x \to -\infty} \frac{x}{\sqrt{4 + x^2}}$$

$$= \lim_{x \to -\infty} \frac{x}{-x \sqrt{\frac{4}{x^2} + 1}}$$

$$= \lim_{x \to -\infty} \frac{-1}{\sqrt{\frac{4}{x^2} + 1}}$$

$$= \frac{-1}{\sqrt{1}} = -1,$$

so there are horizontal asymptotes at $y = 1$ and $y = -1$.

23. $4 - x^2 = 0 \Rightarrow 4 = x^2$ so we have vertical asymptotes at $x = \pm 2$. We have

$$\lim_{x \to \pm\infty} \frac{x}{4 - x^2}$$

$$= \lim_{x \to \pm\infty} \frac{x}{x^2 \left(\frac{4}{x^2} - 1 \right)}$$

$$= \lim_{x \to \pm\infty} \frac{1}{x \left(\frac{4}{x^2} - 1 \right)} = 0.$$

So there is a horizontal asymptote at $y = 0$.
$f(x) \to \infty$ as $x \to -2^-$
$f(x) \to -\infty$ as $x \to -2^+$
$f(x) \to \infty$ as $x \to 2^-$
$f(x) \to \infty$ as $x \to 2^+$

25. The denominator factors: $x^2 - 2x - 3 = (x - 3)(x + 1)$. Since neither $x = 3$ nor $x = -1$ are zeros of the numerator, we see that $f(x)$ has vertical asymptotes at $x = 3$ and $x = -1$.

$f(x) \to -\infty$ as $x \to 3^-$, $f(x) \to \infty$ as $x \to 3^+$, $f(x) \to \infty$ as $x \to -1^-$, $f(x) \to -\infty$ as $x \to -1^+$.

We have

$$\lim_{x \to \pm\infty} \frac{3x^2 + 1}{x^2 - 2x - 3}$$

$$\lim_{x \to \pm\infty} \frac{3 + 1/x^2}{1 - 2/x - 3/x^2} = 3.$$

So there is a horizontal asymptote at $y = 3$.

27. Note that $0 \le 1 - \cos x \le 2$ for all x. For this range of values, the cotangent function is undefined only for 0, so $f(x) = \cot(1 - \cos x)$ will have a vertical asymptote whenever $1 - \cos x = 0$, i.e., whenever $\cos x = 1$. This happens when $x = 2k\pi$ for any integer k. As $f(x)$ approaches any of these asymptotes (from either side), it behaves like $\cot x$ approaching 0 from the right, so $f(x) \to \infty$ as x approaches any of these asymptotes from either side. No horizontal asymptotes.

29. The function is continuous for all x except 0, but

$$\lim_{x \to 0} \frac{4 \sin x}{x} = 4 \lim_{x \to 0} \frac{\sin x}{x} = 4 \cdot 1 = 4$$

so there are no vertical asymptotes. We have

$$\lim_{x \to \infty} \frac{4 \sin x}{x} = 0 \text{ and } \lim_{x \to -\infty} \frac{4 \sin x}{x} = 0$$

since $\dfrac{4 \sin x}{x}$ can be bounded between $-\dfrac{4}{x}$ and $\dfrac{4}{x}$ and

$$\lim_{x \to \infty} \left[-\frac{4}{x} \right] = \lim_{x \to \infty} \left[-\frac{4}{x} \right] = \lim_{x \to \infty} \frac{4}{x}$$

$$= \lim_{x \to -\infty} \frac{4}{x} = 0.$$

Thus, $y = 0$ is a horizontal asymptote.

31. Vertical asymptotes at $x = \pm 2$. The slant asymptote is $y = -x$.

33. Vertical asymptotes at

$$x = \frac{-1 \pm \sqrt{17}}{2}.$$

The slant asymptote is $y = x - 1$.

35. When x is large, the value of the fraction is close to 1.

37. When x is close to -1, the value of the fraction is close to 1.

39. We multiply by

$$\frac{\sqrt{4x^2 - 2x + 1} + 2x}{\sqrt{4x^2 - 2x + 1} + 2x}$$

to get:

$$\lim_{x \to \infty} (\sqrt{4x^2 - 2x + 1} - 2x)$$
$$= \lim_{x \to \infty} \frac{-2x + 1}{\sqrt{4x^2 - 2x + 1} + 2x} \cdot \frac{1/x}{1/x}$$
$$= \lim_{x \to \infty} \frac{-2 + 1/x}{\sqrt{4 - 2/x + 1/x^2} + 2}$$
$$= \frac{-2}{\sqrt{4} + 2} = -\frac{1}{2}.$$

41. $\lim_{x \to \infty} (\sqrt{5x^2 + 4x + 7} - \sqrt{5x^2 + x + 3})$

If we multiply by

$$\frac{\sqrt{5x^2 + 4x + 7} + \sqrt{5x^2 + x + 3}}{\sqrt{5x^2 + 4x + 7} + \sqrt{5x^2 + x + 3}},$$

we get

$$\lim_{x \to \infty} \frac{(5x^2 + 4x + 7) - (5x^2 + x + 3)}{\sqrt{5x^2 + 4x + 7} + \sqrt{5x^2 + x + 3})} =$$
$$\lim_{x \to \infty} \frac{3x + 4}{\sqrt{5x^2 + 4x + 7} + \sqrt{5x^2 + x + 3}} =$$
$$\lim_{x \to \infty} \frac{3 + \frac{4}{x}}{\sqrt{5 + \frac{4}{x} + \frac{7}{x^2}} + \sqrt{5 + \frac{1}{x} + \frac{3}{x^2}}}$$
$$= \frac{3}{2\sqrt{5}} = \frac{3\sqrt{5}}{10}.$$

43. $\lim_{x \to \infty} f(x) = \lim_{x \to 0^+} f(1/x)$ because $1/x \to \infty$ as $x \to 0^+$. $\lim_{x \to -\infty} f(x) = \lim_{x \to 0^-} f(1/x)$ because $1/x \to -\infty$ as $x \to 0^-$.

45. $\lim_{x \to 0^+} \frac{80x^{-0.3} + 60}{2x^{-0.3} + 5} \left(\frac{x^{0.3}}{x^{0.3}} \right)$
$$= \lim_{x \to 0^+} \frac{80 + 60x^{0.3}}{2 + 5x^{0.3}}$$
$$= \frac{80}{2} = 40 \text{ mm}$$

$$\lim_{x \to \infty} \frac{80x^{-0.3} + 60}{2x^{-0.3} + 5} = \frac{60}{5} = 12 \text{ mm}$$

47. $f(x) = \dfrac{80x^{-0.3} + 60}{10x^{-0.3} + 30}$

49. $\lim_{t \to \infty} v_N = \lim_{t \to \infty} \dfrac{Ft}{m} = \infty$

$$\lim_{t \to \infty} v_E = \lim_{t \to \infty} \frac{Fct}{\sqrt{m^2c^2 + F^2t^2}}$$
$$= \lim_{t \to \infty} \frac{Fct}{t\sqrt{\frac{m^2c^2}{t^2} + F^2}}$$
$$= \lim_{t \to \infty} \frac{Fc}{\sqrt{\frac{m^2c^2}{t^2} + F^2}}$$
$$= \frac{Fc}{\sqrt{F^2}} = c$$

51. We must restrict the domain to $v_0 \geq 0$ because the formula makes sense only if the rocket is launched upward. To find v_e, set $19.6R - v_0^2 = 0$. We get $v_0 = \sqrt{19.6R}$. If the rocket is launched with initial velocity $\geq v_e$, it will never return to earth; hence v_e is called the escape velocity.

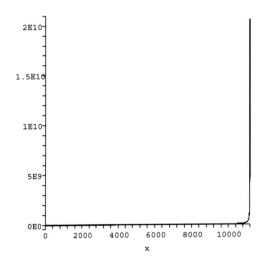

53. Suppose the degree of q is n. If we divide both $p(x)$ and $q(x)$ by x^n, then the new denominator will approach a constant while the new numerator tends to ∞, so there is no horizontal asymptote.

55. When we do long division, we get a remainder of $x + 2$, so the degree of p is one greater than the degree of q.

57. The function $q(x) = -2(x - 3)^2$ satisfies the given conditions.

59. True.

61. False.

63. True.

65. Vertical asymptote at $x = 2$. Horizontal asymptotes at $y = 4$ and $y = 0$.

67. As $t \to \infty$, $\dfrac{1}{t^2 + 1} \to 0$ and $\sin(at)$ oscillates between -1 and 1 for any positive constant a. Thus, $\frac{1}{t^2+1} \sin(at) \to 0$ as $t \to \infty$. In the following graph, we see that suspension system A damps out at about 8.4 seconds, while system B takes about 2.0 seconds to damp out.

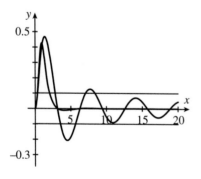

69. $g(x) = \sin x, h(x) = x$ at $a = 0$

1.6 FORMAL DEFINITION OF THE LIMIT

1. **(a)** $|(x^2 + 1) - 1| < 0.1 \Longrightarrow |x^2| < 0.1 \Longrightarrow |x| < \sqrt{0.1}$
Let $\delta = \sqrt{0.1} \approx 0.32$.

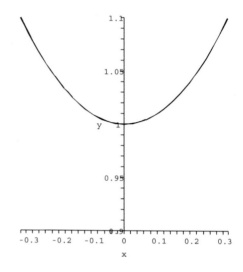

(b) $|(x^2 + 1) - 1| < 0.05 \Longrightarrow |x^2| < 0.05 \Longrightarrow |x| < \sqrt{0.05}$
Let $\delta = \sqrt{0.05} \approx 0.22$.

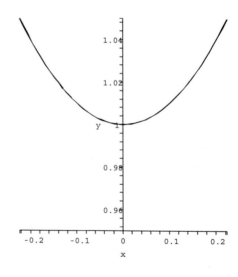

3. **(a)** From the graph, we determine that we can take $\delta = 0.45$, as shown below.

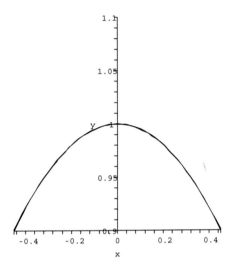

(b) From the graph, we determine that we can take $\delta = 0.31$, as shown below.

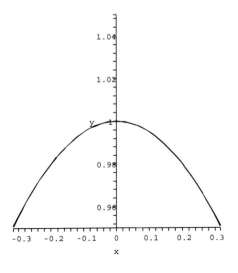

5. **(a)** From the graph, we determine that we can take $\delta = 0.39$, as shown below.

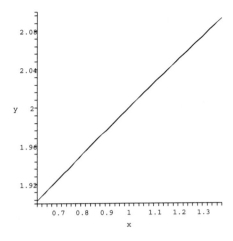

(b) From the graph, we determine that we can take $\delta = 0.19$, as shown below.

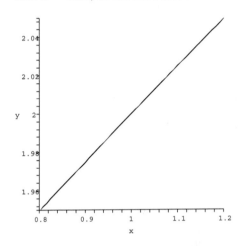

7. **(a)** From the graph, we determine that we can take $\delta = 0.02$, as shown below.

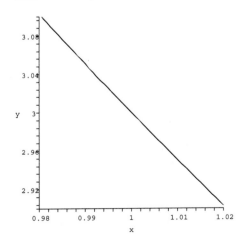

(b) From the graph, we determine that we can take $\delta = 0.009$, as shown below.

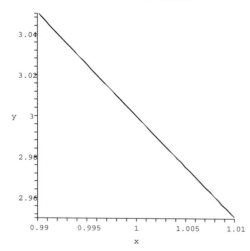

9. We want $|3x - 0| < \varepsilon \Leftrightarrow 3|x| < \varepsilon \Leftrightarrow |x| = |x - 0| < \varepsilon/3$ Take $\delta = \varepsilon/3$.

11. We want $|3x + 2 - 8| < \varepsilon \Leftrightarrow |3x - 6| < \varepsilon \Leftrightarrow 3|x - 2| < \varepsilon \Leftrightarrow |x - 2| < \varepsilon/3$. Take $\delta = \varepsilon/3$.

13. We want $|3 - 4x - (-1)| < \varepsilon \Leftrightarrow |-4x + 4| < \varepsilon \Leftrightarrow 4|-x + 1| < \varepsilon \Leftrightarrow 4|x - 1| < \varepsilon \Leftrightarrow |x - 1| < \varepsilon/4$. Take $\delta = \varepsilon/4$.

15. We want $\left| \dfrac{x^2 + x - 2}{x - 1} - 3 \right| < \varepsilon$. We have

$$\left| \frac{x^2 + x - 2}{x - 1} - 3 \right|$$
$$= \left| \frac{(x + 2)(x - 1)}{x - 1} - 3 \right|$$
$$= |x + 2 - 3|$$
$$= |x - 1|.$$

Take $\delta = \varepsilon$.

17. We want $\left| x^2 - 1 - 0 \right| < \varepsilon$. We have $|x^2 - 1| = |x - 1||x + 1|$. We require that $\delta < 1$, i.e., $|x - 1| < 1$ so $0 < x < 2$ and $|x + 1| < 3$. Then $|x^2 - 1| = |x - 1||x + 1| < 3|x - 1|$. Requiring this to be less than ε gives $|x - 1| < \varepsilon/3$, so $\delta = \min\{1, \varepsilon/3\}$.

19. We want $\left| x^2 - 1 - 3 \right| < \varepsilon$. We have $|x^2 - 4| = |x - 2||x + 2|$. We require that $\delta < 1$, i.e.,

$|x - 2| < 1$ so $1 < x < 3$ and $|x + 2| < 5$. Then $|x^2 - 4| = |x - 2||x + 2| < 5|x - 2|$. Requiring this to be less than ε gives $|x - 2| < \varepsilon/5$, so $\delta = \min\{1, \varepsilon/5\}$.

21. Let $f(x) = mx + b$. Since $f(x)$ is continuous, we know that $\lim\limits_{x \to a} f(x) = ma + b$. So we want to find a δ which forces $|mx + b - (ma + b)| < \varepsilon$. But $|mx + b - (ma + b)| = |mx - ma| = |m||x - a|$. So as long as $|x - a| < \delta = \varepsilon/|m|$, we will have $|f(x) - (ma + b)| < \varepsilon$. This δ clearly does not depend on a. This is due to the fact that $f(x)$ is a linear function, so the slope is constant, which means that the ratio of the change in y to the change in x is constant.

23. For a function $f(x)$ defined on some open interval (c, a) we say

$$\lim_{x \to a^-} f(x) = L$$

if, given any number $\varepsilon > 0$, there is another number $\delta > 0$ such that whenever $x \in (c, a)$ and $a - \delta < x < a$, we have $|f(x) - L| < \varepsilon$.

For a function $f(x)$ defined on some open interval (a, c) we say

$$\lim_{x \to a^+} f(x) = L$$

if, given any number $\varepsilon > 0$, there is another number $\delta > 0$ such that whenever $x \in (a, c)$ and $a < x < a + \delta$, we have $|f(x) - L| < \varepsilon$.

25. As $x \to 1^+$, $x - 1 > 0$ so we compute

$$\frac{2}{x - 1} > 100$$
$$2 > 100(x - 1)$$
$$\frac{2}{100} > x - 1$$

So take $\delta = 2/100 = 0.02$.

27. We look at the graph of $\cot x$ as $x \to 0^+$ and we find that we should take $\delta = 0.0095$.

29. As $x \to 2^-$, $4 - x^2 > 0$ so we compute

$$\frac{2}{\sqrt{4-x^2}} > 100$$

$$2 > 100\sqrt{4-x^2}$$

$$\frac{2}{100} > \sqrt{4-x^2}$$

$$\frac{4}{10,000} > 4 - x^2 = (2-x)(2+x)$$

Take $\delta < 1$ so that $1 < x < 2$ so we have $2 + x < 4$. Then $(2-x)(2+x) < (2-x)4$. Now, if we require $|x-2| < \frac{4}{40,000}$ then $\frac{2}{\sqrt{4-x^2}} > 100$. So let $\delta = \frac{4}{40,000} = 0.0001$.

31. We want M such that if $x > M$,

$$\left| \frac{x^2-2}{x^2+x+1} - 1 \right| < 0.1$$

We have

$$\left| \frac{x^2-2}{x^2+x+1} - 1 \right|$$

$$= \left| \frac{x^2-2-(x^2+x+1)}{x^2+x+1} \right|$$

$$= \left| \frac{-x-3}{x^2+x+1} \right|$$

$$= \left| \frac{x+3}{x^2+x+1} \right|$$

Now, as long as $x > 3$, we have

$$\left| \frac{x+3}{x^2+x+1} \right| < \left| \frac{2x}{x^2+x} \right|$$

$$= \left| \frac{2}{x+1} \right|$$

We want $\left| \frac{2}{x+1} \right| < 0.1$. Since $x \to \infty$, we can take $x > 0$, so we solve $\frac{2}{x+1} < 0.1$ to get $x > 19$, i.e., $M = 19$. Other values of M are possible.

33. We have

$$\left| \frac{x^2+3}{4x^2-4} - \frac{1}{4} \right| = \left| \frac{x^2+3-(x^2-1)}{4x^2-4} \right|$$

$$= \left| \frac{4}{4x^2-4} \right|$$

$$= \left| \frac{1}{x^2-1} \right|$$

Since $x \to -\infty$, we may take $x < -1$ so that $x^2 - 1 > 0$. We now need $\frac{1}{x^2-1} < 0.1$. Solving for x gives $|x| > \sqrt{11} \approx 3.3166$. So we can take $N = -3.4$.

35. We want M such that if $x > M$,

$$\left| \frac{x}{\sqrt{x^2+10}} - 1 \right| < 0.1.$$

Since $\frac{x}{\sqrt{x^2+10}} < 1$ for all x, we have

$$\left| \frac{x}{\sqrt{x^2+10}} - 1 \right| = 1 - \frac{x}{\sqrt{x^2+10}}.$$

$$1 - \frac{x}{\sqrt{x^2+10}} < 0.1$$

$$\frac{x}{\sqrt{x^2+10}} > 0.9$$

$$x > 0.9\sqrt{x^2+10}$$

$$x^2 > 0.81(x^2+10)$$

$$0.19x^2 > 8.1$$

$$x > \sqrt{\frac{8.1}{0.19}} \approx 6.5$$

We take $M = 7$.

37. Let $\varepsilon > 0$ be given and let $M = \sqrt[3]{2/\varepsilon}$. Then if $x > M$,

$$\left| \frac{2}{x^3} \right| < \left| \frac{2}{\left(\sqrt[3]{2/\varepsilon} \right)^3} \right| = \varepsilon.$$

39. Let $\varepsilon > 0$ be given and let $M = \varepsilon^{-1/k}$. Then if $x > M$,

$$\left| \frac{1}{x^k} \right| < \left| \frac{1}{\left(\varepsilon^{-1/k} \right)^k} \right| = \varepsilon.$$

41. Let $\varepsilon > 0$ be given and assume $\varepsilon \le 1/2$. Let $N = -(\frac{1}{\varepsilon} - 2)^{1/2}$. Then if $x < N$,

$$\left| \frac{1}{x^2 + 2} - 3 - (-3) \right| = \left| \frac{1}{x^2 + 2} \right|$$

$$< \left| \frac{1}{\left(-(\frac{1}{\varepsilon} - 2)^{1/2} \right)^2 + 2} \right| = \varepsilon.$$

43. Let $N < 0$ be given and let $\delta = \sqrt[4]{-2/N}$. Then for any x such that $|x + 3| < \delta$,

$$\left| \frac{-2}{(x + 3)^4} \right| > \left| \frac{-2}{(\sqrt[4]{-2/N})^4} \right| = |N|.$$

45. Let $M > 0$ be given and let $\delta = \sqrt{4/M}$. Then for any x such that $|x - 5| < \delta$,

$$\left| \frac{4}{(x - 5)^2} \right| > \left| \frac{4}{(\sqrt{4/M})^2} \right| = |M|.$$

47. We observe that $\lim\limits_{x \to 1^-} f(x) = 2$ and $\lim\limits_{x \to 1^+} f(x) = 4$. For any $x \in (1, 2)$,

$$|f(x) - 2| = |x^2 + 3 - 2| = |x^2 + 1| > 2.$$

So if $\varepsilon \le 2$, there is no $\delta > 0$ to satisfy the definition of limit.

49. We observe that $\lim\limits_{x \to 1^-} f(x) = 2$ and $\lim\limits_{x \to 1^+} f(x) = 4$. For any $x \in (1, \sqrt{2})$,

$$|f(x) - 2| = |5 - x^2 - 2|$$
$$= |3 - x^2| > |3 - (\sqrt{2})^2| = 1.$$

So if $\varepsilon \le 1$, there is no $\delta > 0$ to satisfy the definition of limit. Other values of ε are possible.

51. We want to find, for any given $\varepsilon > 0$, a $\delta > 0$ such that whenever $0 < |r - 2| < \delta$, we have $|2r^2 - 8| < \varepsilon$. We see that

$$|2r^2 - 8| = 2|r^2 - 4| = 2|r - 2||r + 2|.$$

Since we want a radius close to 2, we may take $|r - 2| < 1$ which implies $|r + 2| < 5$ and so

$$|2r^2 - 8| < 10|r - 2|$$

whenever $|r - 2| < 1$. If we then take $\delta = \min\{1, \varepsilon/10\}$, we see that whenever $0 < |r - 2| < \delta$, we have

$$|2r^2 - 8| < 10 \cdot \delta \le 10 \cdot \frac{\varepsilon}{10} = \varepsilon.$$

53. Let $L = \lim\limits_{x \to a} f(x)$. Given any $\varepsilon > 0$, we know there exists $\delta > 0$ such that whenever $0 < |x - a| < \delta$, we have

$$|f(x) - L| < \frac{\varepsilon}{|c|}.$$

Here, we can take $\varepsilon/|c|$ instead of ε because there is such a δ for *any* ε, including $\varepsilon/|c|$. But now we have

$$|c \cdot f(x) - c \cdot L| = |c| \cdot |f(x) - L|$$
$$< |c| \cdot \frac{\varepsilon}{|c|} = \varepsilon.$$

Therefore, $\lim\limits_{x \to a} [c \cdot f(x)] = c \cdot L$, as desired.

55. Let $\varepsilon > 0$ be given. Since $\lim\limits_{x \to a} f(x) = L$, there exists $\delta_1 > 0$ such that whenever $0 < |x - a| < \delta_1$, we have

$$|f(x) - L| < \varepsilon.$$

In particular, we know that

$$L - \varepsilon < f(x).$$

Similarly, since $\lim\limits_{x \to a} h(x) = L$, there exists $\delta_2 > 0$ such that whenever $0 < |x - a| < \delta_2$, we have

$$|h(x) - L| < \varepsilon.$$

In particular, we know that

$$h(x) < L + \varepsilon.$$

Let $\delta = \min\{\delta_1, \delta_2\}$. Then whenever $0 < |x - a| < \delta$, we have

$$L - \varepsilon < f(x) \le g(x) \le h(x) < L + \varepsilon.$$

Therefore

$$|g(x) - L| < \varepsilon$$

and so $\lim\limits_{x \to a} g(x) = L$ as desired.

57. Let $\varepsilon > 0$ be given. If $\lim\limits_{x \to a} f(x) = L$, then there exists $\delta > 0$ such that if $0 < |x - a| < \delta$, then $|f(x) - L| < \varepsilon$. Since $|f(x) - L - 0| = |f(x) - L|$, this is precisely what we need to see that $\lim\limits_{x \to a} [f(x) - L] = 0$.

59. If $2 < x < \sqrt{4.1}$ then $4 < x^2 < 4.1$ so (for $x \in (2, \sqrt{4.1})$), $x^2 - 4 < 4.1 - 4 = 0.1$.

If $\sqrt{3.9} < x < 2$ then $3.9 < x^2 < 4$ so (for $x \in (\sqrt{3.9}, 2)$), $x^2 - 4 > 3.9 - 4 = -0.1$.

For the limit definition, we need to take $\delta = \min\{\delta_1, \delta_2\} = \delta_1$ to ensure that x^2 is within 0.1 of 4 on *both* sides of $x = 2$.

1.7 LIMITS AND LOSS-OF-SIGNIFICANCE ERRORS

1. The limit is $\frac{1}{4}$.

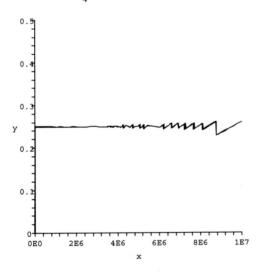

We can rewrite the function as

$$f(x) = x(\sqrt{4x^2 + 1} - 2x) \cdot \frac{\sqrt{4x^2 + 1} + 2x}{\sqrt{4x^2 + 1} + 2x}$$

$$= \frac{x(4x^2 + 1 - 4x^2)}{\sqrt{4x^2 + 1} + 2x}$$

$$= \frac{x}{\sqrt{4x^2 + 1} + 2x}$$

to avoid loss-of-significance errors.

In the table below, the middle column contains values calculated using $f(x) = x(\sqrt{4x^2 + 1} - 2x)$, while the third column contains values calculated using the rewritten $f(x)$.

x	Old $f(x)$	New $f(x)$
1	0.236068	0.236068
10	0.249844	0.249844
100	0.249998	0.249998
1000	0.250000	0.250000
10000	0.250000	0.250000
100000	0.249999	0.250000
1000000	0.250060	0.250000
10000000	0.260770	0.250000
100000000	0.000000	0.250000
1000000000	0.000000	0.250000

3. The limit is 1.

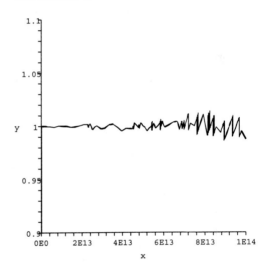

We can rewrite the function as

$$\sqrt{x}(\sqrt{x+4} - \sqrt{x+2}) \cdot \frac{\sqrt{x+4} + \sqrt{x+2}}{\sqrt{x+4} + \sqrt{x+2}}$$

$$= \frac{\sqrt{x}[(x+4) - (x+2)]}{\sqrt{x+4} + \sqrt{x+2}}$$

$$= \frac{2\sqrt{x}}{\sqrt{x+4} + \sqrt{x+2}}$$

to avoid loss-of-significance errors.

In the table below, the middle column contains values calculated using $f(x) = \sqrt{x}\left(\sqrt{x+4}\right.$

$-\sqrt{x+2}\Big)$, while the third column contains values calculated using the rewritten $f(x)$.

x	Old $f(x)$	New $f(x)$
1	0.504017	0.504017
10	0.877708	0.877708
100	0.985341	0.985341
1000	0.998503	0.998503
10000	0.999850	0.999850
100000	0.999985	0.999985
1000000	0.999998	0.999999
10000000	1.000000	1.000000
100000000	1.000000	1.000000
1000000000	1.000000	1.000000
10000000000	1.000000	1.000000
1E+11	0.999990	1.000000
1E+12	1.000008	1.000000
1E+13	0.999862	1.000000
1E+14	0.987202	1.000000
1E+15	0.942432	1.000000
1E+16	0.000000	1.000000
1E+17	0.000000	1.000000

5. The limit is 1.

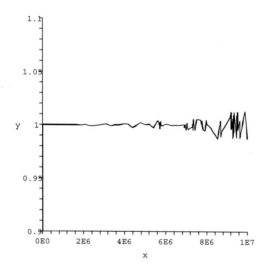

We can multiply $f(x)$ by

$$\frac{\sqrt{x^2+4}+\sqrt{x^2+2}}{\sqrt{x^2+4}+\sqrt{x^2+2}}$$

to rewrite the function as

$$\frac{x[x^2+4-(x^2+2)]}{\sqrt{x^2+4}+\sqrt{x^2+2}}=\frac{2x}{\sqrt{x^2+4}+\sqrt{x^2+2}}$$

to avoid loss-of-significance errors.

In the table below, the middle column contains values calculated using $f(x)=x\left(\sqrt{x^2+4}-\sqrt{x^2+2}\right)$, while the third column contains values calculated using the rewritten $f(x)$.

x	Old $f(x)$	New $f(x)$
1	0.504017	0.504017
10	0.985341	0.985341
100	0.999850	0.999850
1000	0.999998	0.999999
10000	1.000000	1.000000
100000	1.000000	1.000000
1000000	1.000008	1.000000
10000000	0.987202	1.000000
100000000	0.000000	1.000000
1000000000	0.000000	1.000000

7. The limit is 1/6.

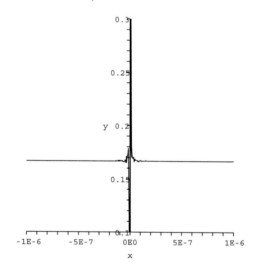

We can rewrite the function as

$$\frac{1-\cos 2x}{12x^2}\cdot\frac{1+\cos 2x}{1+\cos 2x}=\frac{\sin^2 2x}{12x^2(1+\cos 2x)}$$

to avoid loss-of-significance errors.

In the table below, the middle column contains values calculated using $f(x)=\dfrac{1-\cos 2x}{12x^2}$, while the third column contains values calculated using the rewritten $f(x)$. Note that $f(x) = f(-x)$ and so we get the same values when x is negative (which allows us to conjecture the two-sided limit as $x \to 0$).

x	Old $f(x)$	New $f(x)$
1	0.118012	0.118012
0.1	0.166112	0.166112
0.01	0.166661	0.166661
0.001	0.166667	0.166667
0.0001	0.166667	0.166667
0.00001	0.166667	0.166667
0.000001	0.166663	0.166667
0.0000001	0.166533	0.166667
0.00000001	0.185037	0.166667
0.000000001	0	0.166667
1E-10	0	0.166667

9. The limit is $\frac{1}{2}$.

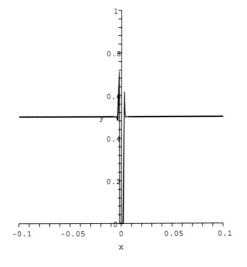

We can rewrite the function as
$$\frac{1 - \cos x^3}{x^6} \cdot \frac{1 + \cos x^3}{1 + \cos x^3}$$
$$= \frac{\sin^2(x^3)}{x^6(1 + \cos x^3)}$$
to avoid loss-of-significance errors.

In the table below, the middle column contains values calculated using $f(x) = \frac{1 - \cos x^3}{x^6}$, while the third column contains values calculated using the rewritten $f(x)$. Note that $f(x) = f(-x)$ and so we get the same values when x

is negative (which allows us to conjecture the two-sided limit as $x \to 0$).

x	Old $f(x)$	New $f(x)$
1	0.459698	0.459698
0.1	0.500000	0.500000
0.01	0.500044	0.500000
0.001	0.000000	0.500000
0.0001	0.000000	0.500000

11. The limit is 2/3.

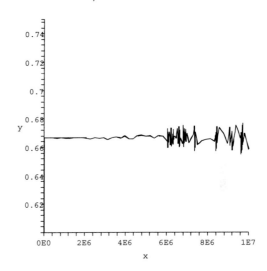

We can multiply $f(x)$ by
$$1 = \frac{g(x)}{g(x)}$$
where
$$g(x) = (x^2 + 1)^{\frac{2}{3}} + (x^2 + 1)^{\frac{1}{3}}(x^2 - 1)^{\frac{1}{3}} + (x^2 - 1)^{\frac{2}{3}}$$
to rewrite the function as
$$\frac{2x^{4/3}}{g(x)}$$
to avoid loss-of-significance errors.

In the table below, the middle column contains values calculated using
$$f(x) = x^{4/3}(\sqrt[3]{x^2 + 1} - \sqrt[3]{x^2 - 1}),$$
while the third column contains values calculated using the rewritten $f(x)$.

x	Old $f(x)$	New $f(x)$
1	1.259921	1.259921
10	0.666679	0.666679
100	0.666667	0.666667
1000	0.666667	0.666667
10000	0.666668	0.666667
100000	0.666532	0.666667
1000000	0.63	0.666667
10000000	2.154435	0.666667
100000000	0.000000	0.666667
1000000000	0.000000	0.666667

13. $\lim\limits_{x \to 1} \dfrac{x^2 + x - 2}{x - 1} = \lim\limits_{x \to 1} \dfrac{(x+2)(x-1)}{x-1} =$
$\lim\limits_{x \to 1}(x+2) = 3$ $\lim\limits_{x \to 1} \dfrac{x^2 + x - 2.01}{x - 1}$ does not
exist, since when x is close to 1, the numerator
is close to $-.01$ (a small but non-zero number)
and the denominator is close to 0.

15. $f(1) = 0$; $g(1) = 0.00159265$
$f(10) = 0$; $g(10) = -0.0159259$
$f(100) = 0$; $g(100) = -0.158593$
$f(1000) = 0$; $g(1000) = -0.999761$

17. $(1.000003 - 1.000001) \times 10^7 = 20$. On a
computer with a 6-digit mantissa, the calculation would be $(1.00000 - 1.00000) \times 10^7 = 0$.

Chapter 1 Review Exercises

1. The slope appears to be 2.

Second Point	m_{sec}
(3, 3)	3
(2.1, 0.21)	2.1
(2.01, 0.0201)	2.01
(1, −1)	1
(1.9, −0.19)	1.9
(1.99, −0.0199)	1.99

3. **(a)** For the x-values of our points here we use
(approximations of) 0, $\frac{\pi}{16}$, $\frac{\pi}{8}$, $\frac{3\pi}{16}$, and $\frac{\pi}{4}$.

Left	Right	Length
(0, 0)	(0.2, 0.2)	0.276
(0.2, 0.2)	(0.39, 0.38)	0.272
(0.39, 0.38)	(0.59, 0.56)	0.262
(0.59, 0.56)	(0.785, 0.71)	0.248
	Total	1.058

(b) For the x-values of our points here we use
(approximations of) 0, $\frac{\pi}{32}$, $\frac{\pi}{16}$, $\frac{3\pi}{32}$, $\frac{\pi}{8}$, $\frac{5\pi}{32}$,
$\frac{3\pi}{16}$, $\frac{7\pi}{32}$, and $\frac{\pi}{4}$.

Left	Right	Length
(0, 0)	(0.1, 0.1)	0.139
(0.1, 0.1)	(0.2, 0.2)	0.138
(0.2, 0.2)	(0.29, 0.29)	0.137
(0.29, 0.29)	(0.39, 0.38)	0.135
(0.39, 0.38)	(0.49, 0.47)	0.132
(0.49, 0.47)	(0.59, 0.56)	0.129
(0.59, 0.56)	(0.69, 0.63)	0.126
(0.69, 0.63)	(0.785, 0.71)	0.122
	Total	1.058

5. Let $f(x) = \dfrac{\tan x^3}{x^2}$.

x	$f(x)$
0.1	0.1
0.01	0.01
0.001	0.001
−0.001	−0.001
−0.01	−0.01
−0.1	−0.1

The limit appears to be 0.

7. Let $f(x) = \dfrac{x + 2}{|x + 2|}$.

x	$f(x)$
−1.9	1
−1.99	1
−1.999	1
−2.1	−1
−2.01	−1
−2.001	−1

$\lim\limits_{x \to -2} \dfrac{x+2}{|x+2|}$ does not exist.

9. Let $f(x) = \dfrac{\sqrt{x^2+4}}{3x+1}$.

x	$f(x)$
-10	-0.3517
-100	-0.3345
-1000	-0.3334
$-10,000$	-0.3333

The limit appears to be $-\frac{1}{3}$.

11. (a) $\lim\limits_{x \to -1^-} f(x) = 1$

(b) $\lim\limits_{x \to -1^+} f(x) = -2$

(c) $\lim\limits_{x \to -1} f(x)$ does not exist

(d) $\lim\limits_{x \to 0} f(x) = 0$

13. $x = -1,\ x = 1$

15. $\lim\limits_{x \to 2} \dfrac{x^2 - x - 2}{x^2 - 4} = \lim\limits_{x \to 2} \dfrac{(x-2)(x+1)}{(x+2)(x-2)} =$

$\lim\limits_{x \to 2} \dfrac{x+1}{x+2} = \dfrac{3}{4}$

17. $\lim\limits_{x \to 0^+} \dfrac{x^2 + x}{\sqrt{x^4 + 2x^2}} = \lim\limits_{x \to 0^+} \dfrac{x(x+1)}{x\sqrt{x^2 + 2}}$

$= \lim\limits_{x \to 0^+} \dfrac{x+1}{\sqrt{x^2 + 2}} = \dfrac{1}{\sqrt{2}}$

but

$\lim\limits_{x \to 0^-} \dfrac{x^2 + x}{\sqrt{x^4 + 2x^2}} = \lim\limits_{x \to 0^-} \dfrac{x(x+1)}{(-x)\sqrt{x^2 + 2}} =$

$\lim\limits_{x \to 0^-} -\dfrac{x+1}{\sqrt{x^2 + 2}} = -\dfrac{1}{\sqrt{2}}$

Since the left and right limits are not equal,

$\lim\limits_{x \to 0} \dfrac{x^2 + x}{\sqrt{x^4 + 2x^2}}$ does not exist

19. $\lim\limits_{x \to 0} (2 + x) \sin(1/x) = \lim\limits_{x \to 0} 2 \sin(1/x)$;

however, since $\lim\limits_{x \to 0} \sin(1/x)$ does not exist, it follows that $\lim\limits_{x \to 0} (2 + x) \sin(1/x)$ also does not exist.

21. $\lim\limits_{x \to 2} f(x) = 5$

23. Multiply the function by

$\dfrac{(1+2x)^{\frac{2}{3}} + (1+2x)^{\frac{1}{3}} + 1}{(1+2x)^{\frac{2}{3}} + (1+2x)^{\frac{1}{3}} + 1}$

to get

$\lim\limits_{x \to 0} \dfrac{\sqrt[3]{1+2x} - 1}{x}$

$= \lim\limits_{x \to 0} \dfrac{2}{(1+2x)^{\frac{2}{3}} + (1+2x)^{\frac{1}{3}} + 1} = \dfrac{2}{3}$,

25. $\lim\limits_{x \to 0} \cot(x^2) = \infty$

27. $\lim\limits_{x \to \infty} \dfrac{x^2 - 4}{3x^2 + x + 1} = \lim\limits_{x \to \infty} \dfrac{x^2 \left(1 - \frac{4}{x^2}\right)}{x^2 \left(3 + \frac{1}{x} + \frac{1}{x^2}\right)}$

$= \lim\limits_{x \to \infty} \dfrac{1 - \frac{4}{x^2}}{3 + \frac{1}{x} + \frac{1}{x^2}} = \dfrac{1}{3}$

29. Since $\lim\limits_{x \to \pi/2} \tan^2 x = \infty$, it follows that

$\lim\limits_{x \to \pi/2} (-\tan^2 x) = -\infty$.

31. $\lim\limits_{x \to -\infty} \dfrac{2x}{x^2 + 3x - 5}$

$= \lim\limits_{x \to -\infty} \dfrac{2x}{x^2 \left(1 + \frac{3}{x} + \frac{5}{x^2}\right)}$

$= \lim\limits_{x \to -\infty} \dfrac{2}{x \left(1 + \frac{3}{x} + \frac{5}{x^2}\right)} = 0$

33. $\lim\limits_{x \to 0^-} \dfrac{\sin x}{|\sin x|} = \lim\limits_{x \to 0^-} \dfrac{\sin x}{-\sin x} = -1$

$\lim\limits_{x \to 0^+} \dfrac{\sin x}{|\sin x|} = \lim\limits_{x \to 0^-} \dfrac{\sin x}{\sin x} = 1$

$\lim\limits_{x \to 0} \dfrac{\sin x}{|\sin x|}$ does not exist.

35. $0 \le \dfrac{x^2}{x^2 + 1} < 1 \Rightarrow -2\,|x| \le \dfrac{2x^3}{x^2 + 1} < 2\,|x|$

$\lim\limits_{x \to 0} -2\,|x| = 0$; $\lim\limits_{x \to 0} 2\,|x| = 0$

By the Squeeze Theorem,

$\lim\limits_{x \to 0} \dfrac{2x^3}{x^2 + 1} = 0$.

37. $f(x) = \dfrac{x-1}{x^2 + 2x - 3} = \dfrac{x-1}{(x+3)(x-1)}$

has a nonremovable discontinuity at $x = -3$ and a removable discontinuity at $x = 1$.

39. $\displaystyle\lim_{x\to 0^-} f(x) = \lim_{x\to 0^-} \sin x = 0$

$\displaystyle\lim_{x\to 0^+} f(x) = \lim_{x\to 0^+} x^2 = 0$

$\displaystyle\lim_{x\to 2^-} f(x) = \lim_{x\to 2^-} x^2 = 4$

$\displaystyle\lim_{x\to 2^+} f(x) = \lim_{x\to 2^+} (4x - 3) = 5$

f has a nonremovable discontinuity at $x = 2$.

41. $f(x) = \dfrac{x+2}{x^2 - x - 6} = \dfrac{x+2}{(x-3)(x+2)}$

continuous on $(-\infty, -2)$, $(-2, 3)$ and $(3, \infty)$.

43. $f(x) = \sin(1 + \cos x)$ is continuous on the interval $(-\infty, \infty)$.

45. $f(x) = \dfrac{x+1}{(x-2)(x-1)}$ has vertical asymptotes at $x = 1$ and $x = 2$.

$\displaystyle\lim_{x\to\pm\infty} \dfrac{x+1}{x^2 - 3x + 2}$

$= \displaystyle\lim_{x\to\pm\infty} \dfrac{x\left(1 + \frac{1}{x}\right)}{x^2\left(1 - \frac{3}{x} + \frac{2}{x^2}\right)}$

$= \displaystyle\lim_{x\to\pm\infty} \dfrac{1 + \frac{1}{x}}{x\left(1 - \frac{3}{x} + \frac{2}{x^2}\right)} = 0$

So $f(x)$ has a horizontal asymptote at $y = 0$.

47. $f(x) = \dfrac{x^2}{x^2 - 1} = \dfrac{x^2}{(x+1)(x-1)}$

has vertical asymptotes at $x = -1$ and $x = 1$.

$\displaystyle\lim_{x\to\pm\infty} \dfrac{x^2}{x^2 - 1} = \lim_{x\to\pm\infty} \dfrac{x^2}{x^2\left(1 - \frac{1}{x^2}\right)}$

$= \displaystyle\lim_{x\to\pm\infty} \dfrac{1}{1 - \frac{1}{x^2}} = \dfrac{1}{1} = 1$

So $f(x)$ has a horizontal asymptote at $y = 1$.

49. $f(x)$ is continuous for all x, so there are no vertical asymptotes.

Performing long division, we find $y = x - 1$ is a slant asymptote.

51. $f(x)$ is discontinuous when $\cos x = 1$, i.e., when $x = 2k\pi$ for any integer k. Thus, there is a vertical asymptote at $x = 2k\pi$ for any integer k.

No horizontal or slant asymptotes.

53. The limit is $\frac{1}{4}$.

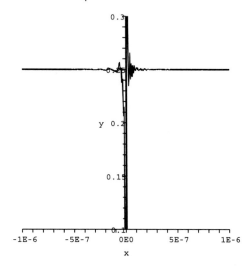

We can rewrite the function as

$\dfrac{1 - \cos x}{2x^2} = \left(\dfrac{1 - \cos x}{2x^2}\right)\left(\dfrac{1 + \cos x}{1 + \cos x}\right) =$

$\dfrac{1 - \cos^2 x}{2x^2(1 + \cos x)} = \dfrac{\sin^2 x}{2x^2(1 + \cos x)}$

to avoid loss-of-significance errors.

In the table below, the middle column contains values calculated using $f(x) = \frac{1 - \cos x}{2x^2}$, while the third column contains values calculated using the rewritten $f(x)$. Note that $f(x) = f(-x)$ and so we get the same values when x is negative (which allows us to conjecture the two-sided limit as $x \to 0$).

x	old $f(x)$	new $f(x)$
1	0.229849	0.229849
0.1	0.249792	0.249792
0.01	0.249998	0.249998
0.001	0.250000	0.250000
0.0001	0.250000	0.250000
0.00001	0.250000	0.250000
0.000001	0.250022	0.250000
0.0000001	0.249800	0.250000
0.00000001	0.000000	0.250000
0.000000001	0.000000	0.250000

55. The limit of θ' as x approaches 0 is 66 radians per second, far faster than the player can maintain focus. From about 9 feet on in to the plate the player can't keep his or her eyes on the ball.

Differentiation

2.1 TANGENT LINES AND VELOCITY

1.

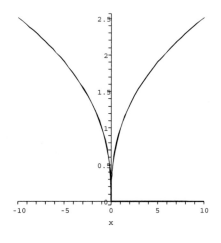

3. The tangent line is vertical and coincides with the y-axis:

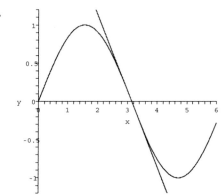

5. At $x = 1$ the slope of the tangent line appears to be about -1.

7. C, B, A, D. At the point labeled C, the slope is very steep and negative. At point B, the slope is zero and at point A, the slope is just more than zero. The slope of the line tangent to point D is large and positive.

9. **(a)** Points $(1, 0)$ and $(2, 6)$. Slope is $\frac{6-0}{1} = 6$.

(b) Points $(2, 6)$ and $(3, 24)$. Slope is $\frac{24-6}{1} = 18$.

(c) Points $(1.5, 1.875)$ and $(2, 6)$. Slope is $\frac{6-1.875}{0.5} = 8.25$.

(d) Points $(2, 6)$ and $(2.5, 13.125)$. Slope is $\frac{13.125-6}{0.5} = 14.25$.

(e) Points $(1.9, 4.959)$ and $(2, 6)$. Slope is $\frac{6-4.959}{0.1} = 10.41$.

(f) Points $(2, 6)$ and $(2.1, 7.161)$. Slope is $\frac{7.161-6}{0.1} = 11.61$.

(g) Slope seems to be approximately 11.

11. **(a)** Points $(1, 0.54)$ and $(2, -0.65)$. Slope is $\frac{-0.65-0.54}{1} = -1.19$.

(b) Points $(2, -0.65)$ and $(3, -0.91)$. Slope is $\frac{-0.91-(-0.65)}{1} = -0.26$.

(c) Points $(1.5, -0.628)$ and $(2, -0.654)$. Slope is $\frac{-0.654-(-0.628)}{0.5} = -0.05$.

(d) Points $(2, -0.65)$ and $(2.5, 1.00)$. Slope is $\frac{1.00-(-0.65)}{0.5} = 3.3$.

(e) Points $(1.9, -0.89)$ and $(2, -0.65)$. Slope is $\frac{-0.65-(-0.89)}{0.1} = 2.4$.

(f) Points $(2, -0.654)$ and $(2.1, -0.298)$. Slope is $\frac{-0.298-(-0.654)}{0.1} = 3.56$.

(g) Slope seems to be approximately 3.

13.

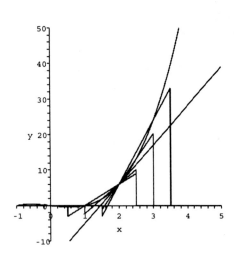

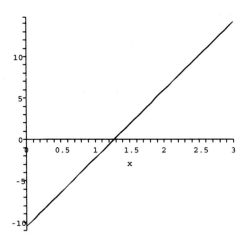

The third secant line is indistinguishable from the tangent line.

15. The sequence of graphs should look like:

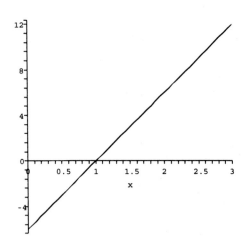

17. Slope is

$$\lim_{h \to 0} \frac{f(1+h) - f(1)}{h}$$

$$= \lim_{h \to 0} \frac{(1+h)^2 - 2 - (-1)}{h}$$

$$= \lim_{h \to 0} \frac{h^2 + 2h}{h} = \lim_{h \to 0} (h + 2) = 2.$$

Tangent line is $y = 2(x - 1) - 1$ or $y = 2x - 3$.

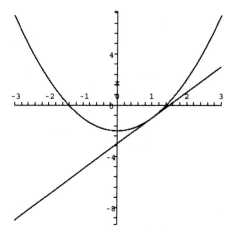

19. Slope is

$$\lim_{h\to 0}\frac{f(-2+h)-f(-2)}{h}$$

$$=\lim_{h\to 0}\frac{(-2+h)^2-3(-2+h)-(10)}{h}$$

$$=\lim_{h\to 0}\frac{4-4h+h^2+6-3h-10}{h}$$

$$=\lim_{h\to 0}\frac{-7h+h^2}{h}=\lim_{h\to 0}(-7+h)=-7.$$

Tangent line is $y=-7(x+2)+10$ or $y=-7x-4.$

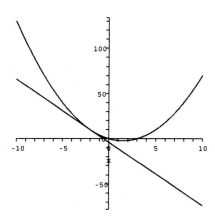

21. Slope is $\lim_{h\to 0}\dfrac{f(1+h)-f(1)}{h}$

$$=\lim_{h\to 0}\frac{\frac{2}{(1+h)+1}-\frac{2}{1+1}}{h}$$

$$=\lim_{h\to 0}\frac{\frac{2}{2+h}-1}{h}$$

$$=\lim_{h\to 0}\frac{\left(\frac{2-(2+h)}{2+h}\right)}{h}$$

$$=\lim_{h\to 0}\frac{\left(\frac{-h}{2+h}\right)}{h}$$

$$=\lim_{h\to 0}\frac{-1}{2+h}=\frac{-1}{2}$$

Tangent line is $y=-\frac{1}{2}(x-1)+1$ or $y=-\frac{x}{2}+\frac{3}{2}.$

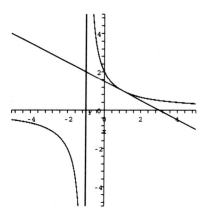

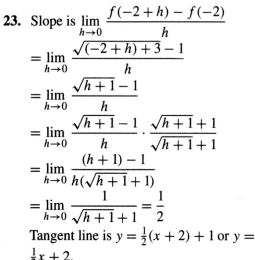

23. Slope is $\lim_{h\to 0}\dfrac{f(-2+h)-f(-2)}{h}$

$$=\lim_{h\to 0}\frac{\sqrt{(-2+h)+3}-1}{h}$$

$$=\lim_{h\to 0}\frac{\sqrt{h+1}-1}{h}$$

$$=\lim_{h\to 0}\frac{\sqrt{h+1}-1}{h}\cdot\frac{\sqrt{h+1}+1}{\sqrt{h+1}+1}$$

$$=\lim_{h\to 0}\frac{(h+1)-1}{h(\sqrt{h+1}+1)}$$

$$=\lim_{h\to 0}\frac{1}{\sqrt{h+1}+1}=\frac{1}{2}$$

Tangent line is $y=\frac{1}{2}(x+2)+1$ or $y=\frac{1}{2}x+2.$

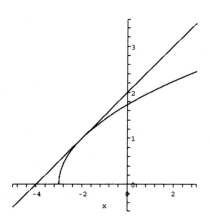

25. Numerical evidence suggests that
$$\lim_{h \to 0^+} \frac{f(1+h) - f(1)}{h} = 1$$
while
$$\lim_{h \to 0^-} \frac{f(1+h) - f(1)}{h} = -1.$$
Since these are not equal, there is no tangent line. A graph makes it apparent that this function has a "corner" at $x = 1$.

27. Numerical evidence suggests that
$$\lim_{h \to 0^+} \frac{f(0+h) - f(0)}{h}$$
$$= \lim_{h \to 0^-} \frac{f(0+h) - f(0)}{h}$$
$$= 0$$
Since the slope of the tangent line from the left equals that from the right and the function appears to be continuous in the graph, we conjecture that the tangent line exists and has slope 0.

29. Looking at the graph, we see that there is a jump discontinuity at $a = 0$. Thus there cannot be a tangent line, as the tangent line from the left would be different from the tangent line from the right.

31. (a) Points (0, 10) and (2, 74). Average velocity is $\frac{74-10}{2} = 32$.

(b) Second point (1, 26). Average velocity is $\frac{74-26}{1} = 48$.

(c) Second point (1.9, 67.76). Average velocity is $\frac{74-67.76}{0.1} = 62.4$.

(d) Second point (1.99, 73.3616). Average velocity is $\frac{74-73.3616}{0.01} = 63.84$.

(e) The instantaneous velocity seems to be 64.

33. (a) Points (0, 0) and (2, $\sqrt{20}$). Average velocity is $\frac{\sqrt{20}-0}{2-0} = 2.236068$.

(b) Second point (1, 3). Average velocity is $\frac{\sqrt{20}-3}{2-1} = 1.472136$.

(c) Second point (1.9, $\sqrt{18.81}$). Average velocity is $\frac{\sqrt{20}-\sqrt{18.81}}{2-1.9} = 1.3508627$.

(d) Second point (1.99, $\sqrt{19.8801}$). Average velocity is $\frac{\sqrt{20}-\sqrt{19.8801}}{2-1.99} = 1.3425375$.

(e) One might conjecture that these numbers are approaching 1.34. The exact limit is $\frac{6}{\sqrt{20}} \approx 1.341641$.

35. (a) Velocity at time $t = 1$ is
$$\lim_{h \to 0} \frac{f(1+h) - f(1)}{h}$$
$$= \lim_{h \to 0} \frac{-16(1+h)^2 + 5 - (-11)}{h}$$
$$= \lim_{h \to 0} \frac{-16 - 32h - 16h^2 + 5 + 11}{h}$$
$$= \lim_{h \to 0} \frac{-32h - 16h^2}{h}$$
$$= \lim_{h \to 0} (-32 - 16h) = -32.$$

(b) Velocity at time $t = 2$ is
$$\lim_{h \to 0} \frac{f(2+h) - f(2)}{h}$$
$$= \lim_{h \to 0} \frac{-16(4 + 4h + h^2) + 5 + 59}{h}$$
$$= \lim_{h \to 0} \frac{-64 - 64h - 16h^2 + 64}{h}$$
$$= \lim_{h \to 0} (-64 - 16h) = -64.$$

37. The slope of the tangent line at $p = 1$ is approximately
$$\frac{-20 - 0}{2 - 0} = -10$$
which means that at $p = 1$, the freezing temperature of water decreases by 10 degrees Celsius per 1 atm increase in pressure. The slope of the tangent line at $p = 3$ is approximately
$$\frac{-11 - (-20)}{4 - 2} = 4.5$$
which means that at $p = 3$, the freezing temperature of water increases by 4.5 degrees Celsius per 1 atm increase in pressure.

39. The hiker reached the top at the highest point on the graph (about 1.75 hours). The hiker was going the fastest on the way up at about 1.5 hours. The hiker was going the fastest on the way down at the point where the tangent line has the least (i.e., most negative) slope, at about 4 hours, at the end of the hike. Where the graph is level, the hiker was either resting, or walking on flat ground.

41. A possible graph of the temperature with respect to time:

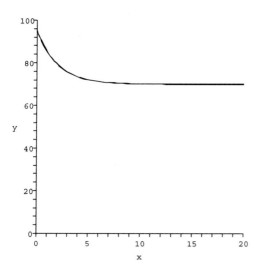

Graph of the rate of change of the temperature:

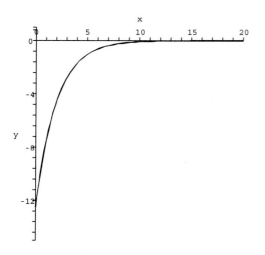

43. (a) To say that

$$\frac{f(4) - f(2)}{2} = 21,034$$

per year is to say that the average rate of change in the bank balance between Jan. 1, 2002 and Jan. 1, 2004 was 21,034 ($ per year).

(b) To say that

$$2[f(4) - f(3.5)] = 25,036$$

(note that $2[f(4) - f(3.5)] = \frac{f(4) - f(3.5)}{1/2}$) per year is to say that the average rate of change between July 1, 2003 and Jan. 1, 2004 was 25,036 ($ per year).

(c) To say that

$$\lim_{h \to 0} \frac{f(4 + h) - f(4)}{h} = \$30,000$$

is to say that the instantaneous rate of change in the balance on Jan. 1, 2004 was 30,000 ($ per year).

45. We are given $\theta(t) = 0.4t^2$. We are advised that θ is measured in radians, and that t is time. Let us assume that t is measured in seconds.

Three rotations corresponds to $\theta = 6\pi$. Proceeding, if $\theta(t) = 6\pi$ then $0.4t^2 = 6\pi$ and solving for t yields $t = \sqrt{15\pi} \approx 6.865$ (seconds).

At that exact moment of time (call it a), the exact angular velocity is

$$\lim_{h \to 0} \frac{\theta(a + h) - \theta(a)}{h}$$
$$= \lim_{h \to 0} \frac{0.4(\sqrt{15\pi} + h)^2 - 6\pi}{h}$$
$$= \lim_{h \to 0} \frac{0.4(15\pi + 2h\sqrt{15\pi} + h^2) - 6\pi)}{h}$$
$$= \lim_{h \to 0} \frac{0.8h\sqrt{15\pi} + 0.4h^2}{h}$$
$$= \lim_{h \to 0} (0.8\sqrt{15\pi} + 0.4h) = 0.8\sqrt{15\pi}$$
$$= \sqrt{9.6\pi} \approx 5.492$$

and the units would be *radians per second*.

47. $v_{avg} = \frac{f(s) - f(r)}{s - r}$
$$= \frac{as^2 + bs + c - (ar^2 + br + c)}{s - r}$$

$$= \frac{a(s^2 - r^2) + b(s - r)}{s - r}$$

$$= \frac{a(s + r)(s - r) + b(s - r)}{s - r} \text{nl} = a(s + r) + b$$

Let $v(r)$ be the velocity at $t = r$. We have

$$v(r) = \lim_{h \to 0} \frac{f(r + h) - f(r)}{h}$$

$$= \lim_{h \to 0}$$

$$\frac{a(r + h)^2 + b(r + h) + c - (ar^2 + bh + c)}{h}$$

$$= \lim_{h \to 0} \frac{a(r^2 + 2rh + h^2) + bh - ar^2}{h}$$

$$= \lim_{h \to 0} \frac{h(2ar + ah + b)}{h}$$

$$= \lim_{h \to 0} (2ar + ah + b) = 2ar + b.$$

So $v(r) = 2ar + b$. The same argument shows that $v(s) = 2as + b$.

Finally, $\dfrac{v(r) + v(s)}{2}$

$$= \frac{(2ar + b) + (2as + b)}{2} \frac{2a(s + r) + 2b}{2}$$

$$= a(s + r) + b = v_{avg}.$$

49. Let $x = h + a$. Then $h = x - a$, and clearly

$$\frac{f(a + h) - f(a)}{h} = \frac{f(x) - f(a)}{x - a}.$$

It is also clear that $x \to a$ if and only if $h \to 0$. Therefore if one of the two limits exists, then so does the other and

$$\lim_{h \to 0} \frac{f(a + h) - f(a)}{h} = \lim_{x \to a} \frac{f(x) - f(a)}{x - a}.$$

51. First, compute the slope of the tangent line. Using the result of #49, it is convenient to assume x is near but not exactly $1/2$, and write

$$\lim_{x \to 1/2} \frac{f(x) - f(1/2)}{x - (1/2)} = \frac{x^2 - (1/4)}{x - (1/2)}$$

$$= \lim_{x \to 1/2} \frac{(x - (1/2))(x + (1/2))}{x - (1/2)}$$

$$= \lim_{x \to 1/2} [x + (1/2)] = 1.$$

Next, we quickly write the equation of the tangent line in point-slope form:

$$y = 1(x - (1/2)) + (1/4) \text{ or } y = x - (1/4).$$

The location of the tree is the point $(x, y) = (1, 3/4)$ and this point is indeed on the tangent line. The tree will be hit if the car gets that far (that being something we have no way of knowing).

2.2 THE DERIVATIVE

1. Using (2.1):

$$f'(1) = \lim_{h \to 0} \frac{f(1 + h) - f(1)}{h}$$

$$= \lim_{h \to 0} \frac{3(1 + h) + 1 - (4)}{h}$$

$$= \lim_{h \to 0} \frac{3h}{h} = \lim_{h \to 0} 3 = 3$$

Using (2.2):

$$\lim_{b \to 1} \frac{f(b) - f(1)}{b - 1}$$

$$= \lim_{b \to 1} \frac{3b + 1 - (3 + 1)}{b - 1}$$

$$= \lim_{b \to 1} \frac{3b - 3}{b - 1}$$

$$= \lim_{b \to 1} \frac{3(b - 1)}{b - 1} = \lim_{b \to 1} 3 = 3$$

3. Using (2.1): Since

$$\frac{f(1 + h) - f(1)}{h} = \frac{\sqrt{3(1 + h) + 1} - 2}{h}$$

$$= \frac{\sqrt{4 + 3h} - 2}{h} \cdot \frac{\sqrt{4 + 3h} + 2}{\sqrt{4 + 3h} + 2}$$

$$= \frac{4 + 3h - 4}{h(\sqrt{4 + 3h} + 2)} = \frac{3h}{h(\sqrt{4 + 3h} + 2)}$$

$$= \frac{3}{\sqrt{4 + 3h} + 2}, \text{ we have}$$

$$f'(1) = \lim_{h \to 0} \frac{f(1 + h) - f(1)}{h}$$

$$= \lim_{h \to 0} \frac{3}{\sqrt{4 + 3h} + 2}$$

$$= \frac{3}{\sqrt{4 + 3(0)} + 2} = \frac{3}{4}.$$

Using (2.2): Since

$$\frac{f(b) - f(1)}{b - 1}$$

$$= \frac{\sqrt{3b + 1} - 2}{b - 1}$$

$$= \frac{(\sqrt{3b + 1} - 2)(\sqrt{3b + 1} + 2)}{(b - 1)(\sqrt{3b + 1} + 2)}$$

$$= \frac{(3b + 1) - 4}{(b - 1)\sqrt{3b + 1} + 2}$$

$$= \frac{3(b - 1)}{(b - 1)\sqrt{3b + 1} + 2}$$

$$= \frac{3}{\sqrt{3b + 1} + 2}, \text{ we have}$$

$$f'(1) = \lim_{b \to 1} \frac{f(b) - f(1)}{b - 1}$$

$$= \lim_{b \to 1} \frac{3}{\sqrt{3b + 1} + 2}$$

$$= \frac{3}{\sqrt{4} + 2} = \frac{3}{4}.$$

5.
$$\lim_{h \to 0} \frac{f(x + h) - f(x)}{h}$$

$$= \lim_{h \to 0} \frac{3(x + h)^2 + 1 - (3(x)^2 + 1)}{h}$$

$$= \lim_{h \to 0} \frac{3x^2 + 6xh + 3h^2 + 1 - (3x^2 + 1)}{h}$$

$$= \lim_{h \to 0} \frac{6xh + 3h^2}{h} = \lim_{h \to 0} (6x + 3h) = 6x$$

7.
$$\lim_{b \to x} \frac{f(b) - f(x)}{b - x}$$

$$= \lim_{b \to x} \frac{\frac{3}{b + 1} - \frac{3}{x + 1}}{b - x}$$

$$= \lim_{b \to x} \frac{\frac{3(x+1) - 3(b+1)}{(b+1)(x+1)}}{b - x}$$

$$= \lim_{b \to x} \frac{-3(b - x)}{(b + 1)(x + 1)(b - x)}$$

$$= \lim_{b \to x} \frac{-3}{(b + 1)(x + 1)} = \frac{-3}{(x + 1)^2}$$

9.
$$\lim_{b \to x} \frac{f(b) - f(x)}{b - x}$$

$$= \lim_{b \to x} \frac{\sqrt{3b + 1} - \sqrt{3x + 1}}{b - x}$$

Multiplying by

$$\frac{\sqrt{3b + 1} + \sqrt{3x + 1}}{\sqrt{3b + 1} + \sqrt{3x + 1}}$$

gives

$$\lim_{b \to x} \frac{(3b + 1) - (3x + 1)}{(b - x)(\sqrt{3b + 1} + \sqrt{3x + 1})}$$

$$= \lim_{b \to x} \frac{3(b - x)}{(b - x)(\sqrt{3b + 1} + \sqrt{3x + 1})}$$

$$= \lim_{b \to x} \frac{3}{(\sqrt{3b + 1} + \sqrt{3x + 1})}$$

$$= \frac{3}{2\sqrt{3x + 1}}$$

11.
$$\lim_{b \to x} \frac{f(b) - f(x)}{b - x}$$

$$= \lim_{b \to x} \frac{b^3 + 2b - 1 - (x^3 + 2x - 1)}{b - x}$$

$$= \lim_{b \to x} \frac{b^3 - x^3 + 2b - 2x}{b - x}$$

$$= \lim_{b \to x} \frac{(b - x)(b^2 + bx + x^2) + 2(b - x)}{b - x}$$

$$= \lim_{b \to x} \frac{(b - x)(b^2 + bx + x^2 + 2)}{b - x}$$

$$= \lim_{b \to x} (b^2 + bx + x^2 + 2)$$

$$= 3x^2 + 2$$

13. The function has negative slope for $x < 0$, positive slope for $x > 0$, and zero slope at $x = 0$. Its slope function (derivative) can only be (c).

15. Here, moving from left to right, the slope goes from negative to positive to negative to positive. Its slope function (derivative) can only be (a).

17. The graph is increasing to the left of the jump and decreasing to the right. The derivative of this function must be (b) which is postive to the left of the jump and negative to the right.

19. The derivative should look like:

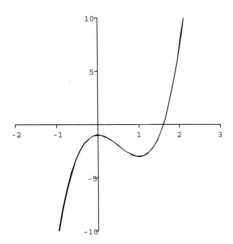

21. The derivative should look like:

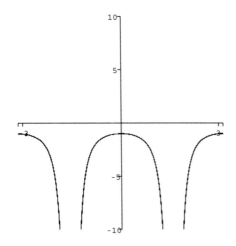

23. One possible graph of $f(x)$:

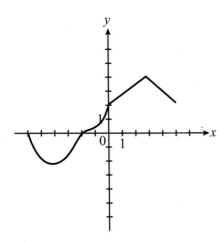

25. $f(x)$ is not differentiable at $x = 0$ or $x = 2$. The graph looks like:

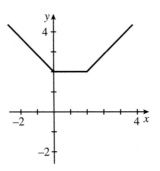

27. $\displaystyle\lim_{h \to 0} \frac{(0 + h)^p - 0^p}{h} = \lim_{h \to 0} \frac{h^p}{h} = \lim_{h \to 0} h^{p-1}$

The last limit does not exist when $p < 1$, equals 1 when $p = 1$, and is 0 when $p > 1$. Thus, $f'(0)$ exists when $p \geq 1$.

29. $\displaystyle\lim_{x \to a} \frac{[f(x)]^2 - [f(a)]^2}{x^2 - a^2}$

$\displaystyle = \lim_{x \to a} \frac{[f(x) - f(a)][f(x) + f(a)]}{(x - a)(x + a)}$

$\displaystyle = \left[\lim_{x \to a} \frac{f(x) - f(a)}{(x - a)}\right]\left[\lim_{x \to a} \frac{f(x) + f(a)}{(x + a)}\right]$

$\displaystyle = f'(a) \cdot \frac{2f(a)}{2a}$

$\displaystyle = \frac{f(a)f'(a)}{a}$

31. We estimate the derivative at $x = 60$ as follows:

$$\frac{3.9 - 2.4}{80 - 40} = \frac{1.5}{40} = 0.0375$$

For every increase of 1 revolution per second of topspin, there is an increase of $0.0375°$ in margin of error.

33. Compute average velocities:

Time Interval	Average Velocity
(1.7, 2.0)	9.0
(1.8, 2.0)	9.5
(1.9, 2.0)	10.0
(2.0, 2.1)	10.0
(2.0, 2.2)	9.5
(2.0, 2.3)	9.0

Our best estimate of the velocity at $t = 2$ is 10.

35. We compile the rate of change in ton-MPG over each of the four two-year intervals for which data is given:

Intervals	Rate of Change
(1992,1994)	$\frac{45.7-44.9}{2} = 0.4$
(1994,1996)	0.4
(1996,1998)	0.4
(1998,2000)	0.2

These rates of change are measured in ton-MPG per year. Either the first or second (they happen to agree) could be used as an estimate for the one-year interval "1994" while only the last is a promising estimate for the one-year interval "2000." The mere fact that all these numbers are positive suggests that efficiency is improving, but the last number being smaller seems to suggest that the rate of improvement is slipping.

37. The left-hand derivative is

$$D_- f(0) = \lim_{h \to 0^-} \frac{f(h) - f(0)}{h}$$

$$= \lim_{h \to 0^-} \frac{2h + 1 - 1}{h} = 2$$

The right-hand derivative is

$$D_+ f(0) = \lim_{h \to 0^+} \frac{f(h) - f(0)}{h}$$

$$= \lim_{h \to 0^+} \frac{3h + 1 - 1}{h} = 3$$

39. $D_+ f(0) = \lim_{h \to 0^+} \frac{f(h) - f(0)}{h}$

$$= \lim_{h \to 0} \frac{k(h) - k(0)}{h} = k'(0).$$

$$D_- f(0) = \lim_{h \to 0^-} \frac{f(h) - f(0)}{h}$$

$$= \lim_{h \to 0^-} \frac{g(h) - k(0)}{h}$$

$$= \lim_{h \to 0} \frac{g(h) - g(0)}{h} = g'(0)$$

$g(0)$ is defined since g is differentiable at 0, and $g(0) = k(0)$ since f is continuous at 0. If $f(x)$ has a jump discontinuity at $x = 0$, it would be because its *left limit* at $x = 0$, namely, $g(0)$, is not the same as the *value* which is $k(0)$. In that case there could be no left derivative (by Theorem 2.1) and one would have to reject the statement $D_- f(0) = g'(0)$.

41. If $f'(x) > 0$ for all x, then the tangent lines all have positive slope, so the function is always sloping up.

43.

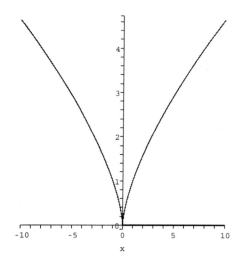

From the graph, we see that $f(x)$ appears to be continuous at $x = 0$, where it has both *limit* and *value* zero. However, when we try to compute its derivative at $x = 0$, we come to the difference quotient

$$\frac{f(0 + h) - f(0)}{h} = \frac{f(h)}{h} = \frac{h^{2/3}}{h} = \frac{1}{h^{1/3}}$$

Clearly this expression has no finite limit as h approaches zero. The numbers get large without bound. We do sometimes say that the vertical line $x = 0$ is the tangent line, but as a line it has no *slope* (just as the function has no derivative).

45. Let $f(x) = -1 - x^2$; then for all x, we have $f(x) \le x$. But at $x = -1$, we find $f(-1) = -2$ and

$$f'(-1) = \lim_{h \to 0} \frac{f(-1+h) - f(-1)}{h}$$

$$= \lim_{h \to 0} \frac{-1 - (-1+h)^2 - (-2)}{h}$$

$$= \lim_{h \to 0} \frac{1 - (1 - 2h + h^2)}{h}$$

$$= \lim_{h \to 0} \frac{2h - h^2}{h} = \lim_{h \to 0} (2 - h) = 2$$

So, $f'(x)$ is not always less than 1.

47. **(a)** meters per second

(b) items per dollar

49. If $f'(t) < 0$, the function $f(t)$ is negatively sloped and decreasing, meaning the stock is losing value with the passing of time. This may be the basis for selling the stock if the current trend is expected to be a long term one.

51. The following sketches are consistent with the hypotheses of infection rate rising, peaking, and returning to zero. We started with the derivative $I'(t)$ (infection rate) and had to think backwards to construct the function $I(t)$. One can see in $I(t)$ the slope increasing up to the time of peak infection rate, thereafter the *slope* decreasing but not the *values*. They merely level off.

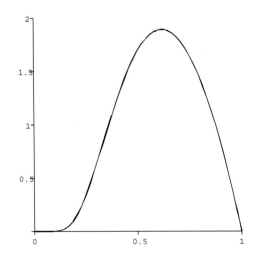

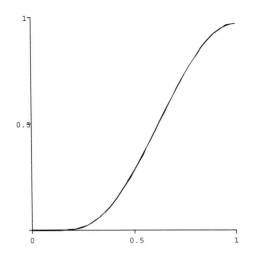

53. Because the curve appears to be bending upward, the slopes of the secant lines (based at $x = 1$ and with upper endpoint beyond 1) will increase with the upper endpoint. This has also the effect that any one of these slopes is greater than the actual derivative. Therefore

$$f'(1) < \frac{f(1.5) - f(1)}{0.5} < \frac{f(2) - f(1)}{1}$$

As to where $f(1)$ fits in this list, it seems necessary to read the graph and come up with estimates of $f(1)$ about 4, and $f(2)$ about 7. That would put the third number in the above list at about 3, comfortably less than $f(1)$.

55. This is a tricky one. It happens that for the function $f(x) = x^2 - x$, the value at $x = 1$ is *zero* ($f(1) = 0$)! Because of this fact,

$$\frac{(1+h)^2 - (1+h)}{h} = \frac{f(1+h) - f(1)}{h}$$

and the answer should be:
$f(x) = x^2 - x$ and $a = 1$.

57. $\displaystyle \lim_{h \to 0} \frac{\left(\frac{1}{2+h}\right) - \left(\frac{1}{2}\right)}{h}$ would be $f'(a)$ for $f(x) = \dfrac{1}{x}$ and $a = 2$.

59. One possible such graph:

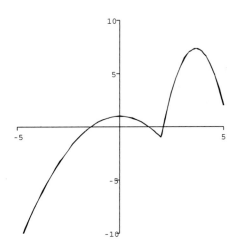

61. We have:

$$f(t) = \begin{cases} 100 & 0 < t \le 20 \\ 100 + 10(t - 20) & 20 < t \le 80 \\ 700 + 8(t - 80) & t > 80 \end{cases}$$

This is another example of a *piecewise linear* function (this one is continuous), and although not differentiable at the transition times $t = 20$ or $t = 80$, elsewhere we have

$$f'(t) = \begin{cases} 0 & 0 < t < 20 \\ 10 & 20 < t < 80 \\ 8 & t > 80 \end{cases}$$

2.3 COMPUTATION OF DERIVATIVES: POWER RULE

1.
$$f'(x) = \frac{d}{dx}(x^3) - \frac{d}{dx}(2x) + \frac{d}{dx}(1)$$
$$= 3x^2 - 2\frac{d}{dx}(x) + 0$$
$$= 3x^2 - 2(1)$$
$$= 3x^2 - 2$$

3.
$$f'(t) = \frac{d}{dt}(3t^3) - \frac{d}{dt}\left(2\sqrt{t}\right)$$
$$= 3\frac{d}{dt}(t^3) - 2\frac{d}{dt}\left(t^{1/2}\right)$$
$$= 3(3t^2) - 2\left(\frac{1}{2}t^{-1/2}\right)$$
$$= 9t^2 - \frac{1}{\sqrt{t}}$$

5.
$$f'(x) = \frac{d}{dx}\left(\frac{3}{x}\right) - \frac{d}{dx}(8x) + \frac{d}{dx}(1)$$
$$= 3\frac{d}{dx}(x^{-1}) - 8\frac{d}{dx}(x) + 0$$
$$= 3(-x^{-2}) - 8(1)$$
$$= -\frac{3}{x^2} - 8$$

7.
$$h'(x) = \frac{d}{dx}\left(\frac{10}{\sqrt{x}}\right) - \frac{d}{dx}(2x)$$
$$= 10\frac{d}{dx}\left(x^{-1/2}\right) - 2\frac{d}{dx}(x)$$
$$= 10\left(-\frac{1}{2}x^{-3/2}\right) - 2(1)$$
$$= -5x^{-3/2} - 2$$
$$= \frac{-5}{x\sqrt{x}} - 2$$

9.
$$f'(s) = \frac{d}{ds}\left(2s^{3/2}\right) - \frac{d}{ds}\left(3s^{-1/3}\right)$$
$$= 2\frac{d}{ds}\left(s^{3/2}\right) - 3\frac{d}{ds}\left(s^{-1/3}\right)$$
$$= 2\left(\frac{3}{2}s^{1/2}\right) - 3\left(-\frac{1}{3}s^{-4/3}\right)$$
$$= 3s^{1/2} + s^{-4/3}$$
$$= 3\sqrt{s} + \frac{1}{\sqrt[3]{s^4}}$$

11.
$$f'(x) = \frac{d}{dx}\left(2\sqrt[3]{x}\right) + \frac{d}{dx}(3)$$
$$= 2\frac{d}{dx}\left(x^{1/3}\right) + 0$$
$$= 2\left(\frac{1}{3}x^{-2/3}\right) = \frac{2}{3}x^{-2/3}$$
$$= \frac{2}{3\sqrt[3]{x^2}}$$

13. $f(x) = x(3x^2 - \sqrt{x}) = 3x^3 - x^{3/2}$ so
$$f'(x) = 3\frac{d}{dx}(x^3) - \frac{d}{dx}\left(x^{3/2}\right)$$
$$= 3(3x^2) - \left(\frac{3}{2}x^{1/2}\right)$$
$$= 9x^2 - \frac{3}{2}\sqrt{x}$$

15.
$$f(x) = \frac{3x^2 - 3x + 1}{2x}$$
$$= \frac{3x^2}{2x} - \frac{3x}{2x} + \frac{1}{2x}$$
$$= \frac{3}{2}x - \frac{3}{2} + \frac{1}{2}x^{-1} \text{ so}$$
$$f'(x) = \frac{d}{dx}\left(\frac{3}{2}x\right) - \frac{d}{dx}\left(\frac{3}{2}\right)$$
$$+ \frac{d}{dx}\left(\frac{1}{2}x^{-1}\right)$$
$$= \frac{3}{2}\frac{d}{dx}(x) - 0 + \frac{1}{2}\frac{d}{dx}(x^{-1})$$
$$= \frac{3}{2}(1) + \frac{1}{2}(-1x^{-2})$$
$$= \frac{3}{2} - \frac{1}{2x^2}$$

17. $f'(x) = \frac{d}{dx}(x^4 + 3x^2 - 2) = 4x^3 + 6x$
$$f''(x) = \frac{d}{dx}(4x^3 + 6x) = 12x^2 + 6$$

19. $f(x) = 2x^4 - 3x^{-1/2}$ so
$$\frac{df}{dx} = 8x^3 + \frac{3}{2}x^{-3/2}$$
$$\frac{d^2f}{dx^2} = 24x^2 - \frac{9}{4}x^{-5/2}$$

21. $f'(x) = 4x^3 + 6x$
$f''(x) = 12x^2 + 6$
$f'''(x) = 24x$
$f^{(4)}(x) = 24$

23.
$$f(x) = \frac{x^2 - x + 1}{\sqrt{x}}$$
$$= x^{3/2} - x^{1/2} + x^{-1/2} \text{ so}$$
$$f'(x) = \frac{d}{dx}\left(x^{3/2} - x^{1/2} + x^{-1/2}\right)$$
$$= \frac{3}{2}x^{1/2} - \frac{1}{2}x^{-1/2} - \frac{1}{2}x^{-3/2}$$
$$f''(x) = \frac{d}{dx}\left(\frac{3}{2}x^{1/2} - \frac{1}{2}x^{-1/2} - \frac{1}{2}x^{-3/2}\right)$$
$$= \frac{3}{4}x^{-1/2} + \frac{1}{4}x^{-3/2} + \frac{3}{4}x^{-5/2}$$
$$f'''(x) = \frac{d}{dx}\left(\frac{3}{4}x^{-1/2} + \frac{1}{4}x^{-3/2} + \frac{3}{4}x^{-5/2}\right)$$
$$= -\frac{3}{8}x^{-3/2} - \frac{3}{8}x^{-5/2} - \frac{15}{8}x^{-7/2}$$

25. $s(t) = -16t^2 + 40t + 10$
$v(t) = s'(t) = -32t + 40$
$a(t) = v'(t) = s''(t) = -32$

27. $s(t) = \sqrt{t} + 2t^2 = t^{1/2} + 2t^2$
$$v(t) = s'(t) = \frac{1}{2}t^{-1/2} + 4t$$
$$a(t) = v'(t) = s''(t) = -\frac{1}{4}t^{-3/2} + 4$$

29. $v(t) = -32t + 40$, $v(1) = 8$, going up.
$a(t) = -32$, $a(1) = -32$, speed decreasing.

31. $v(t) = 20t - 24$, $v(2) = 16$, going up.
$a(t) = 20$, $a(1) = 20$, speed increasing.

33. $f(x) = 4\sqrt{x} - 2x, a = 4$

$f(4) = 4\sqrt{4} - 2(4) = 0$

$f'(x) = \dfrac{d}{dx}\left(4x^{1/2} - 2x\right)$

$\quad = 2x^{-1/2} - 2 = \dfrac{2}{\sqrt{x}} - 2$

$f'(4) = 1 - 2 = -1$

The equation of the tangent line is
$y = -1(x - 4) + 0$ or $y = -x + 4$.

35. $f(x) = x^2 - 2, a = 2, f(2) = 2$
$f'(x) = 2x$
$f'(2) = 4$
The equation of the tangent line is
$y = 4(x - 2) + 2$ or $y = 4x - 6$.

37. $f(x) = x^3 - 3x + 1$
$f'(x) = 3x^2 - 3$
The tangent line to $y = f(x)$ is horizontal
when $f'(x) = 0$: $3x^2 - 3 = 0$
$\Longleftrightarrow 3(x^2 - 1) = 0$
$\Longleftrightarrow 3(x + 1)(x - 1) = 0$
$\Longleftrightarrow x = -1$ or $x = 1$.

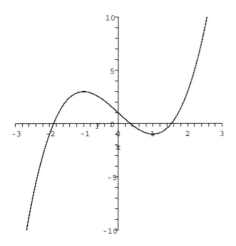

The graph shows that the first is a relative
maximum, the second is a relative minimum.

39. $f(x) = x^{2/3}$
$f'(x) = \dfrac{2}{3}x^{-1/3} = \dfrac{2}{3\sqrt[3]{x}}$
The slope of the tangent line to $y = f(x)$ does
not exist where the derivative is undefined,
which is only when $x = 0$.

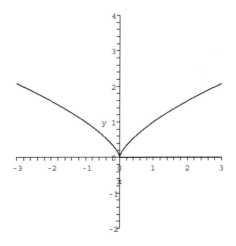

In this case, because the function is continuous, we might say that the tangent line is
the vertical line $x = 0$. The feature at $x = 0$ is
sometimes known as a *cusp*.

41. As regards the (a) function, its derivative
would be negative for all negative x and positive for all positive x. Since no such function
appears among the pictures, this (a) function
has to be the one whose derivative is absent
from the list. There being no f''' in the list, (a)
has to be f''.

This same (a) function is negative for a certain
interval of the form $(-a, a)$, and the (c) function is decreasing on a similar type of interval.
Thus the (a) function (f'') is apparently the
derivative of the (c) function. It follows that
(c) must be f'.

This leaves (b) for f itself, and our identifica-
tions are consistent in every respect.

43. $f(x) = \sqrt{x} = x^{1/2}$
$f'(x) = \dfrac{1}{2}x^{-1/2}$
$f''(x) = \dfrac{1}{2}\left(-\dfrac{1}{2}\right)x^{-3/2}$
$f'''(x) = \left(\dfrac{1}{2}\right)\left(\dfrac{-1}{2}\right)\left(\dfrac{-3}{2}\right)x^{-5/2}$
$f^{(n)}(x) = (-1)^{n-1}\dfrac{\Pi_n}{2^n}x^{-(2n-1)/2}$
in which Π_n is the product of the first $n - 1$
odd integers (starting from 1 and ending at
$2n - 3$).

45. $f(x) = ax^2 + bx + c \Rightarrow f(0) = c$
$f'(x) = 2ax + b \Rightarrow f'(0) = b$
$f''(x) = 2a \Rightarrow f''(0) = 2a$
Given $f''(0) = 3$, we learn $2a = 3$, or $a = 3/2$.
Given $f'(0) = 2$ we learn $2 = b$, and given $f(0) = -2$, we learn $c = -2$. In the end
$f(x) = ax^2 + bx + c = \frac{3}{2}x^2 + 2x - 2$.

47. For $y = \frac{1}{x}$, we have $y' = -\frac{1}{x^2}$. Thus, the slope of the tangent line at $x = a$ is $-\frac{1}{a^2}$.

When $a = 1$, the slope of the tangent line at $(1, 1)$ is -1, and the equation of the tangent line is $y = -x + 2$. The tangent line intersects the axes at $(0, 2)$ and $(2, 0)$. Thus, the area of the triangle is $\frac{1}{2}(2)(2) = 2$.

When $a = 2$, the slope of the tangent line at $(2, \frac{1}{2})$ is $-\frac{1}{4}$, and the equation of the tangent line is $y = -\frac{1}{4}x + 1$. The tangent line intersects the axes at $(0, 1)$ and $(4, 0)$. Thus, the area of the triangle is $\frac{1}{2}(4)(1) = 2$.

In general, the equation of the tangent line is $y = -\left(\frac{1}{a^2}\right)x + \frac{2}{a}$. The tangent line intersects the axes at $(0, \frac{2}{a})$ and $(2a, 0)$. Thus, the area of the triangle is

$$\frac{1}{2}(2a)\left(\frac{2}{a}\right) = 2$$

49. **(a)** $g'(x) = \lim_{h \to 0} \frac{g(x+h) - g(x)}{h}$

$= \lim_{h \to 0} \frac{1}{h}\left[\max_{a \le t \le x+h} f(t) - \max_{a \le t \le x} f(t)\right]$

$= \lim_{h \to 0} \frac{1}{h}\left[f(x+h) - f(x)\right]$

$= f'(x)$

(b) $g'(x) = \lim_{h \to 0} \frac{g(x+h) - g(x)}{h}$

$= \lim_{h \to 0} \frac{1}{h}\left[\max_{a \le t \le x+h} f(t) - \max_{a \le t \le x} f(t)\right]$

$= \lim_{h \to 0} \frac{1}{h}\left[f(a) - f(a)\right]$

$= 0$

51. If $d(t)$ represents the national debt, then $d'(t)$ represents the rate of change of the national

debt. The debt itself, by implication, is increasing and therefore $d'(t) > 0$.

Since the rate of increase has been reduced, this implies $d''(t)$ is being reduced. We cannot conclude anything about the size of $d(t)$.

53. $w(b) = cb^{3/2} nl\, w'(b) = \frac{3c}{2}b^{1/2} = \frac{3c\sqrt{b}}{2}$

$w'(b) > 1$ when

$$\frac{3c\sqrt{b}}{2} > 1, \quad \sqrt{b} > \frac{2}{3c}\, b > \frac{4}{9c^2}.$$

Since c is constant, when b is large enough, b will be greater than $\frac{4}{9c^2}$. After this point, when b increases by 1 unit, the leg width w is increasing by more than 1 unit, so that leg width is increasing faster than body length.

This puts a limitation on the size of land animals since, eventually, the body will not be long enough to accomodate the width of the legs.

55. We can approximate $f'(2000) \approx \frac{9039.5 - 8690.7}{2001 - 1999}$ $= 174.4$. This is the rate of change of the GDP in billions of dollars per year.

To approximate $f''(2000)$, we first estimate $f'(1999) \approx \frac{9016.8 - 8347.3}{2000 - 1998} = 334.75$ and $f'(1998) \approx \frac{8690.7 - 8004.5}{1999 - 1997} = 343.1$.

Since these values are decreasing, $f''(2000)$ is negative. We estimate $f''(2000) \approx \frac{174.4 - 334.75}{2000 - 1999}$ $= -160.35$. This represents the rate of change of the rate of change of the GDP over time. In 2000, the GDP is increasing by a rate of 174.4 billion dollars per year, but this increase is decreasing by a rate of 160.35 billion dollars-per-year per year.

57. Newton's Law states that force equals mass times acceleration. That is, if $F(t)$ is the driving force at time t, then $m \cdot f''(t) = m \cdot a(t) = F(t)$ in which m is the mass, appropriately unitized. The third derivative of the distance function is then $f'''(t) = a'(t) = \frac{1}{m}F'(t)$. It is both the derivative of the acceleration and directly proportional to the rate of change in force. Thus an abrupt change in acceleration

59. Try $f(x) = cx^4$ for some constant c. Then $f'(x) = 4cx^3$ so c must be 1. One possible answer is $f(x) = x^4$.

61. $f'(x) = \sqrt{x} = x^{1/2}$

$f(x) = \dfrac{2}{3}x^{3/2}$ is one possible function.

63. $\displaystyle\lim_{h\to 0} \frac{f(a+h) - 2f(a) + f(a-h)}{h^2}$

$= \displaystyle\lim_{h\to 0} \left[\frac{f(a+h) - f(a)}{h^2} \right.$

$\left. - \frac{[f(a) - f(a-h)]}{h^2} \right]$

$= \displaystyle\lim_{h\to 0} \frac{1}{h} \left[\frac{f(a+h) - f(a)}{h} \right.$

$\left. - \frac{f(a) - f(a-h)}{h} \right]$

$= \displaystyle\lim_{h\to 0} \frac{1}{h} \left[\lim_{h\to 0} \frac{f(a+h) - f(a)}{h} \right.$

$\left. - \lim_{h\to 0} \frac{f(a) - f(a-h)}{h} \right]$

$= \displaystyle\lim_{h\to 0} \frac{1}{h} \left[f'(a) - f'(a-h) \right]$

Now let $k = -h$ in the previous equation, to get

$\displaystyle\lim_{h\to 0} \frac{f(a+h) - 2f(a) + f(a-h)}{h^2}$

$= \displaystyle\lim_{k\to 0} \frac{1}{-k} \left[f'(a) - f'(a+k) \right]$

$= \displaystyle\lim_{k\to 0} \frac{1}{k} \left[f'(a+k) - f'(a) \right]$

$= f''(a)$

2.4 THE PRODUCT AND QUOTIENT RULES

1. $f(x) = (x^2 + 3)(x^3 - 3x + 1)$

$f'(x) = \dfrac{d}{dx}(x^2 + 3) \cdot (x^3 - 3x + 1)$

$\quad + (x^2 + 3) \cdot \dfrac{d}{dx}(x^3 - 3x + 1)$

$\quad = (2x)(x^3 - 3x + 1)$

$\quad + (x^2 + 3)(3x^2 - 3)$

3. $f(x) = (\sqrt{x} + 3x)\left(5x^2 - \dfrac{3}{x}\right)$

$\quad = (x^{1/2} + 3x)(5x^2 - 3x^{-1})$

$f'(x) = \left(\dfrac{1}{2}x^{-1/2} + 3\right)(5x^2 - 3x^{-1})$

$\quad + (x^{1/2} + 3x)(10x + 3x^{-2})$

5. $f(x) = \dfrac{3x-2}{5x+1}$

$f'(x) = \dfrac{\left((5x+1)\frac{d}{dx}(3x-2) - (3x-2)\frac{d}{dx}(5x+1)\right)}{(5x+1)^2}$

$\quad = \dfrac{3(5x+1) - (3x-2)5}{(5x+1)^2}$

$\quad = \dfrac{15x+3 - 15x+10}{(5x+1)^2} = \dfrac{13}{(5x+1)^2}$

7. $f(x) = \dfrac{3x - 6\sqrt{x}}{5x^2 - 2} = \dfrac{3(x - 2x^{1/2})}{5x^2 - 2}$

$f'(x) =$

$3\dfrac{\left((5x^2-2)\frac{d}{dx}(x-2x^{1/2}) - (x-2x^{1/2})\frac{d}{dx}(5x^2-2)\right)}{(5x^2-2)^2}$

$= 3\dfrac{\left((5x^2-2)(1-x^{-1/2}) - (x-2x^{1/2})(10x)\right)}{(5x^2-2)^2}$

9. $f(x) = \dfrac{(x+1)(x-2)}{x^2 - 5x + 1} = \dfrac{x^2 - x - 2}{x^2 - 5x + 1}$

$f'(x) =$

$\dfrac{\left((x^2-5x+1)\frac{d}{dx}(x^2-x-2) - (x^2-x-2)\frac{d}{dx}(x^2-5x+1)\right)}{(x^2-5x+1)^2}$

$= \dfrac{\left((x^2-5x+1)(2x-1) - (x^2-x-2)(2x-5)\right)}{(x^2-5x+1)^2}$

11. We do not recommend treating this one as a quotient, but advise preliminary simplification.

$$f(x) = \frac{x^2 + 3x - 2}{\sqrt{x}}$$

$$= \frac{x^2}{\sqrt{x}} + \frac{3x}{\sqrt{x}} - \frac{2}{\sqrt{x}}$$

$$= x^{3/2} + 3x^{1/2} - 2x^{-1/2}$$

$$f'(x) = \frac{3}{2}x^{1/2} + \frac{3}{2}x^{-1/2} + x^{-3/2}$$

13. We simplify instead of using the product rule.

$$f(x) = x\left(\sqrt[3]{x} + 3\right) = x^{4/3} + 3x$$

$$f'(x) = \frac{4}{3}x^{1/3} + 3$$

15. $f(x) = (x^2 - 1)\frac{x^3 + 3x^2}{x^2 + 2}$

$$f'(x) =$$

$$\frac{d}{dx}(x^2 - 1) \cdot \left(\frac{x^3 + 3x^2}{x^2 + 2}\right)$$

$$+ (x^2 - 1) \cdot \frac{d}{dx}\left(\frac{x^3 + 3x^2}{x^2 + 2}\right)$$

We have

$$\frac{d}{dx}\left(\frac{x^3 + 3x^2}{x^2 + 2}\right) =$$

$$\frac{(x^2 + 2)\frac{d}{dx}(x^3 + 3x^2) - (x^3 + 3x^2)\frac{d}{dx}(x^2 + 2)}{(x^2 + 2)^2}$$

$$= \frac{(x^2 + 2) \cdot (3x^2 + 6x) - (x^3 + 3x^2) \cdot (2x)}{(x^2 + 2)^2}$$

$$= \frac{3x^4 + 6x^2 + 6x^3 + 12x - (2x^4 + 6x^3)}{(x^2 + 2)^2}$$

$$= \frac{x^4 + 6x^2 + 12x}{(x^2 + 2)^2}$$

so $f'(x) =$

$$(2x) \cdot \left(\frac{x^3 + 3x^2}{x^2 + 2}\right) + (x^2 - 1) \cdot \frac{x^4 + 6x^2 + 12x}{(x^2 + 2)^2}$$

17. $\dfrac{d}{dx}\left[f(x)g(x)h(x)\right]$

$$= \frac{d}{dx}\left[(f(x)g(x))\,h(x)\right]$$

$$= (f(x)g(x))\,h'(x) + h(x)\frac{d}{dx}\,(f(x)g(x))$$

$$= (f(x)g(x))\,h'(x)$$

$$+ h(x)\left(f(x)g'(x) + g(x)f'(x)\right)$$

$$= f'(x)g(x)h(x)$$

$$+ f(x)g'(x)h(x) + f(x)g(x)h'(x)$$

In the general case of a product of n functions, the derivative will have n terms to be added, each term a product of all but one of the functions multiplied by the derivative of the remaining function.

19. $f'(x) = \left[\frac{d}{dx}(x^{2/3})\right](x^2 - 2)(x^3 - x + 1)$

$$+ x^{2/3}\left[\frac{d}{dx}(x^2 - 2)\right](x^3 - x + 1)$$

$$+ x^{2/3}(x^2 - 2)\frac{d}{dx}(x^3 - x + 1)$$

$$= \frac{2}{3}x^{-1/3}(x^2 - 2)(x^3 - x + 1)$$

$$+ x^{2/3}(2x)(x^3 - x + 1)$$

$$+ x^{2/3}(x^2 - 2)(3x^2 - 1)$$

21. $h(x) = f(x)g(x)$
$h'(x) = f'(x)g(x) + f(x)g'(x)$

(a) $h(1) = f(1)g(1) = (-2)(1) = -2$

$$h'(1) = f'(1)g(1) + f(1)g'(1)$$

$$= (3)(1) + (-2)(-2) = 7$$

So the equation of the tangent line is
$y = 7(x - 1) - 2$ or $y = 7x + 9$.

(b) $h(0) = f(0)g(0) = (-1)(3) = -3$

$$h'(0) = f'(0)g(0) + f(0)g'(0)$$

$$= (-1)(3) + (-1)(-1)$$

$$= -2$$

So the equation of the tangent line is
$y = -2x - 3$.

23. $h(x) = x^2 f(x)$
$h'(x) = 2xf(x) + x^2 f'(x)$

(a) $h(1) = 1^2 f(1) = -2$

$$h'(1) = 2(1)f(1) + 1^2 f'(1)$$

$$= (2)(-2) + 3 = -1$$

So the equation of the tangent line is
$y = -(x - 1) - 2$ or $y = -x - 1$.

(b) $h(0) = 0^2 f(0) = 0$
$h'(0) = 2(0)f(0) + 0^2 f'(0) = 0$
So the equation of the tangent line is
$y = 0$.

25. The rate at which the quantity Q changes is Q'. Since the amount is said to be "decreasing at a rate of 4%" we have to ask "4% of *what*?" The answer in this type of context is usually 4% of *itself*. In other words, $Q' = -0.04Q$. As for P, the 3% rate of increase would translate as $P' = 0.03P$. By the product rule, with $R = PQ$, we have:

$$R' = (PQ)' = P'Q + PQ'$$
$$= (0.03P)Q + P(-0.04Q)$$
$$= -(0.01)PQ = (-0.01)R.$$

In other words, revenue is decreasing at a rate of 1%.

27. $R' = Q'P + QP'$
At a certain moment of time (call it t_0) we are given $P(t_0) = 20$ ($/item)
$Q(t_0) = 20,000$ (items)
$P'(t_0) = 1.25$ ($/item/year)
$Q'(t_0) = 2,000$ (items/year)
$\Rightarrow R'(t_0) = 2,000(20) + (20,000)1.25$
$\qquad = 65,000$ $/year
So revenue is increasing by $65,000/year at the time t_0.

29. If $u(m) = \dfrac{82.5m - 6.75}{m + 0.15}$ then using the quotient rule,

$$\frac{du}{dm} = \frac{(m + 0.15)(82.5) - (82.5m - 6.75)1}{(m + 0.15)^2}$$

$$= \frac{19.125}{(m + 0.15)^2}$$

which is clearly positive. It seems to be saying that initial ball speed is an increasing function of the mass of the bat. Meanwhile,
$$u'(1) = \frac{19.125}{1.15^2} \approx 14.46$$
$$u'(1.2) = \frac{19.125}{1.35^2} \approx 10.49,$$
which suggests that the rate at which this speed is increasing is decreasing.

31. If $u(m) = \dfrac{14.11}{m + 0.05} = \dfrac{282.2}{20m + 1}$, then

$$\frac{du}{dm} = \frac{(20m + 1) \cdot 0 - 282.2(20)}{(20m + 1)^2}$$

$$= \frac{-5644}{(20m + 1)^2}$$

This is clearly negative, which means that impact speed of the ball is a decreasing function of the weight of the club. It appears that the explanation may have to do with the stated fact that the speed of the club is inversely proportional to its mass. Although the lesson of Example 4.6 was that a heavier club makes for greater ball velocity, that was assuming a fixed club speed, quite a different assumption from this problem.

33. $f'(x) = \lim\limits_{h \to 0} \dfrac{f(x + h) - f(x)}{h}$

$f'(0) = \lim\limits_{h \to 0} \dfrac{f(h) - f(0)}{h}$

$\qquad = \lim\limits_{h \to 0} \dfrac{hg(h) - 0}{h}$

$\qquad = \lim\limits_{h \to 0} \dfrac{hg(h)}{h}$

$\qquad = \lim\limits_{h \to 0} g(h)$

$\qquad = g(0)$

since g is continuous at $x = 0$.

When $g(x) = |x|$, $g(x)$ is continuous but not differentiable at $x = 0$. We have $f(x) = x|x| = \begin{cases} -x^2 & x < 0 \\ x^2 & x \geq 0. \end{cases}$ This is differentiable at $x = 0$.

35. Answers depend on CAS.

37. For any constant k, the derivative of $\sin kx$ is $k \cos kx$.

Graph of $\dfrac{d}{dx}\sin x$:

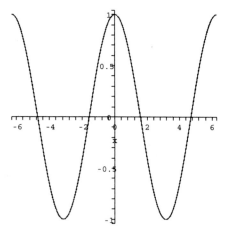

Graph of $\dfrac{d}{dx}\sin 2x$:

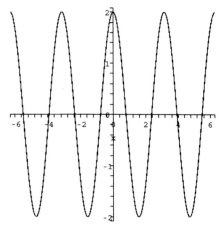

Graph of $\dfrac{d}{dx}\sin 3x$:

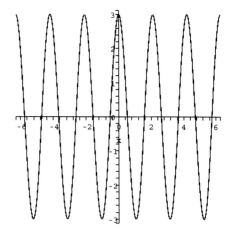

39. CAS answers may vary.

41. If $F(x) = f(x)g(x)$ then

$$F'(x) = f'(x)g(x) + f(x)g'(x) \text{ and}$$
$$F''(x) = f''(x)g(x) + f'(x)g'(x)$$
$$+ f'(x)g'(x) + f(x)g''(x)$$
$$= f''(x)g(x) + 2f'(x)g'(x)$$
$$+ f(x)g''(x)$$
$$F'''(x) = f'''(x)g(x) + f''(x)g'(x)$$
$$+ 2f''(x)g'(x) + 2f'(x)g''(x)$$
$$+ f'(x)g''(x) + f(x)g'''(x)$$
$$= f'''(x)g(x) + 3f''(x)g'(x)$$
$$+ 3f'(x)g''(x) + f(x)g'''(x)$$

One can see obvious parallels to the binomial coefficients as they come from Pascal's Triangle:
$$(a+b)^2 = a^2 + 2ab + b^2$$
$$(a+b)^3 = a^3 + 3a^2b + 3ab^2 + b^3.$$

On this basis, one could correctly predict the pattern of the fourth or any higher derivative.

43. If $g(x) = [f(x)]^2 = f(x)f(x)$, then $g'(x) = f'(x)f(x) + f(x)f'(x)$
$$= 2f(x)f'(x).$$

45.
$$\left(P + \frac{n^2a}{V^2}\right)(V - nb) = nRT$$

$$P + \frac{n^2a}{V^2} = \frac{nRT}{V - nb}$$

$$P = \frac{nRT}{V - nb} - \frac{n^2a}{V^2}$$

$$P'(V) = \frac{-nRT}{(V - nb)^2} + \frac{2n^2a}{V^3}$$

$$P''(V) = \frac{2nRT}{(V - nb)^3} + \frac{6n^2a}{V^4}.$$

Obviously, if $P'(V) = 0$, then
$$\frac{2na}{V^3} = \frac{RT}{(V - nb)^2}(= X)$$
in which X is a temporary name. If $P''(V)$ is *also* zero, then

$$0 = P''(V) = \frac{2nX}{(V-nb)} - \frac{3nX}{V}$$

$$= nX\left[\frac{2}{V-nb} - \frac{3}{V}\right] = \frac{nX(3nb-V)}{V(V-nb)},$$

$$\Rightarrow V = 3nb, \text{ so } V - nb = 2nb, \text{ and}$$

$$X = \frac{2na}{V^3} = \frac{2a}{27n^2b^3}.$$

$$RT = (V-nb)^2 X = 4n^2b^2 X = \frac{8a}{27b},$$

so $T = \dfrac{8a}{27bR}$, and since

$$P = \frac{nRT}{V-nb} - \frac{n^2a}{V^2}, \text{ we have}$$

$$P = \frac{8an}{27b(2nb)} - \frac{n^2a}{9n^2b^2} = \frac{a}{27b^2}.$$

In summary,

$$(T_c, P_c, V_c) = \left(\frac{8a}{27bR}, \frac{a}{27b^2}, 3nb\right)$$

Substitute in the given numbers; in particular $T_c = 647°$ (Kelvin).

47.
$$f(x) = \frac{x^{2.7}}{1+x^{2.7}}$$

$$f'(x) = \frac{\left(1+x^{2.7}\right) \cdot 2.7x^{1.7} - 2.7x^{1.7} \cdot \left(x^{2.7}\right)}{\left(1+x^{2.7}\right)^2}$$

$$= \frac{2.7x^{1.7}}{\left(1+x^{2.7}\right)^2}$$

The fact that $f'(x) > 0$ when $x > 0$ suggest to us that the amount of the enzyme continues to increase as the amount of the activator increases.

49. $\dfrac{d}{dx}\left[x^3 f(x)\right] = 3x^2 \cdot f(x) + x^3 f'(x)$

51. Utilizing $\dfrac{d}{dx}(\sqrt{x}) = \dfrac{1}{2\sqrt{x}}$ (which is a special case of the power rule), we find

$$\frac{d}{dx}\left(\frac{\sqrt{x}}{f(x)}\right) = \frac{f(x)\frac{1}{2\sqrt{x}} - \sqrt{x} f'(x)}{[f(x)]^2}$$

$$= \frac{f(x) - 2xf'(x)}{2\sqrt{x}[f(x)]^2}.$$

2.5 THE CHAIN RULE

1. $f(x) = (x^3 - 1)^2$
Using the chain rule:
$$f'(x) = 2(x^3 - 1)(3x^2) = 6x^2(x^3 - 1)$$
Using the product rule:

$$f(x) = (x^3 - 1)(x^3 - 1)$$
$$f'(x) = (3x^2)(x^3 - 1) + (x^3 - 1)(3x^2)$$
$$= 2(3x^2)(x^3 - 1)$$
$$= 6x^2(x^3 - 1)$$

Using preliminary multiplication:

$$f(x) = x^6 + 2x^3 + 1$$
$$f'(x) = 6x^5 + 6x^2$$
$$= 6x^2(x^3 - 1)$$

3. $f(x) = (x^2 + 1)^3$
Chain rule:
$$f'(x) = 3(x^2 + 1)^2 \cdot 2x = 6x(x^2 + 1)^2$$
Using preliminary multiplication:
$$f(x) = x^6 + 3x^4 + 3x^2 + 1$$
$$f'(x) = 6x^5 + 12x^3 + 6x$$

5. $f(x) = \sqrt{x^2 + 4} = (x^2 + 4)^{1/2}$
$$f'(x) = \frac{1}{2}(x^2 + 4)^{-1/2} \cdot 2x$$

$$= \frac{1}{2\sqrt{x^2 + 4}} \cdot 2x$$

$$= \frac{x}{\sqrt{x^2 + 4}}$$

7. $f(x) = x^5\sqrt{x^3 + 2}$
$$f'(x) = x^5\frac{1}{2\sqrt{x^3 + 2}}3x^2 + 5x^4\sqrt{x^3 + 2}$$

$$= \frac{3x^7 + 10x^4(x^3 + 2)}{2\sqrt{x^3 + 2}}$$

$$= \frac{13x^7 + 20x^4}{2\sqrt{x^3 + 2}}$$

9.

$$f(x) = \frac{x^3}{(x^2 + 4)^2}$$

$$f'(x) = \frac{3x^2(x^2 + 4)^2 - 2(x^2 + 4)(2x)x^3}{(x^2 + 4)^4}$$

$$= \frac{3x^4 + 12x^2 - 4x^4}{(x^2 + 4)^3}$$

$$= \frac{x^2(12 - x^2)}{(x^2 + 4)^3}$$

11.

$$f(x) = \frac{6}{\sqrt{x^2 + 4}} = 6(x^2 + 4)^{-1/2}$$

$$f'(x) = -3(x^2 + 4)^{-3/2} \cdot 2x$$

$$= \frac{-6x}{(x^2 + 4)^{3/2}}$$

13.

$$f(x) = (\sqrt{x} + 3)^{4/3}$$

$$f'(x) = \frac{4(\sqrt{x} + 3)^{1/3}}{3} \cdot \frac{1}{2\sqrt{x}}$$

$$= \frac{2(\sqrt{x} + 3)^{1/3}}{3\sqrt{x}}$$

15.

$$f(x) = \left(\sqrt{x^3 + 2} + 2x\right)^{-2}$$

$$f'(x) =$$

$$-2\left(\sqrt{x^3 + 2} + 2x\right)^{-3}\left[\frac{3x^2}{2\sqrt{x^3 + 2}} + 2\right]$$

$$= -\frac{3x^2 + 4\sqrt{x^3 + 2}}{(\sqrt{x^3 + 2} + 2x)^3 \cdot \sqrt{x^3 + 2}}$$

17.

$$f(x) = \frac{x}{\sqrt{x^2 + 1}}$$

$$f'(x) = \frac{\sqrt{x^2 + 1} - x\left(\frac{1}{2\sqrt{x^2+1}}\right)2x}{x^2 + 1}$$

$$= \frac{1}{(x^2 + 1)\sqrt{x^2 + 1}}$$

19.

$$f(x) = \sqrt{\frac{x}{x^2 + 1}}$$

$$f'(x) = \frac{1}{2\sqrt{\frac{x}{x^2+1}}} \cdot \frac{(x^2 + 1) - 2x^2}{(x^2 + 1)^2}$$

$$= \frac{1 - x^2}{2\sqrt{x}(x^2 + 1)^{3/2}}$$

21.

$$f(x) = \sqrt[3]{x\sqrt{x^4 + 2x\sqrt[4]{\frac{8}{x + 2}}}}$$

$$f(x) = \left(x\left[x^4 + 2x\left(\frac{8}{x + 2}\right)^{1/4}\right]^{1/2}\right)^{1/3}$$

$$f'(x) = \frac{1}{3}\left(x\left[x^4 + 2x\left(\frac{8}{x + 2}\right)^{1/4}\right]^{1/2}\right)^{-2/3} \cdot$$

$$\left(\left[x^4 + 2x\left(\frac{8}{x + 2}\right)^{1/4}\right]^{1/2} + \right.$$

$$+ x\left(\frac{1}{2}\right)\left[x^4 + 2x\left(\frac{8}{x + 2}\right)^{1/4}\right]^{-1/2} \cdot$$

$$\left[4x^3 + 2\left(\frac{8}{x + 2}\right)^{1/4}\right.$$

$$\left.\left. + 2x\left(\frac{1}{4}\right)\left(\frac{8}{x + 2}\right)^{-3/4}\left(\frac{-8}{(x + 2)^2}\right)\right]\right)$$

23.

$$f(x) = \sqrt{x^2 + 16}, \quad a = 3, \quad f(3) = 5$$

$$f'(x) = \frac{1}{2\sqrt{x^2 + 16}}(2x) = \frac{x}{\sqrt{x^2 + 16}}$$

$$f'(3) = \frac{3}{\sqrt{3^2 + 16}} = \frac{3}{5}$$

So the tangent line is $y = \frac{3}{5}(x - 3) + 5$

or $y = \frac{3}{5}x + \frac{16}{5}$.

25. $s(t) = \sqrt{t^2 + 8}$

$$v(t) = s'(t) = \frac{2t}{2\sqrt{t^2 + 8}} = \frac{t}{\sqrt{t^2 + 8}} \text{ m/s}$$

$$v(2) = \frac{2}{\sqrt{12}} = \frac{1}{\sqrt{3}} = \frac{\sqrt{3}}{3} \text{ m/s}$$

27. For higher derivatives, fractional exponents will be required.

$$f(x) = \sqrt{2x + 1} = (2x + 1)^{1/2}$$

$$f'(x) = \frac{1}{2}(2x + 1)^{-1/2} \cdot 2 = (2x + 1)^{-1/2}$$

$$f''(x) = -\frac{1}{2}(2x + 1)^{-3/2}(2)$$

$$= -(2x + 1)^{-3/2}$$

$$f'''(x) = -\left(-\frac{3}{2}\right)(2x + 1)^{-5/2} \cdot 2$$

$$= 3(2x + 1)^{-5/2}$$

$$f^{(4)}(x) = 3\left(-\frac{5}{2}\right)(2x + 1)^{-7/2} \cdot 2$$

$$= -15(2x + 1)^{-7/2}$$

$$f^{(n)}(x) =$$

$$(-1)^{n+1}1 \cdot 3 \cdots (2n - 3)(2x + 1)^{-(2n-1)/2}$$

29. $h'(1) = f'(g(1))g'(1)$
$g(1) = 4$, so $h'(1) = f'(4)g'(1)$.

From the table, we have:

$$f'(4) \approx \frac{2 - (-2)}{5 - 3} = 2, \text{ and}$$

$$g'(1) \approx \frac{6 - 2}{2 - 0} = 2 \text{ so}$$
$$h'(1) \approx 4.$$

31. $k'(3) = g'(f(3))f'(3)$
$f(3) = -2$, so $k'(3) = g'(-2)f'(3)$.

From the table, we have:

$$f'(3) \approx \frac{0 - (-3)}{4 - 2} = \frac{3}{2}, \text{ and}$$

$$g'(-2) \approx \frac{2 - 6}{-1 - (-3)} = -2 \text{ so}$$
$$k'(1) \approx -3.$$

33. $h'(x) = f'(g(x))g'(x)$

$$h'(1) = f'(g(1))g'(1)$$

$$= f'(2) \cdot (-2) = -6$$

35. $f(x) = (x^2 + 3)^2 \cdot 2x$. Recognizing the "$2x$" as the derivative of $x^2 + 3$, we guess $g(x) = c(x^2 + 3)^3$ where c is some constant. $g'(x) = 3c(x^2 + 3)^2 \cdot 2x$, which will be $f(x)$ only if $3c = 1$, so $c = 1/3$, and $g(x) = \dfrac{(x^2 + 3)^3}{3}$.

37. $f(x) = \dfrac{x}{\sqrt{x^2 + 1}}$.
Recognizing the "x" as half the derivative of $x^2 + 1$, and knowing that differentiation throws the square root into the denominator, we guess $g(x) = c\sqrt{x^2 + 1}$ where c is some constant and find that

$$g'(x) = \frac{c}{2\sqrt{x^2 + 1}}(2x)$$

will match $f(x)$ if $c = 1$, so

$$g(x) = \sqrt{x^2 + 1}.$$

39. As a temporary device given *any* f, set $g(x) = f(-x)$. Then by the chain rule,

$$g'(x) = f'(-x)(-1) = -f'(-x).$$

In the even case ($g = f$) this reads $f'(-x) = -f'(x)$ and shows f' is odd. In the odd case ($g = -f$ and therefore $g' = -f'$), this reads $-f'(x) = -f'(-x)$ or $f'(x) = f'(-x)$ and shows f' is even.

41. $\dfrac{d}{dx}f(\sqrt{x}) = f'(\sqrt{x}) \cdot \dfrac{d}{dx}\sqrt{x}$

$$= f'(\sqrt{x}) \cdot \frac{1}{2\sqrt{x}}$$

43.

$$\frac{d}{dx}\left(\frac{1}{1+[f(x)]^2}\right)$$

$$= -\left(\frac{1}{1+[f(x)]^2}\right)^2 \cdot \frac{d}{dx}\left(1+[f(x)]^2\right)$$

$$= -\frac{1}{\left(1+[f(x)]^2\right)^2}\cdot 2f(x)\cdot f'(x)$$

45. $f'(x) = b'(a(x))a'(x)$.
$a(2) = 0$, $b'(0) = -3$, $a'(2) = 2$, so
$f'(2) = -3\cdot 2 = -6$.

47. $f'(x) = c'(a(x))a'(x)$.
$a(-1) = 0$, $c'(0) = -3$, $a'(-1) = -2$,
so $f'(-1) = -3\cdot -2 = 6$.

49. $f(x) = \left(x^3 - 3x^2 + 2x\right)^{1/3}$
$f'(x) =$
$\frac{1}{3}\left(x^3 - 3x^2 + 2x\right)^{-2/3}\cdot\left(3x^2 - 6x + 2\right)$
The derivative of f does not exist at values of x for which

$$0 = x^3 - 3x^2 + 2x$$

$$= x\left(x^2 - 3x + 2\right)$$

$$= x\,(x-1)\,(x-2)$$

Thus, the derivative of f does not exist for $x = 0, 1, 2$. The derivative fails to exist at these points because the tangent lines at these points are vertical.

2.6 DERIVATIVES OF TRIGONOMETRIC FUNCTIONS

1. The peaks and valleys of $\cos(x)$ (e.g., 0, π, 2π, etc.) are matched with the zeros of $\sin(x)$, and the decreasing intervals for $\cos(x)$ (e.g., $[0, \pi]$) correspond to the intervals where $\sin(x)$ is positive, hence where $-\sin(x)$ is negative. These features lend credibility to the notion that $-\sin(x)$ might be the derivative of $\cos(x)$.

3. $f(x) = 4\sin x - x$
$f'(x) = 4\cos x - 1$

5. $f(x) = \tan^3 x - \csc^4 x$
$f'(x) = 3\tan^2 x \sec^2 x$
$\qquad + 4\csc^3 x \csc x \cot x$
$\qquad = 3\tan^2 x \sec^2 x + 4\csc^4 x \cot x$

7. $f(x) = x\cos 5x^2$
$f'(x) = (1)\cos 5x^2 + x(-\sin 5x^2)\cdot 10x$
$\qquad = \cos 5x^2 - 10x^2 \sin 5x^2$

9. $f(x) = \sin(\tan(x^2))$
$f'(x) = \cos(\tan(x^2))\cdot \sec^2(x^2)\cdot 2x$

11.
$$f(x) = \frac{\sin(x^2)}{x^2}$$

$$f'(x) = \frac{x^2\cos(x^2)\cdot 2x - \sin(x^2)\cdot 2x}{x^4}$$

$$= \frac{2x[x^2\cos(x^2) - \sin(x^2)]}{x^4}$$

$$= \frac{2[x^2\cos(x^2) - \sin(x^2)]}{x^3}$$

13. $f(t) = \sin t \sec t = \tan t$
$f'(t) = \sec^2 t$

15.
$$f(x) = \frac{1}{\sin(4x)} = \csc(4x)$$

$f'(x) = -\csc(4x)\cot(4x)\cdot(4)$
$\qquad = -4\csc(4x)\cot(4x)$
$$\qquad = \frac{-4\cos(4x)}{\sin^2(4x)}$$

17. $f(x) = 2\sin x\cos x$
$f'(x) = 2\cos x\cdot\cos x + 2\sin x(-\sin x)$
$\qquad = 2\cos^2 x - 2\sin^2 x$

19. $f(x) = \tan \sqrt{x^2 + 1}$

$f'(x) = (\sec^2 \sqrt{x^2 + 1}) \cdot$

$\left(\dfrac{1}{2}\right)(x^2 + 1)^{-1/2}(2x)$

$= \dfrac{x}{\sqrt{x^2 + 1}} \sec^2 \sqrt{x^2 + 1}$

21. Answers depend on CAS.

23. Answers depend on CAS.

25. $f(x) = \sin 4x, \quad a = \dfrac{\pi}{8},$

$f\left(\dfrac{\pi}{8}\right) = \sin \dfrac{\pi}{2} = 1$

$f'(x) = 4 \cos 4x$

$f'\left(\dfrac{\pi}{8}\right) = 4 \cos \dfrac{\pi}{2} = 0$

So the equation of the tangent line is

$y = 0\left(x - \dfrac{\pi}{8}\right) + 1$ or $y = 1$.

27. $f(x) = \cos x, \quad a = \dfrac{\pi}{2},$

$f\left(\dfrac{\pi}{2}\right) = \cos \dfrac{\pi}{2} = 0$

$f'(x) = -\sin x$

$f'\left(\dfrac{\pi}{2}\right) = -\sin \dfrac{\pi}{2} = -1$

So the equation of the tangent line is

$y = -1\left(x - \dfrac{\pi}{2}\right) + 0$ or $y = -x + \pi/2$.

29. $s(t) = t^2 - \sin(2t), \quad t_0 = 0$

$v(t) = s'(t) = 2t - 2\cos(2t)$

$v(0) = 0 - 2\cos(0) = 0 - 2 = -2$ ft/s

31. $s(t) = \dfrac{\cos t}{t}, \quad t_0 = \pi$

$v(t) = s'(t)$

$= \dfrac{-1}{t^2} \cos t + \dfrac{1}{t}(-\sin t)$

$v(\pi) = -\dfrac{\cos \pi}{\pi^2} - \dfrac{\sin \pi}{\pi}$

$= \dfrac{1}{\pi^2} - \dfrac{1}{\pi}(0) = \dfrac{1}{\pi^2}$ ft/s

33. $f(t) = 4 \sin 3t$

$f'(t) = 12 \cos 3t$

The maximum speed of 12 occurs when the vertical position is zero.

35. $Q(t) = 3 \sin 2t + t + 4$

$I(t) = \dfrac{dQ}{dt} = 6 \cos 2t + 1$ At time $t = 0$,

$I(0) = 7$ amps. At time $t = 1$, $I(1) = 6 \cos 2 + 1 \approx -1.497$ amps.

37. $f(x) = \sin x$

$f'(x) = \cos x$

$f''(x) = -\sin x$

$f'''(x) = -\cos x$

$f^{(4)}(x) = \sin x = f(x)$

$\Rightarrow f^{(75)}(x) = (f^{(72)})^{(3)}(x)$

$= (f^{(18 \cdot 4)})^{(3)}(x)$

$= f'''(x) = -\cos x$

$f^{(150)}(x) = (f^{(148)})^{(2)}(x)$

$= (f^{(37 \cdot 4)})^{(2)}(x)$

$= f''(x) = -\sin x$

39. Since $0 \leq \sin \theta \leq \theta$, we have
$-\theta \leq -\sin \theta \leq 0$ which implies
$-\theta \leq \sin(-\theta) \leq 0$
so for $-\dfrac{\pi}{2} \leq \theta \leq 0$ we have
$\theta \leq \sin \theta \leq 0$.
We also know that
$\lim_{\theta \to 0^-} \theta = 0 = \lim_{\theta \to 0^-} 0$,nl so the Squeeze Theorem implies that
$\lim_{\theta \to 0^-} \sin \theta = 0.$

41. If $f(x) = \cos(x)$, then

$$\frac{f(x+h) - f(x)}{h}$$

$$= \frac{\cos(x+h) - \cos(x)}{h}$$

$$= \frac{\cos x \cos h - \sin x \sin h - \cos x}{h}$$

$$= (\cos x)\frac{(\cos h - 1)}{h} - (\sin x)\left(\frac{\sin h}{h}\right).$$

Taking the limit

$$f'(x) = \lim_{h \to 0} \frac{f(x+h) - f(x)}{h}$$

$$= (\cos x) \cdot \lim_{h \to 0} \frac{\cos h - 1}{h}$$

$$- (\sin x) \cdot \lim_{h \to 0} \frac{\sin h}{h}$$

$$= \cos x \cdot 0 - \sin x \cdot 1$$

$$= -\sin x.$$

43. (a)
$$\lim_{x \to 0} \frac{\sin 3x}{x} = \lim_{x \to 0} \frac{3 \sin 3x}{3x}$$

$$= 3 \cdot \lim_{x \to 0} \frac{\sin(3x)}{(3x)}$$

$$= 3 \cdot 1 = 3$$

(b)
$$\lim_{t \to 0} \frac{\sin t}{4t} = \frac{1}{4} \lim_{t \to 0} \frac{\sin t}{t}$$

$$= \frac{1}{4} \cdot 1 = \frac{1}{4}$$

(c)
$$\lim_{x \to 0} \frac{\cos x - 1}{5x}$$

$$= \frac{1}{5} \lim_{x \to 0} \frac{\cos x - 1}{x} = 0$$

(d) Let $u = x^2$: then $u \to 0$ as $x \to 0$, and

$$\lim_{x \to 0} \frac{\sin x^2}{x^2} = \lim_{u \to 0} \frac{\sin u}{u} = 1$$

45. The function defined by $y = \frac{\sin x}{x}$ is continuous and differentiable on its entire domain.

We need only check that f has these properties at $x = 0$.

In Example 2.4 of Chapter 1, we showed numerically that $\lim_{x \to 0} \frac{\sin x}{x} = 1$. Therefore, since $\lim_{x \to 0} f(x) = f(0)$, f is continuous at 0.

Using a table of values, we can show that

$$\lim_{h \to 0} \frac{f(0+h) - f(0)}{h}$$

$$= \lim_{h \to 0} \frac{\frac{\sin(0+h)}{0+h} - 1}{h}$$

$$= \lim_{h \to 0} \frac{1}{h}\left[\frac{\sin h}{h} - 1\right] = 0.$$

Since the limit exists, f is differentiable at $x = 0$.

47. The function defined by $y' = \frac{x \cos x - \sin x}{x^2}$ is differentiable on its entire domain. We need only check for differentiability of f' at $x = 0$.

Using a table of values, we can show that

$$\lim_{h \to 0} \frac{f'(0+h) - f'(0)}{h}$$

$$= \lim_{h \to 0} \frac{\frac{h \cos h - \sin h}{h^2} - 0}{h}$$

$$= \lim_{h \to 0} \frac{h \cos h - \sin h}{h^3} = -\frac{1}{3}.$$

Since the limit exists, f' is differentiable at $x = 0$. Thus, f'' exists for all x.

The function defined by

$$y'' = \frac{2 \sin x - 2x \cos x - x^2 \sin x}{x^3}$$ is continuous on its entire domain. We need only check for continuity of f'' at $x = 0$.

Using a table of values, we can show that
$$\lim_{x \to 0} \frac{2 \sin x - 2x \cos x - x^2 \sin x}{x^3} = -\frac{1}{3}.$$ The limit above shows that $f''(0) = -\frac{1}{3}$. Since $\lim_{x \to 0} f''(x) = f''(0)$, f'' is continuous at 0.

49. The sketch: $y = x$ and $y = \sin(x)$

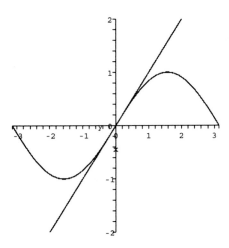

It is not possible visually to either detect or rule out intersections near $x = 0$ (other than zero itself).

We have that $f'(x) = \cos x$, which is less than 1 for $0 < x < 1$. If $\sin x \geq x$ for some x in the interval $(0, 1)$, then there would be a point on the graph of $y = \sin x$ which lies above the line $y = x$, but then (since $\sin x$ is continuous) the slope of the tangent line of $\sin x$ would have to be greater than or equal to 1 at some point in that interval, contradicting $f'(x) < 1$. Since $\sin x < x$ for $0 < x < 1$, we have $-\sin x > -x$ for $0 < x < 1$. Then $-\sin x = \sin(-x)$ so $\sin(-x) > -x$ for $0 < x < 1$, which is the same as saying $\sin x > x$ for $-1 < x < 0$.

Since $-1 \leq \sin x \leq 1$, the only interval on which $y = \sin x$ might intersect $y = x$ is $[-1, 1]$. We know they intersect at $x = 0$ and we just showed that they do not intersect on the intervals $(-1, 0)$ and $(0, 1)$. So the only other points they might intersect are $x = \pm 1$, but we know that $\sin(\pm 1) \neq \pm 1$, so these graphs intersect only at $x = 0$.

51. As seen from the graphs, changing the scale on the x-axis increases the number of oscillations or periods on the display. As the number of periods on the display increase, the graph looks more and more like a bunch of line segments. Its inflection points and concavity are no longer detectable.

2.7 IMPLICIT DIFFERENTIATION

1. Explicitly:

$4y^2 = 8 - x^2$

$y^2 = \frac{8-x^2}{4}$

$y = \pm \frac{\sqrt{8-x^2}}{2}$ (choose plus to fit $(2,1)$)

For $y = \frac{\sqrt{8-x^2}}{2}$,

$y' = \frac{1}{2} \frac{(-2x)}{2\sqrt{8-x^2}} = \frac{-x}{2\sqrt{8-x^2}}$,

$y'(2) = -1/2.$

Implicitly:

$\frac{d}{dx}(x^2 + 4y^2) = \frac{d}{dx}(8)$

$2x + 8y \cdot y' = 0$

$y' = \frac{-2x}{8y} = \frac{-x}{4y}$

at $(2, 1): y' = \frac{-2}{4 \cdot 1} = -\frac{1}{2}$

3. Explicitly:

$y(1 - 3x^2) = \cos x$

$y = \frac{\cos x}{1 - 3x^2}$

$$y'(x) = \frac{(1 - 3x^2)(-\sin x) - \cos x(-6x)}{(1 - 3x^2)^2}$$

$$= \frac{-\sin x + 3x^2 \sin x + 6x \cos x}{(1 - 3x^2)^2}$$

$y'(0) = 0$

Implicitly:

$\frac{d}{dx}(y - 3x^2 y) = \frac{d}{dx}(\cos x)$

$y' - (6xy + 3x^2 y') = -\sin x$

$y'(1 - 3x^2) = 6xy - \sin x$

$y' = \frac{6xy - \sin x}{1 - 3x^2}$

at $(0, 1): y' = 0$ (again).

5. $\dfrac{d}{dx}(x^2y^2 + 3y) = \dfrac{d}{dx}(4x)$

$2xy^2 + x^2 2y \cdot y' + 3y' = 4$

$y'(2x^2y + 3) = 4 - 2xy^2$

$y' = \dfrac{4 - 2xy^2}{2x^2y + 3}$

7. $\dfrac{d}{dx}(\sqrt{xy} - 4y^2) = \dfrac{d}{dx}(12)$

$\dfrac{1}{2\sqrt{xy}} \cdot \dfrac{d}{dx}(xy) - 8y \cdot y' = 0$

$\dfrac{1}{2\sqrt{xy}} \cdot (xy' + y) - 8y \cdot y' = 0$

$(xy' + y) - 16y \cdot y'\sqrt{xy} = 0$

$y'\left(x - 16y\sqrt{xy}\right) = -y$

$y' = \dfrac{-y}{\left(x - 16y\sqrt{xy}\right)} = \dfrac{y}{16y\sqrt{xy} - x}$

9. $x + 3 = 4xy + y^3$

$1 = \dfrac{d}{dx}\left(4xy + y^3\right) = 4(xy' + y) + 3y^2y'$

$1 - 4y = y'(3y^2 + 4x)$

$y' = \dfrac{1 - 4y}{3y^2 + 4x}$

11. $\dfrac{d}{dx}[\cos(x^2y) - \sin y] = \dfrac{d}{dx}(x)$

$-\sin(x^2y)\dfrac{d}{dx}(x^2y) - (\cos y)y' = 1$

$-\sin(x^2y)[2xy + x^2y'] - (\cos y)y' = 1$

$y'[-x^2\sin(x^2y) - \cos y] = 1 + 2xy\sin(x^2y)$

$y' = -\dfrac{1 + 2xy\sin(x^2y)}{x^2\sin(x^2y) + \cos y}$

13. $\dfrac{d}{dx}\left(\sqrt{x + y} - 4x^2\right) = \dfrac{d}{dx}(y)$

$\dfrac{1}{2\sqrt{x + y}} \cdot (1 + y') - 8x = y'$

$y'\left(\dfrac{1}{2\sqrt{x + y}} - 1\right) = \dfrac{-1}{2\sqrt{x + y}} + 8x$

$y'\left(\dfrac{1 - 2\sqrt{x + y}}{2\sqrt{x + y}}\right) = \dfrac{16x\sqrt{x + y} - 1}{2\sqrt{x + y}}$

$y' = \dfrac{16x\sqrt{x + y} - 1}{1 - 2\sqrt{x + y}}$

15. $\dfrac{d}{dx}[\tan 4y - xy^2] = \dfrac{d}{dx}(2x)$

$\sec^2 4y(4y') - (y^2 + 2xyy') = 2$

$y'(4\sec^2 4y - 2xy) = 2 + y^2$

$y' = \dfrac{2 + y^2}{4\sec^2 4y - 2xy}$

17. Rewrite: $x^2 = 4y^3$

Differentiate by x: $2x = 12y^2 \cdot y'$

$y' = \dfrac{2x}{12y^2} = \dfrac{x}{6y^2}$

at $(2, 1)$: $y' = \dfrac{2}{6 \cdot 1^2} = \dfrac{1}{3}$

The equation of the tangent line is

$y = \dfrac{1}{3}(x - 2) + 1$ or $y = \dfrac{1}{3}(x + 1)$.

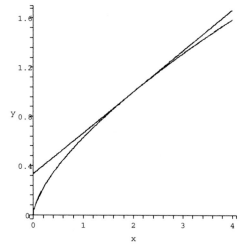

19. This one has $y = 0$ as part of the curve(s), but our point of reference is not on that part, so we can assume y is not zero, cancel it, and come to $x^2y = 4$

$\dfrac{d}{dx}(x^2y) = \dfrac{d}{dx}(4)$

$2xy + x^2 \cdot y' = 0$

$y' = \dfrac{-2y}{x}$

at $(2, 1) : y' = -2/2 = -1$.
The equation of the tangent line is $y = (-1)(x - 2) + 1$ or $y = -x + 3$.

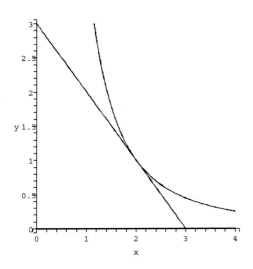

21. $4y^2 = 4x^2 - x^4$
$8yy' = 8x - 4x^3$
$y' = \dfrac{x(2 - x^2)}{2y}$
The slope of the tangent line at $(1, \sqrt{3}/2)$ is

$$m = \dfrac{(1)(2 - 1^2)}{2\left(\frac{\sqrt{3}}{2}\right)}$$

$$= \dfrac{1}{\sqrt{3}} = \dfrac{\sqrt{3}}{3}.$$

The equation of the tangent line is

$$y = \dfrac{\sqrt{3}}{3}(x - 1) + \dfrac{\sqrt{3}}{2}$$

$$y = \dfrac{\sqrt{3}}{3}x + \dfrac{\sqrt{3}}{2} - \dfrac{\sqrt{3}}{3}$$

$$y = \dfrac{\sqrt{3}}{3}x + \dfrac{\sqrt{3}}{6}.$$

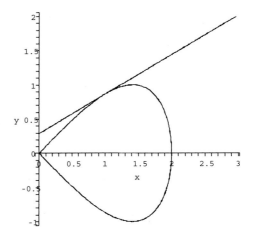

23. $\dfrac{d}{dx}(x^2 + y^3 - 3y) = \dfrac{d}{dx}(4)$
$2x + 3y^2y' - 3y' = 0$
$y'(3y^2 - 3) = -2x$
$y' = \dfrac{2x}{3 - 3y^2}$

Horizontal tangents: From the formula, $y' = 0$ only when $x = 0$. When $x = 0$, we have $0^2 + y^3 - 3y = 4$. Using a CAS to solve this, we find that $y \approx 2.1958$ is a horizontal tangent line, tangent to the curve at the (approximate) point $(0, 2.1958)$.

Vertical tangents: the denominator in y' must be zero. $3 - 3y^2 = 0$
$y^2 = 1$ or $y = \pm 1$.
When $y = 1$ we have
$x^2 + (1)^3 - 3(1) = 4$
$x^2 = 6$ or $x = \pm\sqrt{6} \approx \pm 2.4$.
Also, when $y = -1$, we have
$x^2 + (-1)^3 - 3(-1) = 4$
$x^2 = 2$
$x = \pm\sqrt{2} \approx \pm 1.4$.
Thus, we find 4 vertical tangent lines: $x = -\sqrt{6}$, $x = -\sqrt{2}$, $x = \sqrt{2}$, $x = \sqrt{6}$, tangent to the curve (respectively) at the points $\left(-\sqrt{6}, 1\right)$, $\left(-\sqrt{2}, -1\right)$, $\left(\sqrt{2}, -1\right)$, and $\left(\sqrt{6}, 1\right)$.

25. $\dfrac{d}{dx}(x^2y^2 + 3x - 4y) = \dfrac{d}{dx}(5)$
$x^2 2yy' + 2xy^2 + 3 - 4y' = 0$

Differentiate both sides of this with respect to x:

$$\frac{d}{dx}(x^2 2yy' + 2xy^2 + 3 - 4y') = \frac{d}{dx}(0)$$

$$2(2xyy' + x^2(y')^2 + x^2 yy'')$$
$$+ 2(2xyy' + y^2) - 4y'' = 0$$

$$2xyy' + x^2(y')^2 + x^2 yy''$$
$$+ 2xyy' + y^2 - 2y'' = 0$$

$$4xyy' + x^2(y')^2 + y^2 = y''(2 - x^2 y)$$

$$y'' = \frac{4xyy' + x^2(y')^2 + y^2}{2 - x^2 y}$$

27. $\dfrac{d}{dx}(y^2) = \dfrac{d}{dx}(x^3 - 6x + 4\cos y)$

$2yy' = 3x^2 - 6 - 4\sin y \cdot y'$

Differentiating again with respect to x:

$$2[yy'' + (y')^2]$$
$$= 6x - 4[\sin y \cdot y'' + \cos y \cdot (y')^2],$$
$$yy'' + (y')^2$$
$$= 3x - 2\sin y \cdot y'' - 2\cos y \cdot (y')^2,$$
$$y''(y + 2\sin y) = 3x - [2\cos y + 1](y')^2$$
$$y'' = \frac{3x - [2\cos y + 1](y')^2}{y + 2\sin y}$$

29. $x^2 + y^3 - 2y = 3$

$$y' = \frac{-2x}{3y^2 - 2}$$

If $x = 1.9$, solving for y requires solving the equation $y^3 - 2y + 0.61 = 0$. Using the equation of the tangent line found in Example 7.1, $y = -4x + 9$, $y(1.9) \approx 1.4$.

If $x = 2.1$, solving for y requires solving the equation $y^3 - 2y + 1.41 = 0$. Using the equation of the tangent line found in Example 8.1, $y = -4x + 9$, $y(2.1) \approx 0.6$.

31. Both of the points $(-3, 0)$ and $(0, 3)$ are on the curve:

$0^2 = (-3)^3 - 6(-3) + 9 = -27 + 18 + 9$
$3^2 = (0)^3 - 6(0) + 9 = 9$ The equation of the line through these points has slope

$\dfrac{0 - 3}{-3 - 0} = \dfrac{-3}{-3} = 1$ and y-intercept 3, so $y = x + 3$. This line intersects the curve at: $y^2 = x^3 - 6x + 9$

$(x + 3)^2 = x^3 - 6x + 9$
$x^2 + 6x + 9 = x^3 - 6x + 9$
$x^3 - 12x - x^2 = 0$
$x(x^2 - x - 12) = 0$

Therefore, $x = 0, -3$ or 4 and so the third point is $(4, 7)$.

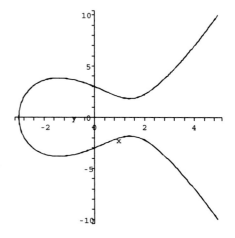

33. $\dfrac{d}{dx}(x^2 y - 2y) = \dfrac{d}{dx}(4)$

$2xy + x^2 y' - 2y' = 0$
$y'(x^2 - 2) = -2xy$
$y' = \dfrac{-2xy}{x^2 - 2}$

The derivative is undefined at $x = \pm\sqrt{2}$, suggesting that there might be vertical tangent lines at these points. Similarly, $y' = 0$ at $y = 0$, suggesting that there might be a horizontal tangent line at this point.

However, plugging $x = \pm\sqrt{2}$ into the original equation gives $0 = 4$, a contradiction which shows that there are no points on this curve with x value $\pm\sqrt{2}$. Likewise, plugging $y = 0$ into the original equation gives $0 = 4$. Again, this is a contradiction which shows that there are no points on the graph with y value of 4.

Sketching the graph, we see that there is a horizontal asymptote at $y = 0$ and vertical asymptotes at $x = \pm\sqrt{2}$.

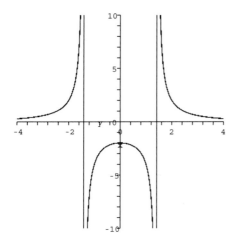

35. If $y_1 = c/x$, then $y_1' = -c/x^2 = -y_1/x$. If $y_2^2 = x^2 + k$, then $2y_2(y_2') = 2x$ and $y_2' = x/y_2$. If we are at a particular point (x_0, y_0) on both graphs, this means $y_1(x_0) = y_0 = y_2(x_0)$ and

$$y_1' \cdot y_2' = \left(\frac{-y_0}{x_0}\right) \cdot \left(\frac{x_0}{y_0}\right) = -1$$

This means that the slopes are negative reciprocals and the curves are orthogonal.

37. For the first type of curve, $y' = 3cx^2$.

For the second type of curve, $2x + 6yy' = 0$, and

$$y' = \frac{-2x}{6y} = \frac{-x}{3y}$$

$$= \frac{-x}{3cx^3} = \frac{-1}{3cx^2}.$$

These are negative reciprocals of each other, so the families of curves are orthogonal.

39. Conjecture: The family of functions $\{y_1 = cx^n\}$ is orthogonal to the family of functions $\{x^2 + ny_2^2 = k\}$ whenever $n \neq 0$.

If $y_1 = cx^n$, then $y_1' = cnx^{n-1} = ny_1/x$. If $ny_2^2 = -x^2 + k$, then $2ny_2(y_2') = -2x$ and $y_2' = -x/ny_2$. If we are at a particular point (x_0, y_0) on both graphs, this means $y_1(x_0) = y_0 = y_2(x_0)$ and

$$y_1' \cdot y_2' = \left(\frac{ny_0}{x_0}\right) \cdot \left(-\frac{x_0}{ny_0}\right) = -1.$$

This means that the slopes are negative reciprocals and the curves are orthogonal.

2.8 THE MEAN VALUE THEOREM

1. $f(x) = x^2 + 1$, $[-2, 2]$
$f(-2) = 5 = f(2)$

As a polynomial, $f(x)$ is continuous on $[-2, 2]$, differentiable on $(-2, 2)$, and the conditions of Rolle's Theorem hold. There exists $c \in (-2, 2)$ such that $f'(c) = 0$. But $f'(c) = 2c$, $\Rightarrow c = 0$.

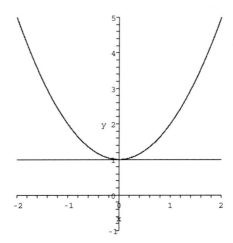

3. $f(x) = x^3 + x^2$, on $[0, 1]$, with $f(0) = 0$, $f(1) = 2$. As a polynomial $f(x)$ is continuous on $[0, 1]$ and differentiable on $(0, 1)$. Since the conditions of the Mean Value Theorem hold there exists a number $c \in (0, 1)$ such that

$$f'(c) = \frac{f(1) - f(0)}{1 - 0} = \frac{2 - 0}{1 - 0} = 2.$$

But $f'(c) = 3c^2 + 2c$.
$\Rightarrow 3c^2 + 2c = 2$,
$3c^2 + 2c - 2 = 0$.

By the quadratic formula

$$c = \frac{-2 \pm \sqrt{2^2 - 4(3)(-2)}}{2(3)}$$

$$= \frac{-2 \pm \sqrt{28}}{6}$$

$$= \frac{-2 \pm 2\sqrt{7}}{6} = \frac{-1 \pm \sqrt{7}}{3}$$

$$\Rightarrow c \approx -1.22 \quad \text{or} \quad c \approx 0.55$$

But since $-1.22 \notin (0, 1)$ we accept only the other alternative:

$$c = \frac{-1 + \sqrt{7}}{3} \approx 0.55$$

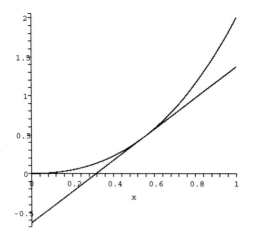

5. $f(x) = \sin x, \left[0, \pi/2\right],$
$f(0) = 0, f(\pi/2) = 1.$
As a trig function, $f(x)$ is continuous on $\left[0, \pi/2\right]$ and differentiable on $(0, \pi/2)$. The conditions of the Mean Value Theorem hold, and there exists $c \in (0, \pi/2)$ such that

$$f'(c) = \frac{f\left(\frac{\pi}{2}\right) - f(0)}{\frac{\pi}{2} - 0}$$

$$= \frac{1 - 0}{\frac{\pi}{2} - 0} = \frac{2}{\pi}.$$

But $f'(c) = \cos(c)$ and c is to be in the first quadrant. From a graphing calculator, we find that $\cos c = \dfrac{2}{\pi}$ when $c \approx 0.881$

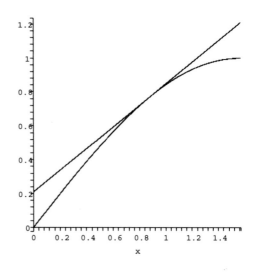

7. If $f'(x) > 0$ for all x then for each (a, b) with $a < b$ we know there exists a $c \in (a, b)$ such that

$$\frac{f(b) - f(a)}{b - a} = f'(c) > 0.$$

$a < b$ makes the denominator positive, and so we must have the numerator also positive, which implies $f(a) < f(b)$.

9. $f'(x) = 3x^2 + 5$. This is positive for all x, so $f(x)$ is increasing.

11. $f'(x) = -3x^2 - 3$. This is negative for all x, so $f(x)$ is decreasing.

13. $f(-1) = -1, f(1) = 1, f(2) = \frac{1}{2}$
$-1 < 1$ and $f(-1) < f(1)$, so $f(x)$ is not a decreasing function. $1 < 2$ and $f(1) > f(2)$, so $f(x)$ is not an increasing function. $f(x)$ is neither an increasing nor a decreasing function.

15. $f'(x) = \frac{1}{2\sqrt{x+1}}$

This is positive for all x in the domain of the function, so $f(x)$ is an increasing function for $x \geq -1$.

17. Let $f(x) = x^3 + 5x + 1$. As a polynomial, $f(x)$ is continuous and differentiable for all x, with $f'(x) = 3x^2 + 5$, which is positive for all x so $f(x)$ is strictly increasing for all x.

Therefore the equation can have at most one solution.

Since $f(x)$ is negative at $x = -1$ and positive at $x = 1$, and $f(x)$ is continuous, there must be a solution to $f(x) = 0$.

19. Let $f(x) = x^4 + 3x^2 - 2$. The derivative is $f'(x) = 4x^3 + 6x$. Since $f'(x) = 0$ has only one solution, by Theorem 8.2 $f(x)$ can have at most two zeros. Since $f(0) = -2$ and $f(-1) = 2 = f(1)$, therefore $f(x) = 0$ has exactly one solution between $x = -1$ and $x = 0$, exactly one solution between $x = 0$ and $x = 1$, and no other solutions.

21. $f(x) = x^3 + ax + b$, $a > 0$. Any cubic (actually any *odd degree*) polynomial heads in opposite directions ($\pm\infty$) as x goes to the oppositely signed infinities, and therefore by the Intermediate Value Theorem has at least one root. For the uniqueness, we look at the derivative, in this case $3x^2 + a$. Because $a > 0$ by assumption, this expression is strictly positive. The function is strictly increasing and can have at most one root.

23. $f(x) = x^5 + ax^3 + bx + c$, $a > 0$, $b > 0$

Here is another odd degree polynomial (see #21) with at least one root.

$f'(x) = 5x^4 + 3ax^2 + b$ is evidently strictly positive because of our assumptions about a, b. Exactly as in #21, there can be at most one root.

25. The average velocity on $[a, b]$ is $\dfrac{s(b) - s(a)}{b - a}$. By the Mean Value Theorem, there exists a $c \in (a, b)$ such that $s'(c) = \dfrac{s(b) - s(a)}{b - a}$. Thus, the instantaneous velocity at $t = c$ is equal to the average velocity between times $t = a$ and $t = b$.

27. Define $h(x) = f(x) - g(x)$. Then h is differentiable because f and g are, and $h(a) = h(b) = 0$. Apply Rolle's theorem to h on $[a, b]$ to conclude that there exists $c \in (a, b)$ such

that $h'(c) = 0$. Thus, $f'(c) = g'(c)$, and so f and g have parallel tangent lines at $x = c$.

29. $f(x) = x^2$
One candidate: $g_0(x) = kx^3$
Because we require $x^2 = g_0'(x) = 3kx^2$, we must have $3k = 1$, $k = 1/3$.

Most general solution:
$g(x) = g_0(x) + c = x^3/3 + c$
where c is an arbitrary constant.

31. Although the obvious first candidate is $g_0(x) = -1/x$, due to the disconnection of the domain by the discontinuity at $x = 0$, we could add *different* constants, one for negative x, another for positive x. Thus the most general solution is:

$$g(x) = \begin{cases} -1/x + a & \text{for } x > 0 \\ -1/x + b & \text{for } x < 0. \end{cases}$$

33. If $g'(x) = \sin x$, then $g(x) = -\cos x + c$ for any constant c.

35. $f(x) = 1/x$ on $[-1, 1]$. We easily see that $f(1) = 1$, $f(-1) = -1$, and $f'(x) = -1/x^2$. If we try to find the c in the interval $(-1, 1)$ for which

$$f'(c) = \frac{f(1) - f(-1)}{1 - (-1)} = \frac{1 - (-1)}{1 - (-1)} = 1,$$

the equation would be $-1/c^2 = 1$ or $c^2 = -1$. There is of course no such c, and the explanation is that the function is not defined for $x = 0 \in (-1, 1)$ and so the function is not continuous.

The hypotheses for the Mean Value Theorem are not fulfilled.

37. $f(x) = \tan x$ on $[0, \pi]$, $f'(x) = \sec^2(x)$. We know the tangent has a massive discontinuity at $x = \pi/2$, so as in #35, we should not be surprised if the Mean Value Theorem does not apply. As applied to the interval $[0, \pi]$ it would say

$$\sec^2(c) = f'(c) = \frac{f(\pi) - f(0)}{\pi - 0}$$

$$= \frac{\tan \pi - \tan 0}{\pi - 0} = 0.$$

But secant = 1/cosine is never 0 in the interval $(-1, 1)$, so no such c exists.

39. If a derivative g' is positive at a single point $x = b$, then $g(x)$ is an increasing function for x sufficiently near b, i.e., $g(x) > g(b)$ for $x > b$ but sufficiently near b. In this problem, we will apply that remark to f' at $x = 0$, and conclude from $f''(0) > 0$ that $f'(x) > f'(0) = 0$ for $x > 0$ but sufficiently small. This being true about the derivative f', it tells us that f itself is increasing on some interval $(0, a)$ and in particular that $f(x) > f(0) = 0$ for $0 < x < a$. On the other side (the negative side) f' is negative, f is decreasing (to zero) and therefore likewise positive. In summary, $x = 0$ is a genuine relative minimum.

41. Consider the function $g(x) = x - \sin(x)$, obviously with $g(0) = 0$ and $g'(x) = 1 - \cos(x)$. If there was ever a point $a > 0$ with $\sin(a) \geq a$, ($g(a) \leq 0$), then by the MVT applied to g on the interval $[0, a]$, there would be a point c ($0 < c < a$) with $g'(c) = \frac{g(a) - g(0)}{a - 0} = \frac{g(a)}{a} \leq 0$.

This would read $1 - \cos(c) = g'(c) \leq 0$ or $\cos(c) \geq 1$. The latter condition is possible only if $\cos(c) = 1$ and $\sin(c) = 0$, in which case c (being positive) would be *at minimum* π. But even in this unlikely case we still would have $\sin(a) \leq 1 < \pi \leq c < a$.

Since $\sin a < a$ for all $a > 0$, we have $-\sin a > -a$ for all $a > 0$, but $-\sin a = \sin(-a)$ so we have $\sin(-a) > -a$ for all $a > 0$. This is the same as saying $\sin a > a$ for all $a < 0$ so in absolute value we have $|\sin a| < |a|$ for all $a \neq 0$.

Thus the only possible solution to the equation $\sin x = x$ is $x = 0$, which we know to be true.

43. $f(x) = \begin{cases} 2x & x \leq 0 \\ 2x - 4 & x > 0 \end{cases}$ $f(x) = 2x - 4$ is continuous and differentiable on $(0, 2)$. Also, $f(0) = 0 = f(2)$. But $f'(x) = 2$ on $(0, 2)$, so there is no c such that $f'(c) = 0$. Rolle's Theorem requires that $f(x)$ be continuous on the closed interval, but we have a jump discontinuity at $x = 0$, which is enough to preclude the applicability of Rolle's.

Chapter 2 Review Exercises

1. $\dfrac{3.4 - 2.6}{1.5 - 0.5} = \dfrac{0.8}{1} = 0.8$

3. $f'(2) = \dfrac{f(2 + h) - f(2)}{h}$

$$= \lim_{h \to 0} \frac{(2 + h)^2 - 2(2 + h) - (0)}{h}$$

$$= \lim_{h \to 0} \frac{4 + 4h + h^2 - 4 - 2h}{h}$$

$$= \lim_{h \to 0} \frac{2h + h^2}{h}$$

$$= \lim_{h \to 0} (2 + h) = 2$$

5. $f'(1) = \lim_{h \to 0} \dfrac{f(1 + h) - f(1)}{h}$

$$= \lim_{h \to 0} \frac{\sqrt{1 + h} - 1}{h}$$

$$= \lim_{h \to 0} \frac{\sqrt{1 + h} - 1}{h} \cdot \frac{\sqrt{1 + h} + 1}{\sqrt{1 + h} + 1}$$

$$= \lim_{h \to 0} \frac{1 + h - 1}{h(\sqrt{1 + h} + 1)}$$

$$= \lim_{h \to 0} \frac{1}{\sqrt{1 + h} + 1} = \frac{1}{2}$$

7.
$$f'(x) = \lim_{h \to 0} \frac{f(x+h) - f(x)}{h}$$

$$= \lim_{h \to 0} \frac{(x+h)^3 + (x+h) - (x^3 + x)}{h}$$

$$= \lim_{h \to 0} \frac{3x^2 h + 3xh^2 + h^3 + h}{h}$$

$$= \lim_{h \to 0} (3x^2 + 3xh + h^2 + 1)$$

$$= 3x^2 + 1$$

9. The point is $(1, 0)$. $y' = 4x^3 - 2$ so the slope at $x = 1$ is 2, and the equation of the tangent line is $y = 2(x - 1) + 0$ or $y = 2x - 2$.

11. The point is $(0, 0)$. $y' = 6 \cos 2x$, so the slope at $x = 0$ is 6, and the equation of the tangent line is $y = 6(x - 0) + 0$ or $y = 6x$.

13. Find the slope to $y - x^2 y^2 = x - 1$ at $(1, 1)$.

$$\frac{d}{dx}(y - x^2 y^2) = \frac{d}{dx}(x - 1)$$

$$y' - 2xy^2 - x^2 2y \cdot y' = 1$$

$$y'(1 - x^2 2y) = 1 + 2xy^2$$

$$y' = \frac{1 + 2xy^2}{1 - 2x^2 y}$$

At $(1, 1)$:
$$y' = \frac{1 + 2(1)(1)^2}{1 - 2(1)^2(1)} = \frac{3}{-1} = -3$$

The equation of the tangent line is $y = -3(x - 1) + 1$ or $y = -3x + 4$.

15. $s(t) = -16t^2 + 40t + 10$

$v(t) = s'(t) = -32t + 40$

$a(t) = v'(t) = -32$

17. $s(t) = 10 \sin 4t$

$v(t) = s'(t)$

$\quad = 40 \cos 4t$

$a(t) = v'(t)$

$\quad = -160 \sin 4t$

19. $v(t) = s'(t) = -32t + 40$

$v(1) = -32(1) + 40 = 8$ ft/s

The ball is rising.

$v(2) = -32(2) + 40 = -24$ ft/s

The ball is falling.

21. (a)
$$m_{\text{sec}} = \frac{f(2) - f(1)}{2 - 1}$$

$$= \frac{\sqrt{3} - \sqrt{2}}{1} \approx 0.318$$

(b)
$$m_{\text{sec}} = \frac{f(1.5) - f(1)}{1.5 - 1}$$

$$= \frac{\sqrt{2.5} - \sqrt{2}}{0.5} \approx 0.334$$

(c)
$$m_{\text{sec}} = \frac{f(1.1) - f(1)}{1.1 - 1}$$

$$= \frac{\sqrt{2.1} - \sqrt{2}}{0.1} \approx 0.349$$

Best estimate for the slope of the tangent line: (c) (approximately 0.349).

23. $f'(x) = 4x^3 - 9x^2 + 2$

25.
$$f'(x) = -\frac{3}{2}x^{-3/2} - 10x^{-3}$$

$$= \frac{-3}{2x\sqrt{x}} - \frac{10}{x^3}$$

27. $f'(t) = 2t(t + 2)^3 + t^2 \cdot 3(t + 2)^2 \cdot 1$

$\quad = 2t(t + 2)^3 + 3t^2(t + 2)^2$

$\quad = t(t + 2)^2(5t + 4)$

29.
$$g'(x) = \frac{(3x^2 - 1) \cdot 1 - x(6x)}{(3x^2 - 1)^2}$$

$$= \frac{3x^2 - 1 - 6x^2}{(3x^2 - 1)^2}$$

$$= -\frac{3x^2 + 1}{(3x^2 - 1)^2}$$

31. $f'(x) = 2x \sin x + x^2 \cos x$

33. $f'(x) = \sec^2 \sqrt{x} \cdot \dfrac{1}{2\sqrt{x}}$

35. $f'(t) = \csc t \cdot 1 + t \cdot (-\csc t \cdot \cot t)$
$= \csc t - t \csc t \cot t$

37. $u(x) = \dfrac{2}{\sqrt{x^2 + 2}} = 2(x^2 + 2)^{-1/2}$

$u'(x) = 2\left[-\dfrac{1}{2}(x^2 + 2)^{-3/2}(2x) \right]$

$= -\dfrac{2x}{\sqrt{(x^2 + 2)^3}}$

39. $f(x) = 3\cos\sqrt{4 - x^2} = 3\cos(4 - x^2)^{1/2}$

$f'(x) = -3\sin(4 - x^2)^{1/2}$

$\left[\dfrac{1}{2}(4 - x^2)^{-1/2}(-2x) \right]$

$= \dfrac{3x \sin\sqrt{4 - x^2}}{\sqrt{4 - x^2}}$

41. $f'(x) = \dfrac{1}{2\sqrt{\sin 4x}} \cdot \cos 4x \cdot 4 = \dfrac{2\cos 4x}{\sqrt{\sin 4x}}$

43. $f'(x) = 2\left(\dfrac{x + 1}{x - 1} \right) \dfrac{d}{dx}\left(\dfrac{x + 1}{x - 1} \right)$

$= 2\left(\dfrac{x + 1}{x - 1} \right) \dfrac{(x - 1) - (x + 1)}{(x - 1)^2}$

$= 2\left(\dfrac{x + 1}{x - 1} \right) \dfrac{-2}{(x - 1)^2}$

$= \dfrac{-4(x + 1)}{(x - 1)^3}$

45. $u'(x)$

$= 4\left[2\sin\left(4 - \sqrt{x}\right)\cos\left(4 - \sqrt{x}\right) \right.$

$\left. \left(-\dfrac{1}{2}x^{-1/2}\right) \right]$

$= -\dfrac{4\sin\left(4 - \sqrt{x}\right)\cos\left(4 - \sqrt{x}\right)}{\sqrt{x}}$

47. The derivative should look roughly like:

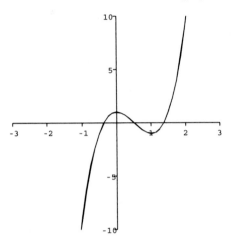

49. $f(x) = x^4 - 3x^3 + 2x^2 - x - 1$
$f'(x) = 4x^3 - 9x^2 + 4x - 1$
$f''(x) = 12x^2 - 18x + 4$

51. $f'(x) = \cos 2x - 2x \sin 2x$
$f''(x) = -2\sin 2x - 2(\sin 2x + 2x \cos 2x) =$
$-4\sin 2x - 4x \cos 2x$
$f'''(x) = -8\cos 2x - 4(\cos 2x - 2x \sin 2x)$
$= -12\cos 2x + 8x \sin 2x$

53. $f(x) = \tan x$
$f'(x) = \sec^2 x$
$f''(x) = 2\sec x \cdot \sec x \tan x$
$= 2\sec^2 x \tan x$

55. $f(x) = \sin 3x$
$f'(x) = \cos 3x \cdot 3 = 3\cos 3x$
$f''(x) = 3(-\sin 3x \cdot 3) = -9\sin 3x$
$f'''(x) = -9\cos 3x \cdot 3 = -27\cos 3x$
$f^{(26)}(x) = -3^{26}\sin 3x$

57. $f(t) = 4\cos 2t$
$v(t) = f'(t) = 4(-\sin 2t) \cdot 2$
$= -8\sin 2t$

(a) The velocity is zero when
$v(t) = -8\sin 2t = 0$, i.e., when
$2t = 0, \pi, 2\pi, \ldots$ so when
$t = 0, \pi/2, \pi, 3\pi/2, \ldots$
$f(t) = 4$ for $t = 0, \pi, 2\pi, \ldots$
$f(t) = 4\cos 2t = -4$ for
$t = \pi/2, 3\pi/2, \ldots$

The position of the spring when the velocity is zero is 4 or −4.

(b) The velocity is a maximum when
$v(t) = -8 \sin 2t = 8$, i.e., when
$2t = 3\pi/2, 7\pi/2, \ldots$ so
$t = 3\pi/4, 7\pi/4, \ldots$
$f(t) = 4 \cos 2t = 0$ for
$t = 3\pi/4, 7\pi/4, \ldots$
The position of the spring when the velocity is at a maximum is zero.

(c) Velocity is at a minimum when
$v(t) = -8 \sin 2t = -8$, i.e., when
$2t = \pi/2, 5\pi/2, \ldots$ so
$t = \pi/4, 5\pi/4, \ldots$
$f(t) = 4 \cos 2t = 0$ for
$t = \pi/4, 5\pi/4, \ldots$
The position of the spring when the velocity is at a minimum is also zero.

59. $\dfrac{d}{dx}(x^2 y - 3y^3) = \dfrac{d}{dx}(x^2 + 1)$

$2xy + x^2 y' - 3 \cdot 3y^2 \cdot y' = 2x$

$y'(x^2 - 9y^2) = 2x - 2xy$

$y' = \dfrac{2x(1 - y)}{x^2 - 9y^2}$

61. $\dfrac{d}{dx}\left(\dfrac{y}{x+1} - 3y\right) = \dfrac{d}{dx} \tan x$

$\dfrac{(x+1)y' - y \cdot (1)}{(x+1)^2} - 3y' = \sec^2 x$

$(x+1)y' - y - 3(x+1)^2 y' = (x+1)^2 \sec^2 x$

$\left[(x+1) - 3(x+1)^2\right] y' = y + (x+1)^2 \sec^2 x$

$y' = \dfrac{y + (x+1)^2 \sec^2 x}{(x+1) - 3(x+1)^2}$

63. When $x = 0$, $-3y^3 = 1$, $y = \dfrac{-1}{\sqrt[3]{3}}$ (call this a).

From our formula (#59), we find $y' = 0$ at this point. To find y'', implicitly differentiate the first derivative (second line in #59):

$2(xy' + y) + (2xy' + x^2 y'')$
$- 9\left[2y(y')^2 + y^2 y''\right] = 2$

At $(0, a)$ with $y' = 0$, we find

$2a - 9a^2 y'' = 2$,

$y'' = \dfrac{-2\sqrt[3]{3}}{9}\left(\sqrt[3]{3} + 1\right)$

Below is a sketch of the graph of $x^2 y - 3y^3 = x^2 + 1$.

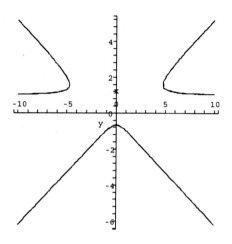

65. $y' = 3x^2 - 12x = 3x(x - 4)$

(a) $y' = 0$ for $x = 0$ $(y = 1)$, and $x = 4$ $(y = -31)$ so there are horizontal tangent lines at $(0, 1)$ and $(4, -31)$.

(b) y' is defined for all x, so there are no vertical tangent lines.

67. $\dfrac{d}{dx}(x^2 y - 4y) = \dfrac{d}{dx}x^2$

$2xy + x^2 y' - 4y' = 2x$

$y'(x^2 - 4) = 2x - 2xy$

$y' = \dfrac{2x - 2xy}{x^2 - 4} = \dfrac{2x(1 - y)}{x^2 - 4}$

(a) $y' = 0$ when $x = 0$ or $y = 1$. At $y = 1$,
$x^2 \cdot 1 - 4 \cdot 1 = x^2$
$x^2 - 4 = x^2$
This is impossible, so there is no x for which $y = 1$. At $x = 0$, $0^2 \cdot y - 4y = 0^2$, so $y = 0$. Therefore, there is a horizontal tangent line at $(0, 0)$.

(b) y' is not defined when $x^2 - 4 = 0$, or $x = \pm 2$. At $x = \pm 2$, $4y - 4y = 4$ so the function is not defined at $x = \pm 2$. There are no vertical tangent lines.

69. $f(x)$ is continuous and differentiable for all x, and $f'(x) = 3x^2 + 7$, which is positive for all x. By Theorem 8.2, if the equation $f(x) = 0$ has two solutions, then $f'(x) = 0$ would have at least one solution, but it has none. Every odd degree polynomial has at least one root, so in this case there is exactly one root.

71. Let $a > 0$. We know that $f(x) = \cos x - 1$ is continuous and differentiable on the interval $(0, a)$. Also $f'(x) = \sin x \leq 1$ for all x. The Mean Value Theorem implies that there exists some c in the interval $(0, a)$ such that

$$f'(c) = \frac{\cos a - 1 - (\cos 0 - 1)}{a - 0}$$

$$= \frac{\cos a - 1}{a}.$$

Since this is equal to $\sin c$ and $\sin c \leq 1$ for any c, we get that

$$\cos a - 1 \leq a$$

as desired. This works for all positive a, but since $\cos x - 1$ is symmetric about the y axis, we get

$$|\cos x - 1| \leq |x|.$$

They are actually equal at $x = 0$.

73. To show that $g(x)$ is continuous at $x = a$, we need to show that the limit as x approaches a of $g(x)$ exists and is equal to $g(a)$. But

$$\lim_{x \to a} g(x) = \lim_{x \to a} \frac{f(x) - f(a)}{x - a},$$

which is the definition of the derivative of $f(x)$ at $x = a$. Since $f(x)$ is differentiable at $x = a$, we know this limit exists and is equal to $f'(a)$, which, in turn, is equal to $g(a)$. Thus $g(x)$ is continuous at $x = a$.

75. $f(x) = x^2 - 2x$ on $[0, 2]$
$f(2) = 0 = f(0)$

If $f'(c) = \dfrac{f(2) - f(0)}{2 - 0} = \dfrac{0 - 0}{2} = 0$ then
$2c - 2 = f'(c) = 0$ so $c = 1$.

77. $f(x) = 3x^2 - \cos x$ One trial: $g_o(x) = kx^3 - \sin x$ $g'_o(x) = 3kx^2 - \cos x$ Need $3k = 3$, $k = 1$, and the general solution is $g(x) = g_o(x) + c = x^3 - \sin x + c$ for c an arbitrary constant.

79. $x = 1$ is to be double root of

$$f(x) = (x^3 + 1) - [m(x - 1) + 2]$$

$$= (x^3 + 1 - 2) - m(x - 1)$$

$$= (x^3 - 1) - m(x - 1)$$

$$= (x - 1)\left[x^2 + x + 1 - m\right]$$

Let $g(x) = x^2 + x + 1 - m$. Then $x = 1$ is a *double* root of f only if $(x - 1)$ is a *factor* of g, in which case $g(1) = 0$. Therefore we require $0 = g(1) = 3 - m$ or $m = 3$. Now $g(x) = x^2 + x - 2 = (x - 1)(x + 2)$, $f(x) = (x - 1)g(x) = (x - 1)^2(x + 2)$ and $x = 1$ is a double root.

The line tangent to the curve $y = x^3 + 1$ at the point $(1, 2)$ has slope $y' = 3x^2 = 3(1) = 3(= m)$. The equation of the tangent line is $y - 2 = 3(x - 1)$ or $y = 3x - 1(= m(x - 1) + 2)$.

Applications of Differentiation

3.1 LINEAR APPROXIMATIONS AND NEWTON'S METHOD

1. $f(x_0) = f(1) = \sqrt{1} = 1$

$f'(x) = \frac{1}{2}x^{-1/2}$

$f'(x_0) = f'(1) = \frac{1}{2}$

So

$L(x) = f(x_0) + f'(x_0)(x - x_0)$

$\quad = 1 + \frac{1}{2}(x - 1)$

$\quad = \frac{1}{2} + \frac{1}{2}x$

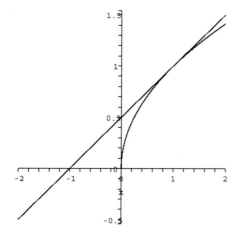

3. $f(x) = \sqrt{2x + 9},\ x_0 = 0$

$f(x_0) = f(0) = \sqrt{2 \cdot 0 + 9} = 3$

$f'(x) = \frac{1}{2}(2x + 9)^{-1/2} \cdot 2 = (2x + 9)^{-1/2}$

$f'(x_0) = f'(0) = (2 \cdot 0 + 9)^{-1/2} = \frac{1}{3}$

So

$L(x) = f(x_0) + f'(x_0)(x - x_0)$

$\quad = 3 + \frac{1}{3}(x - 0)$

$\quad = 3 + \frac{1}{3}x$

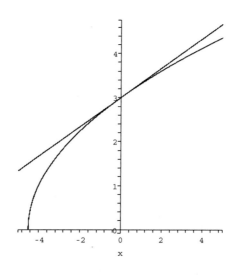

5. $f(x) = \sin 3x,\ x_0 = 0$

$f(x_0) = f(0) = \sin(3 \cdot 0) = \sin 0 = 0$

$f'(x) = 3\cos 3x$

$f'(x_0) = f'(0) = 3\cos(3 \cdot 0) = 3$

$L(x) = f(x_0) + f'(x_0)(x - x_0)$

$\quad = 0 + 3(x - 0)$

$\quad = 3x$

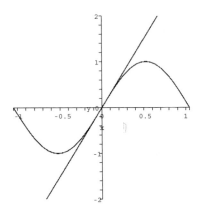

7. (a) $f(0) = g(0) = h(0) = 1$, so all pass through the point (0, 1).

$f'(0) = 2(0 + 1) = 2$,

$g'(0) = 2\cos(2 \cdot 0) = 2$, and

$h'(0) = 2\dfrac{1}{\sqrt{0 + \frac{1}{4}}} = 2$,

so all have slope 2 at $x = 0$.
The linear approximation at $x = 0$ for all three functions is
$L(x) = 1 + 2x$.

(b) Graph of $f(x) = (x + 1)^2$:

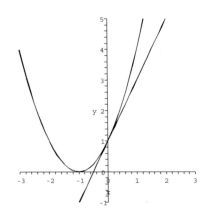

Graph of $g(x) = 1 + \sin(2x)$:

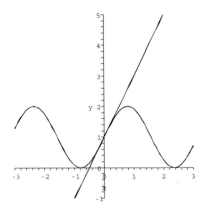

Graph of $h(x) = 2\sqrt{x + \frac{1}{4}}$:

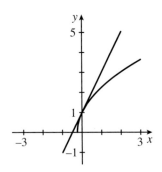

At values close to 0, $g(x) = 1 + \sin(2x)$ seems to have the closest fit with its linear approximation.

9. (a) $f(x) = \sqrt[4]{x},\ x_0 = 16$

$f(16) = \sqrt[4]{16} = 2$

$f'(x) = \dfrac{1}{4}x^{-3/4}$

$f'(16) = \dfrac{1}{4}(16)^{-3/4} = \dfrac{1}{32}$

$L(x) = f(16) + f'(16)(x - 16)$

$\quad = 2 + \dfrac{1}{32}(x - 16)$

$L(16.04) = 2 + \dfrac{1}{32}(0.04) = 2.00125$

(b) $L(16.08) = 2 + \frac{1}{32}(0.08) = 2.0025$

(c) $L(16.16) = 2 + \frac{1}{32}(0.16) = 2.005$

11. (a) $\sqrt[4]{16.04} \approx 2.0012488$

$L(0.04) = 2.00125$

$|2.0012488 - 2.00125|$

$= 0.00000117$

(b) $\sqrt[4]{16.08} \approx 2.0024953$

$L(0.08) = 2.0025$

$|2.0024953 - 2.0025|$

$= 0.00000467$

(c) $\sqrt[4]{16.16} = 2.0049814$

$L(0.16) = 2.005$

$|2.0049814 - 2.005| = 0.0000186$

13. (a)

$$L(x) = f(20) + \frac{18 - 14}{20 - 30}(x - 20)$$

$$L(24) \approx 18 - \frac{4}{10}(24 - 20)$$

$$= 18 - 0.4(4)$$

$$= 16.4 \text{ thousand games}$$

(b)

$$L(x) = f(40) + \frac{14 - 12}{30 - 40}(x - 40)$$

$$f(36) \approx 12 - \frac{2}{10}(36 - 40)$$

$$= 12 - 0.2(-4)$$

$$= 12.8 \text{ thousand games}$$

15. (a)

$$L(x) = f(200) + \frac{142 - 128}{220 - 200}(x - 200)$$

$$L(208) = 128 + \frac{14}{20}(208 - 200)$$

$$= 128 + 0.7(8) = 133.6$$

(b)

$$L(x) = f(240) + \frac{142 - 136}{220 - 240}(x - 240)$$

$$L(232) = 136 - \frac{6}{20}(232 - 240)$$

$$= 136 - 0.3(-8) = 138.4$$

17. The first tangent line intersects the x-axis at a point a little to the right of 1. So x_1 is about 1.25 (very roughly). The second tangent line intersects the x-axis at a point between 1 and

x_1, so x_2 is about 1.1 (very roughly). Newton's Method will converge to the zero at $x = 1$.

19. It wouldn't work because $f'(0) = 0$.

21. $f(x) = x^3 + 3x^2 - 1 = 0$, $x_0 = 1$

$f'(x) = 3x^2 + 6x$

(a)

$$x_1 = x_0 - \frac{f(x_0)}{f'(x_0)}$$

$$= 1 - \frac{1^3 + 3 \cdot 1^2 - 1}{3 \cdot 1^2 + 6 \cdot 1}$$

$$= 1 - \frac{3}{9} = \frac{2}{3}$$

$$x_2 = x_1 - \frac{f(x_1)}{f'(x_1)}$$

$$= \frac{2}{3} - \frac{\left(\frac{2}{3}\right)^3 + 3\left(\frac{2}{3}\right)^2 - 1}{3\left(\frac{2}{3}\right)^2 + 6\left(\frac{2}{3}\right)}$$

$$= \frac{79}{144} \approx 0.5486$$

(b) 0.53209

23. $f(x) = x^4 - 3x^2 + 1 = 0$, $x_0 = 1$

$f'(x) = 4x^3 - 6x$

(a)

$$x_1 = x_0 - \frac{f(x_0)}{f'(x_0)}$$

$$= 1 - \left(\frac{1^4 - 3 \cdot 1^2 + 1}{4 \cdot 1^3 - 6 \cdot 1}\right) = \frac{1}{2}$$

$$x_2 = x_1 - \frac{f(x_1)}{f'(x_1)}$$

$$= \frac{1}{2} - \left(\frac{\left(\frac{1}{2}\right)^4 - 3\left(\frac{1}{2}\right)^2 + 1}{4\left(\frac{1}{2}\right)^3 - 6\left(\frac{1}{2}\right)}\right)$$

$$= \frac{5}{8} = 0.625$$

(b) 0.61803

25. Use $x_{i+1} = x_i - \frac{f(x_i)}{f'(x_i)}$ with
$f(x) = x^3 + 4x^2 - 3x + 1$, and
$f'(x) = 3x^2 + 8x - 3$.

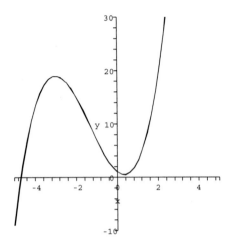

Start with $x_0 = -5$ to find the root near -5:
$x_1 = -4.718750$, $x_2 = -4.686202$,
$x_3 = -4.685780$, $x_4 = -4.685780$

27. Use $x_{i+1} = x_i - \frac{f(x_i)}{f'(x_i)}$ with
$f(x) = x^5 + 3x^3 + x - 1$, and
$f'(x) = 5x^4 + 9x^2 + 1$.

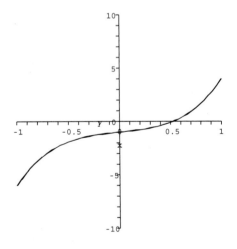

Start with $x_0 = 0.5$ to find the root near 0.5:
$x_1 = 0.526316$, $x_2 = 0.525262$,
$x_3 = 0.525261$, $x_4 = 0.525261$

29. Use $x_{i+1} = x_i - \frac{f(x_i)}{f'(x_i)}$ with
$f(x) = \sin x - x^2 + 1$, and
$f'(x) = \cos x - 2x$

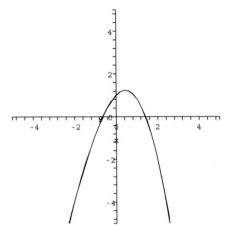

Start with $x_0 = -0.5$ to find the root near -0.5:
$x_1 = -0.644108$, $x_2 = -0.636751$
$x_3 = -0.636733$, $x_4 = -0.636733$

Start with $x_0 = 1.5$ to find the root near 1.5:
$x_1 = 1.413799$, $x_2 = 1.409634$
$x_3 = 1.409624$, $x_4 = 1.409624$

31.
$$x_{n+1} = x_n - \frac{f(x_n)}{f'(x_n)}$$
$$= x_n - \left(\frac{x_n^2 - c}{2x_n} \right)$$
$$= x_n - \frac{x_n^2}{2x_n} + \frac{c}{2x_n}$$
$$= \frac{x_n}{2} + \frac{c}{2x_n}$$
$$= \frac{1}{2} \left(x_n + \frac{c}{x_n} \right)$$

If $x_0 < \sqrt{c}$, then $c/x_0 > \sqrt{c}$, so $x_0 < \sqrt{c} < c/x_0$.

33. $f(x) = x^2 - 11$; $x_0 = 3$; $\sqrt{11} \approx 3.316625$.

35. $f(x) = x^3 - 11$; $x_0 = 2$; $\sqrt[3]{11} \approx 2.22398$.

37. $f(x) = x^{4.4} - 24$; $x_0 = 2$; $\sqrt[4.4]{24} \approx 2.059133$.

39. $f(x) = 4x^3 - 7x^2 + 1 = 0$, $x_0 = 0$
$f'(x) = 12x^2 - 14x$ $x_1 = x_0 - \frac{f(x_0)}{f'(x_0)} = 0 - \frac{1}{0}$
The method fails because $f'(x_0) = 0$. Choose

a different value for x_0 so that $f'(x_0) \neq 0$. Roots are $-0.3454, 0.4362, 1.6592$.

41. $f(x) = x^2 + 1, x_0 = 0$
$f'(x) = 2x$
$x_1 = x_0 - \frac{f(x_0)}{f'(x_0)} = 0 - \frac{1}{0}$
The method fails because $f'(x_0) = 0$. There are no roots.

43. $f(x) = \frac{4x^2 - 8x + 1}{4x^2 - 3x - 7} = 0, x_0 = -1$

Note: $f(x_0) = f(-1)$ is undefined, so Newton's Method fails because x_0 is not in the domain of f. Notice that $f(x) = 0$ only when $4x^2 - 8x + 1 = 0$. So using Newton's Method on $g(x) = 4x^2 - 8x + 1$ with $x_0 = -1$ leads to $x \approx 0.1340$. The other root is $x \approx 1.8660$.

45. (a) With $x_0 = 1.2$,

$x_1 = 0.800000000,$
$x_2 = 0.950000000,$
$x_3 = 0.995652174,$
$x_4 = 0.999962680,$
$x_5 = 0.999999997,$
$x_6 = 1.000000000,$
$x_7 = 1.000000000$

(b) With $x_0 = 2.2$,

$x_0 = 2.200000, x_1 = 2.107692,$
$x_2 = 2.056342, x_3 = 2.028903,$
$x_4 = 2.014652, x_5 = 2.007378,$
$x_6 = 2.003703, x_7 = 2.001855,$
$x_8 = 2.000928, x_9 = 2.000464,$
$x_{10} = 2.000232, x_{11} = 2.000116,$
$x_{12} = 2.000058, x_{13} = 2.000029,$
$x_{14} = 2.000015, x_{15} = 2.000007,$
$x_{16} = 2.000004, x_{17} = 2.000002,$
$x_{18} = 2.000001, x_{19} = 2.000000,$
$x_{20} = 2.000000$

The convergence is much faster with $x_0 = 1.2$.

47. (a) With $x_0 = -1.1$

$x_1 = -1.0507937,$
$x_2 = -1.0256065,$
$x_3 = -1.0128572,$
$x_4 = -1.0064423,$
$x_5 = -1.0032246,$
$x_6 = -1.0016132,$
$x_7 = -1.0008068,$
$x_8 = -1.0004035,$
$x_9 = -1.0002017,$
$x_{10} = -1.0001009,$
$x_{11} = -1.0000504,$
$x_{12} = -1.0000252,$
$x_{13} = -1.0000126,$
$x_{14} = -1.0000063,$

$x_{15} = -1.0000032,$
$x_{16} = -1.0000016,$
$x_{17} = -1.0000008,$
$x_{18} = -1.0000004,$
$x_{19} = -1.0000002,$
$x_{20} = -1.0000001,$
$x_{21} = -1.0000000,$
$x_{22} = -1.0000000$

(b) With $x_0 = 2.1$

$x_0 = 2.100000000,$
$x_1 = 2.006060606,$
$x_2 = 2.000024340,$
$x_3 = 2.000000000,$
$x_4 = 2.000000000$

The rate of convergence in (a) is slower than the rate of convergence in (b).

49. $f(x) = \tan x, f(0) = \tan 0 = 0$

$f'(x) = \sec^2 x, f'(0) = \sec^2 0 = 1$

$L(x) = f(0) + f'(0)(x - 0)$

$\quad = 0 + 1(x - 0) = x$

$L(0.01) = 0.01$

$f(0.01) = \tan 0.01 \approx 0.0100003$

$L(0.1) = 0.1$

$f(0.1) = \tan(0.1) \approx 0.1003$

$L(1) = 1$

$f(1) = \tan 1 \approx 1.557$

51. $f(x) = \sqrt{4 + x}$

$f(0) = \sqrt{4 + 0} = 2$

$f'(x) = \dfrac{1}{2}(4 + x)^{-1/2}$

$f'(0) = \dfrac{1}{2}(4 + 0)^{-1/2} = \dfrac{1}{4}$

$L(x) = f(0) + f'(0)(x - 0) = 2 + \dfrac{1}{4}x$

$L(0.01) = 2 + \dfrac{1}{4}(0.01) = 2.0025$

$f(0.01) = \sqrt{4 + 0.01} \approx 2.002498$

$L(0.1) = 2 + \dfrac{1}{4}(0.1) = 2.025$

$f(0.1) = \sqrt{4 + 0.1} \approx 2.0248$

$L(1) = 2 + \dfrac{1}{4}(1) = 2.25$

$f(1) = \sqrt{4 + 1} \approx 2.2361$

53. If you graph $|\tan x - x|$, you see that the difference is less than 0.01 on the interval $-0.306 < x < 0.306$ (In fact, a slightly larger interval would work as well.)

55. The smallest positive solution of the first equation is 0.132782, and for the second equation the smallest positive solution is 1, so the species modeled by the second equation is certain to go extinct. This is consistent with the models, since the expected number of offspring for the population modeled by

the first equation is 2, while for the second equation it is only 1.

57. The only positive solution is 0.6407.

59. $W(x) = \dfrac{PR^2}{(R + x)^2}, x_0 = 0$

$W'(x) = \dfrac{-2PR^2}{(R+x)^3}$

$L(x) = W(x_0) + W'(x_0)(x - x_0)$

$\quad = \dfrac{PR^2}{(R + 0)^2} + \left(\dfrac{-2PR^2}{(R + 0)^3}\right)(x - 0)$

$\quad = P - \dfrac{2Px}{R}$

$L(x) = 120 - 0.01(120) = P - \dfrac{2Px}{R}$

$\quad = 120 - \dfrac{2 \cdot 120x}{R}$

$0.01 = \dfrac{2x}{R}$

$x = 0.005R = 0.005(20,900,000)$

$\quad = 104,500$ ft

61. (a) As we should expect, when we start with $x_0 = 0.1$, Newton's method converges to 0.

(b) When we start with $x_0 = 1.1$, Newton's method converges to 1.

(c) When we start with $x_0 = 2.1$, Newton's method converges to 2.

3.2 MAXIMUM AND MINIMUM VALUES

1. (a) No absolute extrema.

(b) $f(0) = -1$ is absolute max. There is no absolute minimum (vertical asymptotes at $x = \pm 1$).

(c) No absolute extrema. (They would be at the endpoints, which are not included in the interval.)

3. (a) $f\left(\dfrac{\pi}{2} + 2n\pi\right) = 1$ for any integer n is abs max;

$f\left(\dfrac{3\pi}{2} + 2n\pi\right) = -1$ for any integer n is abs min

(b) $f(0) = 0$ is abs min; $f(\pi/4) = \dfrac{\sqrt{2}}{2}$ is abs max

(c) $f(\pi/2) = 1$ is abs max; there is no abs min, which would occur at both endpoints (not included in the interval).

5. $f(x) = x^2 + 5x - 1$
$f'(x) = 2x + 5$
$2x + 5 = 0$
$x = -5/2$ is a critical number. This is a parabola opening upward, so we have a local minimum.

7. $f(x) = x^3 - 3x + 1$
$f'(x) = 3x^2 - 3$

$3x^2 - 3 = 3(x^2 - 1)$
$\qquad = 3(x+1)(x-1) = 0$

$x = \pm 1$ are critical numbers.

This is a cubic with a positive leading coefficient so $x = -1$ is a local max, $x = 1$ is a local min.

9. $f(x) = x^3 - 3x^2 + 6x$
$f'(x) = 3x^2 - 6x + 6$
$3x^2 - 6x + 6 = 3(x^2 - 2x + 2) = 0$

We can use the quadratic formula to find the roots, which are $x = 1 \pm \sqrt{-1}$. These are imaginary so there are no real critical numbers.

11. $f(x) = x^4 - 3x^3 + 2$
$f'(x) = 4x^3 - 9x^2$
$4x^3 - 9x^2 = x^2(4x - 9) = 0$
$x = 0, 9/4$ are critical numbers $x = 9/4$ is a local min; $x = 0$ is neither a local max nor min.

13. $f(x) = x^{3/4} - 4x^{1/4}$
$f'(x) = \dfrac{3}{4x^{1/4}} - \dfrac{1}{x^{3/4}}$
If $x \neq 0$, $f'(x) = 0$ when $3x^{3/4} = 4x^{1/4}$ $x =$

0, 16/9 are critical numbers. $x = 16/9$ is a local min, $x = 0$ is neither a local max nor min.

15. $f(x) = \sin x \cos x$ on $[0, 2\pi]$
$f'(x) = \cos x \cos x + \sin x(-\sin x)$
$\qquad = \cos^2 x - \sin^2 x$

$\cos^2 x - \sin^2 x = 0$
$\cos^2 x = \sin^2 x$
$\cos x = \pm \sin x$

$x = \pi/4, 3\pi/4, 5\pi/4, 7\pi/4$ are critical numbers.
$x = \pi/4, 5\pi/4$ are local max, $x = 3\pi/4, 7\pi/4$ are local min.

17. $f(x) = \dfrac{x^2 - 2}{x + 2}$
Note that $x = -2$ is not in the domain of f.

$f'(x) = \dfrac{(2x)(x+2) - (x^2-2)(1)}{(x+2)^2}$

$\qquad = \dfrac{2x^2 + 4x - x^2 + 2}{(x+2)^2}$

$\qquad = \dfrac{x^2 + 4x + 2}{(x+2)}$

$f'(x) = 0$ when $x^2 + 4x + 2 = 0$, so the critical numbers are $x = -2 \pm \sqrt{2}$. $x = -2 + \sqrt{2}$ is a local min; $x = -2 + \sqrt{2}$ is a local max.

19. $f(x) = \dfrac{x}{x^2 + 1}$

$f'(x) = \dfrac{1(x^2 + 1) - x(2x)}{(x^2 + 1)^2}$

$\qquad = \dfrac{x^2 + 1 - 2x^2}{(x^2 + 1)^2}$

$\qquad = \dfrac{1 - x^2}{(x^2 + 1)^2}$

$f'(x) = 0$ for $1 - x^2 = 0$, $x = 1, -1$; $f'(x)$ is defined for all x, so $x = 1, -1$ are the critical numbers. $x = -1$ is local min, $x = 1$ is local max.

21. $f(x) = x^{4/3} + 4x^{1/3} + 4x^{-2/3}$
f is not defined at $x = 0$.

$$f'(x) = \frac{4}{3}x^{1/3} + \frac{4}{3}x^{-2/3} - \frac{8}{3}x^{-5/3}$$

$$= \frac{4}{3}x^{-5/3}(x^2 + x - 2)$$

$$= \frac{4}{3}x^{-5/3}(x - 1)(x + 2)$$

$x = -2, 1$ are critical numbers.
$x = -2$ and $x = 1$ are local minima.

23. $f(x) = 2x\sqrt{x+1} = 2x(x+1)^{1/2}$
Domain of f is all $x \geq -1$.

$$f'(x) = 2(x+1)^{1/2} + 2x\left(\frac{1}{2}(x+1)^{-1/2}\right)$$

$$= \frac{2(x+1) + x}{\sqrt{x+1}}$$

$$= \frac{3x + 2}{\sqrt{x+1}}$$

$f'(x) = 0$ for $3x + 2 = 0$, $x = -2/3$.
$f'(x)$ is undefined for $\sqrt{x+1} = 0$, $x = -1$ so
$x = -2/3, -1$ are critical numbers.
$x = -2/3$ is a local min. $x = -1$ is an endpoint
so is neither a local min nor a local max, though
it is a maximum on the interval $[-1, 0)$.

25. Because of the absolute value sign, there may
be critical numbers where the function $x^2 - 1$
changes sign; that is, at $x = \pm 1$. For $x > 1$ and
for $x < -1$, $f(x) = x^2 - 1$ and $f'(x) = 2x$,
so there are no critical numbers on these
intervals. For $-1 < x < 1$, $f(x) = 1 - x^2$ and
$f'(x) = -2x$, so 0 is a critical number. A graph
confirms this analysis and shows there is a
local max at $x = 0$ and local min at $x = \pm 1$.

27. First, let's find the critical numbers for $x < 0$.
In this case,
$f(x) = x^2 + 2x - 1$
$f'(x) = 2x + 2 = 2(x + 1)$
so the only critical number in this interval is
$x = -1$ and it is a local minimum.
Now for $x > 0$, $f(x) = x^2 - 4x + 3$
$f'(x) = 2x - 4 = 2(x - 2)$

so the only critical number is $x = 2$ and it is a
local minimum.

Finally, $x = 0$ is also a critical number, since f
is not continuous and hence not differentiable
at $x = 0$. Indeed, $x = 0$ is a local maximum.

29. $f(x) = x^3 - 3x + 1$
$f'(x) = 3x^2 - 3 = 3(x^2 - 1)$
$f'(x) = 0$ for $x = \pm 1$.

(a) On $[0, 2]$, 1 is the only critical number.
We calculate:
$f(0) = 1$
$f(1) = -1$ is the abs min.
$f(2) = 3$ is the abs max.

(b) On the interval $[-3, 2]$, we have both 1
and -1 as critical numbers. We calculate:
$f(-3) = -17$ is the abs min.
$f(-1) = 3$ is the abs max.
$f(1) = -1$
$f(2) = 3$ is also the abs max.

31. $f(x) = x^{2/3}$
$f'(x) = \frac{2}{3}x^{-1/3} = \frac{2}{3\sqrt[3]{x}}$
$f'(x) \neq 0$ for any x, but $f'(x)$ is undefined for
$x = 0$, so $x = 0$ is critical a number.

(a) On $[-4, -2]$:
$0 \notin [-4, -2]$ so we only look at endpoints.
$f(-4) = \sqrt[3]{16} \approx 2.52$
$f(-2) = \sqrt[3]{4} \approx 1.59$
So $f(-4) = \sqrt[3]{16}$ is the abs max and
$f(-2) = \sqrt[3]{4}$ is the abs min.

(b) On $[-1, 3]$, we have 0 as a critical number.
$f(-1) = 1$
$f(0) = 0$ is the abs min.
$f(3) = 3^{2/3}$ is the abs max.

33. $f(x) = \dfrac{3x^2}{x - 3}$
Note that $x = 3$ is not in the domain of f.

$$f'(x) = \frac{6x(x-3) - 3x^2(1)}{(x-3)^2}$$

$$= \frac{6x^2 - 18x - 3x^2}{(x-3)^2}$$

$$= \frac{3x^2 - 18x}{(x-3)^2}$$

$$= \frac{3x(x-6)}{(x-3)^2}$$

The critical points are $x = 0$, $x = 6$.

(a) On $[-2, 2]$:
$f(-2) = -12/5$
$f(2) = -12$
$f(0) = 0$
Hence abs max is $f(0) = 0$ and abs min is $f(2) = -12$.

(b) On $[2, 8]$, the function is not continuous and in fact has no absolute max or min.

35. $f'(x) = 4x^3 - 6x + 2 = 0$ at about $x = 0.3660, -1.3660$ and at $x = 1$.

(a) $f(-1) = -3$, $f(1) = 1$; $f(0.3660) \approx 1.3481$. The absolute min is $(-1, -3)$ and the absolute max is approximately $(0.3660, 1.3481)$.

(b) The absolute min is approximately $(-1.3660, -3.8481)$ and the absolute max is $(-3, 49)$.

37. $f'(x) = 2x - 3\cos x + 3x \sin x = 0$ at about $x = 0.6371, -1.2269$ and -2.8051.

(a) The absolute min is approximately $(0.6371, -1.1305)$ and the absolute max is approximately $(-1.2269, 2.7463)$.

(b) The absolute min is approximately $(-2.8051, -0.0748)$ and the absolute max is approximately $(-5, 29.2549)$.

39. If an absolute max or min occurs only at the endpoint of a closed interval, then there will be no absolute max or min on the open interval.

29) on $(0, 2)$, $f(1) = -1$ is min, no max. On $(-3, 2)$, $f(-1) = 3$ is max, no min.

30) on $(-3, 1)$, $f(-2) = -14$ is min, no max. On $(-1, 3)$, $f(2) = -14$ is min, no max.

31) on $(-4, -2)$, no max or min. On $(-1, 3)$, $f(0) = 0$ is min, no max.

32) on $(0, 2\pi)$, $f(5\pi/4) = -\sqrt{2}$ is min, $f(\pi/4) = \sqrt{2}$ is max. On $(\pi/2, \pi)$, no max or min.

33) on $(-2, 2)$, $f(0) = 0$ is max; no min. On $(2, 8)$, no max or min

34) on $(0, 1)$, no min or max. On $(-3, 4)$, no max, $f(0) = -2$ is min.

41. On $[-2, 2]$, the absolute maximum is 3 and the absolute minimum doesn't exist.

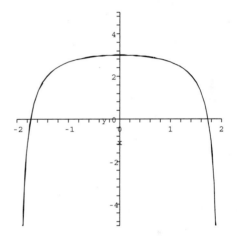

43. On $(-2, 2)$ the absolute maximum is 4 and the absolute minimum is 2.

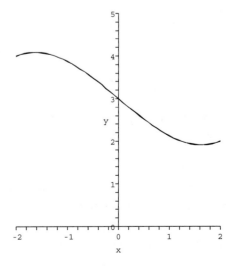

45. You will not be able to construct an example with a continuous function, but there are many examples using a function with a discontinuity, for example, $f(x) = \sec^2 x$.

47. $f(x) = x^3 + cx + 1$
$f'(x) = 3x^2 + c$

We know (perhaps from a pre-calculus course) that for any cubic polynomial with positive leading coefficient, when x is large and positive the value of the polynomial is very large and positive, and when x is large and negative, the value of the polynomial is very large and negative.

Type 1: $c > 0$. There are no critical numbers. As you move from left to right, the graph of f is always rising.

Type 2: $c < 0$ There are two critical numbers $x = \pm\sqrt{-c/3}$. As you move from left to right, the graph rises until we get to the first critical number, then the graph must fall until we get to the second critical number, and then the graph rises again. So the critical number on the left is a local maximum and the critical number on the right is a local minimum.

Type 3: $c = 0$. There is only one critical number, which is neither a local max nor a local min.

49. $f(x) = x^3 + bx^2 + cx + d$
$f'(x) = 3x^2 + 2bx + c$ The quadratic formula says that the critical numbers are

$$x = \frac{-2b \pm \sqrt{4b^2 - 12c}}{6}$$

$$= \frac{-b \pm \sqrt{b^2 - 3c}}{3}.$$

So if $c < 0$, the quantity under the square root is positive and there are two critical numbers. This is like the Type 2 cubics in Exercise 47. We know that as x goes to infinity, the polynomial $x^3 + bx^2 + cx + d$ gets very large and positive, and when x goes to minus infinity, the polynomial is very large but negative. Therefore, the critical number on the left must be a

local max, and the critical number on the right must be a local min.

51. $f(x) = x^4 + cx^2 + 1$
$f'(x) = 4x^3 + 2cx = 2x(2x^2 + c)$
So $x = 0$ is always a critical number.

Case 1: $c \geq 0$. The only solution to $2x(2x^2 + c) = 0$ is $x = 0$, so $x = 0$ is the only critical number. This must be a minimum, since we know that the function $x^4 + cx^2 + 1$ is large and positive when $|x|$ is large (so the graph is roughly U-shaped). We could also note that $f(0) = 1$, and 1 is clearly the absolute minimum of this function if $c \geq 0$.

Case 2: $c < 0$. Then there are two other critical numbers $x = \pm\sqrt{-c/2}$. Now $f(0)$ is still equal to 1, but the value of f at both new critical numbers is less than 1. Hence $f(0)$ is a local max, and both new critical numbers are local minimums.

53. Since f is differentiable on (a, b), it is continuous on the same interval. Since f is decreasing at a and increasing at b, f must have a local minimum for some value c, where $a < c < b$. By Fermat's theorem, c is a critical number for f. Since f is differentiable at c, $f'(c)$ exists, and therefore $f'(c) = 0$.

55. Graph of $f(x) = \dfrac{x^2}{x^2 + 1}$:

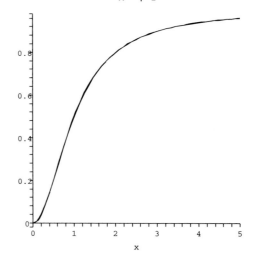

$$f'(x) = \frac{2x(x^2 + 1) - x^2(2x)}{(x^2 + 1)^2}$$

$$= \frac{2x}{(x^2 + 1)^2}$$

$$f''(x) = \frac{2(x^2 + 1)^2 - 2x \cdot 2(x^2 + 1) \cdot 2x}{(x^2 + 1)^4}$$

$$= \frac{2(x^2 + 1)\left[(x^2 + 1) - 4x^2\right]}{(x^2 + 1)^4}$$

$$= \frac{2\left[1 - 3x^2\right]}{(x^2 + 1)^3}$$

$$f''(x) = 0 \text{ for } x = \pm\frac{1}{\sqrt{3}},$$

$$x = -\frac{1}{\sqrt{3}} \notin (0, \infty)$$

$$x = \frac{1}{\sqrt{3}} \text{ is steepest point.}$$

57. $y = x^5 - 4x^3 - x + 10, \ x \in [-2, 2]$

$y' = 5x^4 - 12x^2 - 1$

$x = -1.575, 1.575$ are critical numbers of y. There is a local max at $x = -1.575$, local min at $x = 1.575$. $x = -1.575$ represents the top and $x = 1.575$ represents the bottom of the roller coaster.

$y''(x) = 20x^3 - 24x = 4x(5x^2 - 6) = 0$

$x = 0, \pm\sqrt{6/5}$ are critical numbers of y'. We calculate y' at the critical numbers and at the endpoints $x = \pm 2$:

$y'(0) = -1$

$y'\left(\pm\sqrt{6/5}\right) = -41/5$

$y'(\pm 2) = 31$

So the points where the roller coaster is making the steepest descent are $x = \pm\sqrt{6/5}$, but the steepest part of the roller coast is during the ascents at ± 2.

3.3 INCREASING AND DECREASING FUNCTIONS

1. $y = x^3 - 3x + 2$

$y' = 3x^2 - 3 = 3(x^2 - 1)$

$\quad = 3(x + 1)(x - 1)$

$x = \pm 1$ are critical numbers.
$(x + 1) > 0$ on $(-1, \infty)$, $(x + 1) < 0$ on $(-\infty, -1)$
$(x - 1) > 0$ on $(1, \infty)$, $(x - 1) < 0$ on $(-\infty, 1)$
$3(x + 1)(x - 1) > 0$ on $(1, \infty) \cup (-\infty, -1)$
so y is increasing on $(1, \infty)$ and on $(-\infty, -1)$.
$3(x + 1)(x - 1) < 0$ on $(-1, 1)$, so y is decreasing on $(-1, 1)$.

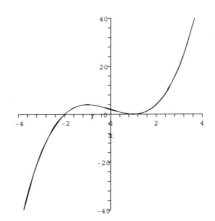

3. $y = x^4 - 8x^2 + 1$

$y' = 4x^3 - 16x = 4x(x^2 - 4)$

$\quad = 4x(x - 2)(x + 2)$

$x = 0, 2, -2$

$4x > 0$ on $(0, \infty)$, $4x < 0$ on $(-\infty, 0)$
$(x - 2) > 0$ on $(2, \infty)$, $(x - 2) < 0$
on $(-\infty, 2)$ $(x + 2) > 0$ on $(-2, \infty)$,
$(x + 2) < 0$ on $(-\infty, -2)$ $4x(x - 2)$
$(x + 2) > 0$ on $(-2, 0) \cup (2, \infty)$, so the function is increasing on $(-2, 0)$ and on $(2, \infty)$.
$4x(x - 2)(x + 2) < 0$ on $(-\infty, -2) \cup (0, 2)$, so y is decreasing on $(-\infty, -2)$ and on $(0, 2)$.

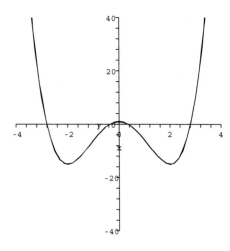

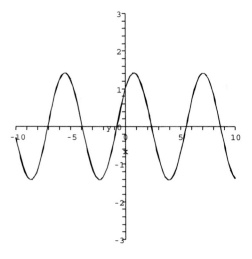

5. $y = (x + 1)^{2/3}$
$y' = \frac{2}{3}(x + 1)^{-1/3} = \frac{2}{3\sqrt[3]{x+1}}$
y' is not defined for $x = -1$.
$\frac{2}{3\sqrt[3]{x + 1}} > 0$ on $(-1, \infty)$, y is increasing.
$\frac{2}{3\sqrt[3]{x + 1}} < 0$ on $(-\infty, -1)$, y is decreasing.

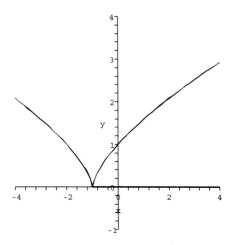

7. $y = \sin x + \cos x$
$y' = \cos x - \sin x = 0$
$\cos x = \sin x$
$x = \pi/4,\ 5\pi/4,\ 9\pi/4$, etc. $\cos x - \sin x >$
0 on $(-3\pi/4, \pi/4) \cup (5\pi/4, 9\pi/4) \cup \ldots$
$\cos x - \sin x < 0$ on $(\pi/4, 5\pi/4) \cup (9\pi/4, 13\pi/4) \cup \ldots$ So $y = \sin x + \cos x$ is decreasing on $(\pi/4, 5\pi/4)$, $(9\pi/4, 13\pi/4)$, etc., and is increasing on $(-3\pi/4, \pi/4)$, $(5\pi/4, 9\pi/4)$, etc.

9. $y = x^4 + 4x^3 - 2$
$y' = 4x^3 + 12x^2 = 4x^2(x + 3)$
Critical numbers are $x = 0$, $x = -3$.
$4x^2(x + 3) > 0$ on $(-3, 0) \cup (0, \infty)$
$4x^2(x + 3) < 0$ on $(-\infty, -3)$
Hence $x = -3$ is a local minimum and $x = 0$ is not an extremum.

11. $y = x^2 - 2x^{2/3} + 2$
$y' = 2x - \frac{4}{3}x^{-1/3}$
y' is undefined for $x = 0$ and $y' = 0$ for $x = \pm\sqrt[4]{\frac{8}{27}} \approx \pm 0.7378$.
$y' < 0$ on $\left(-\infty, -\sqrt[4]{\frac{8}{27}}\right)$
$y' > 0$ on $\left(-\sqrt[4]{\frac{8}{27}}, 0\right)$
$y' < 0$ on $\left(0, \sqrt[4]{\frac{8}{27}}\right)$
$y' > 0$ on $\left(\sqrt[4]{\frac{8}{27}}, \infty\right)$
So $y = x^2 - 2x^{2/3} + 2$ has local minima at $x = \pm\sqrt[4]{\frac{8}{27}}$ and a local maximum at $x = 0$.

13. $y = \dfrac{x}{1 + x^3}$

Note that the function is not defined for $x = -1$.

$$y' = \frac{1(1+x^3) - x(3x^2)}{(1+x^3)^2}$$

$$= \frac{1 + x^3 - 3x^3}{(1+x^3)^2}$$

$$= \frac{1 - 2x^3}{(1+x^3)^2}$$

Critical number is $x = \sqrt[3]{1/2}$. $y' > 0$ on $(-\infty, -1) \cup (-1, \sqrt[3]{1/2})$. $y' < 0$ on $(\sqrt[3]{1/2}, \infty)$. Hence $x = \sqrt[3]{1/2}$ is a local max.

15. $y = \sqrt{x^3 + 3x^2} = (x^3 + 3x^2)^{1/2}$
Domain is all $x \geq -3$.

$$y' = \frac{1}{2}(x^3 + 3x^2)^{-1/2}(3x^2 + 6x)$$

$$= \frac{3x^2 + 6x}{2\sqrt{x^3 + 3x^2}}$$

$$= \frac{3x(x+2)}{2\sqrt{x^3 + 3x^2}}$$

$x = 0, -2, -3$ are critical numbers.
y' undefined at $x = 0, -3$
$y' > 0$ on $(-3, -2) \cup (0, \infty)$
$y' < 0$ on $(-2, 0)$
So $y = \sqrt{x^3 + 3x^2}$ has local max at $x = -2$, local min at $x = 0$. $x = -3$ is an endpoint, and so is not a local extremum.

17. $y = \dfrac{x}{x^2 - 1}$

$$y' = \frac{x^2 - 1 - x(2x)}{(x^2 - 1)^2}$$

$$= -\frac{x^2 + 1}{(x^2 - 1)^2}$$

There are no values of x for which $y' = 0$. There are no critical points, because the values for which y' does not exist (that is, $x = \pm 1$) are not in the domain.

There are vertical asymptotes at $x = \pm 1$, and a horizontal asymptote at $y = 0$. This can be verified by calculating the following limit:

$$\lim_{x \to \pm\infty} \frac{x}{x^2 - 1} = 0.$$

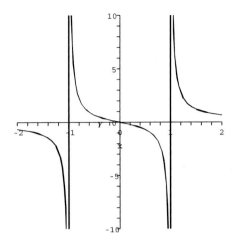

19. $y = \dfrac{x^2}{x^2 - 4x + 3} = \dfrac{x^2}{(x-1)(x-3)}$

Vertical asymptotes $x = 1$, $x = 3$. When $|x|$ is large, the function approaches the value 1, so $y = 1$ is a horizontal asymptote.

$$y' = \frac{2x(x^2 - 4x + 3) - x^2(2x - 4)}{(x^2 - 4x + 3)^2}$$

$$= \frac{2x^3 - 8x^2 + 6x - 2x^3 + 4x^2}{(x^2 - 4x + 3)^2}$$

$$= \frac{-4x^2 + 6x}{(x^2 - 4x + 3)^2}$$

$$= \frac{2x(-2x + 3)}{(x^2 - 4x + 3)^2}$$

$$= \frac{2x(-2x + 3)}{[(x-3)(x-1)]^2}$$

Critical numbers are $x = 0$ (local min) and $x = 3/2$ (local max).

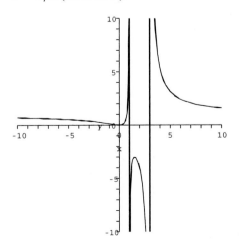

21.

$$y = \frac{x}{\sqrt{x^2 + 1}}$$

$$y' = \frac{\sqrt{x^2 + 1} - x^2/\sqrt{x^2 + 1}}{x^2 + 1}$$

$$= \frac{1}{(x^2 + 1)^{3/2}}$$

The derivative is never zero, so there are no critical points. To verify that there are horizontal asymptotes at $y = \pm 1$:

$$y = \frac{x}{\sqrt{x^2 + 1}}$$

$$= \frac{x}{\sqrt{x^2}\sqrt{1 + \frac{1}{x^2}}}$$

$$= \frac{x}{|x|\sqrt{1 + \frac{1}{x^2}}}$$

Thus, $\lim\limits_{x \to \infty} \dfrac{x}{|x|\sqrt{1 + \frac{1}{x^2}}} = 1$

$\lim\limits_{x \to -\infty} \dfrac{x}{|x|\sqrt{1 + \frac{1}{x^2}}} = -1$

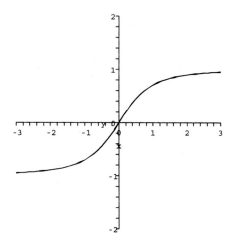

23. $y' = 3x^2 - 26x - 10 = 0$ when

$x = \dfrac{26 \pm \sqrt{796}}{6}$. Local max at $x = -0.3689$;

local min at $x = 9.0356$.

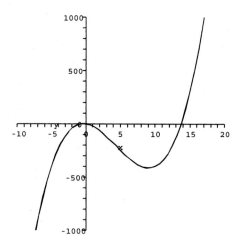

25. $y' = 4x^3 - 45x^2 - 4x + 40$
Local minima at $x = -0.9474$, 11.2599; local max at 0.9374. Global behavior looks like

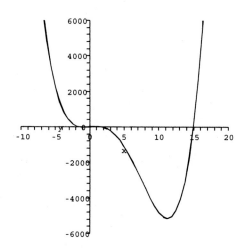

Local behavior of the function near $x = 0$ looks like

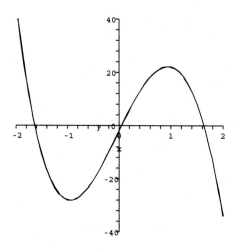

27. $y' = 5x^4 - 600x^2 + 605$ Local minima at $x = -1.0084, 10.9079$; local maxima at $x = -10.9079, 1.0084$. Global behavior looks like

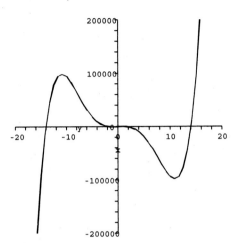

Local behavior of the function near $x = 0$ looks like

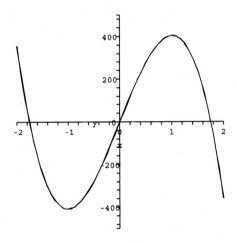

29. One possible graph:

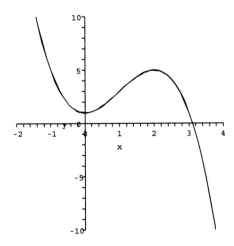

31. One possible graph:

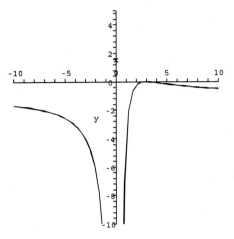

33. The derivative is

$$y' = \frac{-3x^4 + 120x^3 - 1}{(x^4 - 1)^2}.$$

We estimate the critical numbers to be approximately 0.2031 and 39.999. The following graph shows global behavior:

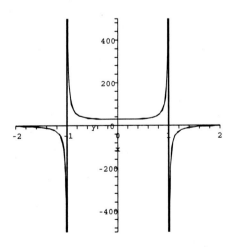

The following graphs show local behavior:

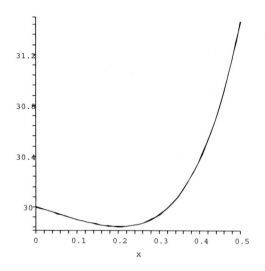

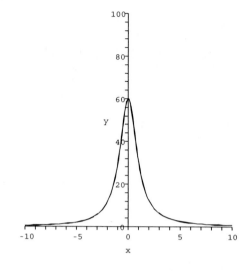

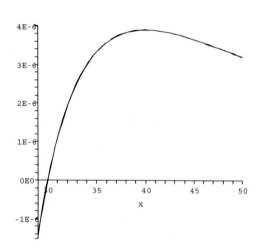

The following graphs show local behavior:

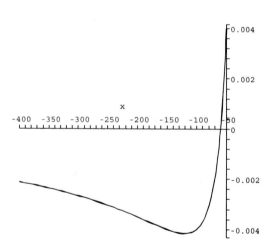

35. The derivative is

$$y' = \frac{-x^2 - 120x + 1}{(x^2 + 1)^2}.$$

We estimate the critical numbers to be approximately 0.008 and -120.008.
The following graph shows global behavior:

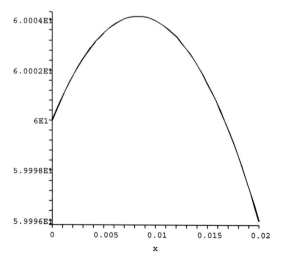

37. $f'(0) = \lim\limits_{x \to 0} \dfrac{f(x) - f(0)}{x - 0}$

$= \lim\limits_{x \to 0} \dfrac{f(x)}{x}$

$= \lim\limits_{x \to 0} \left[1 + 2x \sin\left(\dfrac{1}{x}\right) \right] = 1$

For $x \neq 0$,
$f'(x) =$
$1 + 2\left[2x \sin\left(\frac{1}{x}\right) + x^2 \left(\frac{-1}{x^2}\right) \cos\left(\frac{1}{x}\right) \right]$
$= 1 + 4x \sin\left(\frac{1}{x}\right) - 2\cos\left(\frac{1}{x}\right)$

For values of x close to the origin, the middle term of the derivative is small, and since the last term $-2\cos(1/x)$ reaches its minimum value of -2 in every neighborhood of the origin, f' has negative values on every neighborhood of the origin. Thus, f is not increasing on any neighborhood of the origin.

This conclusion does not contradict Theorem 4.1 because the theorem states that if a function's derivative is positive for all values in an interval, then it is increasing in that interval. In this example, the derivative is not positive throughout any interval containing the origin.

39. f is continuous on $[a, b]$, and $c \in (a, b)$ is a critical number.

(a) If $f'(x) > 0$ for all $x \in (a, c)$ and $f'(x) < 0$ for all $x \in (c, b)$, by Theorem 3.1, f is increasing on (a, c) and decreasing on (c, b), so $f(c) > f(x)$ for all $x \in (a, c)$ and $x \in (c, b)$. Thus $f(c)$ is a local max.

(b) If $f'(x) < 0$ for all $x \in (a, c)$ and $f'(x) > 0$ for all $x \in (c, b)$, by Theorem 3.1, f is decreasing on (a, c) and increasing on (c, b). So $f(c) < f(x)$ for all $x \in (a, c)$ and $x \in (c, b)$. Thus $f(c)$ is a local min.

(c) If $f'(x) > 0$ on (a, c) and (c, b), then $f(c) > f(x)$ for all $x \in (a, c)$ and $f(c) < f(x)$ for all $x \in (c, b)$, so c is not a local extremum.

If $f'(x) < 0$ on (a, c) and (c, b), then $f(c) < f(x)$ for all $x \in (a, c)$ and $f(c) >$

$f(x)$ for all $x \in (c, b)$, so c is not a local extremum.

41. Let $f(x) = 2\sqrt{x}$, $g(x) = 3 - 1/x$.
Then $f(1) = 2\sqrt{1} = 2$, and $g(1) = 3 - 1 = 2$, so $f(1) = g(1)$.
$f'(x) = \dfrac{1}{\sqrt{x}}g'(x) = \dfrac{1}{x^2}$ So $f'(x) > g'(x)$ for

all $x > 1$, and $f(x) = 2\sqrt{x} > 3 - \dfrac{1}{x} = g(x)$ for all $x > 1$.

43. Let $f(x) = \tan x$, $g(x) = x$.
Then $f(0) = \tan 0 = 0$, $g(0) = 0$, so $f(0) = g(0)$. $f'(x) = \sec^2 x$, $\quad g'(x) = 1$. So $f'(x) > g'(x)$ for $0 < x < \frac{\pi}{2}$. Thus $f(x) = \tan x > x = g(x)$ for $0 < x < \frac{\pi}{2}$.

45. TRUE. If $x_1 < x_2$, then $g(x_1) < g(x_2)$ since g is increasing, and then $f(g(x_1)) < f(g(x_2))$ since f is increasing.

47. $s(t) = \sqrt{t + 4} = (t + 4)^{1/2}$
$s'(t) = \frac{1}{2}(t + 4)^{-1/2} = \dfrac{1}{2\sqrt{t+4}} > 0$
So total sales are always increasing at the rate of $\dfrac{1}{2\sqrt{t + 4}}$ thousand dollars per month.

49. If the roots of the derivative are very close together, then the extrema will be very close together and difficult to see on a graph showing global behavior of the function. One function with the given derivative is

$f(x) = \dfrac{1}{3}x^3 - 0.01x + 2.$

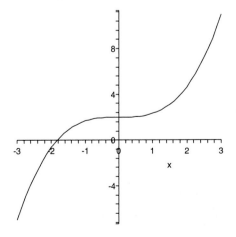

The two extreme points near $x = 0$ are impossible to detect from the graph using a usual scale. To construct a degree 4 polynomial with two hidden extrema near $x = 1$ and another extrema (not hidden) near $x = 0$, we start with a derivative,

$$g'(x) = x(x - 0.9)(x - 1.1)$$
$$= x^3 - 2x^2 + 0.99x.$$

A function with this derivative is

$$g(x) = \frac{1}{4}x^4 - \frac{2}{3}x^3 + \frac{0.99}{2}x^2 - 3$$

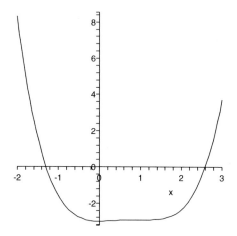

51. $f(x) = x^3 + bx^2 + cx + d$
$f'(x) = 3x^2 + 2bx + c$
$f'(x) \geq 0$ for all x if and only if $(2b)^2 - 4(3)(c) \leq 0$
if and only if $4b^2 \leq 12c$
if and only if $b^2 \leq 3c$.

53. (a) $\mu'(-10) \approx \dfrac{0.0048 - 0.0043}{-12 - (-8)}$

$= \dfrac{0.0005}{-4}$

$= -0.000125$

(b) $\mu'(-6) \approx \dfrac{0.0048 - 0.0043}{-4 - (-8)}$

$= \dfrac{0.0005}{4}$

$= 0.000125$

Whether the warming of the ice due to skating makes it easier or harder depends on the current temperature of the ice. As seen from these examples, the coefficient of friction μ is decreasing when the temperature is $-10°$ and increasing when the temperature is $-6°$.

3.4 CONCAVITY AND THE SECOND DERIVATIVE TEST

1. $f'(x) = 3x^2 - 6x + 4$
$f''(x) = 6x - 6 = 6(x - 1)$
$f''(x) > 0$ on $(1, \infty)$
$f''(x) < 0$ on $(-\infty, 1)$

So f is concave down on $(-\infty, 1)$ and concave up on $(1, \infty)$.

3. $f(x) = x + \frac{1}{x} = x + x^{-1}$
$f'(x) = 1 - x^{-2}$
$f''(x) = 2x^{-3}$
$f''(x) > 0$ on $(0, \infty)$
$f''(x) < 0$ on $(-\infty, 0)$

So f is concave up on $(0, \infty)$ and concave down on $(-\infty, 0)$.

5. $f'(x) = \cos x + \sin x$
$f''(x) = -\sin x + \cos x$
$f''(x) < 0$ on $\dots \left(\frac{\pi}{4}, \frac{5\pi}{4}\right) \cup \left(\frac{9\pi}{4}, \frac{13\pi}{4}\right) \dots$
$f''(x) > 0$ on $\dots \left(-\frac{3\pi}{4}, \frac{\pi}{4}\right) \cup \left(\frac{5\pi}{4}, \frac{9\pi}{4}\right) \dots$
f is concave down on $\dots \left(\frac{\pi}{4}, \frac{5\pi}{4}\right) \cup$
$\left(\frac{9\pi}{4}, \frac{13\pi}{4}\right) \dots$, concave up on $\dots \left(-\frac{3\pi}{4}, \frac{\pi}{4}\right)$
$\cup \left(\frac{5\pi}{4}, \frac{9\pi}{4}\right) \dots$

7. $f'(x) = \frac{4}{3}x^{1/3} + \frac{4}{3}x^{-2/3}$

$f''(x) = \frac{4}{9}x^{-2/3} - \frac{8}{9}x^{-5/3}$

$= \frac{4}{9x^{2/3}}\left(1 - \frac{2}{x}\right)$

The quantity $\dfrac{4}{9x^{2/3}}$ is never negative, so the sign of the second derivative is the same as the sign of $1 - \dfrac{2}{x}$. Hence the function is concave up for $x > 2$ and $x < 0$, and is concave down for $0 < x < 2$.

9. $f(x) = x^4 + 4x^3 - 1$
$f'(x) = 4x^3 + 12x^2 = x^2(4x + 12)$
So the critical numbers are $x = 0$ and $x = -3$.
$f''(x) = 12x^2 + 24x = 12x(x + 2)$.
$f''(0) = 0$ so the second derivative test for $x = 0$ is inconclusive.
$f''(-3) = 36 > 0$ so $x = -3$ is a local minimum.

11. $f(x) = \dfrac{x^2 - 5x + 4}{x}$

$f'(x) = \dfrac{(2x - 5)x - (x^2 - 5x + 4)(1)}{x^2}$

$= \dfrac{x^2 - 4}{x^2}$

So the critical numbers are $x = \pm 2$.

$f''(x) = \dfrac{(2x)(x^2) - (x^2 - 4)(2x)}{x^4} = \dfrac{8x}{x^4}$

$f''(2) = 1 > 0$ so $x = 2$ is a local minimum.
$f''(-2) = -1 < 0$ so $x = -2$ is a local maximum.

13. $y = (x^2 + 1)^{2/3}$

$y' = \dfrac{2}{3}(x^2 + 1)^{-1/3}(2x)$

$= \dfrac{4x(x^2 + 1)^{-1/3}}{3}$

So the only critical number is $x = 0$.

$y'' =$

$\dfrac{4}{3}\left[(x^2 + 1)^{-1/3} + \left(\dfrac{-2x^2}{3}\right)(x^2 + 1)^{-4/3}\right]$

$= \dfrac{4}{3}\dfrac{(x^2 + 1 - \frac{2x^2}{3})}{(x^2 + 1)^{4/3}} = \dfrac{4}{9}\dfrac{(3x^2 + 3 - 2x^2)}{(x^2 + 1)^{4/3}}$

$= \dfrac{4}{9}\dfrac{(x^2 + 3)}{(x^2 + 1)^{4/3}}$

So the function is concave up everywhere, decreasing for $x < 0$, and increasing for $x > 0$. Also $x = 0$ is a local min.

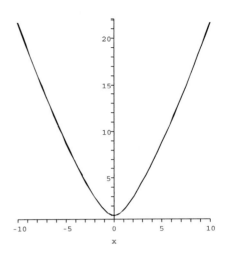

15. $f(x) = \dfrac{x^2}{x^2 - 9}$

$f'(x) = \dfrac{2x(x^2 - 9) - x^2(2x)}{(x^2 - 9)^2}$

$= \dfrac{-18x}{(x^2 - 9)^2}$

$= \dfrac{-18x}{\{(x + 3)(x - 3)\}^2}$

$f''(x) =$

$\dfrac{-18(x^2 - 9)^2 + 18x \cdot 2(x^2 - 9) \cdot 2x}{(x^2 - 9)^4}$

$= \dfrac{54x^2 + 162}{(x^2 - 9)^3}$

$= \dfrac{54(x^2 + 3)}{(x^2 - 9)^3}$

$f'(x) > 0$ on $(-\infty, -3) \cup (-3, 0)$
$f'(x) < 0$ on $(0, 3) \cup (3, \infty)$
$f''(x) > 0$ on $(-\infty, -3) \cup (3, \infty)$
$f''(x) < 0$ on $(-3, 3)$
$f''(0) = \dfrac{162}{(-9)^3}$
f is increasing on $(-\infty, -3) \cup (-3, 0)$, decreasing on $(0, 3) \cup (3, \infty)$, concave up on $(-\infty, -3) \cup (3, \infty)$, concave down on

$(-3, 3)$, $x = 0$ is a local max. f has a horizontal asymptote of $y = 1$ and vertical asymptotes at $x = \pm 3$.

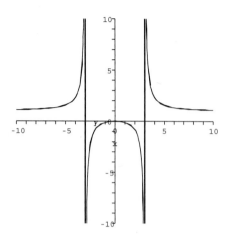

17. $f(x) = x^{3/4} - 4x^{1/4}$

Domain of $f(x)$ is $\{x \mid x \geq 0\}$.

$$f'(x) = \frac{3}{4}x^{-1/4} - x^{-3/4} = \frac{\frac{3}{4}\sqrt{x} - 1}{x^{3/4}}$$

So $x = 0$ and $x = 16/9$ are critical points, but because of the domain we only need to really consider the latter. $f'(1) = -1/4$ so $f(x)$ is decreasing on $(0, 16/9)$.

$f'(4) = \dfrac{0.5}{4^{3/4}} > 0$ so $f(x)$ is increasing on $(16/9, \infty)$.

Thus $x = 16/9$ is the location of a local minimum for $f(x)$.

$$f''(x) = \frac{-3}{16}x^{-5/4} + \frac{3}{4}x^{-7/4}$$

$$= \frac{\frac{-3}{16}\sqrt{x} + \frac{3}{4}}{x^{7/4}}$$

The critical number here is $x = 16$. We find that $f''(x) > 0$ on the interval $(0, 16)$ (so $f(x)$ is concave up on this interval) and $f''(x) < 0$ on the interval $(16, \infty)$ (so $f(x)$ is concave down on this interval).

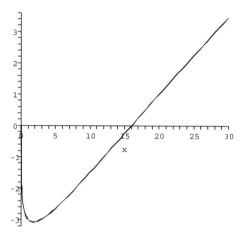

19. The easiest way to sketch this graph is to notice that

$$f(x) = x|x| = \begin{cases} x^2 & x \geq 0 \\ -x^2 & x < 0 \end{cases}$$

Since

$$f'(x) = \begin{cases} 2x & x \geq 0 \\ -2x & x < 0 \end{cases}$$

there is a critical point at $x = 0$. However, it is neither a local maximum nor a local minimum. Since

$$f''(x) = \begin{cases} 2 & x > 0 \\ -2 & x < 0 \end{cases}$$

there is an inflection point at the origin. Note that the second derivative does not exist at $x = 0$.

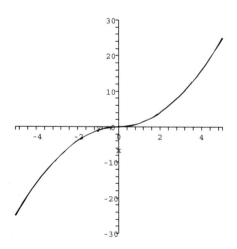

21. $f(x) = x^{1/5}(x+1) = x^{6/5} + x^{1/5}$

$$f'(x) = \frac{6}{5}x^{1/5} + \frac{1}{5}x^{-4/5}$$

$$= \frac{1}{5}x^{-4/5}(6x+1)$$

$$f''(x) = \frac{6}{25}x^{-4/5} - \frac{4}{25}x^{-9/5}$$

$$= \frac{2}{25}x^{-9/5}(3x-2)$$

Note that $f(0) = 0$, and yet the derivatives do not exist at $x = 0$. This means that there is a vertical tangent line at $x = 0$. The first derivative is negative for $x < -1/6$ and positive for $-1/6 < x < 0$ and $x > 0$. The second derivative is positive for $x < 0$ and $x > 2/3$, and negative for $0 < x < 2/3$. Thus, there is a local minimum at $x = -1/6$ and inflection points at $x = 0$ and $x = 2/3$.

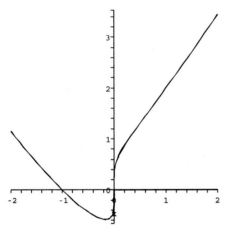

23. $f(x) = x^4 - 26x^3 + x$
$f'(x) = 4x^3 - 78x^2 + 1$
The critical numbers are approximately $-0.1129, 0.1136$ and 19.4993. $f'(-1) < 0$ implies $f(x)$ is decreasing on $(-\infty, -0.1129)$.
$f'(0) > 0$ implies $f(x)$ is increasing on $(-0.1129, 0.1136)$.
$f'(1) < 0$ implies $f(x)$ is decreasing on $(0.1136, 19.4993)$.
$f'(20) > 0$ implies $f(x)$ is increasing on $(19.4993, \infty)$.
Thus $f(x)$ has local minimums at $x = -0.1129$ and $x = 19.4993$ and a local maximum at $x = 0.1136$. nl $f''(x) = 12x^2 - 156x = x(12x -$

156)nl The critical numbers are $x = 0$ and $x = 13$.
$f''(-1) > 0$ implies $f(x)$ is concave up on $(-\infty, 0)$.
$f''(1) < 0$ implies $f(x)$ is concave down on $(0, 13)$.
$f''(20) > 0$ implies $f(x)$ is concave up on $(13, \infty)$.

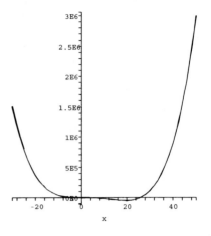

25. $y = \sqrt[3]{2x^2 - 1}$. $y' = \dfrac{4x}{3(2x^2-1)^{2/3}} = 0$ at $x = 0$ and is undefined at $x = \pm\sqrt{1/2}$. $y'' = \dfrac{-4(2x^2+3)}{9(2x^2-1)^{5/3}}$ is never 0, and is undefined where y' is. The function changes concavity at $x = \pm\sqrt{1/2}$, so these are inflection points. The slope does not change at these values, so they are not extrema. The Second Derivative Test shows that $x = 0$ is a minimum.

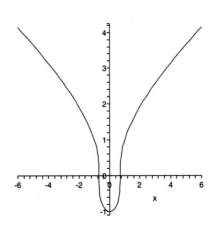

27. $f(x) = x^4 - 16x^3 + 42x^2 - 39.6x + 14$

$f'(x) = 4x^3 - 48x^2 + 84x - 39.6$

$f''(x) = 12x^2 - 96x + 84$

$\qquad = 12(x^2 - 8x + 7)$

$\qquad = 12(x - 7)(x - 1)$

$f'(x) > 0$ on $(0.8952, 1.106) \cup (9.9987, \infty)$
$f'(x) < 0$ on $(-\infty, 0.8952) \cup (1.106, 9.9987)$

$f''(x) > 0$ on $(-\infty, 1) \cup (7, \infty)$
$f''(x) < 0$ on $(1, 7)$
f is increasing on $(0.8952, 1.106)$ and on $(9.9987, \infty)$, decreasing on $(-\infty, 0.8952)$ and on $(1.106, 9.9987)$, concave up on $(-\infty, 1) \cup (7, \infty)$, concave down on $(1, 7)$, $x = 0.8952, 9.9987$ are local min, $x = 1.106$ is local max, $x = 1, 7$ are inflection points.

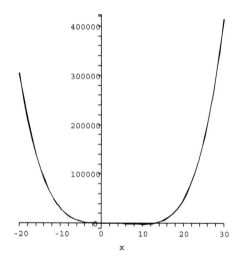

$f''(x) =$

$$\frac{4x\sqrt{x^2 - 4} - (2x^2 - 4)\frac{1}{2}(x^2 - 4)^{-1/2}(2x)}{x^2 - 4}$$

$$= \frac{4x(x^2 - 4) - x(2x^2 - 4)}{(x^2 - 4)^{3/2}}$$

$$= \frac{2x^3 - 12x}{(x^2 - 4)^{3/2}} = \frac{2x(x^2 - 6)}{(x^2 - 4)^{3/2}}$$

$f'(x) > 0$ on $(-\infty, -2) \cup (2, \infty)$
$f''(x) > 0$ on $\left(-\sqrt{6}, -2\right) \cup \left(\sqrt{6}, \infty\right)$
$f''(x) < 0$ on $\left(-\infty, -\sqrt{6}\right) \cup \left(2, \sqrt{6}\right)$
f is increasing on $(-\infty, -2)$ and on $(2, \infty)$, concave up on $\left(-\sqrt{6}, -2\right) \cup \left(\sqrt{6}, \infty\right)$, concave down on $\left(-\infty, -\sqrt{6}\right) \cup \left(2, \sqrt{6}\right)$, $x = \pm\sqrt{6}$ are inflection points.

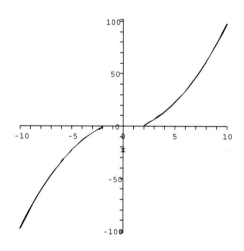

29. $f(x) = x\sqrt{x^2 - 4}$; f undefined on $(-2, 2)$

$f'(x) = \sqrt{x^2 - 4}$

$\qquad + x\left(\dfrac{1}{2}\right)(x^2 - 4)^{-1/2}(2x)$

$\qquad = \sqrt{x^2 - 4} + \dfrac{x^2}{\sqrt{x^2 - 4}}$

$\qquad = \dfrac{2x^2 - 4}{\sqrt{x^2 - 4}}$

31. $f(x)$ is concave up on $(-\infty, -0.5)$ and $(0.5, \infty)$;
$f(x)$ is concave down on $(-0.5, 0.5)$.

33. $f(x)$ is concave up on $(1, \infty)$;
$f(x)$ is concave down on $(-\infty, 1)$.

35. One possible graph:

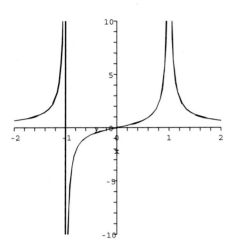

37. One possible graph:

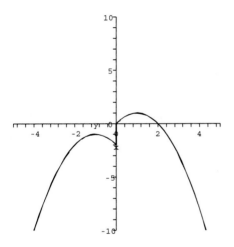

39. $f(x) = ax^3 + bx^2 + cx + d$
$f'(x) = 3ax^2 + 2bx + c$
$f''(x) = 6ax + 2b$
Thus, $f''(x) = 0$ for $x = -b/3a$. Since f'' changes sign at this point, f has an inflection point at $x = -b/3a$. Note that $a \neq 0$.

For the quartic function (where again $a \neq 0$),
$f(x) = ax^4 + bx^3 + cx^2 + dx + e$
$f'(x) = 4ax^3 + 3bx^2 + 2cx + d$
$f''(x) = 12ax^2 + 6bx + 2c$
$\qquad = 2(6ax^2 + 3bx + c)$
The second derivative is zero when

$x = \dfrac{-3b \pm \sqrt{9b^2 - 24ac}}{12a}$

$\quad = \dfrac{-3b \pm \sqrt{3(3b^2 - 8ac)}}{12a}$

There are two distinct solutions to the previous equation (and therefore two inflection points) if and only if $3b^2 - 8ac > 0$.

41. The function has the following properties:
increasing on $(0, \infty)$;
decreasing on $(-\infty, 0)$;
local minimum at $x = 0$;
concave up on $(-\infty, \infty)$;
no inflection points.

43. For exercise 41: increasing on $(-\infty, -1)$ and $(1, \infty)$;
decreasing on $(-1, 1)$;
local maximum at $x = -1$;
local minimum at $x = 1$;
concave up on $(0, \infty)$;
concave down on $(-\infty, 0)$;
inflection point at $x = 0$.

For exercise 42: increasing on $(0, 2)$ and $(2, \infty)$;
decreasing on $(-\infty, 0)$;
local minimum at $x = 0$;
concave up on $(-\infty, 1)$ and $(2, \infty)$;
concave down on $(1, 2)$;
inflection points at $x = 1$ and $x = 2$.

45. We need to know $w'(0)$ to know if the depth is increasing.

47. $s(x) = -3x^3 + 270x^2 - 3600x + 18,000$
$s'(x) = -9x^2 + 540x - 3600$
$s''(x) = -18x + 540 = 0$
$x = 30$. This is a max because the graph of $s'(x)$ is a parabola opening down. So spend $30,000 on advertising to maximize the rate of change of sales. This is also the inflection point of $s(x)$.

49. $C(x) = 0.01x^2 + 40x + 3600$
$\overline{C}(x) = \dfrac{C(x)}{x} = 0.01x + 40 + 3600x^{-1}$
$\overline{C}'(x) = 0.01 - 3600x^{-2} = 0$

$x = 600$. This is a min because $\bar{C}''(x) = 7200x^{-3} > 0$ for $x > 0$, so the graph is concave up. So manufacture 600 units to minimize average cost.

51. $f(x) = x^4 + cx^3$
$f'(x) = 4x^3 + 3cx^2 = x^2(4x + 3c)$
$f''(x) = 12x^2 + 6cx = 6x(2x + c)$

So $x = 0$ is inflection point, and $-\dfrac{3}{4}c$ is local min, $-\dfrac{1}{2}c$ is inflection point.

53. Let $f(x) = -1 - x^2$. Then
$f'(x) = -2x$
$f''(x) = -2$
so f is concave down for all x, but $-1 - x^2 = 0$ has no solution.

55. Since the tangent line points above the sun, the sun appears higher in the sky than it really is.

3.5 OVERVIEW OF CURVE SKETCHING

1. $f(x) = x^3 - 3x^2 + 3x$
$\quad\quad = x(x^2 - 3x + 3)$

The only x-intercept is $x = 0$; the y-intercept is $(0, 0)$.

$f'(x) = 3x^2 - 6x + 3$
$\quad\quad = 3(x^2 - 2x + 1) = 3(x - 1)^2$

$f'(x) > 0$ for all $x \ne 1$, so $f(x)$ is increasing for all $x \ne 1$ and has no local extrema.

$f''(x) = 6x - 6 = 6(x - 1)$

There is an inflection point at $x = 1$: $f(x)$ is concave down on $(-\infty, 1)$ and concave up on $(1, \infty)$.

Finally, $f(x) \to \infty$ as $x \to \infty$ and $f(x) \to -\infty$ as $x \to -\infty$.

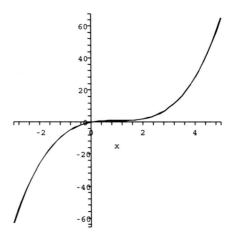

3. $f(x) = x^5 - 2x^3 + 1$

The x-intercepts are $x = 1$ and $x \approx -1.5129$; the y-intercept is $(0, 1)$. $f'(x) = 5x^4 - 6x^2 = x^2(5x^2 - 6)$

The critical numbers are $x = 0$ and $x = \pm\sqrt{6/5}$. Plugging values from each of the intervals into $f'(x)$, we find that $f'(x) > 0$ on $(-\infty, -\sqrt{6/5})$ and $(\sqrt{6/5}, \infty)$ so $f(x)$ is increasing on these intervals. $f'(x) < 0$ on $(-\sqrt{6/5}, 0)$ and $(0, \sqrt{6/5})$ so $f(x)$ is decreasing on these intervals. Thus $f(x)$ has a local maximum at $-\sqrt{6/5}$ and a local minimum at $\sqrt{6/5}$.

$f''(x) = 20x^3 - 12x = 4x(5x^2 - 3)$

The critical numbers are $x = 0$ and $x = \pm\sqrt{3/5}$. Plugging values from each of the intervals into $f''(x)$, we find that $f''(x) > 0$ on $(-\sqrt{3/5}, 0)$ and $(\sqrt{3/5}, \infty)$ so $f(x)$ is concave up on these intervals. $f''(x) < 0$ on $(-\infty, -\sqrt{3/5})$ and $(0, \sqrt{3/5})$ so $f(x)$ is concave down on these intervals. Thus $f(x)$ has inflection points at all three of these critical numbers. Finally, $f(x) \to \infty$ as $x \to \infty$ and $f(x) \to -\infty$ as $x \to -\infty$.

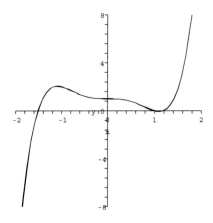

5. $f(x) = x + \dfrac{4}{x} = \dfrac{x^2 + 4}{x}$

This function has no x- or y-intercepts. The domain is $\{x \mid x \neq 0\}$. $f(x)$ has a vertical asymptote at $x = 0$ such that $f(x) \to -\infty$ as $x \to 0^-$ and $f(x) \to \infty$ as $x \to 0^+$.

$$f'(x) = 1 - 4x^{-2} = \dfrac{x^2 - 4}{x^2}$$

The critical numbers are $x = \pm 2$. We find that $f'(x) > 0$ on $(-\infty, -2)$ and $(2, \infty)$ so $f(x)$ is increasing on these intervals. $f'(x) < 0$ on $(-2, 0)$ and $(0, 2)$, so $f(x)$ is decreasing on these intervals. Thus $f(x)$ has a local maximum at $x = -2$ and a local minimum at $x = 2$. $f''(x) = 8x^{-3}$. $f''(x) < 0$ on $(-\infty, 0)$ so $f(x)$ is concave down on this interval and $f''(x) > 0$ on $(0, \infty)$ so $f(x)$ is concave up on this interval, but $f(x)$ has an asymptote (not an inflection point) at $x = 0$.
Finally, $f(x) \to x$ as $x \to -\infty$ and $f(x) \to x$ as $x \to \infty$. $y = x$ is an asymptote.

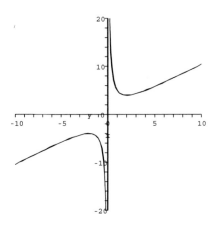

7. $f(x) = \sqrt{x^2 + 1}$
The function is defined for all real numbers. The y-intercept is $(0, 1)$. There are no x-intercepts.

$$f'(x) = \dfrac{1}{2}(x^2 + 1)^{-1/2} 2x = \dfrac{x}{\sqrt{x^2 + 1}}$$

The only critical number is $x = 0$. $f'(x) < 0$ when $x < 0$ and $f'(x) > 0$ when $x > 0$ so $f(x)$ is increasing on $(0, \infty)$ and decreasing on $(-\infty, 0)$. Thus $f(x)$ has a local minimum at $x = 0$.

$$f''(x) = \dfrac{\sqrt{x^2 + 1} - x\left[\frac{1}{2}(x^2+1)^{-1/2} 2x\right]}{x^2 + 1}$$

$$= \dfrac{1}{(x^2 + 1)^{3/2}}$$

Since $f''(x) > 0$ for all x, we see that $f(x)$ is concave up for all x.
$f(x) \to \infty$ as $x \to \pm\infty$.

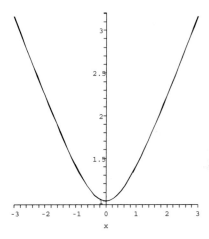

9. $f(x) = \dfrac{4x}{x^2 - x + 1}$
The denominator is never zero, so the domain is all real numbers. $(0, 0)$ is the only x-intercept.

The function has horizontal asymptote at $y = 0$.

$$f'(x) = \dfrac{4(1 - x^2)}{(x^2 - x + 1)^2}$$

There are critical numbers at $x = \pm 1$.

$$f''(x) = \frac{8(x^3 - 3x + 1)}{(x^2 - x + 1)^3}$$

with zeros at approximately $x = -1.8793$, 0.3473, and 1.5321. $f''(x)$ changes sign at these values, so these are inflection points. The Second Derivative Test shows that $x = -1$ is a minimum, and $x = 1$ is a maximum.

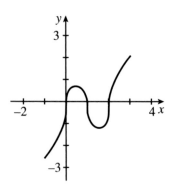

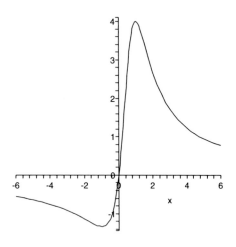

11. $f(x) = (x^3 - 3x^2 + 2x)^{1/3}$
The function is defined for all real numbers. There are x-intercepts at $x = 0$, 1, and 2.

$$f'(x) = \frac{3x^2 - 6x + 2}{3(x^3 - 3x^2 + 2x)^{2/3}}$$

There are critical numbers at $x = \dfrac{3 \pm \sqrt{3}}{3}$, 0, 1, and 2.

$$f''(x) = \frac{-6x^2 + 12x - 8}{9(x^3 - 3x^2 + 2x)^{5/3}}$$

with critical numbers $x = 0$, 1, and 2. $f''(x)$ changes sign at these values, so these are inflection points. The Second Derivative Test shows that $x = \dfrac{3 + \sqrt{3}}{3}$ is a minimum, and $x = \dfrac{3 - \sqrt{3}}{3}$ is a maximum.
$f(x) \to x - 1$ as $x \to -\infty$ and $f(x) \to x - 1$ as $x \to \infty$ so $y = x - 1$ is an asymptote.

13. $f(x) = x^5 - 5x = x(x^4 - 5)$
x-intercepts are $x = 0$ and $x = \pm\sqrt[4]{5}$.
The y-intercept is $(0, 0)$.
$f'(x) = 5x^4 - 5 = 5(x^4 - 1)$
The critical numbers are $x = \pm 1$.
$f''(x) = 20x^3$ so $x = -1$ is a local maximum and $x = 1$ is a local minimum. $f(x)$ is increasing on $(-\infty, -1)$ and $(1, \infty)$ and decreasing on $(-1, 1)$. It is concave up on $(0, \infty)$ and concave down on $(-\infty, 0)$, with an inflection point at $x = 0$.
$f(x) \to -\infty$ as $x \to -\infty$ and $f(x) \to \infty$ as $x \to \infty$.

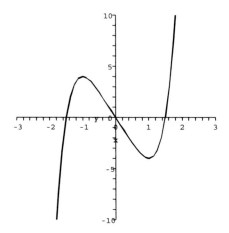

15. $f(x) = \dfrac{2}{x} + \sqrt{\dfrac{1}{x^2} + 9}$

$$f'(x) = -\frac{2}{x^2} - \frac{1}{x^3\sqrt{\frac{1}{x^2} + 9}}$$

$$f''(x) = \frac{4}{x^3} - \frac{1}{x^6\left(\sqrt{\frac{1}{x^2} + 9}\right)^3} + \frac{3}{x^4\sqrt{\frac{1}{x^2} + 9}}$$

$f'(x) < 0$ on $(-\infty, 0) \cup (0, \infty)$
$f''(x) < 0$ on $(-\infty, 0)$
$f''(x) > 0$ on $(0, \infty)$
f is decreasing on $(-\infty, 0)$ and on $(0, \infty)$, concave down on $(-\infty, 0)$, and concave up on $(0, \infty)$. f is undefined at $x = 0$.

$$\lim_{x \to 0^+} f(x) = \infty$$

$$\lim_{x \to 0^-} f(x) = -\infty$$

$$\lim_{x \to \infty} f(x) = \lim_{x \to -\infty} f(x) = 3$$

So f has a vertical asymptote at $x = 0$ and a horizontal asymptote at $y = 3$.

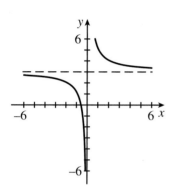

17. $f(x) = (x^3 - 3x^2 + 2x)^{2/3}$
The function is defined for all real numbers. There are x-intercepts at $x = 0$, 1, and 2.
$$f'(x) = \frac{2(3x^2 - 6x + 2)}{3(x^3 - 3x^2 + 2x)^{1/3}}$$
There are critical numbers at $x = \dfrac{3 \pm \sqrt{3}}{3}$, 0, 1, and 2.
$$f''(x) = \frac{18x^4 - 72x^3 + 84x^2 - 24x - 8}{9(x^3 - 3x^2 + 2x)^{4/3}}$$
with critical numbers $x = 0$, 1, and 2 and $x \approx -0.1883$ and 2.1883. $f''(x)$ changes sign at these last two values, so these are inflection points. The Second Derivative Test shows that $x = \dfrac{3 \pm \sqrt{3}}{3}$ are both maxima. Local minima occur at $x = 0$, 1, and 2.
$f(x) \to \infty$ as $x \to \pm\infty$.

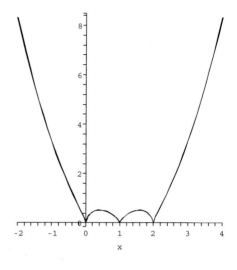

19. $f(x) = \dfrac{x^2 + 1}{3x^2 - 1}$
Note that $x = \pm\sqrt{1/3}$ are not in the domain of the function, but yield vertical asymptotes. There are no x-intercepts.

$$f'(x) = \frac{2x(3x^2 - 1) - (x^2 + 1)(6x)}{(3x^2 - 1)^2}$$
$$= \frac{(6x^3 - 2x) - (6x^3 + 6x)}{(3x^2 - 1)^2}$$
$$= \frac{-8x}{(3x^2 - 1)^2}$$

So the only critical point is $x = 0$.
$f'(x) > 0$ for $x < 0$
$f'(x) < 0$ for $x > 0$
so f is increasing on $(-\infty, -\sqrt{1/3})$ and on $(-\sqrt{1/3}, 0)$; decreasing on $(0, \sqrt{1/3})$ and on $(\sqrt{1/3}, \infty)$. Thus there is a local max at $x = 0$.
$$f''(x) = 8 \cdot \frac{9x^2 + 1}{(3x^2 - 1)^3}$$
$f''(x) > 0$ on $(-\infty, -\sqrt{1/3}) \cup (\sqrt{1/3}, \infty)$
$f''(x) < 0$ on $(-\sqrt{1/3}, \sqrt{1/3})$
Hence f is concave up on $(-\infty, -\sqrt{1/3})$ and on $(\sqrt{1/3}, \infty)$; concave down on $(-\sqrt{1/3}, \sqrt{1/3})$.

Finally, when $|x|$ is large, the function approached $1/3$, so $y = 1/3$ is a horizontal asymptote.

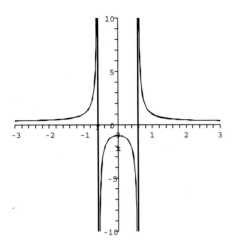

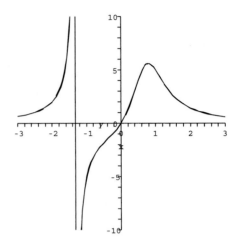

21. $f(x) = \dfrac{5x}{x^3 - x + 1}$

Looking at the graph of $y = x^3 - x + 1$, we see that there is one real root, at approximately -1.325; so the domain of the function is all x except for this one point, and $x \approx -1.325$ will be a vertical asymptote. There is a horizontal asymptote of $y = 0$.

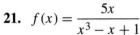

$$f'(x) = 5\frac{1 - 2x^3}{(x^3 - x - 1)^2}$$

The only critical point is $x = \sqrt[3]{1/2}$. By the first derivative test, this is a local max.

$$f''(x) = 10\frac{3x^5 + x^3 - 6x^2 + 1}{(x^3 - x + 1)^3}$$

The numerator of f'' has three real roots, which are approximately $x = -0.39018$, $x = 0.43347$, and $x = 1.1077$. $f''(x) > 0$ on $(-\infty, -1.325) \cup (-0.390, 0.433) \cup (1.108, \infty)$. $f''(x) < 0$ on $(-1.325, -0.390) \cup (0.433, 1.108)$. So f is concave up on $(-\infty, -1.325) \cup (-0.390, 0.433) \cup (1.108, \infty)$ and concave down on $(-1.325, -0.390) \cup (0.433, 1.108)$. Hence $x = -0.39018$, $x = 0.43347$, and $x = 1.1077$ are inflection points.

23. $f(x) = x^2\sqrt{x^2 - 9}$

f is undefined on $(-3, 3)$. There are x-intercepts at $x = -3, 3$.

$$f'(x) =$$

$$2x\sqrt{x^2 - 9} + x^2\left(\frac{1}{2}(x^2 - 9)^{-1/2} \cdot 2x\right)$$

$$= 2x\sqrt{x^2 - 9} + \frac{x^3}{\sqrt{x^2 - 9}}$$

$$= \frac{2x(x^2 - 9) + x^3}{\sqrt{x^2 - 9}}$$

$$= \frac{3x^3 - 18x}{\sqrt{x^2 - 9}} = \frac{3x(x^2 - 6)}{\sqrt{x^2 - 9}}$$

$$= \frac{3x(x + \sqrt{6})(x - \sqrt{6})}{\sqrt{x^2 - 9}}$$

Critical points ± 3. (Note that f is undefined at $x = 0, \pm\sqrt{6}$.)

$$f''(x) = \frac{(9x^2 - 18)\sqrt{x^2 - 9}}{x^2 - 9}$$

$$-\frac{(3x^3 - 18x) \cdot \frac{1}{2}(x^2 - 9)^{-1/2} \cdot 2x}{x^2 - 9}$$

$$= \frac{(9x^2 - 18)(x^2 - 9) - x(3x^3 - 18x)}{(x^2 - 9)^{3/2}}$$

$$= \frac{6x^4 - 81x^2 + 162}{(x^2 - 9)^{3/2}}$$

$f''(x) = 0$ when

$$x^2 = \frac{81 \pm \sqrt{81^2 - 4(6)(162)}}{2(6)}$$

$$= \frac{81 \pm \sqrt{2673}}{12} = \frac{1}{4}(27 \pm \sqrt{297}).$$

So $x \approx \pm 3.325$ or $x \approx \pm 1.562$, but these latter values are not in the domain. So only ± 3.325 are potential inflection points.

$f'(x) > 0$ on $(3, \infty)$

$f'(x) < 0$ on $(-\infty, -3)$

$f''(x) > 0$ on $(-\infty, -3.3) \cup (3.3, \infty)$

$f''(x) < 0$ on $(-3.3, 3) \cup (3, 3.3)$

f is increasing on $(3, \infty)$, decreasing on $(-\infty, -3)$, concave up on $(-\infty, -3.3) \cup (3.3, \infty)$, concave down on $(-3.3, -3) \cup (3, 3.3)$. $x = \pm 3.3$ are inflection points.

Global graph of $f(x)$:

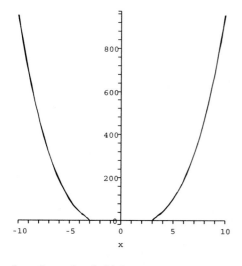

Local graph of $f(x)$:

25.
$$f(x) = \frac{25 - 50\sqrt{x^2 + 0.25}}{x}$$

$$= 25\left(\frac{1 - 2\sqrt{x^2 + 0.25}}{x}\right)$$

$$= 25\left(\frac{1 - \sqrt{4x^2 + 1}}{x}\right)$$

Note that $x = 0$ is not in the domain of the function.

$$f'(x) = 25\left(\frac{1 - \sqrt{4x^2 + 1}}{x^2\sqrt{4x^2 + 1}}\right)$$

We see that there are no critical points. Indeed, $f' < 0$ wherever f is defined. One can verify that

$f''(x) > 0$ on $(0, \infty)$

$f''(x) < 0$ on $(-\infty, 0)$

Hence the function is concave up on $(0, \infty)$ and concave down on $(-\infty, 0)$.

$$\lim_{x \to \infty} \frac{25 - 50\sqrt{x^2 + 0.25}}{x}$$

$$= \lim_{x \to \infty}\left[\frac{25}{x} - \frac{50\sqrt{x^2 + 0.25}}{x}\right]$$

$$= \lim_{x \to \infty}\left[0 - 50\frac{x\sqrt{1 + \frac{0.25}{x^2}}}{x}\right]$$

$$= \lim_{x \to \infty}\left[-50\sqrt{1 + \frac{0.25}{x^2}}\right] = -50$$

$$\lim_{x \to -\infty} \frac{25 - 50\sqrt{x^2 + 0.25}}{x}$$

$$= \lim_{x \to \infty}\left[\frac{25}{x} - \frac{50\sqrt{x^2 + 0.25}}{x}\right]$$

$$= \lim_{x \to -\infty}\left[0 - 50\frac{(-x)\sqrt{1 + \frac{0.25}{x^2}}}{x}\right]$$

$$= \lim_{x \to \infty} 50\sqrt{1 + \frac{0.25}{x^2}} = 50$$

So f has horizontal asymptotes at $y = 50$ and $y = -50$.

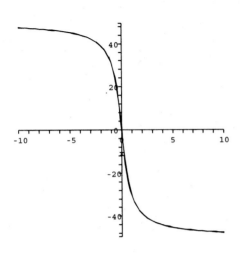

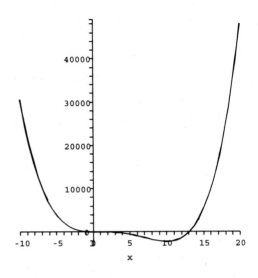

Local graph of $f(x)$:

27. $f(x) = x^4 - 16x^3 + 42x^2 - 39.6x + 14$

$f'(x) = 4x^3 - 48x^2 + 84x - 39.6$

$f''(x) = 12x^2 - 96x + 84$

$\quad = 12(x^2 - 8x + 7)$

$\quad = 12(x - 7)(x - 1)$

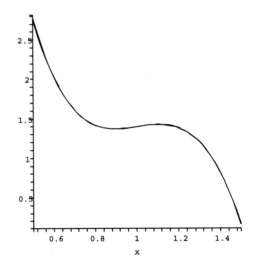

$f'(x) > 0$ on $(0.8952, 1.106) \cup (9.9987, \infty)$

$f'(x) < 0$ on $(-\infty, 0.8952) \cup (1.106, 9.9987)$

$f''(x) > 0$ on $(-\infty, 1) \cup (7, \infty)$

$f''(x) < 0$ on $(1, 7)$

f is increasing on $(0.8952, 1.106)$ and on $(9.9987, \infty)$, decreasing on $(-\infty, 0.8952)$ and on $(1.106, 9.9987)$, concave up on

$(-\infty, 1) \cup (7, \infty)$, concave down on $(1, 7)$, $x = 0.8952, 9.9987$ are local min, $x = 1.106$ is local max, $x = 1, 7$ are inflection points. $f(x) \to \infty$ as $x \to \pm\infty$.

Global graph of $f(x)$:

29. $f(x) = x^4 + cx^2$

$f'(x) = 4x^3 + 2cx$

$f''(x) = 12x^2 + 2c$

$c = 0$: 1 extremum, 0 inflection points

$c < 0$: 3 extrema, 2 inflection points

$c > 0$: 1 extremum, 0 inflection points

$c \to -\infty$: the graph widens and lowers

$c \to +\infty$: the graph narrows

31.
$$f(x) = \frac{x^2}{x^2 + c^2}$$

$$f'(x) = \frac{2c^2 x}{(x^2 + c^2)^2}$$

$$f''(x) = \frac{2c^4 - 6c^2 x^2}{(x^2 + c^2)^3}$$

If $c = 0$: $f(x) = 1$, except that f is undefined at $x = 0$.

$c < 0$, $c > 0$: horizontal asymptote at $y = 1$, local min at $x = 0$, since the derivative changes sign from negative to positive at $x = 0$; also there are inflection points at $x = \pm c/\sqrt{3}$.

As $c \to -\infty$, $c \to +\infty$: the graph widens.

33. When $c = 0$, $f(x) = \sin(0) = 0$.

Since $\sin x$ is an odd function, $\sin(-cx) = -\sin(cx)$. Thus negative values of c give the reflection through the x-axis of their positive counterparts. For large values of c, the graph looks just like $\sin x$, but with a very small period.

35. $f(x) = \frac{x}{x^2 + b}$
$f(0) = 0$
$f(x) > 0$ for $x > 0$

$$\lim_{x \to \infty} \frac{x}{x^2 + b} = \lim_{x \to \infty} \frac{1/x}{1 + b/x^2} = 0$$

$f'(x) = \frac{b - x^2}{(x^2 + b)^2}$, so there is a unique critical point at $x = \sqrt{b}$, which must be the maximum. The bigger b is, the further the max is from the origin. For time since conception, $\sqrt{b}$ represents the most common gestation time. For survival time, $\sqrt{b}$ represents the most common life span.

37. No: Let $f(x) = \frac{x + 1}{x^2 + 1}$. The roots of the denominator are complex, so there are no vertical asymptotes.

No: Let $f(x) = \frac{x^4 - 2x + 3}{x^2 + 1}$. This function goes to ∞ as $x \to \pm\infty$.

39. $f(x) = \frac{3x^2 - 1}{x} = 3x - \frac{1}{x}$

$y = 3x$ is a slant asymptote.

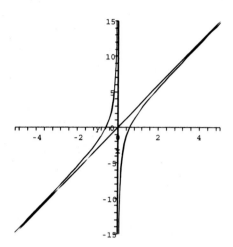

41. $f(x) = \frac{x^3 - 2x^2 + 1}{x^2} = x - 2 + \frac{1}{x^2}$

$y = x - 2$ is a slant asymptote.

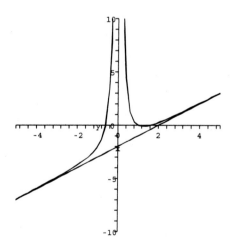

43. $f(x) = \frac{x^4}{x^3 + 1} = x - \frac{x}{x^3 + 1}$

$y = x$ is a slant asymptote.

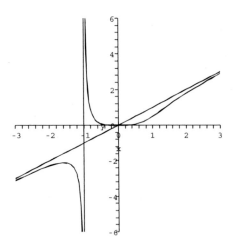

45. One possibility:

$$f(x) = \frac{3x^2}{(x-1)(x-2)}$$

47. One possibility:

$$f(x) = \frac{2x}{\sqrt{(x-1)(x+1)}}$$

3.6 OPTIMIZATION

1. $f(x) = x^2 + 1$ has a minimum at $x = 0$, while $\sin(x^2 + 1)$ has minima where $x^2 + 1 = 3\pi/2 + 2n\pi$.

3.
$$A = xy = 1800$$
$$y = \frac{1800}{x}$$
$$P = 2x + y = 2x + \frac{1800}{x}$$
$$P' = 2 - \frac{1800}{x^2} = 0$$
$$2x^2 = 1800$$
$$x = 30$$
$P'(x) > 0$ for $x > 30$
$P'(x) < 0$ for $0 < x < 30$
So $x = 30$ is min.
$$y = \frac{1800}{x} = \frac{1800}{30} = 60$$

So the dimensions are $30' \times 60'$ and the minimum perimeter is 120 ft.

5.
$$P = 2x + 3y = 120$$
$$3y = 120 - 2x$$
$$y = 40 - \frac{2}{3}x$$
$$A = xy$$
$$A(x) = x\left(40 - \frac{2}{3}x\right)$$
$$A'(x) = 1\left(40 - \frac{2}{3}x\right) + x\left(-\frac{2}{3}\right)$$
$$= 40 - \frac{4}{3}x = 0$$
$$40 = \frac{4}{3}x$$
$$x = 30$$
$A'(x) > 0$ for $0 < x < 30$
$A'(x) < 0$ for $x > 30$
So $x = 30$ is max, $y = 40 - \frac{2}{3} \cdot 30 = 20$.

So the dimensions are $20' \times 30'$.

7.
$$A = xy$$
$$P = 2x + 2y$$
$$2y = P - 2x$$
$$y = \frac{P}{2} - x$$
$$A(x) = x\left(\frac{P}{2} - x\right)$$
$$A'(x) = 1 \cdot \left(\frac{P}{2} - x\right) + x(-1)$$
$$= \frac{P}{2} - 2x = 0$$
$$P = 4x$$
$$x = \frac{P}{4}$$
$A'(x) > 0$ for $0 < x < P/4$
$A'(x) < 0$ for $x > P/4$
So $x = P/4$ is max,
$$y = \frac{P}{2} - x = \frac{P}{2} - \frac{P}{4} = \frac{P}{4}$$

So the dimensions are $\frac{P}{4} \times \frac{P}{4}$. Thus we have a square.

9.
$$d = \sqrt{(x-0)^2 + (y-1)^2}$$
$$y = x^2$$
$$d = \sqrt{x^2 + (x^2-1)^2}$$
$$= (x^4 - x^2 + 1)^{1/2}$$
$$d'(x) = \frac{1}{2}(x^4 - x^2 + 1)^{-1/2}(4x^3 - 2x)$$
$$= \frac{2x(2x^2-1)}{2\sqrt{x^4 - x^2 + 1}} = 0$$

$x = 0, \pm\sqrt{1/2}$;
$f(0) = 1$, $f(\sqrt{1/2}) = 3/4$, $f(-\sqrt{1/2}) = \frac{3}{4}$;
Thus $x = \pm\sqrt{1/2}$ are min, and the points on $y = x^2$ closest to $(0, 1)$ are $(\sqrt{1/2}, 1/2)$ and $(-\sqrt{1/2}, 1/2)$.

11.
$$d = \sqrt{(x-0)^2 + (y-0)^2}$$
$$y = \cos x$$
$$d = \sqrt{x^2 + \cos^2 x}$$
$$d'(x) = \frac{2x - 2\cos x \sin x}{2\sqrt{x^2 + \cos^2 x}} = 0$$
$$x = \cos x \sin x$$
$$x = 0$$

So $x = 0$ is min and the point on $y = \cos x$ closest to $(0, 0)$ is $(0, 1)$.

13. For $(0, 1)$, $(\sqrt{1/2}, 1/2)$ on $y = x^2$, we have $y' = 2x$, $y'(\sqrt{1/2}) = 2 \cdot \sqrt{1/2} = \sqrt{2}$ and
$$m = \frac{\frac{1}{2} - 1}{\sqrt{\frac{1}{2}} - 0} = -\frac{1}{\sqrt{2}}.$$

For $(0, 1)$, $(-\sqrt{1/2}, 1/2)$ on $y = x^2$, we have $y'(-\sqrt{1/2}) = 2(-\sqrt{1/2}) = -\sqrt{2}$ and
$$m = \frac{\frac{1}{2} - 1}{-\sqrt{\frac{1}{2}} - 0} = \frac{1}{\sqrt{2}}.$$

For $(3, 4)$, $(2.06, 4.2436)$ on $y = x^2$, we have $y'(2.06) = 2(2.06) = 4.12$ and
$$m = \frac{4.2436 - 4}{2.06 - 3} = -0.2591 \approx -\frac{1}{4.12}.$$

15.
$$V = l \cdot w \cdot h$$
$$V(x) = (10 - 2x)(6 - 2x) \cdot x, \quad 0 \le x \le 3$$
$$V'(x) = -2(6 - 2x) \cdot x + (10 - 2x)(-2) \cdot x$$
$$+ (10 - 2x)(6 - 2x)$$
$$= 60 - 64x + 12x^2$$
$$= 4(3x^2 - 16x + 15)$$
$$= 0$$
$$x = \frac{16 \pm \sqrt{(-16)^2 - 4 \cdot 3 \cdot 15}}{6}$$
$$= \frac{8}{3} \pm \frac{\sqrt{19}}{3}$$
$$x = \frac{8}{3} + \frac{\sqrt{19}}{3} > 3$$

$V'(x) > 0$ for $x < 8/3 - \sqrt{19}/3$
$V'(x) < 0$ for $x > 8/3 - \sqrt{19}/3$

So $x = \dfrac{8}{3} - \dfrac{\sqrt{19}}{3}$ is a max.

17. Let x be the distance from the connection point to the easternmost development. Then $0 \le x \le 5$.
$$f(x) = \sqrt{3^2 + (5 - x)^2} + \sqrt{4^2 + x^2},$$
$$0 \le x \le 5$$
$$f'(x) = -(9 + (5 - x)^2)^{-1/2}(5 - x)$$
$$+ \frac{1}{2}(16 + x^2)^{-1/2}(2x)$$
$$= \frac{x - 5}{\sqrt{9 + (5 - x)^2}} + \frac{x}{\sqrt{16 + x^2}}$$
$$= 0$$
$$x = \frac{20}{7} \approx 2.857$$
$$f(0) = 4 + \sqrt{34} \approx 9.831$$
$$f\left(\frac{20}{7}\right) = \sqrt{74} \approx 8.602$$
$$f(5) = 3 + \sqrt{41} \approx 9.403$$

So $x = 20/7$ is minimum. The length of new line at this point is approximately 8.6 miles. Since $f(0) \approx 9.8$ and $f(5) \approx 9.4$, the water line should be 20/7 miles west of the second development or equivalently, 15/7 miles east of the first.

19. $C(x) = 5\sqrt{16 + x^2} + 2\sqrt{36 + (8 - x)^2}$

$\quad 0 \le x \le 8$

$C(x) = 5\sqrt{16 + x^2} + 2\sqrt{100 - 16x + x^2}$

$C'(x) = 5\left(\dfrac{1}{2}\right)(16 + x^2)^{-1/2} \cdot 2x$

$\quad + 2\left(\dfrac{1}{2}\right)(100 - 16x + x^2)^{-1/2}(2x - 16)$

$\quad = \dfrac{5x}{\sqrt{16 + x^2}} + \dfrac{2x - 16}{\sqrt{100 - 16x + x^2}}$

$\quad = 0$

$x \approx 1.2529$

$C(0) = 40$

$C(1.2529) \approx 39.0162$

$C(8) \approx 56.7214$

The highway should emerge from the marsh 1.2529 miles east of the bridge. If we build a straight line to the interchange, we have $x = 3.2$.

Since $C(3.2) - C(1.2529) \approx 1.964$, we save $1.964 million.

21. $C(x) = 5\sqrt{16 + x^2} + 3\sqrt{36 + (8 - x)^2}$

$\quad 0 \le x \le 8$

$C'(x) = \dfrac{5x}{\sqrt{16 + x^2}} + \dfrac{3x - 24}{\sqrt{100 - 16x + x^2}}$

Setting $C'(x) = 0$ yields

$x \approx 1.8941$.

$C(0) = 50$

$C(1.8941) \approx 47.8104$

$C(8) \approx 62.7214$

The highway should emerge from the marsh 1.8941 miles east of the bridge. So if we must use the path from exercise 19, the extra cost is $C(1.2529) - C(1.8941)$

$= 48.0452 - 47.8104 = 0.2348$

or about $234.8 thousand.

23. $T(x) = \dfrac{\sqrt{1 + x^2}}{v_1} + \dfrac{\sqrt{1 + (2 - x)^2}}{v_2}$

$T'(x) = \dfrac{1}{v_1} \cdot \dfrac{1}{2}(1 + x^2)^{-1/2} \cdot 2x$

$\quad + \dfrac{1}{v_2}(1 + (2 - x)^2)^{-1/2} \cdot (2 - x)(-1)$

$\quad = \dfrac{x}{v_1\sqrt{1 + x^2}} + \dfrac{x - 2}{v_2\sqrt{1 + (2 - x)^2}}$

Note that

$T'(x) = \dfrac{1}{v_1} \cdot \dfrac{x}{\sqrt{1 + x^2}}$

$\quad - \dfrac{1}{v_2} \cdot \dfrac{(2 - x)}{\sqrt{1 + (2 - x)^2}}$

$\quad = \dfrac{1}{v_1} \sin \theta_1 - \dfrac{1}{v_2} \sin \theta_2$

When $T'(x) = 0$, we have

$\dfrac{1}{v_1} \sin \theta_1 = \dfrac{1}{v_2} \sin \theta_2$

$\dfrac{\sin \theta_1}{\sin \theta_2} = \dfrac{v_1}{v_2}$

25. Cost: $C = 2(2\pi r^2) + 2\pi rh$

Convert from fluid ounces to cubic inches:

$12 \text{ fl oz} = 12 \text{ fl oz} \cdot 1.80469 \text{ in}^3/\text{fl oz}$

$\quad = 21.65628 \text{ in}^3$

Volume: $V = \pi r^2 h$ so

$h = \dfrac{V}{\pi r^2} = \dfrac{21.65628}{\pi r^2}$

$C = 4\pi r^2 + 2\pi r\left(\dfrac{21.65628}{\pi r^2}\right)$

$C(r) = 4\pi r^2 + 43.31256r^{-1}$

$C'(r) = 8\pi r - 43.31256r^{-2}$

$\quad = \dfrac{8\pi r^3 - 43.31256}{r^2}$

$r = \sqrt[3]{\dfrac{43.31256}{8\pi}} = 1.1989''$

when $C'(r) = 0$.

$C'(r) < 0$ on $(0, 1.1989)$

$C'(r) > 0$ on $(1.1989, \infty)$
Thus $r = 1.1989$ minimizes the cost.

$$h = \frac{21.65628}{\pi(1.1989)^2} = 4.7957''$$

27. $V(r) = cr^2(r_0 - r)$

$V'(r) = 2cr(r_0 - r) + cr^2(-1)$

$\quad\quad = 2crr_0 - 3cr^2$

$\quad\quad = cr(2r_0 - 3r)$

$V'(r) = 0$ when $r = 2r_0/3$
$V'(r) > 0$ on $\left(0, 2r_0/3\right)$
$V'(r) < 0$ on $\left(2r_0/3, \infty\right)$
Thus $r = 2r_0/3$ maximizes the velocity.
$r = 2r_0/3 < r_0$, so the windpipe contracts.

29.
$$p(x) = \frac{V^2 x}{(R + x)^2}$$

$$p'(x) = \frac{V^2(R + x)^2 - V^2 x \cdot 2(R + x)}{(R + x)^4}$$

$$\quad\quad = \frac{V^2 R^2 - V^2 x^2}{(R + x)^4}$$

$p'(x) = 0$ when $x = R$
$p'(x) > 0$ on $(0, R)$
$p'(x) < 0$ on (R, ∞)

Thus $x = R$ maximizes the power absorbed.

31. Let r be the radius of the semicircle and $2r$
and w the dimensions of the rectangle. Then
the perimeter is $\pi r + 4r + 2w = 8 + \pi$.

$$w = \frac{8 + \pi - r(\pi + 4)}{2}$$

$$A(r) = \frac{\pi r^2}{2} + 2rw$$

$$\quad\quad = \frac{\pi r^2}{2} + r(8 + \pi - r(\pi + 4))$$

$$\quad\quad = r^2\left(-4 - \frac{\pi}{2}\right) + r(8 + \pi)$$

$A'(r) = -2r\left(4 + \frac{\pi}{2}\right) + (8 + \pi) = 0$
$A'(r) = 0$ when $r = 1$
$A'(r) > 0$ on $(0, 1)$
$A'(r) < 0$ on $(1, \infty)$

Thus $r = 1$ maximizes the area so
$w = \dfrac{8 + \pi - (\pi + 4)}{2} = 2$. The dimensions
of the rectangle are 2×2.

33. $l \times w = 92$, $w = 92/l$

$A(l) = (l + 4)(w + 2)$

$\quad\quad = (l + 4)(92/l + 2)$

$\quad\quad = 92 + 368/l + 2l + 8$

$\quad\quad = 100 + 368l^{-1} + 2l$

$A'(l) = -368l^{-2} + 2$

$\quad\quad = \dfrac{2l^2 - 368}{l^2}$

$A'(l) = 0$ when $l = \sqrt{184} = 2\sqrt{46}$
$A'(l) < 0$ on $(0, 2\sqrt{46})$
$A'(l) > 0$ on $(2\sqrt{46}, \infty)$

So $l = 2\sqrt{46}$ minimizes the total area.
When $l = 2\sqrt{46}$, $w = \dfrac{92}{2\sqrt{46}} = \sqrt{46}$.

For the minimum total area, the printed area
has width $\sqrt{46}$ in. and length $2\sqrt{46}$ in., and the
advertisement has overall width $\sqrt{46} + 2$ in.
and overall length $2\sqrt{46} + 4$ in.

35. Let L represent the length of the ladder. Then
from the diagram, it follows that
$L = a \sec \theta + b \csc \theta$. Therefore,

$$\frac{dL}{d\theta} = a \sec \theta \tan \theta - b \csc \theta \cot \theta$$

$$0 = a \sec \theta \tan \theta - b \csc \theta \cot \theta$$

$$a \sec \theta \tan \theta = b \csc \theta \cot \theta$$

$$\frac{b}{a} = \frac{\sec \theta \tan \theta}{\csc \theta \cot \theta}$$

$$\quad = \frac{1}{\cos \theta} \frac{\sin \theta}{\cos \theta} \frac{\sin \theta}{1} \frac{\sin \theta}{\cos \theta}$$

$$\quad = \tan^3 \theta$$

Thus,
$\tan \theta = \sqrt[3]{b/a} = \sqrt[3]{4/5}$.
Solving $\tan \theta - \sqrt[3]{4/5} = 0$ numerically on the
interval $\left(0, \frac{\pi}{2}\right)$, we have
$\theta \approx 0.748$ rad or 42.87 degrees.

Thus, the length of the longest ladder that can
fit around the corner is approximately

$L = a \sec\theta + b \csc\theta$

$\quad = 5\sec(0.748) + 4\csc(0.748)$

$\quad \approx 12.7\,\text{ft}$

37. Using the result of exercise 36 and solving for b:

$L = (a^{2/3} + b^{2/3})^{3/2}$

$L^{2/3} = a^{2/3} + b^{2/3}$

$b^{2/3} = L^{2/3} - a^{2/3}$

$\quad b = (L^{2/3} - a^{2/3})^{3/2}$

$\quad\quad = (8^{2/3} - 5^{2/3})^{3/2}$

$\quad\quad \approx 1.16\,\text{ft}$

39.

$R(x) = \dfrac{35x - x^2}{x^2 + 35}$

$R'(x) = -35\dfrac{x^2 + 2x - 35}{(x^2 + 35)^2}$

$\quad\quad = -35\dfrac{(x-5)(x+7)}{(x^2+35)^2}$

Hence the only critical number for $x \geq 0$ is $x = 5$ (that is, 5000 items). This must correspond to the absolute maximum, since $R(0) = 0$ and $R(x)$ is negative for $x > 35$. So maximum revenue is $R(5) = 2.5$ (that is, \$2500).

41. $Q'(t)$ is efficiency because it represents the number of additional items produced per unit time.

$Q(t) = -t^3 + 12t^2 + 60t$

$Q'(t) = -3t^2 + 24t + 60$

$\quad\quad = 3(-t^2 + 8t + 20)$

This is the quantity we want to maximize.

$Q''(t) = 3(-2t + 8)$ so the only critical number is $t = 4$ hours. This must be the maximum since the function $Q'(t)$ is a parabola opening down.

43. Let $C(t)$ be the total cost of the tickets. Then

$C(t) = $ (price per ticket)(# of tickets)

$C(t) = (40 - (t - 20))(t)$

$\quad\quad = (60 - t)(t) = 60t - t^2$

for $20 \leq t \leq 50$. Then $C'(t) = 60 - 2t$, so $t = 30$ is the only critical number. This must correspond to the maximum since $C(t)$ is a parabola opening down.

45.

$R = \dfrac{2v^2 \cos^2\theta}{g}(\tan\theta - \tan\beta)$

$R'(\theta) = \dfrac{2v^2}{g}\Big[2\cos\theta(-\sin\theta)(\tan\theta - \tan\beta)$

$\quad\quad\quad + \cos^2\theta \cdot \sec^2\theta\Big]$

$\quad\quad = \dfrac{2v^2}{g}\Big[-2\cos\theta\sin\theta \cdot \dfrac{\sin\theta}{\cos\theta}$

$\quad\quad\quad +2\cos\theta\sin\theta\tan\beta$

$\quad\quad\quad + \cos^2\theta \cdot \dfrac{1}{\cos^2\theta}\Big]$

$\quad\quad = \dfrac{2v^2}{g}\Big[-2\sin^2\theta + \sin(2\theta)\tan\beta + 1\Big]$

$\quad\quad = \dfrac{2v^2}{g}\Big[-2\sin^2\theta + \sin(2\theta)\tan\beta$

$\quad\quad = \quad + (\sin^2\theta + \cos^2\theta)\Big]$

$\quad\quad = \dfrac{2v^2}{g}\Big[\sin(2\theta)\tan\beta$

$\quad\quad\quad + (\cos^2\theta - \sin^2\theta)\Big]$

$\quad\quad = \dfrac{2v^2}{g}\Big[\sin(2\theta)\tan\beta + \cos(2\theta)\Big]$

$R'(\theta) = 0$ when

$\tan\beta = \dfrac{-\cos(2\theta)}{\sin(2\theta)} = -\cot(2\theta)$

$\quad\quad = -\tan\left(\dfrac{\pi}{2} - 2\theta\right)$

$\quad\quad = \tan\left(2\theta - \dfrac{\pi}{2}\right)$

Hence $\beta = 2\theta - \pi/2$, so

$\theta = \dfrac{1}{2}\left(\beta + \dfrac{\pi}{2}\right)$

$\quad = \dfrac{\beta}{2} + \dfrac{\pi}{4} = \dfrac{\beta°}{2} + 45°$

(a) $\beta = 10°, \theta = 50°$

(b) $\beta = 0°, \theta = 45°$

(c) $\beta = -10°, \theta = 40°$

47. The perimeter is $2\pi r + 2l$ where r is the radius of the semicircle and l is the length of the rectangle.

$$2\pi r + 2l = 400$$
$$l = 200 - \pi r$$

The area of the rectangle is $A(r) = 2r(200 - \pi r) = 400r - 2\pi r^2$.

$$A'(r) = 400 - 4\pi r$$

$$A'(r) = 0 \text{ when } r = \frac{400}{4\pi} = \frac{100}{\pi}.$$

This is a maximum since $A''(r) = -4\pi < 0$.

The width of the rectangle is $2r = \dfrac{200}{\pi}$ meters and the length is $200 - \pi r = 100$ meters.

200 meters are used on the straightaways, so 200 meters must also be used on the curves, so that the track is 400 meters.

49. $A = 4xy$

$$\frac{dA}{dx} = 4\left(xy' + y\right)$$

To determine $y' = \dfrac{dy}{dx}$, use the equation for the ellipse:

$$1 = \frac{x^2}{a^2} + \frac{y^2}{b^2}$$

$$0 = \frac{2x}{a^2} + \frac{2yy'}{b^2}$$

$$\frac{2yy'}{b^2} = -\frac{2x}{a^2}$$

$$y' = -\frac{b^2}{a^2}\frac{x}{y}$$

Substituting this expression for y' into the expression for $\dfrac{dA}{dx}$, we get

$$\frac{dA}{dx} = 4(xy' + y)$$

$$= 4\left[x\left(-\frac{b^2}{a^2}\frac{x}{y}\right) + y\right]$$

$$= 4\left(-\frac{b^2}{a^2}\frac{x^2}{y} + y\right)$$

The area is maximized when its derivative is zero:

$$0 = 4\left(-\frac{b^2}{a^2}\frac{x^2}{y} + y\right)$$

$$\frac{b^2}{a^2}\frac{x^2}{y} = y$$

$$\frac{x^2}{a^2} = \frac{y^2}{b^2}$$

Substituting the previous relationship into the equation for the ellipse, we get

$$\frac{x^2}{a^2} = \frac{y^2}{b^2} = \frac{1}{2}$$

and therefore,

$$x = \frac{a}{\sqrt{2}} \quad \text{and} \quad y = \frac{b}{\sqrt{2}}$$

Thus, the maximum area is

$$A = 4\frac{a}{\sqrt{2}}\frac{b}{\sqrt{2}} = 2ab$$

Since the area of the circumscribed rectangle is $4ab$, the required ratio is

$$2ab : \pi ab : 4ab = 1 : \frac{\pi}{2} : 2$$

51. Suppose that $a = b$ in the isoscles triangle, so that

$$A^2 = s(s-a)(s-b)(s-c) = s(s-a)^2(s-c)$$

Since $s = \frac{1}{2}(a + b + c)$, it follows that $s = \frac{1}{2}(2a + c) = a + \frac{c}{2}$, so that $s - a = \frac{c}{2}$. Thus,

$$A^2 = s\left(\frac{c^2}{4}\right)(s-c)$$

$$= \frac{s}{4}\left(sc^2 - c^3\right)$$

Since s is a constant (it's half of the perimeter), we can now differentiate to get

$$2A\frac{dA}{dc} = \frac{s}{4}\left(2sc - 3c^2\right)$$
$$0 = c(2s - 3c)$$

Thus, the area is maximized when $2s - 3c = 0$, which means $c = \frac{2}{3}s$. Solving for a, we get

$$a = s - \frac{c}{2} = s - \frac{s}{3} = \frac{2}{3}s.$$

Thus, the area is maximized when $a = b = c$; in other words the area is maximized when the triangle is equilateral.

The maximum area is

$$A = \sqrt{s(s - c)^3} = \sqrt{s\left(\frac{s}{3}\right)^3}$$

$$= \frac{s^2}{9}\sqrt{3} = \frac{p^2}{36}\sqrt{3}$$

3.7 RELATED RATES

1. $V(t) = \text{(depth)(area)} = \frac{\pi}{48}[r(t)]^2$
(units in cubic feet per min)

$$V'(t) = \frac{\pi}{48}2r(t)r'(t) = \frac{\pi}{24}r(t)r'(t)$$

We are given $V'(t) = \frac{120}{7.5} = 16$.

Hence $16 = \frac{\pi}{24}r(t)r'(t)$ so

$$r'(t) = \frac{(16)(24)}{\pi r(t)}.$$

(a) When $r = 100$,

$$r'(t) = \frac{(16)(24)}{100\pi} = \frac{96}{25\pi}$$
$$\approx 1.2223 \text{ ft/min},$$

(b) When $r = 200$,

$$r'(t) = \frac{(16)(24)}{200\pi} = \frac{48}{25\pi}$$
$$\approx 0.61115 \text{ ft/min}$$

3. From exercise 1,

$$V'(t) = \frac{\pi}{48}2r(t)r'(t) = \frac{\pi}{24}r(t)r'(t),$$

so $\frac{g}{7.5} = \frac{\pi}{24}(100)(0.6) = 2.5\pi$,

so $g = (7.5)(2.5)\pi$
$$= 18.75\pi \approx 58.905 \text{ gal/min}.$$

5. $t =$ hours elapsed since injury
$r =$ radius of the infected area
$A =$ area of the infection
$A = \pi r^2$
$A'(t) = 2\pi r(t) \cdot r'(t)$
When $r = 3$ mm, $r' = 1$ mm/hr,
$A' = 2\pi(3)(1) = 6\pi$ mm^2/hr

7. $V(t) = \frac{4}{3}\pi[r(t)]^3$
$V'(t) = 4\pi[r(t)]^2r'(t) = Ar'(t)$
If $V'(t) = kA(t)$, then
$$r'(t) = \frac{V'(t)}{A(t)} = \frac{kA(t)}{A(t)} = k.$$

9. $10^2 = x^2 + y^2$
$$0 = 2x\frac{dx}{dt} + 2y\frac{dy}{dt}$$
$$\frac{dy}{dt} = -\frac{x}{y}\frac{dx}{dt}$$
$$= -\frac{6}{8}(3)$$
$$= -2.25 \text{ ft/s}$$

11. Let α represent the angle adjacent to θ in the smaller triangle and β the angle adjacent to θ in the larger triangle. Then $\alpha + \beta + \theta = \pi$ so that $\alpha'(t) + \beta'(t) + \theta'(t) = 0$ or $\theta'(t) = -\alpha'(t) - \beta'(t)$. Let x be the base of the smaller triangle. Then $60 - x$ is the base of the larger triangle.

$\tan\alpha = \dfrac{20}{x}$ and $\tan\beta = \dfrac{40}{60-x}$ so that

$$\sec^2\alpha(t)\cdot\alpha'(t) = -\dfrac{20}{[x(t)]^2}\cdot x'(t)\text{ and}$$

$$\sec^2\beta(t)\cdot\beta'(t) = \dfrac{40}{[60-x(t)]^2}\cdot x'(t)\text{ or}$$

$$\alpha'(t) = -\dfrac{20\cos^2\alpha(t)}{[x(t)]^2}x'(t)\text{ and}$$

$$\beta'(t) = \dfrac{40\cos^2\beta(t)}{[60-x(t)]^2}x'(t).$$

Thus,

$$\theta'(t) = \dfrac{20\cos^2\alpha(t)}{[x(t)]^2}x'(t) - \dfrac{40\cos^2\beta(t)}{[60-x(t)]^2}x'(t).$$

Now $x'(t) = -4$ and $x(t) = 30$. When $x = 30$, $\cos\alpha = \dfrac{3}{\sqrt{13}}$ and $\cos\beta = \dfrac{3}{5}$.

So, $\theta'(t) = \dfrac{20\left(\frac{3}{\sqrt{13}}\right)^2}{30^2}(-4) - \dfrac{40\left(\frac{3}{5}\right)^2}{30^2}(-4)$

≈ 0.00246 rad/s.

13. We know $[x(t)]^2 + 4^2 = [s(t)]^2$. Hence $2x(t)x'(t) = 2s(t)s'(t)$, so

$$x'(t) = \dfrac{s(t)s'(t)}{x(t)} = \dfrac{-240s(t)}{x(t)}.$$

When $x = 40$, $s = \sqrt{40^2 + 4^2} = 4\sqrt{101}$, so at that moment

$$x'(t) = \dfrac{(-240)(4\sqrt{101})}{40} = -24\sqrt{101}.$$

So the speed is $24\sqrt{101} \approx 241.2$ mph.

15. If the police car is not moving, then $x'(t) = 0$, but all the other data are unchanged. So

$$d'(t) = \dfrac{x(t)x'(t) + y(t)y'(t)}{\sqrt{[x(t)]^2 + [y(t)]^2}}$$

$$= \dfrac{-(1/2)(50)}{\sqrt{1/4 + 1/16}}$$

$$= \dfrac{-100}{\sqrt{5}} \approx -44.721.$$

This is more accurate.

17. $d'(t) = \dfrac{x(t)x'(t) + y(t)y'(t)}{\sqrt{[x(t)]^2 + [y(t)]^2}}$

$$= \dfrac{-(1/2)(\sqrt{2}-1)(50) - (1/2)(50)}{\sqrt{1/4 + 1/4}}$$

$$= -50.$$

19. $\overline{C}(x) = 10 + \dfrac{100}{x}$

$\overline{C}'(x(t)) = \dfrac{-100}{x^2}\cdot x'(t)$

$\overline{C}'(10) = -1(2) = -2$ dollars per item, so average cost is decreasing at the rate of \$2 per year.

21. From the table, we see that the recent trend is for advertising to increase by \$4000 per year. A good estimate is then $x'(2) \approx 4$ (in units of thousands). From Example 7.4, we have $s'(t) = 60[9 + x(t)]^{-3/2}x'(t)$.

Using our estimate that $x'(2) \approx 4$ and since $x(2) = 24$, we get $s'(24) \approx 60[9 + 24]^{-3/2}(4) \approx 1.266$. Thus, sales are increasing at the rate of approximately \$1266 per year.

23. We have $\tan\theta = \dfrac{x}{2}$, so

$$\dfrac{d}{dt}(\tan\theta) = \dfrac{d}{dt}\left(\dfrac{x}{2}\right)$$

$$\sec^2\theta\cdot\theta' = \dfrac{1}{2}x'$$

$$\theta' = \dfrac{1}{2\sec^2\theta}\cdot x' = \dfrac{x'\cos^2\theta}{2}$$

at $x = 0$, we have $\tan\theta = \dfrac{x}{2} = \dfrac{0}{2}$ so $\theta = 0$ and we have $x' = -130$ft/s so

$$\theta' = \dfrac{(-130)\cdot\cos^2 0}{2} = -65\text{ rad/s.}$$

25. $t =$ number of seconds since launch
$x =$ height of rocket in miles after t seconds
$\theta =$ camera angle in radians after t seconds

$$\tan\theta = \frac{x}{2}$$

$$\frac{d}{dx}(\tan\theta) = \frac{d}{dx}\left(\frac{x}{2}\right)$$

$$\sec^2\theta \cdot \theta' = \frac{1}{2}x'$$

$$\theta' = \frac{\cos^2\theta \cdot x'}{2}$$

When $x = 3$, $\tan\theta = 3/2$, so $\cos\theta = 2/\sqrt{13}$.

$$\theta' = \frac{\left(\frac{2}{\sqrt{13}}\right)^2 (0.2)}{2} \approx 0.03 \text{ rad/s}$$

27. Let θ be the angle between the end of the shadow and the top of the lamppost. Then $\tan\theta = \dfrac{6}{s}$ and $\tan\theta = \dfrac{18}{s+x}$, so

$$\frac{x+s}{18} = \frac{s}{6}$$

$$\frac{d}{dt}\left(\frac{x+s}{18}\right) = \frac{d}{dt}\left(\frac{s}{6}\right)$$

$$\frac{x'+s'}{18} = \frac{s'}{6}$$

$$x' + s' = 3s'$$

$$s' = \frac{x'}{2}$$

Since $x' = 2$, $s' = 2/2 = 1$ ft/s.

29. $P(t)V'(t) + P'(t)V(t) = 0$

$$\frac{P'(t)}{V'(t)} = -\frac{P(t)}{V(t)} = -\frac{c}{V(t)^2}$$

31. Let $r(t)$ be the length of the rope at time t and $x(t)$ be the distance (along the water) between the boat and the dock.

$$r(t)^2 = 36 + x(t)^2$$

$$2r(t)r'(t) = 2x(t)x'(t)$$

$$x'(t) = \frac{r(t)r'(t)}{x(t)} = \frac{-2r(t)}{x(t)}$$

$$= \frac{-2\sqrt{36+x^2}}{x}$$

When $x = 20$, $x' = -2.088$; when $x = 10$, $x' = -2.332$.

33. $f(t) = \dfrac{1}{2L(t)}\sqrt{\dfrac{T}{\rho}} = \dfrac{110}{L(t)}.$

$$f'(t) = \frac{-110}{L(t)^2}L'(t).$$

When $L = 1/2$, $f(t) = 220$ cycles per second. If $L' = -4$ at this time, then $f'(t) = 1760$ cycles per second per second. It will only take 1/8 second at this rate for the frequency to go from 220 to 440, and raise the pitch one octave.

35. When the tank is half full, the radius of the top level of water is 60 feet, which is a maximum. The radius has been increasing to this point ($r'(t) > 0$) and will now begin to decrease ($r'(t) < 0$). So, $r'(t) = 0$.

37. The volume of the conical pile is $V = \frac{1}{3}\pi r^2 h$. Since $h = 2r$, we can write the volume as

$$V = \frac{1}{3}\pi\left(\frac{h}{2}\right)^2 h = \frac{1}{12}\pi h^3$$

Thus,

$$\frac{dV}{dt} = \frac{\pi h^2}{4}\cdot\frac{dh}{dt}$$

$$20 = \frac{\pi 6^2}{4}\cdot\frac{dh}{dt}$$

$$\frac{dh}{dt} = \frac{20}{9\pi}$$

$$\frac{dr}{dt} = \frac{10}{9\pi}$$

3.8 RATES OF CHANGE IN ECONOMICS AND THE SCIENCES

1. The marginal cost function is
$C'(x) = 3x^2 + 40x + 90$.
The marginal cost at $x = 50$ is $C'(50) = 9590$.
The cost of producing the 50th item is $C(50) - C(49) = 9421$.

3. The marginal cost function is
$C'(x) = 3x^2 + 42x + 110$.
The marginal cost at $x = 100$ is $C'(100) = 34,310$. The cost of producing the 100th item is $C(100) - C(99) = 33,990$.

5. $C'(x) = 3x^2 - 60x + 300$
$C''(x) = 6x - 60 = 0$
$x = 10$ is the inflection point because $C''(x)$ changes from negative to positive at this value. After this point, cost rises more sharply.

7. $\overline{C}(x) = C(x)/x = 0.1x + 3 + \dfrac{2000}{x}$

$\overline{C}'(x) = 0.1 - \dfrac{2000}{x^2}$

Critical number is $x = 100\sqrt{2} \approx 141.4$.

$\overline{C}'(x)$ is negative to the left of the critical number and positive to the right, so this must be the minimum.

9. $\overline{C}(x) = C(x)/x = \dfrac{\sqrt{x^3 + 9}}{x} = (x + 9x^{-2})^{1/2}$

$\overline{C}'(x) = \dfrac{1}{2}(x + 9x^{-2})^{1/2}(1 - 18x^{-3})$

Critical number is $x = \sqrt[3]{18}$. $\overline{C}'(x)$ is negative to the left of the critical number and positive to the right, so this must be the minimum.

11. $C(x) = 0.01x^2 + 40x + 3600$

$C'(x) = 0.02x + 40$

$\overline{C}(x) = \dfrac{C(x)}{x} = 0.01x + 40 + \dfrac{3600}{x}$

$C'(100) = 42$

$\overline{C}(100) = 77$

so $C'(100) < \overline{C}(100)$

$\overline{C}(101) = 76.65 < \overline{C}(100)$

13. $\overline{C}'(x) = 0.01 - \dfrac{3600}{x^2} = 0$

so $x = 600$ is min and

$C'(600) = 52$

$\overline{C}(600) = 52$

15. $P(x) = R(x) - C(x)$ When profit is maximized, $P'(x) = R'(x) - C'(x) = 0$, so $R'(x) = C'(x)$.

17. $E = \dfrac{p}{f(p)} f'(p)$

$= \dfrac{p}{200(30 - p)}(-200) = \dfrac{p}{p - 30}$

To solve $\dfrac{p}{p - 30} < -1$, multiply both sides by the negative quantity $p - 30$, to get $p > (-1)(p - 30)$ or $p > 30 - p$, so $2p > 30$, so $15 < p < 30$.

19. $f(p) = 100p(20 - p) = 100(20p - p^2)$

$E = \dfrac{p}{f(p)} f'(p)$

$= \dfrac{p}{100p(20 - p)}(100)(20 - 2p)$

$= \dfrac{20 - 2p}{20 - p}$ or $\dfrac{2p - 20}{p - 20}$

To solve $\dfrac{20 - 2p}{20 - p} < -1$, multiply both sides by the positive quantity $20 - p$ to get $20 - 2p < (-1)(20 - p)$, or $20 - 2p < p - 20$, so $40 < 3p$, so $40/3 < p < 20$.

21. Elasticity of demand at price $p = 15$ is, by definition, the relative change in demand divided by the relative change in price, as price increases from 15 to an amount slightly larger than 15. So if (rel change in demand)/(rel change in price) is less than (-1), then rel change in demand is less than (-1)(rel change in price). This means that demand goes down more than price goes up, so revenue should decrease. (See exercise 23.)

23. $[pf(p)]' < 0$
if and only if $p'f(p) + pf'(p) < 0$
if and only if $f(p) + pf'(p) < 0$
if and only if $pf'(p) < -f(p)$
if and only if $\dfrac{pf'(p)}{f(p)} < -1$.

25. $f(x) = 2x(4 - x)$

$$f'(x) = 2(4 - x) + 2x(-1) = 8 - 4x$$
$$= 4(2 - x) = 0$$

$x = 2$ is a maximum since the graph of $f(x)$ is a downward-opening parabola.

27. $x'(t) = 2x(t)[4 - x(t)] = 0$
$x(t) = 0$, $x(t) = 4$ are critical numbers.
$x'(t) > 0$ for $0 < x(t) < 4$
$x'(t) < 0$ for $x(t) > 4$
So $x(t) = 4$ is the maximum concentration.

$x'(t) = 0.5x(t)[5 - x(t)]$ $x(t) = 0$, $x(t) = 5$
are critical numbers.
$x'(t) > 0$ for $0 < x(t) < 5$
$x'(t) < 0$ for $x(t) > 5$
So $x(t) = 5$ is the maximum concentration.

29. $y'(t) = c \cdot y(t)[K - y(t)]$

$$y(t) = Kx(t)$$

$$y'(t) = Kx'(t)$$

$$Kx'(t) = c \cdot Kx(t)[K - Kx(t)]$$

$$x'(t) = c \cdot Kx(t)[1 - x(t)]$$
$$= rx(t)[1 - x(t)]$$

$$r = cK$$

31. $x'(t) = [a - x(t)][b - x(t)]$ for $x(t) = a$,
$x'(t) = [a - a][b - a] = 0$
So the concentration of product is staying the same.

If $a < b$ and $x(0) = 0$ then $x'(t) > 0$ for $0 < x < a < b$ $x'(t) < 0$ for $a < x < b$ Thus $x(t) = a$ is a maximum.

33. $R(x) = \dfrac{rx}{k + x}$, $x \geq 0$

$$R'(x) = \dfrac{rk}{(k + x)^2}$$

There are no critical numbers. Any possible maximum would have to be at the endpoint $x = 0$, but in fact R is increasing on $[0, \infty)$, so there is no maximum (although as x goes to infinity, R approaches r).

35. $PV^{7/5} = c$

$$\frac{d}{dP}\left(PV^{7/5}\right) = \frac{d}{dP}(c) = 0$$

$$V^{7/5} + \frac{7}{5}PV^{2/5}\frac{dV}{dP} = 0$$

$$V + \frac{7}{5}P\frac{dV}{dP} = 0$$

$$\frac{dV}{dP} = \frac{-5}{7}\frac{V}{P}.$$

But $V^{7/5} = c/P$, so $V = (c/P)^{5/7}$. Hence

$$\frac{dV}{dP} = \frac{-5}{7}\frac{V}{P}$$

$$= \frac{-5}{7}\frac{(c/P)^{5/7}}{P} = \frac{-5c^{5/7}}{7P^{12/7}}.$$

As pressure increases, volume decreases.

37. $m'(x) = 4 - \cos x$, so the rod is less dense at the ends.

39. $m'(x) = 4$, so the rod is homogeneous.

41. The rate of population growth is given by
$f(p) = 4p(5 - p) = 4(5p - p^2)$
$f'(p) = 4(5 - 2p)$,
so the only critical number is $p = 2.5$. Since the graph of f is a parabola opening down, this must be a max.

43.
$$f'(x) = \frac{-64x^{-1.4}(4x^{-0.4} + 15)}{(4x^{-0.4} + 15)^2}$$

$$- \frac{(160x^{-0.4} + 90)(-1.6x^{-1.4})}{(4x^{-0.4} + 15)^2}$$

$$= \frac{-816x^{-1.4}}{(4x^{-0.4} + 15)^2} < 0$$

So $f(x)$ is decreasing. This shows that pupils shrink as light increases.

45. If for some x marginal revenue equals marginal cost, then
$P'(x) = R'(x) - C'(x) = 0$,
so x is a critical number, but it may not be a maximum.

47. If v is not greater than c, the fish will never make any headway.

$$E'(v) = \frac{v(v - 2c)}{(v - c)^2}$$

so the only critical number is $v = 2c$. When $c < v < 2c$, $E'(v) < 0$ and when $v > 2c$, $E'(v) > 0$, so we must have a minimum.

Chapter 3 Review Exercises

1. $f(x) = \cos 3x$, $x_0 = 0$,

$f'(x) = -3 \sin 3x$

$L(x) = f(x_0) + f'(x_0)(x - x_0)$

$\quad = f(0) + f'(0)(x - 0)$

$\quad = \cos(3 \cdot 0) - 3 \sin(3 \cdot 0)(x - 0)$

$\quad = 1$

3. $f(x) = \sqrt[3]{x} = x^{1/3}$, $x_0 = 8$

$f'(x) = \frac{1}{3} x^{-2/3}$

$\quad L(x) = f(x_0) + f'(x_0)(x - x_0)$

$\quad\quad = f(8) + f'(8)(x - 8)$

$\quad\quad = \sqrt[3]{8} + \frac{1}{3}(8)^{-2/3}(x - 8)$

$\quad\quad = 2 + \frac{1}{12}(x - 8)$

$\quad L(7.96) = 2 + \frac{1}{12}(7.96 - 8) \approx 1.99666$

5. From the graph of $f(x) = x^3 + 5x - 1$, there is one root.

$f'(x) = 3x^2 + 5$

Starting with $x_0 = 0$, Newton's method gives $x_1 = 0.2$, $x_2 = 0.198437$, and $x_3 = 0.198437$.

7. $f'(x) = 3x^2 - 3$, so that $f'(1) = 0$.

x_1 is undefined when $x_0 = 1$.

9. $f'(x) = 3x^2 + 6x - 9 = 3(x^2 + 2x - 3)$

$\quad = 3(x + 3)(x - 1)$

So the critical numbers are $x = 1$ and $x = -3$.

$f'(x) > 0$ on $(-\infty, -3) \cup (1, \infty)$

$f'(x) < 0$ on $(-3, 1)$

Hence f is increasing on $(-\infty, -3)$ and on $(1, \infty)$ and f is decreasing on $(-3, 1)$. Thus there is a local max at $x = -3$ and a local min at $x = 1$.

$f''(x) = 3(2x + 2) = 6(x + 1)$

$f''(x) > 0$ on $(-1, \infty)$

$f''(x) < 0$ on $(-\infty, -1)$

Hence f is concave up on $(-1, \infty)$ and concave down on $(-\infty, -1)$, and there is an inflection point at $x = -1$.

11. $f'(x) = 4x^3 - 12x^2 = 4x^2(x - 3)$

$x = 0, 3$ are critical numbers.

$f'(x) > 0$ on $(3, \infty)$

$f'(x) < 0$ on $(-\infty, 0) \cup (0, 3)$

f increasing on $(3, \infty)$, decreasing on $(-\infty, 3)$ so $x = 3$ is a local min.

$f''(x) = 12x^2 - 24x = 12x(x - 2)$

$f''(x) > 0$ on $(-\infty, 0) \cup (2, \infty)$

$f''(x) < 0$ on $(0, 2)$

f is concave up on $(-\infty, 0) \cup (2, \infty)$, concave down on $(0, 2)$ so $x = 0, 2$ are inflection points.

13.

$f'(x) = \dfrac{x^2 - (x - 90)(2x)}{x^4}$

$\quad = \dfrac{-(x - 180)}{x^3}$

$x = 180$ is the only critical number.

$f'(x) < 0$ on $(-\infty, 0) \cup (180, \infty)$

$f'(x) > 0$ on $(0, 180)$

$f(x)$ is decreasing on $(-\infty, 0) \cup (180, \infty)$ and increasing on $(0, 180)$ so $f(x)$ has a local maximum at $x = 180$.

$f''(x) = -\dfrac{x^3 - (x - 180)(3x^2)}{x^6}$

$\quad = -\dfrac{-2x + 540}{x^4}$

$f''(x) < 0$ on $(-\infty, 0) \cup (0, 270)$

$f''(x) > 0$ on $(270, \infty)$ so $x = 270$ is an inflection point.

15. $f'(x) = \dfrac{x^2 + 4 - x(2x)}{(x^2 + 4)^2}$

$= \dfrac{4 - x^2}{(x^2 + 4)^2}$

$x = \pm 2$ are critical numbers.
$f'(x) > 0$ on $(-2, 2)$
$f'(x) < 0$ on $(-\infty, -2) \cup (2, \infty)$
f increasing on $(-2, 2)$, decreasing on
$(-\infty, -2)$ and on $(2, \infty)$, so f has a local min
at $x = -2$ and a local max at $x = 2$. $f''(x) =$
$\dfrac{-2x(x^2 + 4)^2 - (4 - x^2)[2(x^2 + 4) \cdot 2x]}{(x^2 + 4)^4}$

$= \dfrac{2x^3 - 24x}{(x^2 + 4)^3}$

$f''(x) > 0$ on $\left(-\sqrt{12}, 0\right) \cup \left(\sqrt{12}, \infty\right)$

$f''(x) < 0$ on $\left(-\infty, -\sqrt{12}\right) \cup \left(0, \sqrt{12}\right)$

f is concave up on $\left(-\sqrt{12}, 0\right) \cup \left(\sqrt{12}, \infty\right)$,

concave down on $\left(-\infty, -\sqrt{12}\right) \cup \left(0, \sqrt{12}\right)$

so $x = \pm\sqrt{12}, 0$ are inflection points.

17. $f'(x) = 3x^2 + 6x - 9$

$= 3(x + 3)(x - 1)$

$x = -3$, $x = 1$ are critical numbers,
but $x = -3 \notin [0, 4]$.
$f(0) = 0^3 + 3 \cdot 0^2 - 9 \cdot 0 = 0$
$f(4) = 4^3 + 3 \cdot 4^2 - 9 \cdot 4 = 76$
$f(1) = 1^3 + 3 \cdot 1^2 - 9 \cdot 1 = -5$
So $f(4) = 76$ is absolute max on $[0, 4]$, $f(1) =$
-5 is absolute min.

19. $f'(x) = \frac{4}{5}x^{-1/5}$
$x = 0$ is critical number.
$f(-2) = (-2)^{4/5} \approx 1.74$
$f(3) = (3)^{4/5} \approx 2.41$
$f(0) = (0)^{4/5} = 0$
$f(0) = 0$ is absolute min, $f(3) = 3^{4/5}$ is absolute max.

21. $f'(x) = 3x^2 + 8x + 2$

$f'(x) = 0$ when

$x = \dfrac{-8 \pm \sqrt{64 - 24}}{6} = -\dfrac{4}{3} \pm \dfrac{\sqrt{10}}{3}$

$x = -\dfrac{4}{3} - \dfrac{\sqrt{10}}{3}$ is local max,

$x = -\dfrac{4}{3} + \dfrac{\sqrt{10}}{3}$ is local min.

23. $f'(x) = 5x^4 - 4x + 1 = 0$
$x \approx 0.2553, 0.8227$
local min at $x \approx 0.8227$,
local max at $x \approx 0.2553$.

25. One possible graph:

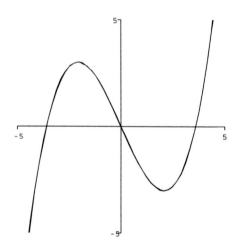

27. $f'(x) = 4x^3 + 12x^2 = 4x^2(x + 3)$
$f''(x) = 12x^2 + 24x = 12x(x + 2)$
$f'(x) > 0$ on $(-3, 0) \cup (0, \infty)$
$f'(x) < 0$ on $(-\infty, -3)$
$f''(x) > 0$ on $(-\infty, -2) \cup (0, \infty)$
$f''(x) < 0$ on $(-2, 0)$
f increasing on $(-3, \infty)$, decreasing on
$(-\infty, -3)$, concave up on $(-\infty, -2) \cup$
$(0, \infty)$, concave down on $(-2, 0)$, local min
at $x = -3$, inflection points at $x = -2, 0$.
$f(x) \to \infty$ as $x \to \pm\infty$.

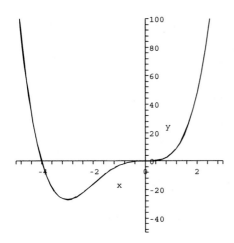

29. $f'(x) = 4x^3 + 4 = 4(x^3 + 1)$
$f''(x) = 12x^2$
$f'(x) > 0$ on $(-1, \infty)$
$f'(x) < 0$ on $(-\infty, -1)$
$f''(x) > 0$ on $(-\infty, 0) \cup (0, \infty)$
f increasing on $(-1, \infty)$, decreasing on $(-\infty, -1)$, concave up on $(-\infty, \infty)$, local min at $x = -1$.
$f(x) \to \infty$ as $x \to \pm\infty$.

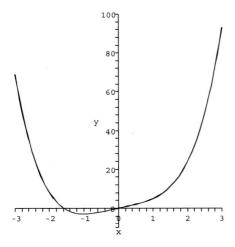

31. $f'(x) = \dfrac{x^2 + 1 - x(2x)}{(x^2 + 1)^2}$

$= \dfrac{1 - x^2}{(x^2 + 1)^2}$

$f''(x) =$
$\dfrac{-2x(x^2 + 1)^2 - (1 - x^2)2(x^2 + 1)2x}{(x^2 + 1)^4}$

$= \dfrac{2x(x^2 - 3)}{(x^2 + 1)^3}$

$f'(x) > 0$ on $(-1, 1)$

$f'(x) < 0$ on $(-\infty, -1) \cup (1, \infty)$
$f''(x) > 0$ on $\left(-\sqrt{3}, 0\right) \cup \left(\sqrt{3}, \infty\right)$
$f''(x) < 0$ on $\left(-\infty, -\sqrt{3}\right) \cup \left(0, \sqrt{3}\right)$
f increasing on $(-1, 1)$, decreasing on $(-\infty, -1)$ and on $(1, \infty)$, concave up on

$\left(-\sqrt{3}, 0\right) \cup \left(\sqrt{3}, \infty\right),$

concave down on

$\left(-\infty, -\sqrt{3}\right) \cup \left(0, \sqrt{3}\right),$

local min at $x = -1$, local max at $x = 1$, inflection points at $0, \pm\sqrt{3}$.

$\lim\limits_{x \to \infty} \dfrac{x}{x^2 + 1} = \lim\limits_{x \to -\infty} \dfrac{x}{x^2 + 1} = 0$

So f has a horizontal asymptote at $y = 0$.

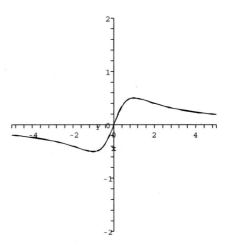

33. $f'(x) = \dfrac{(2x)(x^2 + 1) - x^2(2x)}{(x^2 + 1)^2}$

$= \dfrac{2x}{(x^2 + 1)^2}$

$f''(x) = \dfrac{2(x^2 + 1)^2 - 2x \cdot 2(x^2 + 1)2x}{(x^2 + 1)^4}$

$= \dfrac{2 - 6x^2}{(x^2 + 1)^3}$

$f'(x) > 0$ on $(0, \infty)$
$f'(x) < 0$ on $(-\infty, 0)$
$f''(x) > 0$ on $\left(-\sqrt{\tfrac{1}{3}}, \sqrt{\tfrac{1}{3}}\right)$

$f''(x) < 0$ on $\left(-\infty, -\sqrt{\frac{1}{3}}\right) \cup \left(\sqrt{\frac{1}{3}}, \infty\right)$

f increasing on $(0, \infty)$, decreasing on $(-\infty, 0)$, concave up on

$$\left(-\sqrt{\frac{1}{3}}, \sqrt{\frac{1}{3}}\right),$$

concave down on

$$\left(-\infty, -\sqrt{\frac{1}{3}}\right) \cup \left(\sqrt{\frac{1}{3}}, \infty\right),$$

local min at $x = 0$, inflection points at $x = \pm\sqrt{1/3}$.

$$\lim_{x\to\infty} \frac{x^2}{x^2+1} = \lim_{x\to-\infty} \frac{x^2}{x^2+1} = 1$$

So f has a horizontal asymptote at $y = 1$.

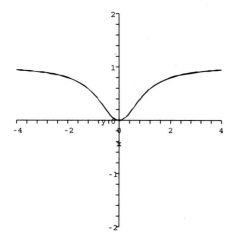

35.
$$f'(x) = \frac{3x^2(x^2-1) - x^3(2x)}{(x^2-1)^2}$$

$$= \frac{x^4 - 3x^2}{(x^2-1)^2}$$

$$f''(x) = \frac{(4x^3 - 6x)(x^2-1)^2}{(x^2-1)^4}$$

$$- \frac{(x^4 - 3x^2)2(x^2-1)2x}{(x^2-1)^4}$$

$$= \frac{2x^3 + 6x}{(x^2-1)^3}$$

$f'(x) > 0$ on $\left(-\infty, -\sqrt{3}\right) \cup \left(\sqrt{3}, \infty\right)$

$f'(x) < 0$ on $\left(-\sqrt{3}, -1\right) \cup (-1, 0) \cup (0, 1) \cup \left(1, \sqrt{3}\right)$

$f''(x) > 0$ on $(-1, 0) \cup (1, \infty)$

$f''(x) < 0$ on $(-\infty, -1) \cup (0, 1)$

f increasing on $(-\infty, -\sqrt{3})$ and on $(\sqrt{3}, \infty)$; decreasing on $(-\sqrt{3}, -1)$ and on $(-1, 1)$ and on $(1, \sqrt{3})$; concave up on $(-1, 0) \cup (1, \infty)$, concave down on $(-\infty, -1) \cup (0, 1)$; $x = -\sqrt{3}$ local max; $x = \sqrt{3}$ local min; $x = 0$ inflection point. f is undefined at $x = -1$ and $x = 1$.

$$\lim_{x\to 1^+} \frac{x^3}{x^2-1} = \infty, \text{ and}$$

$$\lim_{x\to 1^-} \frac{x^3}{x^2-1} = -\infty.$$

$$\lim_{x\to -1^+} \frac{x^3}{x^2-1} = \infty, \text{ and}$$

$$\lim_{x\to -1^-} \frac{x^3}{x^2-1} = -\infty.$$

So f has vertical asymptotes at $x = 1$ and $x = -1$.

By long division, we see that $y = x$ is a slant asymptote.

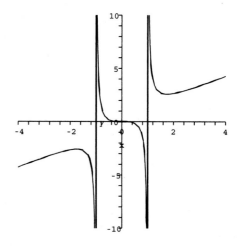

37. $d = \sqrt{(x-2)^2 + (y-1)^2}$
$\quad = \sqrt{(x-2)^2 + (2x^2-1)^2}$

$f(x) = (x - 2)^2 + (2x^2 - 1)^2$

$f'(x) = 2(x - 2) + 2(2x^2 - 1)4x$

$\quad = 16x^3 - 6x - 4$

$f'(x) = 0$ when $x \approx 0.8237$

$f'(x) < 0$ on $(-\infty, 0.8237)$

$f'(x) > 0$ on $(0.8237, \infty)$

So $x \approx 0.8237$ corresponds to the closest point. $y = 2x^2 = 2(0.8237)^2 = 1.3570$
$(0.8237, 1.3570)$ is closest to $(2, 1)$.

39. $C(x) = 6\sqrt{4^2 + (4 - x)^2} + 2\sqrt{2^2 + x^2}$

$C'(x) =$

$6 \cdot \frac{1}{2}[16 + (4 - x)^2]^{-1/2} \cdot 2(4 - x)(-1)$

$+ 2 \cdot \frac{1}{2}(4 + x^2)^{-1/2} \cdot 2x$

$= \dfrac{6(x - 4)}{\sqrt{16 + (4 - x)^2}} + \dfrac{2x}{\sqrt{4 + x^2}} \quad C'(x) = 0$

when $x \approx 2.864$

$C'(x) < 0$ on $(0, 2.864)$

$C'(x) > 0$ on $(2.864, 4)$

So $x \approx 2.864$ gives the minimum cost. Locate highway corner $4 - 2.864 = 1.136$ miles east of point A.

41. Area: $A = 2\pi r^2 + 2\pi rh$
Convert to in^3:

16 fl oz $= 16$ fl oz $\cdot 1.80469$ in^3/fl oz

$\quad = 28.87504$ in^3

Volume: $V = \pi r^2 h$

$h = \dfrac{\text{Vol}}{\pi r^2} = \dfrac{28.87504}{\pi r^2}$

$A(r) = 2\pi \left(r^2 + \dfrac{28.87504}{\pi r} \right)$

$A'(r) = 2\pi \left(2r - \dfrac{28.87504}{\pi r^2} \right)$

$2\pi r^3 = 28.87504$

$r = \sqrt[3]{\dfrac{28.87504}{2\pi}} \approx 1.663$

$A'(r) < 0$ on $(0, 1.663)$

$A'(r) > 0$ on $(1.663, \infty)$

So $r \approx 1.663$ gives the minimum surface area.

$h = \dfrac{28.87504}{\pi(1.663)^2} \approx 3.325$

43. $\rho(x) = m'(x) = 2x$
As you move along the rod to the right, its density increases.

45. The marginal cost function is
$C'(x) = 0.04x + 20$.
$C'(20) = 0.04(20) + 20 = 20.8$
$C(20) - C(19) =$
$0.02(20)^2 + 20(20) + 1800$
$-[0.02(19)^2 + 20(19) + 1800]$
$= 20.78$

Integration

4.1 ANTIDERIVATIVES

1. $\displaystyle\int x^3\,dx = \frac{x^4}{4} + c;\ \frac{x^4}{4},\ \frac{x^4}{4} + 3,\ \frac{x^4}{4} - 2$

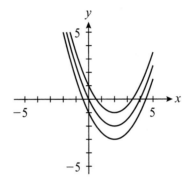

5. $\displaystyle\int (3x^4 - 3x)\,dx = \frac{3}{5}x^5 - \frac{3}{2}x^2 + c$

7. $\displaystyle\int \left(3\sqrt{x} - \frac{1}{x^4}\right)\,dx = \int (3x^{1/2} - x^{-4})\,dx$

$\displaystyle = 2x^{3/2} + \frac{x^{-3}}{3} + c$

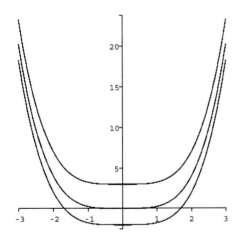

9. $\displaystyle\int \frac{x^{1/3} - 3}{x^{2/3}}\,dx$

$\displaystyle = \int (x^{-1/3} - 3x^{-2/3})\,dx$

$\displaystyle = \frac{3}{2}x^{2/3} - 9x^{1/3} + c$

3. $\displaystyle\int (x - 2)\,dx = \frac{x^2}{2} - 2x + c;\ \frac{x^2}{2} - 2x,$

$\displaystyle \frac{x^2}{2} - 2x + 1,\ \frac{x^2}{2} - 2x - 1$

11. $\displaystyle\int (2\sin x + \cos x)\,dx$

$\displaystyle = -2\cos x + \sin x + c$

13. $\displaystyle\int 2\sec x \tan x \, dx$

$= 2\sec x + c$

15. $\displaystyle\int 5\sec^2 x \, dx = 5\tan x + c$

17. $\displaystyle\int (3\cos x - 2)\, dx = 3\sin x - 2x + c$

19. $\displaystyle\int\left(5x - \frac{3}{x^2}\right) dx = \int (5x - 3x^{-2})\, dx$

$= \dfrac{5}{2}x^2 + 3x^{-1} + c = \dfrac{5}{2}x^2 + \dfrac{3}{x} + c$

21. $\displaystyle\int \frac{x^2 + 4}{x^2}\, dx = \int (1 + 4x^{-2})\, dx$

$= x - 4x^{-1} + c = x - \dfrac{4}{x} + c$

23. $\displaystyle\int x^{1/4}(x^{5/4} - 4)\, dx$

$= \displaystyle\int (x^{3/2} - 4x^{1/4})\, dx$

$= \dfrac{2}{5}x^{5/2} - \dfrac{16}{5}x^{5/4} + c$

25. (a) N/A

(b) $\displaystyle\int (\sqrt{x^3} + 4)\, dx = \dfrac{2}{5}x^{5/2} + 4x + c$

27. (a) N/A

(b) $\displaystyle\int \sec^2 x \, dx = \tan x + c$

29. Use a CAS to find the antiderivative and verify by computing the derivative:

1.9(f) $\displaystyle\int x \sin 2x \, dx$

$= \dfrac{\sin 2x}{4} - \dfrac{x \cos 2x}{2} + c$

Verify:

$\dfrac{d}{dx}\left(\dfrac{\sin 2x}{4} - \dfrac{x \cos 2x}{2}\right)$

$= \dfrac{2\cos 2x}{4} - \dfrac{\cos 2x - 2x \sin 2x}{2}$

$= x \sin 2x$

31. Use a CAS to find antiderivatives and verify by computing the derivatives:

(a) $\displaystyle\int x^2 \sin x \, dx = -x^2 \cos x + 2x \sin x + 2\cos x + c$

$\dfrac{d}{dx}(-x^2 \cos x + 2x \sin x + 2\cos x)$

$= -2x \cos x + x^2 \sin x + 2\sin x + 2x \cos x - 2\sin x = x^2 \sin x$

(b) $\displaystyle\int \frac{\cos x}{\sin^3 x}\, dx = -\frac{1}{2\sin^2 x} + c$

$\dfrac{d}{dx}\left(-\dfrac{1}{2\sin^2 x}\right)$

$= -\dfrac{2\sin^2 x \cdot 0 - 1 \cdot 4\sin x \cos x}{4\sin^4 x}$

$= \dfrac{\cos x}{\sin^3 x}$

(c) $\displaystyle\int \frac{\sin\sqrt{x}}{\sqrt{x}}\, dx = -2\cos\sqrt{x} + c$

$\dfrac{d}{dx}\left(-2\cos\sqrt{3}\right) = 2\sin\sqrt{x} \cdot \dfrac{1}{2\sqrt{x}}$

$= \dfrac{\sin\sqrt{x}}{\sqrt{x}}$

33. Finding the antiderivative,

$f(x) = \dfrac{x^3}{3} + \dfrac{x^2}{2} + c.$

Since $f(0) = 4$, we have

$4 = f(0) = 0 + 0 + c = c.$ Therefore,

$f(x) = \dfrac{x^3}{3} + \dfrac{x^2}{2} + 4.$

35. Finding the antiderivative of f'' gives $f'(x) = 12x + c.$

Since $f'(0) = 2$, we have $2 = f'(0) = c$ and therefore $f'(x) = 12x + 2.$

Finding the antiderivative of $f'(x)$ gives $f(x) = 6x^2 + 2x + c.$

Since $f(0) = 3$, we have $3 = f(0) = c$ and $f(x) = 6x^2 + 2x + 3.$

37. Taking antiderivatives,

$$f''(x) = 3 \sin x + 4x^2$$

$$f'(x) = -3 \cos x + \frac{4}{3}x^3 + c_1$$

$$f(x) = -3 \sin x + \frac{1}{3}x^4 + c_1 x + c_2.$$

39. Taking antiderivatives,

$$f'''(x) = 4 - 2/x^4$$

$$f''(x) = 4x + \frac{2}{3}x^{-3} + c_1$$

$$f'(x) = 2x^2 - \frac{1}{3}x^{-2} + c_1 x + c_2$$

$$f(x) = \frac{2}{3}x^3 + \frac{1}{3}x^{-1} + \frac{c_1}{2}x^2 + c_2 x + c_3$$

41. Position is the antiderivative of velocity,
$s(t) = 3t - 6t^2 + c.$
Since $s(0) = 3$, we have $c = 3$. Thus,
$s(t) = 3t - 6t^2 + 3.$

43. First we find velocity, which is the antiderivative of acceleration,
$v(t) = -3 \cos t + t + c.$
Since $v(0) = 0$ we have $-3 + 0 + c = 0$,
$c = 3$, and $v(t) = -3 \cos t + t + 3$.

Position is the antiderivative of velocity,

$$s(t) = -3 \sin t + \frac{1}{2}t^2 + 3t + c.$$

Since $s(0) = 4$, we have $c = 4$. Thus,

$$s(t) = -3 \sin t + \frac{1}{2}t^2 + 3t + 4.$$

45. The key is to find the velocity and position functions. We start with constant acceleration a, a constant. Then, $v(t) = at + v_0$ where v_0 is the initial velocity. The initial velocity is 30 miles per hour, but we will work in miles per second (30 mph = $\frac{1}{120}$ mi/s). $v(t) = at + \frac{1}{120}.$

We know that the car accelerates to 50 mph ($50\text{mph} = \frac{1}{72}$ mi/s) in 4 seconds, so $v(4) =$

$\frac{1}{72}$. Therefore, $a \cdot 4 + \frac{1}{120} = \frac{1}{72}$ and $a = \frac{1}{720}$ mi/s^2.

So,

$$v(t) = \frac{1}{720}t + \frac{1}{120} \text{ and}$$

$$s(t) = \frac{1}{1440}t^2 + \frac{1}{120}t + s_0$$

where s_0 is the initial position. We can assume the starting position is $s_0 = 0$.

Then, $s(t) = \frac{1}{1440}t^2 + \frac{1}{120}t$ and the distance traveled by the car during the 4 seconds is $s(4) = \frac{2}{45}$ miles ≈ 235 feet.

47. There are many correct answers, but any correct answer will be a vertical shift of this answer.

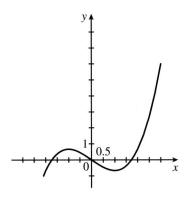

49. All functions that have the derivative shown in exercise 47 are vertical translations of the graph given as the answer for exercise 47.

51. To estimate the acceleration over each interval, we estimate $v'(t)$ by computing the slope of the tangent lines. For example, for the interval $[0, 0.5]$:

$$a \approx \frac{v(0.5) - v(0)}{0.5 - 0} = -31.6 \text{ m/s}^2.$$

Notice, acceleration should be negative since the object is falling.

To estimate the distance traveled over the interval, we estimate the velocity and multiply by the time (distance is rate times time). For an estimate for the velocity, we will use the average of the velocities at the endpoints. For example, for the interval [0, 0.5], the time interval is 0.5 and the velocity is -11.9. Therefore the position changed is $(-11.9)(0.5) = -5.95$ meters. The distance traveled will be 5.95 meters (distance should be positive).

Interval	Acceleration	Distance
[0.0, 0.5]	-31.6	5.95
[0.5, 1.0]	-24.2	12.925
[1.0, 1.5]	-11.6	17.4
[1.5, 2.0]	-3.6	19.3

53. To estimate the speed over the interval, we first approximate the acceleration over the interval by averaging the acceleration at the endpoints of the interval. Then, the velocity will be the acceleration times the length of time. For example, for the interval [0, 0.5] the average acceleration is -0.9 and $v(0.5) = 70 + (-0.9)(0.5) = 69.55$.

And, the distance traveled is the speed times the length of time.

For the time $t = 0.5$, the distance would be $\dfrac{70 + 69.55}{2} \times 0.5 \approx 34.89$ meters.

Time	Speed	Distance
0	70	0
0.5	69.55	34.89
1.0	70.3	69.85
1.5	70.35	105.01
2.0	70.65	140.26

55. We start by taking antiderivatives:
$f'(x) = x^2/2 - x + c_1$
$f(x) = x^3/6 - x^2/2 + c_1 x + c_2$.

Now, we use the data that we are given. We know that $f(1) = 2$ and $f'(1) = 3$, which gives us
$3 = f'(1) = 1/2 - 1 + c_1,$
and
$2 = f(1) = 1/6 - 1/2 + c_1 + c_2.$

Therefore $c_1 = 7/2$ and $c_2 = -7/6$ and the function is
$$f(x) = \frac{x^3}{6} - \frac{x^2}{2} + \frac{7x}{2} - \frac{7}{6}.$$

57. $\dfrac{d}{dx}(\sin x^2) = \cos(x^2) \cdot 2x = 2x \cos x^2$

Therefore, $\displaystyle\int 2x \cos x^2 \, dx = \sin x^2 + c.$

59. $\dfrac{d}{dx}[x^2 \sin 2x]$
$= 2x \sin 2x + x^2 \cos 2x \cdot 2$
$= 2(x \sin 2x + x^2 \cos 2x)$

Therefore,
$$\int \left(x \sin 2x + x^2 \cos 2x \right) dx$$
$$= \frac{1}{2}x^2 \sin 2x + c.$$

61. $\displaystyle\int \frac{x \cos(x^2)}{\sqrt{\sin(x^2)}} dx = \sqrt{\sin(x^2)} + c$

since $\dfrac{d}{dx}\sqrt{\sin x^2} = \dfrac{1}{2}(\sin x^2)^{-1/2} \cos x^2 \cdot 2x.$

63. To derive these formulas, all that needs to be done is to take the derivatives to see that the integrals are correct:

$\dfrac{d}{dx}(\tan x) = \sec^2 x$

$\dfrac{d}{dx}(\sec x) = \sec x \tan x$

4.2 SUMS AND SIGMA NOTATION

1. $\displaystyle\sum_{i=1}^{50} i^2 = \frac{(50)(51)(101)}{6} = 42{,}925$

3. $\displaystyle\sum_{i=1}^{10} \sqrt{i}$

$$= 1 + \sqrt{2} + \sqrt{3} + \sqrt{4} + \sqrt{5} + \sqrt{6}$$
$$+ \sqrt{7} + \sqrt{8} + \sqrt{9} + \sqrt{10}$$
$$\approx 22.47$$

5. $\displaystyle\sum_{i=1}^{6} 3i^2 = 3 + 12 + 27 + 48 + 75 + 108$
$$= 273$$

7. $\displaystyle\sum_{i=6}^{10} (4i + 2)$
$$= (4(6) + 2) + (4(7) + 2) + (4(8) + 2)$$
$$+ (4(9) + 2) + (4(10) + 2)$$
$$= 26 + 30 + 34 + 38 + 42$$
$$= 170$$

9. $\displaystyle\sum_{i=1}^{70} (3i - 1)$
$$= 3 \cdot \sum_{i=1}^{70} i - 70$$
$$= 3 \cdot \frac{70(71)}{2} - 70$$
$$= 7,385$$

11. $\displaystyle\sum_{i=1}^{40} (4 - i^2)$
$$= 160 - \sum_{i=1}^{40} i^2$$
$$= 160 - \frac{(40)(41)(81)}{6}$$
$$= 160 - 22,140$$
$$= -21,980$$

13. $\displaystyle\sum_{i=1}^{100} (i^2 - 3i + 2)$
$$= \sum_{i=1}^{100} i^2 - 3 \cdot \sum_{i=1}^{100} i + 200$$
$$= \frac{(100)(101)(201)}{6} - 3 \cdot \frac{100(101)}{2} + 200$$
$$= 338,350 - 15,150 + 200$$
$$= 323,400$$

15. $\displaystyle\sum_{i=1}^{200} (4 - 3i - i^2)$

$$= 800 - 3 \cdot \sum_{i=1}^{200} i - \sum_{i=1}^{200} i^2$$
$$= 800 - 3 \cdot \frac{200(201)}{2} - \frac{(200)(201)(401)}{6}$$
$$= -2,746,200$$

17. $\displaystyle\sum_{i=3}^{n} (i^2 - 3)$
$$= \sum_{i=3}^{n} i^2 + \sum_{i=0}^{n} (-3)$$
$$= \sum_{i=1}^{n} i^2 - 1^2 - 2^2 + (n + 1)(-3) - 3(-3)$$
$$= \frac{n(n + 1)(2n + 1)}{6} - 3(n + 1) + 4$$

19. $\displaystyle\sum_{i=1}^{n} \frac{1}{n} \left[\left(\frac{i}{n} \right)^2 + 2 \left(\frac{i}{n} \right) \right]$
$$= \frac{1}{n} \left[\sum_{i=1}^{n} \frac{i^2}{n^2} + 2 \sum_{i=1}^{n} \frac{i}{n} \right]$$
$$= \frac{1}{n} \left[\frac{1}{n^2} \sum_{i=1}^{n} i^2 + \frac{2}{n} \sum_{i=1}^{n} i \right]$$
$$= \frac{1}{n} \left[\frac{1}{n^2} \left(\frac{n(n + 1)(2n + 1)}{6} \right) \right.$$
$$\left. + \frac{2}{n} \left(\frac{n(n + 1)}{2} \right) \right]$$
$$= \frac{(n + 1)(2n + 1)}{6n^2} + \frac{n + 1}{n}$$
$$\lim_{n \to \infty} \sum_{i=1}^{n} \frac{1}{n} \left[\left(\frac{i}{n} \right)^2 + 2 \left(\frac{i}{n} \right) \right]$$
$$= \lim_{n \to \infty} \left[\frac{(n + 1)(2n + 1)}{6n^2} + \frac{n + 1}{n} \right]$$
$$= \frac{2}{6} + 1 = \frac{4}{3}$$

21. $\displaystyle\sum_{i=1}^{n} \frac{1}{n} \left[4 \left(\frac{2i}{n} \right)^2 - \left(\frac{2i}{n} \right) \right]$
$$= \frac{1}{n} \left[16 \sum_{i=1}^{n} \frac{i^2}{n^2} - 2 \sum_{i=1}^{n} \frac{i}{n} \right]$$

$$= \frac{1}{n}\left[\frac{16}{n^2}\sum_{i=1}^{n}i^2 - \frac{2}{n}\sum_{i=1}^{n}i\right]$$

$$= \frac{1}{n}\left[\frac{16}{n^2}\left(\frac{n(n+1)(2n+1)}{6}\right)\right.$$

$$\left. - \frac{2}{n}\left(\frac{n(n+1)}{2}\right)\right]$$

$$= \frac{8(n+1)(2n+1)}{3n^2} - \frac{n+1}{n}$$

$$\lim_{n\to\infty}\sum_{i=1}^{n}\frac{1}{n}\left[4\left(\frac{2i}{n}\right)^2 - \left(\frac{2i}{n}\right)\right]$$

$$= \lim_{n\to\infty}\left[\frac{8(n+1)(2n+1)}{3n^2} - \frac{n+1}{n}\right]$$

$$= \frac{16}{3} - 1 = \frac{13}{3}$$

23. $\displaystyle\sum_{i=1}^{n}f(x_i)\Delta x$

$$= \sum_{i=1}^{5}(x_i^2 + 4x_i)\cdot 0.2$$

$$= (0.2^2 + 4(0.2))(0.2) + \ldots$$
$$+ (1^2 + 4)(0.2)$$
$$= (0.84)(0.2) + (1.76)(0.2)$$
$$+ (2.76)(0.2) + (3.84)(0.2)$$
$$+ (5)(0.2)$$
$$= 2.84$$

25. $\displaystyle\sum_{i=1}^{n}f(x_i)\Delta x$

$$= \sum_{i=1}^{10}(4x_i^2 - 2)\cdot 0.1$$

$$= (4(2.1)^2 - 2)(0.1) + \ldots$$
$$+ (4(3)^2 - 2)(0.1)$$
$$= (15.64)(0.1) + (17.36)(0.1)$$
$$+ (19.16)(0.1) + (21.04)(0.1)$$
$$+ (23)(0.1) + (25.04)(0.1)$$
$$+ (27.16)(0.1) + (29.36)(0.1)$$
$$+ (31.64)(0.1) + (34)(0.1)$$
$$= 24.34$$

27. Distance
$$= 50(2) + 60(1) + 70(1/2) + 60(3)$$
$$= 375 \text{ miles.}$$

29. Remember to convert minutes into hours.

Distance
$$= 15\left(\frac{1}{3}\right) + 18\left(\frac{1}{2}\right) + 16\left(\frac{1}{6}\right)$$

$$+ 12\left(\frac{2}{3}\right)$$

$$= 24\frac{2}{3} \text{ miles.}$$

31. On the time interval [0, 0.25], the estimated velocity is the average velocity
$$\frac{120 + 116}{2} = 118 \text{ feet per second. We esti-}$$
mate the distance traveled during the time interval [0, 0.25] to be $(118)(0.25 - 0) = 29.5$ feet.

Altogether, the distance traveled is estimated as
$$(236/2)(0.25) + (229/2)(0.25)$$
$$+ (223/2)(0.25) + (218/2)(0.25)$$
$$+ (214/2)(0.25) + (210/2)(0.25)$$
$$+ (207/2)(0.25) + (205/2)(0.25)$$
$$= 217.75 \text{ feet.}$$

33. We want to prove that
$$\sum_{i=1}^{n}i^3 = \frac{n^2(n+1)^2}{4}$$

is true for all integers $n \geq 1$.

For $n = 1$, we have
$$\sum_{i=1}^{1}i^3 = 1 = \frac{1^2(1+1)^2}{4},$$

as desired. So the proposition is true for $n = 1$.

Next, assume that
$$\sum_{i=1}^{k}i^3 = \frac{k^2(k+1)^2}{4},$$

for some integer $k \geq 1$.

In this case, we have by the induction assumption that for $n = k + 1$,

$$\sum_{i=1}^{n} i^3 = \sum_{i=1}^{k+1} i^3 = \sum_{i=1}^{k} i^3 + (k+1)^3$$

$$= \frac{k^2(k+1)^2}{4} + (k+1)^3$$

$$= \frac{k^2(k+1)^2 + 4(k+1)^3}{4}$$

$$= \frac{(k+1)^2(k^2 + 4k + 4)}{4}$$

$$= \frac{(k+1)^2(k+2)^2}{4}$$

$$= \frac{n^2(n+1)^2}{4}$$

as desired.

35. $\displaystyle\sum_{i=1}^{10} (i^3 - 3i + 1)$

$$= \sum_{i=1}^{10} i^3 - 3\sum_{i=1}^{10} i + 10$$

$$= \frac{100(11)^2}{4} - 3\frac{10(11)}{2} + 10$$

$$= 2870$$

37. $\displaystyle\sum_{i=1}^{100} (i^5 - 2i^2)$

$$= \sum_{i=1}^{100} i^5 - 2\sum_{i=1}^{100} i^2$$

$$= \frac{(100^2)(101^2)[2(100^2) + 2(100) - 1]}{12}$$

$$- 2\frac{100(101)(201)}{6}$$

$$= 171,707,655,800$$

39. $\displaystyle\sum_{i=1}^{n} (ca_i + db_i) = \sum_{i=1}^{n} ca_i + \sum_{i=1}^{n} db_i$

$$= c\sum_{i=1}^{n} a_i + d\sum_{i=1}^{n} b_i$$

41. $\displaystyle\sum_{i=1}^{n} 3\left(\frac{1}{4}\right)^i = \frac{\frac{3}{4} - \frac{3}{4}(\frac{1}{4})^n}{1 - \frac{1}{4}} = 1 - \left(\frac{1}{4}\right)^n$

$$\lim_{n\to\infty}\left[1 - \left(\frac{1}{4}\right)^n\right] = 1 - 0 = 1$$

4.3 AREA

1. **(a)** Evaluation points:
0.125, 0.375, 0.625, 0.875.
Notice that $\Delta x = 0.25$.

$$A_4 = [f(0.125) + f(0.375) + f(0.625)$$
$$+ f(0.875)](0.25)$$
$$= [(0.125)^2 + 1 + (0.375)^2 + 1$$
$$+ (0.625)^2 + 1 + (0.875)^2 + 1](0.25)$$
$$= 1.38125.$$

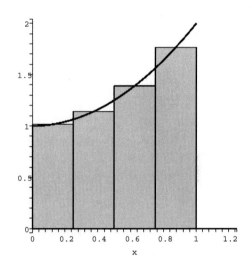

(b) Evaluation points:
0.25, 0.75, 1.25, 1.75.
Notice that $\Delta x = 0.5$.
$$A_4 = [f(0.25) + f(0.75) + f(1.25)$$
$$+ f(1.75)](0.5)$$
$$= [(0.25)^2 + 1 + (0.75)^2 + 1 + (1.25)^2 +$$
$$1$$
$$+ (1.75)^2 + 1](0.5)$$
$$= 4.625.$$

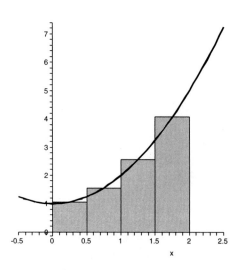

$+ \sin(11\pi/16) + \sin(13\pi/16)$
$+ \sin(15\pi/16)](\pi/8)$
$= 2.0129.$

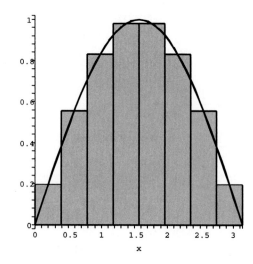

3. **(a)** Evaluation points:
$\pi/8,\ 3\pi/8,\ 5\pi/8,\ 7\pi/8.$
Notice that $\Delta x = \pi/4$.

$A_4 = [f(\pi/8) + f(3\pi/8) + f(5\pi/8)$
$+ f(7\pi/8)](\pi/4)$
$= [\sin(\pi/8) + \sin(3\pi/8) + \sin(5\pi/8)$
$+ \sin(7\pi/8)](\pi/4)$
$= 2.05234.$

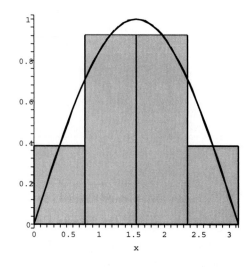

(b) Evaluation points:
$\pi/16,\ 3\pi/16,\ 5\pi/16,\ 7\pi/16,\ 9\pi/16,$
$11\pi/16,\ 13\pi/16,\ 15\pi/16.$
Notice that $\Delta x = \pi/8$.

$A_4 = [f(\pi/16) + f(3\pi/16) + f(5\pi/16)$
$+ f(7\pi/16) + f(9\pi/16) + f(11\pi/16)$
$+ f(13\pi/16) + f(15\pi/16)](\pi/8)$
$= [\sin(\pi/16) + \sin(3\pi/16) + \sin(5\pi/16)$
$+ \sin(7\pi/16) + \sin(9\pi/16)$

5. **(a)** There are 16 rectangles and the evaluation points are given by $c_i = i\,\Delta x$ where i is from 0 to 15.

$$A_{16} = \Delta x \sum_{i=0}^{15} f(c_i)$$

$$= \frac{1}{16} \sum_{i=0}^{15} \left[\left(\frac{i}{16} \right)^2 + 1 \right] \approx 1.3027$$

(b) There are 16 rectangles and the evaluation points are given by $c_i = i\,\Delta x + \frac{\Delta x}{2}$ where i is from 0 to 15.

$$A_{16} = \Delta x \sum_{i=0}^{15} f(c_i)$$

$$= \frac{1}{16} \sum_{i=0}^{15} \left[\left(\frac{i}{16} + \frac{1}{32} \right)^2 + 1 \right]$$

$$\approx 1.3330$$

(c) There are 16 rectangles and the evaluation points are given by $c_i = i\,\Delta x + \Delta x$ where i is from 0 to 15.

$$A_{16} = \Delta x \sum_{i=0}^{15} f(c_i)$$

$$= \frac{1}{16} \sum_{i=0}^{15} \left[\left(\frac{i}{16} + \frac{1}{16} \right)^2 + 1 \right]$$

$$\approx 1.3652$$

7. **(a)** There are 16 rectangles and the evaluation points are the left endpoints which are given by $c_i = 1 + i\,\Delta x$ where i is from 0 to 15.

$$A_{16} = \Delta x \sum_{i=0}^{15} f(c_i)$$

$$= \frac{3}{16} \sum_{i=0}^{15} \sqrt{1 + \frac{3i}{16}} + 2 \approx 6.2663$$

(b) There are 16 rectangles and the evaluation points are the midpoints which are given by $c_i = 1 + i\,\Delta x + \Delta x/2$ where i is from 0 to 15.

$$A_{16} = \Delta x \sum_{i=0}^{15} f(c_i)$$

$$= \frac{3}{16} \sum_{i=0}^{15} \sqrt{1 + \frac{3i}{16} + \frac{3}{32}} + 2$$

$$\approx 6.3340$$

(c) There are 16 rectangles and the evaluation points are the right endpoints which are given by $c_i = 1 + i\,\Delta x$ where i is from 1 to 16.

$$A_{16} = \Delta x \sum_{i=1}^{16} f(c_i)$$

$$= \frac{3}{16} \sum_{i=1}^{16} \sqrt{1 + \frac{3i}{16}} + 2$$

$$\approx 6.4009$$

9. **(a)** There are 50 rectangles and the evaluation points are given by $c_i = i\,\Delta x$ where i is from 0 to 49.

$$A_{50} = \Delta x \sum_{i=0}^{50} f(c_i)$$

$$= \frac{\pi}{100} \sum_{i=0}^{50} \cos\left(\frac{\pi i}{100}\right) \approx 1.0156$$

(b) There are 50 rectangles and the evaluation points are given by $c_i = \frac{\Delta x}{2} + i\,\Delta x$ where i is from 0 to 49.

$$A_{50} = \Delta x \sum_{i=0}^{50} f(c_i)$$

$$= \frac{\pi}{100} \sum_{i=0}^{50} \cos\left(\frac{\pi}{200} + \frac{\pi i}{100}\right)$$

$$\approx 1.00004$$

(c) There are 50 rectangles and the evaluation points are given by $c_i = \Delta x + i\,\Delta x$ where i is from 0 to 49.

$$A_{50} = \Delta x \sum_{i=0}^{50} f(c_i)$$

$$= \frac{\pi}{100} \sum_{i=0}^{50} \cos\left(\frac{\pi}{100} + \frac{\pi i}{100}\right)$$

$$\approx 0.9842$$

11.

n	Left Endpoint	Midpoint	Right Endpoint
10	10.56	10.72	10.56
50	10.662	10.669	10.662
100	10.6656	10.6672	10.6656
500	10.6666	10.6667	10.6666
1000	10.6667	10.6667	10.6667
5000	10.6667	10.6667	10.6667

13.

n	Left Endpoint	Midpoint	Right Endpoint
10	15.48000	17.96000	20.68000
50	17.4832	17.9984	18.5232
100	17.7408	17.9996	18.2608
500	17.9480	17.9999	18.0520
1000	17.9740	17.9999	18.0260
5000	17.9948	17.9999	18.0052

15. $\Delta x = \frac{1}{n}$. We will use right endpoints as evaluation points, $x_i = \frac{i}{n}$.

$$A_n = \sum_{i=1}^{n} f(x_i)\,\Delta x$$

$$= \frac{1}{n} \sum_{i=1}^{n} \left[\left(\frac{i}{n} \right)^2 + 1 \right]$$

$$= \frac{1}{n^3} \sum_{i=1}^{n} i^2 + 1$$

$$= \frac{1}{n^3} \left(\frac{n(n+1)(2n+1)}{6} \right) + 1$$

$$= \frac{8n^2 + 3n + 1}{6n^2}$$

Now to compute the exact area, we take the limit as $n \to \infty$:

$$A = \lim_{n \to \infty} A_n$$

$$= \lim_{n \to \infty} \frac{8n^2 + 3n + 1}{6n^2}$$

$$= \lim_{n \to \infty} \left[\frac{8}{6} + \frac{3}{6n} + \frac{1}{6n^2} \right]$$

$$= \frac{4}{3}$$

17. $\Delta x = \frac{2}{n}$. We will use right endpoints as evaluation points, $x_i = 1 + \frac{2i}{n}$.

$$A_n = \sum_{i=1}^{n} f(x_i) \Delta x$$

$$= \frac{2}{n} \sum_{i=1}^{n} 2 \left(1 + \frac{2i}{n} \right)^2 + 1$$

$$= \frac{2}{n} \sum_{i=1}^{n} \left(\frac{8i^2}{n^2} + \frac{8i}{n} + 3 \right)$$

$$= \frac{16}{n^3} \sum_{i=1}^{n} i^2 + \frac{16}{n^2} \sum_{i=1}^{n} i + 6$$

$$= \frac{16}{n^3} \left(\frac{n(n+1)(2n+1)}{6} \right)$$

$$\quad + \frac{16}{n^2} \left(\frac{n(n+1)}{2} \right) + 6$$

$$= \frac{16n(n+1)(2n+1)}{6n^3} + \frac{16n(n+1)}{2n^2} + 6$$

Now to compute the exact area, we take the limit as $n \to \infty$:

$$A = \lim_{n \to \infty} A_n$$

$$= \lim_{n \to \infty} \left(\frac{16n(n+1)(2n+1)}{6n^3} + \frac{16n(n+1)}{2n^2} + 6 \right)$$

$$= \frac{32}{6} + \frac{16}{2} + 6$$

$$= 19 \frac{1}{3}$$

19. Using left-hand endpoints:
$L_8 = [f(0.0) + f(0.1) + f(0.2) + f(0.3) + f(0.4) + f(0.5) + f(0.6) + f(0.7)](0.1)$
$= (2.0 + 2.4 + 2.6 + 2.7 + 2.6 + 2.4 + 2.0 + 1.4)(0.1) = 1.81$.

Right endpoints:
$R_8 = [f(0.1) + f(0.2) + f(0.3) + f(0.4) + f(0.5) + f(0.6) + f(0.7) + f(0.8)](0.1)$
$= (2.4 + 2.6 + 2.7 + 2.6 + 2.4 + 2.0 + 1.4 + 0.6)(0.1) = 1.67$.

21. Using left-hand endpoints:
$L_8 = [f(1.0) + f(1.1) + f(1.2) + f(1.3) + f(1.4) + f(1.5) + f(1.6) + f(1.7)](0.1)$
$= (1.8 + 1.4 + 1.1 + 0.7 + 1.2 + 1.4 + 1.8 + 2.4)(0.1) = 1.18$.

Right endpoints:
$R_8 = [f(1.1) + f(1.2) + f(1.3) + f(1.4) + f(1.5) + f(1.6) + f(1.7) + f(1.8)](0.1)$
$= (1.4 + 1.1 + 0.7 + 1.2 + 1.4 + 1.8 + 2.4 + 2.6)(0.1) = 1.26$.

23. Let L, M, and R be the values of the Riemann sums with left endpoints, midpoints, and right endpoints. Let A be the area under the curve. Then: $L < M < A < R$.

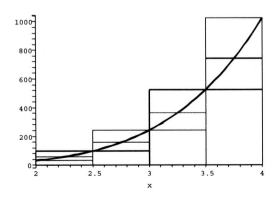

25. Let L, M, and R be the values of the Riemann sums with left endpoints, midpoints, and right endpoints. Let A be the area under the curve. Then: $R < A < M < L$.

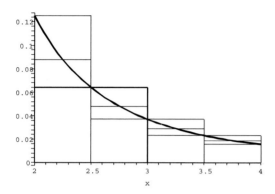

27. There are many possible answers here. One possibility is to use $x = 1/\sqrt{12}$ on $[0, 0.5]$ and $x = \sqrt{7/12}$ on $[0.5, 1]$.

29. We subdivide the interval $[a, b]$ into n equal subintervals. If you are located at $a + (b - a)/n$ (the first right endpoint), then each step of distance Δx takes you to a new right endpoint. To arrive at the ith right endpoint, you have to take $(i - 1)$ steps to the right of distance Δx. Therefore,
$c_i = a + (b - a)/n + (i - 1)\Delta x = a + i\,\Delta x.$

31. We subdivide the interval $[a, b]$ into n equal subintervals. The first evaluation point is $a + \Delta x/2$. From this evaluation point, each step of distance Δx takes you to a new evaluation point. To arrive at the ith evaluation point, you have to take $(i - 1)$ steps to the right of distance Δx. Therefore,
$c_i = a + \Delta x/2 + (i - 1)\Delta x$
$= a + (i - 1/2)\,\Delta x$, for $i = 1, \ldots, n.$

33. $A \approx (0.2 - 0.1)(0.002) + (0.3 - 0.2)(0.004) + (0.4 - 0.3)(0.008) + (0.5 - 0.4)(0.014) + (0.6 - 0.5)(0.026) + (0.7 - 0.6)(0.048) + (0.8 - 0.7)(0.085) + (0.9 - 0.8)(0.144) + (0.95 - 0.9)(0.265) + (0.98 - 0.95)(0.398) + (0.99 - 0.98)(0.568) + (1 - 0.99)(0.736) + 1/2 \cdot [(0.1 - 0)(0.002) + (0.2 - 0.1)(0.004 - 0.002) + (0.3 - 0.2)(0.008 - 0.004) + (0.4 - 0.3)(0.014 - 0.008) + (0.5 - 0.4)$

$(0.026 - 0.014) + (0.6 - 0.5)(0.048 - 0.026) + (0.7 - 0.6)(0.085 - 0.048) + (0.8 - 0.7)(0.144 - 0.085) + (0.9 - 0.8)(0.265 - 0.144) + (0.95 - 0.9)(0.398 - 0.265) + (0.98 - 0.95)(0.568 - 0.398) + (0.99 - 0.98)(0.736 - 0.568)\ (1 - 0.99)(1 - 0.736)]$
≈ 0.092615

The Lorentz curve looks like:

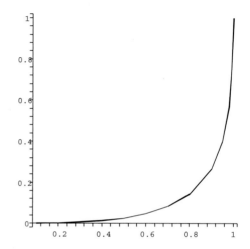

35. $U_4 = \dfrac{2}{4} \displaystyle\sum_{i=1}^{4} \left(\dfrac{i}{2}\right)^2$

$= \dfrac{1}{8} \displaystyle\sum_{i=1}^{4} i^2$

$= \dfrac{1}{8} \left[1^2 + 2^2 + 3^2 + 4^2\right]$

$= \dfrac{30}{8} = 3.75$

$L_4 = \dfrac{2}{4} \displaystyle\sum_{i=1}^{4} \left(\dfrac{i-1}{2}\right)^2$

$= \dfrac{1}{8} \displaystyle\sum_{i=1}^{4} i^2$

$= \dfrac{1}{8} \left[0^2 + 1^2 + 2^2 + 3^2\right]$

$= \dfrac{14}{8} = 1.75$

37. (a) $U_n = \dfrac{2}{n} \displaystyle\sum_{i=1}^{n} \left(\dfrac{2i}{n}\right)^2$

$$= \left(\frac{2}{n}\right)^3 \sum_{i=1}^{n} i^2$$

$$= \left(\frac{2}{n}\right)^3 \frac{n(n+1)(2n+1)}{6}$$

$$= \frac{4}{3}\frac{n(n+1)(2n+1)}{n^3}$$

$$= \frac{4}{3}\left(1+\frac{1}{n}\right)\left(2+\frac{1}{n}\right)$$

$$\lim_{n\to\infty} U_n = \frac{4}{3}(2) = \frac{8}{3}$$

(b) $L_n = \dfrac{2}{n}\displaystyle\sum_{i=1}^{n}\left(\dfrac{2(i-1)}{n}\right)^2$

$$= \left(\frac{2}{n}\right)^3 \sum_{i=1}^{n} (i-1)^2$$

$$= \left(\frac{2}{n}\right)^3 \sum_{i=1}^{n-1} i^2$$

$$= \left(\frac{2}{n}\right)^3 \frac{(n-1)(n)(2n-1)}{6}$$

$$= \frac{4}{3}\frac{(n-1)(n)(2n-1)}{n^3}$$

$$= \frac{4}{3}\left(1-\frac{1}{n}\right)\left(2-\frac{1}{n}\right)$$

$$\lim_{n\to\infty} L_n = \frac{4}{3}(2) = \frac{8}{3}$$

39. **(a)** $U_n = \dfrac{2}{n}\displaystyle\sum_{i=1}^{n}\left[\left(0+\dfrac{2}{n}i\right)^3+1\right]$

$$= \frac{2}{n}\sum_{i=1}^{n}\left[\left(\frac{2i}{n}\right)^3+1\right]$$

$$= \left(\frac{2}{n}\right)^4\sum_{i=1}^{n} i^3 + \frac{2}{n}\sum_{i=1}^{n} 1$$

$$= \frac{2^4}{n^4}\left[\frac{n^2(n+1)^2}{4}\right] + \frac{2}{n}(n)$$

$$= \frac{4(n+1)^2}{n^2} + 2$$

$$= \frac{4(n^2+2n+1)}{n^2} + 2$$

$$= 4\left(1+\frac{2}{n}+\frac{1}{n^2}\right) + 2$$

$$= 6 + \frac{8}{n} + \frac{4}{n^2}$$

$$\lim_{n\to\infty} U_n = 6$$

(b) $L_n = \dfrac{2}{n}\displaystyle\sum_{i=0}^{n-1}\left[\left(0+\dfrac{2}{n}i\right)^3+1\right]$

$$= \frac{2}{n}\sum_{i=0}^{n-1}\left[\left(\frac{2i}{n}\right)^3+1\right]$$

$$= \left(\frac{2}{n}\right)^4\sum_{i=0}^{n-1} i^3 + \frac{2}{n}\sum_{i=1}^{n} 1$$

$$= \frac{2^4}{n^4}\left[\frac{(n-1)^2 n^2}{4}\right] + \frac{2}{n}(n)$$

$$= \frac{4(n-1)^2}{n^2} + 2$$

$$= \frac{4(n^2-2n+1)}{n^2} + 2$$

$$= 4\left(1-\frac{2}{n}+\frac{1}{n^2}\right) + 2$$

$$= 6 - \frac{8}{n} + \frac{4}{n^2}$$

$$\lim_{n\to\infty} L_n = 6$$

4.4 THE DEFINITE INTEGRAL

1. $\displaystyle\int_0^3 (x^3+x)\,dx$

$$= \sum_{i=1}^{n}(c_i^3+c_i)\Delta x = \sum_{i=1}^{n}(c_i^3+c_i)\cdot\frac{3}{n},$$

$$c_i = \frac{x_i+x_{i-1}}{2},\ x_i = \frac{3i}{n}$$

Riemann sum ≈ 24.75

3. $\displaystyle\int_0^{\pi} \sin x^2\,dx$

$$= \sum_{i=1}^{n}\sin c_i^2\,\Delta x = \sum_{i=1}^{n}\sin c_i^2\left(\frac{\pi}{n}\right),$$

$c_i = \dfrac{x_i + x_{i-1}}{2}$, $x_i = \dfrac{i\pi}{n}$

Riemann sum ≈ 0.77

5. For n rectangles, $\Delta x = \dfrac{1}{n}$, $x_i = i\,\Delta x$.

$$R_n = \sum_{i=1}^{n} f(x_i)\,\Delta x$$

$$= \sum_{i=1}^{n} 2x_i\,\Delta x = \frac{1}{n}\sum_{i=1}^{n} 2\left(\frac{i}{n}\right)$$

$$= \frac{2}{n^2}\sum_{i=1}^{n} i = \frac{2}{n^2}\left(\frac{n(n+1)}{2}\right)$$

$$= \frac{(n+1)}{n}$$

To compute the value of the integral, we take the limit as $n \to \infty$,

$$\int_0^1 2x\,dx = \lim_{n\to\infty} R_n$$

$$= \lim_{n\to\infty} \frac{(n+1)}{n} = 1.$$

7. For n rectangles, $\Delta x = 2/n$,
$x_i = i\,\Delta x = 2i/n$.

$$R_n = \sum_{i=1}^{n} f(x_i)\,\Delta x$$

$$= \sum_{i=1}^{n}(x_i^2)\,\Delta x = \frac{2}{n}\sum_{i=1}^{n} 2\left(\frac{2i}{n}\right)^2$$

$$= \frac{2}{n}\sum_{i=1}^{n}\frac{4i^2}{n^2} = \frac{8}{n^3}\sum_{i=1}^{n} i^2$$

$$= \frac{8}{n^3}\left(\frac{n(n+1)(2n+1)}{6}\right)$$

$$= \frac{4(n+1)(2n+1)}{3n^2}$$

To compute the value of the integral, we take the limit as $n \to \infty$,

$$\int_0^2 x^2\,dx = \lim_{n\to\infty} R_n$$

$$= \lim_{n\to\infty} \frac{4(n+1)(2n+1)}{3n^2} = \frac{8}{3}.$$

9. For n rectangles, $\Delta x = 2/n$,
$x_i = 1 + i\,\Delta x = 1 + 2i/n$.

$$R_n = \sum_{i=1}^{n} f(x_i)\,\Delta x$$

$$= \sum_{i=1}^{n}(x_i^2 - 3)\,\Delta x$$

$$= \frac{2}{n}\sum_{i=1}^{n}\left[\left(1 + \frac{2i}{n}\right)^2 - 3\right]$$

$$= \sum_{i=1}^{n}\left(\frac{8i}{n^2} + \frac{8i^2}{n^3} - \frac{4}{n}\right)$$

$$= \frac{8n(n+1)}{2n^2} + \frac{8n(n+1)(2n+1)}{6n^3} - 4$$

To compute the value of the integral, we take the limit as $n \to \infty$,

$$\int_1^3 (x^2 - 3)\,dx = \lim_{n\to\infty} R_n$$

$$= \frac{8}{2} + \frac{16}{6} - 4 = \frac{8}{3}.$$

11. Notice that the graph of $y = 4 - x^2$ is above the x-axis between $x = -2$ and $x = 2$:

$$\int_{-2}^2 (4 - x^2)\,dx.$$

13. Notice that the graph of $y = x^2 - 4$ is below the x-axis between $x = -2$ and $x = 2$. Since we are asked for area and the area in question is below the x-axis, we have to be a bit careful.

$$\int_{-2}^2 [-(x^2 - 4)]\,dx$$

15. $\displaystyle\int_0^\pi \sin x\,dx$

17. $\left|\displaystyle\int_0^1 (x^3 - 3x^2 + 2x)\,dx\right|$

$+ \left|\displaystyle\int_1^2 (x^3 - 3x^2 + 2x)\,dx\right|$

$= \displaystyle\int_0^1 (x^3 - 3x^2 + 2x)\,dx$

$$-\int_1^2 (x^3 - 3x^2 + 2x)\, dx$$

19. The total distance is the total area under the curve whereas the total displacement is the signed area under the curve. In this case, from $t = 0$ to $t = 4$, the function is always positive so the total distance is equal to the total displacement. This means we want to compute the definite integral $\int_0^4 \frac{1}{\sqrt{t^2+1}}\, dt$. The right-hand sum for $n = 1000$ is approximately 2.0932.

So, the total distance traveled is approximately 2.1 and the final position is

$$s(b) \approx s(0) + 2.1 = 0 + 2.1 = 2.1.$$

21. $\displaystyle\int_0^2 f(x)\, dx + \int_2^3 f(x)\, dx$

$$= \int_0^3 f(x)\, dx$$

23. $\displaystyle\int_0^2 f(x)\, dx + \int_2^1 f(x)\, dx$

$$= \int_0^2 f(x)\, dx - \int_1^2 f(x)\, dx$$

$$= \int_0^1 f(x)\, dx$$

25.

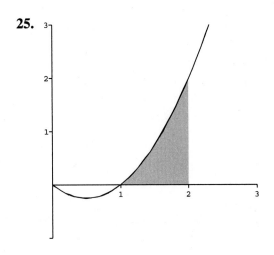

27.

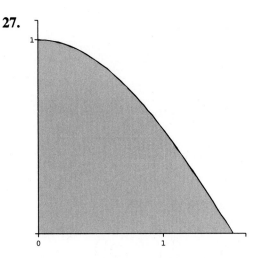

29. The function $f(x) = 3\cos x^2$ is decreasing on $[\pi/3, \pi/2]$. Therefore, on this interval, the maximum occurs at the left endpoint and is $f(\pi/3) = 3\cos(\pi^2/9)$. The minimum occurs at the right endpoint and is $f(\pi/2) = 3\cos(\pi^2/4)$.

Using these to estimate the value of the integral gives the following inequality:

$$\frac{\pi}{6} \cdot \left(3\cos\frac{\pi^2}{4}\right) \le \int_{\pi/3}^{\pi/2} 3\cos x^2\, dx$$

$$\le \frac{\pi}{6} \cdot \left(3\cos\frac{\pi^2}{9}\right)$$

$$-1.23 \le \int_{\pi/3}^{\pi/2} 3\cos x^2\, dx \le 0.72$$

31. The function $f(x) = \sqrt{2x^2 + 1}$ is increasing on $[0, 2]$. Therefore, on this interval, the maximum occurs at the right endpoint and is $f(2) = 3$. The minimum occurs at the left endpoint and is $f(0) = 1$.

Using these to estimate the value of the integral gives the following inequality:

$$(2)(1) \le \int_0^2 \sqrt{x^2 + 1}\, dx \le (2)(3)$$

$$2 \le \int_0^2 \sqrt{x^2 + 1}\, dx \le 6$$

33. We are looking for a value c, such that

$$f(c) = \frac{1}{2 - 0}\int_0^2 3x^2\, dx.$$

Since $\int_0^2 3x^2\,dx = 8$, we want to find c so that

$f(c) = 4$ or, $3c^2 = 4$.

Solving this equation gives $c = \pm\dfrac{2}{\sqrt{3}}$

We are interested in the value that is in the interval $[0, 2]$, so $c = \dfrac{2}{\sqrt{3}}$.

35. $f_{ave} = \dfrac{1}{4}\int_0^4 (2x+1)\,dx$

$= \dfrac{1}{4}\lim_{n\to\infty}\sum_{i=1}^{n}\dfrac{4}{n}\left(\dfrac{8i}{n}+1\right)$

$= \lim_{n\to\infty}\sum_{i=1}^{n}\left(\dfrac{8n(n+1)}{2n^2}+1\right)$

$= 4+1 = 5$

37. $f_{ave} = \dfrac{1}{3-1}\int_1^3 (x^2-1)\,dx$

$= \dfrac{1}{2}\lim_{n\to\infty}\sum_{i=1}^{n}\dfrac{2}{n}\left[\left(1+\dfrac{2i}{n}\right)^2 - 1\right]$

$= \lim_{n\to\infty}\sum_{i=1}^{n}\dfrac{1}{n}\left(\dfrac{4i}{n}+\dfrac{4i^2}{n^2}\right)$

$= \lim_{n\to\infty}\left(\dfrac{4n(n+1)}{2n^2}+\dfrac{4n(n+1)(2n+1)}{6n^3}\right)$

$= 2+\dfrac{4}{3} = \dfrac{10}{3}$

39. This is just a restatement of the Integral Mean Value Theorem.

41. Between $x = 0$ and $x = 2$, the area below the x-axis is much less than the area above the x-axis. Therefore,

$\int_0^2 f(x)\,dx > 0.$

43. Between $x = 0$ and $x = 2$, the area below the x-axis is slightly greater than the area above the x-axis. Therefore,

$\int_0^2 f(x)\,dx < 0.$

45. Imagine the interval $[0, 2]$ is divided into n subintervals. If n is even, then the point $x = 1$ must be one of the boundary points. If we take the midpoint evaluations to approximate Riemann sums for $\int_0^2 f(x)\,dx$ and $\int_0^2 g(x)\,dx$, all the values $f(c_i)$ and $g(c_i)$ are going to be exactly the same for same index number i, since the only difference between $f(x)$ and $g(x)$ occurs at $x = 1$, and $x = 1$ is never going to be one of the c_i's. Thus the approximated Riemann sums for $\int_0^2 f(x)\,dx$ and $\int_0^2 g(x)\,dx$ are going to be the same.

47. $\int_0^4 f(x)\,dx$

$= \int_0^1 f(x)\,dx + \int_1^4 f(x)\,dx$

$= \int_0^1 2x\,dx + \int_1^4 4\,dx$

$\int_0^1 2x\,dx$ is the area of a triangle with base 1 and height 2 and therefore has area $\frac{1}{2}(1)(2) = 1$.

$\int_1^4 4\,dx$ is the area of a rectangle with base 3 and height 4 and therefore has area $(3)(4) = 12$.

Therefore,

$\int_0^4 f(x)\,dx = 1+12 = 13.$

49. Since $b(t)$ represents the birthrate (in births per month), the total number of births from time $t = 0$ to $t = 12$ is given by the integral $\int_0^{12} b(t)\,dt$.

Similarly, the total number of deaths from time $t = 0$ to $t = 12$ is given by the integral $\int_0^{12} a(t)\,dt$.

Of course, the net change in population is the number of birth minus the number of deaths:

Population Change
= Births − Deaths

$$= \int_0^{12} b(t)\, dt - \int_0^{12} a(t)\, dt$$

$$= \int_0^{12} [b(t) - a(t)]\, dt.$$

Next we solve the inequality
$410 - 0.3t > 390 + 0.2t$
$20 > 0.5t$ then $t < 40$ months.

Therefore $b(t) > a(t)$ when $t < 40$ months. The population is increasing when the birthrate is greater than the death rate, which is during the first 40 months. After 40 months, the population is decreasing. The population would reach a maximum at $t = 40$ months.

51. From $PV = 10$ we get $P(V) = 10/V$. By definition,

$$\int_2^4 P(V)\, dV = \int_2^4 \frac{10}{V}\, dV$$

$$= \lim_{n \to \infty} \sum_{i=1}^n \frac{2}{n} \cdot \frac{10}{2 + \frac{2i}{n}}.$$

An estimate of the value of this integral for $n = 100$ is 6.93.

53. $\dfrac{1}{2 - 0} \displaystyle\int_0^2 f(x)\, dx = 5$

$$\int_0^2 f(x)\, dx = 10$$

and

$$\frac{1}{6 - 2} \int_2^6 f(x)\, dx = 11$$

$$\int_2^6 f(x)\, dx = 44.$$

The average value of f over $[0, 6]$ is

$$\frac{1}{6 - 0} \int_0^6 f(x)\, dx$$

$$= \frac{1}{6} \left(\int_0^2 f(x)\, dx + \int_2^6 f(x)\, dx \right)$$

$$= \frac{1}{6}(10 + 44) = 9.$$

55. $\displaystyle\int_0^2 3x\, dx = \frac{1}{2}bh = \frac{1}{2}(2)(6) = 6$

57. $\displaystyle\int_0^2 \sqrt{4 - x^2} = \frac{1}{4}\pi r^2 = \frac{1}{4}\pi \left(2^2\right) = \pi$

59. (a) Average temperature

$$= \frac{1}{24}[3(44) + 3(52) + 3(70) + 3(82)$$

$$+ 3(86) + 3(80) + 3(72) + 3(56)\,]$$

$$= \frac{3}{24}[44 + 52 + 70 + 82 + 86 + 80$$

$$+ 72 + 56\,]$$

$$= \frac{542}{8} = 67.75.$$

(b) Average temperature

$$= \frac{1}{24}[3(46) + 3(44) + 3(52) + 3(70)$$

$$+ 3(82) + 3(86) + 3(80) + 3(72)\,]$$

$$= \frac{3}{24}[46 + 44 + 52 + 70 + 82 + 86$$

$$+ 80 + 72\,]$$

$$= \frac{1}{8}[532] = 66.5.$$

61. Since r is the rate at which items are shipped, rt is the number of items shipped between time 0 and time t. Therefore, $Q - rt$ is the number of items remaining in inventory at time t. Since $Q - rt = 0$ when $t = Q/r$, the formula is valid for $0 \le t \le Q/r$. The average value of $f(t) = Q - rt$ on the time interval $[0, Q/r]$ is

$$\frac{1}{Q/r - 0} \int_0^{Q/r} f(t)\, dt$$

$$= \frac{r}{Q} \int_0^{Q/r} (Q - rt)\, dt$$

$$= \frac{r}{Q} \left(\frac{1}{2}Q \cdot \frac{Q}{r} \right)$$

$$= \frac{r}{Q} \left[\frac{Q^2}{2r} \right] = \frac{Q}{2},$$

since the area under the graph on the interval is a triangle with base $\frac{Q}{2}$ and altitude Q.

63. Delivery is completed in time Q/p, and since in that time Qr/p items are shipped, the inventory when delivery is completed is

$$Q - \frac{Qr}{p} = Q\left(1 - \frac{r}{p}\right).$$

The inventory at any time is given by

$$g(t) = \begin{cases} (p - r)t & \text{for} \quad t \in \left[0, \frac{Q}{p}\right] \\ Q - rt & \text{for} \quad t \in \left[\frac{Q}{p}, \frac{Q}{r}\right] \end{cases}.$$

The graph of g has two linear pieces. The average value of g over the interval $[0, Q/r]$ is the area under the graph (which is the area of a triangle of base Q/r and height $Q(1 - r/p)$) divided by the length of the interval (which is the base of the triangle). Thus the average value of the function is $(1/2)bh$ divided by b, which is

$$(1/2)h = (1/2)Q(1 - r/p).$$

This time the total cost is

$$f(Q) = c_0\frac{D}{Q} + c_c\frac{Q}{2}\left(1 - \frac{r}{p}\right)$$

$$f'(Q) = -\frac{c_0 D}{Q^2} + \frac{c_c(1 - \frac{r}{p})}{2}$$

$$f'(Q) = 0 \text{ gives } \frac{c_0 D}{Q^2} = \frac{c_c}{2}\left(1 - \frac{r}{p}\right)$$

$$Q = \sqrt{\frac{2c_0 D}{c_c(1 - r/p)}}.$$

The order size to minimize the total cost is

$$Q = \sqrt{\frac{2c_0 D}{c_c(1 - r/p)}}.$$

65. The maximum of

$$F(t) = 9 - 10^8(t - 0.0003)^2$$

occurs when $10^8(t - 0.0003)^2$ reaches its minimum, that is, when $t = 0.0003$. At that time
$F(0.0003) = 9$ thousand pounds.

We estimate the value of

$$\int_0^{0.0006} [9 - 10^8(t - 0.0003)^2]\, dt$$

using midpoint sum and $n = 20$, and get $m\,\Delta v \approx 0.00360$ thousand pound-seconds, so $\Delta v \approx 360$ ft per second.

67. Since $f(x) = x^3$ is an odd function, the area under the curve and above the x-axis from 0 to 1 is the same as the area under the x-axis and above the curve for -1 to 0.

You can see that $\int_{-1}^{1} x^3\sqrt{x + 1}\, dx > 0$ by thinking of $\sqrt{x + 1}$ as a weighting factor. Since the weighting factor is always positive, and is greater for $x > 0$ than it is for $x < 0$, the "negative" area to the left of $x = 0$ counts less than the "positive" area to the right of $x = 0$.

4.5 THE FUNDAMENTAL THEOREM OF CALCULUS

1. $\displaystyle\int_0^2 (2x - 3)\, dx$

$$= \left(x^2 - 3x\right)\Big|_0^2 = -2$$

3. $\displaystyle\int_{-1}^1 (x^3 + 2x)\, dx$

$$= \left(\frac{x^4}{4} + x^2\right)\Big|_{-1}^1 = 0$$

5. $\displaystyle\int_0^4 (\sqrt{x} + 3x)\, dx$

$$= \left(\frac{2}{3}x^{\frac{3}{2}} + \frac{3x^2}{2}\right)\Big|_0^4 = \frac{88}{3}$$

7. $\displaystyle\int_0^1 (x\sqrt{x} + x^{1/3})\, dx = \int_0^1 (x^{3/2} + x^{1/3})\, dx$

$$= \left(\frac{2}{5}x^{\frac{5}{2}} + \frac{3}{4}x^{4/3}\right)\Big|_0^1 = \frac{23}{20}$$

9. $\int_0^{\frac{\pi}{4}} (\sec x \tan x)\, dx$

$= \sec x \Big|_0^{\frac{\pi}{4}} = \sqrt{2} - 1$.

11. $\int_{\pi/2}^{\pi} (2\sin x - \cos x)\, dx$

$= (-2\cos x - \sin x) \Big|_{\pi/2}^{\pi}$

$= (-2(-1) - 0) - (0 - 1) = 3$

13. $\int_1^4 \dfrac{x-3}{\sqrt{x}}\, dx$

$= \int_1^4 \left(x^{1/2} - 3x^{-1/2} \right) dx$

$= \dfrac{2}{3} x^{3/2} - 6x^{1/2} \Big|_1^4 = -\dfrac{4}{3}$

15. $\int_0^4 x(x-2)\, dx = \int_0^4 (x^2 - 2x)\, dx$

$= \left(\dfrac{x^3}{3} - x^2 \right) \Big|_0^4 = \dfrac{16}{3}$

17. $\int_0^2 \sqrt{x^2 + 1}\, dx$

$= \lim_{n\to\infty} \sum_{i=1}^{n} \dfrac{2}{n} \sqrt{\left(\dfrac{2i}{n} + 1 \right)}$

Estimating using $n = 20$, we get the Riemann sum ≈ 2.96.

19. $\int_1^4 \dfrac{x^2}{x^2 + 4}\, dx$

$= \lim_{n\to\infty} \sum_{i=1}^{n} \dfrac{3}{n} \dfrac{(1 + (3i/n)^2)}{(1 + 3i/n)^2 + 4}$

Estimating using $n = 20$, we get the Riemann sum ≈ 1.71.

21. $\int_0^{\pi/4} \dfrac{\sin x}{\cos^2 x}\, dx$

$= \int_0^{\pi/4} \tan x \sec x\, dx$

$= \sec x \Big|_0^{\pi/4} = \sqrt{2} - 1$

23. $f'(x) = x^2 - 3x + 2$

25. $f'(x) = \left(\cos(3x^2) + 1 \right) \dfrac{d}{dx}(x^2)$

$= \left(\cos(3x^2) + 1 \right)(2x)$

27. $f(x) = \int_x^{-1} \sqrt{t^2 + 1}\, dt = -\int_{-1}^{x} \sqrt{t^2 + 1}\, dt$

$f'(x) = -\sqrt{x^2 + 1}$

29. $y'(x) = \sin \sqrt{x^2 + \pi^2}$

At the point in question, $y(0) = 0$ and $y'(0) = \sin \pi = 0$.

Therefore, the tangent line has slope 0 and passes through the point $(0, 0)$. The equation of this line is $y = 0$.

31. $y'(x) = \cos(\pi x^3)$

At the point in question, $y(2) = 0$ and $y'(2) = \cos 8\pi = 1$.

Therefore, the tangent line has slope 1 and passes through the point $(2, 0)$.

The equation of this line is $y = x - 2$.

33. $f'(x) = x^2 - 3x + 2$

Setting $f'(x) = 0$ we get $(x - 1)(x - 2) = 0$, $x = 1, 2$.

$f'(x) \begin{cases} > 0 & \text{when } t < 1 \text{ or } t > 2 \\ < 0 & \text{when } 1 < t < 2 \end{cases}$

$f(1) = \int_0^1 (t^2 - 3t + 2)\, dt = (t^3/3 - 3t^2/2 +$

$2t) \Big|_0^1 = \dfrac{5}{6}$

$f(2) = \int_0^2 (t^2 - 3t + 2)\, dt$

$= (t^3/3 - 3t^2/2 + 2t) \Big|_0^2 = \dfrac{2}{3}$

Hence $f(x)$ has a local maximum at the point $(1, 5/6)$ and a local minimum at the point $(2, 2/3)$.

35. The graph of $y = 4 - x^2$ is above the x-axis over the interval $[-2, 2]$.

$$\int_{-2}^{2} (4 - x^2)\, dx$$

$$= (4x - \frac{x^3}{3}) \Big|_{-2}^{2}$$

$$= \frac{32}{3}$$

37. The graph of $y = x^2$ is above the x-axis over the interval $[0, 2]$.

$$\int_{0}^{2} x^2\, dx = \frac{x^3}{3} \Big|_{0}^{2} = \frac{8}{3}$$

39. The graph of $y = \sin x$ is above the x-axis over the interval $[0, \pi]$.

$$\int_{0}^{\pi} \sin x\, dx = -\cos x \Big|_{0}^{\pi} = 2$$

41. If you look at the graph of $1/x^2$, it is obvious that there is positive area between the curve and the x-axis over the interval $[-1, 1]$. In addition to this, there is a vertical asymptote in the interval that we are integrating over which should alert us to a possible problem.

The problem is that $1/x^2$ is not continuous on $[-1, 1]$ (the discontinuity occurs at $x = 0$) and that continuity is one of the conditions in the Fundamental Theorem of Calculus, Part I (Theorem 5.1).

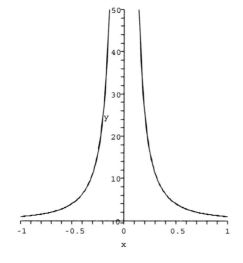

43. $s(t) = 40t + \cos t + c$,
$s(0) = 0 + \cos 0 + c = 2$ so therefore $c = 1$
and
$s(t) = 40t + \cos t + 1$.

45. $v(t) = 4t - \dfrac{t^2}{2} + c_1$,
$v(0) = c_1 = 8$ so therefore $c_1 = 8$ and

$$v(t) = 4t - \frac{t^2}{2} + 8.$$

$$s(t) = 2t^2 - \frac{t^3}{6} + 8t + c_2,$$

$s(0) = c_2 = 0$ so therefore $c_2 = 0$ and

$$s(t) = 2t^2 - \frac{t^3}{6} + 8t.$$

47. $\omega(t) = 10t + c_1$ and $\omega(0) = 0$ gives that $c_1 = 0$
and hence $\omega(t) = 10t$.
$\omega(0.8) = 8$ rad/s.
$v(0.8) = 3(8) = 24$ ft/s.
$\theta(t) = 5t^2 + c_2$ and $\theta(0) = 0$, so $c_2 = 0$
and $\theta(t) = 5t^2$ $\theta(0.8) = 5(0.8^2) = 3.2$ rad.

49. $f_{ave} = \dfrac{1}{3-1} \displaystyle\int_{1}^{3} (x^2 - 1)\, dx$

$$= \frac{1}{2} \left(\frac{x^3}{3} - x \right) \Big|_{1}^{3} = \frac{10}{3}$$

51. $f_{ave} = \dfrac{1}{1-0} \displaystyle\int_{0}^{1} (2x - 2x^2)\, dx$

$$= \left(x^2 - \frac{2x^3}{3} \right) \Big|_{0}^{1} = \frac{1}{3}$$

53. $f_{ave} = \dfrac{1}{\pi/2 - 0} \displaystyle\int_{0}^{\pi/2} \cos x\, dx$

$$= \frac{2}{\pi} (\sin x) \Big|_{0}^{\pi/2} = \frac{2}{\pi}$$

55. $\displaystyle\int_{0}^{3} f(x)\, dx < \int_{0}^{2} f(x)\, dx$

$$< \int_{0}^{1} f(x)\, dx$$

57. Using the Fundamental Theorem of Calculus, it follows that an antiderivative of $\sin\sqrt{x^2+1}$ is $\int_a^x \sin\sqrt{t^2+1}\,dt$ where a is a constant.

59. The next shipment must arrive when the inventory is zero. This occurs at time T:

$$f(t) = Q - r\sqrt{t}$$
$$f(T) = 0 = Q - r\sqrt{T}$$
$$r\sqrt{T} = Q$$
$$T = \frac{Q^2}{r^2}.$$

The average value of f on $[0, T]$ is

$$\frac{1}{T}\int_0^T f(t)\,dt$$
$$= \frac{1}{T}\int_0^T (Q - rt^{1/2})\,dt$$
$$= \frac{1}{T}\left[Qt - \frac{2}{3}rt^{3/2}\right]_0^T$$
$$= \frac{1}{T}\left[QT - \frac{2}{3}rT^{3/2}\right]$$
$$= Q - \frac{2}{3}r\sqrt{T}$$
$$= Q - \frac{2}{3}r\frac{Q}{r}$$
$$= \frac{Q}{3}$$

61. When $a < 2$ or $a > 2$, f is continuous. Using the Fundamental Theorem of Calculus,

$$\left[\lim_{x\to a} F(x)\right] - F(a)$$
$$= \lim_{x\to a}[F(x) - F(a)]$$
$$= \lim_{x\to a}\left[\int_0^x f(t)\,dt - \int_0^a f(t)\,dt\right]$$
$$= \lim_{x\to a}\left[\int_a^x f(t)\,dt\right]$$
$$= 0.$$

When $a = 2$,

$$\lim_{x\to a^-}\left[\int_a^x f(t)\,dt\right]$$
$$= \lim_{x\to 2^-}\left[\int_2^x t\,dt\right]$$

$$= \lim_{x\to 2^-}\left[\frac{t^2}{2}\right]_2^x$$
$$= \lim_{x\to 2^-}\left[\frac{x^2}{2} - \frac{2^2}{2}\right]$$
$$= 0 \text{ and}$$
$$\lim_{x\to a^+}\left[\int_a^x f(t)\,dt\right]$$
$$= \lim_{x\to 2^+}\left[\int_2^x (t+1)\,dt\right]$$
$$= \lim_{x\to 2^+}\left[\frac{t^2}{2} + t\right]_2^x$$
$$= \lim_{x\to 2^+}\left[\frac{x^2}{2} + x - \frac{2^2}{2} - 2\right] = 0.$$

Thus, for all values of a,

$$\left[\lim_{x\to a} F(x)\right] - F(a) = 0$$
$$\lim_{x\to a} F(x) = F(a).$$

Thus, F is continuous for all x. However, $F'(2)$ does not exist, which is shown as follows:

$$F'(2) = \lim_{h\to 0}\frac{F(2+h) - F(2)}{h}$$
$$= \lim_{h\to 0}\frac{1}{h}\left[\int_0^{2+h} f(t)\,dt - \int_0^2 f(t)\,dt\right]$$
$$= \lim_{h\to 0}\frac{1}{h}\int_2^{2+h} f(t)\,dt$$

We'll show that this limit does not exist by showing that the left and right limits are different. The right limit is

$$\lim_{h\to 0^+}\frac{1}{h}\int_2^{2+h} f(t)\,dt$$
$$= \lim_{h\to 0^+}\frac{1}{h}\int_2^{2+h}(t+1)\,dt$$
$$= \lim_{h\to 0^+}\frac{1}{h}\left[\frac{t^2}{2} + t\right]_2^{2+h}$$
$$= \lim_{h\to 0^+}\frac{1}{h}\left[\frac{(2+h)^2}{2} + 2 + h - \frac{2^2}{2} - 2\right]$$
$$= \lim_{h\to 0^+}\frac{1}{h}\left[\frac{h^2 + 4h + 4}{2} + 2 + h - 4\right]$$

$$= \lim_{h \to 0^+} \frac{1}{h} \left[\frac{h^2}{2} + 3h \right]$$

$$= \lim_{h \to 0^+} \left[\frac{h}{2} + 3 \right]$$

$$= 3.$$

The left limit is $\displaystyle \lim_{h \to 0^-} \frac{1}{h} \int_2^{2+h} f(t)\, dt$

$$= \lim_{h \to 0^-} \frac{1}{h} \int_2^{2+h} t\, dt$$

$$= \lim_{h \to 0^-} \frac{1}{h} \left[\frac{t^2}{2} \right]_2^{2+h}$$

$$= \lim_{h \to 0^-} \frac{1}{h} \left[\frac{(2+h)^2}{2} - \frac{2^2}{2} \right]$$

$$= \lim_{h \to 0^-} \frac{1}{h} \left[\frac{h^2 + 4h + 4}{2} - 2 \right]$$

$$= \lim_{h \to 0^-} \left[\frac{h}{2} + 2 \right]$$

$$= 2.$$

Thus, $F'(2)$ does not exist. This result does not contradict the Fundamental Theorem of Calculus, because in this situation, $f(x)$ is not continuous, and thus The Fundamental Theorem of Calculus does not apply.

63. $g(x) = \displaystyle \int_0^x \left[\int_0^u f(t)\, dt \right] du$

$g'(x) = \displaystyle \int_0^x f(t)\, dt$

$g''(x) = f(x)$

A zero of f corresponds to a zero of the second derivative of g (possibly an inflection point of g).

65. The integral in part (a) is improper, because the integrand has an asymptote at one of the limits of integration. The Fundamental Theorem of Calculus applies to the integrals in parts (b) and (c).

67. **(a)** $\displaystyle \lim_{n \to \infty} \frac{1}{n} \left[\sin \frac{\pi}{n} + \sin \frac{2\pi}{n} + \cdots + \sin \pi \right]$

$$= \int_0^1 \sin(\pi x)\, dx$$

$$= -\frac{1}{\pi} \cos(\pi x) \Big|_0^1$$

$$= -\frac{1}{\pi} (\cos \pi - \cos 0)$$

$$= -\frac{1}{\pi} (-1 - 1)$$

$$= \frac{2}{\pi}$$

(b) $\displaystyle \lim_{n \to \infty} \frac{2}{n}$

$$\left[\frac{1}{(1 + 2/n)^2} + \frac{1}{(1 + 4/n)^2} + \cdots + \frac{1}{3^2} \right]$$

$$= \int_1^3 \frac{1}{x^2}\, dx$$

$$= -x^{-1} \Big|_1^3$$

$$= \frac{2}{3}$$

69. Let $F(x) = \displaystyle \int_{a(x)}^{b(x)} f(t)\, dt$,

$G(x) = \displaystyle \int_0^{b(x)} f(t)\, dt$,

$H(x) = \displaystyle \int_0^{a(x)} f(t)\, dt$. Then

$F(x) = G(x) - H(x)$.

$G'(x) = f(b(x)) b'(x)$

$H'(x) = f(a(x)) a'(x)$.

$F'(x) = G'(x) - H'(x)$

$\quad = f(b(x)) b'(x) - f(a(x)) a'(x)$

4.6 INTEGRATION BY SUBSTITUTION

1. Let $u = x^3 + 2$ and then $du = 3x^2\, dx$ and

$$\int x^2 \sqrt{x^3 + 2}\, dx = \frac{1}{3} \int u^{1/2}\, du$$

$$= \frac{2}{9} u^{3/2} + c = \frac{2}{9} (x^3 + 2)^{3/2} + c.$$

3. Let $u = \sqrt{x} + 2$ and then $du = \frac{1}{2}x^{-1/2}\,dx$ and

$$\int \frac{(\sqrt{x}+2)^3}{\sqrt{x}}\,dx = 2\int u^3\,du$$

$$= \frac{2}{4}u^4 + c = \frac{1}{2}(\sqrt{x}+2)^4 + c.$$

5. Let $u = x^4 + 3$ and then $du = 4x^3\,dx$ and

$$\int x^3\sqrt{x^4+3}\,dx = \frac{1}{4}\int u^{1/2}\,du$$

$$= \frac{1}{6}u^{3/2} + c = \frac{1}{6}(x^4+3)^{3/2} + c.$$

7. Let $u = \cos x$ and then $du = -\sin x\,dx$ and

$$\int \frac{\sin x}{\sqrt{\cos x}}\,dx = -\int \frac{du}{\sqrt{u}}$$

$$= -2\sqrt{u} + c = -2\sqrt{\cos x} + c.$$

9. Let $u = x^3$ and then $du = 3x^2\,dx$ and

$$\int x^2 \cos x^3\,dx = \frac{1}{3}\int \cos u\,du$$

$$= \frac{1}{3}\sin u + c = \frac{1}{3}\sin x^3 + c.$$

11. Let $u = \tan x$. Then $du = \sec^2 x\,dx$.

$$\int \sec^2 x \cos(\tan x)\,dx$$

$$= \int \cos u\,du = \sin u + c = \sin(\tan x) + c.$$

13. Let $u = \sqrt{x} = x^{1/2}$. Then $du = \frac{1}{2}x^{-1/2}\,dx = \frac{1}{2\sqrt{x}}\,dx$.

$$\int \frac{\cos\sqrt{x}}{\sqrt{x}}\,dx = 2\int \cos u\,du = 2\sin u + c = \sin\sqrt{x} + c.$$

15. Let $u = \sin 3x + 1$. Then $du = 3\cos 3x\,dx$.

$$\int \frac{\cos 3x}{(\sin 3x + 1)^3}\,dx = \frac{1}{3}\int u^{-3}\,du$$

$$= -\frac{1}{6}u^{-2} + c$$

$$= -\frac{1}{6}(\sin 3x + 1)^{-2} + c.$$

17. Let $u = \sqrt{x} + 1$ and then $du = \frac{1}{2\sqrt{x}}\,dx$ and

$$\int \frac{1}{\sqrt{x}\left(\sqrt{x}+1\right)^2}\,dx = 2\int u^{-2}\,du$$

$$= -2u^{-1} + c = -2\left(\sqrt{x}+1\right)^{-1} + c.$$

19. Let $u = \frac{2}{x} + 1$. Then $du = -\frac{2}{x^2}\,dx$.

$$\int \frac{4}{x^2(2/x+1)^3}\,dx = -2\int u^{-3}\,du = u^{-2} +$$

$$= (2/x + 1)^{-2} + c$$

21. Let $u = 3 - x$. Then $du = -\,dx$ and $x = 3 - u$.

$$\int \frac{9x^2}{(3-x)^4}\,dx = -9\int (3-u)^2 u^{-4}\,du$$

$$= -9\int (9u^{-4} - 6u^{-3} + u^{-2})\,du$$

$$= -9\left(-3u^{-3} + 3u^{-2} - u^{-1}\right) + c$$

$$= \frac{27}{(3-x)^3} - \frac{27}{(3-x)^2} + \frac{9}{3-x} + c$$

23. Let $u = 4x - 1$. Then $du = 4\,dx$.

$$\int \frac{\sqrt{4x^5 - x^4}}{x^2}\,dx = \int \sqrt{4x-1}\,dx$$

$$= \frac{1}{4}\int u^{1/2}\,du$$

$$= \frac{1}{4}\left(\frac{2}{3}u^{3/2}\right) + c = \frac{1}{6}(4x-1)^{3/2} + c.$$

25. Let $u = x + 7$. Then $du = dx$ and $x = u - 7$.

$$\int \frac{2x+3}{(x+7)^3}\,dx = \int [2(u-7)+3]\,u^{-3}\,du$$

$$= \int (2u^{-2} - 11u^{-3})\,du$$

$$= -2u^{-1} + \frac{11}{2}u^{-2} + c$$

$$= \frac{11}{2(x+7)^2} - \frac{2}{x+7} + c$$

27. Let $u = \sqrt{1 + \sqrt{x}}$ and then

$$(u^2 - 1)^2 = x$$

$$2(u^2 - 1)(2u)\,du = dx \text{ and}$$

$$\int \frac{1}{\sqrt{1+\sqrt{x}}}\,dx$$

$$= \int \frac{4u(u^2-1)}{u}\,du$$

$$= 4 \int (u^2-1)\,du$$

$$= 4 \left(\frac{u^3}{3} - u\right) + c$$

$$= \frac{4}{3}(1+\sqrt{x})^{3/2} - 4(1+\sqrt{x})^{1/2} + c$$

29. Let $u = x^2 + 1$ and then
$du = 2x\,dx$, $u(0) = 1$, $u(2) = 5$

$$\int_0^2 x\sqrt{x^2+1}\,dx = \frac{1}{2}\int_1^5 \sqrt{u}\,du$$

$$= \frac{1}{2}\cdot\frac{2}{3}u^{3/2}\Big|_1^5 = \frac{1}{3}\left(\sqrt{125}-1\right)$$

$$= \frac{5}{3}\sqrt{5} - \frac{1}{3}$$

31. Let $u = x^2 + 1$ and then $du = 2x\,dx$, $u(-1) = 2 = u(1)$ and

$$\int_{-1}^1 \frac{x}{(x^2+1)^{1/2}}\,dx$$

$$= \frac{1}{2}\int_2^2 u^{-1/2}\,du = 0$$

33. Let $u = x^3$. Then $du = 3x^2\,dx$, $u(0) = 0$ and $u(1) = 1$.

$$\int_0^1 \cos x^3\,dx = \frac{1}{3}\int_0^1 \cos u\,du$$

$$= \frac{1}{3}\sin u\Big|_0^1 = \frac{1}{3}\sin 1$$

35. $\displaystyle \int_1^4 \frac{x-1}{\sqrt{x}}\,dx = \int_1^4 \left(x^{1/2} - x^{-1/2}\right)dx$

$$= \left(\frac{2}{3}x^{3/2} - 2x^{1/2}\right)\Big|_1^4$$

$$= \left(\frac{16}{3}-4\right) - \left(\frac{2}{3}-2\right) = \frac{8}{3}$$

37. (a) $\displaystyle \int_0^\pi \sin x^2\,dx \approx 0.77$ using midpoint evaluation with $n \geq 40$.

(b) Let $u = x^2$ and then $du = 2x\,dx$ and

$$\int_0^\pi x\sin x^2\,dx = \frac{1}{2}\int_0^{\pi^2} \sin u\,du$$

$$= \frac{1}{2}(-\cos u)\Big|_0^{\pi^2}$$

$$= -\frac{1}{2}\cos \pi^2 + \frac{1}{2}$$

$$\approx 0.95134$$

39. (a) $\displaystyle \int_0^2 \frac{4x^2}{(x^2+1)^2}\,dx \approx 1.414$ using right endpoint evaluation with $n \geq 50$.

(b) Let $u = x^2 + 1$ and $du = 2x\,dx$

$$\int_0^2 \frac{4x}{(x^2+1)^2}\,dx = \int_1^5 2\frac{1}{u^2}\,du$$

$$= \int_1^5 2u^{-2}\,du$$

$$= -2u^{-1}\Big|_1^5$$

$$= \frac{8}{5}$$

41. $\displaystyle \frac{1}{2}\int_0^4 f(u)\,du$

43. $\displaystyle \int_0^1 f(u)\,du$

45. $\displaystyle \int_{-a}^a f(x)\,dx$

$$= \int_{-a}^0 f(x)\,dx + \int_0^a f(x)\,dx$$

Let $u = -x$ and $du = -dx$ in the first integral. Then,

$$\int_{-a}^a f(x)\,dx$$

$$= -\int_a^0 f(-u)\,du + \int_0^a f(x)\,dx$$

$$= \int_0^a f(-u)\,du + \int_0^a f(x)\,dx$$

If f is even, then $f(-u) = f(u)$, and so

$$\int_{-a}^a f(x)\,dx$$

$$= \int_0^a f(u)\,du + \int_0^a f(x)\,dx$$

$$= \int_0^a f(x)\,dx + \int_0^a f(x)\,dx$$

$$= 2\int_0^a f(x)\,dx$$

If f is odd, then $f(-u) = -f(u)$, and so

$$\int_{-a}^a f(x)\,dx$$

$$= -\int_0^a f(u)\,du + \int_0^a f(x)\,dx$$

$$= -\int_0^a f(x)\,dx + \int_0^a f(x)\,dx$$

$$= 0$$

47. Let $u = 10 - x$, so that $du = -dx$. Then,

$$I = \int_0^{10} \frac{\sqrt{x}}{\sqrt{x} + \sqrt{10-x}}\,dx$$

$$= -\int_{x=0}^{x=10} \frac{\sqrt{10-u}}{\sqrt{10-u} + \sqrt{u}}\,du$$

$$= -\int_{u=10}^{u=0} \frac{\sqrt{10-u}}{\sqrt{10-u} + \sqrt{u}}\,du$$

$$= \int_{u=0}^{u=10} \frac{\sqrt{10-u}}{\sqrt{10-u} + \sqrt{u}}\,du$$

$$I = \int_{x=0}^{x=10} \frac{\sqrt{10-x}}{\sqrt{10-x} + \sqrt{x}}\,dx$$

The last equation follows from the previous one because u and x are dummy variables of integration. Now note that

$$\frac{\sqrt{x}}{\sqrt{x} + \sqrt{10-x}}$$

$$= \frac{\sqrt{x} + \sqrt{10-x} - \sqrt{10-x}}{\sqrt{x} + \sqrt{10-x}}$$

$$= 1 - \frac{\sqrt{10-x}}{\sqrt{x} + \sqrt{10-x}}$$

Thus,

$$\int_0^{10} \frac{\sqrt{x}}{\sqrt{x} + \sqrt{10-x}}\,dx$$

$$= \int_0^{10} \left[1 - \frac{\sqrt{10-x}}{\sqrt{x} + \sqrt{10-x}} \right]\,dx$$

$$= \int_0^{10} 1\,dx - \int_0^{10} \frac{\sqrt{10-x}}{\sqrt{x} + \sqrt{10-x}}\,dx$$

$$I = \int_0^{10} 1\,dx - I$$

$$2I = 10$$

$$I = 5$$

49. Let $u = 6 - x$, so that $du = -dx$. Then,

$$I = \int_2^4 \frac{\sin^2(9-x)}{\sin^2(9-x) + \sin^2(x+3)}\,dx$$

$$= -\int_4^2 \frac{\sin^2(u+3)}{\sin^2(u+3) + \sin^2(9-u)}\,du$$

$$= \int_2^4 \frac{\sin^2(u+3)}{\sin^2(u+3) + \sin^2(9-u)}\,du$$

$$= \int_2^4 \frac{\sin^2(x+3)}{\sin^2(x+3) + \sin^2(9-x)}\,dx$$

$$= \int_2^4 \left[1 - \frac{\sin^2(9-x)}{\sin^2(x+3) + \sin^2(9-x)} \right]\,dx$$

$$I = \int_2^4 1\,dx - I$$

$$2I = 2$$

$$I = 1$$

51. Let $6 - u = x + 4$; that is, let $u = 2 - x$, so that $du = -dx$.

Then,

$$I = \int_0^2 \frac{f(x+4)}{f(x+4) + f(6-x)}\,dx$$

$$= -\int_2^0 \frac{f(6-u)}{f(6-u) + f(u+4)}\,du$$

$$= \int_0^2 \frac{f(6-u)}{f(6-u) + f(u+4)}\,du$$

$$= \int_0^2 \frac{f(6-x)}{f(6-x) + f(x+4)} \, dx$$

$$= \int_0^2 \frac{f(6-x) + f(x+4) - f(x+4)}{f(6-x) + f(x+4)} \, dx$$

$$= \int_0^2 \left[1 - \frac{f(x+4)}{f(6-x) + f(x+4)} \right] dx$$

$$I = \int_0^2 1 \, dx - I$$

$$2I = 2$$

$$I = 1$$

53. Let $u = x^{1/6}$, so that
$du = (1/6)x^{-5/6}dx$, which means
$6u^5 du = dx$.

Thus,

$$\int \frac{1}{\sqrt{x} + \sqrt[3]{x}} dx$$

$$= 6 \int \frac{u^5}{u^3 + u^2} \, du$$

$$= 6 \int \frac{u^3}{u+1} \, du.$$

55. The point is that if we let $u = x^4$, then we get
$x = \pm u^{1/4}$, and so we need to pay attention to
the sign of u and x. A safe way is to solve the
original indefinite integral in terms of x, and
then solve the definite integral using boundary
points in terms of x.

$$\int_{-2}^1 4x^4 dx = \int_{x=-2}^{x=1} u^{1/4} \, du$$

$$= \frac{4}{5} u^{5/4} \Big|_{x=-2}^{x=1}$$

$$= \frac{4}{5} x^5 \Big|_{x=-2}^{x=1}$$

$$= \frac{4}{5}(1^5 - (-2)^5)$$

$$= \frac{4}{5}(1 - (-32))$$

$$= \frac{4(33)}{5} = \frac{132}{5}$$

57. $V(t) = V_p \sin(2\pi f t)$

$$V^2(t) = V_p^2 \sin^2(2\pi f t)$$

$$= V_p^2 \left(\frac{1}{2} - \frac{1}{2} \cos(4\pi f t) \right)$$

$$= \frac{V_p^2}{2}(1 - \cos(4\pi f t))$$

$$\text{rms} = \sqrt{f \int_0^{1/f} V^2(t) \, dt}$$

$$= \sqrt{f \int_0^{1/f} \frac{V_p^2}{2}(1 - \cos(4\pi f t)) \, dt}$$

$$= \frac{V_p \sqrt{f}}{\sqrt{2}} \sqrt{\int_0^{1/f} (1 - \cos(4\pi f t)) \, dt}$$

$$= \frac{V_p \sqrt{f}}{\sqrt{2}} \sqrt{\left(t - \frac{\sin(4\pi f t)}{4\pi f} \right) \Big|_0^{1/f}}$$

$$= \frac{V_p \sqrt{f}}{\sqrt{2}} \sqrt{\frac{1}{f}} = \frac{V_p}{\sqrt{2}}$$

4.7 NUMERICAL INTEGRATION

1. Midpoint Rule:

$$\int_0^1 \left(x^2 + 1 \right) dx$$

$$\approx \frac{1}{4} \left[f \left(\frac{1}{8} \right) + f \left(\frac{3}{8} \right) + f \left(\frac{5}{8} \right) \right. $$
$$\left. + f \left(\frac{7}{8} \right) \right]$$

$$= \frac{85}{64}$$

Trapezoidal Rule:

$$\int_0^1 \left(x^2 + 1 \right) dx$$

$$\approx \frac{1-0}{2(4)} \left[f(0) + 2f \left(\frac{1}{4} \right) + 2f \left(\frac{1}{2} \right) \right.$$
$$\left. + 2f \left(\frac{3}{4} \right) + f(1) \right]$$

$$= \frac{43}{32}$$

Simpson's Rule:

$$\int_0^1 \left(x^2 + 1\right)\, dx$$

$$= \frac{1-0}{3(4)}\left[f(0) + 4f\left(\frac{1}{4}\right) + 2f\left(\frac{1}{2}\right) \right.$$

$$\left. + 4f\left(\frac{3}{4}\right) + f(1)\right]$$

$$= \frac{4}{3}$$

3. Midpoint Rule:

$$\int_1^3 \frac{1}{x}\, dx$$

$$\approx \frac{3-1}{4}\left[f\left(\frac{5}{4}\right) + f\left(\frac{7}{4}\right) + f\left(\frac{9}{4}\right) \right.$$

$$\left. + f\left(\frac{11}{4}\right)\right]$$

$$= \frac{1}{2}\left(\frac{4}{5} + \frac{4}{7} + \frac{4}{9} + \frac{4}{11}\right)$$

$$= \frac{3776}{3465}$$

Trapezoidal Rule:

$$\int_1^3 \frac{1}{x}\, dx$$

$$\approx \frac{3-1}{2(4)}\left[f(1) + 2f\left(\frac{3}{2}\right) + 2f(2) \right.$$

$$\left. + 2f\left(\frac{5}{2}\right) + f(3)\right]$$

$$= \frac{1}{4}\left(1 + \frac{4}{3} + 1 + \frac{4}{5} + \frac{1}{3}\right)$$

$$= \frac{67}{60}$$

Simpson's Rule:

$$\int_1^3 \frac{1}{x}\, dx$$

$$\approx \frac{3-1}{3(4)}\left[f(1) + 4f\left(\frac{3}{2}\right) + 2f(2) \right.$$

$$\left. + 4f\left(\frac{5}{2}\right) + f(3)\right]$$

$$= \frac{1}{6}\left(1 + \frac{8}{3} + 1 + \frac{8}{5} + \frac{1}{3}\right)$$

$$= \frac{11}{10}$$

5. (a) Left endpoints:

$$\int_0^2 f(x)\, dx$$

$$\approx \frac{2-0}{4}\left[f(0) + f(0.5) + f(1) \right.$$

$$\left. + f(1.5)\right]$$

$$= \frac{1}{2}(1 + 0.25 + 0 + 0.25)$$

$$= 0.75$$

(b) Midpoint Rule:

$$\int_0^2 f(x)\, dx$$

$$\approx \frac{2-0}{4}\left[f(0.25) + f(0.75) \right.$$

$$\left. + f(1.25) + f(1.75)\right]$$

$$= \frac{1}{2}(0.6 + 0.1 + 0.1 + 0.6)$$

$$= 0.7$$

(c) Trapezoidal Rule:

$$\int_0^2 f(x)\, dx$$

$$\approx \frac{2-0}{2(4)}\left[f(0) + 2f(0.5) + 2f(1) \right.$$

$$\left. + 2f(1.5) + f(2)\right]$$

$$= \frac{1}{4}(1 + 0.5 + 0 + 0.5 + 1)$$

$$= 0.75$$

7.

n	Midpoint	Trapezoidal	Simpson
10	0.5538	0.5889	0.5660
20	0.5629	0.5713	0.5655
50	0.5652	0.5666	0.5657

9.

n	Midpoint	Trapezoidal	Simpson
10	51.7073	51.8311	51.7554
20	51.7381	51.7692	51.7485
50	51.7468	51.7518	51.7485

n	Simpson	ES_n
10	0	0
20	0	0
40	0	0
80	0	0

11. The exact value of this integral is

$$\int_0^1 5x^4 \, dx = x^5 \Big|_0^1 = 1 - 0 = 1$$

n	Midpoint	EM_n
10	0.99168	8.3×10^{-3}
20	0.99792	2.1×10^{-3}
40	0.99948	5.2×10^{-4}
80	0.99987	1.3×10^{-4}

n	Trapezoidal	ET_n
10	1.01665	1.6×10^{-2}
20	1.00417	4.1×10^{-3}
40	1.00104	1.0×10^{-3}
80	1.00026	2.6×10^{-4}

n	Simpson	ES_n
10	1.000066	6.6×10^{-5}
20	1.0000041	4.2×10^{-6}
40	1.00000026	2.6×10^{-7}
80	1.00000016	1.6×10^{-8}

13. The exact value of this integral is

$$\int_0^\pi \cos x \, dx = \sin x \Big|_0^\pi = 0.$$

n	Midpoint	EM_n
10	0	0
20	0	0
40	0	0
80	0	0

n	Trapezoidal	ET_n
10	0	0
20	0	0
40	0	0
80	0	0

15. If you double n, the error in the Midpoint Rule is divided by 4, the error in the Trapezoidal Rule is divided by 4, and the error in Simpson's Rule is divided by 16.

17. Midpoint Rule:

$$\pi \approx \frac{1}{4} \left[\frac{4}{1 + (1/8)^2} + \frac{4}{1 + (3/8)^2} + \frac{4}{1 + (5/8)^2} + \frac{4}{1 + (7/8)^2} \right] \approx 3.1468$$

Trapezoidal Rule:

$$\pi \approx \frac{1}{8} \left[\frac{4}{1 + 0^2} + 2 \cdot \frac{4}{1 + (1/4)^2} + 2 \cdot \frac{4}{1 + (1/2)^2} + 2 \cdot \frac{4}{1 + (3/4)^2} \frac{4}{1 + 1^2} \right]$$
$$\approx 3.1312$$

Simpson's Rule:

$$\pi \approx \frac{1}{12} \left[\frac{4}{1 + 0^2} + 4 \cdot \frac{4}{1 + (1/4)^2} + 2 \cdot \frac{4}{1 + (1/2)^2} + 4 \cdot \frac{4}{1 + (3/4)^2} \frac{4}{1 + 1^2} \right]$$
$$\approx 3.1416$$

19. Midpoint Rule:
$\sin 1 \approx 0.843666$

Trapezoidal Rule:
$\sin 1 \approx 0.837084$

Simpson's Rule:
$\sin 1 \approx 0.841489$

21.
$$f(x) = \frac{4}{1+x^2} = 4(1+x^2)^{-1},$$

$$f'(x) = -4(1+x^2)^{-2}(2x) = -8x(1+x^2)^{-2},$$

$$f''(x) = -8(1+x^2)^{-2} + 16x(1+x^2)^{-3}(2x)$$

$$= \frac{-8(1+x^2) + 32x^2}{(1+x^2)^3} = \frac{24x^2 - 8}{(1+x^2)^3}$$

A graph of $y = f''(x)$ shows that $|f''(x)| \le 8$ on the interval $[0, 1]$.

(a) $|EM_4| \le 8\dfrac{(1-0)^3}{24(4)^2} \approx 0.021$

(b) $|ET_4| \le 8\dfrac{(1-0)^3}{12(4)^2} \approx 0.042$

23. From exercise 21, we know that $K = 8$.

(a)
$$8\frac{(1-0)^3}{24n^2} \le 10^{-7}$$

$$\frac{1}{3} \cdot 10^7 \le n^2$$

$$n \ge \sqrt{\frac{1}{3} \cdot 10^7} \approx 1825.7$$

Take $n = 1826$.

(b)
$$8\frac{(1-0)^3}{12n^2} \le 10^{-7}$$

$$\frac{2}{3} \cdot 10^7 \le n^2$$

$$n \ge \sqrt{\frac{2}{3} \cdot 10^7} \approx 2581.99$$

Take $n = 2582$.

25. $f(x) = 5x^4$, $f'(x) = 20x^3$, $f''(x) = 60x^2$, $f'''(x) = 120x$, and $f^{(4)}(x) = 120$. Thus, $|f''(x)| \le 60$ and $|f^{(4)}(x)| \le 120$ on the interval $[0, 1]$.

Midpoint Rule ($n = 80$):

$$|EM_{80}| \le 60\frac{(1-0)^3}{24(80)^2} \approx 0.000391$$

From exercise 11, the actual error is 0.00013.

Trapezoidal Rule ($n = 80$):

$$|ET_{80}| \le 60\frac{(1-0)^3}{12(80)^2} \approx 0.000781$$

From exercise 11, the actual error is 0.00026.

Simpson's Rule ($n = 80$):

$$|ES_{80}| \le 120\frac{(1-0)^5}{180(80)^4} \approx 1.63 \times 10^{-8}$$

From exercise 11, the actual error is 1.6×10^{-8}.

27. Trapezoidal Rule:

$$\int_0^2 f(x)\,dx$$

$$\approx \frac{2-0}{2(8)}\big[f(0) + 2f(0.25) + 2f(0.5)$$

$$+ 2f(0.75) + 2f(1) + 2f(1.25)$$

$$+ 2f(1.5) + 2f(1.75) + f(2)\big]$$

$$= \frac{1}{8}\,[4.0 + 9.2 + 10.4 + 9.6 + 10$$

$$+ 9.2 + 8.8 + 7.6 + 4.0\;]$$

$$= 9.1$$

Simpson's Rule:

$$\int_0^2 f(x)\,dx$$

$$\approx \frac{2-0}{3(8)}\big[f(0) + 4f(0.25) + 2f(.05)$$

$$+ 4f(0.75) + 2f(1) + 4f(1.25) + 2f(1.5)$$

$$+ 4\,4f(1.75) + f(2)\big]$$

$$= \frac{1}{12}(4.0 + 18.4 + 10.4 + 19.2 + 10$$

$$+ 18.4 + 8.8 + 15.2 + 4.0)$$

$$\approx 9.033$$

29. Simpson's Rule:

$$\int_0^{120} f(x)\,dx$$

$$\approx \frac{120-0}{3(12)}[f(0) + 4f(10) + 2f(20)$$

$$+ 4f(30) + 2f(40) + 4f(50) + 2f(60)$$

$$+ 4f(70) + 2f(80) + 4f(90)$$

$$+ 2f(100) + 4f(110) + f(120)]$$

$$= \frac{10}{3}(56 + 216 + 116 + 248 + 116 + 232$$
$$+ 124 + 224 + 104 + 192 + 80 + 128 +$$
$$22) \approx 6193 \text{ square feet}$$

31. Simpson's Rule:

$$\int_0^{12} v(t)\, dt$$

$$\approx \frac{12 - 0}{3(12)}[f(0) + 4f(1) + 2f(2)$$
$$+ 4f(3) + 2f(4) + 4f(5) + 2f(6)$$
$$+ 4f(7) + 2f(8) + 4f(9) + 2f(10)$$
$$+ 4f(11) + f(12)]$$

$$= \frac{1}{3}(40 + 168 + 80 + 176 + 96 + 200$$
$$+ 92 + 184 + 84 + 176 + 80 + 168$$
$$+ 42) \approx 529 \text{ feet}$$

33. Simpson's Rule: $\int_0^{2.4} f(x)\, dx$

$$\approx \frac{2.4 - 0}{3(12)}[f(0) + 4f(0.2) + 2f(0.4)$$
$$+ 4f(0.6) + 2f(0.8) + 4f(1) + 2f(1.2)$$
$$+ 4f(1.4) + 2f(1.6) + 4f(1.8) + 2f(2)$$
$$+ 4f(2.2) + f(2.4)]$$

$$= \frac{1}{15}(0 + 0.8 + 0.8 + 4 + 3.2 + 8 + 4.4$$
$$+ 8 + 3.2 + 4.8 + 1.2 + 0.8 + 0)$$
$$\approx 2.6 \text{ liters}$$

35. (a) Midpoint Rule:

$$M_n < \int_a^b f(x)\, dx$$

(b) Trapezoidal Rule:

$$T_n > \int_a^b f(x)\, dx$$

(c) Simpson's Rule: not enough information.

37. (a) Midpoint Rule:

$$M_n > \int_a^b f(x)\, dx$$

(b) Trapezoidal Rule:

$$T_n < \int_a^b f(x)\, dx$$

(c) Simpson's Rule:
not enough information.

39. (a) Midpoint Rule:

$$M_n < \int_a^b f(x)\, dx$$

(b) Trapezoidal Rule:

$$T_n > \int_a^b f(x)\, dx$$

(c) Simpson's Rule:

$$S_n = \int_a^b f(x)\, dx$$

41. $\frac{1}{2}(R_L + R_R)$

$$= \frac{1}{2}\left[\sum_{i=0}^{n-1} f(x_i)\Delta x + \sum_{i=1}^{n} f(x_i)\Delta x\right]$$

$$= \frac{\Delta x}{2}\left[f(x_0) + \sum_{i=1}^{n-1} f(x_i)\right.$$

$$+ \left.\sum_{i=1}^{n-1} f(x_i) + f(x_n)\right]$$

$$= \frac{b - a}{2n}\left[f(x_0) + 2\sum_{i=1}^{n-1} f(x_i) + f(x_n)\right]$$

$$= T_n$$

43. (a) $\int_{-1}^{1} x\, dx = 0$

$$\left(-\frac{1}{\sqrt{3}}\right) + \left(\frac{1}{\sqrt{3}}\right) = 0$$

(b) $\int_{-1}^{1} x^2\, dx = \frac{2}{3}$

$$\left(-\frac{1}{\sqrt{3}}\right)^2 + \left(\frac{1}{\sqrt{3}}\right)^2 = \frac{2}{3}$$

(c) $\int_{-1}^{1} x^3\, dx = 0$

$$\left(-\frac{1}{\sqrt{3}}\right)^3 + \left(\frac{1}{\sqrt{3}}\right)^3 = 0$$

45. Simpson's Rule is not applicable because $\dfrac{\sin x}{x}$ is not defined at $x = 0$.

$$L = \lim_{x \to 0} \frac{\sin x}{x} = 1$$

The two functions $f(x)$ and $\dfrac{\sin x}{x}$ differ only at one point, $x = 0$, so

$$\int_0^\pi f(x)\,dx = \int_0^\pi \frac{\sin x}{x}\,dx.$$

We can now apply Simpson's Rule with $n = 2$:

$$\int_0^\pi f(x)\,dx$$

$$\approx \frac{\pi}{6}\left(1 + \frac{4\sin(\pi/2)}{\pi/2} + \frac{\sin \pi}{\pi}\right)$$

$$= \frac{\pi}{2}\left(\frac{1}{3} + \frac{8}{3\pi}\right)$$

$$\approx \frac{\pi}{2} \cdot 1.18$$

47. Let I be the exact integral. Then we have

$$T_n - I \approx -2(M_n - I)$$
$$T_n - I \approx 2I - 2M_n$$
$$T_n + 2M_n \approx 3I$$
$$\frac{T_n}{3} + \frac{2}{3}M_n \approx I$$

Chapter 4 Review Exercises

1. $\displaystyle\int (4x^2 - 3)\,dx = \frac{4}{3}x^3 - 3x + c$

3. $\displaystyle\int \frac{4}{\sqrt{x}}\,dx = 4\frac{x^{1/2}}{1/2} + c = 8\sqrt{x} + c$

5. $\displaystyle\int 2\sin 4x\,dx = -\frac{1}{2}\cos 4x + c$

7. $\displaystyle\int (x - \cos 4x)\,dx = \frac{x^2}{2} - \frac{1}{4}\sin 4x + c$

9. $\displaystyle\int \frac{x^2 + 4x}{x}\,dx = \int (x + 4)\,dx$
$$= \frac{x^2}{2} + 4x + c$$

11. $\displaystyle\int x\left(1 - \frac{1}{x}\right)dx = \int (x - 1)\,dx$
$$= \frac{1}{2}x^2 - x + c$$

13. Let $u = x^2 + 4$, then $du = 2x\,dx$ and
$$\int x\sqrt{x^2 + 4}\,dx$$
$$= \frac{1}{2}\int u^{1/2}\,du = \frac{1}{3}u^{3/2} + c$$
$$= \frac{1}{3}(x^2 + 4)^{3/2} + c.$$

15. Let $u = x^3$, $du = 3x^2\,dx$.
$$\int 6x^2 \cos x^3\,dx = 2\int \cos u\,du$$
$$= 2\sin u + c = 2\sin x^3 + c$$

17. Let $u = \dfrac{1}{x} = x^{-1}$. Then $du = -x^{-2}\,dx = -\dfrac{1}{x^2}\,dx$.
$$\int \frac{\sin(1/x)}{x^2}\,dx = -\int \sin u\,du = \cos u + c = \cos(1/x) + c$$

19. Let $u = \tan x$. Then $du = \sec^2 x\,dx$.
$$\int \tan x \sec^2 x\,dx = \int u\,du = \frac{1}{2}u^2 + c = \frac{1}{2}\tan^2 x + c$$

21. $f(x) = \displaystyle\int (3x^2 + 1)\,dx = x^3 + x + c$
$$f(0) = c = 2$$
$$f(x) = x^3 + x + 2$$

23. $s(t) = \displaystyle\int (-32t + 10)\,dt$
$$= -16t^2 + 10t + c$$
$$s(0) = c = 2$$
$$s(t) = -16t^2 + 10t + 2$$

25. $\displaystyle\sum_{i=1}^{6}(i^2 + 3i)$

$$= (1^2 + 3 \cdot 1) + (2^2 + 3 \cdot 2) + (3^2 + 3 \cdot 3)$$
$$+ (4^2 + 3 \cdot 4) + (5^2 + 3 \cdot 5) + (6^2 + 3 \cdot 6)$$
$$= 4 + 10 + 18 + 28 + 40 + 54$$
$$= 154$$

27. $\displaystyle\sum_{i=1}^{100}(i^2 - 1)$

$$= \sum_{i=1}^{100} i^2 - \sum_{i=1}^{100} 1$$

$$= \frac{100(101)(201)}{6} - 100$$

$$= 338{,}250$$

29. $\displaystyle\frac{1}{n^3}\sum_{i=1}^{n}(i^2 - i)$

$$= \frac{1}{n^3}\left(\sum_{i=1}^{n} i^2 - \sum_{i=1}^{n} i\right)$$

$$= \frac{1}{n^3}\left(\frac{n(n+1)(2n+1)}{6} - \frac{n(n+1)}{2}\right)$$

$$= \frac{(n+1)(2n+1)}{6n^2} - \frac{n+1}{2n^2}$$

$$\lim_{n\to\infty}\frac{1}{n^3}\sum_{i=1}^{n}(i^2 - i)$$

$$= \lim_{n\to\infty}\left(\frac{(n+1)(2n+1)}{6n^2} - \frac{n+1}{2n^2}\right)$$

$$= \frac{2}{6} - 0 = \frac{1}{3}$$

31. Riemann sum $= \displaystyle\frac{2}{8}\sum_{i=1}^{8} c_i^2 = 2.65625$

33. Riemann sum $= \displaystyle\frac{3}{8}\sum_{i=1}^{8}\sqrt{c_i + 1} \approx 4.668$

35. **(a)** Left endpoints:

$$\int_0^{1.6} f(x)\,dx$$

$$\approx \frac{1.6 - 0}{8}(f(0) + f(0.2) + f(0.4)$$
$$+ f(0.6) + f(0.8) + f(1) + f(1.2)$$
$$+ f(1.4))$$

$$= \frac{1}{5}(1 + 1.4 + 1.6 + 2 + 2.2 + 2.4$$
$$+ 2 + 1.6)$$
$$= 2.84$$

(b) Right endpoints:

$$\int_0^{1.6} f(x)\,dx$$

$$\approx \frac{1.6 - 0}{8}(f(0.2) + f(0.4) + f(0.6)$$
$$+ f(0.8) + f(1) + f(1.2) + f(1.4)$$
$$+ f(1.6))$$

$$= \frac{1}{5}(1.4 + 1.6 + 2 + 2.2 + 2.4$$
$$+ 2 + 1.6 + 1.4)$$
$$= 2.92$$

(c) Trapezoidal Rule: $\displaystyle\int_0^{1.6} f(x)\,dx$

$$\approx \frac{1.6 - 0}{2(8)}[f(0) + 2f(0.2) + 2f(0.4)$$
$$+ 2f(0.6) + 2f(0.8) + 2f(1)$$
$$+ 2f(1.2) + 2f(1.4) + f(1.6)]$$
$$= 2.88$$

(d) Simpson's Rule:

$$\int_0^{1.6} f(x)\,dx$$

$$\approx \frac{1.6 - 0}{3(8)}[f(0) + 4f(0.2) + 2f(0.4)$$
$$+ 4f(0.6) + 2f(0.8) + 4f(1)$$
$$+ 2f(1.2) + 4f(1.4) + f(1.6)]$$
$$\approx 2.907$$

37. See Example 7.9.
Simpson's Rule is expected to be most accurate.

39. We will compute the area A_n of n rectangles using right endpoints. In this case $\Delta x = \frac{1}{n}$ and $x_i = \frac{i}{n}$

$$A_n = \sum_{i=1}^{n} f(x_i)\Delta x = \frac{1}{n}\sum_{i=1}^{n} f\left(\frac{i}{n}\right)$$

$$= \frac{1}{n} \sum_{i=1}^{n} 2 \cdot \left(\frac{i}{n}\right)^2$$

$$= \frac{2}{n^3} \sum_{i=1}^{n} i^2$$

$$= \left(\frac{2}{n^3}\right) \frac{n(n+1)(2n+1)}{6}$$

$$= \frac{(n+1)(2n+1)}{3n^2}$$

Now, to find the integral, we take the limit:

$$\int_0^1 x^2 \, dx$$

$$= \lim_{n \to \infty} A_n$$

$$= \lim_{n \to \infty} \frac{(n+1)(2n+1)}{3n^2}$$

$$= \frac{2}{3}$$

41. Area $= \displaystyle\int_0^3 (3x - x^2) \, dx$

$$= \left(\frac{3x^2}{2} - \frac{x^3}{3}\right)\Big|_0^3 = \frac{9}{2}$$

43. The velocity is always positive, so distance traveled is equal to change in position.

$$\text{Dist} = \int_1^2 (40 - 10t) \, dt$$

$$= (40t - 5t^2)\Big|_1^2 = 25$$

45. $f_{ave} = \displaystyle\frac{1}{\pi/2 - 0} \int_0^{\pi/2} \cos x \, dx$

$$= \frac{2}{\pi} \sin x \Big|_0^{\pi/2} = \frac{2}{\pi}$$

47. $\displaystyle\int_0^2 (x^2 - 2) \, dx = \left(\frac{x^3}{3} - 2x\right)\Big|_0^2 = -\frac{4}{3}$

49. $\displaystyle\int_0^{\pi/2} \sin 2x \, dx = -\frac{1}{2} \cos 2x \Big|_0^{\pi/2} = 1$

51. $\displaystyle\int_0^{10} (1 - \sqrt{t}) \, dt = \left[t - \frac{2}{3} t^{3/2}\right]\Big|_0^{10}$

$$= 10 - \frac{20}{3}\sqrt{10}$$

53. Let $u = x^3 - 1$. Then $du = 3x^2 \, dx$, $u(0) = -1$, and $u(2) = 7$.

$$\int_0^2 x^2 (x^3 - 1)^3 \, dx = \frac{1}{3} \int_{-1}^7 u^3 \, du = \frac{1}{12} u^4 \Big|_{-1}^7$$

$$= 200$$

55. $\displaystyle\int_0^2 x\sqrt{x^2 + 4} \, dx$

$$= \left(\frac{1}{2} \cdot \frac{2}{3} \cdot (x^2 + 4)^{3/2}\right)\Big|_0^2$$

$$= \frac{16\sqrt{2} - 8}{3}$$

57. $\displaystyle\int_1^2 \left(x + \frac{1}{x}\right) dx = \int_1^2 (x^2 + 2 + x^{-2}) \, dx$

$$= \left[\frac{x^3}{3} + 2x - \frac{1}{x}\right]\Big|_1^2 = \frac{29}{6}$$

59. $f'(x) = \sin x^2 - 2$

61. (a) Midpoint Rule:

$$\int_0^1 \sqrt{x^2 + 4} \, dx$$

$$\approx \frac{1 - 0}{4} \left[f\left(\frac{1}{8}\right) + f\left(\frac{3}{8}\right) \right. $$
$$\left. + f\left(\frac{5}{8}\right) + f\left(\frac{7}{8}\right) \right]$$

$$\approx 2.079$$

(b) Trapezoidal Rule:

$$\int_0^1 \sqrt{x^2 + 4} \, dx$$

$$\approx \frac{1 - 0}{2(4)} \left[f(0) + 2f\left(\frac{1}{4}\right) \right.$$

$$\left. + 2f\left(\frac{1}{2}\right) + 2f\left(\frac{3}{4}\right) \right.$$

$$+ f(1) \Big]$$

$$\approx 2.083$$

(c) Simpson's Rule: $\displaystyle\int_0^1 \sqrt{x^2 + 4}\, dx$

$$\approx \frac{1-0}{3(4)} \left[f(0) + 4f\left(\frac{1}{4}\right) \right.$$

$$\left. +2f\left(\frac{1}{2}\right) + 4f\left(\frac{3}{4}\right) + f(1) \right]$$

$$\approx 2.080$$

63.

n	Midpoint	Trapezoid	Simpson's
20	2.08041	2.08055	2.08046
40	2.08045	2.08048	2.08046

Applications of the Definite Integral

5.1 AREA BETWEEN CURVES

1.
$$\text{Area} = \int_1^3 \left[x^3 - (x^2 - 1) \right] dx$$
$$= \left(\frac{x^4}{4} - \frac{x^3}{3} + x \right) \Bigg|_1^3$$
$$= \left(\frac{81}{4} - \frac{27}{3} + 3 \right) - \left(\frac{1}{4} - \frac{1}{3} + 1 \right)$$
$$= \frac{160}{12} = \frac{40}{3}$$

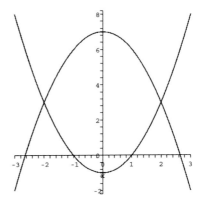

3.
$$\text{Area} = \int_{-2}^0 \left[x^4 - (x - 1) \right] dx$$
$$= \left(\frac{1}{5} x^5 - \frac{1}{2} x^2 + x \right)_{-2}^0$$
$$= 0 - \left(-\frac{32}{5} - 2 - 2 \right) = \frac{52}{5}$$

5.
$$\text{Area} = \int_{-2}^2 \left[7 - x^2 - \left(x^2 - 1 \right) \right] dx$$
$$= \left(8x - \frac{2x^3}{3} \right) \Bigg|_{-2}^2$$
$$= \left(16 - \frac{16}{3} \right) - \left(-16 + \frac{16}{3} \right)$$
$$= \frac{64}{3}$$

7.
$$\text{Area} = \int_{-1}^2 (3x + 2 - x^3) \, dx$$
$$= \left(\frac{3}{2} x^2 + 2x - \frac{1}{4} x^4 \right) \Bigg|_{-1}^2$$
$$= (6 + 4 - 4) - \left(\frac{3}{2} - 2 - \frac{1}{4} \right)$$
$$= \frac{27}{4}$$

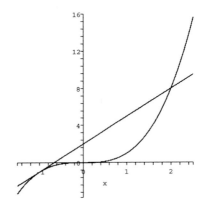

9.

$$\text{Area} = \int_{-1}^{1} (2 - x^2 - |x|) \, dx$$

$$= \int_{-1}^{0} (2 - x^2 + x) \, dx + \int_{0}^{1} (2 - x^2 - x) \, dx$$

$$= \left(2x - \frac{1}{3}x^3 + \frac{1}{2}x^2 \right) \Big|_{-1}^{0}$$

$$+ \left(2x - \frac{1}{3}x^3 - \frac{1}{2}x^2 \right) \Big|_{0}^{1}$$

$$= 0 - \left(-2 + \frac{1}{3} + \frac{1}{2} \right) + \left(2 - \frac{1}{3} - \frac{1}{2} \right)$$

$$- 0 = \frac{7}{3}$$

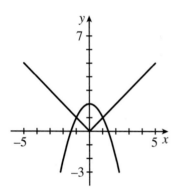

11.

$$\text{Area} = \int_{-2}^{0} \left[x - (x^2 - 6) \right] \, dx$$

$$= \left(\frac{1}{2}x^2 - \frac{1}{3}x^3 + 6x \right) \Big|_{-2}^{3}$$

$$= \left(\frac{9}{2} - 9 + 18 \right) - \left(2 + \frac{8}{3} - 12 \right) = \frac{125}{6}$$

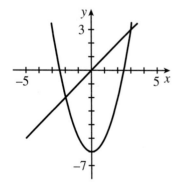

13.

$$\text{Area} = \int_{-1}^{1.35321} (2 + x - x^4) \, dx$$

$$= \left(2x + \frac{1}{2}x^2 - \frac{1}{5}x^5 \right) \Big|_{-1}^{1.35321}$$

$$= 4.01449$$

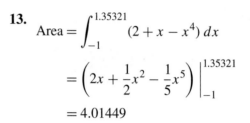

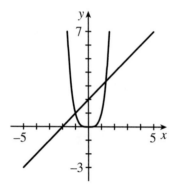

15.

$$\text{Area} = \int_{0}^{0.8767} \left(\sin x - x^2 \right) \, dx$$

$$= \left(-\cos x - \frac{x^3}{3} \right) \Big|_{0}^{0.8767}$$

$$\approx 0.135698$$

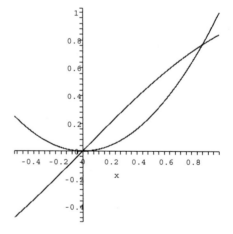

17.

$$\text{Area} = \int_{0}^{1} [(2 - y) - y] \, dy$$

$$= \int_{0}^{1} [2 - 2y] \, dy$$

$$= \left(2y - y^2 \right) \Big|_{0}^{1}$$

$$= 1 - 0 = 1$$

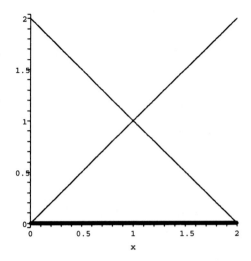

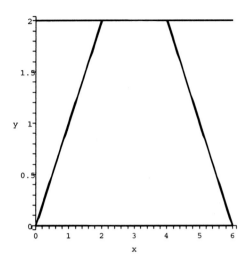

19.
$$\text{Area} = \int_0^1 [x - (-x)]\, dx$$
$$= 2\int_0^1 x\, dx = x^2 \Big|_0^1$$
$$= 1 - 0 = 1$$

23.
$$A = \frac{1}{b-a}\int_a^b f(x)\, dx$$
$$= \frac{1}{3-0}\int_0^3 x^2\, dx = \left(\frac{1}{3}\cdot\frac{x^3}{3}\right)\Big|_0^3$$
$$= \frac{27}{9} - 0 = 3$$

Relative to the interval [0, 3], the inequality $x^2 < 3$ holds only on the subinterval $[0, \sqrt{3})$. We find

$$\int_0^{\sqrt{3}} \left(3 - x^2\right)\, dx$$
$$= \left(3x - \frac{x^3}{3}\right)\Big|_0^{\sqrt{3}}$$
$$= (3\sqrt{3} - \sqrt{3}) - (0 - 0)$$
$$= 2\sqrt{3}, \text{ whereas}$$
$$\int_{\sqrt{3}}^3 \left(x^2 - 3\right)\, dx$$
$$= \left(\frac{x^3}{3} - 3x\right)\Big|_{\sqrt{3}}^3$$
$$= (9 - 9) - (\sqrt{3} - 3\sqrt{3})$$
$$= 2\sqrt{3}, \text{ the same.}$$

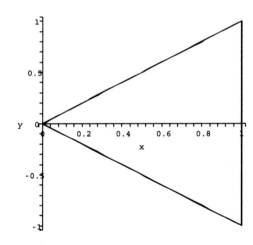

21.
$$\text{Area} = \int_0^2 [(6 - y) - y]\, dy$$
$$= \int_0^2 (6 - 2y)\, dy$$
$$= \left(6y - y^2\right)\Big|_0^2$$
$$= (12 - 4) - (0 - 0)$$
$$= 8$$

25. $f(4) = 16.1 + 1.4(4) = 21.7$
$g(4) = 19.7 + -0.5(4) = 21.7$

21.7 represents the consumption rate (million barrels per year) at time $t = 4$ (1/1/74).

$$\int_4^{10} [f(t) - g(t)] \, dt$$

$$= \int_4^{10} [(16.1 + 1.4t) - (19.7 + 0.5t)] \, dt$$

$$= \int_4^{10} (0.9t - 3.6) \, dt$$

$$= (0.45t^2 - 3.6t) \Big|_4^{10}$$

$$= 16.2 \text{ million barrels}$$

27. $b(0) = d(0) = 2$ and $b'(t) = 0.1 > 0.06 = d'(t)$, so $b(t) \geq d(t)$ for $t \geq 0$. The area between the graphs represents the difference between the number of births and the number of deaths.

$$\int_0^{10} [b(t) - d(t)] \, dt$$

$$= \int_0^{10} [(2 + 0.1t) - (2 + 0.06t)] \, dt$$

$$= \int_0^{10} 0.04t \, dt$$

$$= 0.02t^2 \Big|_0^{10}$$

$$= 200 \text{ million}$$

29.
$$\int_0^{0.4} f_c(x) \, dx \approx \frac{0.4}{3(4)} \{ f_c(0) + 4 f_c(0.1)$$
$$+ 2 f_c(0.2) + 4 f_c(0.3) + f_c(0.4) \} = 291.67$$
$$\int_0^{0.4} f_e(x) \, dx \approx \frac{0.4}{3(4)} \{ f_e(0) + 4 f_e(0.1)$$
$$+ 2 f_e(0.2) + 4 f_e(0.3) + f_e(0.4) \} = 102.33$$
$$\frac{\int_0^{0.4} f_c(x) - \int_0^{0.4} f_e(x)}{\int_0^4 f_c(x)} \approx \frac{291.67 - 102.33}{291.67}$$
$$\approx 0.6492$$

$1 - 0.6492 = 0.3508$, so the proportion of energy retained is about 35.08%.

31.
$$\int_0^3 f_s(x) \, dx \approx \frac{3}{3(4)} \{ f_s(0) + 4 f_s(0.75)$$
$$+ 2 f_s(1.5) + 4 f_s(2.25) + f_s(3) \} = 860$$
$$\int_0^3 f_r(x) \, dx \approx \frac{3}{3(4)} \{ f_r(0) + 4 f_r(0.75)$$
$$+ 2 f_r(1.5) + 4 f_r(2.25) + f_r(3) \} = 800$$
$$1 - \left(\frac{860 - 800}{860} \right) \approx 0.9302$$

Energy returned by the tendon is 93.02%.

33. These integrals represent the difference in the distances traveled by the two runners over the time periods in question. If it was a race, the first runner would be ahead by 2 miles after a time of π and the second runner would have caught up to the first runner by the time $t = 2\pi$.

$$\int_0^\pi [f(t) - g(t)] \, dt$$

$$= \int_0^\pi [10 - (10 - \sin t)] \, dt = (-\cos t) \Big|_0^\pi = 2$$

$$\int_0^{2\pi} [f(t) - g(t)] \, dt$$

$$= \int_0^{2\pi} [10 - (10 - \sin t)] \, dt = (-\cos t) \Big|_0^{2\pi} = 0$$

35. Without formulae or tables, only rough or qualitative estimates are possible.

Time	1	2	3	4	5
Amount	397	403	401	412	455

37. In this setup, p is price and q is quantity. We find that $D(q) = S(q)$ only if

$$10 - \frac{q}{40} = 2 + \frac{q}{120} + \frac{q^2}{1200}$$

$$12{,}000 - 30q = 2400 + 10q + q^2$$

$$q^2 + 40q - 9600 = 0$$

$$(q - 80)(q + 120) = 0$$

Within the range of the graph, the only solution is at $q = 80$. Thus $q^* = 80$ and $p^* = D(q^*) = S(q^*) = 8$.

Consumer surplus, as an area, is that part of the picture below the D curve, above $p = p^*$, and to the left of $Q = q^*$.

Numerically in this case the consumer surplus is

$$\int_0^{q^*} D(q)\, dq - p^* q^*$$

$$= \int_0^{80} \left(10 - \frac{q}{40}\right) dq - 80(8)$$

$$= \left(10q - \frac{q^2}{80}\right)\bigg|_0^{80} - 640$$

$$= 800 - 80 - 640 = 80$$

The units are dollars (q counting items, p in dollars per item).

39. The curves, meeting as they do at 2 and 5, represent the derivatives C' and R'. The area (*a*) between the curves over the interval $[0, 2]$ is the loss resulting from the production of the first 2000 items. The area (*b*) between the curves over the interval $[2, 5]$ is the profit resulting from the production of the next 3000 items. The area (*c*), as the *sum* of the two previous (call it $(a) + (b)$), is *without meaning*. However, the *difference* $(b) - (a)$ would be the total profit on the first 5000 items, or, if negative, would represent the loss. The area (*d*) between the curves over the interval $[5, 6]$ represents the loss attributable to the (unprofitable) production of the next thousand items after the first 5000.

41. Assume $a < 0$ so that $ax^2 + bx + c > mx + n$ on $[a, b]$.

Area $= \int_A^B [(ax^2 + bx + c) - (mx + n)]\, dx$

Since the curves intersect at $x = A$ and $x = B$, A and B are both solutions of

$ax^2 + bx + c = mx + n$. So

$$a \cdot A^2 + b \cdot A + c = m \cdot A + n$$
$$a \cdot B^2 + b \cdot B + c = m \cdot B + n.$$

Subtracting the second equation from the first and then dividing by $A - B$, we have

$$a(A^2 - B^2) + b(A - B) = m(A - B)$$
$$a(A + B) + b = m.$$

Thus,

$$a \cdot A^2 + b \cdot A + c = [a(A + B) + b] \cdot A + n$$
$$- a \cdot A^2 + a \cdot AB + b \cdot A + n.$$

So, $n = c - a \cdot AB$.

Area $= \displaystyle\int_A^B [(ax^2 + bx + c) - ((a(A + B) + b)$

$$x + c - a \cdot AB)]\, dx$$

$$= \int_A^B [(ax^2 - a(A + B)x + a \cdot AB)]\, dx$$

$$= a\left[\frac{1}{3}x^3 - \frac{A + B}{2}x^2 + ABx\right]\bigg|_A^B$$

$$= \frac{a}{6}(2B^3 - 3AB^2 - 3B^3 + 6AB^2 - 2A^3$$
$$+ 3A^3 + 3A^2B - 6A^2B)$$

$$= \frac{a}{6}(A^3 - 3A^2B + 3AB^2 - B^3)$$

$$= \frac{a}{6}(A - B)^3 = -\frac{a}{6}(B - A)^3$$

$$= \frac{|a|}{6}(B - A)^3, \text{ since } a < 0.$$

The case for $a > 0$ is similar.

43. Let the upper parabola be
$y = y_1 = qx^2 + v + h$
and let the lower be
$y = y_2 = px^2 + v$. They are to meet at $x = w/2$, so we must have
$qw^2/4 + h = pw^2/4$, hence
$h = (p - q)w^2/4$ or $(q - p)w^2 = -4h$.

Using symmetry, the area between the curves is given by the integral

$$2 \int_0^{w/2} (y_1 - y_2)\, dx$$

$$= 2 \int_0^{w/2} [h + (q - p)x^2]\, dx$$

$$= 2[hw/2 + (q - p)w^3/24]$$

$$= w[h + (q - p)w^2/12]$$

$$= w[h - 4h/12] = (2/3)wh.$$

45. Solving for x in $x - x^2 = L$, we get

$$x = \frac{1 \pm \sqrt{1 - 4L}}{2}$$

$$A_1 = \int_0^{(1-\sqrt{1-4L})/2} [L - (x - x^2)]\, dx$$

$$= \left(Lx - \frac{x^2}{2} + \frac{x^3}{3} \right) \Big|_0^{(1-\sqrt{1-4L})/2}$$

$$A_2 = \int_{(1-\sqrt{1-4L})/2}^{(1+\sqrt{1-4L})/2} [(x - x^2) - L]\, dx$$

$$= \left(\frac{x^2}{2} - \frac{x^3}{3} - Lx \right) \Big|_{(1-\sqrt{1-4L})/2}^{(1+\sqrt{1-4L})/2}.$$

By setting $A_1 = A_2$, we get the final answer:

$$L = \frac{3}{16}.$$

5.2 VOLUME: SLICING, DISKS, AND WASHERS

1.
$$V = \int_{-1}^3 A(x)\, dx = \int_{-1}^3 (x + 2)\, dx$$

$$= \left(\frac{x^2}{2} + 2x \right) \Big|_{-1}^3$$

$$= \left(\frac{9}{2} + 6 \right) - \left(\frac{1}{2} - 2 \right)$$

$$= 12$$

3.
$$V = \pi \int_0^2 (4 - x)^2\, dx$$

$$= -\frac{\pi}{3} (4 - x)^3 \Big|_0^2$$

$$= -\frac{\pi}{3}(8 - 64) = \frac{56\pi}{3}$$

5.
$$V = \int_0^{60} \pi x^2\, dy$$

$$= \pi \int_0^{60} 60[60 - y]\, dy$$

$$= 60\pi \left(60y - \frac{y^2}{2} \right) \Big|_0^{60}$$

$$= 60\pi \left[60^2 - \frac{60^2}{2} \right]$$

$$= \frac{60^3 \pi}{2} = 108{,}000\pi \ \text{ft}^3$$

7. $f(0) = 750, f(500) = 0$

$$f(x) = -\tfrac{3}{2}x + 750$$

$$V = \int_0^{500} \left(-\frac{3}{2}x + 750 \right)^2 dx$$

$$= -\frac{2}{9} \left(-\frac{3}{2}x + 750 \right)^3 \Big|_0^{500}$$

$$= -\frac{2}{9}(0 - 750^3) = 93{,}750{,}000 \ \text{ft}^3$$

which is one-third the area of the base times the height.

9. $f(0) = 3, f(30) = \dfrac{1}{2} \left(6 \ \text{in.} = \dfrac{1}{2} \ \text{ft} \right)$

$$f(x) = -\frac{1}{12}x + 3$$

$$V = \int_0^{30} \left(-\frac{1}{12}x + 3 \right)^2 dx$$

$$= -\frac{12}{3} \left(-\frac{1}{12}x + 3 \right)^3 \Big|_0^{30}$$

$$= -\frac{12}{3} \left(\frac{1}{8} - 27 \right) = \frac{215}{2} \ \text{ft}^3$$

11. The key to evaluating the integral is to simplify $\sin^2 \frac{x}{2}$ using the identity $\sin^2 \alpha = \frac{1}{2}(1 - \cos 2\alpha)$.

$$V = \pi \int_0^{2\pi} \left(4 + \sin \frac{x}{2}\right)^2 dx$$

$$= \pi \int_0^{2\pi} \left(16 + 8 \sin \frac{x}{2} + \sin^2 \frac{x}{2}\right) dx$$

$$= \pi \int_0^{2\pi} \left(16 + 8 \sin \frac{x}{2} + \frac{1}{2}(1 - \cos x)\right) dx$$

$$= \pi \left(16x - 16 \cos \frac{x}{2} + \frac{1}{2}x - \frac{1}{2} \sin x\right) \Big|_0^{2\pi}$$

$$= 33\pi^2 + 32\pi \text{ in}^3$$

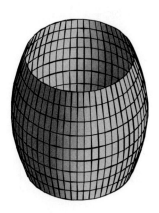

13.
$$V = \int_0^1 A(x)\, dx$$

$$\approx \frac{1}{3(10)} [A(0) + 4A(0.1) + 2A(0.2)$$
$$+ 4A(0.3) + 2A(0.4) + 4A(0.5)$$
$$+ 2A(0.6) + 4A(0.7) + 2A(0.8)$$
$$+ 4A(0.9) + A(1.0)]$$

$$= \frac{7.4}{30} \approx 0.2467 \text{ cm}^3$$

15.
$$V = \int_0^2 A(x)\, dx$$

$$\approx \frac{2}{3(4)} [A(0) + 4A(0.5) + 2A(1)$$
$$+ 4A(1.5) + A(2)]$$

$$= 2.5 \text{ ft}^3$$

17. (a)
$$V = \pi \int_0^2 (2 - x)^2\, dx$$

$$= -\pi \left(\frac{(2-x)^3}{3}\right) \Big|_0^2$$

$$= \frac{8\pi}{3}$$

(b)
$$V = \pi \int_0^2 \left[3^2 - \{3 - (2 - x)\}^2\right] dx$$

$$= \pi \int_0^2 \left[9 - \{1 + x\}^2\right] dx$$

$$= \pi \left[9x \Big|_0^2 - \frac{(1+x)^3}{3} \Big|_0^2\right]$$

$$= \pi \left[18 - \frac{3^3 - 1^3}{3}\right] = \frac{28\pi}{3}$$

19. (a)
$$V = \pi \int_0^2 (y^2)^2 dy = \pi \int_0^2 y^4 dy$$

$$= \pi \left(\frac{y^5}{5}\right) \Big|_0^2 = \frac{32\pi}{5}$$

(b)
$$V = \pi \int_0^2 (4)^2 dy$$

$$- \pi \int_0^2 (4 - y^2)^2 dy$$

$$= \pi \int_0^2 (-y^4 + 8y^2) dy$$

$$= \pi \left(-\frac{y^5}{5} + \frac{8y^3}{3}\right) \Big|_0^2$$

$$= \pi \left[\left(-\frac{32}{5} + \frac{64}{3}\right) - (0 + 0)\right]$$

$$= \frac{224\pi}{15}$$

21. (a)

$$V = \int_0^2 \pi(2)^2\, dy + \int_2^{\sqrt{6}}$$

$$\pi[2^2 - (y^2 - 4)^2]\, dy$$

$$= 4\pi y \big|_0^2 + \int_2^{\sqrt{6}} \pi(-y^4 + 8y^2 - 12)\, dy$$

$$= 8\pi + \pi \left(-\frac{1}{5}y^5 + \frac{8}{3}y^3 - 12y\right)\Big|_2^{\sqrt{6}}$$

$$= \pi \left(\frac{256}{15} - \frac{16}{5}\sqrt{6}\right) \approx 28.99$$

(b)

$$V = \int_0^2 \pi(2 + \sqrt{x+4})^2\, dx$$

$$- \int_0^2 \pi(2)^2\, dx$$

$$= \pi \int_0^2 (4\sqrt{x+4} + x + 4)\, dx$$

$$= \pi \left(\frac{8}{3}(x+4)^{3/2} + \frac{1}{2}x^2 + 4x\right)\Big|_0^2$$

$$= \frac{2\pi}{3}(24\sqrt{6} - 17)$$

23. (a) Use a numerical method to approximate the integral.

$$V = \pi \int_0^1 \left(\sqrt{\frac{x}{x^2+2}}\right)^2 dx$$

$$= \pi \int_0^1 \frac{x}{x^2+2}\, dx$$

$$\approx 0.637$$

(b) Use a numerical method to approximate the integral.

$$V = \pi \int_0^1 \left[3^2 - \left(3 - \sqrt{\frac{x}{x^2+2}}\right)^2\right] dx$$

$$= \pi \int_0^1 \left(6\sqrt{\frac{x}{x^2+2}} - \frac{x}{x^2+2}\right) dx$$

$$\approx 7.4721$$

25. (a)

$$V = \pi \int_0^3 (3-y)^2\, dy$$

$$= \pi \int_0^3 \left(9 - 6y + y^2\right) dy$$

$$= \pi \left(9y - 3y^2 + \frac{y^3}{3}\right)\Big|_0^3$$

$$= 9\pi$$

(b)

$$V = \pi \int_0^3 (3-x)^2\, dx$$

$$= \pi \int_0^3 \left(9 - 6x + x^2\right) dx$$

$$= \pi \left(9x - 3x^2 + \frac{x^3}{3}\right)\Big|_0^3$$

$$= 9\pi$$

(c)

$$V = \pi \int_0^3 \left[(3)^2 - (3 - (3-x))^2\right] dx$$

$$= \pi \int_0^3 [9 - x^2]\, dx$$

$$= \pi \left(9x - \frac{x^3}{3}\right)\Big|_0^3$$

$$= 18\pi$$

(d)

$$V = \pi \int_0^3 \left[((3-x) + 3)^2 - 3^2\right] dx$$

$$= \pi \int_0^3 \left[(6-x)^2 - 9\right] dx$$

$$= \pi \int_0^3 \left(27 - 12x + x^2\right) dx$$

$$= \pi \left(27x - 6x^2 + \frac{x^3}{3}\right)\Big|_0^3$$

$$= 36\pi$$

(e)
$$V = \pi \int_0^3 \left[(3)^2 - (3 - (3 - y))^2 \right] dy$$
$$= \pi \int_0^3 \left(9 - y^2 \right) dy$$
$$= \pi \left(9y - \frac{y^3}{3} \right) \Big|_0^3$$
$$= 18\pi$$

(f)
$$V = \pi \int_0^3 [((3 - y) + 3)^2 - (3)^2] \, dy$$
$$= \pi \int_0^3 \left[(6 - y)^2 - 9 \right] dy$$
$$= \pi \int_0^3 (27 - 12y + y^2) \, dy$$
$$= \pi \left(27y - 6y^2 + \frac{y^3}{3} \right) \Big|_0^3$$
$$= 36\pi$$

27. (a)
$$V = \int_0^1 \pi \left[1^2 - \left(\sqrt{y} \right)^2 \right] dy$$
$$= \pi \int_0^1 (1 - y) \, dy$$
$$= \pi \left(y - \frac{y^2}{2} \right) \Big|_0^1$$
$$= \frac{\pi}{2}$$

(b)
$$V = \int_0^1 \pi \left(x^2 \right)^2 dx$$
$$= \pi \frac{x^5}{5} \Big|_0^1 = \frac{\pi}{5}$$

(c)
$$V = \int_0^1 \pi \left(1 - \sqrt{y} \right)^2 dy$$
$$= \pi \int_0^1 \left(1 - 2y^{1/2} + y \right) dy$$
$$= \pi \left(y - \frac{4}{3} y^{3/2} + \frac{y^2}{2} \right) \Big|_0^1$$
$$= \frac{\pi}{6}$$

(d)
$$V = \int_0^1 \pi (1)^2 \, dx - \int_0^1 \pi \left(1 - x^2 \right)^2 dx$$
$$= \pi \int_0^1 \left(2x^2 - x^4 \right) dx$$
$$= \pi \left(\frac{2}{3} x^3 - \frac{x^5}{5} \right) \Big|_0^1$$
$$= \frac{7\pi}{15}$$

(e)
$$V = \int_0^1 \pi (2)^2 \, dy - \int \pi \left(1 + \sqrt{y} \right)^2 dy$$
$$= \pi \int_0^1 \left(3 - 2y^{1/2} - y \right) dy$$
$$= \pi \left(3y - \frac{4}{3} y^{3/2} - \frac{y^2}{2} \right) \Big|_0^1$$
$$= \frac{7\pi}{6}$$

(f)
$$V = \int_0^1 \pi \left(x^2 + 1 \right)^2 dx$$
$$- \int_0^1 \pi (1)^2 \, dx$$
$$= \pi \int_0^1 \left(x^4 + 2x^2 \right) dx$$
$$= \pi \left(\frac{x^5}{5} + \frac{2}{3} x^3 \right) \Big|_0^1$$
$$= \frac{13\pi}{15}$$

29.

$$V = \pi \int_0^h \left(\sqrt{\frac{y}{a}}\right)^2 dy$$

$$= \frac{\pi}{a} \int_0^h y \, dy = \frac{\pi h^2}{2a}$$

The volume of a cylinder of height h and radius $\sqrt{h/a}$ is

$$h \cdot \pi (\sqrt{h/a})^2 = \frac{\pi h^2}{a}.$$

31. We can choose either x or y to be our integration variable,

$$V = \pi \int_{-1}^1 1 \, dx = \pi x \Big|_{-1}^1 = 2\pi$$

33. The line connecting the two points $(0, 1)$ and $(1, -1)$ has equation $y = -2x + 1$, or $x = \frac{1-y}{2}$

$$V = \int_{-1}^1 \pi \left(\frac{1-y}{2}\right)^2 dy$$

$$= \pi \left(\frac{y}{4} - \frac{y^2}{4} + \frac{y^3}{12}\right) \Big|_{-1}^1$$

$$= \frac{2\pi}{3}$$

35.

$$V = \pi \int_{-r}^r \left(\sqrt{r^2 - y^2}\right)^2 dy$$

$$= \pi \int_{-r}^r (r^2 - y^2) \, dy$$

$$= \pi \left(r^2 y - \frac{y^3}{3}\right) \Big|_{-r}^r$$

$$= \frac{4}{3}\pi r^3$$

37. If we compute the two volumes using disks parallel to the base, we have identical cross sections, so the volumes are the same.

39. **(a)** If each of these line segments is the base of square, then the cross-sectional area is evidently

$$A(x) = 4(1 - x^2).$$

The volume would be

$$V_a = \int_{-1}^1 A(x) \, dx$$

$$= 2 \int_0^1 A(x) \, dx$$

$$= 8 \left(x - \frac{x^3}{3}\right) \Big|_0^1 = \frac{16}{3}$$

(b) These segments I_x cannot be the literal "bases" of circles, because circles "sit" on a single point of tangency. They could however be diameters. Assuming so, the cross sectional area would be "$\pi/2$ times radius-squared" or $\pi(1 - x^2)/2$. The resulting volume would be $\pi/8$ times the previous case, or $2\pi/3$. (The resulting solid is a hemisphere of radius 1, so the volume is $\frac{1}{2}$ that of a sphere of radius 1.)

41. The distance across the region for any x, $-1 \le x \le 1$, is $(2 - x^2) - x^2 = 2(1 - x^2)$.

(a) $A(x) = 4(1 - x^2)^2 = 4(1 - 2x^2 + x^4)$

The volume is

$$V = 2 \int_0^1 A(x) \, dx$$

$$= 8 \left(x - \frac{2x^3}{3} + \frac{x^5}{5}\right) \Big|_0^1$$

$$= 8 \left(1 - \frac{2}{3} + \frac{1}{5}\right) = \frac{64}{15}.$$

(b) The radius of the semicircular cross section for x, $-1 \le x \le 1$, is half the distance across the region, $1 - x^2$, so $A(x) = \frac{\pi}{2}(1 - x^2)^2$.

$$V = \int_{-1}^{1} \frac{\pi}{2}(1-x^2)^2\,dx$$

$$= \frac{\pi}{2}\int_{-1}^{1}(1-2x^2+x^4)\,dx$$

$$= \frac{\pi}{2}\left[x - \frac{2}{3}x^3 + \frac{1}{5}x^5\right]\Big|_{-1}^{1}$$

$$= \frac{8\pi}{15}$$

(c) The base of each equilateral triangle is $2(1-x^2)$ and the altitude is $\sqrt{3}(1-x^2)$, so

$$A(x) = \frac{1}{2}\left[2(1-x^2)\right]\left[\sqrt{3}(1-x^2)\right] = \sqrt{3}(1-x^2)^2.$$

$$V = \int_{-1}^{1}\sqrt{3}(1-x^2)^2\,dx$$

$$= \sqrt{3}\int_{-1}^{1}(1-2x^2+x^4)\,dx$$

$$= \sqrt{3}\left[x - \frac{2}{3}x^3 + \frac{1}{5}x^5\right]\Big|_{-1}^{1}$$

$$= \frac{16\sqrt{3}}{15}$$

43. We must estimate $\pi \int_0^3 [f(x)]^2\,dx$.

The given table can be extended to give these respective values for $f(x)^2$: 4, 1.44, 0.81, 0.16, 1.0, 1.96, 2.56. Simpson's approximation to the integral would be

$$\frac{3}{(3)(6)}\{4 + 4(1.44) + 2(0.81)$$

$$+4(0.16) + 2(1.0) + 4(1.96) + 2.56\}$$

The sum in the braces is 24.42, and this must be multiplied by $\pi/6$ giving a final answer of 12.786.

45. Let $h(t)$ be the height of the water at time t. The height is always increasing, so h is an increasing function for all t in its domain. Initially, h increases quickly since the vase is narrow at the bottom, but h grows more slowly as the vase widens, so the graph starts out concave down. As the top of the water reaches the point where the vase begins to narrow, h will increase more quickly again, so the graph will be concave up. Finally, as the top of the water reaches the point where the vase begins to widen again, the graph will switch back to concave down. In summary, the graph of h is increasing with two inflection points.

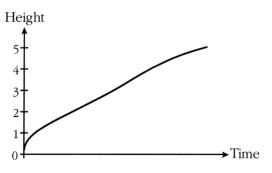

5.3 VOLUMES BY CYLINDRICAL SHELLS

1. Radius of a shell: $r = 2 - x$
Height of a shell: $h = x^2$

$$V = \int_{-1}^{1} 2\pi(2-x)x^2\,dx$$

$$= 2\pi\left(\frac{2x^3}{3} - \frac{x^4}{4}\right)\Big|_{-1}^{1}$$

$$= \frac{8\pi}{3}$$

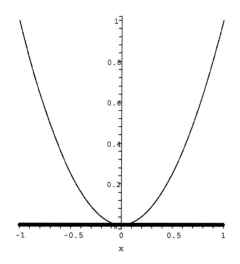

3. Radius of a shell: $r = x$
Height of a shell: $h = 2x$

$$V = \int_0^1 2\pi x(2x)\,dx$$

$$= \frac{4\pi}{3}x^3\Big|_0^1 = \frac{4\pi}{3}$$

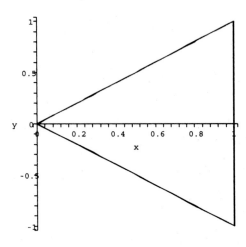

5. Radius of a shell: $r = 3 - y$
Height of a shell: $h = 2y$

$$V = \int_0^2 2\pi(3 - y)(2y)\,dy$$

$$= 4\pi\left(\frac{3}{2}y^2 - \frac{1}{3}y^3\right)\Big|_0^2 = \frac{40\pi}{3}$$

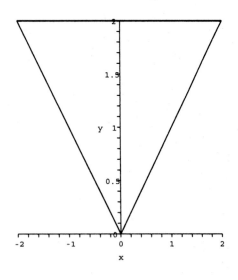

7. Radius of a shell: $r = y$
Height of a shell: $h = \sqrt{1 - (y - 1)^2}$

$$V = \int_0^2 2\pi y\sqrt{1 - (y - 1)^2}\,dy$$

$$= \int_{-1}^1 2\pi(u + 1)\sqrt{1 - u^2}\,du$$

$$= \int_{-1}^1 2\pi u\sqrt{1 - u^2}\,du$$

$$+ \int_{-1}^1 2\pi\sqrt{1 - u^2}\,du$$

The value of the first integral is 0. The integral $\int_{-1}^1 \sqrt{1 - u^2}\,du$ is the area of a semicircle of radius 1, so $\int_{-1}^1 \sqrt{1 - u^2}\,du = \frac{1}{2}\pi$.

$$V = 0 + \frac{1}{2}\pi \cdot 2\pi = \pi^2$$

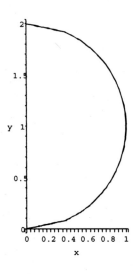

9. $$V = \int_{-1}^1 2\pi(x + 2)\left((2 - x^2) - x^2\right)dx$$

$$= 2\pi\int_{-1}^1 \left(4 + 2x - 4x^2 - 2x^3\right)dx$$

$$= 2\pi\left(4x + x^2 - \frac{4x^3}{3} - \frac{x^4}{2}\right)\Big|_{-1}^1$$

$$= \frac{32\pi}{3}$$

11.
$$V = \int_{-2}^{2} 2\pi(y+2)(4-y^2)\, dy$$

$$= \int_{-2}^{2} (4y - y^3 + 8 - 2y^2)\, dy$$

$$= 2\pi \left(2y^2 - \frac{1}{4}y^4 + 8y - \frac{2}{3}y^3 \right) \Big|_{-2}^{2}$$

$$= \frac{128\pi}{3}$$

13.
$$V = \int_{-1}^{2} 2\pi(2-x)(x-(x^2-2))\, dx$$

$$= 2\pi \int_{-1}^{2} \left(4 - 3x^2 + x^3 \right) dx$$

$$= 2\pi \left(4x - x^3 + \frac{x^4}{4} \right) \Big|_{-1}^{2}$$

$$= \frac{27\pi}{2}$$

15.
$$V = \int_{-2}^{4} 2\pi(5-y)[9-(y-1)^2]\, dy$$

$$= 2\pi \int_{-2}^{4} (y^3 - 7y^2 + 2y + 40)\, dy$$

$$= 2\pi \left(\frac{y^4}{4} - \frac{7y^3}{3} + y^2 + 40y \right) \Big|_{-2}^{4}$$

$$= 288\pi$$

17. (a)
$$V = \int_{2}^{4} 2\pi(y)\,(y-(4-y))\, dy$$

$$= 2\pi \int_{2}^{4} \left(2y^2 - 4y \right) dy$$

$$= 2\pi \left(\frac{2y^3}{3} - 2y^2 \right) \Big|_{2}^{4}$$

$$= \frac{80\pi}{3}$$

(b)
$$V = \int_{2}^{4} \pi[y^2 - (4-y)^2]\, dy$$

$$= \pi \int_{2}^{4} (8y - 16)\, dy$$

$$= \pi(4y^2 - 16y) \Big|_{2}^{4}$$

$$= 16\pi$$

(c)
$$V = \int_{2}^{4} \pi\,(4-(4-y))^2\, dy$$

$$\qquad - \int_{2}^{4} \pi(4-y)^2\, dy$$

$$= \pi \int_{2}^{4} y^2\, dy$$

$$\qquad - \pi \int_{2}^{4} (16 - 8y + y^2)\, dy$$

$$= \pi \int_{2}^{4} (-16 + 8y)\, dy$$

$$= \pi\,(-16y + 4y^2) \Big|_{2}^{4} = 16\pi$$

(d)
$$V = \int_{2}^{4} 2\pi(4-y)\,(y-(4-y))\, dy$$

$$= 2\pi \int_{2}^{4} \left(-2y^2 + 12y - 16 \right) dy$$

$$= 2\pi \left(-\frac{2y^3}{3} + 6y^2 - 16y \right) \Big|_{2}^{4}$$

$$= \frac{16\pi}{3}$$

19. (a) Method of shells.
$$V = \int_{-2}^{3} 2\pi(3-x)[x-(x^2-6)]\, dx$$

$$= \int_{-2}^{3} 2\pi(x^3 - 4x^2 - 3x + 18)\, dx$$

$$= 2\pi \left(\frac{1}{4}x^4 - \frac{4}{3}x^3 - \frac{3}{2}x^2 + 18x \right) \Big|_{-2}^{3}$$

$$= \frac{625\pi}{6}$$

(b) Method of washers.

$$V = \int_{-2}^{3} \pi \left[(3 - (x^2 - 6))^2 - (3 - x)^2 \right] dx$$

$$= \pi \int_{-2}^{3} \left[(9 - x^2)^2 - (3 - x)^2 \right] dx$$

$$= \pi \int_{-2}^{3} (72 + 6x - 19x^2 + x^4) \, dx$$

$$= \pi \left(72x + 3x^2 - \frac{19}{3}x^3 + \frac{1}{5}x^5 \right) \Big|_{-2}^{3}$$

$$= \frac{625\pi}{3}$$

(c) Method of shells.

$$V = \int_{-2}^{3} 2\pi(3 + x)[x - (x^2 - 6)] \, dx$$

$$= \int_{-2}^{3} 2\pi(-x^3 - 2x^2 + 9x + 18) \, dx$$

$$= 2\pi \left(-\frac{1}{4}x^4 - \frac{2}{3}x^3 + \frac{9}{2}x^2 + 18x \right) \Big|_{-2}^{3}$$

$$= \frac{875\pi}{6}$$

(d) Method of washers.

$$V = \int_{-2}^{3} \pi[(6 + x)^2 - (x^2)^2] \, dx$$

$$= \int_{-2}^{3} \pi(-x^4 + x^2 + 12x + 36) \, dx$$

$$= \pi \left(-\frac{1}{5}x^5 + \frac{1}{3}x^3 + 6x^2 + 36x \right) \Big|_{-2}^{3}$$

$$= \frac{500\pi}{3}$$

21. (a)

$$V \approx 2\pi \int_{-0.89}^{0.89} (2 - x)$$

$$\cdot (\cos x - x^4) \, dx$$

$$\approx 16.723$$

(b)

$$V \approx \pi \int_{-0.89}^{0.89} [(2 - x^4)^2$$

$$- (2 - \cos x)^2] \, dx$$

$$\approx 12.635$$

(c)

$$V \approx \pi \int_{-0.89}^{0.89} [(\cos x)^2 - (x^4)^2] \, dx$$

$$\approx 4.088$$

(d)

$$V \approx 2\pi \int_{0}^{0.89} x(\cos x - x^4) \, dx$$

$$\approx 1.497$$

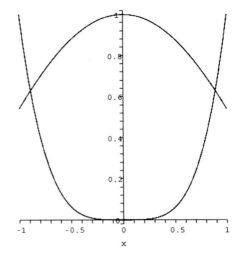

23. (a)

$$V = \int_{0}^{1} \pi(2 - x)^2 \, dx$$

$$- \int_{0}^{1} \pi \left(x^2 \right)^2 \, dx$$

$$= \pi \int_{0}^{1} (x^2 - 4x + 4) \, dx$$

$$- \pi \int_{0}^{1} x^4 \, dx$$

$$= \pi \int_{0}^{1} (-x^4 + x^2 - 4x + 4) \, dx$$

$$= \pi \left(-\frac{x^5}{5} + \frac{x^3}{3} - 2x^2 + 4x \right) \Big|_{0}^{1}$$

$$= \frac{32\pi}{15}$$

(b)

$$V = \int_0^1 2\pi x \left(2 - x - x^2\right) dx$$

$$= 2\pi \int_0^1 \left(2x - x^2 - x^3\right) dx$$

$$= 2\pi \left(x^2 - \frac{x^3}{3} - \frac{x^4}{4}\right) \Big|_0^1$$

$$= \frac{5\pi}{6}$$

(c)

$$V = \int_0^1 2\pi (1 - x)(2 - x - x^2) dx$$

$$= 2\pi \int_0^1 \left(x^3 - 3x + 2\right) dx$$

$$= 2\pi \left(\frac{x^4}{4} - \frac{3x^2}{2} + 2x\right) \Big|_0^1$$

$$= \frac{3\pi}{2}$$

(d)

$$V = \int_0^1 \pi \left(2 - x^2\right)^2 dx$$

$$\quad - \int_0^1 \pi \left(2 - (2 - x)\right)^2 dx$$

$$= \pi \int_0^1 (x^4 - 4x^2 + 4) dx$$

$$\quad - \pi \int_0^1 x^2 dx$$

$$= \pi \int_0^1 (x^4 - 5x^2 + 4) dx$$

$$= \pi \left(\frac{x^5}{5} - \frac{5x^3}{3} + 4x\right) \Big|_0^1$$

$$= \frac{38\pi}{15}$$

25. (a)

$$V = 2\pi \int_0^1 y(2 - y - y^2) dy$$

$$= 2\pi \int_0^1 (-y^3 - y^2 + 2y) dy$$

$$= 2\pi \left(-\frac{y^4}{4} - \frac{y^3}{3} + y^2\right) \Big|_0^1$$

$$= \frac{5\pi}{6}$$

(b)

$$V = 2\pi \int_0^1 [(2 - y)^2 - (y^2)^2] dy$$

$$= 2\pi \int_0^1 (-y^4 + y^2 - 4y + 4) dy$$

$$= 2\pi \left(-\frac{y^5}{5} + \frac{y^3}{3} - 2y^2 + 4y\right) \Big|_0^1$$

$$= \frac{64\pi}{15}$$

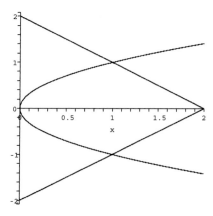

27. Axis of revolution: x-axis
Region bounded by:

$$y = 2x - x^2, \ y = 0$$

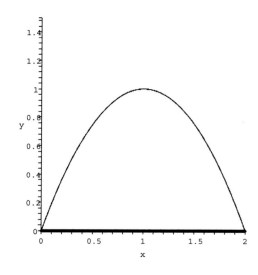

29. Axis of revolution: y-axis
Region bounded by: $x = \sqrt{y}$, $x = y$

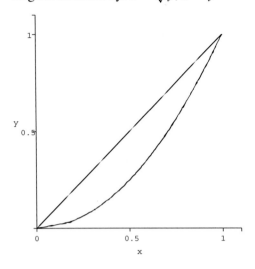

31. Axis of revolution: y-axis
Region bounded by: $y = x$, $y = x^2$

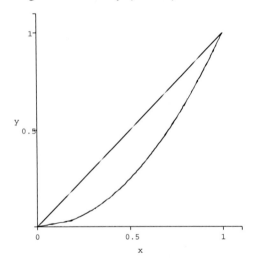

33. If the r-interval $[0, R]$ is partitioned by points r_i, the circular band

$$\left\{ r_i^2 \leq x^2 + y^2 \leq r_{i+1}^2 \right\}$$

has approximate area $c(r_i)\Delta r_i$ (length times thickness). The limit of the sum of these areas is

$$A = \lim \sum_{i=1}^{n} c(r_i)\Delta r_i = \int_0^R c(r)\,dr.$$

Because we know that $c(r) = 2\pi r$, we can evaluate the integral, getting

$$2\pi \frac{r^2}{2}\bigg|_0^R = \pi R^2.$$

35. The volume that we are looking for is twice the volume of a shell with radius x and height $\sqrt{1 - x^2}$. In other words, the bead is mathematically the solid formed up from revolving the region bounded by $y = \sqrt{1 - x^2}$, $x = 1/2$, and the x-axis around the y-axis. Therefore

$$V = 2 \cdot \int_{1/2}^1 2\pi x \sqrt{1 - x^2}\,dx.$$

Let $u = 1 - x^2$, $du = -2x\,dx$, and

$$V = 4\pi \int_{1/2}^1 x\sqrt{1 - x^2}\,dx$$

$$= -\frac{1}{2}\, 4\pi \int_{3/4}^0 u^{1/2}\,du$$

$$= 2\pi \cdot \frac{2}{3} u^{3/2}\bigg|_0^{3/4}$$

$$= \frac{\sqrt{3}\pi}{2}\,\mathrm{cm}^3.$$

37. $$V = \int_0^1 x(1 - x^2)\,dx$$

$$= \int_0^1 (x - x^3)\,dx$$

$$= \left(\frac{x^2}{2} - \frac{x^4}{4}\right)\bigg|_0^1 = \frac{1}{4}$$

$$V_1 = \int_c^1 x(1 - x^2)\,dx$$

$$= \left(\frac{x^2}{2} - \frac{x^4}{4}\right)\bigg|_c^1 = \frac{1}{4} - \frac{c^2}{2} + \frac{c^4}{4}$$

We want

$$V - V_1 = \frac{1}{10}V$$

Then

$$\frac{c^2}{2} - \frac{c^4}{4} = \frac{1}{40}$$
$$c \approx 0.2265$$

5.4 ARC LENGTH AND SURFACE AREA

1. For $n = 2$, the evaluation points are 0, 0.5, 1.

$$s \approx s_1 + s_2$$

$$= \sqrt{(0 - 0.5)^2 + [f(0) - f(0.5)]^2}$$

$$+ \sqrt{(1 - 0.5)^2 + [f(1) - f(0.5)]^2}$$

$$= \sqrt{0.5^2 + 0.5^4} + \sqrt{0.5^2 + 0.75^2}$$

$$\approx 1.4604$$

For $n = 4$, the evaluation points are 0, 0.25, 0.5, 0.75, 1.

$$s \approx \sum_{i=1}^{4} s_i \approx 1.4743$$

3. For $n = 2$, the evaluation points are $0, \pi/2, \pi$.

$$s \approx s_1 + s_2$$

$$= \sqrt{(\pi/2)^2 + [\cos(\pi/2) - \cos 0]^2}$$

$$+ \sqrt{(\pi/2)^2 + [\cos \pi - \cos(\pi/2)]^2}$$

$$\approx 3.7242$$

For $n = 4$, the evaluation points are $0, \pi/4, \pi/2, 3\pi/4, \pi$.

$$s \approx \sum_{i=1}^{4} s_i \approx 3.7901$$

5. This is a straight line segment from $(0, 1)$ to $(2, 5)$. As such, its length is

$$s = \sqrt{(5 - 1)^2 + (2 - 0)^2}$$

$$= \sqrt{20} = 2\sqrt{5}.$$

7. $y'(x) = 6x^{1/2}$, the arc length integrand is $\sqrt{1 + (y')^2} = \sqrt{1 + 36x}$.

Let $u = 1 + 36x$, then

$$s = \int_1^2 \sqrt{1 + 36x}\, dx$$

$$= \int_{37}^{73} \sqrt{u} \left(\frac{du}{36} \right)$$

$$= \frac{2}{3(36)} u^{3/2} \Big|_{37}^{73}$$

$$= \frac{1}{54}(73\sqrt{73} - 37\sqrt{37})$$

$$\approx 7.3824$$

9.

$$y'(x) = \frac{x^{1/2}}{2} - \frac{x^{-1/2}}{2}$$

$$= \frac{1}{2} \left(\sqrt{x} - \frac{1}{\sqrt{x}} \right),$$

$$1 + (y')^2 = 1 + \frac{1}{4} \left(x - 2 + \frac{1}{x} \right)$$

$$= \frac{1}{4} \left(x + 2 + \frac{1}{x} \right)$$

$$= \left[\frac{1}{2} \left(\sqrt{x} + \frac{1}{\sqrt{x}} \right) \right]^2 .$$

$$s = \int_1^4 \sqrt{1 + (y')^2}$$

$$= \frac{1}{2} \int_1^4 \left(\sqrt{x} + \frac{1}{\sqrt{x}} \right) dx$$

$$= \frac{x^{3/2}}{3} \Big|_1^4 + \sqrt{x} \Big|_1^4$$

$$= \frac{7}{3} + 1 = \frac{10}{3}$$

11.

$$s = \int_{-1}^1 \sqrt{1 + \left(3x^2 \right)^2}\, dx$$

$$= \int_{-1}^1 \sqrt{1 + 9x^4}\, dx \approx 3.0957$$

13. $s = \int_0^2 \sqrt{1 + (2 - 2x)^2}\, dx \approx 2.9579$

15.
$$s = \int_0^\pi \sqrt{1 + (-\sin x)^2}\, dx$$
$$= \int_0^\pi \sqrt{1 + \sin^2 x}\, dx \approx 3.8202$$

17. $s = \int_0^\pi \sqrt{1 + (x \sin x)^2}\, dx = 4.6984$

19.
$$y' = \frac{1}{5}x$$
$$s \int_{-10}^{10} \sqrt{1 + \left(\frac{1}{5}x\right)^2}\, dx \approx 29.58 \text{ ft}$$

21. For $x = \pm 10$, $y = 15$. For $x = 0$, $y = 10$. The sag is $15 - 10 = 5$ feet.

23. $y = 0$ when $x = 0$ and when $x = 60$, so the punt traveled 60 yards horizontally.

$$y'(x) = 4 - \frac{2}{15}x = \frac{2}{15}(30 - x).$$

This is zero only when $x = 30$, at which point the punt was $(30)^2/15 = 60$ yards high.

$$s = \int_0^{60} \sqrt{1 + \left(4 - \frac{2}{15}x\right)^2}\, dx$$
$$\approx 139.4 \text{ yards}$$
$$v = \frac{s}{4 \sec} = \frac{139.4 \text{ yards}}{4 \sec} \cdot \frac{3 \text{ feet}}{1 \text{ yard}}$$
$$= 104.55 \text{ ft/s}$$

25.
$$\frac{d}{dx} \sqrt{2} \int_0^x \sqrt{1 - \frac{\sin^2 u}{2}}\, du$$
$$= \frac{\sqrt{2}}{\sqrt{2}} \sqrt{2 - \sin^2 x}$$
$$= \sqrt{1 + \cos^2 x}$$

27. The antiderivatives returned by some CAS still include an integral, indicating that the CAS can't find an antiderivative in closed form (see exercise 28). In this case, numerical integration is the method of choice.

29.
$$S = 2\pi \int_0^1 x^2 \sqrt{1 + (2x)^2}\, dx$$
$$\approx 3.8097$$

31.
$$s = 2\pi \int_0^2 (2x - x^2)\sqrt{1 + (2 - 2x)^2}\, dx$$
$$\approx 10.9655$$

33.
$$s = 2\pi \int_0^{\pi/2} \cos x \sqrt{1 + \sin^2 x}\, dx$$
$$\approx 7.2118$$

35.
$$L_1 = \int_{-\pi/6}^{\pi/6} \sqrt{1 + \cos^2 x}\, dx \approx 1.44829$$

L_2

$$= \sqrt{\left(\sin\frac{\pi}{6} - \sin\left(-\frac{\pi}{6}\right)\right)^2 + \left(\frac{2\pi}{6}\right)^2}$$
$$\approx 1.44797$$

Hence

$$\frac{L_2}{L_1} = \frac{1.44797}{1.44829} \approx 0.9998.$$

37.
$$L_1 = \int_3^5 \sqrt{1 + (2x)^2}\, dx \approx 16.12716$$
$$L_2 = \sqrt{2^2 + (5^2 - 3^2)^2} \approx 16.12452$$

Hence

$$\frac{L_2}{L_1} \approx 0.99984.$$

39.

$$s_1 = \int_0^1 \sqrt{1 + (6x^5)^2}\, dx$$

$$= \int_0^1 \sqrt{1 + 36x^{10}}\, dx \approx 1.672$$

$$s_2 = \int_0^1 \sqrt{1 + (8x^7)^2}\, dx$$

$$= \int_0^1 \sqrt{1 + 64x^{14}}\, dx \approx 1.720$$

$$s_3 = \int_0^1 \sqrt{1 + (10x^9)^2}\, dx$$

$$= \int_0^1 \sqrt{1 + 100x^{18}}\, dx \approx 1.754$$

As $n \to \infty$, the length approaches 2, since one can see that the graph of $y = x^n$ on $[0, 1]$ approaches a path consisting of the horizontal line segment from $(0, 0)$ to $(1, 0)$ followed by the vertical line segment from $(1, 0)$ to $(1, 1)$.

41. $y_1 = x^4,\ y_1' = 4x^3$
$y_2 = x^2,\ y_2' = 2x$

Since both are increasing for positive x, y_1 is "steeper" (y_2 is "flatter") if and only if $y_1' > y_2'$, i.e.,

$$4x^3 > 2x,\ x^2 > \frac{1}{2}, \quad x > \sqrt{\frac{1}{2}}$$

$y = x^6$ is steeper than $y = x^4$ if and only if $6x^5 > 4x^3$, $x^2 > \frac{2}{3}$, $x > \sqrt{\frac{2}{3}}$.

43.

$$S = \int_{-1}^1 2\pi \sqrt{1 - y^2} \sqrt{1 + \left(\frac{y}{\sqrt{1 - y^2}}\right)^2}\, dy$$

$$= \int_{-1}^1 2\pi \sqrt{1 - y^2} \frac{1}{\sqrt{1 - y^2}}\, dy$$

$$= \int_{-1}^1 2\pi\, dy = 4\pi$$

45. $6\pi : 4\pi : (\sqrt{5} + 1)\pi = 3 : 2 : \tau$

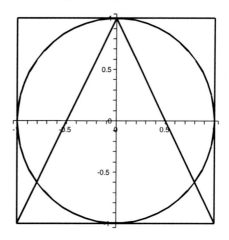

5.5 PROJECTILE MOTION

1. $y(0) = 80,\ y'(0) = 0$

3. $y(0) = 60,\ y'(0) = 10$

5. The initial conditions are
$y(0) = 30$ and $y'(0) = 0$.
We want to find $y'(t)$ when $y(t) = 0$.

We start with the equation $y''(t) = -32$. Integrating gives $y'(t) = -32t + c_1$. From the initial velocity, we have $0 = y'(0) = -32(0) + c_1$, and so $y'(t) = -32t$.

Integrating again gives $y(t) = -16t^2 + c_2$. From the initial position, we have $30 = y(0) = -16(0) + c_2$ and so $y(t) = -16t^2 + 30$.

Solving $y(t) = 0$ gives $t = \pm\sqrt{\frac{15}{8}}$. The positive solution is the solution we are interested in. This is the time when the diver hits the water. The diver's velocity is therefore

$$y'\left(\sqrt{\frac{15}{8}}\right) = -32\sqrt{\frac{15}{8}} \quad \approx -43.8 \text{ ft/sec}$$

7. If an object is dropped (time zero, zero initial velocity) from an initial height of y_0, then the impact moment is $t_0 = \sqrt{y_0}/4$ and the impact velocity (ignoring possible negative sign) is

$$v_{\text{impact}} = 32t_0 = 8\sqrt{y_0}.$$

Therefore if the object is dropped from 30 ft, the impact velocity is

$8\sqrt{30} \approx 43.8178$ feet per second.

If dropped from 120 ft, impact velocity is

$8\sqrt{120} \approx 87.6356$ feet per second.

From 3000 ft, impact velocity is

$8\sqrt{3000} \approx 438.178$ feet per second.

From a height of hy_0, the impact velocity is

$$8\sqrt{hy_0} = 8\sqrt{h}\sqrt{y_0} = \sqrt{h}\left(8\sqrt{y_0}\right),$$

which is to say that impact velocity increases by a factor of $\sqrt{h}$ when initial height increases by a factor of h.

9. See the solution to exercise 7. We are told that $t_0 = 4$, hence

$t_0 = \frac{\sqrt{y_0}}{4}$, $4 = \frac{\sqrt{y_0}}{4}$, $16 = \sqrt{y_0}$, $y_0 = 256$. The height of the cliff is 256 feet.

11. With y the height at time t, we go back to
$y'' = -32$,
$v(t) = y' = -32t + v(0)$, and
$y = -16t^2 + tv(0) + y(0)$.

In this case, with launch at 64 feet per second from ground level, we have $v(0) = 64$ and $y(0) = 0$, hence
$y = -16t^2 + 64t = -16(t-2)^2 + 64$
(completing the square), and $v(t) = -32(t - 2)$.

This is zero at $t = 2$, which is when he peaks. At time 2, $y = 64$. He remains in the air until the next time $y = 0$, which is at $t = 4$. We see that $v(4) = -64$, which is, except for sign, the same as his launch velocity.

13. Let v_0 be the initial velocity. Then $h''(t) = -32$, $h'(t) = -32t + v_0$ and $h(t) = -16t^2 + v_0 t$. h is maximized when $h'(t) = 0$.

$0 = h'(t) = -32t + v_0$ yields $t = \dfrac{v_0}{32}$.

The height at this time is

$$h\left(\frac{v_0}{32}\right) = -16\left(\frac{v_0}{32}\right)^2 + v_0\left(\frac{v_0}{32}\right) = \frac{v_0^2}{64}.$$

So, $\dfrac{v_0^2}{64} = \dfrac{5}{3}$ since 20 in. $= \dfrac{5}{3}$ ft. This gives

$v_0 = 8\sqrt{\frac{5}{3}}$ ft/sec.

This is considerably less than Michael Jordan's initial velocity of about 17 feet per second, but the difference in velocity is not as dramatic as in height (20 inches to 54 inches).

15. If the initial conditions are

$y(0) = H$ and $y'(0) = 0$,

integrating $y''(t) = -32$ gives
$y'(t) = -32t + c_1$.
The initial condition gives
$y'(t) = -32t + v_0 = -32t$.
Integrating gives
$y(t) = -16t^2 + c_2$. The initial condition gives
$y(t) = -16t^2 + H$.

The impact occurs when $y(t_0) = 0$ or when $t_0 = \sqrt{H}/4$.

Therefore the impact velocity is
$y'(t_0) = -32t_0 = -8\sqrt{H}$.

17. $d'' = 9.8$ m/s^2 $= 980$ cm/s^2,
$d'(0) = 0$, $d(0) = 0$
$d' = 980t$, $d = 490t^2$
So, $t = \sqrt{\dfrac{d}{490}} \approx 0.045\sqrt{d}$.

19. The starting point is $y'' = -9.8$,
$y'(0) = 98\sin(\pi/3) = 49\sqrt{3}$. We get
$y(t) = -4.9t^2 + ty'(0)$
$= -4.9t\left(t - [v(0)/4.9]\right)$
$= -4.9t\left(t - 10\sqrt{3}\right)$

The flight time is $10\sqrt{3}$ seconds. As to the horizontal range, we have $x'(t) = 98\cos(\pi/3) = 49$. Therefore $x(t) = 49t$ and in this case, the horizontal range is $49(10\sqrt{3}) = 490\sqrt{3} \approx 849$ meters. The horizontal range is the same as in Example 5.4.

21. With 6° replacing 7° in Example 5.5, we have
$y'(0) = 176 \sin(-6°) \approx -18.4$ ft/s, so
$y'(t) = -32t - 18.40$ and $y(t) = -16t^2 - 18.40t + 10$.
$x'(0) = 176 \cos(-6°) \approx 175.04$ ft/s, so
$x'(t) = 175.04$ and $x(t) = 175.04t$.
$x(t) = 39$ when $t \approx 0.2228$ second.
$y(0.2228) \approx 5$ feet, so the ball clears the net.
$y(t) = 0$ when $-16t^2 - 18.40t + 10 = 0$ or
when $t \approx 0.4026$ (ignoring the negative solution).
$x(0.4026) \approx 70$, so that the ball lands past the service line; the serve is not in. Let θ be the angle with $\theta > 0$. Then $y(t) = -16t^2 - (176 \sin\theta)t + 10$ and $x(t) = (176 \cos\theta)t$.

When $x(t) = 39$, $t = \frac{39}{176 \cos\theta}$. To clear the net, $y\left(\frac{39}{176 \cos\theta}\right) > 3$, so $-16\left(\frac{39}{176 \cos\theta}\right)^2$
$-176 \sin\theta \left(\frac{39}{176 \cos\theta}\right) + 10 > 3$ or $-\frac{1521}{1936}$
$\sec^2\theta - 39 \tan\theta + 10 > 3$. Thus, θ can be at most 9° and the ball still clear the net (solve the inequality graphically). When $y(t) = 0$, $t = \frac{-176 \sin\theta + \sqrt{176^2 \sin^2\theta + 640}}{32} = T$. Solving $x(T) \leq 60$, we get that θ must be at least 7.7°.

23. Let $(x(t), y(t))$ be the trajectory. In this case
$y(0) = 6, x(0) = 0$
$y'(0) = 0, x'(0) = 130$
$x''(t) = 0, x'(t) = 130$
$x(t) = 130t$

This is 60 at time $t = 6/13$. Meanwhile,

$y''(t) = -32, y'(t) = -32t$
$y(t) = -16t^2 + 6$
$y\left(\frac{6}{13}\right) = -16\left(\frac{6}{13}\right)^2 + 6 = \frac{438}{169}$
$y\left(\frac{6}{13}\right) \approx 2.59$ ft

25. Let $(x(t), y(t))$ be the trajectory. In this case 5° is converted to $\pi/36$ radians.

$y(0) = 5, x(0) = 0$
$y'(0) = 120 \sin\frac{\pi}{36} \approx 10.46$
$x'(0) = 120 \cos\frac{\pi}{36} \approx 119.54$
$x''(0) = 0$
$x'(t) = 119.54$
$x(t) = 119.54t$

This is 120 when
$t = 120/119.54 \approx 1.00385$ second.

Meanwhile,

$y''(t) = -32$
$y'(t) = -32t + 10.46$
$y(t) = -16t^2 + 10.46t + 5$
$y(1.00385) = -16(1.00385)^2$
$\qquad + 10.46(1.00385) + 5$
$y(1.00385) \approx -0.62$ ft

The ball hits the ground before reaching first base.

27. (a) Assuming that the ramp height h is the same as the height of the cars, this problem is asking for the initial speed v_0 required to achieve a horizontal flight distance of 125 feet from a launch angle of 30° above the horizontal. We may assume $x(0) = 0$, $y(0) = h$, and we find

$y'(0) = v_0 \sin\frac{\pi}{6} = \frac{v_0}{2}$
$x'(0) = v_0 \cos\frac{\pi}{6} = \frac{\sqrt{3}}{2}v_0$
$y''(t) = -32, x''(t) = 0$
$y'(t) = -32t + \frac{v_0}{2}, x'(t) = \frac{\sqrt{3}}{2}v_0$
$y(t) = -16t^2 + \frac{v_0}{2}t + h,$
$x(t) = \frac{\sqrt{3}}{2}v_0t.$

$x(t)$ will be 125 if $t = 250/\left(\sqrt{3}v_0\right)$ at which time we require that y be h. Therefore

$$-16\left(\frac{250}{\sqrt{3}v_0}\right)^2 + \frac{v_0}{2}\left(\frac{250}{\sqrt{3}v_0}\right) = 0.$$

$$v_0 = \sqrt{\frac{8000}{\sqrt{3}}} \approx 68 \text{ ft/s.}$$

(b) With an angle of $45° = \pi/4$, the equations become

$$y'(0) = v_0 \sin\frac{\pi}{4} = \frac{v_0}{\sqrt{2}}$$

$$x'(0) = v_0 \cos\frac{\pi}{4} = \frac{v_0}{\sqrt{2}}$$

$$y''(t) = -32, \quad x''(t) = 0$$

$$y'(t) = -32t + \frac{v_0}{\sqrt{2}}, \quad x'(t) = \frac{v_0}{\sqrt{2}}$$

$$y(t) = -16t^2 + \frac{v_0 t}{\sqrt{2}} + h,$$

$$x(t) = \frac{v_0 t}{\sqrt{2}}$$

where h is the height of the ramp.
We now solve $x(t) = 125$ which gives

$$t_0 = t = \frac{125\sqrt{2}}{v_0}.$$

At this distance, we want the car to be at a height h to clear the cars. This gives the equation $y(t_0) = h$, or

$$-16\left(\frac{125\sqrt{2}}{v_0}\right)^2$$

$$+ \frac{125v_0\sqrt{2}}{v_0\sqrt{2}} + h = h.$$

Solving for v_0 gives

$$v_0 = 20\sqrt{10} \approx 63.24 \text{ ft/s.}$$

29. (a) In this case with

$$\theta_0 = 0 \text{ and } \omega = 1$$

$$x''(t) = -25\sin(4t)$$

$$x'(0) = x(0) = 0$$

$$x'(t) = \frac{25}{4}\cos 4t - \frac{25}{4}$$

$$x(t) = \frac{25}{16}\sin 4t - \frac{25}{4}t$$

(b) With $\theta_0 = \frac{\pi}{2}$ and $\omega = 1$

$$x''(t) = -25\sin\left(4t + \frac{\pi}{2}\right)$$

$$x'(0) = x(0) = 0$$

$$x'(t) = \frac{25}{4}\cos\left(4t + \frac{\pi}{2}\right)$$

$$x(t) = \frac{25}{16}\sin\left(4t + \frac{\pi}{2}\right) - \frac{25}{16}$$

31. From exercise 5, time of impact is

$$t = \frac{\sqrt{30}}{4} \text{ seconds.}$$

$2\frac{1}{2}$ somersaults corresponds to 5π radians of revolution. Therefore the average angular velocity is

$$\frac{5\pi}{\sqrt{30}/4} = \frac{20\pi}{\sqrt{30}} \approx 11.5 \text{ rad/sec.}$$

33. Let $(x(t), y(t))$ be the trajectory of the center of the basketball. We are assuming that $y(0) = 6$, $x(0) = 0$, the angle of launch θ of the shot is $52°$ ($\theta = \frac{13\pi}{45}$ in radians) and the initial speed is 25 feet per second. Therefore

$$y'(0) = 25\sin\frac{13\pi}{45} \approx 19.70$$

$$x'(0) = 25\cos\frac{13\pi}{45} \approx 15.39$$

$$y''(t) = -32, \quad x''(t) = 0$$

$$y'(t) = -32t + 19.70, \quad x'(t) = 15.39$$

$$y(t) = -16t^2 + 19.70t + 6,$$

$$x(t) = 15.39t.$$

x will be 15 when t is about $15/15.39 \approx$ 0.9746 sec, at which time y will be about

$$-16(0.9746)^2 + 19.70(0.9746) + 6 \approx 10$$

In other words, the center of the ball is at position (15, 10) and the shot is good.

More generally, with unknown θ, the number 19.70 is replaced by $25 \sin \theta$, while the number 15.39 is replaced by $25 \cos \theta$. y will be exactly 10 if

$$-16t^2 + 25t \sin \theta + 6 = 10$$

$$t = \frac{25 \sin \theta + \sqrt{625 \sin^2 \theta - 256}}{32}$$

$x = 25t \cos \theta$.

As a function of θ, this last expression is too complicated to use calculus (easily) to maximize and minimize it on the θ-interval (48°, 57°), but quick spreadsheet calculations give these values:

(Observe that x is not a monotonic function of θ in this range. It takes its maximum when θ is between 52.4 and 52.5 degrees. The evidence is overwhelming that all the shots will be good.)

θ Degrees	t Seconds	x Feet
48.0	0.8757	14.6484
49.0	0.9021	14.7958
50.0	0.9274	14.9024
51.0	0.9516	14.9710
52.0	0.9748	15.0038
52.1	0.9771	15.0051
52.2	0.9793	15.0062
52.3	0.9816	15.0069
52.4	0.9838	15.0073
52.5	0.9861	15.0073
52.6	0.9883	15.0070
52.7	0.9905	15.0064
52.8	0.9928	15.0054
52.9	0.9950	15.0042
53.0	0.9972	15.0026
54.0	1.0187	14.9690
55.0	1.0394	14.9044
56.0	1.0594	14.8100
57.0	1.0787	14.6869

35. $85° = \frac{17}{36}\pi$ radians.

$$x'(0) = 100 \cos\left(\frac{17}{36}\pi\right) \approx 8.72$$

$$y'(0) = 100 \sin\left(\frac{17}{36}\pi\right) \approx 99.62$$

$$x''(0) = -20$$
$$y''(0) = 0$$
$$y(t) = 99.62t$$
$$x(t) = -10t^2 + 8.72t$$
$$y(t_0) = 90 \text{ when } t_0 \approx 0.903$$
$$x(t_0) = x(0.903) \approx -0.29$$

The ball just barely gets into the goal.

37. Let $(x(t), y(t))$ be the trajectory of the plane. Some of our data is in feet, so we will take $g = -32$ in this problem. We have

$$y''(t) - 32$$
$$y'(t) = -32t + y'(0)$$
$$y(t) = -16t^2 + y'(0)t + y(0)$$
$$x'(t) = c$$
$$x(t) = ct + x(0)$$

Solving for t, we have

$$\frac{1}{c}(x - x(0)) = t.$$

Substituting this expression for t in $y(t)$, we have

$$y - y(0)$$
$$= -16\left[\frac{1}{c}(x - x(0))\right]^2$$
$$+ y'(0)\left[\frac{1}{c}(x - x(0))\right]$$

Hence the path is a parabola.

Turning to the question of the duration of weightlessness, we can assume $x(0) = 0$, and we know that $y'(t) = 0$ when $y - y(0) = 2500$. For this unknown time t_1 (the moment when y' is zero), we have $0 = -32t_1 + y'(0)$. Therefore $t_1 = y'(0)/32$, and

$$2500 = y(t_1) - y(0)$$

$$= -16\left[\frac{y'(0)}{32}\right]^2 + y'(0)\left[\frac{y'(0)}{32}\right]$$

$$= \frac{y'(0)^2}{64},$$

hence $y'(0)^2 = 64(2500)$, $y'(0) = 8(50) = 400$, and $t_1 = 400/32 = 25/2$.

We now know that

$$y - y(0) = -16t^2 + 400t$$

for all t. The second time (t_2) that $y(t) = y(0)$ (after time zero) occurs when

$$t = 400/16 = 25 \text{ seconds.}$$

This is the duration of the weightless experience. Note that $t_2 = 2t_1$. The plane must pull out of the dive soon after this time.

39. Let $y(t)$ be the height of the first ball at time t, and let v_{0y} be the initial velocity. We can assume $y(0) = 0$.

As usual, we have

$y'' = -32$, $y' = -32t + v_{0y}$,
$y = -16t^2 + tv_{0y}$.

The second return to height zero is at time $t = v_{0y}/16$. If this is to be $5/2$, then $v_{0y} = 40$. But the maximum occurs at time

$$v_{0y}/32 = 5/4,$$

at which time the height ($y(5/4)$) is

$$-16(25/16) + 40(5/4) = 25 \text{ feet.}$$

For eleven balls, the difference is that the second return to zero is to be at time $11/4$, hence $v_{0y} = 44$, and the maximum height is 30.25.

41. Let v_0 be the initial velocity.

$$v_{0y} = v_0 \sin\left(\frac{\pi}{2} - \alpha\right) \text{ and } v_{0x} = v_0 \cos\left(\frac{\pi}{2} - \alpha\right)$$

$$\sin\left(\frac{\pi}{2} - \alpha\right) = \frac{v_{0y}}{v_0} \text{ and } \cos\left(\frac{\pi}{2} - \alpha\right) = \frac{v_{0x}}{v_0}$$

$$\sin\alpha = \frac{v_0}{v_{0y}} \text{ and } \cos\alpha = \frac{v_0}{v_{0x}}$$

$$\tan\alpha = \frac{\sin\alpha}{\cos\alpha} = \frac{v_{0x}}{v_{0y}}$$

From exercise 16, $h = \dfrac{v_{0y}^2}{64}$ is the maximum height.

From exercise 40, $w = \dfrac{v_{0x}v_{0y}}{16}$ is the horizontal distance.

$$w = r\left(\frac{v_{0y}^2}{64}\right)\left(\frac{v_{0x}}{v_{0y}}\right) = 4h\tan\alpha$$

43. We must use the result

$$\Delta\alpha \approx \frac{\Delta\omega}{4h} \text{ from exercise 42.}$$

With $h = 25$ from exercise 39 (10 balls) and $\omega = 1$, we get $\Delta\alpha$ about $1/100 = 0.01$ radians or about $0.6°$.

45. The initial conditions are

$x'(0) = 220\cos(9.3°) \approx 217.1$, $x(0) = 0$
$y'(0) = 220\sin(9.3°) \approx 35.55$, $y(0) = 0$

Integrating $x''(t) = 0$ and $y''(t) = -32$ and using the initial conditions gives

$x'(t) = 217.1$
$x(t) = 217.1t$
$y'(t) = -32t + 35.55$
$y(t) = -16t^2 + 35.55t$

Solving for $y(t) = 0$ gives $t = 0$ and $t \approx 2.222$. Plugging the second value into $x(t)$ gives

$x(2.222) = 482.4$ ft.

47. On Earth:
$x(t) = (60\cos25°)t \approx 54.378t$
$y(t) = -16t^2 + (60\sin25°)t \approx -16t^2 + 25.357t$

The ball hits the ground when $t = \dfrac{25.357}{16} \approx$ 1.58 sec.

$x(1.58) \approx 86.18$ ft

On the Moon:

$x(t) = (60 \cos 25°)t \approx 54.378t$

$y(t) = -2.6t^2 + (60 \sin 25°)t \approx -2.6t^2 + 25.357t$

The ball hits the ground when $t = \dfrac{25.357}{16} \approx$ 9.75 sec.

$x(9.75) \approx 530.34$ ft

49. 28 mph ≈ 41.07 ft/sec

Using the result of exercise 16, the maximum height is $\dfrac{41.07^2}{64} \approx 26.35$ ft.

51. Let $h(t)$ be the height of the target t seconds after the paint ball is fired and h_0 its height when the paint ball is fired. Assuming $h'(0) = 0$, we have $h(t) = -16t^2 + h_0$.

Let $x(t)$ and $y(t)$ represent the path of the paint ball with initial velocity v_0.

The angle α from the horizontal to the target at $t = 0$ satisfies $\tan \alpha = \dfrac{h_0}{20}$, so that

$$\cos \alpha = \frac{20}{\sqrt{20^2 + h_0^2}} \text{ and } \sin \alpha = \frac{h_0}{\sqrt{20^2 + h_0^2}}.$$

$$x(t) = \frac{20v_0}{\sqrt{20^2 + h_0^2}}t \text{ and } y(t)$$

$$= -16t^2 + \frac{h_0 v_0}{\sqrt{20^2 + h_0^2}}t$$

$$x(t) = 20 \text{ when } t = \frac{\sqrt{20^2 + h_0^2}}{v_0}$$

So when $x(t) = 20$,

$$h(t) = -16\left(\frac{\sqrt{20^2 + h_0^2}}{v_0}\right)^2 + h_0, \text{ and}$$

$$y(t) = -16\left(\frac{\sqrt{20^2 + h_0^2}}{v_0}\right)^2$$

$$+ \frac{h_0 v_0}{\sqrt{20^2 + h_0^2}}\left(\frac{\sqrt{20^2 + h_0^2}}{v_0}\right)$$

$$= -16\left(\frac{\sqrt{20^2 + h_0^2}}{v_0}\right)^2 + h_0.$$

Since $h(t) = y(t)$ when $x(t) = 20$, the paint ball hits the target.

53. (a) The speed at the bottom is given by

$$\frac{1}{2}mv^2 = mgH, \ v = \sqrt{2gH}.$$

(b) Use the result from (a).

$$v = \sqrt{2gH} = \sqrt{2 \cdot 16g} = 4\sqrt{2g}$$
$$= 4\sqrt{2 \cdot 32} = 32 \text{ ft/s}$$

(c) At half way down,

$$\frac{1}{2}mv^2 + mg8 = mg16,$$

$$v = \sqrt{2 \cdot (16 - 8)g} = 4\sqrt{g}$$

$$= 4\sqrt{32} \approx 22.63 \text{ ft/s}.$$

(d) At half way down, the slope of the line tangent to $y = x^2$ is $2 \cdot \sqrt{8} = 4\sqrt{2}$.

Hence we know that

$$\frac{v_y}{v_x} = 4\sqrt{2}.$$

At the same time,

$$(v_y)^2 + (v_x)^2 = (4\sqrt{g})^2$$

$$v_x^2 = \frac{16g}{33}$$

$$v_x = 4\sqrt{\frac{g}{33}} \approx 3.939 \text{ ft/s}$$

$$v_y = 16\sqrt{\frac{2g}{33}} \approx 22.282 \text{ ft/s}$$

5.6 APPLICATIONS OF INTEGRATION TO PHYSICS AND ENGINEERING

1. We first determine the value of the spring constant k. We convert to feet so that our units of work is foot-pounds.

$$5 = F(1/3) = \frac{k}{3} \text{ and so } k = 15.$$

$$W = \int_0^{1/2} F(x)\, dx$$

$$= \int_0^{1/2} 15x\, dx = \frac{15}{8} \text{ foot-pounds.}$$

3. The force is constant (250 pounds) and the distance is 20/12 feet, so the work is

$$W = Fd = (250)(20/12)$$
$$= 1250/3 \text{ foot-pounds.}$$

5. If x is between 0 and 30,000 feet, then the weight of the rocket at altitude x is $10,000 - \frac{1}{15}x$.

Therefore the work is

$$\int_0^{30,000} \left(10,000 - \frac{x}{15}\right) dx$$

$$= \left(10,000x - \frac{x^2}{30}\right)\Big|_0^{30,000}$$

$$= 270,000,000 \text{ ft-lb.}$$

7.
$$W = \int_0^1 800x(1-x)\, dx$$

$$= \left(400x^2 - \frac{800}{3}x^3\right)\Big|_0^1$$

$$= \frac{400}{3} \text{ mile-lb}$$

$$= 704,000 \text{ ft-lb}$$

9.
$$W = \int_0^{100} 62.4\pi$$

$$\cdot (100x - x^2)(200 + x)\, dx$$

$$= 62.4\pi \int_0^{100} \left(20,000x - 100x^2 - x^3\right) dx$$

$$\approx 8,168,140,899 \text{ ft-lb}$$

11. The difference between this exercise and Example 6.3 is that here the integral runs from 10 to 20 rather than from 0 to 20.

$$W = \int_{10}^{20} 62.4\pi x(20-x)^2\, dx$$

$$= 62.4\pi \left[400\frac{x^2}{2} - 40\frac{x^3}{3} + \frac{x^4}{4}\right]_{10}^{20}$$

$$= 62.4\pi \left[\frac{40,000}{3} - \frac{27,500}{3}\right]$$

$$= 62.4\pi \left(\frac{12,500}{3}\right) \approx 816,814 \text{ ft-lb}$$

13.
$$W = \int_0^{10} ax\, dx = \frac{100a}{2}$$

$$W_1 = \int_0^c ax\, dx = \frac{ac^2}{2}$$

$$W_1 = \frac{W}{2} \text{ gives } \frac{ac^2}{2} = \frac{1}{2}\frac{100a}{2}$$

$$c = \sqrt{50} \approx 7.07 \text{ feet}$$

The answer is greater than 5 feet because the deeper the laborer digs, the more distance it is required for him to lift the dirt out of the hole.

15. We estimate the integral using Simpson's Rule:

$$J = \int_0^{0.0008} F(t)\,dt$$

$$\approx \frac{0.0008}{3(8)}[0 + 4(1000) + 2(2100)$$

$$+ 4(4000) + 2(5000) + 4(5200)$$

$$+ 2(2500) + 4(1000) + 0]$$

$$\approx 2.133$$

$$2.13 = J = m\,\Delta v = .01\Delta v$$

$$\Delta v = 213$$

The velocity after impact is therefore $213 - 100 = 113$ ft/sec.

17. We compute the impulse using Simpson's Rule.

$$J \approx \frac{0.6}{3(6)}[0 + 4(8000) + 2(16{,}000)$$

$$+ 4(24{,}000) + 2(15{,}000) + 4(9000)$$

$$+ 0]$$

$$\approx 7533.3$$

$$7533.3 = J = m\,\Delta v = 200\Delta v$$

$$\Delta v = 37.7 \text{ ft/sec}$$

Since the velocity after the crash is zero, this number is the estimated original velocity.

19. $f'(t)$ is zero at $t = 3$, and the maximum thrust is $f(3) = 11$.

It is implicit in the drawing that the thrust is zero after time 6. Therefore the impulse is

$$\int_0^6 \frac{66t}{t^2 + 9}\,dt \approx 53.11.$$

21. $m = \int_0^6 \left(\frac{x}{6} + 2\right) dx = 15$

$$M = \int_0^6 x\left(\frac{x}{6} + 2\right) dx = 48$$

Therefore,

$$\bar{x} = \frac{M}{m} = \frac{48}{15} = \frac{16}{5} = 3.2$$

So the center of mass is to the right of $x = 3$.

23.
$$m = \int_{-3}^{27} \left(\frac{1}{46} + \frac{x + 3}{690}\right)^2 dx$$

$$= \frac{690}{3}\left(\frac{1}{46} + \frac{x + 3}{690}\right)^3 \Big|_{-3}^{27}$$

$$\approx 0.0614 \text{ slugs}$$

$$\approx 31.5 \text{ oz}$$

25.
$$M = \int_{-3}^{27} x\left(\frac{1}{46} + \frac{x + 3}{690}\right)^2 dx$$

$$\approx 1.0208$$

$$\bar{x} = \frac{M}{m} = \frac{1.0208}{0.0614} \approx 16.6 \text{ in.}$$

This is 3 inches less than the bat of Example 6.5, a reflection of the translation three inches to the left on the number line.

27.
$$m = \int_0^{30} 0.00468\left(\frac{3}{16} + \frac{x}{60}\right) dx$$

$$\approx 0.0614 \text{ slug}$$

$$M = \int_0^{30} 0.00468x\left(\frac{3}{16} + \frac{x}{60}\right) dx$$

$$\approx 1.0969$$

Weight $= m(32)(16) = 31.4$ oz

$$\bar{x} = \frac{M}{m} = \frac{1.0969}{0.0614} \approx 17.86 \text{ in.}$$

29. Area of the base is $\frac{1}{2}(3 + 1) = 2$.

Area of the body is $1 \times 4 = 4$.

Area of the tip is $\frac{1}{2}(1 \times 1) = \frac{1}{2}$.

Base:

$$m = \rho A = 2\rho$$

$$M = \int_0^1 \rho x(3 - 2x)\,dx = \frac{5}{6}\rho$$

$$\bar{x} = \frac{M}{m} = \frac{5\rho/6}{2\rho} = \frac{5}{12}$$

Body:

$$m = \rho A = 4\rho$$

$$M = \int_1^5 \rho x \, dx = 12\rho$$

$$\bar{x} = \frac{M}{m} = \frac{12\rho}{4\rho} = 3$$

Tip:

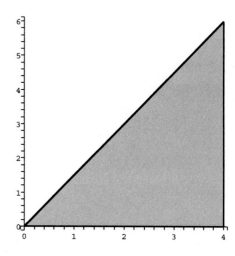

$$m = \rho A = \frac{1}{2}\rho$$

$$M = \int_5^6 \rho x (6 - x) \, dx = \frac{8}{3}$$

$$\bar{x} = \frac{M}{m} = \frac{8\rho/3}{\rho/2} = \frac{16}{3}$$

31. The x-coordinate of the centroid is the same as the center of mass from $x = 0$ to $x = 4$ with density $\rho(x) = \dfrac{3}{2}x$, hence

$$\bar{x} = \frac{M}{m} = \frac{\int_0^4 3/2 \cdot x^2 \, dx}{\int_0^4 3/2 \cdot x \, dx} = \frac{32}{12} = \frac{8}{3}$$

The y-coordinate of the centroid is the same as the center of mass from $y = 0$ to $y = 6$ with density $\rho(y) = 4 - \dfrac{2}{3}y$, hence

$$\bar{y} = \frac{M}{m} = \frac{\int_0^6 \left(4y - \frac{2}{3}y^2\right) dy}{\int_0^6 \left(4 - \frac{2}{3}y\right) dy} = \frac{24}{12} = 2$$

So the center of the given triangle is the point $(8/3, 2)$.

33. This time the x-coordinate of the centroid is obviously $x = 0$, so the question remains to find the y-coordinate.

This is the same as finding the center of mass from $y = 0$ to $y = 4$ with density $\rho(y) = 2\sqrt{4 - y}$, hence

$$m = \int_0^4 2\sqrt{4 - y} \, dy = -2 \cdot \frac{2}{3}(4 - y)^{3/2} \Big|_0^4 = \frac{32}{3}$$

$$M = \int_0^4 2y\sqrt{4 - y} \, dy = -\int_4^0 2(4 - u)u^{1/2} \, du$$

$$= 2\int_0^4 (4u^{1/2} - u^{3/2}) \, du$$

$$= 2\left(\frac{8}{3}u^{3/2} - \frac{2}{5}u^{5/2}\right)\Big|_0^4$$

$$= \frac{256}{15}$$

$$\bar{y} = \frac{M}{m} = \frac{256/15}{32/3} = \frac{8}{5} = 1.6$$

So the centroid is the point $(0, 8/5)$.

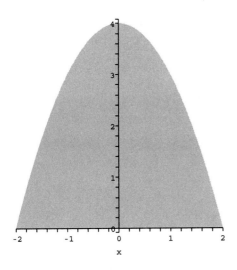

35. With x the depth, the horizontal width is a linear function of x, given by $x + 40$. Hence,

$$F = \int_0^{60} 62.4x(x + 40)\, dx$$

$$= 62.4 \left(\frac{x^3}{3} + 20x^2 \right) \Big|_0^{60}$$

$$= 8{,}985{,}600 \text{ lb}.$$

37. Let x be the vertical deviation above the center of the window, the horizontal width of the window is given by $2\sqrt{25 - x^2}$, depth of water $40 + x$, and hydrostatic force

$$62.4 \int_{-5}^{5} (x + 40)2\sqrt{25 - x^2}\, dx$$
$$\approx 196{,}035 \text{ pounds}.$$

39. Assume that the center of the circular window descends to 1000 feet.

$$F = \int_0^{0.5} 62.4(999.75 + x)$$

$$\cdot 2\sqrt{(0.25)^2 - (0.25 - x)^2}\, dx$$

$$= \int_0^{0.5} 124.8(999.75 + x)\sqrt{0.5x - x^2}\, dx$$

$$\approx 12{,}252 \text{ lb}$$

41. With $h(t)$ the height of the vaulter at time t (time measured from the peak of his vault)

$$h(0) = 20,\ h'(0) = 0,\ h''(0) = -32$$

$$h'(t) = -32t,\ h(t) = -16t^2 + 20$$

$$h(t) = 0 \text{ for } t = \sqrt{\frac{20}{16}} = \frac{\sqrt{5}}{2}$$

$$v = h'\left(\frac{\sqrt{5}}{2} \right) = -16\sqrt{5}$$

$$\approx -35.7 \text{ ft/s}$$

$$\frac{1}{2}mv^2 = \frac{1}{2}\left(\frac{200}{32} \right)(16^2 \cdot 5)$$

$$= 4000 \text{ ft-lb}$$

43. $(100 \text{ tons})(20 \text{ miles/hr})$

$$= \frac{(100 \cdot 2000 \text{ lbs})(20 \cdot 5280 \text{ ft})}{3600 \text{ sec}}$$

$$\approx 5{,}866{,}667 \text{ ft-lb/s}$$

$$= \frac{5{,}866{,}667}{550} \text{ hp}$$

$$\approx 10{,}667 \text{ hp}$$

45. The bat in exercise 23 models the bat of Example 6.5 choked up 3 in.

From Example 6.5:

$$f(x) = \left(\frac{1}{46} + \frac{x}{690} \right)^2;$$

$$\int_0^{30} f(x) \cdot x^2\, dx \approx 27.22.$$

From exercise 23:

$$f(x) = \left(\frac{1}{46} + \frac{x + 3}{690} \right)^2;$$

$$\int_{-3}^{27} f(x) \cdot x^2\, dx \approx 20.54.$$

Reduction in moment:
$$\frac{27.22 - 20.54}{27.22} \approx 24.5\%$$

47. $\displaystyle \int_{-a}^{a} 2\rho x^2 b \sqrt{1 - \frac{x^2}{a^2}}\, dx = \frac{1}{4}\rho\pi a^3 b$

49. Using the formula in exercise 48, we find that the moments are 1323.8ρ for the wooden racket, 1792.9ρ for the mid-sized racket, and 2361.0ρ for the oversized racket. The ratios are

$$\frac{\text{mid}}{\text{wood}} \approx 1.35, \quad \frac{\text{over}}{\text{wood}} \approx 1.78.$$

9.
$$A = \int_0^6 [(10 + 2t) - (4 + t)]dt$$

$$= \int_0^6 (6 + t)dt = \left(6t + \frac{t^2}{2}\right)\Big|_0^6 = 54$$

Population $= 10,000 + 54 = 10,054$

Chapter 5 Review Exercises

1.
$$\text{Area} = \int_0^\pi \left(x^2 + 2 - \sin x\right) dx$$

$$= \left(\frac{x^3}{3} + 2x + \cos x\right)\Big|_0^\pi$$

$$= \frac{\pi^3}{3} + 2\pi - 2$$

3.
$$\text{Area} = \int_0^1 \left[x^3 - \left(2x^2 - x\right)\right] dx$$

$$= \left(\frac{x^4}{4} - \frac{2}{3}x^3 + \frac{x^2}{2}\right)\Big|_0^1 = \frac{1}{12}$$

5. Solving $x^2 - 2 = 2 - x^2$, we get $2x^2 = 4$, so $x = \pm\sqrt{2}$.

$$\text{Area} = \int_{-\sqrt{2}}^{\sqrt{2}} \left[(2 - x^2) - (x^2 - 2)\right] dx$$

$$= \int_{-\sqrt{2}}^{\sqrt{2}} (4 - 2x^2) dx$$

$$= \left(4x - \frac{2}{3}x^3\right)\Big|_{-\sqrt{2}}^{\sqrt{2}} = \frac{16\sqrt{2}}{3}$$

7.
$$\text{Area} = \int_0^1 x^2 dx + \int_1^2 (2 - x) dx$$

$$= \frac{x^3}{3}\Big|_0^1 + \left(2x - \frac{x^2}{2}\right)\Big|_1^2$$

$$= \frac{1}{3} + (4 - 2) - \left(2 - \frac{1}{2}\right) = \frac{5}{6}$$

11.
$$V = \int_0^2 \pi(3 + x)^2 dx$$

$$= \pi \int_0^2 (9 + 6x + x^2) dx$$

$$= \pi \left(9x + 3x^2 + \frac{x^3}{3}\right)\Big|_0^2$$

$$= \frac{98\pi}{3}$$

13. Use Simpson's Rule:

$$V \approx \frac{0.4}{3}[0.4 + 4(1.4) + 2(1.8) + 4(2.0) + 2(2.1) + 4(1.8) + 2(1.1) + 4(0.4) + 0] \approx 4.373$$

15. (a)
$$V = \int_{-2}^2 \pi(4)^2 dx - \int_{-2}^2 \pi(x^2)^2 dx$$

$$= \pi \int_{-2}^2 (16 - x^4) dx$$

$$= \pi \left(16x - \frac{x^5}{5}\right)\Big|_{-2}^2$$

$$= \frac{256\pi}{5}$$

(b)
$$V = \int_0^4 \pi(\sqrt{y})^2 dy = \pi \int_0^4 y\, dy$$

$$= \frac{\pi y^2}{2}\Big|_0^4 = 8\pi$$

(c)
$$V = \int_{-2}^{2} 2\pi(2-x)(4-x^2)\,dx$$

$$= 2\pi \int_{-2}^{2} (8 - 4x - 2x^2 + x^3)\,dx$$

$$= 2\pi \left(8x - 2x^2 - \frac{2}{3}x^3 + \frac{1}{4}x^4\right)\Big|_{-2}^{2}$$

$$= \frac{128\pi}{3}$$

(d)
$$V = \int_{-2}^{2} \pi(6)^2\,dx$$

$$- \int_{-2}^{2} \pi(x^2+2)^2\,dx$$

$$= \pi \int_{-2}^{2} (-x^4 - 4x^2 + 32)\,dx$$

$$= \pi\left(-\frac{x^5}{5} - \frac{4x^3}{3} + 32x\right)\Big|_{-2}^{2}$$

$$= \frac{1408\pi}{15}$$

17. (a)
$$V = \int_{0}^{1} 2\pi y((2-y) - y)\,dy$$

$$= 2\pi \int_{0}^{1} (2y - 2y^2)\,dy$$

$$= 2\pi\left(y^2 - \frac{2y^3}{3}\right)\Big|_{0}^{1}$$

$$= \frac{2\pi}{3}$$

(b)
$$V = \int_{0}^{1} \pi(2-y)^2\,dy$$

$$- \int_{0}^{1} \pi(y)^2\,dy$$

$$= \pi \int_{0}^{1} (4 - 4y)\,dy$$

$$= \pi\,(4y - 2y^2)\Big|_{0}^{1} = 2\pi$$

(c)
$$V = \int_{0}^{1} \pi((2-y) + 1)^2\,dy$$

$$- \int_{0}^{1} \pi(y+1)^2\,dy$$

$$= \pi \int_{0}^{1} (9 - 6y + y^2)\,dy$$

$$- \pi \int_{0}^{1} (y^2 + 2y + 1)\,dy$$

$$= \pi \int_{0}^{1} (8 - 8y)\,dy$$

$$= \pi\,(8y - 4y^2)\Big|_{0}^{1} = 4\pi$$

(d)
$$V = \int_{0}^{1} 2\pi(4-y)((2-y) - y)\,dy$$

$$= 2\pi \int_{0}^{1} (8 - 10y + 2y^2)\,dy$$

$$= 2\pi\left(8y - 5y^2 + \frac{2y^3}{3}\right)\Big|_{0}^{1}$$

$$= \frac{22\pi}{3}$$

19. $s = \int_{-1}^{1} \sqrt{1 + (4x^3)^2}\,dx \approx 3.2$

21. $y = (x+1)^{1/2}$, $y' = \frac{1}{2}(x+1)^{-1/2}$

$$= \frac{1}{2\sqrt{x+1}},\ (y')^2 = \frac{1}{4(x+1)}$$

$$s = \int_{0}^{3} \sqrt{1 + \frac{1}{4(x+1)}}\,dx \approx 3.1678$$

23. $S = \int_{0}^{1} 2\pi(1-x^2)\sqrt{1+4x^2}\,dx$
≈ 5.483

25. $h''(t) = -32$
$h(0) = 64,\ h'(0) = 0$
$h'(t) = -32t$
$h(t) = -16t^2 + 64$

This is zero when $t = 2$, at which time $h'(2) = -32(2) = -64$ feet per second.

27. $y''(t) = -32$, $x''(t) = 0$,

$y(0) = 0$, $x(0) = 0$

$y'(0) = 48 \sin\left(\dfrac{\pi}{9}\right)$

$x'(0) = 48 \cos\left(\dfrac{\pi}{9}\right)$

$y'(0) \approx 16.42$, $x'(0) \approx 45.11$

$y'(t) = -32t + 16.42$

$y(t) = -16t^2 + 16.42t$

This is zero at $t = 1.026$. The time of flight is 1.026 seconds. Meanwhile,

$x'(t) \equiv 45.11$

$x(t) = 45.11t$

$x(1.026) = 45.11(1.026) \approx 46.3$ ft

This is the horizontal range.

29. $y(0) = 6$, $x(0) = 0$

$y'(0) = 80 \sin\left(\dfrac{2\pi}{45}\right) \approx 11.13$,

$x'(0) = 80 \cos\left(\dfrac{2\pi}{45}\right) \approx 79.22$

$y''(t) = -32$, $x''(t) = 0$

$y'(t) = -32t + 11.13$

$y(t) = -16t^2 + 11.13t + 6$

$x'(t) = 79.22$

$x(t) = 79.22t$

This is 120 (40 yards) when t is about 1.51. At this time, the vertical height (if still in flight) would be
$y(1.51) = -16(1.51)^2 + 11.13(1.51) + 6$
$= -13.6753$.

Since this is negative, we conclude the ball is not still in flight, has hit the ground, and was not catchable.

31. $h''(t) = -32$

$h'(0) = v_0$

$h(0) = 0$

$h'(t) = -32t + v_0$

This is zero at $t = v_0/32$.

$h\left(\dfrac{v_0}{32}\right) = -16\left(\dfrac{v_0^2}{32^2}\right) + \dfrac{v_0^2}{32} = \dfrac{v_0^2}{64}$

If this is to be 128, then clearly v_0 must be

$\sqrt{(64)(128)} = 64\sqrt{2}$ ft/sec.

Impact speed from ground to ground is the same as launch speed, which can be verified by first finding the time t of return to ground:
$-16t^2 + v_0 t = 0$

$t = v_0/16$

and then compiling

$h'\left(v_0/16\right) = -32(v_0/16) + v_0 = -v_0$.

33. $F = kx$, $60 = k \cdot 1$, $k = 60$

$W = \displaystyle\int_0^{2/3} 60x \, dx = 30x^2 \Big|_0^{2/3}$

$= \dfrac{30 \cdot 4}{9} = \dfrac{40}{3}$ ft-lb

35.
$m = \displaystyle\int_0^4 \left(x^2 - 2x + 8\right) dx$

$= \left(\dfrac{x^3}{3} - x^2 + 8x\right)\Big|_0^4 = \dfrac{112}{3}$

$M = \displaystyle\int_0^4 x\left(x^2 - 2x + 8\right) dx$

$= \displaystyle\int_0^4 \left(x^3 - 2x^2 + 8x\right) dx$

$= \left(\dfrac{x^4}{4} - \dfrac{2x^3}{3} + 4x^2\right)\Big|_0^4 = \dfrac{256}{3}$

$\bar{x} = \dfrac{M}{m} = \dfrac{\frac{256}{3}}{\frac{112}{3}} = \dfrac{256}{112} = \dfrac{16}{7}$

Center of mass is greater than 2 because the object has greater density on the right side of the interval [0, 4].

37.
$$F = \int_0^{80} 62.4x(60 + x)\, dx$$

$$= 62.4 \int_0^{80} (60x + x^2)\, dx$$

$$= 62.4\left(30x^2 + \frac{x^3}{3}\right)\Big|_0^{80}$$

$$\approx 22{,}630{,}400 \text{ lb}$$

39. $J \approx \dfrac{0.0008}{3(8)}\{0 + 4(800) + 2(1600)$

$+ 4(2400) + 2(3000) + 4(3600)$
$+ 2(2200) + 4(1200) + 0\} = 1.52$

$J = m\Delta v$

$1.52 = 0.01\Delta v$
$\Delta v = 152 \text{ ft/s}$
$152 - 120 = 32 \text{ ft/s}$

Exponentials, Logarithms, and Other Transcendental Functions

6.1 THE NATURAL LOGARITHM

1. $\ln 4 = \displaystyle\int_1^4 \frac{dt}{t}$

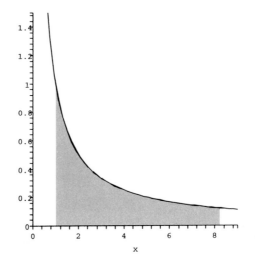

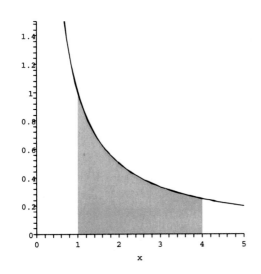

3. $\ln 8.2 = \displaystyle\int_1^{8.2} \frac{dt}{t}$

5. (a) Simpson's Rule with $n = 32$:

$$\ln 4 = \int_1^4 \frac{dx}{x} \approx 1.386296874$$

(b) Simpson's Rule with $n = 64$:

$$\ln 4 = \int_1^4 \frac{dx}{x} \approx 1.386294521$$

7. The table below and the graph of $y = \dfrac{\ln x}{\csc x}$ both suggest that $\displaystyle\lim_{x \to 0^+} \frac{\ln x}{\csc x} = 0.$

x	$\dfrac{\ln x}{\csc x}$
0.1	−0.2299
0.01	−0.0461
0.001	−0.0069
0.0001	−0.0009
0.00001	−0.0001

9. The table below and the graph of
$$y = \frac{1}{\ln(x+1)} - \frac{1}{x} \text{ both suggest that}$$

$$\lim_{x \to 0^+} \left[\frac{1}{\ln(x+1)} - \frac{1}{x} \right] = \frac{1}{2}.$$

x	$\dfrac{1}{\ln(x+1)} - \dfrac{1}{x}$
0.1	0.4921
0.01	0.4992
0.001	0.4999
0.0001	0.5000

11. $\dfrac{d}{dx} \ln 4x^2 = \dfrac{1}{4x^2} \dfrac{d}{dx} 4x^2 = \dfrac{1}{4x^2}(8x) = \dfrac{2}{x}$

13. $[\ln(\cos x)]' = \dfrac{1}{\cos x}(-\sin x) = -\tan x$

15. $\dfrac{d}{dx} \ln \sin(\ln(\cos x^3))$

$= \cos(\ln(\cos x^3)) \dfrac{d}{dx} \ln(\cos x^3)$

$= \cos(\ln(\cos x^3)) \dfrac{1}{\cos x^3} \dfrac{d}{dx} \cos x^3$

$= \cos(\ln(\cos x^3)) \dfrac{1}{\cos x^3}(-\sin x^3) \dfrac{d}{dx} x^3$

$= -3x^2 \tan x^3 \cos(\ln(\cos x^3))$

17. $\ln \sqrt{2} + 3 \ln 2 = \ln 2^{1/2} + 3 \ln 2$

$= \dfrac{1}{2} \ln 2 + 3 \ln 2$

$= \dfrac{7}{2} \ln 2$

19. $2 \ln 3 - \ln 9 + \ln \sqrt{3} = 2 \ln 3 - \ln 3^2 + \ln 3^{1/2}$

$= 2 \ln 3 - 2 \ln 3 + \dfrac{1}{2} \ln 3$

$= \dfrac{1}{2} \ln 3$

21. Use $u = x^2 + 1$, then $du = 2x \, dx$.

$$\int \frac{2x}{x^2+1} dx = \int \frac{1}{u} du$$

$$= \ln |u| + c$$

$$= \ln \left| x^2 + 1 \right| + c$$

$$= \ln(x^2 + 1) + c$$

since $x^2 + 1 > 0$.

23. Use $u = \cos 2x$, then $du = -2 \sin 2x \, dx$.

$$\int \tan 2x \, dx = \int \frac{\sin 2x}{\cos 2x} dx = \int \frac{\sin 2x}{\cos 2x} dx$$

$$= -\frac{1}{2} \int \frac{1}{u} du$$

$$= -\frac{1}{2} \ln |u| + c$$

$$= -\frac{1}{2} \ln |\cos x| + c$$

25. Use $u = \ln x$, then $du = \dfrac{1}{x} dx$.

$$\int \frac{1}{x \ln x} dx = \int \frac{1}{\ln x} \frac{1}{x} dx$$

$$= \int \frac{1}{u} du$$

$$= \ln |u| + c$$

$$= \ln |\ln x| + c$$

27. Use $u = \ln x + 1$, then $du = \dfrac{1}{x} dx$.

$$\int \frac{(\ln x + 1)^2}{x} dx = \int (\ln x + 1)^2 \frac{1}{x} dx$$

$$= \int u^2 du$$

$$= \frac{1}{3} u^3 + c$$

$$= \frac{1}{3}(\ln x + 1)^3 + c$$

29. Let $u = \ln x$. Then $du = \dfrac{1}{x} \, dx$, $u(1) = 0$ and

$u(2) = \ln 2$.

$$\int_1^2 \frac{\sqrt{\ln x}}{x} \, dx = \int_1^2 (\ln x)^{1/2} \frac{1}{x} \, dx$$

$$= \int_0^{\ln 2} u^{1/2} \, du = \frac{2}{3} u^{3/2} \Big|_0^{\ln 2} = \frac{2}{3} (\ln 2)^{3/2}$$

31. $\dfrac{d}{dx} \left[\ln \sqrt{x^2 + 1} \right] = \dfrac{d}{dx} \left[\ln(x^2 + 1)^{1/2} \right]$

$$= \frac{d}{dx} \left[\frac{1}{2} \ln(x^2 + 1) \right]$$

$$= \frac{1}{2} \frac{1}{x^2 + 1} 2x$$

$$= \frac{x}{x^2 + 1}$$

33. $\dfrac{d}{dx} \left[\ln \dfrac{x^4}{x^5 + 1} \right] = \dfrac{d}{dx} \left[\ln x^4 - \ln(x^5 + 1) \right]$

$$= \frac{d}{dx} \left[\ln x^4 \right] - \frac{d}{dx} \left[\ln(x^5 + 1) \right]$$

$$= \frac{4}{x} - \frac{5x^4}{x^5 + 1}$$

35. $\ln \dfrac{1}{x - 1} = \ln 1 - \ln(x - 1) = -\ln(x - 1)$

As $x \to 1^+$, $x - 1 \to 0^+$, so $\ln(x - 1) \to -\infty$.

Thus, $\displaystyle\lim_{x \to 1^+} \ln \dfrac{1}{x - 1} = \lim_{x \to 1^+} [-\ln(x - 1)]$

$= \infty$.

37. $\displaystyle\lim_{x \to \infty} \ln(x + 1)^{4/\ln(x+1)} =$

$$\lim_{x \to \infty} \left[\frac{4}{\ln(x + 1)} \ln(x + 1) \right]$$

$$= \lim_{x \to \infty} 4 = 4$$

39. $f(x) = x^{\sin x}$

$\ln f(x) = \ln x^{\sin x} = (\sin x) \ln x$

$\dfrac{f'(x)}{f(x)} = \cos x \cdot \ln x + \sin x \cdot \dfrac{1}{x}$

$$f'(x) = f(x) \left[\cos x \ln x + \frac{\sin x}{x} \right]$$

$$= x^{\sin x} \left[\cos x \ln x + \frac{\sin x}{x} \right]$$

41. $f(x) = (\sin x)^x$

$\ln f(x) = \ln(\sin x)^x = x \ln(\sin x)$

$$\frac{f'(x)}{f(x)} = 1 \cdot \ln(\sin x) + x \cdot \frac{\cos x}{\sin x}$$

$$f'(x) = f(x) \left[\ln(\sin x) + x \cot x \right]$$

$$= (\sin x)^x \left[\ln(\sin x) + x \cot x \right]$$

43.

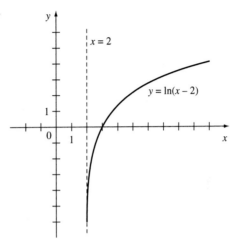

Increasing and concave down on $(2, \infty)$.

45.

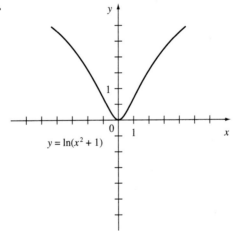

Increasing on $(0, \infty)$, decreasing on $(-\infty, 0)$.
Concave up on $(-1, 1)$,
concave down on $(-\infty, -1) \cup (1, \infty)$.

47.

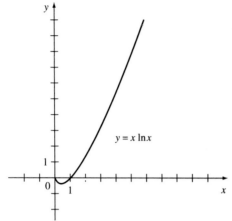

$y = x \ln x$

Decreasing on $(0, 0.368)$, increasing on $(0.368, \infty)$. Concave up on $(0, \infty)$.

49.
$$\ln \frac{a}{b} = \ln a \cdot \frac{1}{b}$$
$$= \ln a + \ln \frac{1}{b}$$
$$= \ln a + \ln b^{-1}$$
$$= \ln a - \ln b$$

51.
$$s(x) = x^2 \ln \frac{1}{x} = x^2 \ln x$$
$$s'(x) = -x^2 \frac{1}{x} \ln x \cdot 2x$$
$$= -x - 2x \ln x$$
$$= -x(1 + 2 \ln x)$$

The critical values are $x = 0$ and $x \approx 0.607$. The function is undefined for $x = 0$ and the derivative changes from $+$ to $-$ at $x \approx 0.607$. Therefore, there is a maximum at $x \approx 0.607$.

53. First substitution:
$$u = \ln \sqrt{x}, \; du = \frac{1}{\sqrt{x}} \cdot \frac{1}{2\sqrt{x}} dx = \frac{1}{2x} dx$$
$$\int \frac{1}{x \ln \sqrt{x}} dx = 2 \int \frac{1}{\ln \sqrt{x}} \cdot \frac{1}{2x} dx =$$
$$2 \int \frac{1}{u} du = 2 \ln |u| + c_1 = 2 \ln |\ln \sqrt{x}| + c_1$$

Second substitution: $u = \ln x, \; du = \frac{1}{x} dx$
$$\int \frac{1}{x \ln \sqrt{x}} dx =$$

$$2 \int \frac{1}{x \ln x} dx = 2 \int \frac{1}{u} du = 2 \ln |u| + c_2 = 2 \ln |\ln x| + c_2$$

$$2 \ln |\ln \sqrt{x}| + c_1 = 2 \ln \left| \frac{1}{2} \ln x \right| + c_1$$

$$= 2[\ln \frac{1}{2} + \ln |\ln x|] + c_1$$

$$= 2 \ln \frac{1}{2} + 2 \ln |\ln x| + c_1$$

$$= 2 \ln |\ln x| + c_2$$

where $c_2 = 2 \ln \frac{1}{2} + c_1$.

55. $T = \frac{-1}{c} \ln \left(1 - c \cdot \frac{b - 1}{v0}\right)$

$b = 300, a = 0, v_0 = 125, c = 0.1$

$$T = \frac{-1}{0.1} \ln \left(1 - 0.1 \cdot \frac{300 - 0}{125}\right)$$
$$= 2.744 \text{ sec}$$
$$T(x) = -10 \ln(1 - 0.0008(300 - x))$$
$$-10 \ln(1 - 0.0008x) + 0.1$$
$$T'(x) =$$
$$-10 \left(\frac{0.0008}{0.76 + 0.0008x} - \frac{0.0008}{1 - 0.0008x}\right) = 0$$
$$0.0008(1 - 0.0008x)$$
$$= 0.0008(0.76 + 0.0008x)$$
$$T'(x) = 0 \text{ when } x = \frac{1 - 0.76}{0.0016} = 150 \text{ft.}$$
$$T'(x) < \text{ on } (0, 150)$$
$$T'(x) > 0 \text{ on } (150, 300)$$

Hence $x = 150$ minimizes the total time.
$$T(150) = -10 \ln(1 - 0.0008(300 - 150))$$
$$-10 \ln(1 - 0.0008(150)) + 0.1 = 2.656 \text{ sec.}$$
So the relay is faster.

If the delay is 0.2 sec, the relay takes longer.

57. $T(x) = -10 \ln \left(1 - 0.1 \frac{300 - x}{125}\right)$

$$- 10 \ln \left(1 - 0.1 \frac{x}{100}\right) + 0.1$$

$$= -10(\ln(1 - 0.0008(300 - x))$$

$$- 10 \ln(1 - 0.001x) + 0.01$$

$$T'(x) =$$

$$-10 \left(\frac{0.0008}{0.76 + 0.0008x} - \frac{0.001}{1 - 0.001x}\right) = 0$$

$0.0008(1 - 0.001x) = 0.001(0.76 + 0.0008x)$

$T'(x) = 0$ when $x = 25$ ft.

$T'(x) < 0$ on $(0, 25)$

$T'(x) > 0$ on $(25, 300)$

Hence $x = 25$ minimizes the total time.

$$T(25) = -10\ln(1 - 0.0008(300 - 25))$$
$$-10\ln(1 - 0.001(25)) + 0.1$$
$$= 2.838 \text{ sec.}$$

So the relay takes longer. Without the delay, the relay would take 2.738 sec, so a delay of $2.744 - 2.738 = 0.006$ sec makes the two times equal.

59. Let $y = c + \ln \dfrac{f}{1 - f} = c + \ln f - \ln(1 - f)$.

Then $y' = \dfrac{1}{f} + \dfrac{1}{1 - f} = \dfrac{1}{1 - f} \cdot \dfrac{1}{f(1 - f)} = \dfrac{1}{f - f^2}$

and $y'' = \dfrac{-(1 - 2f)}{(f - f^2)^2} = \dfrac{2f - 1}{f^2(1 - f^2)}$.

$y'' = 0$ when $f = \dfrac{1}{2}$, $y'' < 0$ on $\left(0, \frac{1}{2}\right)$ and

$y'' > 0$ on $\left(\dfrac{1}{2}, 1\right)$, so y' has a minimum at

$f = \dfrac{1}{2}$.

$\displaystyle\lim_{f \to 1} y = \lim_{f \to 1} (c + \ln f - \ln(1 - f)) = \infty$

since the first term approaches c, the second term approaches 0, and the third term approaches $-\infty$.

61. Use the right endpoints of each subinterval as evaluation points. Then the rectangles will be completely contained in the region under the graph of $y = \frac{1}{x}$, so that the sum of their areas will be less than the area under the graph, $\ln(n)$; $\displaystyle\lim_{n \to \infty} \ln(n) = \infty$.

6.2 INVERSE FUNCTIONS

1. $f(x) = x^5$ and $g(x) = x^{1/5}$

$f(g(x)) = f(x^{1/5})$

$= (x^{1/5})^5 = x^{(5/5)} = x$

$g(f(x)) = g(x^5) = (x^5)^{1/5}$

$= x^{(5/5)} = x$

3. $f(x) = 2x^3 + 1$ and $g(x) = \sqrt[3]{\dfrac{x - 1}{2}}$

$f(g(x)) = 2\left(\sqrt[3]{\dfrac{x - 1}{2}}\right)^3 + 1$

$= 2\left(\dfrac{x - 1}{2}\right) + 1 = x$

$g(f(x)) = \sqrt[3]{\dfrac{f(x) - 1}{2}}$

$= \sqrt[3]{\dfrac{2x^3 + 1 - 1}{2}}$

$= \sqrt[3]{x^3} = x$

5. The function is one-to-one since $f(x) = x^3$ is one-to-one. To find the inverse function, write

$$y = x^3 - 2$$
$$y + 2 = x^3$$
$$\sqrt[3]{y + 2} = x$$

So $f^{-1}(x) = \sqrt[3]{x + 2}$.

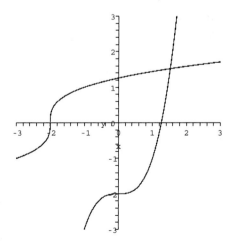

7. The function $y = x^5$ is one-to-one and hence so is $f(x) = x^5 - 1$. To find a formula for the inverse, write

$$y = x^5 - 1$$
$$y + 1 = x^5$$
$$\sqrt[5]{y + 1} = x$$

So $f^{-1}(x) = \sqrt[5]{x + 1}$.

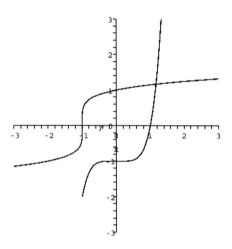

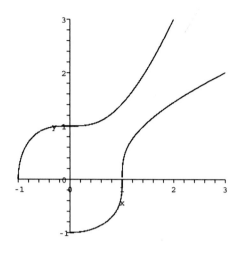

9. The function is not one-to-one since it is an even function $(f(-x) = f(x))$. In particular, $f(2) = 18 = f(-2)$.

11. Here, the natural domain requires that the radicand (the object inside the radical) be nonnegative. Hence $x \geq -1$ is required, while all function values are nonnegative. Therefore the inverse, if defined at all, will be defined only for nonnegative numbers. Sometimes one can determine the existence of an inverse in the process of trying to find its formula. This is an example: Write

$$y = \sqrt{x^3 + 1}$$
$$y^2 = x^3 + 1$$
$$y^2 - 1 = x^3$$
$$\sqrt[3]{y^2 - 1} = x$$

The left side is a formula for $f^{-1}(y)$, good for $y \geq 0$. Therefore, $f^{-1}(x) = \sqrt[3]{x^2 - 1}$ whenever $x \geq 0$.

13. (a) Since $f(0) = -1$, $f^{-1}(-1) = 0$.
$$f'(x) = 3x^2 + 4$$
$$(f^{-1})'(-1) = \frac{1}{3(0)^2 + 4} = \frac{1}{4}$$

(b) Since $f(1) = 4$, $f^{-1}(4) = 1$.
$$(f^{-1})'(4) = \frac{1}{3(1)^2 + 4} = \frac{1}{7}$$

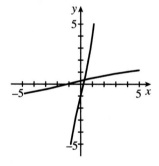

15. (a) Since $f(-1) = -5$, $f^{-1}(-5) = -1$.
$$f'(x) = 5x^4 + 9x^2 + 1$$
$$(f^{-1})'(-5) = \frac{1}{5(-1)^4 + 9(-1)^2 + 1}$$
$$= \frac{1}{15}$$

(b) Since $f(1) = 5$, $f^{-1}(5) = 1$.
$$(f^{-1})'(-5) = \frac{1}{5(1)^4 + 9(1)^2 + 1} = \frac{1}{15}$$

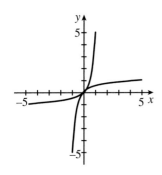

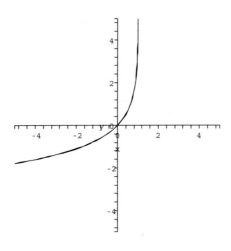

17. (a) Since $f(2) = 4$, $f^{-1}(4) = 2$.

$$f'(x) = \frac{1}{2}(x^3 + 2x + 4)^{-1/2}(3x^2 + 2)$$

$$f'(2) = \frac{7}{4}$$

$$(f^{-1})'(4) = \frac{1}{f'(2)} = \frac{4}{7}$$

(b) Since $f(0) = 2$, $f^{-1}(2) = 0$.

$$f'(0) = \frac{1}{2}$$

$$(f^{-1})'(2) = \frac{1}{f'(0)} = 2$$

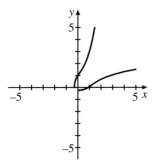

19. Reflect the graph across the line $y = x$.

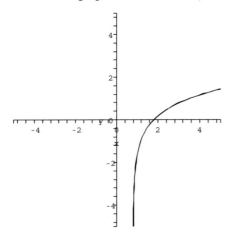

21. Reflect the graph across the line $y = x$.

23. Since 23 is halfway between 20 ($= f(2)$) and 26 ($= f(3)$), the x-value for $y = 23$ should be halfway between 2 and 3, i.e., $f^{-1}(23)$ is estimated linearly by 2.5.

Since the graph is increasing and concave down, this estimate is too high.

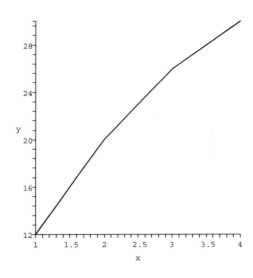

25. Since 5 is three-quarters of the way from 2 ($= f(3)$) to 6 ($= f(2)$), the x-value should be three-quarters of the way from 3 to 2, i.e., $f^{-1}(5)$ is estimated linearly by 2.25.

Since the graph is decreasing and concave up, this estimate is too high.

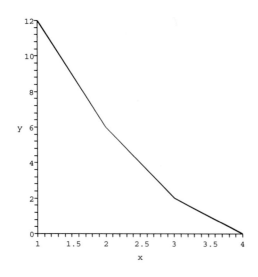

27. If $f(x) = x^3 - 5$, then the horizontal line test is passed, so $f(x)$ is one-to-one.

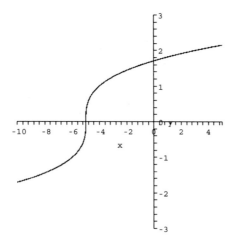

29. The function $f(x) = x^3 + 2x - 1$ passes the horizontal line test and is invertible.

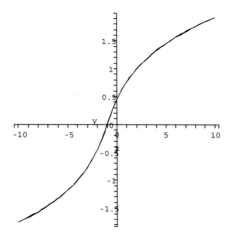

31. Not one-to-one. Fails horizontal line test.

33. If $f(x) = \dfrac{1}{x+1}$, then the horizontal line test is passed, so $f(x)$ is one-to-one.

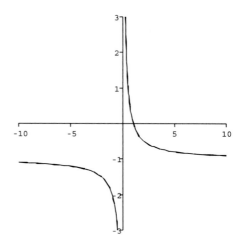

35. If $f(x) = \dfrac{x}{x+4}$, then the horizontal line test is passed so $f(x)$ is one-to-one.

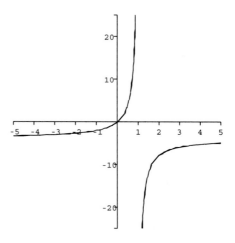

37. $f(g(x)) = (g(x))^2 = \left(\sqrt{x}\right)^2 = x$

$g(f(x)) = \sqrt{f(x)} = \sqrt{x^2} = |x|$.

Because $x \geq 0$, the absolute value is the same as x. Thus these functions (both defined only when $x \geq 0$) are inverses.

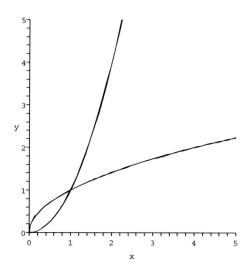

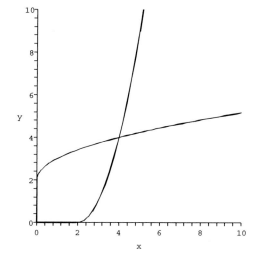

39. With $f(x) = x^2$ defined only for $x \leq 0$, (shown in the graph in quadrant II) the horizontal line test is passed. The formula for the inverse function g is $g(x) = -\sqrt{x}$ shown in quadrant IV and defined only for $x \geq 0$.

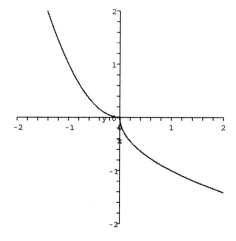

41. The graph of $y = (x - 2)^2$ is a parabola with vertex at $(2, 0)$. If we take only the right half $\{x \geq 2\}$ (shown below as the lower right graph) the horizontal line test is passed, and the formula for the inverse function g is $g(x) = 2 + \sqrt{x}$ defined only for $x \geq 0$ and shown below as the upper left graph.

43. In the first place, for $f(x)$ to be defined, the radicand must be nonnegative, i.e., $0 \leq x^2 - 2x = x(x - 2)$, which entails either $x \leq 0$ or $x \geq 2$. One can restrict the domain to either of these intervals and have an invertible function. Taking the latter for convenience, the inverse will be found as follows:

$$y = \sqrt{x^2 - 2x}$$
$$y^2 = x^2 - 2x = x^2 - 2x + 1 - 1$$
$$= (x - 1)^2 - 1$$
$$y^2 + 1 = (x - 1)^2$$
$$\sqrt{y^2 + 1} = \pm(x - 1)$$

With $x \geq 2$ and the left side nonnegative, we must choose the plus sign. We can then write $x = 1 + \sqrt{y^2 + 1}$.

The right side is now a formula for $f^{-1}(y)$, seemingly good for any y, but we recall from the original formula (as a radical) that y must be nonnegative. We summarize the conclusion:

$$f^{-1}(x) = 1 + \sqrt{x^2 + 1}, \quad (x \geq 0).$$

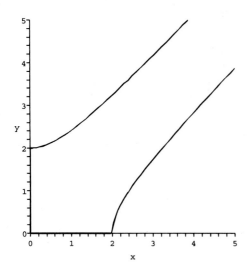

45. The function $\sin(x)$ is increasing and one-to-one on the interval

$$-\frac{\pi}{2} \leq x \leq \frac{\pi}{2}.$$

One does not "find" the inverse in the sense of solving the equation $y = \sin(x)$ and obtaining a formula. It is done only in theory or as a graph. The name of the inverse is the "arcsin" function $(y = \arcsin(x))$, and some of its properties are developed in the next section.

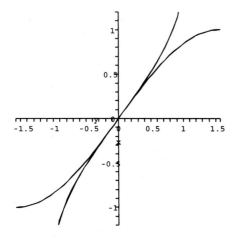

47. Income more often than not rises and falls over time, so the function is unlikely to be one-to-one. In short, income functions *usually* do not have inverses.

49. During an interval of free fall following a drop, the height is decreasing with time and an inverse exists. After impact, if there is a bounce then some of the heights are repeated and the function is no longer one-to-one on the expanded time interval.

51. Two three-dimensional shapes with congruent profiles will cast identical shadows if the congruent profiles face the light source. Such objects need not be fully identical in shape. (For an example, think of a sphere and a hemisphere with the flat side of the latter facing the light.) The shadow as a function of shape is not one-to-one and does not have an inverse.

53. The usual meaning of a "ten percent cut in salary" is that the new salary is 90% of the old. Thus after a ten percent raise the salary is 1.1 times the original, and after a subsequent ten percent cut, the salary is 90% of the raised salary, or 0.9 times 1.1 times the original salary. The combined effect is 99% of the original, and therefore the ten percent raise and the ten percent cut are not inverse operations.

The 10%-raise function is $y = f(x) = (1.1)x$, and the inverse relation is $x = y/(1.1) = (0.90909\ldots)y$. Thus $f^{-1}(x) = (0.90909)x$ and in the language of cuts, this is a pay cut of fractional value $1 - 0.90909\ldots = 0.090909\ldots$ or $9.0909\ldots$ percent.

55. From exercise 13, the slope is $\frac{1}{7}$. The point is $(4, 1)$, so the equation of the tangent line to the graph of $f^{-1}(x)$ is $y = \frac{1}{7}(x - 4) + 1$.

57. Quadratic fit via a CAS gives the function $f(x) = 7x^2 - 13.8x + 19$. This is equal to 30 when $x \approx 2.58$.

6.3 THE EXPONENTIAL FUNCTION

1.

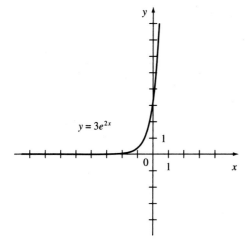

$y = 3e^{2x}$

3.

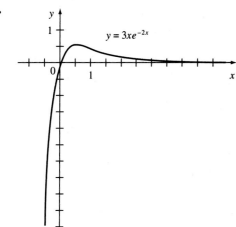

$y = 3xe^{-2x}$

5.

$y = e^{1/x}$

7. $\dfrac{d}{dx}\left[4e^{3x}\right] = 4e^{3x}(3) = 12e^{3x}$

9. $\dfrac{d}{dx}\left[\dfrac{e^{4x}}{x}\right] = \dfrac{(4e^{4x})x - e^{4x}(1)}{x^2} = \dfrac{(4x-1)e^{4x}}{x^2}$

11. $\dfrac{d}{dx}\left[2e^{x^3 3x}\right] = 2e^{x^3 - 3x}(3x^2 - 3)$

$$= 2(3x^2 - 3)e^{x^3 - 3x}$$

13. $\dfrac{d}{dx}\sqrt{e^{2x^2}}$

$$= \dfrac{1}{2}(e^{2x^2})^{-1/2}\dfrac{d}{dx}e^{2x^2} = \dfrac{1}{2}(e^{2x^2})^{-1/2}e^{2x^2}\dfrac{d}{dx}2x^2$$

$$= \dfrac{1}{2}(e^{2x^2})^{-1/2}e^{2x^2}(4x) = 2x\sqrt{e^{2x^2}} = 2xe^{2x^2}$$

15. $f(x) = e^{\ln x^2}$

$$f'(x) = e^{\ln x^2} \cdot \dfrac{d}{dx}\ln x^2$$

$$e^{\ln x^2} \cdot \dfrac{2}{x} = 2x$$

Much easier if one noticed at the outset that $f(x) = x^2$.

17. $f(x) = \ln\sqrt{4e^{3x}} = \dfrac{1}{2}\left[\ln(4 \cdot e^{3x})\right]$

$$= \dfrac{1}{2}[\ln 4 + \ln e^{3x}] = \dfrac{\ln 4 + 3x}{2}$$

$$f'(x) = \dfrac{3}{2}$$

19. Let $u = 3x$, then $du = 3\,dx$.

$$\int e^{3x}\,dx = \dfrac{1}{3}\int e^{3x}3\,dx$$

$$= \dfrac{1}{3}\int e^u\,du$$

$$= \dfrac{1}{3}e^u + c$$

$$= \dfrac{1}{3}e^{3x} + c$$

21. Let $u = x^2$, then $du = 2x\,dx$.

$$\int x\,e^{x^2}\,dx = \frac{1}{2}\int e^{x^2}2x\,dx$$

$$= \frac{1}{2}\int e^u\,du$$

$$= \frac{1}{2}e^u + c$$

$$= \frac{1}{2}e^{x^2} + c$$

23. Let $u = \cos x$, then $du = -\sin x\,dx$.

$$\int \sin x e^{\cos x}\,dx = -\int e^{\cos x}(-\sin x)\,dx$$

$$= -\int e^u\,du$$

$$= -e^u + c$$

$$= -e^{\cos x} + c$$

25. Let $u = \frac{1}{x}$, then $du = -\frac{1}{x^2}\,dx$.

$$\int \frac{e^{1/x}}{x^2}\,dx = -\int e^{1/x}\left(-\frac{1}{x^2}\right)\,dx$$

$$= -\int e^u\,du$$

$$= -e^u + c$$

$$= -e^{1/x} + c$$

27. $$\int (1 + e^x)^2\,dx = \int (1 + 2e^x + e^{2x})\,dx$$

$$= x + 2e^x + \frac{1}{2}e^{2x} + c$$

29. $$\int \ln e^{x^2}\,dx = \int x^2\,dx = \frac{1}{3}x^3 + c$$

31. $$\int_0^1 e^{3x}\,dx = \frac{1}{3}\left[e^{3x}\right]_0^1 = \frac{1}{3}(e^3 - 1)$$

$$= \frac{1}{3}e^3 - \frac{1}{3}$$

33. $\int_{-2}^2 xe^{-x^2}\,dx = -\frac{1}{2}\left[e^{-x^2}\right]_{-2}^2 = -\frac{1}{2}(e^{-4} - e^{-4}) = 0$

35. $f(x) = xe^{-2x}$
$f'(x) = xe^{-2x}(-2) + e^{-2x} = e^{-2x}(1 - 2x)$

Critical numbers: $e^{-2x}(1 - 2x) = 0$
$e^{-2x} = 0$ or $1 - 2x = 0$
none $x = \frac{1}{2}$

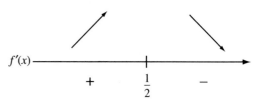

Therefore, $f\left(\frac{1}{2}\right) = \frac{1}{2}e^{-1}$ is a maximum.

$$f''(x) = e^{-2x}(-2) + (1 - 2x)e^{-2x}(-2)$$

$$= -2e^{-2x}(2 - 2x)$$

$$= -4e^{-2x}(1 - x)$$

Inflection points: $-4e^{-2x}(1 - x) = 0$

$e^{-2x} = 0$ or $1 - x = 0$
none $x = 1$

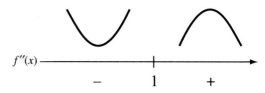

Therefore, $f(1) = e^{-2}$ is an inflection point.

37. $f(x) = x^2 e^{-2x}$

$$f'(x) = x^2 e^{-2x}(-2) + e^{-2x}(2x)$$

$$= 2xe^{-2x}(1 - x)$$

critical numbers: $2xe^{-2x}(1 - x) = 0$

$x = 0$ or $e^{-2x} = 0$ or $1 - x = 0$
none $x = 1$

Therefore, $f(0) = 0$ is a minimum and $f(1) = e^{-2}$ is a maximum.

$$f''(x) = 2e^{-2x}(1 - 2x) + 2(x - x^2)e^{-2x}(-2)$$

$$= 2e^{-2x}(2x^2 - 4x + 1)$$

Inflection points: $2e^{-2x}(2x^2 - 4x + 1) = 0$

$e^{-2x} = 0$ or $2x^2 - 4x + 1 = 0$

none $x = 1 \pm \frac{\sqrt{2}}{2}$

$f''(x)$

+ $1 - \frac{\sqrt{2}}{2}$ − $1 + \frac{\sqrt{2}}{2}$ +

Therefore,

$$f\left(1 - \frac{\sqrt{2}}{2}\right) \approx 0.048 \text{ and } f\left(1 + \frac{\sqrt{2}}{2}\right) \approx 0.096$$

are inflection points.

39. As in the proof of part (i) of Theorem 1.1, we make use of the rules of logarithms.

(ii) To show $\dfrac{e^r}{e^s} = e^{r-s}$, note that

$$\ln \frac{e^r}{e^s} = \ln e^r - \ln e^s = r - s = \ln e^{r-s}.$$

Therefore, $\dfrac{e^r}{e^s} = e^{r-s}$.

(iii) To show $(e^r)^t = e^{r \cdot t}$, note that
$\ln(e^r)^t = t \ln(e^r) = t \cdot r = \ln e^{t \cdot r}$.
Therefore, $(e^r)^t = e^{r \cdot t}$.

41. $x(t) = \dfrac{6}{2e^{-8t} + 1} = 6(2e^{-8t} + 1)^{-1}$

$x'(t) = -6(2e^{-8t} + 1)^{-2} \cdot (-16e^{-8t})$

$= \dfrac{96e^{-8t}}{(2e^{-8t} + 1)^2}$

Since $e^{-8t} > 0$ for any t, both numerator and denominator are positive, so that $x'(t) > 0$. Then, since $x(t)$ is an increasing function with a limiting value of 6 (as t goes to infinity), the concentration never exceeds (indeed, never reaches) the value of 6.

43. $f(x) = e^{-x^2/2}$

$f'(x) = e^{-x^2/2} \cdot (-2x/2)$

$= -xe^{-x^2/2}$

$f''(x) = -\left[x \cdot -xe^{-x^2/2} + 1 \cdot e^{-x^2/2}\right]$

$= e^{-x^2/2}(x^2 - 1)$

This will be zero only when $x = \pm 1$.

45.

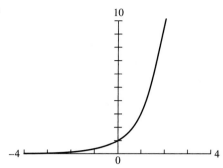

47.

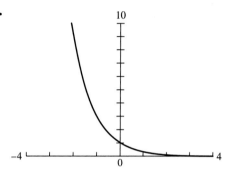

49. For $b > 1$, the graph of $y = b^x$ is increasing and concave up. For $0 < b < 1$, the graph of $y = b^x$ is decreasing and concave up.

51. $\dfrac{d}{dx}\left[3^{2x}\right] = (\ln 3)3^{2x}(2) = 2(\ln 3)3^{2x}$

53. $\dfrac{d}{dx}\left[3^{x^2}\right] = (\ln 3)3^{x^2}(2x) = 2x(\ln 3)3^{x^2}$

55. $\dfrac{d}{dx}\left[\log_4 x^2\right] = \dfrac{1}{\ln 4}\dfrac{1}{x^2}(2x) = \dfrac{2}{x \ln 4}$

57. $\displaystyle\int 2^x dx = \dfrac{1}{\ln 2}2^x + c$

59. Let $u = x^2$, then $du = 2x \, dx$.

$$\int x \, 2^{x^2} \, dx = \frac{1}{2}\int 2^{x^2} \cdot 2x \, dx$$

$$= \frac{1}{2}\int 2^u \, du$$

$$= \frac{1}{2}\frac{1}{\ln 2}2^u + c$$

$$= \frac{1}{2 \ln 2}2^{x^2} + c$$

61. For small x we approximate e^x by $x + 1$.

$$\frac{L(e^{2\pi d/L} - e^{-2\pi d/L})}{e^{2\pi d/L} + e^{-2\pi d/L}}$$

$$\approx \frac{L\left[\left(1 + \frac{2\pi d}{L}\right) - \left(1 - \frac{2\pi d}{L}\right)\right]}{\left(1 + \frac{2\pi d}{L}\right) + \left(1 - \frac{2\pi d}{L}\right)}$$

$$\approx \frac{L\left(\frac{4\pi d}{L}\right)}{2} = 2\pi d$$

$$f(d) \approx \frac{4.9}{\pi} \cdot 2\pi d = 9.8d$$

63. With $t = 90$ and $r = 1/25$, we have

$$P(n) = \frac{3.6^n}{n!} e^{-3.6}.$$

We compute P for the first few values of n:

n	P
0	$e^{-3.6}$
1	$3.6e^{-3.6}$
2	$6.48e^{-3.6}$
3	$7.776e^{-3.6}$
4	$6.9984e^{-3.6}$

Once $n > 3$, the values of P will decrease as n increases. This is due to the fact that to get $P(n + 1)$ from $P(n)$, we multiply $P(n)$ by $3.6/(n + 1)$. Since $n > 3$, $3.6/(n + 1) < 1$ and so $P(n + 1) < P(n)$. Thus we see from the table that P is maximized at $n = 3$. It makes sense that P would be maximized at $n = 3$ because

$$(90 \text{ mins}) \left(\frac{1}{25} \text{ goals/min}\right) = 3.6 \text{ goals}.$$

65. $W(t) = a \cdot e^{-be^{-t}}$

as $t \to \infty$, $-be^{-t} \to 0$, so $W(t) \to a$.

$W'(t) = a \cdot e^{-be^{-t}} \cdot be^{-t}$

as $t \to \infty$, $be^{-t} \to 0$, so $W'(t) \to 0$.

$W''(t) = (a \cdot e^{-be^{-t}} \cdot be^{-t}) \cdot be^{-t}$

$\qquad + (a \cdot e^{-be^{-t}}) \cdot (-be^{-t})$

$\qquad = a \cdot e^{-be^{-t}} \cdot be^{-t} \left[be^{-t} - 1\right]$

$W''(t) = 0$ when $be^{-t} = 1$

$$e^{-t} = b^{-1}$$
$$-t = \ln b^{-1}$$
$$t = \ln b$$

$$W'(\ln b) = a \cdot e^{-be^{-\ln b}} \cdot be^{-\ln b}$$

$$= a \cdot e^{-b\left(\frac{1}{b}\right)} \cdot b \cdot \frac{1}{b} = ae^{-1}$$

Maximum growth rate is ae^{-1} when $t = \ln b$.

67. For $a = 70$, $b = 0.2$,

$$f(t) = \frac{70}{1 + 3e^{-0.2t}} = 70(1 + 3e^{-0.2t})^{-1}$$

$$f(2) = \frac{70}{1 + 3e^{-0.2 \cdot 2}} \approx 23$$

$$f'(t) = -70(1 + 3e^{-0.2t})^{-2}(3e^{-0.2t})(-0.2)$$

$$= \frac{42e^{-0.2t}}{(1 + 3e^{-0.2t})^2}$$

$$f'(2) = \frac{42e^{-0.2 \cdot 2}}{(1 + 3e^{-0.2 \cdot 2})^2} \approx 3.105$$

This says that at time $t = 2$ hours, the rate at which the spread of the rumor is increasing is about 3% of the population per hour.

$$\lim_{t \to \infty} f(t) = \frac{70}{1 + 0} = 70$$

so 70% of the population will eventually hear the rumor.

69.

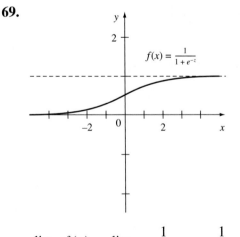

$$\lim_{x \to \infty} f(x) = \lim_{x \to \infty} \frac{1}{1 + e^{-x}} = \frac{1}{1 + 0} = 1$$

$$\lim_{x \to \infty} f(x) = \lim_{x \to \infty} \frac{1}{1 + e^{-x}} = \frac{1}{1 + \infty} = 0$$

If $g(x)$ is $f(x)$ rounded to the nearest integer, then $g(x) = \begin{cases} 0, & x < 0 \\ 1, & x \geq 0 \end{cases}$ and the threshold for switching from "off" to "on" is $x = 0$. To move the threshold to $x = 4$, redefine the function as $f(x) = \dfrac{1}{1 + e^{-(x-4)}}$.

71. We solve the equation $\int_0^3 2ke^{-kx}\, dx = 1$ for k.

$$\int_0^3 2ke^{-kx}\, dx = -2 \left[e^{-kx}\right]_0^3 = 2(1 - e^{-3k})$$

$$2(1 - e^{-3k}) = 1$$

$$1 - e^{-3k} = \frac{1}{2}$$

$$e^{-3k} = \frac{1}{2}$$

$$-3k = \ln \frac{1}{2}$$

$$k = -\frac{1}{3} \ln \frac{1}{2} = \frac{1}{3} \ln 2$$

The pdf $f(x)$ can be written in simpler form:

$$f(x) = 2ke^{-kx} = 2 \left(\frac{1}{3} \ln 2\right) e^{-\left(\frac{1}{3} \ln 2\right)x}$$

$$= \left(\frac{2}{3} \ln 2\right) 2^{-x/3}.$$

The probability a given cell is between 1 and 2 days old is given by

$$\int_1^2 \left(\frac{2}{3} \ln 2\right) 2^{-x/3}\, dx$$

$$= \left(\frac{2}{3} \ln 2\right) \cdot \frac{1}{\ln 2} (-3) \left[2^{-x/3}\right]_1^2$$

$$= -2(2^{-2/3} - 2^{-1/3})$$

$$= 2^{2/3} - 2^{1/3}$$

$$\approx 0.327, \text{ or about } 32.7\%.$$

73. For $h > 0$, $h = \ln e^h = \int_1^{e^h} \frac{1}{x}\, dx$.

Since $\frac{1}{x}$ is continuous on $[1, e^h]$, the Integral Mean Value Theorem (Theorem 4.4.4) implies there is a number $\bar{x} \in (1, e^h)$ for which

$$\frac{1}{\bar{x}} = \frac{1}{e^h - 1} \int_1^{e^h} \frac{1}{x}\, dx. \text{ Hence,}$$

$$\frac{e^h - 1}{\bar{x}} = \int_1^{e^h} \frac{1}{x}\, dx = h, \text{ and this implies}$$

$$\frac{e^h - 1}{h} = \bar{x}. \text{ So } \lim_{h \to 0^+} \frac{e^h - 1}{h} = \lim_{h \to 0^+} \bar{x}. \text{ As}$$

$h \to 0^+$, $e^h \to 1^+$ and since $\bar{x} \in (1, e^h)$ it must be that $\bar{x} \to 1$. Hence $\lim_{h \to 0^+} \dfrac{e^h - 1}{h} = \lim_{h \to 0^+} \bar{x} = 1$. A similar argument works for $h < 0$, but with $h < 0$, $e^h < 1$, so $h = \ln e^h =$

$$\int_1^{e^h} \frac{1}{x}\, dx = -\int_{e^h}^1 \frac{1}{x}\, dx. \text{ Now, the Integral}$$

Mean Value Theorem implies there is a number $\bar{x} \in (e^h, 1)$ for which

$$-\frac{1}{\bar{x}} = \frac{1}{1 - e^h} \int_{e^h}^1 -\frac{1}{x}\, dx. \text{ Hence,}$$

$$\frac{1 - e^h}{\bar{x}} = \int_{e^h}^1 -\frac{1}{x}\, dx = h, \text{ and we have}$$

$$\frac{e^h - 1}{h} = \bar{x}. \text{ As } h \to 0^-, e^h \to 1^- \text{ and since}$$

$\bar{x} \in (e^h, 1)$ it must be that $\bar{x} \to 1$. Hence $\lim_{h \to 0^-} \dfrac{e^h - 1}{h} = \lim_{h \to 0^-} \bar{x} = 1$. We conclude that $\lim_{h \to 0} \dfrac{e^h - 1}{h}$ exists and equals 1.

75. The table suggests that $\lim_{n \to \infty} \dfrac{\ln(1 + 1/n)}{1/n} = 1$.

n	$\dfrac{\ln(1 + 1/n)}{1/n}$
10	0.9531
100	0.9950
1000	0.9995
10,000	1.0000

77. Let $f(x) = \ln x$. Then $f^{-1}(x) = e^x$. The conditions of Theorem 2.3 are satisfied, so

$$\frac{d}{dx} e^x = \frac{1}{f'(e^x)} = \frac{1}{1/e^x} = e^x.$$

6.4 THE INVERSE TRIGONOMETRIC FUNCTIONS

1.

(a) On the unit circle, the angle $\theta \in \left[-\dfrac{\pi}{2}, \dfrac{\pi}{2} \right]$ for which $\sin \theta = 0$ is $\theta = 0$. Hence, $\sin^{-1} 0 = 0$.

(b) On the unit circle, the angle $\theta \in \left[-\dfrac{\pi}{2}, \dfrac{\pi}{2} \right]$ for which $\sin \theta = -\dfrac{1}{2}$ is $\theta = -\dfrac{\pi}{6}$. Hence, $\sin^{-1}\left(-\dfrac{1}{2} \right) = -\dfrac{\pi}{6}$.

(c) On the unit circle, the angle $\theta \in \left[-\dfrac{\pi}{2}, \dfrac{\pi}{2} \right]$ for which $\sin \theta = -1$ is $\theta = -\dfrac{\pi}{2}$. Hence, $\sin^{-1}(-1) = -\dfrac{\pi}{2}$.

3.

(a) On the unit circle, the angle $\theta \in \left(-\dfrac{\pi}{2}, \dfrac{\pi}{2} \right)$ for which $\tan \theta = 1$ is $\theta = \dfrac{\pi}{4}$. Hence, $\tan^{-1} 1 = \dfrac{\pi}{4}$.

(b) On the unit circle, the angle $\theta \in \left(-\dfrac{\pi}{2}, \dfrac{\pi}{2} \right)$ for which $\tan \theta = 0$ is $\theta = 0$. Hence, $\tan^{-1} 0 = 0$.

(c) On the unit circle, the angle $\theta \in \left(-\dfrac{\pi}{2}, \dfrac{\pi}{2} \right)$ for which $\tan \theta = -1$ is $\theta = -\dfrac{\pi}{4}$. Hence, $\tan^{-1}(-1) = -\dfrac{\pi}{4}$.

5.

(a) On the unit circle, the angle $\theta \in \left[0, \dfrac{\pi}{2} \right) \cup \left(\dfrac{\pi}{2}, \pi \right]$ for which $\sec \theta = 1$ is $\theta = 0$. Hence, $\sec^{-1} 1 = 0$.

(b) On the unit circle, the angle $\theta \in \left[0, \dfrac{\pi}{2} \right) \cup \left(\dfrac{\pi}{2}, \pi \right]$ for which $\sec \theta = 2$ is $\theta = \dfrac{\pi}{3}$. Hence, $\sec^{-1} 2 = \dfrac{\pi}{3}$.

(c) On the unit circle, the angle $\theta \in \left[0, \dfrac{\pi}{2} \right) \cup \left(\dfrac{\pi}{2}, \pi \right]$ for which $\sec \theta = \sqrt{2}$ is $\theta = \dfrac{\pi}{4}$. Hence, $\sec^{-1} \sqrt{2} = \dfrac{\pi}{4}$.

7. Let $\sin^{-1} x = \theta$, then $\cos(\sin^{-1} x) = \cos \theta$.

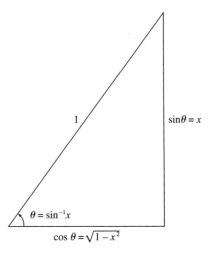

$\cos \theta = \sqrt{1 - x^2}, \ 1 \le x \le 1$

9. Let $\sec^{-1} x = \theta$, then $\tan(\sec^{-1} x) = \tan \theta$.

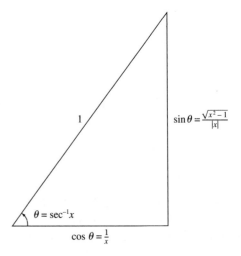

$\tan\theta = \sqrt{x^2 - 1}$, $x \geq 1$, or
$\tan\theta = -\sqrt{x^2 1}$, $x \leq -1$.

11. Let $\cos^{-1}\dfrac{1}{2} = \theta$, then $\sin\left(\cos^{-1}\dfrac{1}{2}\right) = \sin\theta$.

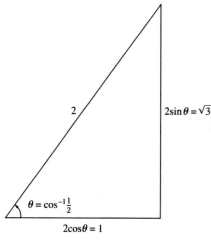

2

$2\sin\theta = \sqrt{3}$

$\theta = \cos^{-1}\dfrac{1}{2}$

$2\cos\theta = 1$

$\sin\theta = \dfrac{\sqrt{3}}{2}$

13. Let $\cos^{-1}\dfrac{3}{5} = \theta$, then $\tan\left(\cos^{-1}\dfrac{3}{5}\right) = \tan\theta$.

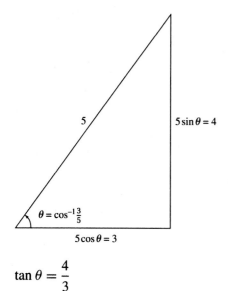

5

$5\sin\theta = 4$

$\theta = \cos^{-1}\dfrac{3}{5}$

$5\cos\theta = 3$

$\tan\theta = \dfrac{4}{3}$

15. The range of the inverse cosine is $0 \leq y \leq \pi$, and $\dfrac{\pi}{8}$ is in this range and has the same cosine as $-\dfrac{\pi}{8}$, so $\cos^{-1}(\cos(-\pi/8)) = \pi/8$.

17. Expanding by the double angle formula,
$\sin\left(2\sin^{-1}(x/3)\right)$

$$= 2\sin\left(\sin^{-1}(x/3)\right)\cos\left(\sin^{-1}(x/3)\right)$$

$$= 2 \cdot \frac{x}{3} \cdot \sqrt{1 - \left(\frac{x^2}{3}\right)}$$

$$= \frac{2x}{9}\sqrt{9 - x^2} \text{ for } |x| \leq 3.$$

19.

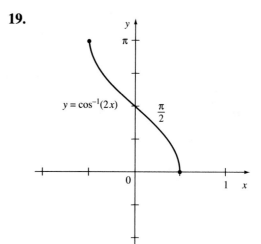

$y = \cos^{-1}(2x)$

21.

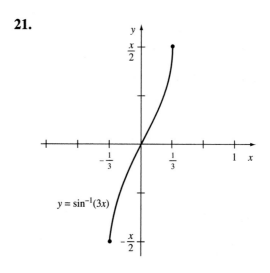

$y = \sin^{-1}(3x)$

23.

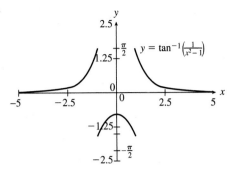

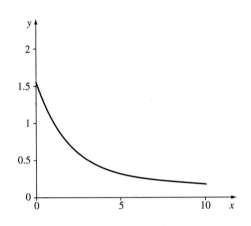

25. $\cos(2x) = 0$

$2x = \dfrac{\pi}{2}n$

$x = \dfrac{\pi}{4}n$, n any odd integer

27. $\sin(3x) = 0$

$3x = 0 + \pi n$

$x = 0 + \dfrac{\pi}{3}n = \dfrac{\pi}{3}n$, n any integer

29. $\tan(5x) = 1$

$5x = \dfrac{\pi}{4} + \pi n$

$x = \dfrac{\pi}{20} + \dfrac{\pi}{5}n$, n any integer

31. $\sec(2x) = 2$

$2x = \dfrac{\pi}{3} + 2\pi n,\ -\dfrac{\pi}{3} + 2\pi n$

$x = \dfrac{\pi}{6} + \pi n,\ -\dfrac{\pi}{6} + \pi n$, n any integer

33. $y = \csc^{-1} x$ if and only if $\csc y = x$ and

$y \in \left[-\dfrac{\pi}{2}, 0\right) \cup \left(0, \dfrac{\pi}{2}\right]$

$y = \cot^{-1} x$ if and only if $\cot y = x$ and

$y \in (0, \pi)$.

35. Since x is measured in feet, use $12x$ so that both measurements are in inches.

$\tan A = \dfrac{20}{12x}$

$A = \tan^{-1}\left(\dfrac{20}{12x}\right) = \tan^{-1}\left(\dfrac{5}{3x}\right)$

6.5 THE CALCULUS OF THE INVERSE TRIGONOMETRIC FUNCTIONS

1. $\dfrac{d}{dx}\left[\sin^{-1}(3x^2)\right] = \dfrac{1}{\sqrt{1 - 9x^4}}\dfrac{d}{dx}(3x^2)$

$= \dfrac{6x}{\sqrt{1 - 9x^4}}$

3. $\dfrac{d}{dx}\left[\sec^{-1} x^2\right] = \dfrac{1}{|x^2|\sqrt{(x^2)^2 - 1}}\dfrac{d}{dx}(x^2)$

$= \dfrac{2}{x\sqrt{x^4 - 1}}$

5. $\dfrac{d}{dx}\left[x \cos^{-1}(2x)\right] = x\dfrac{-1}{\sqrt{1 - (2x)^2}}2 + \cos^{-1}(2x)$

$= \dfrac{-2x}{\sqrt{1 - 4x^2}} + \cos^{-1}(2x)$

7. $\dfrac{d}{dx}\left[\cos^{-1}(\sin x)\right] = \dfrac{-1}{\sqrt{1 - (\sin x)^2}}\cos x$

$= -\dfrac{\cos x}{\sqrt{\cos^2 x}}$

$= -\dfrac{\cos x}{|\cos x|} = \pm 1$

9. $\dfrac{d}{dx}\left[\tan^{-1}(\sec x)\right] = \dfrac{1}{1 + (\sec x)^2} \cdot (\sec x \tan x)$

$= \dfrac{\sec x \tan x}{1 + \sec^2 x}$

37. $I_1 = \int_0^1 \sqrt{1 - x^2}\, dx$ is one fourth of the area of a circle with radius 1, so

$$\int_0^1 \sqrt{1 - x^2}\, dx = \frac{\pi}{4}.$$

$$I_2 = \int_0^1 \sqrt{1 + x^2}\, dx = \arctan x \Big|_0^1$$

$$= \arctan 1 - \arctan 0 = \tfrac{\pi}{4}$$

n	$S_n(\sqrt{1 - x^2})$	$S_n\left(\frac{1}{1 + x^2}\right)$
4	0.77090	0.78539
8	0.78030	0.78539

The second integral $\int \dfrac{1}{1 + x^2}\, dx$ provides a better algorithm for estimating π.

39. The function $\tan^{-1} x$ is continuous and differentiable everywhere, so for any $a \neq 0$ we can apply the Mean Value Theorem to get $\frac{\tan^{-1} a - \tan^{-1} 0}{a - 0} = \frac{1}{1 + c^2}$ for some c between 0 and a. Taking absolute values, we get $\left| \frac{\tan^{-1} a}{a} \right| = \left| \frac{1}{1 + c^2} \right| < 1$, so $|\tan^{-1} a| < |a|$ for $a \neq 0$. This means that the only solution to $\tan^{-1} x = x$ is $x = 0$.

41. The linear approximation for the inverse tangent function at $x = 0$ is

$$f(x) \approx f(0) + f'(0)(x - 0)$$
$$\tan^{-1}(x) \approx \tan^{-1}(0) + \tfrac{1}{1 + 0^2}(x - 0)$$
$$\tan^{-1}(x) \approx x$$

Using this approximation,

$$\phi = \tan^{-1}\left(\frac{3[1 - d/D] - w/2}{D - d} \right)$$

$$\phi \approx \frac{3[1 - d/D] - w/2}{D - d}$$

If $d = 0$, then $\phi \approx \frac{3 - w/2}{D}$. Thus, if w or D increases, then ϕ decreases.

43. We find the derivative of $f(t)$:

$$f'(t) = \frac{a^2 + t^2 - t(2t)}{(a^2 + t^2)^2}$$
$$= \frac{a^2 - t^2}{(a^2 + t^2)^2}.$$

The denominator is always positive, while the numerator is positive when $a^2 > t^2$, i.e., when $a > t$. We now find the derivative of $\theta(x)$:

$$\theta'(x) = \frac{1}{1 + \left(\frac{29.25}{x} \right)^2} \left(\frac{-29.25}{x^2} \right)$$
$$- \frac{1}{1 + \left(\frac{10.75}{x} \right)^2} \left(\frac{-10.75}{x^2} \right)$$
$$= \frac{-29.25}{x^2 + (29.25)^2} + \frac{10.75}{x^2 + (10.75)^2}.$$

We consider each of the two terms of the last line above as instances of $f(t)$, the first as $-f(29.25)$ and the second as $f(10.75)$. Now, for any given x where $x \geq 30$, this x is our a in $f(t)$ and since $a = x$ is greater than 29.25 and greater than 10.75, $f(t)$ is increasing for these two t values and this value of a. Thus $f(29.25) > f(10.75)$. This means that

$$\theta'(x) = -f(29.25) + f(10.75) < 0$$

(where $a = x$) and so $\theta(x)$ is decreasing for $x \geq 30$. Since $\theta(x)$ is decreasing for $x \geq 30$, the announcers would be wrong to suggest that the angle increases by backing up 5 yards when the team is between 50 and 60 feet away from the goal post.

45. Let $f(x) = \cos x$ on the interval $[0, \pi]$. Then $f^{-1}(x) = \cos^{-1} x$ and all the conditions of the theorem are satisfied, so

$$\frac{d}{dx} \cos^{-1} x = \frac{1}{-\sin(\cos^{-1} x)} = \frac{-1}{\sqrt{1 - x^2}}$$

since $\sin(\cos^{-1} x) = \sqrt{1 - x^2}$.

6.6 THE HYPERBOLIC FUNCTIONS

1.

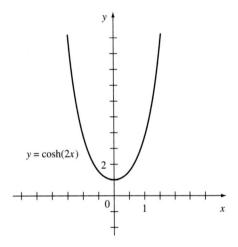

$y = \cosh(2x)$

3.

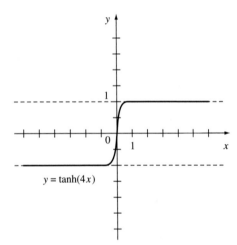

$y = \tanh(4x)$

5.

$y = \cosh(2x)\sinh(2x)$

7. $\dfrac{d}{dx}[\cosh 4x] = \sinh 4x \dfrac{d}{dx}(4x) = 4\sinh 4x$

9. $\dfrac{d}{dx}\left[\sinh \sqrt{x}\right] = \cosh \sqrt{x}\dfrac{d}{dx}\left(\sqrt{x}\right)$

$= \dfrac{\cosh \sqrt{x}}{2\sqrt{x}}$

11. $\dfrac{d}{dx}\left[\cosh^{-1} 2x\right] = \dfrac{1}{\sqrt{(2x)^2 - 1}}\dfrac{d}{dx}(2x) =$

$\dfrac{2}{\sqrt{4x^2 - 1}}$

13. $\dfrac{d}{dx}\left[x^2 \sinh 2x\right] = x^2 \cosh 2x \cdot 2 + \sinh 2x \cdot 2x$

$= 2x^2 \cosh 2x + 2x \sinh 2x$

15. $\dfrac{d}{dx}[\tanh 4x] = \text{sech}^2 4x \dfrac{d}{dx}4x$

$= 4\,\text{sech}^2 4x$

17. $\displaystyle\int \cosh 6x\, dx$

Let $u = 6x$. Then $du = 6\, dx$.

$\dfrac{1}{6}\displaystyle\int \cosh u\, du = \dfrac{1}{6}\sinh u + c = \dfrac{1}{6}\sinh 6x + c$

19. $\displaystyle\int \tanh 3x\, dx = \int \dfrac{\sinh 3x}{\cosh 3x}\, dx$

Let $u = \cosh 3x$. Then $du = \sinh 3x \cdot 3\, dx$.

$\dfrac{1}{3}\displaystyle\int \dfrac{1}{u}\, du = \dfrac{1}{3}\ln |u| + c = \dfrac{1}{3}\ln(\cosh 3x) + c$

21. $\displaystyle\int_0^1 \dfrac{e^{4x} - e^{-4x}}{2}\, dx = \int_0^1 \sinh 4x\, dx$

$= \dfrac{1}{4}[\cosh 4x]_0^1$

$= \dfrac{1}{4}(\cosh 4 - 1)$

23. $\displaystyle\int \cos x \sinh(\sin x)\, dx$

Let $u = \sin x$. Then $du = \cos x\, dx$.

$\displaystyle\int \sinh u\, du = \cosh u + c = \cosh(\sin x) + c$

25. $\displaystyle\int_0^1 \cosh x\, e^{\sinh x}\, dx$

Let $u = \sinh x$. Then $du = \cosh x\, dx$.

$$\int_a^b e^u \, du = \left[e^u \right]_a^b = \left[e^{\sinh x} \right]_0^1 = e^{\sinh(1)} - 1$$

27. $\dfrac{d}{dx}[\cosh x] = \dfrac{d}{dx}\left[\dfrac{e^x + e^{-x}}{2} \right]$

$= \dfrac{e^x - e^x}{2}$

$= \sinh x$

$\dfrac{d}{dx}[\tanh x] = \dfrac{d}{dx}\left[\dfrac{\sinh x}{\cosh x} \right]$

$\qquad = \dfrac{\cosh x \frac{d}{dx}[\sinh x] - \sinh x \frac{d}{dx}[\cosh x]}{\cosh^2 x}$

$\qquad = \dfrac{\cosh^2 x - \sinh^2 x}{\cosh^2 x}$

$\qquad = \dfrac{1}{\cosh^2 x}$

$\qquad = \operatorname{sech}^2 x$

29. First, $e^x > e^{-x}$ if $x > 0$ and $e^x < e^{-x}$ if $x < 0$.

Since $\sinh x = \dfrac{e^x - e^{-x}}{2}$, we have that

$e^x - e^{-x} > 0$ if $x > 0$ and
$e^x - e^{-x} < 0$ if $x < 0$.

Therefore, $\sinh x > 0$ if $x > 0$ and $\sinh x < 0$ if $x < 0$.

31. $\dfrac{d}{dx}[\cosh x] = \sinh x$
Critical values:

$\sinh x = 0$
$\qquad x = 0$

$\dfrac{dy}{dx}$:

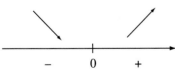

$\qquad\quad - \qquad 0 \qquad +$

There is a minimum of $\cosh(0) = 1$ at $x = 0$. $y = \cosh x$ is decreasing on $(-\infty, 0)$ and increasing on $(0, \infty)$.

$\dfrac{d^2}{dx^2}[\cosh x] = \dfrac{d}{dx}[\sinh x] = \cosh x$

Points of inflection:
$\cosh x = 0$, no solution

There are no points of inflection. The graph is concave up on $(-\infty, \infty)$.

33. If $y = \cosh^{-1} x$ then $x = \cosh y$ and
$x = \dfrac{e^y + e^{-y}}{2}$.

Also $\sinh y = \dfrac{e^y - e^{-y}}{2}$.

$e^y = \cosh y + \sinh y$

$e^y = \cosh y + \sqrt{\sinh^2 y}$

$e^y = \cosh y + \sqrt{\cosh^2 y - 1}$

$e^y = x + \sqrt{x^2 - 1}$

$y = \cosh^{-1} x = \ln\left(x + \sqrt{x^2 - 1} \right)$

35. For the function
$f(x) = 10 \cosh\left(\dfrac{x}{10} \right)$, $-20 \le x \le 20$, the amount of sag will occur at the minimum value of the function. Find the critical values.

$f'(x) = \sinh\left(\dfrac{x}{10} \right) = 0$

$\qquad x = 0$

The lowest point of the cable is at
$f(0) = 10 \cosh(0) = 10$
The highest point of the cable is at
$f(20) = 10 \cosh(2) \approx 37.62$.
Therefore, the amount of sag is
$37.62 - 10 \approx 27.62$ feet.

37. $f(x) = 15 \cosh\left(\dfrac{x}{15} \right)$, $-25 \le x \le 25$

$f'(x) = \sinh\left(\dfrac{x}{15} \right)$

The critical value is $x = 0$.
The lowest point of the cable is at
$f(0) = 15 \cosh(0) = 15$.
The highest point of the cable is at
$f(25) = 15 \cosh\left(\dfrac{25}{15} \right) \approx 41.13$.

Therefore, the amount of sag is
$41.13 - 15 = 26.13$ feet.

39. $f(x) = a \cosh\left(\dfrac{x}{a}\right), \ -b \le x \le b$

$f'(x) = \sinh\left(\dfrac{x}{a}\right)$

The critical value is $x = 0$.
The lowest point of the cable is at
$f(0) = a \cosh(0) = a$.
The highest point of the cable is at

$f(b) = a \cosh\left(\dfrac{b}{a}\right)$.

Therefore, the amount of sag is

$f(b) - f(0) = a \cosh\left(\dfrac{b}{a}\right) - a$.

$L = \displaystyle\int_{-b}^{b} \sqrt{1 + \left[\sinh\left(\dfrac{x}{a}\right)\right]^2}\, dx$

$= \displaystyle\int_{-b}^{b} \cosh\left(\dfrac{x}{a}\right) dx$

$= a\left[\sinh\left(\dfrac{x}{a}\right)\right]_{-b}^{b}$

$= a\left[\sinh\left(\dfrac{b}{a}\right) - \sinh\left(-\dfrac{b}{a}\right)\right]$

Since $y = \sinh x$ is an odd function (see exercise 40),

$\sinh\left(-\dfrac{b}{a}\right) = -\sinh\left(\dfrac{b}{a}\right)$. So,

$L = a\left[\sinh\left(\dfrac{b}{a}\right) + \sinh\left(\dfrac{b}{a}\right)\right]$

$= 2a \sinh\left(\dfrac{b}{a}\right)$.

41. $\cosh x + \sinh x = \dfrac{e^x + e^{-x}}{2} + \dfrac{e^x - e^{-x}}{2} = e^x$

43. $\displaystyle\lim_{x \to \infty} \dfrac{e^x - e^{-x}}{e^x + e^{-x}} = 1$ since both e^{-x} terms tend to 0.

$\displaystyle\lim_{x \to \infty} \dfrac{e^x - e^{-x}}{e^x + e^{-x}} = \lim_{x \to \infty} \dfrac{e^{-x} - e^{-x}}{e^{-x} + e^x} = -1$.

45. **(a)** $mv'(t) = -mg + kv^2 = k\left(v^2 - \dfrac{mg}{k}\right)$

Thus, $\dfrac{v'(t)}{v^2 - mg/k} = \dfrac{k}{m}$.

(b) $\dfrac{1}{v^2 - a^2}\dfrac{dv}{dt} = \dfrac{k}{m}$, where $a = \sqrt{\dfrac{mg}{k}}$

$\dfrac{dv}{v^2 - a^2} = \dfrac{dv}{2a}\left(\dfrac{1}{v - a} - \dfrac{1}{v + a}\right) = \dfrac{k}{m}dt$

$\dfrac{dv}{2a}\left(\dfrac{1}{v - a} - \dfrac{1}{v + a}\right) = \dfrac{k}{m}dt$

Integrating,

$\dfrac{1}{2a}\ln\left|\dfrac{v - a}{v + a}\right| = \dfrac{k}{m}t + C$.
v is negative and less than a in absolute value, so

$\ln\left(\dfrac{a - v}{a + v}\right) = \dfrac{2ak}{m}t + C'$.
Exponentiating,

$\dfrac{a - v}{a + v} = ce^{\frac{2ak}{m}}$, where $c = e^{C'}$ is a postive constant.

Solving for v and using $\dfrac{2ak}{m} = 2\sqrt{\dfrac{kg}{m}}$, we get

$v(t) = -a\dfrac{ce^{2\sqrt{kg/m}t} - 1}{ce^{2\sqrt{kg/m}t} + 1}$, which is the answer given in part (c).

(d) $v(0) = 0$ implies that
$ce^{2\sqrt{kg/m}(0)} - 1 = 0$ or $c = 1$.

(e) Let $x = \sqrt{\dfrac{kg}{m}}t$ in

$\tanh x = \dfrac{e^{2x} - 1}{e^{2x} + 1}$.

(f) The fraction in part (c) tends to 1, so the limiting velocity is $-a$.

47. For the first skydiver:
terminal velocity, -80 m/s;
distance in 2 seconds, 19.41 m;
distance in 4 seconds, 75.45 m.
For the second skydiver:

terminal velocity, -40 m/s;
distance in 2 seconds, 18.86 m;
distance in 4 seconds, 68.35 m.

49. For an initial velocity $v_0 = 2000$, we set the derivative of the velocity equal to 0 and solve the resulting equation in a CAS. The maximum acceleration of -9.797 occurs at about 206 seconds.

51.

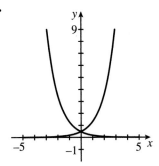

As $x \to \infty$, $e^{-x} \to 0$, so $\displaystyle\lim_{x \to \infty} \sinh x = \frac{e^x}{2}$

and $\displaystyle\lim_{x \to \infty} \cosh x = \frac{e^x}{2}$. As $x \to -\infty$, $e^x \to 0$,

so $\displaystyle\lim_{x \to -\infty} \sinh x = \frac{-e^{-x}}{2}$ and $\displaystyle\lim_{x \to -\infty} \cosh x = \frac{e^{-x}}{2}$.

Chapter 6 Review Exercises

1. $\dfrac{d}{dx}\left[\ln(x^3 + 5)\right] = \dfrac{1}{x^3 + 5}\dfrac{d}{dx}(x^3 + 5) = \dfrac{3x^2}{x^3 + 5}$

3. $\dfrac{d}{dx}\left[\ln\sqrt{x^4 + x}\right] = \dfrac{d}{dx}\left[\dfrac{1}{2}\ln(x^4 + x)\right]$

$= \dfrac{1}{2}\dfrac{1}{x^4 + x}\dfrac{d}{dx}(x^4 + x)$

$= \dfrac{4x^3 + 1}{2(x^4 + x)}$

5. $\dfrac{d}{dx}\left[e^{-x^2}\right] = e^{-x^2}\dfrac{d}{dx}[-x^2] = -2xe^{-x^2}$

7. $\dfrac{d}{dx}\left[4^{x^3}\right] = (\ln 4)4^{x^3}\dfrac{d}{dx}[x^3] = 3x^2(\ln 4)4^{x^3}$

9. $\dfrac{d}{dx}\left[\sin^{-1} 2x\right] = \dfrac{1}{\sqrt{1 - (2x)^2}}\dfrac{d}{dx}[2x]$

$= \dfrac{2}{\sqrt{1 - 4x^2}}$

11. $\dfrac{d}{dx}\left[\tan^{-1}(\cos 2x)\right]$

$= \dfrac{1}{1 + (\cos 2x)^2}\dfrac{d}{dx}[\cos 2x]$

$= -\dfrac{2\sin 2x}{1 + \cos^2 2x}$

13. $\dfrac{d}{dx}\left[\cosh\sqrt{x}\right] = \sinh\sqrt{x}\dfrac{d}{dx}\left[\sqrt{x}\right]$

$= \dfrac{\sinh\sqrt{x}}{2\sqrt{x}}$

15. $\dfrac{d}{dx}\left[\sinh^{-1} 3x\right] = \dfrac{1}{\sqrt{1 + (3x)^2}}\dfrac{d}{dx}[3x]$

$= \dfrac{3}{\sqrt{1 + 9x^2}}$

17. $\displaystyle\int \dfrac{x^2}{x^3 + 4}dx$

Let $u = x^3 + 4$. Then $du = 3x^2\, dx$.

$\dfrac{1}{3}\displaystyle\int\dfrac{1}{u}du = \dfrac{1}{3}\ln|u| + c = \dfrac{1}{3}\ln\left|x^3 + 4\right| + c$

19. $\displaystyle\int_0^1 \dfrac{x}{x^2 + 1}dx$

Let $u = x^2 + 1$. Then $du = 2x\, dx$. If $x = 1$,

then $u = 2$. If $x = 0$, then $u = 1$. $\dfrac{1}{2}\displaystyle\int_1^2\dfrac{1}{u}du = $

$\dfrac{1}{2}\left[\ln|u|\right]_1^2 = \dfrac{1}{2}\ln 2$

21. $\displaystyle\int \dfrac{\sin(\ln x)}{x}dx$

Let $u = \ln x$. Then $du = \dfrac{1}{x}dx$.

$\displaystyle\int \sin(u)\, du = -\cos(u) + c = -\cos(\ln x) + c$

23. $\displaystyle\int e^{-4x}\, dx$

Let $u = -4x$. Then $du = -4\,dx$.

$$-\frac{1}{4}\int e^u\,du = -\frac{1}{4}e^u + c = -\frac{1}{4}e^{-4x} + c$$

25. $\displaystyle\int \frac{e^{\sqrt{x}}}{\sqrt{x}}\,dx$

Let $u = \sqrt{x}$. Then $du = \dfrac{1}{2\sqrt{x}}\,dx$.

$$2\int e^u\,du = 2e^u + c = 2e^{\sqrt{x}} + c$$

27. $\displaystyle\int_0^2 e^{3x}\,dx$

Let $u = 3x$. Then $du = 3\,dx$. If $x = 2$, then $u = 6$. If $x = 0$, then $u = 0$.

$$\frac{1}{3}\int_0^6 e^u\,du = \frac{1}{3}\left[e^u\right]_0^6 = \frac{1}{3}(e^6 - 1)$$

29. $\displaystyle\int 3^{4x}\,dx$

Let $u = 4x$. Then $du = 4\,dx$.

$$\frac{1}{4}\int 3^u\,du = \frac{1}{4}\frac{1}{\ln 3}3^u + c = \frac{1}{4\ln 3}3^{4x} + c$$

31. $\displaystyle\int \frac{3}{x^2+4}\,dx = \frac{3}{4}\int \frac{1}{\left(\frac{x}{2}\right)^2 + 1}\,dx$

Let $u = \dfrac{x}{2}$. Then $du = \dfrac{1}{2}\,dx$.

$$2\cdot\frac{3}{4}\int \frac{1}{u^2+1}\,du = \frac{3}{2}\tan^{-1}\left(\frac{x}{2}\right) + c$$

33. $\displaystyle\int \frac{x^2}{\sqrt{1-x^6}}\,dx = \int \frac{x^2}{\sqrt{1-(x^3)^2}}\,dx$

Let $u = x^3$. Then $du = 3x^2\,dx$.

$$\frac{1}{3}\int \frac{1}{\sqrt{1-u^2}}\,du = \frac{1}{3}\sin^{-1}(u) + c$$
$$= \frac{1}{3}\sin^{-1}(x^3) + c$$

35. $\displaystyle\int \frac{9x}{x^2\sqrt{x^4-1}}\,dx = \int \frac{9x}{x^2\sqrt{(x^2)^2-1}}\,dx$

Let $u = x^2$. Then $du = 2x\,dx$.

$$\frac{9}{2}\int \frac{1}{u\sqrt{u^2-1}}\,du = \frac{9}{2}\sec^{-1}(u) + c$$
$$= \frac{9}{2}\sec^{-1}(x^2) + c$$

37. $\displaystyle\int \frac{4}{\sqrt{1+x^2}}\,dx = 4\sinh^{-1}x + c$

39. $\displaystyle\int \cosh 4x\,dx = \frac{1}{4}\sinh 4x + c$

41. $\dfrac{d}{dx}[x^3 - 1] = 3x^2$

Since $3x^2 \geq 0$ for all x, the function $f(x) = x^3 - 1$ is increasing on $(-\infty, \infty)$. The function is one-to-one.

$$y = x^3 - 1$$
$$y + 1 = x^3$$
$$(y+1)^{1/3} = x$$

Therefore, $f^{-1}(x) = (x+1)^{1/3}$.

43. $\dfrac{d}{dx}\left[e^{2x^2}\right] = 4xe^{2x^2}$

Since $4xe^{2x^2} < 0$ on $(-\infty, 0)$ and $4xe^{2x^2} > 0$ on $(0, \infty)$, the function is not one-to-one. There is no inverse.

45. (a) $f(x) = x^5 + 2x^3 - 1$
$f'(x) = 5x^4 + 6x^2$

If $a = f(x) = 2$ then $x = 1$. $(f^{-1})'(2) =$
$$\frac{1}{f'(1)} = \frac{1}{11}$$

(b)

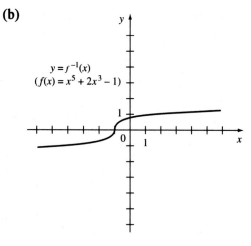

$y = f^{-1}(x)$
$(f(x) = x^5 + 2x^3 - 1)$

47. (a) $f(x) = \sqrt{x^3 + 4x}$

$$f'(x) = \frac{3x^2 + 4}{2\sqrt{x^3 + 4x}}$$

If $a = f(x) = 4$ then $x = 2$.

$$(f^{-1})'(4) = \frac{1}{f'(2)} = \frac{1}{\dfrac{3(2)^2 + 4}{2\sqrt{(2)^3 + 4(2)}}}$$

$$= \frac{8}{16} = \frac{1}{2}$$

(b)

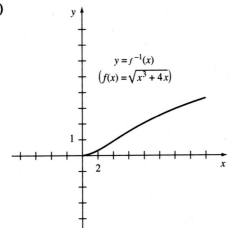

$y = f^{-1}(x)$

$\left(f(x) = \sqrt{x^3 + 4x}\right)$

49. On the unit circle, the angle $\theta \in \left[-\dfrac{\pi}{2}, \dfrac{\pi}{2}\right]$ for which $\sin\theta = 1$ is $\theta = \dfrac{\pi}{2}$. Hence, $\sin^{-1} 1 = \dfrac{\pi}{2}$.

51. On the unit circle, the angle $\theta \in \left(-\dfrac{\pi}{2}, \dfrac{\pi}{2}\right)$ for which $\tan\theta = -1$ is $\theta = -\dfrac{\pi}{4}$. Hence, $\tan^{-1}(-1) = -\dfrac{\pi}{4}$.

53. $\sin(\sec^{-1} 2) = \sin\dfrac{\pi}{3} = \dfrac{\sqrt{3}}{2}$

55. $\sin^{-1}\left(\sin\left(\dfrac{3\pi}{4}\right)\right) = \sin^{-1}\left(\dfrac{\sqrt{2}}{2}\right) = \dfrac{\pi}{4}$

57. $\sin 2x = 1$

$$2x = \frac{\pi}{2} + 2\pi n$$

$$x = \frac{\pi}{4} + \pi n, \text{ for any integer } n$$

59.

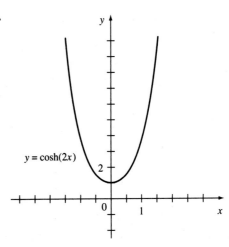

$y = \cosh(2x)$

61.

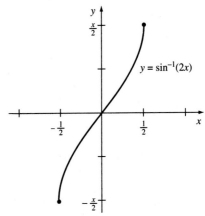

$y = \sin^{-1}(2x)$

63.

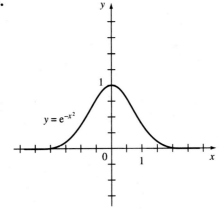

$y = e^{-x^2}$

65. $f(x) = 20\cosh\left(\dfrac{x}{20}\right)$, $-25 \le x \le 25$

$$f'(x) = \sinh\frac{x}{20}$$

The critical value is $x = 0$.

The lowest point of the cable is at $x = 0$,

$f(0) = 20 \cosh(0) = 20$.

The highest point of the cable is at $x = \pm 25$,

$$f(\pm 25) = 20 \cosh\left(+\frac{25}{20}\right) \approx 37.77.$$

Therefore, the amount of sag is
$37.77 - 20 = 17.77$ feet.

67. For any positive integer n, $\displaystyle\lim_{x \to \infty} \frac{e^x}{x^n} = \infty$, so e^x dominates.

69. Let θ_1 be the angle from the horizontal to the upper line segment defining θ and let θ_2 be the angle from the horizontal to the lower line segment defining θ. Then the length of the side opposite θ_2 is $\dfrac{H - P}{2}$ while the length of the side opposite θ_1 is $\dfrac{H + P}{2}$. Then

$$\theta(x) = \theta_1 - \theta_2$$

$$= \tan^{-1}\left(\frac{H + P}{2x}\right)$$

$$- \tan^{-1}\left(\frac{H - P}{2x}\right)$$

and so

$$\theta'(x) = \frac{1}{1 + \left(\frac{H+P}{2x}\right)^2}\left(-\frac{H + P}{2x^2}\right)$$

$$- \frac{1}{1 + \left(\frac{H-P}{2x}\right)^2}\left(-\frac{H - P}{2x^2}\right).$$

We set this equal to 0:

$$0 = \frac{-2(H + P)}{4x^2 + (H + P)^2} + \frac{2(H - P)}{4x^2 + (H - P)^2}$$

and solve for x:

$$\frac{2(H + P)}{4x^2 + (H + P)^2} = \frac{2(H - P)}{4x^2 + (H - P)^2}$$

$$8x^2(H + P) - 8x^2(H - P)$$

$$= 2(H - P)(H + P)^2$$

$$- 2(H + P)(H - P)^2$$

$$8x^2(2P) = 2(H - P)(H + P)(2P)$$

$$x^2 = \frac{H^2 - P^2}{4}$$

$$x = \frac{\sqrt{H^2 - P^2}}{2}.$$

Integration Techniques

7.1 REVIEW OF FORMULAS AND TECHNIQUES

1. $\displaystyle\int \sin 6x\, dx = -\frac{1}{6}\cos x + c$

3. $\displaystyle\int \sec 2x \tan 2x\, dx = \frac{1}{2}\sec 2x + c$

5. $\displaystyle\int e^{3-2x}\, dx = -\frac{1}{2}e^{3-2x} + c$

7. Let $u = 1 + x^{2/3}$, $du = \frac{2}{3}x^{-1/3}\, dx$.

$\displaystyle\int \frac{4}{x^{1/3}(1+x^{2/3})}\, dx = 4\left(\frac{3}{2}\right)\int u^{-1}\, du$

$\qquad = 6\ln|u| + c = 6\ln\left(1 + x^{2/3}\right) + c$

9. Let $u = \sqrt{x}$, $du = \frac{1}{2\sqrt{x}}\, dx$.

$\displaystyle\int \frac{\sin\sqrt{x}}{\sqrt{x}}\, dx = 2\int \sin u\, du$

$\qquad = -2\cos u + c = -2\cos\sqrt{x} + c$

11. Let $u = \sin x$, $du = \cos x\, dx$.

$\displaystyle\int_0^\pi \cos x\, e^{\sin x}\, dx = \int_0^0 e^u\, du = 0$

13. Note that $\dfrac{\sin x}{\cos^2 x} = \dfrac{1}{\cos x} \cdot \dfrac{\sin x}{\cos x}$

$\qquad = \sec x \tan x.$

$\displaystyle\int_{-\pi/4}^0 \sec x \tan x\, dx$

$\qquad = \sec x \Big|_{-\pi/4}^0 = 1 - \sqrt{2}$

15. $\displaystyle\frac{3}{16+x^2}\, dx = \frac{3}{4}\tan^{-1}\frac{x}{4} + c$

17. Let $u = x^3$, $du = 3x^2\, dx$.

$\displaystyle\int \frac{x^2}{1+x^6}\, dx = \frac{1}{3}\int \frac{1}{1+u^2}\, du$

$\qquad = \frac{1}{3}\tan^{-1}u + c = \frac{1}{3}\tan^{-1}x^3 + c$

19. $\displaystyle\frac{1}{\sqrt{4-x^2}}\, dx = \sin^{-1}\frac{x}{2} + c$

21. Let $u = x^2$, $du = 2x\, dx$.

$\displaystyle\int \frac{x}{\sqrt{1-x^4}}\, dx = \frac{1}{2}\int \frac{1}{\sqrt{1-u^2}}\, du$

$\qquad = \frac{1}{2}\sin^{-1}u + c = \frac{1}{2}\sin^{-1}x^2 + c$

23. $\displaystyle\int \frac{4}{5+2x+x^2}\, dx = 4\int \frac{1}{4+(x+1)^2}\, dx$

$\qquad = 2\tan^{-1}\left(\frac{x+1}{2}\right) + c$

25. $\int \dfrac{4x}{5+2x+x^2}\,dx$

$= \int \dfrac{4x+4}{5+2x+x^2}\,dx - \int \dfrac{4}{5+2x+x^2}\,dx$

$= 2\ln|4+(x+1)^2| - 2\tan^{-1}\left(\dfrac{x+1}{2}\right) + c$

27. $\int (x^2+4)^2\,dx = \int (x^4+8x^2+16)\,dx$

$= \dfrac{x^5}{5} + \dfrac{8}{3}x^3 + 16x + c$

29. $\int \dfrac{1}{\sqrt{3-2x-x^2}}\,dx = \int \dfrac{1}{\sqrt{4-(x+1)^2}}\,dx$

$= \sin^{-1}\left(\dfrac{x+1}{2}\right) + c$

31. $\int \dfrac{1+x}{1+x^2}\,dx$

$= \int \dfrac{1}{1+x^2}\,dx + \dfrac{1}{2}\int \dfrac{2x}{1+x^2}\,dx$

$= \tan^{-1} x + \dfrac{1}{2}\ln\left(1+x^2\right) + c$

33. $\displaystyle\int_{-2}^{-1} e^{\ln(x^2+1)}\,dx = \int_{-2}^{-1} (x^2+1)\,dx$

$= \left(\dfrac{x^3}{3}+x\right)\bigg|_{-2}^{-1} = \dfrac{10}{3}$

35. $\displaystyle\int_{3}^{4} x\sqrt{x-3}\,dx$

$= \int_{3}^{4} (x-3+3)\sqrt{x-3}\,dx$

$= \int_{3}^{4} (x-3)^{3/2}\,dx + 3\int_{3}^{4}(x-3)^{1/2}\,dx$

$= \dfrac{2}{5}(x-3)^{5/2}\bigg|_{3}^{4} + 3\cdot\dfrac{2}{3}(x-3)^{3/2}\bigg|_{3}^{4}$

$= \dfrac{12}{5}$

37. Substituting $u = e^x$,

$\displaystyle\int_{0}^{2} \dfrac{e^x}{1+e^{2x}}\,dx = \tan^{-1}e^x \bigg|_{0}^{2}$

$= \tan^{-1}e^2 - \dfrac{\pi}{4}.$

39. $\displaystyle\int_{1}^{4} \dfrac{x^2+1}{\sqrt{x}}\,dx$

$= \int_{1}^{4} x^{3/2}\,dx + \int_{1}^{4} x^{-1/2}\,dx$

$= \dfrac{2}{5}x^{5/2}\bigg|_{1}^{4} + 2x^{1/2}\bigg|_{1}^{4} = \dfrac{72}{5}$

41. $\int \dfrac{5}{3+x^2}\,dx = \dfrac{5}{\sqrt{3}}\arctan\dfrac{x}{\sqrt{3}} + c$

$\int \dfrac{5}{3+x^3}\,dx$: N/A

43. $\int \ln x\,dx$: N/A

Substituting $u = \ln x$,

$\int \dfrac{\ln x}{2x}\,dx = \dfrac{1}{2}\int u\,du$

$= \dfrac{1}{4}u^2 + c = \dfrac{1}{4}(\ln x)^2 + c.$

45. $\int e^{-x^2}\,dx$: N/A

Substituting $u = -x^2$,

$\int xe^{-x^2}\,dx = -\dfrac{1}{2}e^{-x^2} + c.$

47. In this solution, we use polynomial division to rewrite $\dfrac{x^2}{x^2+1}$ as $1 - \dfrac{1}{x^2+1}$.

$$\int_0^2 f(x)\,dx$$

$$= \int_0^1 \frac{x}{x^2+1}\,dx + \int_1^2 \frac{x^2}{x^2+1}\,dx$$

$$= \frac{1}{2}\ln|x^2+1| \ \Big|_0^1 + \int_1^2 \left(1 - \frac{1}{x^2+1}\right)\,dx$$

$$= \frac{1}{2}\ln 2 + \left(x - \tan^{-1}x\right) \ \Big|_1^2$$

$$= \frac{\ln 2}{2} + 1 + \frac{\pi}{4} - \tan^{-1}2$$

49. $\displaystyle \int \frac{4x+1}{2x^2+4x+10}\,dx$

$$= \int \frac{4x+4}{2x^2+4x+10}\,dx - \int \frac{3}{2x^2+4x+10}\,dx$$

$$= \ln|2x^2+4x+10| - \frac{3}{2}\int \frac{1}{(x+1)^2+4}\,dx$$

$$= \ln|2x^2+4x+10| - \frac{3}{4}\tan^{-1}\left(\frac{x+1}{2}\right) + c$$

7.2 INTEGRATION BY PARTS

1. Let $u = x, dv = \cos x\,dx$ so that
$du = dx, v = \sin x.$

$$\int x\cos x\,dx = x\sin x - \int \sin x\,dx$$

$$= x\sin x + \cos x + c$$

3. Let $u = x,\ dv = e^{2x}\,dx$ so that
$du = dx,\ v = \dfrac{1}{2}e^{2x}.$

$$\int xe^{2x}\,dx = \frac{1}{2}xe^{2x} - \int \frac{1}{2}e^{2x}\,dx$$

$$= \frac{1}{2}xe^{2x} - \frac{1}{4}e^{2x} + c$$

5. Let $u = \ln x,\ dv = x^2\,dx,$
$du = \dfrac{1}{x}\,dx, v = \dfrac{1}{3}x^3.$

$$\int x^2 \ln x\,dx = \frac{1}{3}x^3 \ln x - \int \frac{1}{3}x^3 \cdot \frac{1}{x}\,dx$$

$$= \frac{1}{3}x^3 \ln x - \frac{1}{3}\int x^2\,dx$$

$$= \frac{1}{3}x^3 \ln x - \frac{1}{9}x^3 + c$$

7. Let $u = x^2,\ dv = e^{-3x}\,dx,$
$du = 2x\,dx, v = -\dfrac{1}{3}e^{-3x}.$

$$I = \int x^2 e^{-3x}\,dx$$

$$= -\frac{1}{3}x^2 e^{-3x} - \int \left(-\frac{1}{3}e^{-3x}\right) \cdot 2x\,dx$$

$$= -\frac{1}{3}x^2 e^{-3x} + \frac{2}{3}\int xe^{-3x}\,dx$$

Let $u = x,\ dv = e^{-3x}\,dx,\ du = dx,\ v = -\dfrac{1}{3}e^{-3x}.$

$$I = -\frac{1}{3}x^2 e^{-3x}$$

$$+ \frac{2}{3}\left[-\frac{1}{3}xe^{-3x} - \int \left(-\frac{1}{3}e^{-3x}\right)\,dx\right]$$

$$= -\frac{1}{3}x^2 e^{-3x} - \frac{2}{9}xe^{-3x} + \frac{2}{9}\int e^{-3x}\,dx$$

$$= -\frac{1}{3}x^2 e^{-3x} - \frac{2}{9}xe^{-3x} - \frac{2}{27}e^{-3x} + c$$

9. Let $I = \displaystyle\int e^x \sin 4x\,dx.$

$$u = e^x,\ dv = \sin 4x\,dx$$

$$du = e^x\,dx, v = -\frac{1}{4}\cos 4x$$

$$I = -\frac{1}{4}e^x \cos 4x - \int \left(-\frac{1}{4}\cos 4x\right)e^x\,dx$$

$$= -\frac{1}{4}e^x \cos 4x + \frac{1}{4}\int e^x \cos 4x\,dx$$

Using integration by parts again, this time let
$u = e^x, dv = \cos 4x\,dx, du = e^x\,dx,$

$$v = \frac{1}{4}\sin 4x.$$

$$I = -\frac{1}{4}e^x \cos 4x$$

$$+ \frac{1}{4}\left(\frac{1}{4}e^x \sin 4x - \int \frac{1}{4}(\sin 4x)e^x \, dx\right)$$

$$I = -\frac{1}{4}e^x \cos 4x + \frac{1}{16}e^x \sin 4x - \frac{1}{16}I$$

$$\frac{17}{16}I = -\frac{1}{4}e^x \cos 4x + \frac{1}{16}e^x \sin 4x + c_1$$

$$I = -\frac{4}{17}e^x \cos 4x + \frac{1}{17}e^x \sin 4x + c$$

11. Let $I = \int \cos x \cos 2x \, dx$ and $u = \cos x$,

$dv = \cos 2x \, dx$, $du = -\sin x \, dx$,

$v = \frac{1}{2}\sin 2x$.

$$I = \frac{1}{2}\cos x \sin 2x - \int \frac{1}{2}\sin 2x(-\sin x) \, dx$$

$$= \frac{1}{2}\cos x \sin 2x + \frac{1}{2}\int \sin x \sin 2x \, dx$$

Let $u = \sin x$, $dv = \sin 2x \, dx$, $du = \cos x \, dx$,

$v = -\frac{1}{2}\cos 2x$.

$$I = \frac{1}{2}\cos x \sin 2x + \frac{1}{2}\left[-\frac{1}{2}\cos 2x \sin x \right.$$

$$\left. - \int \left(-\frac{1}{2}\cos 2x\right)\cos x \, dx\right]$$

$$= \frac{1}{2}\cos x \sin 2x - \frac{1}{4}\cos 2x \sin x + \frac{1}{4}I$$

$$\frac{3}{4}I = \frac{1}{2}\cos x \sin 2x - \frac{1}{4}\cos 2x \sin x + c_1$$

$$I = \frac{2}{3}\cos x \sin 2x - \frac{1}{3}\cos 2x \sin x + c$$

13. Let $u = x$, $dv = \sec^2 x \, dx$,

$du = dx$, $v = \tan x$.

$$\int x \sec^2 x \, dx = x \tan x - \int \tan x \, dx$$

$$= x \tan x + \ln |\cos x| + c$$

15. Let $u = (\ln x)^2$, $dv = dx$,

$du = 2\frac{\ln x}{x} \, dx$, $v = x$.

$$I = \int (\ln x)^2 \, dx$$

$$= x(\ln x)^2 - \int x \cdot 2\frac{\ln x}{x} \, dx$$

$$= x(\ln x)^2 - 2\int \ln x \, dx$$

Integration by parts again:

$u = \ln x$, $dv = dx$

$du = \frac{1}{x} \, dx$, $v = x$

$$I = x(\ln x)^2 - 2\left[x \ln x - \int x \cdot \frac{1}{x} \, dx\right]$$

$$= x(\ln x)^2 - 2x \ln x + 2\int dx$$

$$= x(\ln x)^2 - 2x \ln x + 2x + c$$

17. Let $u = \ln(\sin x)$, $dv = \cos x \, dx$,

$du = \frac{1}{\sin x} \cdot \cos x \, dx$, $v = \sin x$.

$$I = \int \cos x \ln(\sin x) \, dx$$

$$= \sin x \ln(\sin x)$$

$$- \int \sin x \cdot \frac{1}{\sin x} \cdot \cos x \, dx$$

$$= \sin x \ln(\sin x) - \int \cos x \, dx$$

$$= \sin x \ln(\sin x) - \sin x + c$$

19. Let $u = x$, $dv = \sin 2x \, dx$,

$du = dx$, $v = -\frac{1}{2}\cos 2x$.

$$\int_0^1 x \sin 2x \, dx$$

$$= -\frac{1}{2}x \cos 2x \Big|_0^1 - \int_0^1 \left(-\frac{1}{2}\cos 2x\right) dx$$

$$= -\frac{1}{2}(1\cos 2 - 0 \cos 0)$$

$$+ \frac{1}{2}\int_0^1 \cos 2x \, dx$$

$$= -\frac{1}{2}\cos 2 + \frac{1}{2}\left(\frac{1}{2}\sin 2x\right)\Big|_0^1$$

$$= -\frac{1}{2}\cos 2 + \frac{1}{4}(\sin 2 - \sin 0)$$

$$= -\frac{1}{2}\cos 2 + \frac{1}{4}\sin 2$$

21. Let $u = x$, $dv = \cos \pi x\, dx$,

$$du = dx, v = \frac{1}{\pi}\sin \pi x.$$

$$\int_0^1 x \cos \pi x\, dx$$

$$= \frac{1}{\pi}x \sin \pi x \Big|_0^1 - \int_0^1 \frac{1}{\pi}\sin \pi x\, dx$$

$$= \frac{1}{\pi}(\sin \pi - 0) + \frac{1}{\pi^2}\cos \pi x \Big|_0^1$$

$$= 0 + \frac{1}{\pi^2}(\cos \pi - \cos 0) = -\frac{2}{\pi^2}$$

23. Let $u = x$, $dv = \sin \pi x\, dx$,

$$du = dx, v = -\frac{1}{\pi}\cos \pi x.$$

$$\int_0^1 x \sin \pi x\, dx$$

$$= -\frac{1}{\pi}x \cos \pi x \Big|_0^1 - \int_0^1 \left(-\frac{1}{\pi}\cos \pi x\right) dx$$

$$= -\frac{1}{\pi}(\cos \pi - 0) + \frac{1}{\pi^2}\sin \pi x \Big|_0^1$$

$$= -\frac{1}{\pi}(-1) + \frac{1}{\pi^2}(\sin \pi - \sin 0) = \frac{1}{\pi}$$

25. Let $u = \ln x$, $dv = dx$,

$$du = \frac{1}{x}dx, v = x.$$

$$\int_1^{10} \ln x\, dx = x \ln x \Big|_1^{10} - \int_1^{10} x \cdot \frac{1}{x}dx$$

$$= 10 \ln 10 - 1 \ln 1 - \int_1^{10} dx$$

$$= 10 \ln 10 - 0 - x \Big|_1^{10}$$

$$= 10 \ln 10 - 9$$

27. Let $u = \cos^{-1} x$, $dv = dx$,

$$du = -\frac{1}{\sqrt{1 - x^2}}dx, v = x.$$

$$I = \int \cos^{-1} x\, dx$$

$$= x \cos^{-1} x - \int x\left(-\frac{1}{\sqrt{1 - x^2}}\right) dx$$

$$= x \cos^{-1} x + \int \frac{x}{\sqrt{1 - x^2}}dx$$

Substituting $u = 1 - x^2$, $du = -2x\, dx$,

$$I = x \cos^{-1} x + \int \frac{1}{\sqrt{u}}\left(-\frac{1}{2}du\right)$$

$$= x \cos^{-1} x - \frac{1}{2}\int u^{-1/2}du$$

$$= x \cos^{-1} x - \frac{1}{2}\cdot 2u^{1/2} + c$$

$$= x \cos^{-1} x - \sqrt{1 - x^2} + c.$$

29. Substituting $u = \sqrt{x}$, $du = \frac{1}{2\sqrt{x}}dx$,

we have $dx = 2\sqrt{x}\, du = 2u\, du$.

$$I = \int \sin \sqrt{x}\, dx = 2\int u \sin u\, du$$

$$= 2(-u \cos u + \sin u) + c$$

$$= 2(-\sqrt{x}\cos \sqrt{x} + \sin \sqrt{x}) + c,$$

where we use the result of Example 2.1 to

evaluate $\int u \sin u\, du$.

31. Let $u = \sin(\ln x)$, $dv = dx$,

$$du = \cos(\ln x)\frac{dx}{x}, v = x.$$

$$I = \int \sin(\ln x)\, dx$$

$$= x \sin(\ln x) - \int \cos(\ln x)\, dx$$

Integration by parts again:

$u = \cos(\ln x), dv = dx$

$du = -\sin(\ln x)\frac{dx}{x}, v = x$

$$\int \cos(\ln x)\, dx$$

$$= x\cos(\ln x) + \int \sin(\ln x)\, dx$$

$I = x\sin(\ln x) - x\cos(\ln x) - I$

$2I = x\sin(\ln x) - x\cos(\ln x) + c_1$

$I = \dfrac{1}{2}x\sin(\ln x) - \dfrac{1}{2}x\cos(\ln x) + c$

33. Let $u = e^{2x}, du = 2e^{2x}\, dx$.

$$I = \int e^{6x}\sin(e^{2x})\, dx$$

$$= \frac{1}{2}\int (e^{2x})^2 \sin(e^{2x})(2e^{2x})\, dx$$

$$= \frac{1}{2}\int u^2 \sin u\, du$$

Let $v = u^2, dw = \sin u\, du$,

$dv = 2u\, du, w = -\cos u$.

$$I = \frac{1}{2}\left(-u^2\cos u + 2\int u\cos u\, du\right)$$

$$= -\frac{1}{2}u^2\cos u + \int u\cos u\, du$$

$$= -\frac{1}{2}u^2\cos u + (u\sin u + \cos u) + c$$

$$= -\frac{1}{2}e^{4x}\cos(e^{2x}) + e^{2x}\sin(e^{2x})$$

$$+ \cos(e^{2x}) + c,$$

where we use the result of exercise 1 to evaluate $\int u\cos u\, du$.

35. In this solution, we use the reduction formula (2.4) twice. Let $u = \sqrt[3]{x} = x^{1/3}$,

$du = \frac{1}{3}x^{-2/3}\, dx, 3u^2\, du = dx$.

$$I = \int e^{\sqrt[3]{x}}\, dx = 3\int u^2 e^u\, du$$

$$= 3\left(u^2 e^u - 2\int u e^u\, du\right)$$

$$= 3u^2 e^u - 6\left(u e^u - \int e^u\, du\right)$$

$$= 3u^2 e^u - 6u e^u + 6e^u + c$$

Hence,

$$\int_0^8 e^{\sqrt[3]{x}}\, dx = \int_0^2 3u^2 e^u\, du$$

$$= \left(3u^2 e^u - 6u e^u + 6e^u\right)\Big|_0^2$$

$$= 6e^2 - 6.$$

37. n times. Each integration reduces the power of x by 1.

39. Let $u = \cos^{n-1} x, dv = \cos x\, dx$,

$du = (n-1)(\cos^{n-2} x)(-\sin x)\, dx$,

$v = \sin x$.

$$\int \cos^n x\, dx$$

$$= \sin x \cos^{n-1} x$$

$$\quad - \int (\sin x)(n-1)(\cos^{n-2} x)(-\sin x)\, dx$$

$$= \sin x \cos^{n-1} x$$

$$\quad + \int (n-1)(\cos^{n-2} x)(\sin^2 x)\, dx$$

$$= \sin x \cos^{n-1} x$$

$$\quad + \int (n-1)(\cos^{n-2} x)(1 - \cos^2 x)\, dx$$

$$= \sin x \cos^{n-1} x$$

$$\quad + \int (n-1)(\cos^{n-2} x - \cos^n x)\, dx$$

Thus

$$\int \cos^n x \, dx$$

$$= \sin x \cos^{n-1} x + \int (n-1) \cos^{n-2} x \, dx$$

$$\quad - (n-1) \int \cos^n x \, dx.$$

$$n \int \cos^n x \, dx = \sin x \cos^{n-1} x$$

$$\quad + (n-1) \int \cos^{n-2} x \, dx$$

$$\int \cos^n x \, dx$$

$$= \frac{1}{n} \sin x \cos^{n-1} x + \frac{n-1}{n} \int \cos^{n-2} x \, dx$$

41. $\int x^3 e^x \, dx$

$$= x^3 e^x - 3 \int x^2 e^x \, dx$$

$$= x^3 e^x - 3 \left[x^2 e^x - 2 \int x e^x \, dx \right]$$

$$= x^3 e^x - 3x^2 e^x + 6 \left[x e^x - \int e^x \, dx \right]$$

$$= e^x (x^3 - 3x^2 + 6x - 6) + c$$

43. $\int \cos^3 x \, dx$

$$= \frac{1}{3} \cos^2 x \sin x + \frac{2}{3} \int \cos x \, dx$$

$$= \frac{1}{3} \cos^2 x \sin x + \frac{2}{3} \sin x + c$$

45. The indefinite integral $\int x^4 e^x \, dx$ was evaluated in Example 2.6.

$$\int_0^1 x^4 e^x \, dx$$

$$= e^x (x^4 - 4x^3 + 12x^2 - 24x + 24) \Big|_0^1$$

$$= 9e - 24$$

47. $\int_0^{\pi/2} \sin^5 x \, dx$

$$= -\frac{1}{5} \sin^4 x \cos x \Big|_0^{\pi/2} + \frac{4}{5} \int_0^{\pi/2} \sin^3 x \, dx$$

$$= 0 + \frac{4}{5} \left(-\frac{1}{3} \sin^2 x \cos x - \frac{2}{3} \cos x \right) \Big|_0^{\pi/2}$$

$$= \frac{8}{15}$$

49. m even:

$$\int_0^{\pi/2} \sin^m x \, dx$$

$$= \frac{(m-1)(m-3)\ldots 1}{m(m-2)\ldots 2} \cdot \frac{\pi}{2}$$

m odd:

$$\int_0^{\pi/2} \sin^m x \, dx$$

$$= \frac{(m-1)(m-3)\ldots 2}{m(m-2)\ldots 3}$$

51. First column: each row is the derivative of the previous row;

Second column: each row is the antiderivative of the previous row.

53.

	$\cos x$	
x^4	$\sin x$	$+$
$4x^3$	$-\cos x$	$-$
$12x^2$	$-\sin x$	$+$
$24x$	$\cos x$	$-$
24	$\sin x$	$+$

$$\int x^4 \cos x \, dx$$

$$= x^4 \sin x + 4x^3 \cos x - 12x^2 \sin x$$

$$\quad - 24x \cos x + 24 \sin x + c$$

55.

	e^{2x}	
x^4	$e^{2x}/2$	$+$
$4x^3$	$e^{2x}/4$	$-$
$12x^2$	$e^{2x}/8$	$+$
$24x$	$e^{2x}/16$	$-$
24	$e^{2x}/32$	$+$

$$\int x^4 e^{2x}\,dx$$

$$= \left(\tfrac{x^4}{2} - x^3 + \tfrac{3x^2}{2} - \tfrac{3x}{2} + \tfrac{3}{4}\right) e^{2x} + c$$

57.

	e^{-3x}	
x^3	$-e^{-3x}/3$	$+$
$3x^2$	$e^{-3x}/9$	$-$
$6x$	$-e^{-3x}/27$	$+$
6	$e^{-3x}/81$	$-$

$$\int x^3 e^{-3x}\,dx$$

$$= \left(-\tfrac{x^3}{3} - \tfrac{x^2}{3} - \tfrac{2x}{9} - \tfrac{2}{27}\right) e^{-3x} + c$$

59. (a) Use the identity

$$\cos A \cos B$$
$$= \frac{1}{2}[\cos(A - B) + \cos(A + B)].$$

This identity gives

$$\int_{-\pi}^{\pi} \cos(mx)\cos(nx)\,dx$$

$$= \int_{-\pi}^{\pi} \frac{1}{2}[\cos((m-n)x)$$
$$+ \cos((m+n)x)]\,dx$$

$$= \frac{1}{2}\left[\frac{\sin((m-n)x)}{m-n}\right.$$
$$\left. + \frac{\sin((m+n)x)}{m+n}\right]\Bigg|_{-\pi}^{\pi}$$

$$= 0.$$

(b) Use the identity

$$\sin A \sin B$$
$$= \frac{1}{2}[\cos(A - B) - \cos(A + B)].$$

This identity gives

$$\int_{-\pi}^{\pi} \sin(mx)\sin(nx)\,dx$$

$$= \int_{-\pi}^{\pi} \frac{1}{2}[\cos((m-n)x)$$
$$- \cos((m+n)x)]\,dx$$

$$= \frac{1}{2}\left[\frac{\sin((m-n)x)}{m-n}\right.$$
$$\left. - \frac{\sin((m+n)x)}{m+n}\right]\Bigg|_{-\pi}^{\pi}$$

$$= 0.$$

61. In Example 2.5, we added the constant of integration when we added an integral to both sides. If we do that here, we have $0 = -1 + c$ rather than $0 = -1$.

63. Let $u = \ln x,\ dv = e^x\,dx,$

$$du = \frac{dx}{x},\ v = e^x.$$

$$\int e^x \ln x\,dx = e^x \ln x - \int \frac{e^x}{x}\,dx$$

$$\int e^x \ln x\,dx + \int \frac{e^x}{x}\,dx = e^x \ln x + c$$

$$\int e^x \left(\ln x + \frac{1}{x}\right)\,dx = e^x \ln x + c$$

65. The quotient rule says

$$\frac{d}{dx}\left(\frac{f(x)}{g(x)}\right) = \frac{f'(x)g(x) - f(x)g'(x)}{g^2(x)}$$

$$= \frac{f'(x)}{g(x)} - \frac{f(x)g'(x)}{g^2(x)}.$$

In the form of antiderivatives, we get

$$\int \left[\frac{f'(x)}{g(x)} - \frac{f(x)g'(x)}{g^2(x)}\right]\,dx = \frac{f(x)}{g(x)}$$

$$\int \frac{f'(x)}{g(x)}\,dx - \int \frac{f(x)g'(x)}{g^2(x)}\,dx = \frac{f(x)}{g(x)}$$

$$\int \frac{f'(x)}{g(x)}\,dx = \frac{f(x)}{g(x)} + \int \frac{f(x)g'(x)}{g^2(x)}\,dx$$

7.3 TRIGONOMETRIC TECHNIQUES OF INTEGRATION

1. Let $u = \sin x$, $du = \cos x\, dx$.

$$\int \cos x \sin^4 x\, dx = \int u^4\, du$$

$$= \frac{1}{5}u^5 + c = \frac{1}{5}\sin^5 x + c$$

3. Let $u = \sin x$, $du = \cos x\, dx$.

$$\int_0^{\pi/4} \cos x \sin^3 x\, dx = \int_0^{1/\sqrt{2}} u^3\, du$$

$$= \frac{1}{4}u^4 \Big|_0^{1/\sqrt{2}} = \frac{1}{4 \cdot (\sqrt{2})^4} = \frac{1}{16}$$

5. Let $u = \cos x$, $du = -\sin x\, dx$.

$$\int_0^{\pi/2} \cos^2 x \sin x\, dx = \int_1^0 u^2(-du)$$

$$= \left(-\frac{1}{3}u^3\right) \Big|_1^0 = \frac{1}{3}$$

7. $\displaystyle \int \cos^2 x\, dx = \frac{1}{2}\int (1 + \cos 2x)\, dx$

$$= \frac{1}{2}x + \frac{1}{4}\sin 2x + c$$

9. Let $u = \sec x$, $du = \sec x \tan x\, dx$.

$$\int \tan x \sec^3 x\, dx$$

$$= \int \tan x \sec x \sec^2 x\, dx = \int u^2\, du$$

$$= \frac{1}{3}u^3 + c = \frac{1}{3}\sec^3 x + c$$

11. Let $u = \tan x$, $du = \sec^2 x\, dx$.

$$\int_0^{\pi/4} \tan^4 x \sec^4 x\, dx$$

$$= \int_0^{\pi/4} \tan^4 x \sec^2 x \sec^2 x\, dx$$

$$= \int_0^{\pi/4} \tan^4 x(1 + \tan^2 x) \sec^2 x\, dx$$

$$= \int_0^1 u^4(1 + u^2)\, du$$

$$= \int_0^1 (u^4 + u^6)\, du = \frac{u^5}{5} + \frac{u^7}{7} \Big|_0^1 = \frac{12}{35}$$

13. $\displaystyle \int \cos^2 x \sin^2 x\, dx$

$$= \int \frac{1}{2}(1 + \cos 2x) \cdot \frac{1}{2}(1 - \cos 2x)\, dx$$

$$= \frac{1}{4}\int (1 - \cos^2 2x)\, dx$$

$$= \frac{1}{4}\int \left[1 - \frac{1}{2}(1 + \cos 4x)\right] dx$$

$$= \frac{1}{4}\left(\frac{1}{2}x - \frac{1}{8}\sin 4x\right) + c$$

$$= \frac{1}{8}x - \frac{1}{32}\sin 4x + c$$

15. Let $u = \cos x$, $du = -\sin x\, dx$.

$$\int_{-\pi/3}^0 \sqrt{\cos x} \sin^3 x\, dx$$

$$= \int_{-\pi/3}^0 \sqrt{\cos x}(1 - \cos^2 x) \sin x\, dx$$

$$= \int_{1/2}^1 \sqrt{u}(1 - u^2)(-du)$$

$$= \int_{1/2}^1 (u^{5/2} - u^{1/2})du$$

$$= \frac{2}{7}u^{7/2} - \frac{2}{3}u^{3/2} \Big|_{1/2}^1$$

$$= \frac{25}{168}\sqrt{2} - \frac{8}{21}$$

17. Let $x = 3 \sin \theta 4$, $4 - \dfrac{\pi}{2} < \theta < \dfrac{\pi}{2}$,

$dx = 3 \cos \theta \, d\theta$.

$$\int \frac{1}{x^2 \sqrt{9 - x^2}} \, dx = \int \frac{3 \cos \theta}{9 \sin^2 \theta \cdot 3 \cos \theta} d\theta$$

$$= \frac{1}{9} \int \csc^2 \theta \, d\theta = -\frac{1}{9} \cot \theta + c$$

By drawing a diagram, we see that if $x = \sin \theta$,

then $\cot \theta = \dfrac{\sqrt{9 - x^2}}{x}$. Thus, the integral $=$

$-\dfrac{\sqrt{9 - x^2}}{9x} + c$.

19. This is the area of a quarter of a circle of radius 2.

$$\int_0^2 \sqrt{4 - x^2} \, dx = \pi$$

21. Let $x = 3 \sec \theta$, $dx = 3 \sec \theta \tan \theta \, d\theta$.

$$I = \int \frac{x^2}{\sqrt{x^2 - 9}} \, dx$$

$$= \int \frac{27 \sec^2 \theta \sec \theta \tan \theta}{\sqrt{9 \sec^2 \theta - 9}} d\theta$$

$$= \int 9 \sec^3 \theta \, d\theta$$

Use integration by parts. Let $u = \sec \theta$ and $dv = \sec^2 \theta \, d\theta$. This gives

$$\int \sec^3 \theta \, d\theta$$

$$= \sec \theta \tan \theta - \int \sec \theta \tan^2 \theta \, d\theta$$

$$= \sec \theta \tan \theta - \int \sec \theta (\sec^2 \theta - 1) \, d\theta$$

$$= \sec \theta \tan \theta + \int \sec \theta \, d\theta - \int \sec^3 \theta \, d\theta.$$

$$2 \int \sec^3 \theta \, d\theta$$

$$= \sec \theta \tan \theta + \int \sec \theta \, d\theta$$

We divide by 2 and use Example 3.8 for $\int \sec \theta \, d\theta$.

$$\int \sec^3 \theta \, d\theta$$

$$= \frac{1}{2} \sec \theta \tan \theta + \frac{1}{2} \ln |\sec \theta + \tan \theta| + c$$

Putting all these together and using $\sec \theta = \dfrac{x}{3}$,

$\tan \theta = \dfrac{\sqrt{x^2 - 9}}{3}$:

$$\int \frac{x^2}{\sqrt{x^2 - 9}} \, dx = \int 9 \sec^3 \theta \, d\theta$$

$$= \frac{9}{2} \sec \theta \tan \theta + \frac{9}{2} \ln |\sec \theta + \tan \theta| + c$$

$$= \frac{9}{2} \left(\frac{x}{3} \right) \left(\frac{\sqrt{x^2 - 9}}{3} \right)$$

$$+ \frac{9}{2} \ln \left| \frac{x}{3} + \frac{\sqrt{x^2 - 9}}{3} \right| + c$$

$$= \frac{x \sqrt{x^2 - 9}}{2} + \frac{9}{2} \ln \left| \frac{x + \sqrt{x^2 - 9}}{3} \right| + c$$

23. Let $x = 2 \sec \theta$, $dx = 2 \sec \theta \tan \theta \, d\theta$.

$$\int \frac{2}{\sqrt{x^2 - 4}} \, dx = \int \frac{4 \sec \theta \tan \theta}{2 \tan \theta} d\theta$$

$$= 2 \int \sec \theta \, d\theta$$

$$= 2 \ln |\sec \theta + \tan \theta| + c$$

$$= 2 \ln \left| \frac{x}{2} + \frac{\sqrt{x^2 - 4}}{2} \right| + c$$

$$= 2 \ln \left| x + \sqrt{x^2 - 4} \right| + c$$

25. Let $x = 3 \tan \theta$, $dx = 3 \sec^2 \theta \, d\theta$.

$$\int \frac{x^2}{\sqrt{x^2 + 9}} \, dx$$

$$= \int \frac{27 \tan^2 \theta \sec^2 \theta}{\sqrt{9 \tan^2 \theta + 9}} d\theta$$

$$= \int 9 \tan^2 \theta \sec \theta \, d\theta$$

$$= 9 \int (\sec^2 \theta - 1) \sec \theta \, d\theta$$

$$= 9 \int \sec^3 \theta \, d\theta - 9 \int \sec \theta \, d\theta$$

$$= \frac{9}{2} \sec \theta \tan \theta - \frac{9}{2} \ln |\sec \theta + \tan \theta| + c$$

$$= \frac{9}{2} \left(\frac{\sqrt{x^2 + 9}}{3} \right) \left(\frac{x}{3} \right)$$

$$\quad - \frac{9}{2} \ln \left| \frac{\sqrt{x^2 + 9}}{3} + \frac{x}{3} \right| + c$$

$$= \frac{x \sqrt{x^2 + 9}}{2} - \frac{9}{2} \ln \left| \frac{x + \sqrt{x^2 + 9}}{3} \right| + c$$

27. Let $x = 4 \tan \theta$, $dx = 4 \sec^2 \theta \, d\theta$.

$$\int \sqrt{x^2 + 16} \, dx$$

$$= \int \sqrt{16 \tan^2 \theta + 16} \cdot 4 \sec^2 \theta \, d\theta$$

$$= 16 \int \sec^3 \theta \, d\theta$$

$$= 16 \left(\frac{1}{2} \sec \theta \tan \theta + \frac{1}{2} \int \sec \theta \, d\theta \right)$$

$$= 8 \sec \theta \tan \theta + 8 \int \sec \theta \, d\theta$$

$$= 8 \sec \theta \tan \theta + 8 \ln |\sec \theta + \tan \theta| + c$$

$$= \frac{1}{2} x \sqrt{16 + x^2}$$

$$\quad + 8 \ln \left| \frac{1}{4} \sqrt{16 + x^2} + \frac{x}{4} \right| + c$$

29. Let $u = x^2 + 8$, $du = 2x \, dx$.

$$\int_0^1 x \sqrt{x^2 + 8} \, dx = \frac{1}{2} \int_8^9 u^{1/2} \, du$$

$$= \frac{1}{3} u^{3/2} \Big|_8^9 = \frac{27 - 16\sqrt{2}}{3}$$

31. Using $u = \tan x$ gives

$$\int \tan x \sec^4 x \, dx$$

$$= \int \tan x (1 + \tan^2 x) \sec^2 x \, dx$$

$$= \int u(1 + u^2) \, du$$

$$= \int (u + u^3) \, du = \frac{1}{2} u^2 + \frac{1}{4} u^4 + c$$

$$= \frac{1}{2} \tan^2 x + \frac{1}{4} \tan^4 x + c.$$

Using $u = \sec x$ gives

$$\int \tan x \sec^4 x \, dx$$

$$= \int \tan x \sec x \sec^3 x \, dx$$

$$= \int u^3 \, du = \frac{1}{4} u^4 + c$$

$$= \frac{1}{4} \sec^4 x + c.$$

33. This is using integration by parts followed by substitution.

$u = \sec^{n-2} x$, $dv = \sec^2 x \, dx$

$du = (n - 2) \sec^{n-2} x \tan x \, dx$, $v = \tan x$

$$I = \int \sec^n x \, dx = \sec^{n-2} x \tan x$$

$$\quad - (n - 2) \int \sec^{n-2} x (\sec^2 x - 1) \, dx$$

$$= \sec^{n-2} x \tan x$$

$$\quad - (n - 2) \int (\sec^n x - \sec^{n-2} x) \, dx$$

$$= \sec^{n-2} x \tan x - (n - 2) I$$

$$\quad + (n - 2) \int \sec^{n-2} x \, dx$$

$$(n - 1) I$$

$$= \sec^{n-2} x \tan x + (n - 2) \int \sec^{n-2} x \, dx$$

$$I = \frac{\sec^{n-2} x \tan x}{n - 1} + \frac{n - 2}{n - 1} \int \sec^{n-2} x \, dx$$

35. The average power

$$= \frac{1}{\frac{2\pi}{\omega}} \int_0^{2\pi/\omega} RI^2 \cos^2(\omega t)\, dt$$

$$= \frac{\omega RI^2}{2\pi} \int_0^{2\pi/\omega} \frac{1}{2} [1 + \cos(2\omega t)]\, dt$$

$$= \frac{\omega RI^2}{4\pi} \left[t + \frac{1}{2\omega} \sin(2\omega t) \right]\Big|_0^{2\pi/\omega}$$

$$= \frac{\omega RI^2}{4\pi} \left[\frac{2\pi}{\omega} + \frac{1}{2\omega} \sin\left(\frac{4\omega\pi}{\omega}\right) - 0 \right]$$

$$= \frac{1}{2} RI^2$$

37. Using a CAS we get

(Ex 3.2) $\displaystyle \int \cos^4 x \sin^3 x\, dx$

$$= -\frac{1}{7} \sin^2 x \cos^5 x - \frac{2}{35} \cos^5 x + c$$

(Ex 3.3) $\displaystyle \int \sqrt{\sin x} \cos^5 x\, dx$

$$= \frac{2}{11} \sin^{11/2} x - \frac{4}{7} \sin^{7/2} x$$

$$+ \frac{2}{3} \sin^{3/2} x + c$$

(Ex 3.5) $\displaystyle \int \cos^4 x\, dx$

$$= \frac{1}{4} \cos^3 x \sin x + \frac{3}{8} \cos x \sin x$$

$$+ \frac{3}{8} x + c$$

(Ex 3.6) $\displaystyle \int \tan^3 x \sec^3 x\, dx$

$$= 1/5 \frac{\sin^4 x}{\cos^5 x} + 1/15 \frac{\sin^4 x}{\cos^3 x}$$

$$- 1/15 \frac{\sin^4 x}{\cos x} - 1/15 \sin^2 x \cos x$$

$$- 2/15 \cos x + c$$

Obviously my CAS used different techniques. The answers given by the book are simpler.

39. $\displaystyle -\frac{1}{7} \sin^2 x \cos^5 x - \frac{2}{35} \cos^5 x$

$$= -\frac{1}{7} (1 - \cos^2 x) \cos^5 x - \frac{2}{35} \cos^5 x$$

$$= \frac{1}{7} \cos^7 x - \frac{1}{5} \cos^5 x$$

The conclusion is $c = 0$.

7.4 INTEGRATION OF RATIONAL FUNCTIONS USING PARTIAL FRACTIONS

1. $\displaystyle \frac{x-5}{x^2-1} = \frac{x-5}{(x+1)(x-1)}$

$$= \frac{A}{x+1} + \frac{B}{x-1}$$

$$x - 5 = A(x-1) + B(x+1)$$

$$x = -1 : -6 = -2A; \ A = 3$$

$$x = 1 : -4 = 2B; \ B = -2$$

$$\frac{x-5}{x^2-1} = \frac{3}{x+1} - \frac{2}{x-1}$$

$$\int \frac{x-5}{x^2-1}\, dx = \int \left(\frac{3}{x+1} - \frac{2}{x-1} \right) dx$$

$$= 3 \ln |x+1| - 2 \ln |x-1| + c$$

3. $\displaystyle \frac{6x}{x^2-x-2} = \frac{6x}{(x-2)(x+1)}$

$$= \frac{A}{x-2} + \frac{B}{x+1}$$

$$6x = A(x+1) + B(x-2)$$

$$x = 2 : 12 = 3A; \ A = 4$$

$$x = -1 : -6 = -3B; \ B = 2$$

$$\frac{6x}{x^2-x-2} = \frac{4}{x-2} + \frac{2}{x+1}$$

$$\int \frac{6x}{x^2-x-2}\, dx$$

$$= \int \left(\frac{4}{x-2} + \frac{2}{x+1} \right) dx$$

$$= 4 \ln |x-2| + 2 \ln |x+1| + c$$

5.
$$\frac{-x+5}{x^3-x^2-2x} = \frac{-x+5}{x(x-2)(x+1)}$$

$$= \frac{A}{x} + \frac{B}{x-2} + \frac{C}{x+1}$$

$$-x+5 = A(x-2)(x+1) + Bx(x+1)$$
$$+ cx(x-2)$$

$$x=0: 5=-2A: A=-\frac{5}{2}$$

$$x=2: 3=6B: B=\frac{1}{2}$$

$$x=-1: 6=3C: C=2$$

$$\frac{-x+5}{x^3-x^2-2x} = -\frac{5/2}{x} + \frac{1/2}{x-2} + \frac{2}{x+1}$$

$$\int \frac{-x+5}{x^3-x^2-2x} \, dx$$

$$= \int \left(-\frac{5/2}{x} + \frac{1/2}{x-2} + \frac{2}{x+1} \right) dx$$

$$= -\frac{5}{2} \ln|x| + \frac{1}{2} \ln|x-2|$$
$$+ 2 \ln|x+1| + c$$

7.
$$\frac{x^3+x+2}{x^2+2x-8} = x-2 + \frac{13x-14}{x^2+2x-8}$$

$$= x-2 + \frac{13x-14}{(x+4)(x-2)}$$

$$= x-2 + \frac{A}{x+4} + \frac{B}{x-2}$$

$$13x-14 = A(x-2) + B(x+4)$$

$$x=-4: -66=-6A; A=11$$

$$x=2: 12=6B; B=2$$

$$\frac{x^3+x+2}{x^2+2x-8} = x-2 + \frac{11}{x+4} + \frac{2}{x-2}$$

$$\int \frac{x^3+x+2}{x^2+2x-8} \, dx$$

$$= \int \left(x-2 + \frac{11}{x+4} + \frac{2}{x-2} \right) dx$$

$$= \frac{x^2}{2} - 2x + 11 \ln|x+4|$$
$$+ 2 \ln|x-2| + c$$

9.
$$\frac{5x-23}{6x^2-11x-7} = \frac{5x-23}{(2x+1)(3x-7)}$$

$$= \frac{A}{2x+1} + \frac{B}{3x-7}$$

$$5x-23 = A(3x-7) + B(2x+1)$$

$$x=-\frac{1}{2}: -\frac{51}{2} = -\frac{17}{2}A; A=3$$

$$x=\frac{7}{3}: -\frac{34}{3} = \frac{17}{3}B; B=-2$$

$$\frac{5x-23}{6x^2-11x-7} = \frac{3}{2x+1} - \frac{2}{3x-7}$$

$$\int \frac{5x-23}{6x^2-11x-7} \, dx$$

$$= \int \left(\frac{3}{2x+1} - \frac{2}{3x-7} \right) dx$$

$$= \frac{3}{2} \ln|2x+1| - \frac{2}{3} \ln|3x-7| + c$$

11.
$$\frac{x-1}{x^3+4x^2+4x} = \frac{x-1}{x(x+2)^2}$$

$$= \frac{A}{x} + \frac{B}{x+2} + \frac{C}{(x+2)^2}$$

$$x-1 = A(x+2)^2 + Bx(x+2) + Cx$$

$$x=0: -1=4A; A=-\frac{1}{4}$$

$$x=-2: -3=-2C; C=\frac{3}{2}$$

$$x=1: 0=9A+3B+C; B=\frac{1}{4}$$

$$\frac{x-1}{x^3+4x^2+4x}$$

$$= -\frac{1/4}{x} + \frac{1/4}{x+2} + \frac{3/2}{(x+2)^2}$$

$$\int \frac{x-1}{x^3+4x^2+4x} \, dx$$

$$= \int \left(-\frac{1/4}{x} + \frac{1/4}{x+2} + \frac{3/2}{(x+2)^2} \right) dx$$

$$= -\frac{1}{4} \ln|x| + \frac{1}{4} \ln|x+2| - \frac{3}{2(x+2)} + c$$

13. $\dfrac{x+4}{x^3+3x^2+2x} = \dfrac{x+4}{x(x+2)(x+1)}$

$= \dfrac{A}{x} + \dfrac{B}{x+2} + \dfrac{C}{x+1}$

$x+4 = A(x+2)(x+1) + Bx(x+1)$
$\qquad + Cx(x+2)$

$x=0 : 4 = 2A; \ A = 2$

$x=-2 : 2 = 2B; \ B = 1$

$x=-1 : 3 = -C; \ C = -3$

$\dfrac{x+4}{x^3+3x^2+2x} = \dfrac{2}{x} + \dfrac{1}{x+2} - \dfrac{3}{x+1}$

$\displaystyle\int \dfrac{x+4}{x^3+3x^2+2x}\,dx$

$= \displaystyle\int \left(\dfrac{2}{x} + \dfrac{1}{x+2} - \dfrac{3}{x+1} \right) dx$

$= 2\ln|x| + \ln|x+2| - 3\ln|x+1| + c$

15. $\dfrac{x+2}{x^3+x} = \dfrac{x+2}{x(x^2+1)}$

$= \dfrac{A}{x} + \dfrac{Bx+C}{x^2+1}$

$x+2 = A(x^2+1) + (Bx+C)x$

$= Ax^2 + A + Bx^2 + Cx$

$= (A+B)x^2 + Cx + A$

$A=2; \ C=1; \ B=-2$

$\dfrac{x+2}{x^3+x} = \dfrac{2}{x} + \dfrac{-2x+1}{x^2+1}$

$\displaystyle\int \dfrac{x+2}{x^3+x}\,dx$

$= \displaystyle\int \left(\dfrac{2}{x} + \dfrac{-2x+1}{x^2+1} \right) dx$

$= \displaystyle\int \left(\dfrac{2}{x} - \dfrac{2x}{x^2+1} + \dfrac{1}{x^2+1} \right) dx$

$= 2\ln|x| - \ln(x^2+1) + \tan^{-1}x + c$

17. $\dfrac{4x-2}{16x^4-1} = \dfrac{4x-2}{(4x^2+1)(2x+1)(2x-1)}$

$= \dfrac{Ax+B}{4x^2+1} + \dfrac{C}{2x+1} + \dfrac{D}{2x-1}$

$4x-2 = (Ax+B)(2x+1)(2x-1)$
$\qquad + C(2x-1)(4x^2+1)$
$\qquad + D(2x+1)(4x^2+1)$

$A=-4; \ B=1; \ C=1; \ D=0$

$\dfrac{4x-2}{16x^4-1} = \dfrac{-2x+1}{4x^2+1} + \dfrac{1}{2x+1}$

$\displaystyle\int \dfrac{4x-2}{16x^4-1}\,dx$

$= \displaystyle\int \left(\dfrac{-2x+1}{4x^2+1} + \dfrac{1}{2x+1} \right) dx$

$= \displaystyle\int \left(-\dfrac{1}{4}\dfrac{8x}{4x^2+1} + \dfrac{1}{4x^2+1} + \dfrac{1}{2x+1} \right) dx$

$= -\dfrac{1}{4}\ln|4x^2+1| + \dfrac{1}{2}\tan^{-1}(2x)$

$\qquad + \dfrac{1}{2}\ln|2x+1| + c$

19. $\dfrac{4x^2-7x-17}{6x^2-11x-10}$

$= \dfrac{2}{3} + \dfrac{1}{3}\dfrac{x-31}{(2x-5)(3x+2)}$

$= \dfrac{2}{3} + \dfrac{1}{3}\left[\dfrac{A}{2x-5} + \dfrac{B}{3x+2} \right]$

$x-31 = A(3x+2) + B(2x-5)$

$x=\dfrac{5}{2} : -\dfrac{57}{2} = \dfrac{19}{2}A, \ A=-3;$

$x=-\dfrac{2}{3} : -\dfrac{95}{3} = -\dfrac{19}{3}B, \ B=5;$

$\dfrac{4x^2-7x-17}{6x^2-11x-10}$

$= \dfrac{2}{3} + \dfrac{1}{3}\left[\dfrac{-3}{2x-5} + \dfrac{5}{3x+2} \right]$

$\displaystyle\int \dfrac{4x^2-7x-17}{6x^2-11x-10}\,dx$

$= \displaystyle\int \left(\dfrac{2}{3} - \dfrac{1}{2x-5} + \dfrac{5/3}{3x+2} \right) dx$

$= \dfrac{2}{3}x - \dfrac{1}{2}\ln|2x-5| + \dfrac{5}{9}\ln|3x+2| + c$

21. $\dfrac{2x+3}{x^2+2x+1} = \dfrac{2x+3}{(x+1)^2}$

$= \dfrac{A}{x+1} + \dfrac{B}{(x+1)^2}$

$2x+3 = A(x+1) + B$

$x = -1 : B = 1; A = 2$

$\dfrac{2x+3}{x^2+2x+1} = \dfrac{2}{x+1} + \dfrac{1}{(x+1)^2}$

$\displaystyle\int \dfrac{2x+3}{x^2+2x+1}\, dx$

$= \displaystyle\int \left(\dfrac{2}{x+1} + \dfrac{1}{(x+1)^2} \right)\, dx$

$= 2\ln|x+1| - \dfrac{1}{x+1} + c$

23. $\dfrac{x^3-4}{x^3+2x^2+2x} = 1 + \dfrac{-2x^2-2x-4}{x(x^2+2x+2)}$

$= 1 + \dfrac{A}{x} + \dfrac{Bx+C}{x^2+2x+2}$

$-2x^2 - 2x - 4 = A(x^2+2x+2)$

$\quad + (Bx+C)x$

$= (A+B)x^2 + (2A+C)x + 2A$

$A = -2; B = 0; C = 2$

$\dfrac{x^3-4}{x^3+2x^2+2x}$

$= 1 + \dfrac{-2}{x} + \dfrac{2}{x^2+2x+2}$

$\displaystyle\int \dfrac{x^3-4}{x^3+2x^2+2x}\, dx$

$= \displaystyle\int \left(1 + \dfrac{-2}{x} + \dfrac{2}{(x+1)^2+1} \right)\, dx$

$= x - 2\ln|x| + 2\tan^{-1}(x+1) + c$

25. $\dfrac{x^3+x}{3x^2+2x+1}$

$= \dfrac{x}{3} - \dfrac{2}{9} + \dfrac{1}{9}\dfrac{10x+2}{3x^2+2x+1}$

This is already the PFD, since $3x^2 + 2x + 1$ has no real roots.

$\displaystyle\int \dfrac{x^3+x}{3x^2+2x+1}\, dx$

$= \displaystyle\int \left(\dfrac{x}{3} - \dfrac{2}{9} + \dfrac{1}{9}\dfrac{10x+2}{3x^2+2x+1} \right)\, dx$

$= \displaystyle\int \left(\dfrac{x}{3} - \dfrac{2}{9} + \dfrac{15}{9}\dfrac{6x+2}{3\,3x^2+2x+1} \right.$

$\left. -\dfrac{14}{9}\dfrac{1}{3}\dfrac{1}{3(x+1/3)^2+2/3} \right)\, dx$

$= \dfrac{x^2}{6} - \dfrac{2}{9}x + \dfrac{5}{27}\ln(3x^2+2x+1)$

$\quad - \dfrac{2\sqrt{2}}{27}\tan^{-1}\left(\dfrac{3x+1}{\sqrt{2}} \right) + c$

27. $\dfrac{4x^2+3}{x^3+x^2+x} = \dfrac{A}{x} + \dfrac{Bx+C}{x^2+x+1}$

$4x^2 + 3 = A(x^2+x+1) + (Bx+C)x$

$= Ax^2 + Ax + A + Bx^2 + Cx$

$A = 3; C = -3; B = 1$

$\dfrac{4x^2+3}{x^3+x^2+x} = \dfrac{3}{x} + \dfrac{x-3}{x^2+x+1}$

$\displaystyle\int \dfrac{4x^2+3}{x^3+x^2+x}\, dx$

$= \displaystyle\int \left(\dfrac{3}{x} + \dfrac{x-3}{x^2+x+1} \right)\, dx$

$= \displaystyle\int \left(\dfrac{3}{x} + \dfrac{x+1/2}{x^2+x+1} - \dfrac{7/2}{x^2+x+1} \right)\, dx$

$= 3\ln|x| + \dfrac{1}{2}\ln|x^2+x+1|$

$\quad - \dfrac{7}{\sqrt{3}}\tan^{-1}\left(\dfrac{2x+1}{\sqrt{3}} \right) + c$

29. $\dfrac{3x^3+1}{x^3-x^2+x-1}$

$= 3 + \dfrac{3x^2-3x+4}{x^3-x^2+x-1}$

$= 3 + \dfrac{3x^2-3x+4}{(x^2+1)(x-1)}$

$= 3 + \dfrac{Ax+B}{x^2+1} + \dfrac{C}{x-1}$

$3x^2 - 3x + 4 = (Ax + B)(x - 1) + C(x^2 + 1)$

$= Ax^2 - Ax + Bx - B + Cx^2 + C$

$x = 1 : 4 = 2C; C = 2$

$A + C = 3 : A = 1$

$-A + B = -3 : B = -2$

$\dfrac{3x^3 + 1}{x^3 - x^2 + x - 1} = 3 + \dfrac{x - 2}{x^2 + 1} + \dfrac{2}{x - 1}$

$\displaystyle\int \dfrac{3x^3 + 1}{x^3 - x^2 + x - 1}\, dx$

$= \displaystyle\int \left(3 + \dfrac{x - 2}{x^2 + 1} + \dfrac{2}{x - 1}\right) dx$

$= \displaystyle\int \left(3 + \dfrac{x}{x^2 + 1} - \dfrac{2}{x^2 + 1} + \dfrac{2}{x - 1}\right) dx$

$= 3x + \dfrac{1}{2}\ln(x^2 + 1) - 2\tan^{-1} x$

$\quad + 2\ln|x - 1| + c$

31. $A + C = 0$, and since $A = 0$ we have $C = 0$.

$B + D = 2$, and since $B = 4$ we have $D = -2$.

$\dfrac{4x^2 + 2}{(x^2 + 1)^2} = \dfrac{4}{x^2 + 1} + \dfrac{-2}{(x^2 + 1)^2}$

33. $\dfrac{2x^2 + 4}{(x^2 + 4)^2} = \dfrac{Ax + B}{x^2 + 4} + \dfrac{Cx + D}{(x^2 + 4)^2}$

$2x^2 + 4 = (Ax + B)(x^2 + 4) + Cx + D$

$\quad = Ax^3 + 4Ax + Bx^2 + 4B + Cx + D$

$A = 0; B = 2$

$4A + C = 0 : C = 0$

$4B + D = 4 : D = -4$

$\dfrac{2x^2 + 4}{(x^2 + 4)^2} = \dfrac{2}{x^2 + 4} - \dfrac{4}{(x^2 + 4)^2}$

35. $\dfrac{4x^2 + 3}{(x^2 + x + 1)^2} = \dfrac{Ax + B}{x^2 + x + 1} + \dfrac{Cx + D}{(x^2 + x + 1)^2}$

$4x^2 + 3$

$= (Ax + B)(x^2 + x + 1) + Cx + D$

$= Ax^3 + Ax^2 + Ax + Bx^2 + Bx + B + Cx + D$

$A = 0$

$A + B = 4 : B = 4$

$A + B + C = 0 : C = -4$

$B + D = 3 : D = -1$

$\dfrac{4x^2 + 3}{(x^2 + x + 1)^2}$

$= \dfrac{4}{x^2 + x + 1} - \dfrac{4x + 1}{(x^2 + x + 1)^2}$

37. Let $u = x^3 + 1, du = 3x^2\, dx$.

$\displaystyle\int \dfrac{3}{x^4 + x}\, dx = \int \dfrac{3x^2}{x^3(x^3 + 1)}\, dx$

$= \displaystyle\int \dfrac{1}{(u - 1)u}\, du$

$= \displaystyle\int \left(\dfrac{1}{u - 1} - \dfrac{1}{u}\right) du$

$= \ln|u - 1| - \ln|u| + c$

$= \ln\left|\dfrac{u - 1}{u}\right| + c$

$= \ln\left|\dfrac{x^3}{x^3 + 1}\right| + c$

On the other hand, we can let

$u = \dfrac{1}{x}, du = -\dfrac{1}{x^2}\, dx.$

$\displaystyle\int \dfrac{3}{x^4 + x}\, dx = -\int \dfrac{3u^2}{1 + u^3}\, du$

$= -\ln|1 + u^3| + c$

$= -\ln|1 + 1/x^3| + c$

To see that the two answers are equivalent, note that

$\ln\left|\dfrac{x^3}{x^3 + 1}\right| = -\ln\left|\dfrac{x^3 + 1}{x^3}\right|$

$= -\ln|1 + 1/x^3|.$

7.5 INTEGRATION TABLES AND COMPUTER ALGEBRA SYSTEMS

1. Use # 7 in the table.

$$\int \frac{x}{(2+4x)^2}\,dx$$

$$= \frac{2}{16(2+4x)} + \frac{1}{16}\ln|2+4x| + c$$

$$= \frac{1}{8(2+4x)} + \frac{1}{16}\ln|2+4x| + c$$

3. Substitute $u = 1 + e^x$ and $du = e^x\,dx$.

$$\int e^{2x}\sqrt{1+e^x}\,dx = \int (u-1)\sqrt{u}\,du$$

$$= \int (u^{3/2} - u^{1/2})\,du$$

$$= \frac{2}{5}u^{5/2} - \frac{2}{3}u^{3/2} + c$$

$$= \frac{2}{5}(1+e^x)^{5/2} - \frac{2}{3}(1+e^x)^{3/2} + c$$

$$= \frac{2}{15}(1+e^x)^{3/2}\left[3(1+e^x) - 5\right] + c$$

$$= \frac{2}{15}(1+e^x)(3e^x - 2) + c$$

5. Substitute $u = 2x$ and then use #26 in the table.

$$\int \frac{x^2}{\sqrt{1+4x^2}}\,dx$$

$$= \frac{1}{8}\int \frac{u^2}{\sqrt{1+u^2}}\,du$$

$$= \frac{1}{8}\left[\frac{u}{2}\sqrt{1+u^2}\right.$$

$$\left. - \frac{1}{2}\ln|u + \sqrt{1+u^2}|\right] + c$$

$$= \frac{1}{8}x\sqrt{1+4x^2}$$

$$- \frac{1}{16}\ln|2x + \sqrt{1+4x^2}| + c$$

7. Substitute $u = x^3$ and then use #30 in the table.

$$\int x^8\sqrt{4-x^6}\,dx = \int u^2\sqrt{4-u^2}\cdot\frac{1}{3}\,du$$

$$= \frac{1}{3}\int u^2\sqrt{4-u^2}\,du$$

$$= \frac{1}{3}\left[\frac{u}{8}(2u^2 - 4)\sqrt{4-u^2} + \frac{16}{8}\sin^{-1}\frac{u}{2}\right] + c$$

$$= \frac{1}{24}x^3(2x^6 - 4)\sqrt{4-x^6}$$

$$+ \frac{2}{3}\sin^{-1}\frac{x^3}{2} + c$$

$$\int_0^1 x^8\sqrt{4-x^6}\,dx = \frac{\pi}{9} - \frac{\sqrt{3}}{12}$$

9. Substitute $u = e^x$ and then use #25 in the table.

$$\int \frac{e^x}{\sqrt{e^{2x}+4}}\,dx = \int \frac{1}{\sqrt{u^2+4}}\,du$$

$$= \ln(u + \sqrt{4+u^2}) + c$$

$$= \ln(e^x + \sqrt{4+e^{2x}}) + c$$

$$\int_0^{\ln 2} \frac{x^x}{\sqrt{e^{2x}+4}}\,dx = \ln\left(\frac{2\sqrt{2}+2}{1+\sqrt{5}}\right).$$

11. Substitute $u = x - 3$ and then use #32 in the table.

$$\int \frac{\sqrt{6x - x^2}}{(x-3)^2}\,dx$$

$$= \int \frac{\sqrt{(u+3)(6-(u+3))}}{u^2}\,du$$

$$= \int \frac{\sqrt{9-u^2}}{u^2}\,du$$

$$= -\frac{1}{u}\sqrt{9-u^2} - \sin^{-1}\frac{u}{3} + c$$

$$= -\frac{1}{x-3}\sqrt{9-(x-3)^2}$$

$$- \sin^{-1}\left(\frac{x-3}{3}\right) + c$$

13. Use #85 in the table.

$$\int \tan^6 x \, dx = \frac{1}{5} \tan^5 x - \int \tan^4 x \, dx$$

$$= \frac{1}{5} \tan^5 x - \left[\frac{1}{3} \tan^3 x - \int \tan^2 x \, dx \right]$$

$$= \frac{1}{5} \tan^5 x - \frac{1}{3} \tan^3 x + \tan x - x + c$$

15. Substitute $u = \sin x$ and then use #16a in the table.

$$\int \frac{\cos x}{\sin x \sqrt{4 + \sin x}} \, dx = \int \frac{1}{u\sqrt{4 + u}} \, du$$

$$= \frac{1}{\sqrt{4}} \ln \left| \frac{\sqrt{4 + u} - 2}{\sqrt{4 + u} + 2} \right| + c$$

$$= \frac{1}{2} \ln \left| \frac{\sqrt{4 + \sin x} - 2}{\sqrt{4 + \sin x} + 2} \right| + c$$

17. Substitute $u = x^2$ and then use #62 in the table.

$$\int x^3 \cos x^2 \, dx = \frac{1}{2} \int u \cos u \, du$$

$$= \frac{1}{2}(\cos u + u \sin u) + c$$

$$= \frac{1}{2} \cos x^2 + \frac{1}{2} x^2 \sin x^2 + c$$

19. Substitute $u = \cos x$ and then use #18 in the table.

$$\int \frac{\sin x \cos x}{\sqrt{1 + \cos x}} \, dx = -\int \frac{u}{\sqrt{1 + u}} \, du$$

$$= -\frac{2}{3}(u - 2)\sqrt{1 + u} + c$$

$$= -\frac{2}{3}(\cos x - 2)\sqrt{1 + \cos x} + c$$

21. Substitute $u = \sin x$ and then use #26 in the table.

$$\int \frac{\sin^2 x \cos x}{\sqrt{\sin^2 x + 4}} \, dx = \int \frac{u^2}{\sqrt{u^2 + 4}} \, du$$

$$= \frac{u}{2}\sqrt{4 + u^2} - \frac{4}{2} \ln(u + \sqrt{4 + u^2}) + c$$

$$= \frac{1}{2} \sin x \sqrt{4 + \sin^2 x}$$

$$\qquad - 2 \ln(\sin x + \sqrt{4 + \sin^2 x}) + c$$

23. Substitute $u = -\frac{2}{x^2}$.

$$\int \frac{e^{-2/x^2}}{x^3} \, dx = \frac{1}{4} \int e^u \, du$$

$$= \frac{1}{4} e^u + c = \frac{1}{4} e^{-2/x^2} + c$$

25. Use #50 in the table.

$$\int \frac{x}{\sqrt{4x - x^2}} \, dx$$

$$= -\sqrt{4x - x^2} + 2 \cos^{-1}\left(\frac{2 - x}{2} \right) + c$$

27. Substitute $u = e^x$ and then use #95 in the table.

$$\int e^x \tan^{-1}(e^x) \, dx = \int \tan^{-1} u \, du$$

$$= u \tan^{-1} u - \frac{1}{2} \ln(1 + u^2) + c$$

$$= e^x \tan^{-1} e^x - \frac{1}{2} \ln(1 + e^{2x}) + c$$

29. Answer depends on CAS used.

31. Any answer is wrong because the integrand is undefined for all $x \neq 1$.

33. Answer depends on CAS used.

35. Answer depends on CAS used.

7.6 INDETERMINATE FORMS AND L'HÔPITAL'S RULE

1.
$$\lim_{x \to -2} \frac{x + 2}{x^2 - 4}$$

$$= \lim_{x \to -2} \frac{x + 2}{(x + 2)(x - 2)}$$

$$= \lim_{x \to -2} \frac{1}{x - 2} = -\frac{1}{4}$$

3.
$$\lim_{x\to\infty} \frac{3x^2 + 2}{x^2 - 4}$$

$$= \lim_{x\to\infty} \frac{3 + \frac{2}{x^2}}{1 - \frac{4}{x^2}}$$

$$= \frac{3}{1} = 3$$

5. $\lim_{x\to 0} \dfrac{e^{2x} - 1}{x}$ is type $\frac{0}{0}$;

we apply L'Hôpital's Rule to get

$$\lim_{x\to 0} \frac{2e^{2x}}{1} = \frac{2}{1} = 2.$$

7. $\lim_{x\to 0} \dfrac{\tan^{-1} x}{\sin x}$ is type $\frac{0}{0}$;

we apply L'Hôpital's Rule to get

$$\lim_{x\to 0} \frac{1/\left(1 + x^2\right)}{\cos x} = \lim_{x\to 0} \frac{1}{1} = 1.$$

9. $\lim_{x\to \pi} \dfrac{\sin 2x}{\sin x}$ is type $\frac{0}{0}$;

we apply L'Hôpital's Rule to get

$$\lim_{x\to \pi} \frac{2\cos 2x}{\cos x} = \frac{2(1)}{-1} = -2.$$

11. $\lim_{x\to 0} \dfrac{\sin x - x}{x^3}$ is type $\frac{0}{0}$;

we apply L'Hôpital's Rule thrice to get

$$= \lim_{x\to 0} \frac{\cos x - 1}{3x^2} = \lim_{x\to 0} \frac{-\sin x}{6x}$$

$$= \lim_{x\to 0} \frac{-\cos x}{6} = \frac{-1}{6}.$$

13.
$$\lim_{x\to 1} \frac{\sqrt{x} - 1}{x - 1}$$

$$= \lim_{x\to 1} \frac{\sqrt{x} - 1}{x - 1} \frac{\sqrt{x} + 1}{\sqrt{x} + 1}$$

$$= \lim_{x\to 1} \frac{x - 1}{(x - 1)\left(\sqrt{x} + 1\right)}$$

$$= \lim_{x\to 1} \frac{1}{\sqrt{x} + 1} = \frac{1}{2}$$

15. $\lim_{x\to\infty} \dfrac{x^3}{e^x}$ is type $\frac{\infty}{\infty}$;

we apply L'Hôpital's Rule thrice to get

$$\lim_{x\to\infty} \frac{3x^2}{e^x} = \lim_{x\to\infty} \frac{6x}{e^x}$$

$$= \lim_{x\to\infty} \frac{6}{e^x} = 0.$$

17. $\lim_{x\to 0} \dfrac{x\cos x - \sin x}{x\sin^2 x}$ is type $\frac{0}{0}$;

we apply L'Hôpital's Rule twice to get

$$\lim_{x\to 0} \frac{\cos x - x\sin x - \cos x}{\sin^2 x + 2x\sin x\cos x}$$

$$= \lim_{x\to 0} \frac{-x\sin x}{\sin x\,(\sin x + 2x\cos x)}$$

$$= \lim_{x\to 0} \frac{-x}{\sin x + 2x\cos x}$$

$$= \lim_{x\to 0} \frac{-1}{\cos x + 2\cos x - 2x\sin x}$$

$$= -\frac{1}{3}.$$

19. $\lim_{x\to 1} \dfrac{\sin \pi x}{x - 1}$ is type $\frac{0}{0}$;

we apply L'Hôpital's Rule to get

$$\lim_{x\to 1} \frac{\pi \cos \pi x}{1} = \frac{\pi(-1)}{1} = -\pi.$$

21. $\lim_{x\to\infty} \dfrac{\ln x}{x^2}$ is type $\frac{\infty}{\infty}$;

we apply L'Hôpital's Rule to get

$$\lim_{x\to\infty} \frac{1/x}{2x} = \lim_{x\to\infty} \frac{1}{2x^2} = 0.$$

23. $\lim_{x\to\infty} xe^{-x} = \lim_{x\to\infty} \dfrac{x}{e^x}$ is type $\frac{\infty}{\infty}$;

we apply L'Hôpital's Rule to get

$$\lim_{x\to\infty} \frac{1}{e^x} = 0.$$

25. As x approaches 1 from below, $\ln x$ is a small negative number. Hence $\ln(\ln x)$ is undefined, so the limit is undefined.

27. $\lim\limits_{x\to 0^+} x \ln x = \lim\limits_{x\to 0^+} \dfrac{\ln x}{1/x}$ is type $\dfrac{\infty}{\infty}$;

we apply L'Hôpital's Rule to get

$$\lim_{x\to 0^+} \frac{1/x}{-1/x^2} = \lim_{x\to 0^+}(-x) = 0.$$

29. $\lim\limits_{x\to 0^+} \dfrac{\ln x}{\cot x}$ is type $\dfrac{\infty}{\infty}$;

we apply L'Hôpital's Rule to get

$$\lim_{x\to 0^+} \frac{1/x}{-\csc^2 x}$$

$$= \lim_{x\to 0^+} \frac{-\sin^2 x}{x}$$

$$= \lim_{x\to 0^+} \left(-\sin x \frac{\sin x}{x}\right) = (0)(1) = 0.$$

31. $\lim\limits_{x\to\infty}\left(\sqrt{x^2+1} - x\right)$

$$= \lim_{x\to\infty}\left(\left(\sqrt{x^2+1} - x\right)\frac{\sqrt{x^2+1}+x}{\sqrt{x^2+1}+x}\right)$$

$$= \lim_{x\to\infty}\left(\frac{x^2+1-x^2}{\sqrt{x^2+1}+x}\right)$$

$$= \lim_{x\to\infty}\frac{1}{\sqrt{x^2+1}+x} = 0.$$

33. Let $y = \left(1 + \dfrac{1}{x}\right)^x$. Then $\ln y = x \ln\left(1 + \dfrac{1}{x}\right)$. Then

$$\lim_{x\to\infty} \ln y = \lim_{x\to\infty} x \ln\left(1 + \frac{1}{x}\right)$$

$$= \lim_{x\to\infty} \frac{\ln\left(1 + \frac{1}{x}\right)}{1/x}$$

$$= \lim_{x\to\infty} \frac{\frac{1}{1+\frac{1}{x}}\left(\frac{-1}{x^2}\right)}{-1/x^2}$$

$$= \lim_{x\to\infty} \frac{1}{1 + \frac{1}{x}} = 1.$$

Hence

$$\lim_{x\to\infty} y = \lim_{x\to\infty} e^{\ln y} = e.$$

35. $\lim\limits_{x\to 0^+}\left(\dfrac{1}{\sqrt{x}} - \dfrac{\sqrt{x}}{\sqrt{x+1}}\right)$

$$= \lim_{x\to 0^+}\left(\frac{\sqrt{x+1} - (\sqrt{x})^2}{\sqrt{x}\sqrt{x+1}}\right)$$

$$= \lim_{x\to 0^+}\left(\frac{\sqrt{x+1} - x}{\sqrt{x}\sqrt{x+1}}\right)$$

$$= \infty.$$

37. Let $y = (1/x)^x$. Then $\ln y = x \ln(1/x)$. Then

$$\lim_{x\to 0^+} \ln y = \lim_{x\to 0^+} x \ln(1/x)$$

$$= \lim_{x\to 0^+} [-x \ln x] = 0,$$

by exercise 27. Thus

$$\lim_{x\to 0^+} y = \lim_{x\to 0^+} e^{\ln y} = 1.$$

39. L'Hôpital's Rule does not apply. As $x \to 0$, the numerator gets close to 1 and the denominator is small and positive. Hence the limit is ∞.

41. L'Hôpital's Rule does not apply. As $x \to 0$, the numerator is small and positive while the denominator goes to $-\infty$. Hence the limit is 0. Also $\lim\limits_{x\to 0} \dfrac{2x}{2/x}$, which equals $\lim\limits_{x\to 0} x^2$, is not of the form $\dfrac{0}{0}$ so L'Hôpital's Rule doesn't apply here either.

43. Starting with $\lim\limits_{x\to 0}\dfrac{\sin 3x}{\sin 2x}$, we cannot "cancel sin" to get $\lim\limits_{x\to 0}\dfrac{3x}{2x}$. We can cancel the x's in the last limit to get the final answser of $3/2$. The first step is likely to give a correct answer because the linear approximation of $\sin 3x$ is $3x$, and the linear approximation of $\sin 2x$ is $2x$. The linear approximations are better the closer x is to zero, so the limits are likely to be the same.

45. (a)

$$\lim_{x\to 0}\frac{\sin x^2}{x^2} = \lim_{x\to 0}\frac{2x \cos x^2}{2x}$$

$$= \lim_{x\to 0} \cos x^2 = 1,$$

which is the same as $\lim\limits_{x\to 0}\dfrac{\sin x}{x}$.

(b) $\lim\limits_{x\to 0} \dfrac{1-\cos x^2}{x^4}$

$= \lim\limits_{x\to 0} \dfrac{2x \sin x^2}{4x^3} = \lim\limits_{x\to 0} \dfrac{\sin x^2}{2x^2}$

$= \dfrac{1}{2} \lim\limits_{x\to 0} \dfrac{\sin x^2}{x^2}$

$= (1/2)(1) = 1/2$ (by part (a)),

while

$\lim\limits_{x\to 0} \dfrac{1-\cos x}{x^2} = \lim\limits_{x\to 0} \dfrac{\sin x}{2x}$

$= \dfrac{1}{2} \lim\limits_{x\to 0} \dfrac{\sin x}{x}$

$= \dfrac{1}{2}(1) = \dfrac{1}{2}$

so both of these limits are the same.

47. $\lim\limits_{x\to 0} \dfrac{\sin kx^2}{x^2}$

$= \lim\limits_{x\to 0} \dfrac{2kx \cos kx^2}{2x}$

$= \lim\limits_{x\to 0} k \cos kx^2 = k(1) = k$

Since k can be any real number, the value of a limit with indeterminate form $\frac{0}{0}$ can be any real number.

49. $\lim\limits_{x\to\infty} e^x = \lim\limits_{x\to\infty} x^n = \infty$

$\lim\limits_{x\to\infty} \dfrac{e^x}{x^n} = \infty$ since n applications of L'Hôpital's Rule yields

$\lim\limits_{x\to\infty} \dfrac{e^x}{n!} = \infty.$

Hence, e^x dominates x^n.

51. $\lim\limits_{x\to 0} \dfrac{e^{cx}-1}{x} = \lim\limits_{x\to 0} \dfrac{ce^{cx}}{1} = c$

53. If $x \to 0$, then $x^2 \to 0$, so if $\lim\limits_{x\to 0} \dfrac{f(x)}{g(x)} = L,$

then $\lim\limits_{x\to 0} \dfrac{f(x^2)}{g(x^2)} = L.$ If $a \neq 0$ or 1, then

$\lim\limits_{x\to a} \dfrac{f(x)}{g(x)}$ involves the behavior of the quo-

tient near a, while $\lim\limits_{x\to a} \dfrac{f(x^2)}{g(x^2)}$ involves the behavior of the quotient near the different point a^2 and $a^2 \neq a$.

55. $\lim\limits_{\omega\to 0} \dfrac{2.5(4\omega t - \sin 4\omega t)}{4\omega^2}$

$= \lim\limits_{\omega\to 0} \dfrac{2.5(4t - 4t \cos 4\omega t)}{8\omega}$

$= \lim\limits_{\omega\to 0} \dfrac{2.5(16t^2 \sin 4\omega t)}{8} = 0$

57. (a) $f(x) = \dfrac{(x+1)(2+\sin x)}{x(2+\cos x)}$

   (b) $f(x) = \dfrac{x}{e^x}$

   (c) $f(x) = \dfrac{3x+1}{x-7}$

   (d) $f(x) = \dfrac{3-8x}{1+2x}$

59. The area of triangular region 1 is

$(1/2)(\text{base})(\text{height})$

$= (1/2)(1-\cos\theta)(\sin\theta).$

Let P be the center of the circle. The area of region 2 equals the area of sector APC minus the area of triangle APB. The area of the sector is $\theta/2$, while the area of triangle APB is

$(1/2)(\text{base})(\text{height})$

$= (1/2)(\cos\theta)(\sin\theta).$

Hence the area of region 1 divided by the area of region 2 is

$\dfrac{(1/2)(1-\cos\theta)(\sin\theta)}{\theta/2 - (1/2)(\cos\theta)(\sin\theta)}$

$= \dfrac{(1-\cos\theta)(\sin\theta)}{\theta - \cos\theta \sin\theta}$

$= \dfrac{\sin\theta - \cos\theta \sin\theta}{\theta - \cos\theta \sin\theta}$

$= \dfrac{\sin\theta - (1/2)\sin 2\theta}{\theta - (1/2)\sin 2\theta}$

Then

$$\lim_{\theta \to 0} \frac{\sin \theta - (1/2) \sin 2\theta}{\theta - (1/2) \sin 2\theta}$$

$$= \lim_{\theta \to 0} \frac{\cos \theta - \cos 2\theta}{1 - \cos 2\theta}$$

$$= \lim_{\theta \to 0} \frac{-\sin \theta + 2 \sin 2\theta}{2 \sin 2\theta}$$

$$= \lim_{\theta \to 0} \frac{-\cos \theta + 4 \cos 2\theta}{4 \cos 2\theta}$$

$$= \frac{-1 + 4(1)}{4(1)} = \frac{3}{4}$$

7.7 IMPROPER INTEGRALS

1. Improper, function not defined at $x = 0$

3. Not improper, function continuous on entire interval

5. Improper, function not defined at $x = 0$

7.
$$\int_0^1 x^{-1/3} \, dx = \lim_{R \to 0^+} \int_R^1 x^{-1/3} \, dx$$

$$= \lim_{R \to 0^+} \frac{3}{2} x^{2/3} \Big|_R^1$$

$$= \lim_{R \to 0^+} \frac{3}{2}(1 - R^{2/3}) = \frac{3}{2}$$

9.
$$\int_1^\infty x^{-4/5} \, dx = \lim_{R \to \infty} \int_1^R x^{-4/5} \, dx$$

$$= \lim_{R \to \infty} 5x^{1/5} \Big|_1^R = \lim_{R \to \infty} (5R^{1/5} - 5) = \infty$$

So the original integral diverges.

11.
$$\int_0^1 \frac{1}{\sqrt{1-x}} \, dx = \lim_{R \to 1^-} \int_0^R \frac{1}{\sqrt{1-x}} \, dx$$

$$= \lim_{R \to 1^-} \left[-2\sqrt{1-x} \right]_0^R$$

$$= \lim_{R \to 1^-} \left[-2(\sqrt{1-R} - 1) \right]$$

$$= 2$$

The integral converges to 2.

13.
$$\int_0^1 \ln x \, dx = \lim_{R \to 0^+} \int_R^1 \ln x \, dx$$

$$= \lim_{R \to 0^+} (x \ln x - x) \Big|_R^1$$

$$= \lim_{R \to 0^+} (-1 - R \ln R + R)$$

$$= -1 - \lim_{R \to 0^+} \frac{\ln R}{1/R}$$

$$= -1 - \lim_{R \to 0^+} \frac{1/R}{-1/R^2}$$

$$= -1 + \lim_{R \to 0^+} R = -1$$

15.
$$\int_0^3 \frac{2}{x^2 - 1} \, dx = \int_0^1 \frac{2}{x^2 - 1} \, dx + \int_1^3 \frac{2}{x^2 - 1} \, dx$$

$$\int_0^1 \frac{2}{x^2 - 1} \, dx$$

$$= \lim_{R \to 1^-} \int_0^R \left(\frac{1}{x-1} - \frac{1}{x+1} \right) \, dx$$

$$= \lim_{R \to 1^-} \left[\ln|x-1| - \ln|x+1| \right]_0^R$$

$$= \lim_{R \to 1^-} \left[(\ln|x-1| - \ln|x+1|) - 0 \right]$$

$$= -\infty$$

Since $\displaystyle\int_0^1 \frac{2}{x^2 - 1} \, dx$ diverges, $\displaystyle\int_0^3 \frac{2}{x^2 - 1} \, dx$ also diverges.

17.
$$\int_0^\infty xe^x \, dx = \lim_{R \to \infty} \int_0^R xe^x \, dx$$

$$= \lim_{R \to \infty} (xe^x - e^x) \Big|_0^R$$

$$= \lim_{R \to \infty} [e^R(R - 1) + 1] = \infty$$

So the original integral diverges.

19. Substitute $u = 3x$ and then use #102 in the Table of Integrals.

$$I = \int_{-\infty}^{1} x^2 e^{3x}\, dx = \frac{1}{27} \int_{-\infty}^{3} u^2 e^u\, du$$

$$= \frac{1}{27} \lim_{R\to-\infty} (u^2 e^u - 2u e^u + 2e^u) \Big|_{R}^{3}$$

$$= \frac{5}{27} e^3 - \frac{1}{27} \lim_{R\to-\infty} e^R (R^2 - 2R + 2)$$

But $\displaystyle\lim_{R\to-\infty} e^R(R^2 - 2R + 2)$

$$= \lim_{R\to\infty} e^{-R}(R^2 + 2R + 2)$$

$$= \lim_{R\to\infty} \frac{R^2 + 2R + 2}{e^R} = 0.$$

Hence, $I = \dfrac{5}{27} e^3$.

21. $$\int_{-\infty}^{\infty} \frac{1}{x^2}\, dx = \int_{-\infty}^{-1} \frac{1}{x^2}\, dx + \int_{-1}^{0} \frac{1}{x^2}\, dx$$

$$+ \int_{0}^{1} \frac{1}{x^2}\, dx + \int_{1}^{\infty} \frac{1}{x^2}\, dx$$

$$\int_{-1}^{0} \frac{1}{x^2}\, dx = \lim_{R\to 0^-} \int_{-1}^{R} x^{-2}\, dx$$

$$= \lim_{R\to 0^-} \left[-\frac{1}{x} \right]_{-1}^{R}$$

$$= \lim_{R\to 0^-} \left[-\frac{1}{R} - 1 \right]$$

$$= \infty$$

Since $\displaystyle\int_{-1}^{0} \frac{1}{x^2}\, dx$ diverges, $\displaystyle\int_{-\infty}^{\infty} \frac{1}{x^2}\, dx$ also diverges.

23. $$\int_{-\infty}^{\infty} \frac{1}{1+x^2}\, dx$$

$$= \int_{-\infty}^{0} \frac{1}{1+x^2}\, dx + \int_{0}^{\infty} \frac{1}{1+x^2}\, dx$$

$$= \lim_{R\to-\infty} \int_{R}^{0} \frac{1}{1+x^2}\, dx$$

$$+ \lim_{R\to\infty} \int_{0}^{R} \frac{1}{1+x^2}\, dx$$

$$= \lim_{R\to-\infty} \tan^{-1} x \big|_{R}^{0} + \lim_{R\to\infty} \tan^{-1} x \big|_{0}^{R}$$

$$= \lim_{R\to-\infty} (\tan^{-1} 0 - \tan^{-1} R)$$

$$+ \lim_{R\to\infty} (\tan^{-1} R - \tan^{-1} 0)$$

$$= 0 - \left(-\frac{\pi}{2} \right) + \frac{\pi}{2} - 0 = \pi$$

25. $$\int_{0}^{\pi} \cot x\, dx = \int_{0}^{\pi/2} \cot x\, dx + \int_{\pi/2}^{\pi} \cot x\, dx$$

$$\int_{0}^{\pi/2} \cot x\, dx = \lim_{R\to 0^+} \int_{R}^{\pi/2} \frac{\cos x}{\sin x}\, dx$$

$$= \lim_{R\to 0^+} \ln|\sin x| \Big|_{R}^{\pi/2}$$

$$= \ln|\sin(\pi/2)| - \lim_{R\to 0^+} \ln|\sin R|$$

$$= \infty$$

So the original integral diverges.

27. $$\int_{0}^{2} \frac{x}{x^2 - 1}\, dx$$

$$= \int_{0}^{1} \frac{x}{x^2 - 1}\, dx + \int_{1}^{2} \frac{x}{x^2 - 1}\, dx$$

$$\int_{0}^{1} \frac{x}{x^2 - 1}\, dx = \lim_{R\to 1^-} \int_{0}^{R} \frac{x}{x^2 - 1}\, dx$$

$$= \lim_{R\to 1^-} \frac{1}{2} \ln|x^2 - 1| \Big|_{0}^{R}$$

$$= \lim_{R\to 1^-} \left(\frac{1}{2} \ln|R^2 - 1| - \frac{1}{2} \ln|-1| \right)$$

$$= -\infty$$

So the original integral diverges.

29. $\int_0^1 \frac{2}{\sqrt{1-x^2}}\,dx = \lim_{R\to 1^-}\int_0^R \frac{2}{\sqrt{1-x^2}}\,dx$

$= \lim_{R\to 1^-} 2\sin^{-1}x \Big|_0^R$

$= \lim_{R\to 1^-} 2(\sin^{-1}R - \sin^{-1}0)$

$= 2\left(\frac{\pi}{2} - 0\right) = \pi$

31. Substitute $u = \sqrt{x}$.

$\int \frac{1}{\sqrt{x}e^{\sqrt{x}}}\,dx = \int 2e^{-u}\,du$

Hence, $\int_0^\infty \frac{1}{\sqrt{x}e^{\sqrt{x}}}\,dx$

$= \lim_{R\to 0^+}\int_R^1 \frac{1}{\sqrt{x}e^{\sqrt{x}}}\,dx$

$+ \lim_{R\to\infty}\int_1^R \frac{1}{\sqrt{x}e^{\sqrt{x}}}\,dx$

$= \lim_{R\to 0^+}\left(\frac{-2}{e^{\sqrt{x}}}\Big|_R^1\right) + \lim_{R\to\infty}\left(\frac{-2}{e^{\sqrt{x}}}\Big|_1^R\right)$

$= \lim_{R\to 0^+}\left(\frac{-2}{e} + \frac{2}{e^{\sqrt{R}}}\right) + \lim_{R\to\infty}\left(\frac{2}{e} - \frac{2}{e^{\sqrt{R}}}\right)$

$= 2 + 0 = 2.$

33. $\int_0^\infty \cos x\,dx = \lim_{R\to\infty}\int_0^R \cos x\,dx$

$= \lim_{R\to\infty} \sin x \Big|_0^R = \lim_{R\to\infty}(\sin R - \sin 0)$

$= \lim_{R\to\infty} \sin R$

Since the last limit does not exist, the integral diverges.

35. Converges if and only if $n < 1$.

$\int_0^1 x^{-n}\,dx = \lim_{R\to 0^+}\int_R^1 x^{-n}\,dx$

$= \lim_{R\to 0^+} \frac{x^{-n+1}}{-n+1}\Big|_R^1$

$= \lim_{R\to 0^+}\left(\frac{1}{-n+1} - \frac{R^{-n+1}}{-n+1}\right)$

This limit exists only if $-n + 1 > 0$ or $n < 1$. Therefore the integral converges if and only if $n < 1$.

37. $\int_0^\infty xe^{cx}\,dx$ converges for $c < 0$.

$\int_{-\infty}^0 xe^{cx}\,dx$ converges for $c > 0$.

39. $0 < \frac{x}{1+x^3} < \frac{x}{x^3} = \frac{1}{x^2}$

$\int_1^\infty \frac{1}{x^2}\,dx = \lim_{R\to\infty}\int_1^R \frac{1}{x^2}\,dx$

$= \lim_{R\to\infty}\left(-\frac{1}{x}\right)\Big|_1^R$

$= \lim_{R\to\infty}\left(-\frac{1}{R} + 1\right) = 1$

So $\int_1^\infty \frac{x}{1+x^3}\,dx$ converges.

41. $\frac{x}{x^{3/2}-1} > \frac{x}{x^{3/2}} = \frac{1}{x^{1/2}} > 0$

$\int_2^\infty x^{-1/2}\,dx = \lim_{R\to\infty}\int_2^R x^{-1/2}\,dx$

$= \lim_{R\to\infty} 2\sqrt{x}\Big|_2^R$

$= \lim_{R\to\infty}(2\sqrt{R} - 2\sqrt{2}) = \infty$

So $\int_2^\infty \frac{x}{x^{3/2}-1}\,dx$ diverges.

43. $0 < \frac{3}{x+e^x} < \frac{3}{e^x}$

$\int_0^\infty \frac{3}{e^x}\,dx = \lim_{R\to\infty}\int_0^R \frac{3}{e^x}\,dx$

$= \lim_{R\to\infty}\left(-\frac{3}{e^x}\right)\Big|_0^R$

$= \lim_{R\to\infty}\left(-\frac{3}{e^R} + 3\right) = 3$

So $\int_0^\infty \frac{3}{x+e^x}\,dx$ converges.

45. $\dfrac{\sin^2 x}{1+e^x} \le \dfrac{1}{1+e^x} < \dfrac{1}{e^x}$

$\displaystyle\int_0^\infty \dfrac{1}{e^x}\, dx = \lim_{R\to\infty} \int_0^R \dfrac{1}{e^x}\, dx$

$= \displaystyle\lim_{R\to\infty} (-e^{-x})\ \bigg|_0^R$

$= \displaystyle\lim_{R\to\infty} (-e^{-R} + 1) = 1$

So $\displaystyle\int_0^\infty \dfrac{\sin^2 x}{1+e^x}\, dx$ converges.

47. $\dfrac{x^2 e^x}{\ln x} > e^x$

$\displaystyle\int_2^\infty e^x\, dx = \lim_{R\to\infty} \int_2^R e^x\, dx$

$= \displaystyle\lim_{R\to\infty} e^x\ \bigg|_2^R$

$= \displaystyle\lim_{R\to\infty} (e^R - e^2) = \infty$

So $\displaystyle\int_2^\infty \dfrac{x^2 e^x}{\ln x}\, dx$ diverges.

49. Let $u = \ln 4x,\ dv = x\, dx,$

$du = \dfrac{dx}{x},\ v = \dfrac{x^2}{2}.$

$\displaystyle\int x \ln 4x\, dx = \dfrac{1}{2} x^2 \ln 4x - \dfrac{1}{2} \int x\, dx$

$= \dfrac{1}{2} x^2 \ln 4x - \dfrac{x^2}{4} + c$

$I = \displaystyle\int_0^1 x \ln 4x\, dx = \lim_{R\to 0^+} \int_R^1 x \ln 4x\, dx$

$= \displaystyle\lim_{R\to 0^+} \left(\dfrac{1}{2} x^2 \ln 4x - \dfrac{x^2}{4}\right)\ \bigg|_R^1$

$= \dfrac{1}{2} \ln 4 - \dfrac{1}{4} - \displaystyle\lim_{R\to 0^+} \left(\dfrac{1}{2} R^2 \ln 4R - \dfrac{R^2}{4}\right)$

$= \dfrac{1}{2} \ln 4 - \dfrac{1}{4} - \dfrac{1}{2} \displaystyle\lim_{R\to 0^+} R^2 \ln 4R$

$\displaystyle\lim_{R\to 0^+} R^2 \ln 4R = \lim_{R\to 0^+} \dfrac{\ln 4R}{R^{-2}}$

$= \displaystyle\lim_{R\to 0^+} \dfrac{R^{-1}}{-2R^{-3}} = \lim_{R\to 0^+} \dfrac{R^2}{-2} = 0$

Hence, $I = \dfrac{1}{2} \ln 4 - \dfrac{1}{4}.$

51. The volume is finite:

$V = \pi \displaystyle\int_1^\infty \dfrac{1}{x^2}\, dx = \pi \lim_{R\to\infty} \int_1^R \dfrac{1}{x^2}\, dx$

$= \pi \displaystyle\lim_{R\to\infty} -\dfrac{1}{x}\ \bigg|_1^R$

$= \pi \displaystyle\lim_{R\to\infty} \left(-\dfrac{1}{R} + 1\right) = \pi.$

The surface area is infinite:

$S = 2\pi \displaystyle\int_1^\infty \dfrac{1}{x} \sqrt{1 + \dfrac{1}{x^4}}\, dx$

$\dfrac{1}{x} \sqrt{1 + \dfrac{1}{x^4}} > \dfrac{1}{x}$

and $\displaystyle\int_1^\infty \dfrac{1}{x}\, dx = \lim_{R\to\infty} \int_1^R \dfrac{1}{x}\, dx$

$= \displaystyle\lim_{R\to\infty} \ln|x|\ \bigg|_1^R = \lim_{R\to\infty} \ln R = \infty.$

53. True; we have seen that if $\displaystyle\int_0^\infty f(x)\, dx$ converges, then $\displaystyle\lim_{x\to\infty} f(x) = 0.$

55. False; consider $f(x) = \dfrac{1}{\sqrt{x}}.$

57. $I_p = \displaystyle\int_0^1 x^{-p}\, dx = \lim_{R\to 0^+} \int_R^1 x^{-p}\, dx$

$= \displaystyle\lim_{R\to 0^+} \dfrac{x^{-p+1}}{-p+1}\ \bigg|_R^1$

$= \displaystyle\lim_{R\to 0^+} \dfrac{1 - R^{-p+1}}{-p+1}$

We need $p < 1$ for the above limit to converge. If this is the case,

$$I_p = \frac{1}{-p+1},$$

and for such p

$$\int_0^1 (1-x)^{-p} \, dx$$

$$= \lim_{R \to 1^-} \int_0^R (1-x)^{-p} \, dx$$

$$= \lim_{R \to 1^-} \frac{-(1-x)^{-p+1}}{-p+1} \Big|_0^R$$

$$= \lim_{R \to 1^-} \frac{1 - (1-R)^{-p+1}}{-p+1}$$

$$= \frac{1}{-p+1} = I_p.$$

59. Substitute $u = \sqrt{k}x$.

$$\int_{-\infty}^{\infty} e^{-kx^2} \, dx = \frac{1}{\sqrt{k}} \int_{-\infty}^{\infty} e^{-u^2} \, du$$

$$= \frac{\sqrt{\pi}}{\sqrt{k}}$$

61. (a)

$$\int_0^{\infty} ke^{-2x} \, dx = \lim_{R \to \infty} \int_0^R ke^{-2x} \, dx$$

$$= -\frac{k}{2} \lim_{R \to \infty} e^{-2x} \Big|_0^R$$

$$= -\frac{k}{2} \lim_{R \to \infty} (e^{-2R} - 1) = \frac{k}{2} = 1.$$

So $k = 2$.

(b)

$$\int_0^{\infty} ke^{-4x} \, dx = \lim_{R \to \infty} \int_0^R ke^{-4x} \, dx$$

$$= -\frac{k}{4} \lim_{R \to \infty} e^{-4x} \Big|_0^R$$

$$= -\frac{k}{4} \lim_{R \to \infty} (e^{-4R} - 1) = \frac{k}{4} = 1.$$

So $k = 4$.

(c) If $r > 0$:

$$\int_0^{\infty} ke^{-rx} \, dx = \lim_{R \to \infty} \int_0^R ke^{-rx} \, dx$$

$$= -\frac{k}{r} \lim_{R \to \infty} e^{-rx} \Big|_0^R$$

$$= -\frac{k}{r} \lim_{R \to \infty} (e^{-rR} - 1) = \frac{k}{r} = 1.$$

So $k = r$.

If $r \le 0$:

The integral $\int_0^{\infty} ke^{-rx} \, dx$

diverges for any nonzero value of k, so there is no value of k to make the function a pdf.

63. From exercise 61 (c), we know that this r has to be positive.

Substitute $u = rx$.

$$\mu = \int_0^{\infty} xf(x) \, dx = \int_0^{\infty} rxe^{-rx} \, dx$$

$$= \lim_{R \to \infty} \int_0^R rxe^{-rx} \, dx$$

$$= \frac{1}{r} \lim_{R \to \infty} \int_0^R ue^{-u} \, du$$

$$= \frac{1}{r} \lim_{R \to \infty} e^{-u}(-u - 1) \Big|_0^R$$

$$= \lim_{R \to \infty} \frac{-R-1}{e^R} + \frac{1}{r}$$

$$= 0 + \frac{1}{r} = \frac{1}{r}.$$

65. The probability is given by the formula

$$\int_{\mu}^{\infty} re^{-rx} \, dx = \int_{\frac{1}{r}}^{\infty} re^{-rx} \, dx$$

$$= \lim_{R \to \infty} \int_{\frac{1}{r}}^R re^{-rx} \, dx = \lim_{R \to \infty} (-e^{-rx}) \Big|_{\frac{1}{r}}^R$$

$$= \lim_{R \to \infty} (e^{-rR} + e^{-1}) = e^{-1} < \frac{1}{2}.$$

The mean is different from the median, so it is not odd that the probability we just found is not equal to 1/2.

67. Improper because $\tan(\pi/2)$ is not defined.

The two integrals are equal,

$$\int_0^{\pi/2} \frac{1}{1+\tan x}\, dx = \int_0^{\pi/2} f(x)\, dx,$$

because the two integrands only differ at one value of x, and that except for this value, everything is proper.

$$g(x) = \begin{cases} \frac{\tan x}{1+\tan x} & \text{if } 0 \le x < \frac{\pi}{2} \\ 1 & \text{if } x = \frac{\pi}{2} \end{cases}$$

69. (a) Substitute $u = \dfrac{\pi}{2} - x$.

$$\int_0^{\pi/2} \ln(\sin x)\, dx$$

$$= \lim_{R\to 0^+} \int_R^{\pi/2} \ln(\sin x)\, dx$$

$$= \lim_{R\to \pi/2^-} \left[-\int_R^0 \ln(\sin(\pi/2 - u))\, du \right]$$

$$= \lim_{R\to \pi/2^-} \int_0^R \ln(\cos u)\, du$$

$$= \int_0^{\pi/2} \ln(\cos x)\, dx$$

(b)

$$2 \int_0^{\pi/2} \ln(\sin x)\, dx$$

$$= \int_0^{\pi/2} \ln(\cos x)\, dx + \int_0^{\pi/2} \ln(\sin x)\, dx$$

$$= \int_0^{\pi/2} [\ln(\cos x) + \ln(\sin x)]\, dx$$

$$= \int_0^{\pi/2} \ln(\sin x \cos x)\, dx$$

$$= \int_0^{\pi/2} [\ln(\sin(2x)) - \ln 2]\, dx$$

$$= \int_0^{\pi/2} \ln(\sin(2x))\, dx - \frac{\pi}{2} \ln 2$$

$$= \frac{1}{2} \int_0^{\pi} \ln(\sin x)\, dx - \frac{\pi}{2} \ln 2$$

by substituting x for $2x$.

Hence,

$$2 \int_0^{\pi/2} \ln(\sin x)\, dx$$

$$= \frac{1}{2} \int_0^{\pi} \ln(\sin x)\, dx - \frac{\pi}{2} \ln 2.$$

(c) On the other hand, we notice that the graph of $\sin x$ is symmetric over the interval $[0, \pi]$ across the line $x = \pi/2$, hence

$$\int_0^{\pi} \ln(\sin x)\, dx = 2 \int_0^{\pi/2} \ln(\sin x)\, dx.$$

(d) Then

$$\frac{1}{2} \int_0^{\pi} \ln(\sin x)\, dx = \int_0^{\pi/2} \ln(\sin x)\, dx.$$

So we get

$$\int_0^{\pi/2} \ln(\sin x)\, dx = -\frac{\pi}{2} \ln 2.$$

71. Use integration by parts twice. First time,

let $u = -\dfrac{1}{2}x^3$, $dv = -2xe^{-x^2}\, dx$.

Second time,

let $u = -\dfrac{1}{2}x$, $dv = -2xe^{-x^2}\, dx$.

$$\int x^4 e^{-x^2}\, dx$$

$$= -\frac{1}{2}x^3 e^{-x^2} + \int \frac{3}{2}x^2 e^{-x^2}\, dx$$

$$= -\frac{1}{2}x^3 e^{-x^2}$$

$$+ \frac{3}{2}\left(-\frac{1}{2}xe^{-x^2} + \frac{1}{2}\int e^{-x^2}\, dx \right)$$

$$= -\frac{1}{2}x^3 e^{-x^2} - \frac{3}{4}xe^{-x^2} + \frac{3}{4}\int e^{-x^2}\, dx$$

Putting integration limits to all the above, and realizing that when taking limits to $\pm\infty$, all multiples of e^{-x^2} as shown in above will go to 0 (we have seen this a lot of times before). Then we get

$$\int_{-\infty}^{\infty} x^4 e^{-x^2}\, dx = \frac{3}{4} \int_{-\infty}^{\infty} e^{-x^2}\, dx = \frac{3}{4}\sqrt{\pi}.$$

This means when $n = 2$, the statement
$$\int_{-\infty}^{\infty} x^{2n} e^{-x^2}\, dx = \frac{3 \cdot 1}{2^n} \sqrt{\pi}$$
is true. (We can also check that the case for $n = 1$ is correct.)

For general n, supposing that the statement is true for all $m < n$, then integration by parts gives

$$\int x^{2n} e^{-x^2}\, dx$$
$$= -\frac{1}{2} x^{2n-1} e^{-x^2} + \frac{2n-1}{2} \int x^{2n-2} e^{-x^2}\, dx$$

and hence

$$\int_{-\infty}^{\infty} x^{2n} e^{-x^2}\, dx$$
$$= \frac{2n-1}{2} \int_{-\infty}^{\infty} x^{2n-2} e^{-x^2}\, dx$$
$$= \frac{2n-1}{2} \cdot \frac{(2n-3) \cdots 3 \cdot 1}{2^{n-1}} \sqrt{\pi}$$
$$= \frac{(2n-1) \cdots 3 \cdot 1}{2^n} \sqrt{\pi}.$$

73. Substitute $u = 0.001x$.

$$\mu = \lim_{R \to \infty} \int_0^R 0.001 x e^{-0.001x}\, dx$$
$$= \lim_{R \to \infty} \int_0^R 1000 u e^{-u}\, du$$
$$= \lim_{R \to \infty} 1000 e^{-u}(-u - 1)\, \Big|_0^R$$
$$= \lim_{R \to \infty} 1000 \left[e^{-R}(-R - 1) + 1 \right]$$
$$= 1000$$

75.
$$\int_0^{30} \frac{1}{40} e^{-x/40}\, dx = -e^{-x/40}\, \Big|_0^{30}$$
$$= 1 - e^{-30/40} = 1 - e^{-3/4}$$
$$P(x > 30) = 1 - \left(1 - e^{-3/4}\right) = e^{-3/4}$$

$$\int_0^{35} \frac{1}{40} e^{-x/40}\, dx = -e^{-x/40}\, \Big|_0^{35}$$
$$= 1 - e^{-35/40} = 1 - e^{-7/8}$$
$$P(x > 35) = 1 - \text{ above } = e^{-7/8}$$

$$\int_0^{40} \frac{1}{40} e^{-x/40}\, dx = -e^{-x/40}\, \Big|_0^{40}$$
$$= 1 - e^{-40/40} = 1 - e^{-1}$$
$$P(x > 40) = 1 - \text{ above } = e^{-1}$$

$$\int_0^{45} \frac{1}{40} e^{-x/40}\, dx = -e^{-x/40}\, \Big|_0^{45}$$
$$= 1 - e^{-45/40} = 1 - e^{-9/8}$$
$$P(x > 45) = 1 - \text{ above } = e^{-9/8}$$

Hence,

$$P(x > 35 | x > 30) = \frac{P(x > 35)}{P(x > 30)}$$
$$= \frac{e^{-7/8}}{e^{-3/4}} = e^{-1/8} \approx 0.8825.$$

$$P(x > 40 | x > 35) = \frac{P(x > 40)}{P(x > 35)}$$
$$= \frac{e^{-1}}{e^{-7/8}} = e^{-1/8} \approx 0.8825$$

$$P(x > 45 | x > 40) = \frac{P(x > 45)}{P(x > 40)}$$
$$= \frac{e^{-9/8}}{e^{-1}} = e^{-1/8} \approx 0.8825$$

77. Since

$$\int_0^A c e^{-cx}\, dx = -e^{-cx}\, \Big|_0^A = 1 - e^{-cA},$$
$$P(x > m + n | x > n) = \frac{P(x > m + n)}{P(x > n)}$$
$$= \frac{1 - \int_0^{m+n} c e^{-cx}\, dx}{1 - \int_0^m c e^{-cx}\, dx\, dx} = \frac{e^{-c(m+n)}}{e^{-cn}}$$
$$= e^{-cm}.$$

79. (a) For $x \geq 0$,

$$F_1(x) = \int_{-\infty}^{x} f_1(t)\, dt$$

$$= \int_{0}^{x} f_1(t)\, dt$$

$$= \int_{0}^{x} 2e^{-2t}\, dt = -e^{-2t}\, \Big|_{0}^{x}$$

$$= 1 - e^{-2x}.$$

$$\Omega_1(r) = \frac{\int_{r}^{\infty}[1 - F_1(x)]\, dx}{\int_{-\infty}^{r} F_1(x)\, dx}$$

$$= \frac{\int_{r}^{\infty} e^{-2x}\, dx}{\int_{0}^{r}(1 - e^{-2x})\, dx}$$

$$= \frac{\frac{1}{2}e^{-2r}}{r + \frac{1}{2}e^{-2r} - \frac{1}{2}} = \frac{e^{-2r}}{2r + e^{-2r} - 1}$$

(b) For $0 \leq x \leq 1$,

$$F_2(x) = \int_{-\infty}^{x} f_2(t)\, dt$$

$$= \int_{0}^{x} f_2(t)\, dt$$

$$= \int_{0}^{x} 1\, dt = t\, \Big|_{0}^{x} = x.$$

$$\Omega_2(r) = \frac{\int_{r}^{\infty}[1 - F_2(x)]\, dx}{\int_{-\infty}^{r} F_2(x)\, dx}$$

$$= \frac{\int_{r}^{1}(1 - x)\, dx}{\int_{0}^{r} x\, dx}$$

$$= \frac{\frac{1}{2} - r + \frac{r^2}{2}}{\frac{r^2}{2}} = \frac{1 - 2r + r^2}{r^2}$$

(c)
$$\mu_1 = \int_{-\infty}^{\infty} x f_1(x)\, dx$$

$$= \int_{0}^{\infty} 2x e^{-2x}\, dx$$

$$= \lim_{R \to \infty} \int_{0}^{R} 2x e^{-2x}\, dx$$

$$= \lim_{R \to \infty} e^{-2x}(-x - 1/2)\, \Big|_{0}^{R}$$

$$= \lim_{R \to \infty} e^{-2R}(R + 1/2) + \frac{1}{2} = \frac{1}{2}$$

$$\mu_2 = \int_{-\infty}^{\infty} x f_2(x)\, dx$$

$$= \int_{0}^{1} x\, dx = \frac{x^2}{2}\, \Big|_{0}^{1} = \frac{1}{2}$$

$\mu_1 = \mu_2$ and when $r = 1/2$

$$\Omega_1(1/2) = \frac{e^{-2 \cdot 1/2}}{2 \cdot 1/2 + e^{-2 \cdot 1/2} - 1} = 1$$

$$\Omega_2(1/2) = \frac{1 - 2 \cdot 1/2 + (1/2)^2}{(1/2)^2} = 1$$

(d) The graph of $f_2(x)$ is more stable than that of $f_1(x)$.

$f_1(x) > f_2(x)$ for $0 < x < 0.34$

and $f_1(x) < f_2(x)$ for $x > 1$.

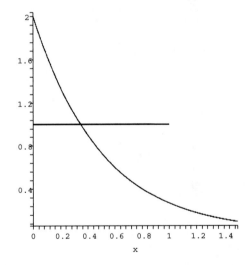

(e)
$$\Omega_1(r) = 1 - \frac{2r - 1}{e^{-2r} + 2r - 1}$$

$$\Omega_2 r = 1 - \frac{2r - 1}{r^2}$$

and $r^2 - (e^{-2r} + 2r - 1)$

$$= e^{-2r} + (r - 1)^2 > 0$$

This means

when $r < 1/2$, $\Omega_1(r) < \Omega_2(r)$

when $r > 1/2$, $\Omega_1(r) > \Omega_2(r)$.

In terms of this example, we see that the riskier investment is only disadvantageous when r small, and will be better when r large.

7.8 PROBABILITY

1. $f(x) = 4x^3 \geq 0$ for $0 \leq x \leq 1$ and

$$\int_0^1 4x^3\, dx = x^4 \Big|_0^1 = 1 - 0 = 1$$

3. $f(x) = x + 2x^3 \geq 0$ for $0 \leq x \leq 1$

and $\int_0^1 (x + 2x^3)\, dx = \dfrac{x^2}{2} + \dfrac{x^4}{2} \Big|_0^1 = 1$

5. $f(x) = \dfrac{1}{2} \sin x \geq 0$ over $[0, \pi]$

and $\int_0^\pi \dfrac{1}{2} \sin x\, dx = -\dfrac{1}{2} \cos x \Big|_0^\pi = 1.$

7. We solve for c:

$$1 = \int_0^1 cx^3\, dx = \dfrac{c}{4}$$

which gives $c = 4$.

9. We solve for c:

$$1 = \int_0^1 ce^{-4x}\, dx = -\dfrac{c}{4}(e^{-4} - 1)$$

which gives $c = \dfrac{4}{1 - e^{-4}}$.

11. We solve for c:

$$1 = \int_0^2 2ce^{-cx}\, dx = (-2e^{-cx}) \Big|_0^2 = 2 - 2e^{-2c}$$

which gives $e^{2c} = 2$ or $c = \dfrac{1}{2} \ln 2 \approx 0.347.$

13. $P(70 \leq x \leq 72)$

$$= \int_{70}^{72} \dfrac{0.4}{\sqrt{2\pi}} e^{-0.08(x-68)^2}\, dx$$

$$\approx 0.157$$

15. $P(84 \leq x \leq 120)$

$$= \int_{84}^{120} \dfrac{0.4}{\sqrt{2\pi}} e^{-0.08(x-68)^2}\, dx$$

$$\approx 7.77 \times 10^{-11}$$

17. $P\left(0 \leq x \leq \dfrac{1}{4}\right) = \int_0^{1/4} 6e^{-6x}\, dx$

$$= -e^{-6x} \Big|_0^{1/4}$$

$$= (-e^{-3/2} + 1) \approx 0.77687$$

19. $P(1 \leq x \leq 2) = \int_1^2 6e^{-6x}\, dx$

$$= -e^{-6x} \Big|_1^2$$

$$= (-e^{-12} + e^{-6}) \approx 0.00247$$

21. $P(0 \leq x \leq 1) = \int_0^1 4xe^{-2x}\, dx$

$$\approx 0.594$$

23. Mean:

$$\int_0^{10} x(4xe^{-2x})\, dx \approx 0.9999995$$

25. (a) Mean:

$$\mu = \int_a^b xf(x)\, dx = \int_0^1 3x^3\, dx$$

$$= \dfrac{3}{4} = 0.75$$

(b) Median, we must solve for m:

$$\dfrac{1}{2} = \int_a^m f(x)\, dx$$

$$= \int_0^m 3x^2\, dx = m^3$$

which gives $m = \dfrac{1}{\sqrt[3]{2}} \approx 0.7937.$

27. (a) Mean:

$$\mu = \int_a^b xf(x)\,dx$$

$$= \int_0^\pi \frac{1}{2}x \sin x\,dx$$

$$= \frac{1}{2}(\sin x - x \cos x)\Big|_0^\pi$$

$$= \frac{\pi}{2} \approx 1.57$$

(b) Median, we must solve for m:

$$\frac{1}{2} = \int_a^m f(x)\,dx$$

$$= \int_0^m \sin x\,dx = \frac{1}{2}(1 - \cos m)$$

which gives

$$m = \cos^{-1}(0) = \frac{\pi}{2} \approx 1.57.$$

29. (a) Mean:

$$\mu = \int_a^b xf(x)\,dx$$

$$= \frac{\ln 3}{2}\int_0^3 xe^{-kx}\,dx$$

$$\approx 1.23$$

(b) Median, we must solve for m:

$$\frac{1}{2} = \int_a^m f(x)\,dx$$

$$= \frac{\ln 3}{2}\int_0^m e^{-kx}\,dx$$

$$= \frac{\ln 3}{2}\left(-\frac{3}{\ln 3}e^{-kx}\right)\Big|_0^m$$

which gives

$$m = 3\frac{\ln 3 - \ln 2}{\ln 3} \approx 1.1072.$$

31. $f(x) = ce^{-4x}$, $[0, b]$, $b > 0$

$$1 = \int_0^b ce^{-4x}\,dx$$

$$= -\frac{c}{4}e^{-4x}\Big|_0^b = -\frac{c}{4}\left(e^{-4b} - 1\right)$$

$$c = \frac{4}{1 - e^{-4b}}$$

As $b \to \infty$, $c \to 4$.

33. $f(x) = ce^{-6x}$, $[0, b]$, $b > 0$

$$1 = \int_0^b ce^{-6x}\,dx$$

$$= \frac{-c}{6}e^{-6x}\Big|_0^b = -\frac{c}{6}\left(e^{-6b} - 1\right)$$

$$c = \frac{6}{1 - e^{-6b}}$$

As $b \to \infty$, $c \to 6$.

$$\mu = \int_0^b xce^{-6x}\,dx$$

$$= \frac{ce^{-6c}}{36}(-6x - 1)\Big|_0^b$$

$$= \frac{ce^{-6b}}{36}(-6b - 1) + \frac{c}{36}$$

As $b \to \infty$, $\mu \to \frac{c}{36} = \frac{1}{6}$.

35. (a) $P(h \le 3)$

$$= P(0) + P(1) + P(2) + P(3)$$

$$= \frac{1}{256} + \frac{8}{256} + \frac{28}{256} + \frac{56}{256}$$

$$= \frac{93}{256}$$

(b) $P(h > 4)$

$$= P(5) + P(6) + P(7) + P(8)$$

$$= \frac{56}{256} + \frac{28}{256} + \frac{8}{256} + \frac{1}{256}$$

$$= \frac{93}{256}$$

(c) $P(0) + P(8) = \dfrac{1}{256} + \dfrac{1}{256} = \dfrac{1}{128}$

(d) $P(1) + P(3) + P(5) + P(7)$

$$= \frac{8}{256} + \frac{56}{256} + \frac{56}{256} + \frac{8}{256}$$

$$= \frac{1}{2}$$

37. (a) $P(4/3) + P(4/2) + P(4/1)$
$+ P(4/0) = 0.1659 + 0.2073$
$+ 0.2073 + 0.1296 = 0.7101$

(b) $P(0/4) + P(1/4) + P(2/4)$
$+ P(3/4) = 0.0256 + 0.0615$
$+ 0.0922 + 0.1106 = 0.2899$

(c) $P(0/4) + P(4/0) = 0.0256$
$+ 0.1296 = 0.1552$

(d) $P(2/4) + P(3/4) + P(4/3)$
$+ P(4/2) = 0.0922 + 0.1106$
$+ 0.1659 + 0.2073 = 0.576$

39. About c:

$$1 = \int_{1}^{100} cp^{-2}\,dp = -cp^{-1}\Big|_{1}^{100}$$

$$= -c\left(\frac{1}{100} - 1\right).$$

Therefore, $c = \dfrac{100}{99}$.

The requested probability is

$$\int_{60}^{70} cp^{-2}\,dp = -cp^{-1}\Big|_{60}^{70}$$

$$= -\frac{100}{99}\left(\frac{1}{70} - \frac{1}{60}\right) \approx 0.0024.$$

41. $f(x) = \dfrac{0.4}{\sqrt{2\pi}}e^{-0.08(x-68)^2}$

$f'(x) = \dfrac{-0.064}{\sqrt{2\pi}}(x - 68)e^{-0.08(x-68)^2}$

$f''(x) = \dfrac{-0.064}{\sqrt{2\pi}}e^{-0.08(x-68)^2}$

$\cdot \left(1 - 0.16(x - 68)^2\right)$

The second derivative is zero when

$$x - 68 = \pm 1/\sqrt{0.16} = \pm 1/0.4 = \pm 5/2.$$

Thus the standard deviation is $\dfrac{5}{2}$.

43. $f'(p) = \dfrac{n!}{m!(n - m)!}\Big[mp^{m-1}(1 - p)^{n-m}$

$- (n - m)p^m(1 - p)^{n-m-1}\Big]$

$f'(p) = 0$ when $p = \dfrac{m}{n}$ and

$$f'(p) \begin{cases} > 0 & \text{if } p < m/n \\ < 0 & \text{if } p > m/n \end{cases}$$

Hence $f(p)$ is maximized when $p = \dfrac{m}{n}$.

In common senses, in order for an event to happen m times in n tries, the probability of the event itself should be about m/n.

45. Suppose the statement is not true. Then there must be a game before which the player's winning percentage is smaller than 75% and after which the player's winning percentage is greater than 75%. Then there are integers a and b (note that $a \geq m$, $b \geq n$, and $a - b = m - n$), such that

$$\frac{a}{b} < \frac{3}{4} \quad \text{and} \quad \frac{a+1}{b+1} > \frac{3}{4}.$$

Then

$4a < 3b$, and $4a + 4 > 3b + 3$

$3b + 4 > 4a + 4 > 3b + 3$.

But there is no integer between the two numbers $3b + 4$ and $3b + 3$, and thus such a situation will never happen. Thus there must be a game after which the player's winning percentage is exactly 75%.

47. The probability of a $2k$-goal game ending in a $k - k$ tie is

$$f(2k) = \frac{(2k) \cdots (k+1)}{(k) \cdots (1)} p^k (1-p)^k$$

$f(2k) < f(2k - 2)$ for general k.

$$\frac{f(2k)}{f(2k-2)} = 2\frac{2k-1}{k} p(1-p)$$

Here $\dfrac{2k-1}{k} = 2 - \dfrac{1}{k} < 2$.

On the other hand,

$$\left(p - \frac{1}{2}\right)^2 \geq 0, \quad p^2 - p + \frac{1}{4} \geq 0$$

$$p - p^2 \leq \frac{1}{4}, \quad p(1-p) \leq \frac{1}{4}$$

Now we get

$$\frac{f(2k)}{f(2k-2)} = 2\frac{2k-1}{k} p(1-p)$$

$$< 2 \cdot 2 \cdot \frac{1}{4} = 1.$$

So $f(2k) < f(2k - 2)$. In other words, the probability of a tie is decreasing as the number of goals increases.

49. To find the maximum, we take the derivative and set it equal to zero:

$$f'(x) = -2ax(bx - 1)(bx + 1)e^{-b^2x^2} = 0.$$

This gives critical numbers

$$x = 0, \pm\frac{1}{b}.$$

Since this will be a pdf for the interval $[0, 4m]$, we only have to check that there is a maximum at $\dfrac{1}{b}$. An easy check shows that $f'(x) > 0$ for $x < \dfrac{1}{b}$ and $f'(x) < 0$ for $x > \dfrac{1}{b}$. Therefore there is a maximum at $x = \dfrac{1}{b}$ (the most common speed).

To find a in terms of m, we want the total probability equal to 1. Since $m = \dfrac{1}{b}$, we also make the substitution $b = \dfrac{1}{m}$.

$$1 = \int_0^{4m} ax^2 e^{-x^2/m^2} \, dx$$

Solving for a gives

$$a = \left(\int_0^{4m} x^2 e^{-x^2/m^2} \, dx\right)^{-1}.$$

Note: this integral is not expressible in terms of elementary functions, so we will leave it like this. Using a CAS, one can find that

$$a \approx 2.2568 m^{-3}.$$

Chapter 7 Review Exercises

1. Substitute $u = \sqrt{x}$.

$$\int \frac{e^{\sqrt{x}}}{\sqrt{x}} \, dx = 2 \int e^u \, du$$

$$= 2e^u + c = 2e^{\sqrt{x}} + c$$

3. Use the table of integrals.

$$\int \frac{x^2}{\sqrt{1-x^2}} \, dx$$

$$= -\frac{1}{2}x\sqrt{1-x^2} + \frac{1}{2}\sin^{-1}x + c$$

5. Use integration by parts twice.

$$\int x^2 e^{-3x} \, dx$$

$$= -\frac{1}{3}x^2 e^{-3x} + \frac{2}{3} \int x e^{-3x} \, dx$$

$$= -\frac{1}{3}x^2 e^{-3x}$$

$$+ \frac{2}{3}\left(-\frac{1}{3}xe^{-3x} + \frac{1}{3}\int e^{-3x} \, dx\right)$$

$$= -\frac{1}{3}x^2 e^{-3x} - \frac{2}{9}xe^{-3x} - \frac{2}{27}e^{-3x} + c$$

7. Substitute $u = x^2$.

$$\int \frac{x}{1+x^4}\, dx = \frac{1}{2} \int \frac{du}{1+u^2}$$

$$= \frac{1}{2} \tan^{-1} u + c = \frac{1}{2} \tan^{-1} x^2 + c$$

9. $\displaystyle \int \frac{x^3}{4+x^4}\, dx = \frac{1}{4} \ln(4 + x^4) + c$

11. $\displaystyle \int e^{2\ln x}\, dx = \int x^2\, dx = \frac{x^3}{3} + c$

13. Use integration by parts.

$$\int_0^1 x \sin 3x\, dx$$

$$= -\frac{1}{3} x \cos 3x \Big|_0^1 + \frac{1}{3} \int_0^1 \cos 3x\, dx$$

$$= -\frac{1}{3} \cos 3 + \frac{1}{9} \sin 3x \Big|_0^1$$

$$= -\frac{1}{3} \cos 3 + \frac{1}{9} \sin 3$$

15. Use the table of integrals.

$$\int_0^{\pi/2} \sin^4 x\, dx$$

$$= -\frac{1}{4} \sin^3 x \cos x \Big|_0^{\pi/2}$$

$$+ \frac{3}{4} \left(\frac{x}{2} - \frac{1}{2} \sin x \cos x \right) \Big|_0^{\pi/2}$$

$$= \frac{3\pi}{16}$$

17. Use integration by parts.

$$\int_{-1}^1 x \sin \pi x\, dx$$

$$= -\frac{1}{\pi} x \cos \pi x \Big|_{-1}^1 + \frac{1}{\pi} \int_{-1}^1 \cos \pi x\, dx$$

$$= \frac{2}{\pi} + \frac{1}{\pi^2} \sin \pi x \Big|_{-1}^1 = \frac{2}{\pi}$$

19. Use integration by parts.

$$\int_1^2 x^3 \ln x\, dx = \frac{x^4}{4} \ln x \Big|_1^2 - \frac{1}{4} \int_1^2 x^3\, dx$$

$$= 4 \ln 2 - \frac{x^4}{16} \Big|_1^2 = 4 \ln 2 - \frac{15}{16}$$

21. Substitute $u = \sin x$.

$$\int \cos x \sin^2 x\, dx = \int u^2\, du$$

$$= \frac{u^3}{3} + c = \frac{\sin^3 x}{3} + c$$

23. Substitute $u = \sin x$.

$$\int \cos^3 x \sin^3 x\, dx = \int (1 - u^2) u^3\, du$$

$$= \frac{u^4}{4} - \frac{u^6}{6} + c = \frac{3 \sin^4 x - 2 \sin^6 x}{12} + c$$

25. Substitute $u = \tan x$.

$$\int \tan^2 x \sec^4 x\, dx = \int u^2 (1 + u^2)\, du$$

$$= \frac{u^3}{3} + \frac{u^5}{5} + c = \frac{5 \tan^3 x + 3 \tan^5 x}{15} + c$$

27. Substitute $u = \sin x$.

$$\int \sqrt{\sin x} \cos^3 x\, dx$$

$$= \int u^{1/2}(1 - u^2)\, du$$

$$= \frac{2}{3} u^{3/2} - \frac{2}{7} u^{7/2} + c$$

$$= \frac{2}{3} \sin^{3/2} x - \frac{2}{7} \sin^{7/2} x + c$$

29. Complete the square.

$$\int \frac{2}{8 + 4x + x^2}\, dx$$

$$= \int \frac{2}{(x+2)^2 + 2^2}\, dx$$

$$= \tan^{-1} \left(\frac{x+2}{2} \right) + c$$

31. Use the table of integrals.

$$\int \frac{2}{x^2\sqrt{4-x^2}}\, dx = -\frac{\sqrt{4-x^2}}{2x} + c$$

33. Substitute $u = 9 - x^2$.

$$\int \frac{x^3}{\sqrt{9-x^2}}\, dx = -\frac{1}{2}\int \frac{(9-u)}{u^{1/2}}\, du$$

$$= -\frac{9}{2}\int u^{-1/2}\, du + \frac{1}{2}\int u^{1/2}\, du$$

$$= -9u^{1/2} + \frac{1}{3}u^{3/2} + c$$

$$= -9(9-x^2)^{1/2} + \frac{1}{3}(9-x^2)^{3/2} + c$$

$$= \sqrt{9-x^2}\left(-9 + \frac{1}{3}(9-x^2)\right) + c$$

$$= \sqrt{9-x^2}\left(-6 - \frac{x^2}{3}\right) + c$$

35. Substitute $u = x^2 + 9$.

$$\int \frac{x^3}{\sqrt{x^2+9}}\, dx = \frac{1}{2}\int (u-9)u^{-1/2}\, du$$

$$= \frac{1}{3}u^{3/2} - 9u^{1/2} + c$$

$$= \frac{1}{3}(x^2+9)^{3/2} - 9(x^2+9)^{1/2} + c$$

$$= \sqrt{x^2+9}\left[\frac{1}{3}(x^2+9) - 9\right] + c$$

$$= \sqrt{x^2+9}\left(\frac{x^2}{3} - 6\right) + c$$

37. Use the method of PFD.

$$\int \frac{x+4}{x^2+3x+2}\, dx$$

$$= \int \left(\frac{3}{x+1} + \frac{-2}{x+2}\right) dx$$

$$= 3\ln|x+1| - 2\ln|x+2| + c$$

39. Use the method of PFD.

$$\int \frac{4x^2+6x-12}{x^3-4x}\, dx$$

$$= \int \left(\frac{3}{x} + \frac{-1}{x+2} + \frac{2}{x-2}\right) dx$$

$$= 3\ln|x| - \ln|x+2| + 2\ln|x-2| + c$$

41. Use the table of integrals.

$$\int e^x \cos 2x\, dx$$

$$= \frac{(\cos 2x + 2\sin 2x)e^x}{5} + c$$

43. Substitute $u = x^2 + 1$.

$$\int x\sqrt{x^2+1}\, dx = \frac{1}{2}\int u^{1/2}\, du$$

$$= \frac{1}{3}u^{3/2} + c = \frac{1}{3}(x^2+1)^{3/2} + c$$

45.
$$\frac{4}{x^2-3x-4} = \frac{A}{x+1} + \frac{B}{x-4}$$
$$4 = A(x-4) + B(x+1)$$
$$x = -1 : 4 = -5A, \text{ so } A = -\frac{4}{5}.$$
$$x = 4 : 4 = 5B, \text{ so } B = \frac{4}{5}.$$
$$\frac{4}{x^2-3x-4} = \frac{-4/5}{x+1} + \frac{4/5}{x-4}$$

47.
$$\frac{-6}{x^3+x^2-2x} = \frac{A}{x} + \frac{B}{x-1} + \frac{C}{x+2}$$
$$-6 = A(x-1)(x+2) + Bx(x+2)$$
$$+ Cx(x-1)$$
$$A = 3; B = -2; C = -1$$
$$\frac{-6}{x^3+x^2-2x} = \frac{3}{x} + \frac{-2}{x-1} + \frac{-1}{x+2}$$

49.
$$\frac{x-2}{x^2+4x+4} = \frac{A}{x+2} + \frac{B}{(x+2)^2}$$
$$x - 2 = A(x+2) + B$$
$$A = 1; B = -4$$
$$\frac{x-2}{x^2+4x+4} = \frac{1}{x+2} + \frac{-4}{(x+2)^2}$$

51. Substitute $u = e^{2x}$.

$$\int e^{3x}\sqrt{4 + e^{2x}}\,dx$$

$$= \int e^{2x}\sqrt{4e^{2x} + e^{4x}}\,dx$$

$$= \frac{1}{2}\int \sqrt{4u + u^2}\,du$$

$$= \frac{1}{2}\int \sqrt{(u+2)^2 - 4}\,du$$

$$= \frac{1}{4}(u+2)\sqrt{(u+2)^2 - 4}$$

$$\quad - \ln|(u+2) + \sqrt{(u+2)^2 - 4}| + c$$

$$= \frac{(e^{2x}+2)\sqrt{4e^{2x} + e^{4x}}}{4}$$

$$\quad - \ln\left|(e^{2x}+2) + \sqrt{4e^{2x} + e^{4x}}\right| + c$$

53. $\int \sec^4 x\,dx$

$$= \frac{1}{3}\sec^2 x \tan x + \frac{2}{3}\int \sec^2 x\,dx$$

$$= \frac{1}{3}\sec^2 x \tan x + \frac{2}{3}\tan x + c$$

55. Substitute $u = 3 - x$.

$$\int \frac{4}{x(3-x)^2}\,dx = -4\int \frac{1}{(3-u)u^2}\,du$$

$$= \frac{4}{9}\ln\left|\frac{3-u}{u}\right| + \frac{4}{3u} + c$$

$$= \frac{4}{9}\ln\left|\frac{x}{3-x}\right| + \frac{4}{3(3-x)} + c$$

57.

$$\int \frac{\sqrt{9+4x^2}}{x^2}\,dx = \int \frac{2\sqrt{\frac{9}{4}+x^2}}{x^2}\,dx$$

$$= 2\left(\frac{-\sqrt{\frac{9}{4}+x^2}}{x} + \ln\left|x + \sqrt{9/4+x^2}\right|\right) + c$$

$$= -\frac{\sqrt{9+4x^2}}{x} + 2\ln\left|x + \sqrt{\frac{9}{4}+x^2}\right| + c$$

59. $\int \frac{\sqrt{4-x^2}}{x}\,dx$

$$= \sqrt{4-x^2} - 2\ln\left|\frac{2+\sqrt{4-x^2}}{x}\right| + c$$

61. Substitute $u = x^2 - 1$.

$$\int_0^1 \frac{x}{x^2-1}\,dx = \int_{-1}^0 \frac{du}{2u}$$

$$= \lim_{R\to 0^-}\int_{-1}^R \frac{du}{2u} = \lim_{R\to 0^-}\frac{1}{2}\ln|u|\,\Big|_{-1}^0$$

This limit does not exist, so the integral diverges.

63. $\int_1^\infty \frac{3}{x^2}\,dx = \lim_{R\to\infty}\int_1^R \frac{3}{x^2}\,dx$

$$= \lim_{R\to\infty} -\frac{3}{x}\Big|_1^R = \lim_{R\to\infty} -\frac{3}{R} + 3 = 3$$

65. $\int_0^\infty \frac{4}{4+x^2}\,dx = \lim_{R\to\infty}\int_0^R \frac{4}{4+x^2}\,dx$

$$= \lim_{R\to\infty} 2\tan^{-1}\frac{x}{2}\Big|_0^R$$

$$= \lim_{R\to\infty} 2\tan^{-1} R = \pi$$

67. $\int_0^2 \frac{3}{x^2}\,dx = \lim_{R\to 0^+}\int_R^2 \frac{3}{x^2}\,dx$

$$= \lim_{R\to 0^+} -\frac{3}{x}\Big|_R^2 = \infty$$

So the original integral diverges.

69. $\lim_{x\to 1}\frac{x^3-1}{x^2-1}$ is type $\frac{0}{0}$;

L'Hôpital's Rule gives

$$\lim_{x\to 1}\frac{3x^2}{2x} = \frac{3}{2}.$$

71. $\lim_{x\to\infty}\frac{e^{2x}}{x^4+2}$ is type $\frac{\infty}{\infty}$;

applying L'Hôpital's Rule four times gives

$$\lim_{x \to \infty} \frac{2e^{2x}}{4x^3}$$

$$= \lim_{x \to \infty} \frac{4e^{2x}}{12x^2} = \lim_{x \to \infty} \frac{8e^{2x}}{24x}$$

$$= \lim_{x \to \infty} \frac{16e^{2x}}{24} = \infty.$$

73.

$$L = \lim_{x \to 2^+} \left| \frac{x+1}{x-2} \right|^{\sqrt{x^2-4}}$$

$$\ln L = \lim_{x \to 2^+} \left(\sqrt{x^2 - 4} \, \ln \left| \frac{x+1}{x-2} \right| \right)$$

$$= \lim_{x \to 2^+} \left(\frac{\ln \left| \frac{x+1}{x-2} \right|}{(x^2 - 4)^{-1/2}} \right)$$

$$= \lim_{x \to 2^+} \left(\frac{\left| \frac{x-2}{x+1} \right| \frac{-3}{(x-2)^2}}{-x(x^2-4)^{-3/2}} \right)$$

$$= \lim_{x \to 2^+} \left(\frac{3(x^2-4)^{3/2}}{x(x+1)(x-2)} \right)$$

$$= \lim_{x \to 2^+} \left(\frac{3(x-2)^{1/2}(x+2)^{3/2}}{x(x+1)} \right)$$

$$\ln L = 0$$
$$L = 1$$

75.

$$\lim_{x \to 0^+} (\tan x \ln x) = \lim_{x \to 0^+} \left(\frac{\ln x}{\cot x} \right)$$

$$= \lim_{x \to 0^+} \left(\frac{1/x}{-\csc^2 x} \right)$$

$$= -\lim_{x \to 0^+} \left(\frac{\sin^2 x}{x} \right)$$

$$= -\lim_{x \to 0^+} \left(\frac{\sin x}{x} \sin x \right)$$

$$= (-1)(0) = 0$$

77. $f(x) = x + 2x^3$ on $[0, 1]$
$f(x) \geq 0$ for $0 \leq x \leq 1$ and

$$\int_0^1 \left(x + 2x^3 \right) dx = \left(\frac{x^2}{2} + \frac{x^4}{2} \right) \Big|_0^1 = 1$$

79.

$$1 = \int_1^2 \frac{c}{x^2} \, dx = \frac{-c}{x} \Big|_1^2 = \frac{-c}{2} + c = \frac{c}{2}$$

Therefore, $c = 2$.

81. (a)

$$P(x < 0.5) = \int_0^{0.5} 4e^{-4x} \, dx$$

$$= -e^{-4x} \Big|_0^{0.5} = 1 - e^{-2} \approx 0.865$$

(b)

$$P(0.5 \leq x \leq 1) = \int_{0.5}^1 4e^{-4x} \, dx$$

$$= -e^{-4x} \Big|_{0.5}^1 = -e^{-4} + e^{-2} \approx 0.117$$

83. (a)

$$\mu = \int_0^1 x \left(x + 2x^3 \right) dx$$

$$= \frac{x^3}{3} + \frac{2x^5}{5} \Big|_0^1 = \frac{11}{15} \approx 0.7333$$

(b) $\dfrac{1}{2} = \displaystyle\int_0^c \left(x + 2x^3 \right) dx$

$$= \frac{x^2}{2} + \frac{x^4}{2} \Big|_0^c = \frac{c^2}{2} + \frac{c^4}{2}$$

Therefore $c^2 + c^4 = 1$,

$$c = \sqrt{\frac{-1 + \sqrt{5}}{2}} \approx 0.786.$$

85. If $c(t) = R$, then the total amount of dye is

$$\int_0^T c(t) \, dt = \int_0^T R \, dt = RT.$$

If $c(t) = 3te^{2Tt}$, then we can use integration by parts to get

$$\int_0^T 3te^{2Tt}\, dt$$

$$= \frac{3t}{2T}e^{2Tt}\Big|_0^T - \int_0^T \frac{3}{2T}e^{2Tt}\, dt$$

$$= \frac{3}{2}e^{2T^2} - \frac{3}{4T^2}e^{2Tt}\Big|_0^T$$

$$= \frac{3}{2}e^{2T^2} - \frac{3}{4T^2}e^{2T^2} + \frac{3}{4T^2}.$$

Since $R = c(T) = 3Te^{2T^2}$, the cardiac output is

$$\frac{RT}{\int_0^T c(t)\, dt} = \frac{RT}{\frac{3}{2}e^{2T^2} - \frac{3}{4T^2}e^{2T^2} + \frac{3}{4T^2}}$$

$$= \frac{4RT^3}{6T^4 e^{2T^2} - 3e^{2T^2} + 3}.$$

87.

$$f_{n,\text{ave}} = \frac{1}{e^n} \int_0^{e^n} \ln x\, dx$$

$$= \frac{1}{e^n} \lim_{R \to 0^+} \int_R^{e^n} \ln x\, dx$$

$$= \frac{1}{e^n} \lim_{R \to 0^+} (x \ln x - x)\Big|_R^{e^n}$$

$$= \frac{1}{e^n} \lim_{R \to 0^+} \left(ne^n - e^n - R\ln R + R\right)$$

$$= n - 1$$

89.

$$R(t) = P(x > t) = 1 - \int_0^t ce^{-cx}\, dx$$

$$= 1 - (-e^{-cx})\Big|_0^t = 1 - (1 - e^{-ct}) = e^{-ct}$$

Hence,

$$\frac{f(t)}{R(t)} = \frac{ce^{-ct}}{e^{-ct}} = c.$$

91. We use a CAS to see that

$$\int_{90}^{100} \frac{1}{\sqrt{450\pi}} e^{-(x-100)^2/450}\, dx$$
$$\approx 24.75\%.$$

Using a CAS, we find that

$$P(x \le 134) \approx 0.988$$

and

$$P(x \le 135) \approx 0.990.$$

Somebody being called a genius needs to have an IQ score of at least 135.

First-Order Differential Equations

8.1 MODELING WITH DIFFERENTIAL EQUATIONS

1. Exponential growth with $k = 4$ so we can use Equation (1.4) to arrive at the general solution of $y = Ae^{4t}$. The initial condition gives $2 = A$ so the solution is $y = 2e^{4t}$.

3. Exponential growth with $k = -3$ so we can use Equation (1.4) to arrive at the general solution of $y = Ae^{-3t}$. The initial condition gives $5 = A$ so the solution is $y = 5e^{-3t}$.

5. Exponential growth with $k = 2$ so we can use Equation (1.4) to arrive at the general solution of $y = Ae^{2t}$. The initial condition gives $2 = Ae^2$, $A = \dfrac{2}{e^2}$ so the solution is $y = \dfrac{2}{e^2}e^{2t}$ or $y = 2e^{2(t-1)}$.

7. Integrating gives the general solution $y = -3t + c$. The initial condition gives $3 = c$ and so the solution is $y = -3t + 3$.

9. The equation for population must be $y(t) = 100e^{kt}$.

We know that in 4 hours, the population doubles, so

$$200 = y(4) = 100e^{k4}.$$

Solving for k gives $k = (\ln 2)/4$ and $y(t) = 100e^{t(\ln 2)/4}$.

To determine when the population reaches 6000, we solve $y(t) = 6000$ or

$$6000 = 100e^{t(\ln 2)/4}.$$

Solving gives

$$t = \frac{4 \ln 60}{\ln 2} \approx 23.628 \text{ hours.}$$

11. The equation for population must be $y(t) = 400e^{kt}$.

We know that in 1 hour, the population is 800, so $800 = y(1) = 400e^k$. Solving for k gives $k = \ln 2$.

$$y(t) = 400e^{t \ln 2}$$

After 10 hours, the population is

$$y(10) = 400e^{10 \ln 2} = 409{,}600 \text{ cells.}$$

13. The equation for population is $y = Ae^{0.44t}$

The initial population is A, so we want to find when the population is $2A$. So we solve the equation

$$2A = Ae^{0.44t}$$

which gives solution

$$t = \frac{\ln 2}{0.44} \approx 1.5753 \text{ hours.}$$

15. With t measured in minutes, and $y = Ae^{kt} = 10^8 e^{kt}$ on the time interval $(0, T)$ (during which no treatment is given), the condition on T is that 10% of the population at time T (surviving after the treatment) will be the same as the initial population.

In other words, $10^8 = (0.1)10^8 e^{kT}$. This gives $e^{kT} = 10$ and $T = \ln(10)/k$.

Since the doubling time is 20 minutes, $2 \cdot 10^8 = 10^8 e^{k \cdot 20}$, $2 = e^{20k}$, and $k = \ln(2)/20$.

$$T = \frac{\ln(10)}{k} = \frac{\ln(10)}{\ln(2)/20}$$

$$= \frac{20 \ln(10)}{\ln(2)} \approx 66.44 \text{ minutes}.$$

17. Given $y(t) = Ae^{rt}$, the doubling time t_d obeys $2A = Ae^{rt_d}$, $2 = e^{rt_d}$. $rt_d = \ln 2$, $t_d = \dfrac{\ln 2}{r}$ as desired.

19. We apply the formula of exercise 18 to get half-life equal to

$$\frac{\ln 2}{1.3863} \approx 0.500 \text{ day}.$$

21. Using the formula in exercise 18, we have $3 = -(\ln 2)/r$ and therefore $r = -(\ln 2)/3$.

Thus the formula for amount of substance is

$$y(t) = Ae^{-t(\ln 2)/3}.$$

The initial condition gives $A = 0.4$ and so

$$y(t) = 0.4e^{-t(\ln 2)/3}.$$

Solving for t in the equation

$$0.01 = 0.4e^{-t(\ln 2)/3} \text{ gives}$$

$$t = \frac{3 \ln(40)}{\ln(2)} \approx 15.97 \text{ hours}.$$

Thus the amount will drop below $0.01\,\mathrm{mg}$ after 15.97 hours.

23. Using the formula in exercise 18, we find the decay constant is

$$r = -\frac{\ln 2}{28}.$$

The formula for the amount of substance is

$$y(t) = Ae^{rt}.$$

After 50 years,

$$y(50) = Ae^{50r} \approx 0.29A.$$

Thus, this is about 29% of the original amount of strontium-90.

25. The half-life is 5730 years, so

$$r = -\frac{\ln 2}{5730}.$$

Solving for t in

$$y(t) = 0.20A = Ae^{rt} \text{ gives}$$

$$t = \frac{5730 \ln(5)}{\ln(2)} \approx 13{,}305 \text{ years}.$$

27. Newton's Law of Cooling gives

$$y(t) = Ae^{kt} + T_a \text{ with } T_a = 70.$$

We have $y(0) = 200$ so

$$200 = A + 70 \text{ and } A = 130.$$

We have $y(1) = 180$ so

$$180 = y(1) = 130e^k + 70 \text{ and}$$

$$k = \ln\left(\frac{110}{130}\right).$$

The temperature will be 120 when
$120 = y(t) = 130e^{\ln(110/130)t} + 70$ and

$$t = \frac{\ln(5/13)}{\ln(11/13)} \approx 5.720 \text{ minutes}.$$

29. Using Newton's Law of Cooling

$$y = Ae^{kt} + T_a$$

with $T_a = 70$, $y(0) = 50$, we get

$$50 = Ae^0 + 70, \ A = -20$$

so that $y(t) = -20e^{kt} + 70$.

Since after two minutes the temperature is 56 degrees, $56 = -20e^{k2} + 70$.

$$e^{2k} = \frac{14}{20} = 0.7$$

$$2k = \ln 0.7, \ k = \frac{1}{2}\ln 0.7$$

Therefore, $y(t) = -20e^{(\ln 0.7)t/2} + 70$.

31. Using Newton's Law of Cooling with ambient temperature 70 degrees, initial temperature 60 degrees, and with time t (in minutes) elapsed since 10:07, we have

$y(t) = Ae^{kt} + 70,$

$60 = Ae^{0k} + 70 = A + 70,$

$A = -10,$ and

$y(t) = -10e^{kt} + 70$ (for the martini).

Two minutes later, its temperature is 61 degrees. Hence

$61 = -10e^{k2} + 70, \ e^{2k} = \dfrac{9}{10}$

$2k = \ln \dfrac{9}{10}, \ k = \dfrac{1}{2} \ln \dfrac{9}{10} = \dfrac{1}{2} \ln 0.9$

Therefore, $y(t) = -10e^{\left(\frac{1}{2} \ln 0.9\right)t} + 70.$

The temperature is 40 degrees at elapsed time t only if

$40 = -10e^{\left(\frac{1}{2} \ln 0.9\right)t} + 70$

$t = \dfrac{2 \ln 3}{\ln 0.9} \approx -20.854$

or about 21 minutes *before* 10:07 p.m. The time was 9:46 p.m.

33. With t the time elapsed since serving, with the ambient temperature 68 degrees, and if the temperature is 160 degrees when $t = 20$, then

$y(t) = Ae^{kt} + T_a,$

$160 = Ae^{k\cdot20} + 68Ae^{20k} = 92.$

After 22 minutes the temperature is 158 degrees.

$158 = Ae^{22k} + 68, \ Ae^{22k} = 90$

$e^{2k} = \dfrac{Ae^{22k}}{Ae^{20k}} = \dfrac{90}{92}, \ k = \dfrac{1}{2} \ln \dfrac{90}{92}$

Therefore,

$y(t) = Ae^{\frac{1}{2}\left(\ln \frac{90}{92}\right)t} + 68.$

Using the first set of numbers:

$Ae^{20 \cdot \frac{1}{2} \ln \frac{90}{92}} = 92$

$A = \dfrac{92}{e^{10 \ln \frac{90}{92}}} \approx 114.615$

$y(t) = 114.615e^{\frac{1}{2}\left(\ln \frac{90}{92}\right)t} + 68$

The serving temperature was

$y(0) = 114.615e^0 + 68 = 182.615$ degrees.

35. Annual:

$A = 1000(1 + 0.08)^1 = \$1080.00$

Monthly:

$A = 1000 \left(1 + \dfrac{0.08}{12}\right)^{12} \approx \$1083.00$

Daily:

$A = 1000 \left(1 + \dfrac{0.08}{365}\right)^{365} \approx \$1083.28$

Continuous:

$A = 1000e^{(0.8)1} \approx \$1083.29$

37. Person A:

$A = 10{,}000e^{0.12 \cdot 20} = \$110{,}231.76$

Person B:

$B = 20{,}000e^{0.12 \cdot 10} = \$66{,}402.34$

39. Let t be the number of years after 1985. Then, assuming continuous compounding at rate r,

$9800 = 34e^{r \cdot 10}, \ e^{10r} = \dfrac{9800}{34}.$

$r = \dfrac{1}{10} \ln \left(\dfrac{9800}{34}\right) \approx 0.566378$

Therefore,

$A = 34e^{\frac{1}{10} \ln \left(\frac{9800}{34}\right)t} = 34 \left(\dfrac{9800}{34}\right)^{t/10}.$

In 2005, $t = 20$ and

$A = 34 \left(\dfrac{9800}{34}\right)^2 = \$2{,}824{,}705.88.$

41. The problem with comparing tax rates for the income bracket $[16K, 20K]$ over a thirteen year time interval, is that due to inflation, the persons in this income bracket in 1988 have less purchasing power than those in the same bracket in 1975, and a lower tax rate may or may not compensate. To quantify and illustrate, assume a 5.5% annual inflation rate. This would translate into a loss of purchasing power amounting to

$$1/(1.055)^{13} \approx 1/(2.006) \approx 1/2$$

which is essentially to say that in terms of comparable purchasing power, the income bracket $[16K, 20K]$ in 1988 corresponds to an income bracket of $[8K, 10K]$ in 1975. One should then go back and look at the tax rate for the latter bracket in 1975. Only if that tax rate exceeds the 1988 rate (15%) for the bracket $[16K, 20K]$ should one consider that taxes have genuinely gone down.

43. $T_1 = 30,000 \cdot 0.15 + (40,000 - 30,000) \cdot 0.28 = \$7300$

$T_2 = 30,000 \cdot 0.15 + (42,000 - 30,000) \cdot 0.28 = \$7860$

$T_1 + 0.05T_1 = \$7665$

The tax T_2 on the new salary is greater than the adjusted tax $(1.05T_1)$ on the old salary.

45. With a constant depreciation rate of 10%, the value of the $40,000 item after ten years would be

$40,000(e^{-(0.1)10}) = 40,000e^{-1}$

$\approx \$14,715.18$

and after twenty years

$40,000(e^{-(0.1)20}) = 40,000e^{-2}$

$\approx \$5413.41.$

By the straight line method, assuming a value of zero after 20 years, the value would be $20,000 after 10 years.

47. Over an interval of elapsed time Δt, a drop from 1000 to 500 would be represented by a slope of $\dfrac{500}{\Delta t}$ while a drop from 10 to 5 would be represented by a slope of $\dfrac{5}{\Delta t}$.

A drop from ln 1000 to ln 500 would result in a slope of

$$\frac{\ln 1000 - \ln 500}{\Delta t} = \frac{\ln \frac{1000}{500}}{\Delta t} = \frac{\ln 2}{\Delta t}$$

while a drop from ln 10 to ln 5 would result in a slope of

$$\frac{\ln 10 - \ln 5}{\Delta t} = \frac{\ln \frac{10}{5}}{\Delta t} = \frac{\ln 2}{\Delta t}.$$

So the slopes of the logarithmic drops are the same. If a population were changing at a constant percentage rate, the graph of population versus time would appear exponential while the graph of the logarithm of population versus time would appear linear.

49. Fitting a line to the first two data points on the plot of time vs. the natural log of the population $(y = \ln(P(x)))$ produces the linear function

$y = 1.468x + 0.182$

which is equivalent to fitting the original data to the exponential function

$P(x) = e^{1.468x + 0.182}$ or

$P(x) = 1.200e^{1.468x}.$

51. As in exercise 49, we let x denote time and $y = \ln P$.

We pick the second and fourth data points to fit a line to (any two data points are fine to use but will give slightly different answers). In this case, the points are

$(1, \ln 15)$ $(3, \ln 33)$

The equation of the line connecting these two points is

$\ln P = y = 0.394x + 2.314.$

Exponentiating this equation gives

$P = e^y = e^{0.394x + 2.314} = 10.115e^{0.394x}.$

53. As in exercise 49, we let x denote time (with $x = 0$ corresponding to the year 1960) and let $y = \ln P$.

Looking at the graph of the modified data, we decide to use the first and last data points. In this case, the points are

$(0, \ln 7.5)$ $(30, \ln 1.6)$

The equation of the line connecting these two points is

$\ln P = y = -0.0515x + 2.0149.$

Exponentiating this equation gives

$$P = e^y = e^{-0.0515x + 2.0149}$$
$$= 7.5e^{-0.0515x}.$$

55. With known conclusion $y = Ae^{-rt}$, $A = 150$, $t = 24$, and $r = \ln(2)/t_h$, we find that with $t_h = 31$ we get

$$y = 150(1/2)^{(24/31)} = 87.7$$

and with $t_h = 46$ we get

$$y = 150(1/2)^{(24/46)} = 104.5.$$

The difference is about 17 mg, at 11% not a dramatically large percentage of the original dose of 150 mg. If one had expected the two numbers to be proportional to the half-lives, one would have expected the difference to be about 48% ($46/31 = 1.48$) and would definitely consider the 11% to be far less than anticipated.

57. In this case, with $t_h = 4$, $A = 1$, $y = Ae^{-rt}$, and $r = \ln(2)/4$, one finds $y = (1/2)^{(t/4)}$.

The curve is a typical exponential, declining from a value of 1 at $t = 0$ to

$$1/2^6 = 1/64 \approx 0.016 \text{ at } t = 24.$$

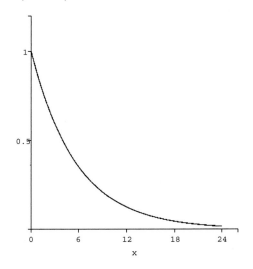

59. With r the rate of continuous compounding, the value of an initial amount X after t years is Xe^{rt}. If the goal is P, then the relation is $P = Xe^{rt}$ or $X = Pe^{-rt}$.

With $r = 0.08$, $t = 10$, $P = 10,000$, we find

$$X = 10,000e^{-0.8} = \$4493.29.$$

61. Here $r = 0.05$ and $T = 3$.

Person A:

$$\int_0^3 60,000e^{-0.05t}\, dt$$

$$= -1,200,000e^{-0.05t}\,\Big|_0^3$$

$$= \$167,150.43$$

Person B:

$$\int_0^3 (60,000 + 3000t)e^{-0.05t}\, dt$$

$$= (-1,200,000e^{-0.05t} - 60,000te^{-0.05t}$$

$$\quad - 1,200,000e^{-0.05t})\,\Big|_0^3$$

$$= \$179,373.42$$

Person C:

$$\int_0^3 60,000e^{0.05t} \cdot e^{-0.05t}\, dt$$

$$= \int_0^3 60,000\, dt = \$180,000$$

63. The comparison is to be made between three years of accumulation of $1,000,000 versus the accumulation of four annual payments of $280,000 at times 0, 1, 2, 3, then the respective figures are

$$1,000,000(1.08)^3 = 1,259,712 \text{ versus}$$

$$280,000(1.08^3 + 1.08^2 + 1.08 + 1)$$
$$= 1,261,711.$$

One should take the annuity.

8.2 SEPARABLE DIFFERENTIAL EQUATIONS

1. Separable.

$$\frac{y'}{\cos y} = 3x + 1$$

3. Not separable.

5. Separable.

$$y' = y(x^2 + \cos x)$$

$$\frac{y'}{y} = x^2 + \cos x$$

7. Not separable.

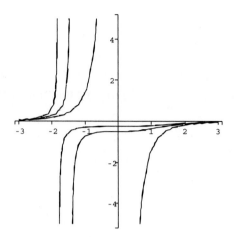

9. $\dfrac{1}{y}y' = x^2 + 1$

$$\int \frac{1}{y}\,dy = \int (x^2 + 1)\,dx$$

$$\ln|y| = \frac{x^3}{3} + x + c$$

$$y = e^{x^3/3 + x + c}$$

$$y = Ae^{x^3/3 + x}$$

13. $yy' = \dfrac{6x^2}{1+x^3}$

$$\int y\,dy = \int \frac{6x^2}{1+x^3}\,dx$$

$$\frac{1}{2}y^2 = 2\ln|1+x^3| + c$$

$$y = \pm\sqrt{4\ln|1+x^3| + c}$$

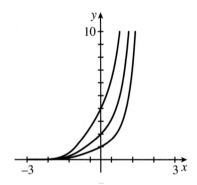

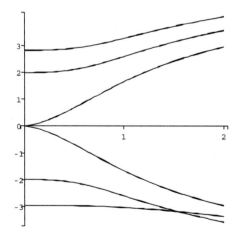

11. $\dfrac{1}{y^2}y' = 2x^2$

$$\int \frac{1}{y^2}\,dy = \int 2x^2\,dx$$

$$-\frac{1}{y} = \frac{2x^3}{3} + c$$

$$y = -\frac{1}{2x^3/3 + c}$$

15. $ye^{-y}y' = 2xe^{-x}$

$$\int ye^{-y}\,dy = \int 2xe^{-x}dx$$

Integrating by parts,

$$\int xe^{-x}\,dx = -xe^{-x} - e^{-x} + c.$$

Therefore,

$$-ye^{-y} - e^{-y} = -2xe^{-x} - 2e^{-x} + c.$$

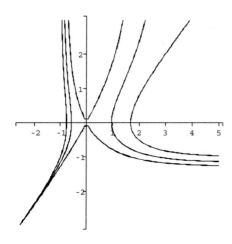

17. $\dfrac{1}{y^2 - y} y' = 1$

$$\int \frac{1}{y^2 - y}\, dy = \int 1\, dx$$

Integrating by partial fraction decomposition,

$$\int \frac{1}{y^2 - y}\, dy = \int \left(\frac{1}{y - 1} - \frac{1}{y} \right) dy$$
$$= \ln|y - 1| - \ln|y| + c.$$

Therefore,

$$\ln\left(\frac{y - 1}{y} \right) = x + c$$

$$\frac{y - 1}{y} = ke^x$$

$$y = \frac{1}{1 - ke^x}$$

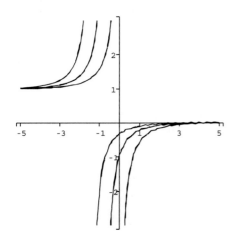

19. $\dfrac{1}{y} y' = \dfrac{x}{1 + x^2}$

$$\int \frac{1}{y}\, dy = \int \frac{x}{1 + x^2}\, dx$$

$$\ln|y| = \frac{1}{2} \ln|1 + x^2| + c$$

$$y = e^{\frac{1}{2} \ln|1 + x^2| + c} = k\sqrt{1 + x^2}$$

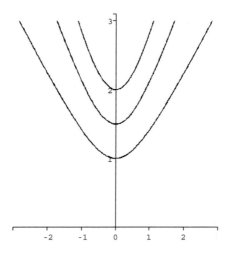

21. $\sec^2 y\, y' = \dfrac{1}{4x - 3}$

$$\int \sec^2 y\, dy = \int \frac{1}{4x - 3}\, dx$$

$$\tan y = \frac{1}{4} \ln|4x - 3| + c$$

$$y = \tan^{-1}\left[\frac{1}{4} \ln|4x - 3| + c \right]$$

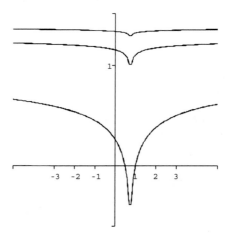

23. $\dfrac{y'}{y} = 3(x+1)^2$

$\ln y = (x+1)^3 + c$

$y = ke^{(x+1)^3}$

Using the initial condition,

$1 = ke, \; k = \dfrac{1}{e}$

$y = \dfrac{1}{e}e^{(x+1)^3}$

25. $yy' = 4x^2$

$\dfrac{y^2}{2} = \dfrac{4x^3}{3} + c$

Using the initial condition,

$\dfrac{2^2}{2} = c = 2$

$\dfrac{y^2}{2} = \dfrac{4x^3}{3} + 2$

27. $\dfrac{y'}{y} = \dfrac{4}{x+3}$

$\ln|y| = 4\ln|x+3| + c$

$|y| = k(|x+3|)^4$

Using the initial condition,

$|1| = k(1)^4$

$k = 1$

$|y| = (|x+3|)^4 = (x+3)^4$

29. $\cos y \, y' = 4x$

$\sin y = 2x^2 + c$

Using the initial condition,

$0 = \sin(0) = y(0) = 0 + c = c$

$\sin y = 2x^2$

31. For this problem we have $M = 2$ and $k = 3$. Using these and the initial condition, we solve for A.

$1 = \dfrac{2Ae^{3(2)(0)}}{1 + Ae^{3(2)(0)}} = \dfrac{2A}{1+A}$

$A = 1$

$y = \dfrac{2e^{6t}}{1 + e^{6t}}$

33. For this problem we have $M = 5$ and $k = 2$. Using these and the initial condition, we solve for A.

$4 = \dfrac{5Ae^{10(0)}}{1 + Ae^{10(0)}}$

$= \dfrac{5A}{1+A}$

$A = 4$

$y = \dfrac{20e^{10t}}{1 + 4e^{10t}}$

35. For this problem we have $M = 1$ and $k = 1$. Using these and the initial condition, we solve for A.

$\dfrac{3}{4} = \dfrac{Ae^{(0)}}{1 + Ae^{(0)}} = \dfrac{A}{1+A}$

$A = 3$

$y = \dfrac{3e^t}{1 + 3e^t}$

37. Substituting $r = Mk$ in

$y' = ry\left(1 - \dfrac{y}{M}\right)$

we get

$y' = Mky\left(1 - \dfrac{y}{M}\right) = ky(M - y)$

$\dfrac{1}{y(M-y)}y' = k$

Adapting the solution

$y = \dfrac{MAe^{Mkt}}{1 + Ae^{Mkt}}$ in (2.6) with $r = Mk$,

we find $y = \dfrac{MAe^{rt}}{1 + Ae^{rt}}$.

In this case with $r = 0.71$, $M = 8 \times 10^7$, and $y(0) = 2 \times 10^7$, we find

$$2 \times 10^7 = y(0) = \frac{8 \times 10^7 A}{1 + A}.$$

Therefore $\dfrac{A}{1+A} = \dfrac{2}{8} = \dfrac{1}{4}$, $A = 1/3$, and after routine simplification we find

$$y(t) = \frac{(8 \times 10^7)e^{0.71t}}{3 + e^{0.71t}}.$$

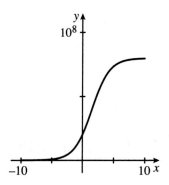

39. Keeping the absolute values, we have

$$\left| \frac{y}{M - y} \right| = Ae^{Mkt} \text{ with } A > 0.$$

If $y > M$, the ratio is negative, and the resolution is $\dfrac{y}{M - y} = -Ae^{Mkt}$. This further resolves as $y = -MAe^{Mkt} + yAe^{Mkt}$, which eventually becomes

$$y = \frac{MAe^{Mkt}}{Ae^{Mkt} - 1} = \frac{MAe^{rt}}{Ae^{rt} - 1}.$$

41. Starting from

$$y = \sqrt[3]{x^3 + \frac{21}{2}x^2 + 9x + 3c}$$

with $y(0) = 0$, we have $c = 0$.
Therefore,

$$y = \sqrt[3]{x^3 + \frac{21}{2}x^2 + 9x}.$$

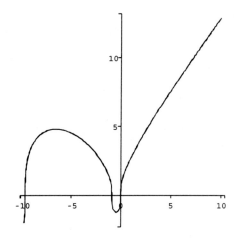

43. Given $y' = \dfrac{x^2 + 7x + 3}{y^2}$, note that $y'(x)$ does not exist for a given x if $y(x) = 0$.
We see that $y(x) = 0$ if

$$-3c = x^3 + \left(\frac{21}{2} \right) x^2 + 9x.$$

The cubic polynomial on the right, call it $h(x)$, has its derivative given by

$$h'(x) = 3x^2 + 21x + 9 = 3(x^2 + 7x + 3),$$

and the roots of $h'(x)$ are

$$x_1 = \frac{-7 - \sqrt{37}}{2} \approx -6.5414$$

$$x_2 = \frac{-7 + \sqrt{37}}{2} \approx -0.4586$$

The effect is that $h(x)$ has a relative maximum at x_1 and a relative minimum at x_2, and so the equation $-3c = h(x)$ has three solutions when $-3c$ lies between the relative minimum and the relative maximum, i.e., if $h(x_2) < -3c < h(x_1)$, or when

$$\frac{-h(x_1)}{3} < c < \frac{-h(x_2)}{3}.$$

Therefore,

$$c_1 = -\left(\frac{217 + 37\sqrt{37}}{12} \right) \approx -36.84$$

$$c_2 = \frac{-217 + 37\sqrt{37}}{12} \approx 0.67185$$

45. With $h(x)$ as in exercise 43, note that when $c = c_2$, $3c_2 = -h(x_2)$ and

$$y = \sqrt[3]{x^3 + \frac{21}{2}x^2 + 9x + 3c_2}$$

$$= \sqrt[3]{h(x) - h(x_2)}$$

$$= \sqrt[3]{(x - x_2)\left[x^2 + \left(x_2 + \frac{21}{2}\right)x + \left(x_2^2 + \frac{21}{2}x_2 + 9\right)\right]}.$$

$$\frac{d}{dx}(y^3) = 3y^2 y' = \frac{d}{dx}(h(x) + 3c_2) = h'(x)$$

Thus,

$$y' = \frac{h'(x)}{3y^2} = \frac{(x - x_1)(x - x_2)}{3y^2}$$

$$= \frac{(x - x_1)(x - x_2)}{3(h(x) - h(x_2))^{2/3}}$$

$$= \frac{(x - x_1)(x - x_2)^{1/3}}{3\left[x^2 + \left(x_2 + \frac{21}{2}\right)x + \left(x_2^2 + \frac{21}{2}x_2 + 9\right)\right]^{2/3}}.$$

By looking at the discriminant of the quadratic in the denominator, we see that there are two roots, i.e., two points with vertical tangent lines.

47. Let A be the accumulated value at time t and d be the amount of the deposits made yearly, then A satisfies

$$A' = 0.06A + d.$$

This differential equation separates to

$$\frac{A'}{0.06A + d} = 1$$

and integrates to

$$\frac{\ln(0.06A + d)}{0.06} = t + c$$

or

$$0.06A + d = ke^{0.06t}.$$

At time $t = 0$, A is the unknown initial investment P, hence $k = 0.06P + 2000$, and so $0.06A + 2000 = (0.06P + 2000)e^{0.06t}$. If we want $A = 1,000,000$ at $t = 20$, we must have

$$62,000 = (0.06P + 2000)e^{1.2}$$

$$P = \frac{62,000e^{-1.2} - 2000}{0.06} \approx \$277,900.69$$

The interest rate makes a considerable difference.

49. The amount owed is increasing at a rate of $0.08A$ because of the interest and decreasing at the rate of $12P$ because of the payments. When the mortgage begins at $t = 0$, the amount owed is \$150,000. Thus the differential equation is $A' = 0.08A - 12P$ with the initial condition $A(0) = 150,000$. Solving this differential equation:

$$\frac{A'}{0.08A - 12P} = 1$$

$$\frac{\ln(0.08A - 12P)}{0.08} = t + c$$

$$0.08A - 12P = ke^{0.08t}$$

Using the initial condition gives

$$k = 12,000 - 12P.$$

We set $A(30) = 0$ and solve for P:

$$-12P = (12,000 - 12P)e^{2.4}$$

$$P = \frac{12,000e^{2.4}}{12(e^{2.4} - 1)} \approx \$1099.77$$

Total amount paid:

$(30)(12)(1099.77) = \$395,917.20$

Total interest:

$395,917.20 - 150,000 = \$245,917.20$

51. Reworking exercise 49:

$$A'(t) = 0.08A(t) - 12P$$

$$A(0) = 150,000$$

where P is the payment made each month. Solving this differential equation:

$$\frac{A'}{0.08A - 12P} = 1$$

$$0.08A - 12P = ke^{0.08t}$$

$$k = 12,000 - 12P$$

We set $A(15) = 0$ and solve for P:

$$-12P = (12,000 - 12P)e^{1.2}$$

$$P = \frac{12,000e^{1.2}}{12(e^{1.2} - 1)} \approx \$1431.01$$

The monthly payments are increased by about $330.

Total amount paid:

$(15)(12)(141.01) = \$257,581.80$

The total amount is decreased by $138,335.40.

Total interest:

$257,582 - 150,000 = \$107,582$

53. Starting with $A' = 0.08A + 10,000$ with the initial condition $A(0) = 0$:

$$\frac{A'}{0.08A + 10,000} = 1$$

$$\frac{\ln(0.08A + 10,000)}{0.08} = t + c$$

$$0.08A + 10,000 = ke^{0.08t}$$

Solving gives

$$0.08A + 10,000 = 10,000e^{0.08t}.$$

At time $t = 10$ we have

$$A = \frac{10,000(e^{0.8} - 1)}{0.08} = \$153,192.62.$$

This would be the amount in his fund at age 40, and it would accumulate in the next 25 years to

$$153,192.62e^{(0.08)25} = \$1,131,948.86.$$

55. Following the conditions of exercise 53, replacing the 8% by an unknown rate r, we come after 10 years of payment and 25 additional years of accumulation to

$$10,000\frac{(e^{10r} - 1)}{r}e^{25r}.$$

For contrast, if the payment rate 10,000 is replaced by 20,000, and the payment interval of 10 years is replaced by 25 years, we come to an accumulation after the twenty-five years of

$$20,000\frac{(e^{25r} - 1)}{r}.$$

Equating the two expressions leads to

$$2(e^{25r} - 1) = e^{35r} - e^{25r} \text{ or}$$

$$3e^{25r} - 2 = e^{35r}.$$

The equation can only be solved with the help of some form of technology, but the answer of r about 0.105 (10.5%) can at least be checked.

57. When the given numbers are substituted for the given symbols, the differential equation becomes

$$x' = (0.4 - x)(0.6 - x) - 0.625x^2$$

$$= \frac{3}{8}x^2 - x + \frac{6}{25}$$

$$= \frac{3}{8}\left(x - \frac{12}{5}\right)\left(x - \frac{4}{15}\right).$$

When separated it takes the form

$$\frac{x'}{(x - g)(x - f)} = r$$

in which $g = 12/5$, $f = 4/15 < g$, and $r = 3/8$.

By partial fractions we find

$$\frac{1}{(x - g)(x - f)}$$

$$= \frac{1}{(g - f)}\left\{\frac{1}{(x - g)} - \frac{1}{(x - f)}\right\}$$

and after integration we find

$$\frac{1}{(g - f)}\ln\left|\frac{x - g}{x - f}\right| = rt + c_1$$

or in this case with

$$g - f = (36/15) - (4/15) = 32/15,$$

$$\ln\left|\frac{x - 12/5}{x - 4/15}\right| = \frac{32}{15}\left(\frac{3}{8}t + c_1\right)$$

$$= \frac{4}{5}t + c_2 \left(c_2 = \frac{32}{15}c_1\right).$$

Using the given initial condition $x = 0.2 = 1/5$ when $t = 0$, we find

$c_2 = \ln|(11/5)/(1/15)| = \ln(33)$,

$$\ln\left|\frac{x - 12/5}{33(x - 4/15)}\right| = \frac{4}{5}t \text{ and}$$

$$\frac{x - 12/5}{33(x - 4/15)} = \pm e^{\frac{4}{5}t} = e^{\frac{4}{5}t}$$

(the choice of sign is + since the left side is 1 when $x = 1/5$).

Concluding the algebra we find

$$\frac{5x - 12}{11(15x - 4)} = e^{\frac{4}{5}t},$$

$$5x - 12 = 11(15x - 4)e^{\frac{4}{5}t},$$

$$x = \frac{12 - 44e^{\frac{4}{5}t}}{5 - 11(15)e^{\frac{4}{5}t}} = \frac{4}{5}\left(\frac{3 - 11e^{\frac{4}{5}t}}{1 - 33e^{\frac{4}{5}t}}\right),$$

and it is apparent that

$$x \to \frac{4}{15} \text{ as } t \to \infty.$$

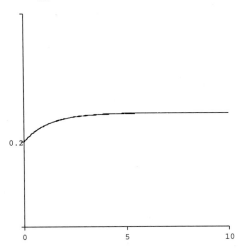

59. After beginning

$$x' = 0.6(0.5 - x)(0.6 - x) - 0.4x(0 + x)$$
$$= 0.6(0.3 - 1.1x + x^2) - 0.4x^2$$
$$= 0.2x^2 - 0.66x + 0.18$$
$$= \frac{1}{5}\left(x^2 - \frac{33}{10}x + \frac{9}{10}\right)$$
$$= \frac{1}{5}(x - 3)\left(x - \frac{3}{10}\right).$$

The parameters g, f, r are respectively

3, 3/10, 1/5. We jump ahead to

$$\ln\left|\frac{x - 3}{x - 3/10}\right| = \frac{27}{10}\left(\frac{t}{5} + c_1\right)$$
$$= \frac{27}{50}t + c_2.$$

In this case with $x = 0.2 = 1/5$ when $t = 0$, we find

$$c_2 = \ln\left|\frac{(1/5) - 3}{(1/5) - (3/10)}\right| = \ln 28,$$

$$\frac{x - 3}{28(x - 3/10)} = \pm e^{\frac{27}{50}t} = e^{\frac{27}{50}t},$$

and the conclusion is

$$5(x - 3) = 14(10x - 3)e^{\frac{27}{50}t},$$

$$x = \frac{15 - 42e^{\frac{27}{50}t}}{5 - 140e^{\frac{27}{50}t}} = \frac{42 - 15e^{-\frac{27}{50}t}}{140 - 5e^{-\frac{27}{50}t}}$$

$$= \frac{3}{5}\left(\frac{14 - 5e^{-\frac{27}{50}t}}{28 - e^{-\frac{27}{50}t}}\right).$$

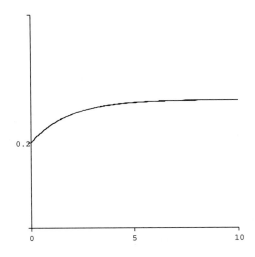

61. (a) This time we are given
$$x' = r(a - x)(b - x) \text{ with } r = 0.4, a = 6, b = 8, \text{ and } x(0) = 0.$$

$$\ln\left|\frac{x - b}{x - a}\right| = (b - a)rt + c_2,$$

in this case

$$\ln\left|\frac{x - 8}{x - 6}\right| = \frac{4}{5}t + c_2,$$

and without using the initial condition,

$$\frac{x-8}{x-6} = ke^{\frac{4}{3}t},$$

$$x - 8 = (x - 6)ke^{\frac{4}{3}t},$$

$$x = \frac{8 - 6ke^{\frac{4}{3}t}}{1 - ke^{\frac{4}{3}t}} = \frac{6k - 8e^{-\frac{4}{3}t}}{k - e^{-\frac{4}{3}t}}.$$

With $x(0) = 0$ we have $k = 4/3$ and

$$x = 24\frac{1 - e^{-\frac{4}{3}t}}{4 - 3e^{-\frac{4}{3}t}}.$$

It is now apparent that
$x \to 6$ as $t \to \infty$, and this makes perfect sense again because of the molecular exchange: the reaction continues as long as there is any "free" A and B, but in this case there is less A initially ($a = 6$), and the process essentially stops when the A is essentially gone, i.e., when $x \to a(= 6)$.

(b) The solution to the differential equation is (with $a = 6$, $b = 8$ and $r = 0.6$):

$$\ln\left|\frac{x-8}{x-6}\right| = \frac{6}{5}t + c_2$$

or

$$x = \frac{6ke^{6t/5} - 8}{ke^{6t/5} - 1}.$$

We use the initial condition $x(0) = 0$, which gives $k = \frac{8}{3}$ and therefore

$$x = \frac{8e^{6t/5} - 8}{\frac{4}{3}e^{6t/5} - 1} = \frac{24\left(e^{6t/5} - 1\right)}{4e^{6t/5} - 3}.$$

Again, it is clear that $x \to 6$ as $t \to \infty$.

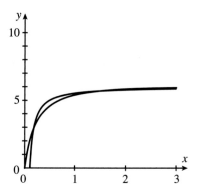

When $r = 0.6$, the amount of substance X approaches the limiting value more quickly.

63. We have

$$y' = 0.025y(8 - y) - 0.2$$
$$= -0.025(y^2 - 8y + 8)$$
$$= -\frac{1}{40}(y - b)(y - a),\ \text{in which}$$
$$b = 4 + \sqrt{8},\ a = 4 - \sqrt{8}.$$

This leads to

$$\ln\left|\frac{y-b}{y-a}\right| = -\frac{1}{40}\left(2\sqrt{8}\right)t + c_2$$

and with $y(0) = 8$ we have

$$\ln\left|\frac{8-b}{8-a}\right| = c_2,$$

$$\ln\left|\frac{(y-b)(8-a)}{(y-a)(8-b)}\right| = \frac{-t\sqrt{8}}{20},$$

$$\frac{y-b}{y-a} = \frac{8-b}{8-a}e^{-\frac{t\sqrt{8}}{20}}.$$

We can see that as $t \to \infty$ the right side goes to zero, hence also the left side, and hence

$$y \to b = 4 + \sqrt{8} \approx 6.828427.$$

This represents an eventual fish population of about 682,800.

65. The equilibrium solutions are the algebraic solutions to the quadratic equation

$$0.025P(8 - P) - R = 0,\ \text{or}$$
$$P^2 - 8P + 40R = 0.$$

In the process of studying exercise 63 ($R = 0.2$) we found it convenient to factor the left side (P was y at the time) and the roots were $b = 4 + \sqrt{8}$ and $a = 4 - \sqrt{8}$.

In exercise 64, the corresponding equation ($R = 0.6$) would be

$$0 = P^2 - 8P + 40R = P^2 - 8P + 24.$$

But this equation has no real roots, hence no equilibrium populations.

67. $P' = 0.05P(8 - P) - 0.6$

$$= -\frac{1}{20}(P^2 - 8P + 12)$$

$$= -\frac{1}{20}(P - 6)(P - 2)$$

Following well-established procedures, we come to

$$\ln\left|\frac{P - 6}{P - 2}\right| = -\frac{1}{5}t + c_2,$$

$$\frac{P - 6}{P - 2} = Ae^{-\frac{t}{5}},$$

We learn from this relation that the ratio $(P - 6)/(P - 2)$ never changes sign, always negative if the initial condition has $P(0)$ in the interval $(2, 6)$. Clearly the exponential approaches zero as $t \to \infty$ and P approaches 6. This last conclusion is true even if $P(0) > 6$.

If on the other hand $0 \le P(0) < 2$, the ratio is forever positive, and we find eventually

$$\frac{P - 6}{P - 2} = \frac{P(0) - 6}{P(0) - 2}e^{-t/5}.$$

Here the right side is a positive decreasing function of t and so must be the left side. The effect is that P itself is decreasing (not obvious) and reaches the value zero when

$$e^{-t/5} = 3\frac{P(0) - 2}{P(0) - 6} \text{ or when}$$

$$t = 5\ln\frac{6 - P(0)}{3[2 - P(0)]} = 5\ln\frac{6 - P(0)}{6 - 3P(0)}.$$

In the ratio inside the (second) logarithm, the numerator is clearly greater than the denominator, which is itself positive. This indicates that there is some moment of positive time after which the population is zero and no further activity occurs.

69. The differential equation is

$$r'(t) = k[r(t) - S].$$

This separates as $\dfrac{r'}{r - S} = k$, and solves as

$\ln(r - S) = kt + c.$

In this case $S = 1000$, $r(0) = 14,000$, and $r(4) = 8000$.

Putting $t = 0$, we see that the constant c is $\ln 13,000$,

$$\ln\frac{r - 1000}{13,000} = kt,$$

and putting $t = 4$,

$$\ln\frac{7000}{13,000} = \ln\frac{7}{13} = 4k.$$

Assembling the available information, we find

$$\ln\frac{r - 1000}{13,000} = kt$$

$$= \frac{t}{4}(4k) = \frac{t}{4}\ln\frac{7}{13}, \text{ and}$$

$$r = 1000 + 13,000\left(\frac{7}{13}\right)^{t/4},$$

or equivalently

$r = 1 + 13e^{-0.15476t}$ thousands.

71. If $P' = kP^{1.1}$, we separate as $P^{-1.1}P' = k$, and get

$$\frac{-10}{P^{1/10}} = \frac{P^{-0.1}}{(-0.1)} = kt + c$$

$$= kt - \frac{10}{P(0)^{1/10}}$$

$$P = \left(-\frac{kt}{10} + \frac{1}{P(0)^{1/10}}\right)^{-10}.$$

We see that P approaches infinity as t approaches $\dfrac{10}{kP(0)^{1/10}}$.

73. From the differential equation, with $z = y'/y$, we find $z = k(M - y)$. This is a line in the (y, z)-plane. The z-intercept is M and the slope is $-k$.

75. $\dfrac{dv}{dt} = 32 - \dfrac{4v^2}{10}$

$$= -\frac{4}{10}\left(v - \sqrt{80}\right)\left(v + \sqrt{80}\right)$$

We see that here $k = -4/10$ is a negative number. The parameters b, a $(b > a)$ are $b = \sqrt{80}$, $a = -\sqrt{80}$.

The solution is

$$\ln \left| \frac{v - \sqrt{80}}{v + \sqrt{80}} \right| = 2kt\sqrt{80} + c_2.$$

Given that $v(0) = 0$, we find $c_2 = 0$ and

$$\frac{\sqrt{80} - v}{\sqrt{80} + v} = e^{2kt\sqrt{80}}.$$

Because $k < 0$, the right-hand side goes to zero as t goes to infinity. Therefore $v \to \sqrt{80}$. This is the terminal velocity.

77. If $y' = ky(M - y)$, then by the product rule

$$y'' = k[y'(M - y) - yy'] = ky'[M - 2y].$$

This will be zero when $y' = 0$ or when $y = M/2$. $y' = 0$ at the equilibrium solutions $y = 0$ and $y = M$. Since these are constant functions, they have no points of inflection, thus the only inflection point is when $y = M/2$.

79.

$$y'' = -ay' \ln \left(\frac{y}{b} \right) - ay'$$

$$= -ay' \left[\ln \left(\frac{y}{b} \right) + 1 \right]$$

Thus we have $y'' = 0$ if and only if $y' = 0$ or if $y = \dfrac{b}{e}$.

Notice that $y' = 0$ can occur when $y = 0$ or when $y = b$. But, as in exercise 77, these are equilibrium solutions—any solution with $y' = 0$ is a constant and cannot have any inflection points.

8.3 DIRECTION FIELDS AND EULER'S METHOD

1.

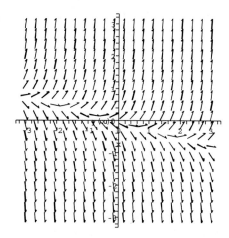

3.

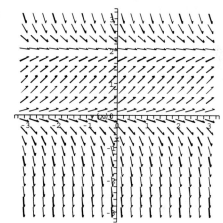

5.

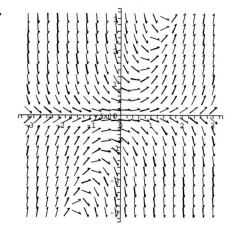

7. Field C.

9. Field D.

11. Field A.

13. For $h = 0.1$:

n	x_n	$y(x_n)$	$f(x_n, y_n)$
0	0.0	1	0
1	0.1	1	0.2
2	0.2	1.02	0.408
3	0.3	1.0608	0.63648
10	1.0	2.334633363	4.669266726
20	2.0	29.49864319	117.9945728

For $h = 0.05$:

n	x_n	$y(x_n)$	$f(x_n, y_n)$
0	0.00	1	0
1	0.05	1	0.10
2	0.10	1.0050	0.201000
3	0.15	1.01505000	0.3045150000
20	1.00	2.510662314	5.021324627
40	2.00	39.09299941	156.3719977

15. First for $h = 0.1$:

n	x_n	$y(x_n)$	$f(x_n, y_n)$
0	0.0	1	3
1	0.1	1.3	3.51
2	0.2	1.651	3.878199
3	0.3	2.0388199	3.998493015
10	1.0	3.847783602	0.58569576
20	2.0	3.999018724	0.00392403

Then for $h = 0.05$:

n	x_n	$y(x_n)$	$f(x_n, y_n)$
0	0.00	1	3
1	0.05	1.15	3.2775
2	0.10	1.313875	3.529232484
3	0.15	1.490336624	3.740243243
20	1.00	3.818763109	0.69210075
40	2.00	3.997787406	0.00884548

17. For $h = 0.1$:

n	x_n	$y(x_n)$	$f(x_n, y_n)$
0	0.0	3	−1
1	0.1	2.9	−0.995162582
2	0.2	2.800483742	−0.9817529887
3	0.3	2.702308443	−0.9614902222
10	1.0	2.094283524	−0.7264040831
20	2.0	1.527574224	−0.3922389411

For $h = 0.05$:

n	x_n	$y(x_n)$	$f(x_n, y_n)$
0	0.0	3	−1
1	0.05	2.95	−0.9987705755
2	0.10	2.900061471	−0.9952240532
3	0.15	2.850300269	−0.9895922921
20	1.00	2.099000947	−0.7311215058
40	2.00	1.534517021	−0.3991817376

19. For $h = 0.1$:

n	x_n	$y(x_n)$	$f(x_n, y_n)$
0	0.0	1	1.0
1	0.1	1.10	1.095445115
2	0.2	1.209544512	1.187242398
3	0.3	1.328268751	1.276036344
10	1.0	2.395982932	1.842819289
20	2.0	4.568765343	2.562960269

For $h = 0.05$:

n	x_n	$y(x_n)$	$f(x_n, y_n)$
0	0.00	1	1
1	0.05	1.05	1.048808848
2	0.10	1.102440442	1.096558454
3	0.15	1.157268365	1.143358371
20	1.00	2.420997836	1.849593965
40	2.00	4.620277218	2.572989937

21. (a) The exact solution to exercise 13 is

$$y(x) = e^{x^2}$$

$$y(1) \approx 2.718281828$$

$$y(2) \approx 54.59815003$$

(b) The exact solution to exercise 14 is

$$y(x) = \sqrt{x^2 + 4}$$
$$y(1) \approx 2.236067977$$
$$y(2) \approx 2.828427125$$

23.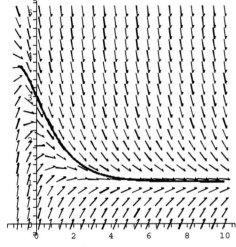

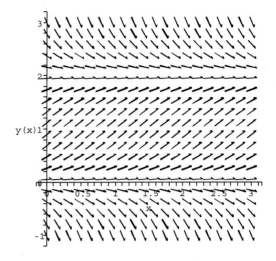

27. Equilibrium solutions come from $y' = 0$, which only occurs when $y = 0$ or $y = \pm 1$. From the direction field, $y = 0$ and $y = -1$ are seen to be unstable equilibria and $y = 1$ is seen to be a stable equilibrium.

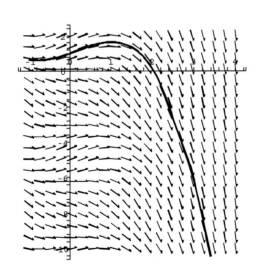

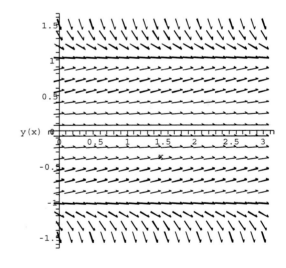

25. Equilibrium solutions come from $y' = 0$, which only occurs when $y = 0$ or $y = 2$. From the direction field, $y = 0$ is seen to be an unstable equilibrium and $y = 2$ is seen to be a stable equilibrium.

29. Equilibrium solutions come from $y' = 0$, which only occurs when $y = 1$. From the direction field, $y = 1$ is seen to be a stable equilibrium.

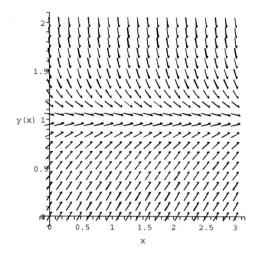

31. The equilibrium solutions are the constant solutions to the DE. If indeed g is a certain constant k, then

$$0 = g' = -k + 3k^2/(1 + k^2)$$
$$= -k(k^2 - 3k + 1)/(k^2 + 1).$$

Thus $k = 0$ is clearly one solution, while the two roots of the quadratic in the numerator are also solutions. These are the numbers

$$a = \frac{3 - \sqrt{5}}{2} \approx 0.3820 \text{ and}$$

$$b = \frac{3 + \sqrt{5}}{2} \approx 2.6810.$$

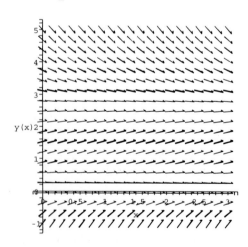

Of the three, 0 and b are stable, while a is unstable. Since $0 < a$, with 0 being stable and a being unstable, $g(t)$ will decrease to 0 when $0 < g(0) < a$, i.e., $g' < 0$. Similarly, when

$g(0) > b$, $g(t)$ will decrease to b, and again $g' < 0$.

Since $a < b$, with a being unstable and b being stable, $g(t)$ will increase to b when $a < g(0) < b$, and $g' > 0$. Thus,

$$\lim_{t \to \infty} g(t) = 0 \text{ if } 0 \le g(0) < a, \text{ while}$$

$$\lim_{t \to \infty} g(t) = b \text{ if } a < g(0).$$

As the problem evolves, g depends not only on time t, but on a certain real parameter x. We could write $g = g_x(t)$, and the dependence on x is through the initial condition:

$$g_x(0) = \frac{3}{2} + \frac{3 \sin(x)}{2}.$$

With x restricted to the interval $[0, 4\pi]$ (4π being about 12.5664), the event $g_x(0) < a$, or equivalently $\lim_{t \to \infty} g_x(t) = 0$, which corresponds to a black-stripe zone, occurs when x lies in one of the two intervals (3.9827, 5.4421) or (10.2658, 11.7253). More precisely, these are the intervals with endpoints

$$\frac{3\pi}{2} \pm \cos^{-1}\left(\frac{\sqrt{5}}{3}\right) \quad \text{and}$$

$$\frac{7\pi}{2} \pm \cos^{-1}\left(\frac{\sqrt{5}}{3}\right).$$

33. Using Euler's method:

For $h = 0.1$:

n	x_n	$y(x_n)$	$f(x_n, y_n)$
0	0.00	3.0000	8.0000
1	0.10	3.8000	13.4400
2	0.20	5.1440	25.4607
3	0.30	7.6901	58.1372
4	0.40	13.5038	181.3525
5	0.50	31.6390	1000.0295

For $h = 0.05$:

n	x_n	$y(x_n)$	$f(x_n, y_n)$
0	0.00	3.0000	8.0000
1	0.05	3.4000	10.5600
2	0.10	3.9280	14.4292
3	0.15	4.6495	20.6175
4	0.20	5.6803	31.2662
10	0.50	218.1215	47,576.0009

For $h = 0.01$:

n	x_n	$y(x_n)$	$f(x_n, y_n)$
0	0.00	3.0000	8.0000
1	0.01	3.0800	8.4864
9	0.09	3.9396	14.5203
10	0.10	4.0848	15.6855
20	0.20	6.5184	41.4900
21	0.21	6.9333	47.0711
30	0.30	15.8434	250.0139

x	Exact
0.000	3.0000
0.100	4.1374
0.200	6.8713
0.300	21.4869
0.400	−18.7351
0.500	−6.5688

35. The vertical asymptote in the solution occurs when the denominator vanishes, which is when $e^{2x} = 2$. So the vertical asymptote is at

$$x = \ln(2)/2 \approx 0.3466.$$

The field diagram cannot give any forewarning of the vertical asymptote. Dependent as the field equations are only on y, they can only hint at things that likewise depend on y. The location of the vertical asymptote is by its very nature an x-measurement.

In this case where the actual x-value does not enter the calculations, the Euler process merely generates the numbers in the recursive sequence $y_n = hy_{n-1}^2 + y_{n-1} - h$ subject to an initial condition of $y_o = 3$. The numbers in

such a sequence will increase to infinity, with growth rate depending on h. The simultaneous determination of x_n through the law $x_n = hn$ has nothing to do with the geometry of the solution to the differential equation. "Jumping over the asymptote" is the pseudo-event that happens when n passes from below $\dfrac{0.3466}{h}$ to above; it has no special relation to the Euler y-numbers, and no relation whatever to the solution of the differential equation.

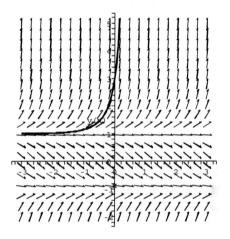

37. The general solution is

$$y = \frac{x^3}{3} - 2x^2 + 2x + c.$$

Using the initial condition $y(3) = 0$ gives $y(0) = c = 3$ and therefore

$$y = \frac{x^3}{3} - 2x^2 + 2x + 3.$$

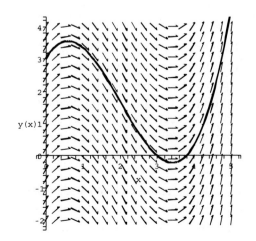

39. Using a CAS gives $y(0) \approx -5.55$.

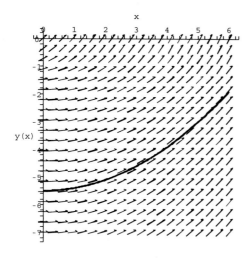

8.4 SYSTEMS OF FIRST-ORDER DIFFERENTIAL EQUATIONS

1. Equilibrium points are those that satisfy $x'(t) = 0$ and $y'(t) = 0$. Substituting into the equations, we have

$$0 = 0.2x - 0.2x^2 - 0.4xy$$
$$0 = -0.1y + 0.2xy$$

$$0 = x(0.2 - 0.2x - 0.4y)$$
$$0 = y(-0.1 + 0.2x)$$

$$x = 0 \text{ or } 0.2 - 0.2x - 0.4y = 0$$
$$y = 0 \text{ or } x = 0.5$$

The equilibrium points are $(0, 0)$, corresponding to the case where there are no predators or prey; $(1, 0)$, corresponding to the case where there are prey but no predators; and $(0.5, 0.25)$, corresponding to both populations being constant, with two times as many prey as predators.

3. Equilibrium points are those that satisfy $x'(t) = 0$ and $y'(t) = 0$. Substituting into the equations, we have

$$0 = 0.3x - 0.1x^2 - 0.2xy$$
$$0 = -0.2y + 0.1xy$$

$$0 = x(0.3 - 0.1x - 0.2y)$$
$$0 = y(-0.2 + 0.1x)$$

$$x = 0 \text{ or } 0.3 - 0.1x - 0.2y = 0$$
$$y = 0 \text{ or } x = 2$$

The equilibrium points are $(0, 0)$, corresponding to the case where there are no predators or prey; $(3, 0)$, corresponding to the case where there are prey but no predators; and $(2, 0.5)$, corresponding to both populations being constant, with four times as many prey as predators.

5. Equilibrium points are those that satisfy $x'(t) = 0$ and $y'(t) = 0$. Substituting into the equations, we have

$$0 = 0.2x - 0.1x^2 - 0.4xy$$
$$0 = -0.3y + 0.1xy$$

$$0 = x(0.2 - 0.1x - 0.4y)$$
$$0 = y(-0.3 + 0.1x)$$

$$x = 0 \text{ or } 0.2 - 0.1x - 0.4y = 0$$
$$y = 0 \text{ or } x = 3$$

The equilibrium points are $(0, 0)$, corresponding to the case where there are no predators or prey; $(2, 0)$, corresponding to the case where there are prey but no predators; and $(3, -0.25)$, corresponding to a negative number of predators, which is impossible.

7. In exercise 1,

$$\frac{dy}{dx} = \frac{-0.1y + 0.2xy}{0.2x - 0.2x^2 - 0.4xy}$$

From the following phase portrait, we observe that $(0, 0)$ is an unstable equilibrium, $(1, 0)$ is an unstable equilibrium, and $(0.5, 0.25)$ is a stable equilibrium.

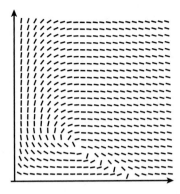

9. In exercise 5,

$$\frac{dy}{dx} = \frac{-0.3y + 0.1xy}{0.2x - 0.1x^2 - 0.4xy}$$

From the following phase portrait, we observe that $(0, 0)$ is an unstable equilibrium, and $(2, 0)$ is a stable equilibrium.

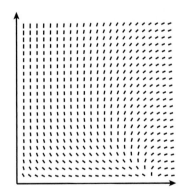

11. The point $(0, 0)$ is an unstable equilibrium.

13. The point $(0.5, 0.5)$ is a stable equilibrium.

15. The point $(1, 0)$ is a stable equilibrium.

17. Equilibrium points are those that satisfy $x'(t) = 0$ and $y'(t) = 0$. Substituting into the equations, we have

$$0 = 0.3x - 0.2x^2 - 0.1xy$$
$$0 = 0.2y - 0.1y^2 - 0.1xy$$

$$0 = x(0.3 - 0.2x - 0.1y)$$
$$0 = y(0.2 - 0.1y - 0.1x)$$

$$x = 0 \text{ or } 0.3 - 0.2x - 0.1y = 0$$
$$y = 0 \text{ or } 0.2 - 0.1y - 0.1x = 0$$

$$x = 0 \text{ or } 2x + y = 3$$
$$y = 0 \text{ or } x + y = 2$$

The equilibrium points are $(0, 0)$, corresponding to the case where neither species exists; $(0, 2)$, corresponding to the case where species Y exists but species X does not; $(1.5, 0)$, corresponding to the case where species X exists but species Y does not; and $(1, 1)$, corresponding to both species existing, with as many of species Y as species X.

19. Equilibrium points are those that satisfy $x'(t) = 0$ and $y'(t) = 0$. Substituting into the equations, we have

$$0 = 0.3x - 0.2x^2 - 0.2xy$$
$$0 = 0.2y - 0.1y^2 - 0.2xy$$

$$0 = x(0.3 - 0.2x - 0.2y)$$
$$0 = y(0.2 - 0.1y - 0.2x)$$

$$x = 0 \text{ or } 0.3 - 0.2x - 0.2y = 0$$
$$y = 0 \text{ or } 0.2 - 0.1y - 0.2x = 0$$

$$x = 0 \text{ or } x + y = 1.5$$
$$y = 0 \text{ or } 2x + y = 2$$

The equilibrium points are $(0, 0)$, corresponding to the case where neither species exists; $(0, 2)$, corresponding to the case where species Y exists but species X does not; $(1.5, 0)$, corresponding to the case where species X exists but species Y does not; and $(0.5, 1)$, corresponding to both species existing, with twice as many of species Y as species X.

21. Equilibrium points are those that satisfy $x'(t) = 0$ and $y'(t) = 0$. Substituting into the equations, we have

$$0 = 0.2x - 0.2x^2 - 0.1xy$$

$$0 = 0.1y - 0.1y^2 - 0.2xy$$

$$0 = x(0.2 - 0.2x - 0.1y)$$

$$0 = y(0.1 - 0.1y - 0.2x)$$

$$x = 0 \text{ or } 0.2 - 0.2x - 0.1y = 0$$

$$y = 0 \text{ or } 0.1 - 0.1y - 0.2x = 0$$

$$x = 0 \text{ or } 2x + y = 2$$

$$y = 0 \text{ or } 2x + y = 1$$

The equilibrium points are (0, 0), corresponding to the case where neither species exists;
(0, 1), corresponding to the case where species
Y exists but species X does not; and (1, 0), corresponding to the case where species X exists
but species Y does not.

23. This is because both species are hurt by the
presence of the other.

25. In exercise 17,

$$\frac{dy}{dx} = \frac{0.2y - 0.1y^2 - 0.1xy}{0.3x - 0.2x^2 - 0.1xy}$$

From the following phase portrait, we observe
that (0, 0) is an unstable equilibrium, (0, 2) is
an unstable equilibrium, (1.5, 0) is an unstable
equilibrium, and (1, 1) is a stable equilibrium.

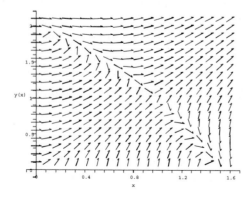

27. (a)

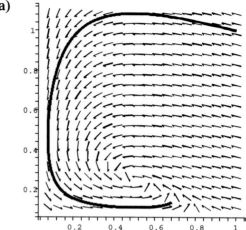

(b)

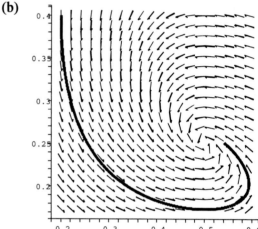

(c)

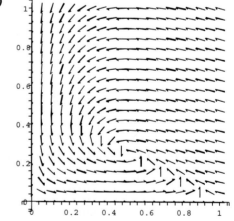

29. Write $u = y$, $v = y'$. We then have

$$u' = v$$

$$v' = -2xv - 4u + 4x^2$$

31. Write $u = y$, $v = y'$. We then have

$$u' = v$$
$$v' = (\cos x)v - xu^2 + 2x$$

33. Write $u_1 = y$, $u_2 = y'$, and $u_3 = y''$.

$$u_1' = u_2$$
$$u_2' = u_3$$
$$u_3' = -2xu_3 + 4u_2 - 2u_1 + x^2$$

35. Write $u_1 = y$, $u_2 = y'$, $u_3 = y''$, and $u_4 = y'''$.

$$u_1' = u_2$$
$$u_2' = u_3$$
$$u_3' = u_4$$
$$u_4' = 2u_4 - xu_2 + 2 - e^x$$

37. An approximate solution is

$x(1) \approx 0.253718$, $y(1) \approx 0.167173$.

x	x_n	y_n
0	0.2	0.2
1	0.2048	0.1964
2	0.2097201152	0.1928742272
3	0.2147629013	0.1894212388
5	0.2252268589	0.1827279868
10	0.2537179002	0.1671729954

39. Equilibrium points are those that satisfy $x'(t) = 0$ and $y'(t) = 0$. Substituting into the equations, we have

$$0 = (x^2 - 4)(y^2 - 9)$$
$$0 = x^2 - 2xy$$

$$0 = (x + 2)(x - 2)(y + 3)(y - 3)$$
$$0 = x(x - 2y)$$

$x = 2$, $x = -2$, $y = 3$, $y = -3$
$x = 0$, $x = 2y$

The equilibrium points are $(2, 1)$, $(-2, -1)$, $(6, 3)$, $(-6, -3)$, $(0, 3)$, $(0, -3)$.

41. Equilibrium points are those that satisfy $x'(t) = 0$ and $y'(t) = 0$. Substituting into the

equations, we have

$$0 = (2 + x)(y - x)$$
$$0 = (4 - x)(x + y)$$

$x = -2$ or $x = y$

$x = 4$ or $x = -y$

The equilibrium points are $(0, 0)$, $(-2, 2)$, $(4, 4)$.

43. For equilibrium solutions, set $x' = y' = 0$ to get

$$0 = 0.4x - 0.1x^2 - 0.2xy$$
$$0 = -0.5y + 0.1xy$$

$$0 = 0.1x(4 - x - 2y)$$
$$0 = 0.1y(-5 + x)$$

Equilibrium points are $(0, 0)$, $(4, 0)$. (Note that $x = 5$ leads to a negative value of y, which is impossible.) Neither of these solutions has nonzero values for both populations, so the species cannot coexist.

Now suppose that the death rate of species Y is D instead of 0.5, and let us search for equilibrium solutions where both population values are nonzero. The equations are now

$$x' = 0.4x - 0.1x^2 - 0.2xy$$
$$y' = -Dy + 0.1xy$$

where $D > 0$.

$$0 = 0.1x(4 - x - 2y)$$
$$0 = 0.1y(x - 10D)$$

Since we are searching for nonzero solutions,

$$0 = 4 - x - 2y$$
$$0 = x - 10D$$

Solving the second equation gives $x = 10D$, and substituting this expression into the first equation gives

$$0 = 4 - 10D - 2y$$
$$0 = 2 - 5D - y$$
$$y = 2 - 5D$$

The equilibrium solution for y will be positive provided that $2 - 5D > 0$, which means that $D < 0.4$.

45. Assume that all coefficients are positive. The equations that define equilibrium are

$$0 = x(b - cx - k_1 y)$$
$$0 = y(-d + k_2 x)$$

For the species to coexist, both x and y must be nonzero, and so the equations reduce to

$$0 = b - cx - k_1 y$$
$$0 = -d + k_2 x$$

Solving the second equation, we get $x = \dfrac{d}{k_2}$.
Substituting the result into the first equation,

$$0 = b - c\frac{d}{k_2} - k_1 y$$

$$k_1 y = b - \frac{cd}{k_2} = \frac{bk_2 - cd}{k_2}$$

$$y = \frac{bk_2 - cd}{k_1 k_2}$$

Thus, $y > 0$ if and only if $bk_2 - cd > 0$, which is equivalent to $bk_2 > cd$.

Chapter 8 Review Exercises

1. We separate variables and integrate.

$$\frac{1}{y} y' = 2$$

$$\int \frac{1}{y}\, dy = \int 2\, dx$$

$$\ln |y| = 2x + c$$

$$y = ke^{2x}$$

The initial condition gives $3 = k$ so the solution is

$$y = 3e^{2x}$$

3. We separate variables and integrate.

$$yy' = 2x$$

$$\int y\, dy = \int 2x\, dx$$

$$\frac{y^2}{2} = x^2 + c$$

$$y = \sqrt{2x^2 + c}$$

The initial condition gives us $2 = \sqrt{c}$, $c = 4$, so the solution is

$$y = \sqrt{2x^2 + 4}.$$

5. We separate variables and integrate.

$$\frac{1}{\sqrt{y}} y' = \sqrt{x}$$

$$\int y^{-1/2}\, dy = \int x^{1/2}\, dx$$

$$2y^{1/2} = \frac{2}{3} x^{3/2} + c$$

$$y = \left(\frac{x^{3/2}}{3} + c \right)^2$$

The initial condition gives us

$$4 = \left(\frac{1}{3} + c \right)^2, c = \tfrac{5}{3}, \text{ so the solution is}$$

$$y = \left(\frac{x^{3/2}}{3} + \frac{5}{3} \right)^2.$$

7. With t measured in hours, we have $y = Ae^{kt}$, $A = y(0) = 10^4$.

If the doubling time is 2, then $2 = e^{2k}$, $k = \ln(2)/2$, and $y = 10^4 e^{t \ln(2)/2} = 10^4 2^{t/2}$.

To reach $y = 10^6$ at a certain unknown time t, we need

$$2^{t/2} = 100,$$

$$t = \frac{2 \ln(100)}{\ln(2)} \approx 13.3 \text{ hours.}$$

9. With t measured in hours, x in milligrams, we get

$$x = 2 \left(\frac{1}{2} \right)^{t/2} = \frac{2}{2^{t/2}}.$$

To get to $x = 0.1$ at a certain unknown time t, we need

$$2^{t/2} = \frac{2}{0.1} = 20,$$

$$t = \frac{2\ln(20)}{\ln(2)} \approx 8.64 \text{ hours.}$$

11. The equation for the doubling time t_d in this case is

$$2 = e^{0.08t_d}, \text{ hence}$$

$$t_d = \frac{\ln(2)}{0.08} \approx 8.66 \text{ years.}$$

13. Newton's Law of Cooling is $y(t) = Ae^{kt} + T_a$. Here, $T_a = 68$ and $y(0) = 180$, so $A = 112$ and $y(t) = 112e^{kt} + 68$. Since $y(1) = 176$, $\frac{108}{112} = e^k$ and $k = \ln\frac{108}{112} = \ln\frac{27}{28}$.

$$y(t) = 112e^{\ln(27/28)t} + 68$$

Set $y(t) = 120$ and solve for t.

$$120 = 112e^{\ln(27/28)t} + 68$$

$$\frac{52}{112} = e^{\ln(27/28)t}$$

$$\ln\frac{52}{112} = t\ln\frac{27}{28}$$

$$t = \frac{\ln(13/28)}{\ln(27/28)} \approx 21.10 \text{ minutes}$$

15. $\dfrac{y'}{y} = 2x^3$

$$\ln|y| = \frac{x^4}{2} + c$$

$$y = Ae^{x^4/2}$$

17. $(y^2 + y)y' = \dfrac{4}{1 + x^2}$

$$\int (y^2 + y)\, dy = \int \frac{4}{1 + x^2}\, dx$$

$$\frac{y^3}{3} + \frac{y^2}{2} = 4\tan^{-1}x + c$$

It is impossible (without using a CAS) to write out the explicit formula of y in terms of x.

19. Equilibrium solutions occur where $y' = 0$, which occurs when $y = 0$ and $y = 2$. $y = 0$ is unstable and $y = 2$ is stable, which can be seen by drawing the direction field.

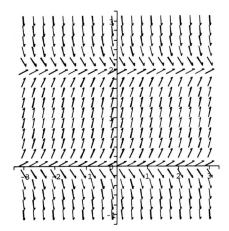

21. Equilibrium solutions occur where $y' = 0$, which occurs when $y = 0$, and it is stable.

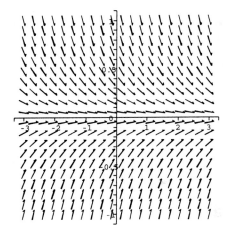

23.

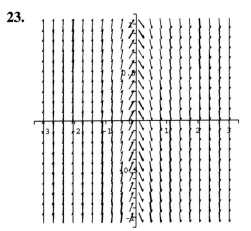

25.

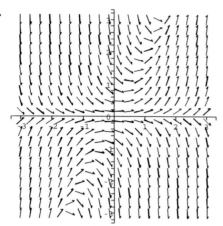

27. The differential equation is

$$x' = (0.3 - x)(0.4 - x) - 0.25x^2$$

$$= 0.12 - 0.7x + 0.75x^2$$

$$= \frac{3}{4}\left(x^2 - \frac{14}{15}x + \frac{4}{25}\right)$$

$$= \frac{3}{4}(x - r)(x - s), \text{ in which}$$

$$r = \frac{7 + \sqrt{13}}{15}, \; s = \frac{7 - \sqrt{13}}{15},$$

$$r - s = \frac{2\sqrt{13}}{15}.$$

When separated it takes the form

$$\frac{x'}{(x - r)(x - s)} = k$$

in which $k = 3/4$.

By partial fractions we find

$$\frac{1}{(x - r)(x - s)}$$

$$= \frac{1}{(r - s)}\left\{\frac{1}{(x - r)} - \frac{1}{(x - s)}\right\}$$

and after integration we find

$$\frac{1}{(r - s)}\ln\left|\frac{x - r}{x - s}\right| = kt + c_1$$

or in this case

$$\ln\left|\frac{x - r}{x - s}\right| = \frac{2\sqrt{13}}{15}\left(\frac{3}{4}t + c_1\right)$$

$$= wt + c_2$$

$$\left(w = \frac{\sqrt{13}}{10} \approx 0.36056, \; c_2 = \frac{2\sqrt{13}}{15}c_1\right).$$

Using the initial condition $x(0) = c$, we find $c_2 = \ln|(c - r)/(c - s)|$,

$$\ln\left|\frac{(c - s)(x - r)}{(c - r)(x - s)}\right| = wt \text{ and}$$

$$\frac{x - r}{x - s} = \pm\frac{c - r}{c - s}e^{wt} = \frac{c - r}{c - s}e^{wt},$$

$$x = \frac{s(r - c)e^{wt} + r(c - s)}{(r - c)e^{wt} + (c - s)}$$

$$= \frac{r\left(\dfrac{c - s}{r - c}\right)e^{-wt} + s}{\left(\dfrac{c - s}{r - c}\right)e^{-wt} + 1}.$$

The choice of sign is $+$ since the left side of the middle equation is $(c - r)/(c - s)$ when $t = 0$ and $x = c$. The last expression is one of many possible ways to normalize. It is apparent that $x \to s \approx 0.22630$ as $t \to \infty$.

Numerically, when $c = 0.1$, this comes to

$$x = \frac{0.22630 - 0.14710e^{-0.36056t}}{1 - 0.20805e^{-0.36056t}}, \text{ and the}$$

graph looks like

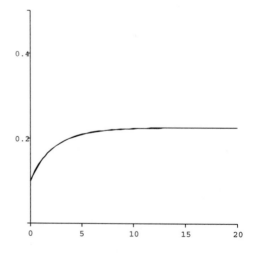

When $c = 0.4$, this comes to

$$x = \frac{0.22630 + 0.39999e^{-0.36056t}}{1 + 0.56574e^{-0.36056t}}, \text{ and the}$$

graph looks like

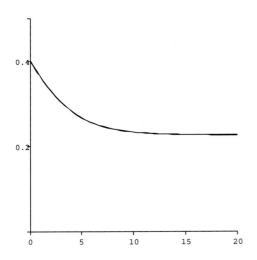

This is impossible since the concentration x will not decrease.

29. The DE $\dfrac{x'}{(x-a)^2} = r$ integrates to

$$\frac{-1}{x-a} = rt + c.$$

$$x - a = \frac{-1}{rt + c}$$

$$x = a - \frac{1}{rt + c}$$

Thus,

$$x(0) = a - \frac{1}{c} \quad \text{and} \quad c = \frac{1}{a - x(0)}.$$

$$x(t) = a - \frac{1}{rt - \frac{1}{a - x(0)}} = a - \frac{a - x(0)}{(a - x(0))rt + 1}.$$

Since $a > x(0)$, it is clear that $0 < x(t) < a$.

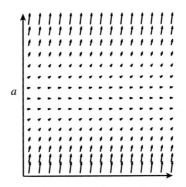

From the direction field, it appears that the limiting amount, $\lim\limits_{t \to \infty} x(t) = a$. Further simplification of the equation for $x(t)$ leads to

$$x(t) = \frac{x(0) + t(ra^2 - arx(0))}{1 + t(ra - rx(0))}$$

$$= \frac{x(0)t^{-1} + a(ar - rx(0))}{t^{-1} + (ra - rx(0))}$$

from which it is clear that the limiting amount is a.

31. With A the amount in the account at time t the DE is $A'(t) = 0.10A + 20{,}000$ with an IC of $A(0) = 100{,}000$. The DE separates and integrates easily, yielding

$$10 \ln |0.10A + 20{,}000| = t + c$$

$$c = 10 \ln(30{,}000), \text{ so}$$

$$0.10A + 20{,}000 = 30{,}000e^{t/10}.$$

If the fortune is to reach $1{,}000{,}000$ at time t, we must have

$$120{,}000 = 30{,}000e^{t/10}.$$

$$\frac{t}{10} = \ln \frac{12}{3} = \ln(4),$$

$$t = 10 \ln(4) \approx 13.86 \text{ years.}$$

33. It is a predator-prey model.

For equilibrium solutions, set $x' = y' = 0$ to get

$$0 = 0.1x - 0.1x^2 - 0.2xy$$

$$0 = -0.1y + 0.1xy$$

which are equivalent to

$0 = 0.1x(1 - x - 2y)$

$0 = 0.1y(-1 + x)$

$x = 0$ or $x = 1 - 2y$

$y = 0$ or $x = 1$.

The equilibrium solutions are $(0, 0)$ (no prey or predators) and $(1, 0)$ (prey but no predators).

35. It is a competing species model.

For equilibrium solutions, set $x' = y' = 0$ to get

$0 = 0.5x - 0.1x^2 - 0.2xy$

$0 = 0.4y - 0.1y^2 - 0.2xy$

which are equivalent to

$0 = 0.1x(5 - x - 2y)$

$0 = 0.1y(4 - y - 2x)$

$x = 0$ or $x + 2y = 5$

$y = 0$ or $2x + y = 4$.

The equilibrium solutions are $(0, 0)$ (none of either species), $(0, 4)$ (none of first species, some of second), $(5, 0)$ (some of first species, none of second), and $(1, 2)$ (twice as many of second species as first species).

37. In exercise 33,

$$\frac{dy}{dx} = \frac{-0.1y + 0.1xy}{0.1x - 0.1x^2 - 0.2xy}.$$

From the direction field, we see that $(0, 0)$ is unstable and $(1, 0)$ is stable.

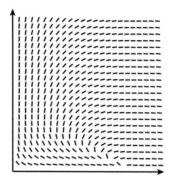

39. Write $u = y$, $v = y'$, then

$u' = v$

$v' = 4x^2v - 2u + 4xu - 1$

Infinite Series

9.1 SEQUENCES OF REAL NUMBERS

1. $1, \dfrac{3}{4}, \dfrac{5}{9}, \dfrac{7}{16}, \dfrac{9}{25}, \dfrac{11}{36}$

3. $4, 2, \dfrac{2}{3}, \dfrac{1}{6}, \dfrac{1}{30}, \dfrac{1}{180}$

5. **(a)** $\lim\limits_{n \to \infty} \dfrac{1}{n^3} = 0$

(b) Given any $\varepsilon > 0$, choose $N \ge \sqrt[3]{\dfrac{1}{\varepsilon}}$. Then for $n > N$,

$$\left| \dfrac{1}{n^3} - 0 \right| = \dfrac{1}{n^3} < \varepsilon.$$

Thus, the sequence converges to 0.

(c)

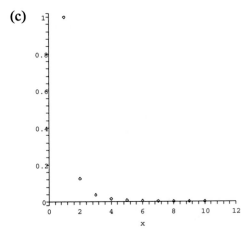

7. **(a)** $\lim\limits_{n \to \infty} \dfrac{n}{n+1} = \lim\limits_{n \to \infty} \dfrac{1}{1 + \frac{1}{n}} = 1$

(b) Given any $\varepsilon > 0$, choose $N \ge \dfrac{1}{\varepsilon}$. Then for $n > N$,

$$\left| \dfrac{n}{n+1} - 1 \right| = \dfrac{1}{n+1} < \dfrac{1}{n} < \varepsilon.$$

Thus, the sequence converges to 1.

(c)

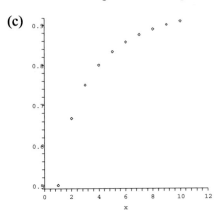

9. **(a)** $\lim\limits_{n \to \infty} \dfrac{2}{\sqrt{n}} = 0$

(b) Given any $\varepsilon > 0$, choose $N \ge 4/\varepsilon^2$. Then for $n > N$,

$$\left| \dfrac{2}{\sqrt{n}} - 0 \right| = \dfrac{2}{\sqrt{n}} < \varepsilon.$$

Thus, the sequence converges to 0.

(c)

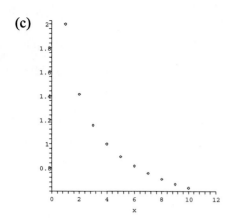

11. $\lim\limits_{n\to\infty}\dfrac{3n^2+1}{2n^2-1}=\lim\limits_{n\to\infty}\dfrac{3+\frac{1}{n^2}}{2-\frac{1}{n^2}}=\dfrac{3}{2}$

The sequence converges to $\dfrac{3}{2}$.

13. $\lim\limits_{n\to\infty}\dfrac{n^2+1}{n+1}=\lim\limits_{n\to\infty}\dfrac{n+\frac{1}{n}}{1+\frac{1}{n}}=\infty$

The sequence diverges.

15. The sequence approaches two different values, $\pm 1/3$; therefore, it diverges.

17. $\lim\limits_{n\to\infty}\left|(-1)^n\dfrac{n+2}{n^2+4}\right|=\lim\limits_{n\to\infty}\dfrac{n+2}{n^2+4}$

$=\lim\limits_{n\to\infty}\dfrac{\frac{1}{n}+\frac{2}{n^2}}{1+\frac{4}{n^2}}=0$

Hence, $\lim\limits_{n\to\infty}\dfrac{(-1)^2(n+2)}{n^2+4}=0$.

The sequence converges to 0.

19. $\lim\limits_{x\to\infty}\dfrac{x}{e^x}=\lim\limits_{x\to\infty}\dfrac{1}{e^x}$ by l'Hôpital's Rule and

$\lim\limits_{x\to\infty}\dfrac{1}{e^x}=0$, so by Theorem 1.2 $\lim\limits_{n\to\infty}\dfrac{n}{e^n}=0$.

The sequence converges to 0.

21. $\lim\limits_{n\to\infty}\dfrac{e^n+2}{e^{2n}-1}=\lim\limits_{n\to\infty}\dfrac{\frac{1}{e^n}+\frac{2}{e^{2n}}}{1-\frac{1}{e^{2n}}}=0$

The sequence converges to 0.

23. $\lim\limits_{x\to\infty}\dfrac{x2^x}{3^x}=\lim\limits_{x\to\infty}\dfrac{x}{\frac{3^x}{2^x}}=\lim\limits_{x\to\infty}\dfrac{x}{\left(\frac{3}{2}\right)^x}$

$=\lim\limits_{x\to\infty}\dfrac{1}{\left(\frac{3}{2}\right)^x\ln\frac{3}{2}}=0,$ by l'Hôpital's

Rule, since $\lim\limits_{x\to\infty}\left(\dfrac{3}{2}\right)^x=\infty$.

Hence, $\lim\limits_{n\to\infty}\dfrac{n2^n}{3^n}=0$, by Theorem 1.2.
The sequence converges to 0.

25. Since $\dfrac{n!}{2^n}=\dfrac{1}{2}\cdot\dfrac{2}{2}\cdots\dfrac{n}{2}\ge\dfrac{1}{2}\cdot\dfrac{n}{2}=\dfrac{n}{4}$ and

$\lim\limits_{n\to\infty}\dfrac{n}{4}=\infty$, we have $\lim\limits_{n\to\infty}\dfrac{n!}{2^n}=\infty$.
The sequence diverges.

27. Since $\lim\limits_{n\to\infty}\dfrac{2n+1}{n}=2,$

so $\lim\limits_{n\to\infty}[\ln(2n+1)-\ln n]$

$=\lim\limits_{n\to\infty}\ln\dfrac{2n+1}{n}=\ln 2.$

The sequence converges to $\ln 2$.

29. $-1\le\cos n\le 1\Rightarrow\dfrac{-1}{n^2}\le\dfrac{\cos n}{n^2}\le\dfrac{1}{n^2}$

for all n, and $\lim\limits_{n\to\infty}\dfrac{-1}{n^2}=\lim\limits_{n\to\infty}\dfrac{1}{n^2}=0,$

so by the Squeeze Theorem,

$\lim\limits_{n\to\infty}\dfrac{\cos n}{n^2}=0.$

31. $0\le|a_n|=\dfrac{1}{ne^n}\le\dfrac{1}{n}$ for $n\ge 1$ and $\lim\limits_{n\to\infty}\dfrac{1}{n}=0,$
so by the Squeeze Theorem and Corollary 1.1,

$\lim\limits_{n\to\infty}\dfrac{(-1)^n}{ne^n}=0.$

33. $\dfrac{a_{n+1}}{a_n}=\left(\dfrac{n+4}{n+3}\right)\cdot\left(\dfrac{n+2}{n+3}\right)$

$=\dfrac{n^2+6n+8}{n^2+6n+9}<1$ for all n, so

$a_{n+1}<a_n$ for all n, so the sequence is decreasing.

35. $\dfrac{a_{n+1}}{a_n} = \left(\dfrac{e^{n+1}}{n+1}\right) \cdot \left(\dfrac{n}{e^n}\right) = \dfrac{e \cdot n}{n+1} > 1$

for all n, so $a_{n+1} > a_n$ for all n, so $\{a_n\}_{n=1}^{\infty}$ is increasing.

37. $\dfrac{a_{n+1}}{a_n} = \left(\dfrac{2^{n+1}}{(n+2)!}\right) \cdot \left(\dfrac{(n+1)!}{2^n}\right)$

$= \dfrac{2}{n+2} < 1$ for all n, so $a_{n+1} < a_n$

for all n, so the sequence is decreasing.

39. $|a_n| = \left|\dfrac{3n^2 - 2}{n^2 + 1}\right| = \dfrac{3n^2 - 2}{n^2 + 1} 4$

$< \dfrac{3n^2}{n^2 + 1} < \dfrac{3n^2}{n^2} = 3$

41. $|a_n| = \left|\dfrac{\sin(n^2)}{n+1}\right| \le \dfrac{1}{n+1} < \dfrac{1}{2}$ for $n > 1$. For

$n = 1, a_1 = \dfrac{\sin 1}{2} \approx 0.4207 < \dfrac{1}{2}$. Thus, $|a_n| <$

$\dfrac{1}{2}$ for all $n \ge 1$.

43. $a_{1000} \approx 7.374312390$,
$e^2 \approx 7.389056099$
$b_{1000} \approx 0.135064522$,
$e^{-2} \approx 0.135335283$.
$\lim\limits_{n \to \infty} a_n = e^2$ and $\lim\limits_{n \to \infty} b_n = e^{-2}$.

45. If side $s = 12''$ and the diameter $D = \dfrac{12}{n}$, then
the number of disks that fit along one side is
$\dfrac{12}{\left(\frac{12}{n}\right)} = n$. Thus, the total number of disks is

$\dfrac{12}{\left(\frac{12}{n}\right)} \cdot \dfrac{12}{\left(\frac{12}{n}\right)} = n \cdot n = n^2$.

$a_n =$ wasted area in box with n^2 disks

$= 12 \cdot 12 - n^2 \left(\dfrac{6}{n}\right)^2 \pi$

$= 144 - 36\pi \approx 30.9 \text{ in}^2$

47. $a_0 = 3.049$,

$a_1 = 3.049 + 0.005(3.049)^{2.01} = 3.096$

$a_2 = 3.096 + 0.005(3.096)^{2.01} = 3.144$

$a_3 = 3.144 + 0.005(3.144)^{2.01} = 3.194$

$a_{10} = 3.594 <$ population in 1970

$a_{20} = 4.376 <$ population in 1980

$a_{30} = 5.589 >$ population in 1990

Estimated population in 2035 is
$a_{75} = 4.131 \times 10^{53}$ billion.

49. In the 3rd month, only the adult rabbits have
newborns, so $a_3 = 2 + 1 = 3$. In the 4th month,
only the 2 pairs of adult rabbits from a_2 can
have newborns, so $a_4 = 3 + 2 = 5$.

In general, $a_n = a_{n-1} + a_{n-2}$.

51. For the first 8 terms we get

$a_1 = 1$,

$a_2 = 2.5$,

$a_3 = 2.05$,

$a_4 = 2.000609756$,

$a_5 = 2.000000093$,

$a_6 = 2.000000000$,

$a_7 = 2.000000000$,

$a_8 = 2.000000000$.

The equation $L = \dfrac{1}{2}\left(L + \dfrac{4}{L}\right)$ has two solutions, $L = 2$ and $L = -2$. Since the terms of
the sequence are positive, we discard the negative solution. Thus the limit must be $L = 2$.

53. Use induction to show that $a_{n+1} > a_n$.

First note that $a_1^2 = 2$, and since $a_2^2 = 2 + \sqrt{2}$,
it follows that $a_2^2 > a_1^2$, so $a_2 > a_1$. Thus, the
statement is true for $n = 1$.

Now assume that the statement is true for
$n = k$ (that is, $a_{k+1} > a_k$), and show that the
statement is true for $n = k + 1$. First note that

$$a_{k+2} = \sqrt{2 + \sqrt{a_{k+1}}}$$

$$a_{k+2}^2 = 2 + \sqrt{a_{k+1}}$$

$$a_{k+2}^2 - 2 = \sqrt{a_{k+1}}$$

$$\left(a_{k+2}^2 - 2\right)^2 = a_{k+1}$$

The previous equation is also valid if we replace k by $k-1$: $\left(a_{k+1}^2 - 2\right)^2 = a_k$. Since $a_{k+1} > a_k$, it follows that $a_{k+1} > \left(a_{k+1}^2 - 2\right)^2$ and therefore $\left(a_{k+2}^2 - 2\right)^2 > \left(a_{k+1}^2 - 2\right)^2$,

$$a_{k+2}^2 - 2 > a_{k+1}^2 - 2,$$

$$a_{k+2}^2 > a_{k+1}^2, \; a_{k+2} > a_{k+1}.$$

Thus, by induction, a_n is increasing.

Now we'll prove that $a_n < 2$ by induction. First note that $a_1 < 2$, and assume that $a_k < 2$. Then

$$a_{k+1} = \sqrt{2 + \sqrt{a_k}}$$

$$a_{k+1}^2 - 2 = \sqrt{a_k}, \; \left(a_{k+1}^2 - 2\right)^2 = a_k$$

$$\left(a_{k+1}^2 - 2\right)^2 < 2, \; a_{k+1}^2 - 2 < \sqrt{2}$$

$$a_{k+1}^2 < 2 + \sqrt{2}, \; a_{k+1}^2 < 4$$

$$a_{k+1} < 2$$

Thus, by induction, $a_n < 2$.

Since a_n is increasing and bounded above by 2, a_n converges. To estimate the limit, we'll approximate the solution of $x = \sqrt{2 + \sqrt{x}}$:

$$x^2 = 2 + \sqrt{x}, \; (x^2 - 2)^2 = x$$
$$x^4 - 4x^2 + 4 = x, \; 0 = x^4 - 4x^2 - x + 4$$
$$0 = (x - 1)(x^3 + x^2 - 3x - 4)$$

Since $a_n > \sqrt{2}$, it follows that $x \neq 1$. Therefore, $0 = x^3 + x^2 - 3x - 4$. Using a CAS, the solution is $x \approx 1.8312$.

55.
$$p > 1 \Rightarrow \lim_{n \to \infty} \frac{1}{p^n} = 0$$

$$p = 1 \Rightarrow \lim_{n \to \infty} \frac{1}{p^n} = 1$$

$$0 < p < 1 \Rightarrow \lim_{n \to \infty} \frac{1}{p^n} \text{ does not exist}$$

$$-1 < p < 0 \Rightarrow \lim_{n \to \infty} \frac{1}{p^n} \text{ does not exist}$$

$$p = -1 \Rightarrow \lim_{n \to \infty} \frac{1}{p^n} \text{ does not exist}$$

$$p < -1 \Rightarrow \lim_{n \to \infty} \frac{1}{p^n} = 0$$

Therefore, the sequence $a_n = 1/p^n$ converges for $p < -1$ and $p \geq 1$.

57.
$$a_n = \frac{1}{n^2}(1 + 2 + 3 + \cdots + n)$$

$$= \frac{1}{n^2}\left(\frac{n(n+1)}{2}\right) = \frac{n+1}{2n}$$

$$= \frac{1}{2}\left(1 + \frac{1}{n}\right)$$

$$\lim_{n \to \infty} a_n = \frac{1}{2}\lim_{n \to \infty}\left(1 + \frac{1}{n}\right) = \frac{1}{2}$$

Thus, the sequence a_n converges to $1/2$. Note that

$$\int_0^1 x \, dx = \lim_{n \to \infty}\left(\sum_{k=1}^n \frac{1}{n}\frac{k}{n}\right).$$

Therefore the sequence a_n converges to $\int_0^1 x \, dx$.

59. Begin by joining the centers of circles C_1 and C_2 with a line segment. The length of this line segment is the sum of the radii of the two circles, which is $r_1 + r_2$. Thus, the square of the length of the line segment is $(r_1 + r_2)^2$. Now, the coordinates of the centers of the circles are (c_1, r_1) and (c_2, r_2). Using the formula for the distance between two points, the square of the length of the line segment joining the two centers is $(c_2 - c_1)^2 + (r_2 - r_1)^2$. Equating the two expressions, we get $(c_2 - c_1)^2 + (r_2 - r_1)^2 = (r_1 + r_2)^2$.

Expanding and simplifying this relationship, we get

$$(c_2 - c_1)^2 = (r_1 + r_2)^2 - (r_2 - r_1)^2$$
$$= r_1^2 + 2r_1r_2 + r_2^2 - (r_2^2 - 2r_1r_2 + r_1^2)$$
$$= 4r_1r_2.$$
$$|c_2 - c_1| = 2\sqrt{r_1r_2}$$

The same reasoning applied to the other two pairs of centers yields analogous results. Without going through the motions again, you can simply take the results above and first replace all subscripts "2" by "3" to get the results for circles C_1 and C_3. Then take these new results and replace all subscripts "1" by "2" to get the results for circles C_2 and C_3. The results are

$$(c_3 - c_1)^2 + (r_3 - r_1)^2 = (r_1 + r_3)^2$$
$$|c_3 - c_1| = 2\sqrt{r_1r_3}$$
$$(c_3 - c_2)^2 + (r_3 - r_2)^2 = (r_2 + r_3)^2$$
$$|c_3 - c_2| = 2\sqrt{r_2r_3}$$

Finally, $|c_1 - c_2| = |c_1 - c_3| + |c_3 - c_2|$
$$2\sqrt{r_1r_2} = 2\sqrt{r_1r_3} + 2\sqrt{r_2r_3}$$
$$\sqrt{r_1r_2} = \sqrt{r_3}\left(\sqrt{r_1} + \sqrt{r_2}\right)$$
$$\sqrt{r_3} = \frac{\sqrt{r_1r_2}}{\sqrt{r_1} + \sqrt{r_2}}$$

61. Let (x_0, y_0) be one point of intersection of the circle and parabola. The distance between the two points $(0, c)$ and (x_0, y_0), where $y_0 = x_0^2$, is r.

$$\sqrt{x_0^2 + (x_0^2 - c)^2} = r$$
$$x_0^2 + x_0^4 - 2cx_0^2 + c^2 = r^2$$
$$y_0^2 + (1 - 2c)y_0 + (c^2 - r^2) = 0$$

We want the solution y_0 to the above equation to be unique, so that
$$(1 - 2c)^2 - 4(c^2 - r^2) = 0$$
$$1 - 4c + 4c^2 - 4c^2 + 4r^2 = 0$$
$$1 - 4c + 4r^2 = 0, \quad c = \frac{1}{4} + r^2$$

9.2 INFINITE SERIES

1. $\sum_{k=0}^{\infty} 3\left(\frac{1}{5}\right)^k$ is a geometric series with $a = 3$ and $|r| = \frac{1}{5} < 1$, so it converges to $\frac{3}{1 - 1/5} = \frac{15}{4}$.

3. $\sum_{k=0}^{\infty} \frac{1}{2}\left(-\frac{1}{3}\right)^k$ is a geometric series with $a = \frac{1}{2}$ and $|r| = \frac{1}{3} < 1$, so it converges to $\frac{1/2}{1 - (-1/3)} = \frac{3}{8}$.

5. $\sum_{k=0}^{\infty} \frac{1}{2}(3)^k$ is a geometric series with $|r| = 3 > 1$, so it diverges.

7. Using partial fractions,
$$S_n = \sum_{k=1}^{n} \frac{4}{k(k+2)} = \sum_{k=1}^{n}\left(\frac{2}{k} - \frac{2}{k+2}\right)$$
$$= \left(2 - \frac{2}{3}\right) + \left(1 - \frac{2}{4}\right) + \left(\frac{2}{3} - \frac{2}{5}\right)$$
$$+ \cdots + \left(\frac{2}{n} - \frac{2}{n+2}\right)$$
$$= 2 + 1 - \frac{2}{n+1} - \frac{2}{n+2}$$
$$= 3 - \frac{4n+6}{n^2 + 3n + 2}$$
and
$$\lim_{n\to\infty} S_n$$
$$= \lim_{n\to\infty}\left(3 - \frac{4n+6}{n^2 + 3n + 2}\right) = 3.$$

Thus, the series converges to 3.

9. $\lim\limits_{k\to\infty} \dfrac{3k}{k+4} = 3 \neq 0$, so by the kth-Term Test for Divergence, the series diverges.

11. $\sum\limits_{k=1}^{\infty} \dfrac{2}{k} = 2\sum\limits_{k=1}^{\infty} \dfrac{1}{k}$ and from Example 2.7, $\sum\limits_{k=1}^{\infty} \dfrac{1}{k}$ diverges, so $2\sum\limits_{k=1}^{\infty} \dfrac{1}{k}$ diverges.

13. Using partial fractions,

$$S_n = \sum_{k=1}^{n} \frac{2k+1}{k^2(k+1)^2}$$

$$= \sum_{k=1}^{n} \left[\frac{1}{k^2} - \frac{1}{(k+1)^2}\right]$$

$$= \left(1 - \frac{1}{4}\right) + \left(\frac{1}{4} - \frac{1}{9}\right) + \left(\frac{1}{9} - \frac{1}{16}\right)$$

$$+ \cdots + \left(\frac{1}{(n-1)^2} - \frac{1}{n^2}\right)$$

$$+ \left(\frac{1}{n^2} - \frac{1}{(n+1)^2}\right)$$

$$= 1 - \frac{1}{(n+1)^2}$$

and

$$\lim_{n\to\infty} S_n = \lim_{n\to\infty}\left(1 - \frac{1}{(n+1)^2}\right) = 1.$$

Thus the series converges to 1.

15. $\sum\limits_{k=2}^{\infty} \dfrac{1}{3^k}$ is a geometric series with

$a = \dfrac{1}{9}$ and $|r| = \dfrac{1}{3} < 1$, so it converges to

$\dfrac{1/9}{1 - 1/3} = \dfrac{1}{6}.$

17. $\sum\limits_{k=0}^{\infty} \left(\dfrac{1}{2^k} - \dfrac{1}{k+1}\right) = \sum\limits_{k=0}^{\infty} \dfrac{1}{2^k} - \sum\limits_{k=0}^{\infty} \dfrac{1}{k+1}.$

The first series is a convergent geometric series but the second series is the divergent harmonic series, so the original series diverges.

19. $\sum\limits_{k=2}^{\infty} \left(\dfrac{2}{3^k} + \dfrac{1}{2^k}\right) = \sum\limits_{k=2}^{\infty} \dfrac{2}{3^k} + \sum\limits_{k=2}^{\infty} \dfrac{1}{2^k}.$

The first series is a geometric series with

$a = \dfrac{2}{9}$ and $|r| = \dfrac{1}{3} < 1$, so it converges to

$\dfrac{2/9}{1 - 1/3} = \dfrac{1}{3}.$

The second series is a geometric series with

$a = \dfrac{1}{4}$ and $|r| = \dfrac{1}{2} < 1$, so it converges to

$\dfrac{1/4}{1 - 1/2} = \dfrac{1}{2}.$

Thus

$$\sum_{k=0}^{\infty} \left(\frac{2}{3^k} + \frac{1}{2^k}\right) = \frac{1}{3} + \frac{1}{2} = \frac{5}{6}.$$

21. $\lim\limits_{k\to\infty} |a_k| = \lim\limits_{k\to\infty} \dfrac{3k}{k+1} = 3 \neq 0$

So by the kth-Term Test for Divergence, the series diverges.

23. The series appears to converge.

n	$\sum\limits_{k=1}^{n} \dfrac{1}{k^2}$
2	1.250000000
4	1.423611111
8	1.527422052
16	1.584346533
32	1.614167263
64	1.629430501
128	1.637152005
256	1.641035436
512	1.642982848
1024	1.643957981

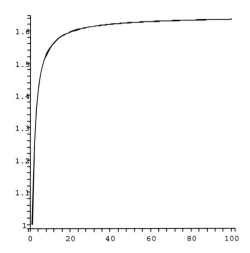

$$L = \sum_{k=1}^{\infty} a_k = \sum_{k=1}^{m-1} a_k + \sum_{k=m}^{\infty} a_k$$

$$= S_{m-1} + \sum_{k=m}^{\infty} a_k.$$

So $\displaystyle\sum_{k=m}^{\infty} a_k = L - S_{m-1}$, and thus converges.

29. Assume $\displaystyle\sum_{k=1}^{\infty} a_k$ converges to A and $\displaystyle\sum_{k=1}^{\infty} b_k$ converges to B. Then the sequences of partial sums converge, and letting

$$S_n = \sum_{k=1}^{n} a_k \text{ and } T_n = \sum_{k=1}^{n} b_k,$$

we have $\displaystyle\lim_{n \to \infty} S_n = A$ and $\displaystyle\lim_{n \to \infty} T_n = B$.

Let $Q_n = \displaystyle\sum_{k=1}^{n} (a_k + b_k)$, the sequence of partial sums for $\displaystyle\sum_{k=1}^{\infty} (a_k + b_k)$. Since S, T, and Q are all finite sums,

$$Q_n = S_n + T_n.$$

Then by Theorem 1.1(i),

$$A + B = \lim_{n \to \infty} S_n + \lim_{n \to \infty} T_n = \lim_{n \to \infty} (S_n + T_n)$$

$$= \lim_{n \to \infty} Q_n = \sum_{k=1}^{\infty} (a_k + b_k).$$

The proofs for $\displaystyle\sum_{k=1}^{\infty} (a_k - b_k)$ and $\displaystyle\sum_{k=1}^{\infty} c a_k$ are similar.

25. The series appears to converge.

n	$\displaystyle\sum_{k=1}^{n} \frac{1}{k!}$
2	4.500000000
4	5.125000000
8	5.154836310
16	5.154845485
32	5.154845485
1024	5.154845485

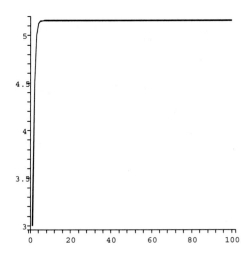

31. Let $S_n = \displaystyle\sum_{k=1}^{n} \frac{1}{k}$. Then

$$S_1 = 1 \text{ and } S_2 = 1 + \frac{1}{2} = \frac{3}{2}.$$

Since $S_8 > \dfrac{5}{2}$, we have

$$S_{16} = S_8 + \frac{1}{9} + \frac{1}{10} + \cdots + \frac{1}{16}$$

$$> S_8 + 8 \left(\frac{1}{16} \right)$$

$$= S_8 + \frac{1}{2} > \frac{5}{2} + \frac{1}{2} = 3. \text{ So } S_{16} > 3.$$

27. Assume $\displaystyle\sum_{k=1}^{\infty} a_k$ converges to L. Then for any m,

$$S_{32} = S_{16} + \frac{1}{17} + \frac{1}{18} + \cdots + \frac{1}{32}$$

$$> S_{16} + 16\left(\frac{1}{32}\right)$$

$$= S_{16} + \frac{1}{2} > 3 + \frac{1}{2} = \frac{7}{2}. \text{ So } S_{32} > \frac{7}{2}.$$

If $n = 64$, then $S_{64} > 4$. If $n = 256$, then $S_{256} > 5$. If $n = 4^{m-1}$, then $S_n > m$.

33. $0.9 + 0.09 + 0.009 + \cdots = \sum\limits_{k=0}^{\infty} 0.9(0.1)^k$

which is a geometric series with $a = 0.9$ and $|r| = 0.1 < 1$, so it converges to $\dfrac{0.9}{1 - 0.1} = 1$.

35. $0.18\overline{1818} = \dfrac{18}{100} + \dfrac{18}{10,000} + \cdots$

$$= 18 \sum_{k=1}^{\infty} \frac{1}{100^k}.$$

This is a geometric series with $a = \dfrac{18}{100}$ and $|r| = \dfrac{1}{100}$, so the sum is

$$\frac{18/100}{1 - 1/100} = \frac{2}{11}.$$

37. The amount of overhang is

$\sum\limits_{k=0}^{n-1} \dfrac{L}{2(n-k)}$. So if $n = 8$, then

$$\sum_{k=0}^{7} \frac{L}{2(8-k)} \approx 1.3589L.$$

When $n = 4$,

$$L \sum_{k=0}^{n-1} \frac{1}{2(n-k)} = L \sum_{k=0}^{3} \frac{1}{2(4-k)}$$
$$= 1.0417L > L.$$

$$\lim_{n\to\infty} \sum_{k=0}^{n-1} \frac{L}{2(n-k)} = \lim_{n\to\infty} \sum_{k=1}^{n} \frac{L}{2k}$$

$$= \frac{L}{2} \lim_{n\to\infty} \sum_{k=1}^{n} \frac{1}{k} = \infty$$

39. If $0 < 2r < 1$, then

$$1 + 2r + (2r)^2 + \cdots = \sum_{k=0}^{\infty} (2r)^k$$

is a geometric series with $a = 1$ and $|2r| = 2r < 1$, so it converges to $\dfrac{1}{1 - 2r}$.

If $r = \dfrac{1}{1000}$, then

$$1 + 0.002 + 0.000004 + \cdots$$

$$= \frac{1}{1 - 2\left(\frac{1}{1000}\right)}$$

$$= \frac{500}{499} = 1.002004008 \ldots .$$

41. $p^2 + 2p(1-p)p^2 + \left[2p(1-p)\right]^2 p^2 + \cdots$

$$= \sum_{k=0}^{\infty} p^2 \left[2p(1-p)\right]^k$$

is a geometric series with $a = p^2$ and $|r| = 2p(1-p) < 1$ because $2p(1-p)$ is a probability and therefore must be between 0 and 1. So the series converges to

$$\frac{p^2}{1 - \left[2p(1-p)\right]} = \frac{p^2}{1 - 2p + 2p^2}.$$

If $p = 0.6$,

$$\frac{0.6^2}{1 - 2(0.6) + 2(0.6)^2} \approx 0.692 > 0.6.$$

If $p > \dfrac{1}{2}, \dfrac{p^2}{1 - 2p(1-p)} > p$.

43. $d + de^{-r} + de^{-2r} + \cdots = \sum\limits_{k=0}^{\infty} d(e^{-r})^k$

which is a geometric series with $a = d$ and $\left|e^{-r}\right| = e^{-r} < 1$ if $r > 0$.

So $\sum\limits_{k=0}^{\infty} d(e^{-r})^k = \dfrac{d}{1 - e^{-r}}$.

If $r = 0.1$,

$$\sum_{k=0}^{\infty} d(e^{-0.1})^k = \frac{d}{1 - e^{-0.1}} = 2,$$

so $d \approx 2(1 - 0.905) \approx 0.19$.

45. $\displaystyle\sum_{k=0}^{\infty} 100{,}000 \left(\frac{3}{4}\right)^k$

$= \dfrac{100{,}000}{1 - 3/4} = \$400{,}000$

$\displaystyle\sum_{k=0}^{\infty} 100{,}000(0.6)^k = \dfrac{100{,}000}{1 - 0.6} = \$250{,}000$

There is a reduction of \$150,000.

47. $\displaystyle\sum_{k=1}^{\infty} \frac{1}{k}$ and $\displaystyle\sum_{k=1}^{\infty} \frac{-1}{k}$

49. The geometric series

$1 + r + r^2 + r^3 + \cdots$

converges to $S = \dfrac{1}{1-r}$ provided that $-1 < r < 1$.

But $-1 < r$ implies that $-r < 1$, and so $1 - r < 2$, and therefore $\dfrac{1}{1-r} > \dfrac{1}{2}$, which means that

$S > \dfrac{1}{2}$.

51. Since $\displaystyle\sum a_k$ converges, $\displaystyle\lim_{k\to\infty} a_k = 0$, by Theorem 2.2. Therefore, $\displaystyle\lim_{k\to\infty} \dfrac{1}{a_k}$ does not exist. In particular, $\displaystyle\lim_{k\to\infty} \dfrac{1}{a_k} \neq 0$.

Therefore, $\displaystyle\sum \dfrac{1}{a_k}$ diverges, by the converse of Theorem 2.2.

53. Since $0 < p < 1$, therefore $-1 < -p < 0$, and $0 < 1 - p < 1$. Thus, the given series is geometric with common ratio $r = 1 - p$, and so converges:

$\displaystyle\sum_{n=1}^{\infty} p(1-p)^{n-1} = p \sum_{n=1}^{\infty} r^{n-1}$

$= p\dfrac{1}{1-r} = \dfrac{p}{1-(1-p)} = \dfrac{p}{p} = 1.$

The sum represents the probability that you eventually win a game.

9.3 THE INTEGRAL TEST AND COMPARISON TESTS

1. $\displaystyle\sum_{k=1}^{\infty} \frac{4}{\sqrt[3]{k}}$ is a divergent p-series, because

$p = \dfrac{1}{3} < 1.$

3. $\displaystyle\sum_{k=4}^{\infty} \frac{1}{k^{11/10}}$ is a convergent p-series, since

$p = \dfrac{11}{10} > 1.$

5. Using the Limit Comparison Test, let $a_k = \dfrac{k+1}{k^2 + 2k + 3}$ and $b_k = \dfrac{1}{k}$, so $\displaystyle\lim_{k\to\infty} \dfrac{a_k}{b_k}$

$= \displaystyle\lim_{k\to\infty} \left(\dfrac{k+1}{k^2 + 2k + 3}\right)\left(\dfrac{k}{1}\right)$

$= \displaystyle\lim_{k\to\infty} \dfrac{k^2 + k}{k^2 + 2k + 3} = 1 > 0$, and since $\displaystyle\sum_{k=3}^{\infty} \dfrac{1}{k}$ is the divergent harmonic series,

$\displaystyle\sum_{k=3}^{\infty} \dfrac{k+1}{k^2 + 2k + 3}$ diverges.

7. Using the Limit Comparison Test, let $a_k = \dfrac{4}{2 + 4k}$ and $b_k = \dfrac{1}{k}$, so $\displaystyle\lim_{k\to\infty} \dfrac{a_k}{b_k}$

$= \displaystyle\lim_{k\to\infty} \left(\dfrac{4}{2 + 4k}\right)\left(\dfrac{k}{1}\right) = \displaystyle\lim_{k\to\infty} \dfrac{4k}{2 + 4k} =$

$1 > 0$, and since $\displaystyle\sum_{k=1}^{\infty} \dfrac{1}{k}$ is the divergent harmonic series, $\displaystyle\sum_{k=1}^{\infty} \dfrac{4}{2 + 4k}$ diverges.

9. Let $f(x) = \dfrac{2}{x \ln x}$. Then f is continuous and positive on $[2, \infty]$ and

$f'(x) = \dfrac{-2(1 + \ln x)}{x^2(\ln x)^2} < 0$

for $x \in [2, \infty)$, so f is decreasing. So we can use the Integral Test,

$$\int_2^\infty \frac{2}{x \ln x}\, dx = \lim_{R \to \infty} \int_2^R \frac{2}{x \ln x}\, dx$$

$$= 2 \lim_{R \to \infty} [\ln(\ln x)]_2^R$$

$$= 2 \lim_{R \to \infty} [\ln(\ln R) - \ln(\ln 2)]$$

$$= \infty$$

so $\displaystyle\sum_{k=2}^\infty \frac{2}{k \ln k}$ diverges.

11. Using the Limit Comparison Test, let $a_k = \dfrac{2k}{k^3+1}$ and $b_k = \dfrac{1}{k^2}$.

Then

$$\lim_{k \to \infty} \frac{a_k}{b_k} = \lim_{k \to \infty} \left(\frac{2k}{k^3+1}\right) \cdot \left(\frac{k^2}{1}\right)$$

$$= \lim_{k \to \infty} \frac{2k^3}{k^3+1} = 2 > 0$$

and $\displaystyle\sum_{k=1}^\infty \frac{1}{k^2}$ is a convergent p-series ($p = 2 > 1$), so $\displaystyle\sum_{k=1}^\infty \frac{2k}{k^3+1}$ converges.

13. Let $f(x) = \dfrac{e^{1/x}}{x^2}$.

Then f is continuous and positive on $[1, \infty)$ and

$$f'(x) = \frac{-e^{1/x}(1+2x)}{x^4} < 0$$

for all $x \in [1, \infty)$, so f is decreasing. Therefore, we can use the Integral Test.

$$\int_1^\infty \frac{e^{1/x}}{x^2}\, dx = \lim_{R \to \infty} \int_1^R \frac{e^{1/x}}{x^2}\, dx$$

$$= \lim_{R \to \infty} \left.-e^{1/x}\right|_1^R = \lim_{R \to \infty} \left(e - e^{1/R}\right)$$

$$= e - 1$$

So the series $\displaystyle\sum_{k=1}^\infty \frac{e^{1/k}}{k^2}$ converges, and so does $\displaystyle\sum_{k=3}^\infty \frac{e^{1/k}}{k^2}$.

15. Let $f(x) = \dfrac{e^{-\sqrt{x}}}{\sqrt{x}}$. Then f is continuous and positive on $[1, \infty)$ and

$$f'(x) = \frac{-(\sqrt{x} - 1)}{2x^{3/2}e^{\sqrt{x}}} < 0$$

for $x \in [1, \infty)$. So f is decreasing. Therefore, we can use the Integral Test.

$$\int_1^\infty \frac{e^{-\sqrt{x}}}{\sqrt{x}}\, dx = \lim_{R \to \infty} \int_1^R \frac{e^{-\sqrt{x}}}{\sqrt{x}}\, dx$$

$$= \lim_{R \to \infty} \left[-2e^{-\sqrt{x}}\right]_1^R = \lim_{R \to \infty} \left[\frac{2}{e} - \frac{2}{e^{\sqrt{R}}}\right]$$

$$= \frac{2}{e}.$$

So $\displaystyle\sum_{k=1}^\infty \frac{e^{\sqrt{k}}}{\sqrt{k}}$ converges.

17. Using the Limit Comparison Test, let $a_k = \dfrac{2k^2}{k^{5/2}+2}$ and $b_k = \dfrac{1}{\sqrt{k}}$. Since

$$\lim_{k \to \infty} \frac{a_k}{b_k} = \lim_{k \to \infty} \frac{2k^2}{k^{5/2}+2} \frac{\sqrt{k}}{1}$$

$$= \lim_{k \to \infty} \frac{2k^{5/2}}{k^{5/2}+2} = 2 > 0$$

and $\displaystyle\sum_{k=1}^\infty \frac{1}{\sqrt{k}}$ is a divergent p-series, so

$$\sum_{k=1}^\infty \frac{2k^2}{k^{5/2}+2} \text{ diverges.}$$

19. Let $a_k = \dfrac{4}{\sqrt{k^3+1}}$ and $b_k = \dfrac{1}{k^{3/2}}$. Since

$$\lim_{k \to \infty} \frac{a_k}{b_k} = \lim_{k \to \infty} \frac{4}{\sqrt{k^3+1}} \frac{k^{3/2}}{1}$$

$$= \lim_{k \to \infty} \frac{4}{\sqrt{1+k^{-3}}} = 4 > 0 \text{ and } \sum_{k=1}^\infty \frac{1}{k^{3/2}} \text{ is}$$

a convergent p-series, $\displaystyle\sum_{k=0}^\infty \frac{4}{\sqrt{k^3+1}}$ converges, by the Limit Comparison Test.

21. Let $f(x) = \dfrac{\tan^{-1} x}{1 + x^2}$ which is continuous and positive on $[1, \infty)$ and

$$f'(x) = \frac{1 - 2x \tan^{-1} x}{(1 + x^2)^2} < 0 \text{ for } x \in [1, \infty),$$

so f is decreasing. So we can use the Integral Test.

$$\int_1^\infty \frac{\tan^{-1} x}{1 + x^2}\, dx = \lim_{R \to \infty} \int_1^R \frac{\tan^{-1} x}{1 + x^2}\, dx$$

$$= \lim_{R \to \infty} \frac{1}{2}(\tan^{-1} x)^2 \Big|_1^R$$

$$= \lim_{R \to \infty} \left[\frac{1}{2}(\tan^{-1} R)^2 - \frac{1}{2}(\tan^{-1} 1)^2 \right]$$

$$= \frac{1}{2}\left(\frac{\pi}{2}\right)^2 - \frac{1}{2}\left(\frac{\pi}{4}\right)^2 = \frac{3\pi^2}{32},$$

so $\displaystyle\sum_{k=1}^\infty \frac{\tan^{-1} k}{1 + k^2}$ converges.

23. Since $\left| \cos^2 k \right| \le 1$, $\dfrac{1}{\cos^2 k} > 1$, so by the kth-Term Test for Divergence, $\displaystyle\sum_{k=1}^\infty \frac{1}{\cos^2 k}$ diverges.

25. Let $f(x) = \dfrac{\ln x}{x}$, which is continuous and positive on $[2, \infty)$ and

$$f'(x) = \frac{1 - \ln x}{x^2} < 0$$

for $x \in [3, \infty)$, so f is decreasing on $[3, \infty)$. Therefore, we can use the Integral Test on $[3, \infty)$.

$$\int_3^\infty \frac{\ln x}{x}\, dx = \lim_{R \to \infty} \int_3^R \frac{\ln x}{x}\, dx$$

$$= \lim_{R \to \infty} \frac{(\ln x)^2}{2} \Big|_3^R$$

$$= \lim_{R \to \infty} \left[\frac{(\ln R)^2}{2} - \frac{(\ln 3)^2}{2} \right]$$

$$= \infty$$

So $\displaystyle\sum_{k=3}^\infty \frac{\ln k}{k}$ diverges and so does $\displaystyle\sum_{k=2}^\infty \frac{\ln k}{k}$.

27. Using the Limit Comparison Test, let $a_k = \dfrac{k^4 + 2k - 1}{k^5 + 3k^2 + 1}$ and $b_k = \dfrac{1}{k}$.

Then $\displaystyle\lim_{k \to \infty} \frac{a_k}{b_k} = \lim_{k \to \infty} \left(\frac{k^4 + 2k - 1}{k^5 + 3k^2 + 1} \right) \left(\frac{k}{1} \right)$

$$= \lim_{k \to \infty} \frac{k^5 + 2k^2 - k}{k^5 + 3k^2 + 1} = 1 > 0.$$

Since $\displaystyle\sum_{k=1}^\infty \frac{1}{k}$ is the divergent harmonic series, $\displaystyle\sum_{k=1}^\infty \frac{k^4 + 2k1}{k^5 + 3k^2 + 1}$ diverges, and so does $\displaystyle\sum_{k=4}^\infty \frac{k^4 + 2k1}{k^5 + 3k^2 + 1}$.

29. $\displaystyle\lim_{k \to \infty} \frac{k + 1}{k + 2} = 1 \ne 0$, so by the kth-Term Test for Divergence, $\displaystyle\sum_{k=3}^\infty \frac{k + 1}{k + 2}$ diverges.

31. Using the Limit Comparison Test, let $a_k = \dfrac{k + 1}{k^3 + 2}$ and $b_k = \dfrac{1}{k^2}$. Then $\displaystyle\lim_{k \to \infty} \frac{a_k}{b_k} =$

$$\lim_{k \to \infty} \left(\frac{k + 1}{k^3 + 2} \right) \left(\frac{k^2}{1} \right) = \lim_{k \to \infty} \frac{k^3 + k^2}{k^3 + 2} = 1 >$$

0 and $\displaystyle\sum_{k=1}^\infty \frac{1}{k^2}$ is a convergent p-series ($p = 2 > 1$), so $\displaystyle\sum_{k=8}^\infty \frac{k + 1}{k^3 + 2}$ converges.

33. $\dfrac{1}{(k + 1)\sqrt{k} + k\sqrt{k + 1}} < \dfrac{1}{k\sqrt{k} + k\sqrt{k}}$

$$0 < \frac{1}{(k + 1)\sqrt{k} + k\sqrt{k + 1}} < \frac{1}{2k^{3/2}}$$

The series $\displaystyle\sum_{k=1}^\infty \frac{1}{2k^{3/2}}$ converges, as it is a p-series with $p > 1$. Thus, by the Comparison Test, the original series converges.

35. We are only concerned with what happens to the terms as $k \to \infty$, so the first few terms don't matter.

37. Since $\displaystyle\lim_{k \to \infty} \frac{a_k}{b_k} = 0$, there exists N, so that for all $k > N$, $\left| \dfrac{a_k}{b_k} - 0 \right| < 1$. Since a_k and

b_k are positive, $\dfrac{a_k}{b_k} < 1$, so $a_k < b_k$. Thus, since $\displaystyle\sum_{k=1}^{\infty} b_k$ converges, $\displaystyle\sum_{k=1}^{\infty} a_k$ converges by the Comparison Test.

39. Case 1: Suppose that

$$\lim_{k\to\infty} \frac{|a_k|}{|b_k|} = 0.$$

Then there exists a natural number N such that, for all $k \geq N$, $|a_k| < |b_k|$. Thus, for all $k \geq N$, $|a_k b_k| < b_k^2$. Thus, by the Comparison Test, $\displaystyle\sum_{k=1}^{\infty} |a_k b_k|$ converges.

Case 2: Suppose that

$$\lim_{k\to\infty} \frac{|b_k|}{|a_k|} = 0.$$

Using the same reasoning as in Case 1, there exists a natural number N such that, for all $k \geq N$, $|b_k| < |a_k|$. Thus, for all $k \geq N$, $|a_k b_k| < a_k^2$. Thus, by the Comparison Test, $\displaystyle\sum_{k=1}^{\infty} |a_k b_k|$ converges.

The only other possibility is that $\displaystyle\lim_{k\to\infty} \frac{|a_k|}{|b_k|} = L$, where $L > 0$. Then it follows that $\displaystyle\lim_{k\to\infty} \frac{a_k^2}{|a_k b_k|} = L$. Thus, by the Limit Comparison Test, $\displaystyle\sum_{k=1}^{\infty} |a_k b_k|$ converges.

41. If $p \leq 1$, then the series diverges because $\dfrac{1}{k(\ln k)^p} \geq \dfrac{1}{k \ln k}$ (at least for $k > 2$) and $\displaystyle\sum_{k=2}^{\infty} \frac{1}{k \ln k}$ diverges (from Exercise 9). If $p > 1$, then let $f(x) = \dfrac{1}{x(\ln x)^p}$. f is continuous and positive on $[2, \infty)$, and

$$f'(x) = \frac{-(\ln(x))^{-p-1}(p + \ln x)}{x^2} < 0$$

so f is decreasing. Thus, we can use the Integral Test.

$$\int_2 \frac{1}{x(\ln x)^p}\, dx$$

$$= \lim_{R\to\infty} \int_2^R \frac{1}{x(\ln x)^p}\, dx$$

$$= \lim_{R\to\infty} \left.\frac{1}{(1-p)(\ln x)^{p-1}}\right|_2^R$$

$$= \lim_{R\to\infty} \left[\frac{1}{(1-p)(\ln R)^{p-1}} - \frac{1}{(1-p)(\ln 2)^{p-1}}\right]$$

$$= -\frac{1}{(1-p)(\ln 2)^{p-1}},$$

so $\displaystyle\sum_{k=2}^{\infty} \frac{1}{k(\ln k)^p}$ converges when $p > 1$.

43. If $p \leq 1$, then $k^p \leq k$ for all $k \geq 1$, so $\dfrac{1}{k^p} \geq \dfrac{1}{k}$ for all $k \geq 1$, and $\dfrac{\ln k}{k^p} \geq \dfrac{\ln k}{k} \geq \dfrac{1}{k}$ for all $k > 2$, and $\displaystyle\sum_{k=2}^{\infty} \frac{1}{k}$ diverges.

So by the Comparison Test, $\displaystyle\sum_{k=2}^{\infty} \frac{\ln k}{k^p}$ diverges.

If $p > 1$, then let $f(x) = \dfrac{\ln x}{x^p}$. Then f is continuous and positive on $[2, \infty)$ and

$$f'(x) = \frac{x^{p-1}(1 - p \ln x)}{x^2 p} < 0$$

for $x > 3$, so f is decreasing. Thus, we can use the Integral Test on $[3, \infty)$.

$$\int_3^\infty \frac{\ln x}{x^p}\, dx = \lim_{R\to\infty} \int_3^R \frac{\ln x}{x^p}\, dx$$

$$= \lim_{R\to\infty} \left.\frac{\ln x}{(1-p)x^{p-1}} - \frac{1}{(1-p)^2 x^{p-1}}\right|_3^R$$

$$= \lim_{R\to\infty} \left[\frac{\ln R}{(1-p)R^{p-1}} - \frac{1}{(1-p)^2 R^{p-1}}\right.$$

$$\left. - \frac{\ln 2}{(1-p)2^{p-1}} + \frac{1}{(1-p)^2 2^{p-1}}\right]$$

$$= \frac{1}{(1-p)^2 2^{p-1}} - \frac{\ln 2}{(1-p)2^{p-1}}$$

because

$$\lim_{R \to \infty} \frac{\ln R}{(1-p)R^{p-1}} = \lim_{R \to \infty} \frac{\frac{1}{R}}{-(1-p)^2 R^{p-2}}$$

$$= \lim_{R \to \infty} \frac{-1}{(1-p)^2 R^{p-1}} = 0$$

by l'Hôpital's Rule and

$$\lim_{R \to \infty} \frac{1}{(1-p)^2 R^{p-1}} = 0$$

because $p > 1$.

Thus by the Integral Test, $\sum_{k=R}^{\infty} \dfrac{\ln k}{k^p}$ converges when $p > 1$.

45. $R_{100} \leq \displaystyle\int_{100}^{\infty} \frac{1}{x^4}\, dx = \frac{1}{3 \cdot 100^3} \approx 3.33 \times 10^{-7}$

47. $R_{50} \leq \displaystyle\int_{50}^{\infty} \frac{6}{x^8}\, dx = \frac{6}{7 \cdot 50^7} \approx 1.097 \times 10^{-12}$

49. $R_{40} \leq \displaystyle\int_{40}^{\infty} xe^{-x^2}\, dx = \frac{1}{2}e^{-40^2} = \frac{1}{2}e^{-1600}$

51. $R_n \leq \displaystyle\int_{n}^{\infty} \frac{3}{x^4}\, dx = \frac{1}{n^3} < 10^{-6}$

So $n^3 > 10^6$ or $n > 10^2$. Thus, $n = 101$.

53. $R_n \leq \displaystyle\int_{n}^{\infty} xe^{-x^2}\, dx = \frac{1}{2}e^{-n^2} < 10^{-6}$.

Taking the natural logarithm, we need $n > \sqrt{\ln 500{,}000}$ or $n \geq 4$.

55. **(a)** can't tell

    (b) converges

    (c) converges

    (d) can't tell

57. $1 + \dfrac{1}{3} + \dfrac{1}{5} + \dfrac{1}{7} + \cdots = \displaystyle\sum_{k=0}^{\infty} \frac{1}{2k+1}$

Using the Limit Comparison Test, let $a_k = \dfrac{1}{2k+1}$ and $b_k = \dfrac{1}{k}$. Then

$$\lim_{k \to \infty} \frac{a_k}{b_k} = \lim_{k \to \infty} \left(\frac{1}{2k+1}\right)\left(\frac{k}{1}\right)$$

$$= \lim_{k \to \infty} \frac{k}{2k+1} = \frac{1}{2} > 0 \text{ and since } \sum_{k=1}^{\infty} \frac{1}{k} \text{ is the}$$

divergent harmonic series, then $\displaystyle\sum_{k=1}^{\infty} \frac{1}{2k+1}$

diverges.

59. If $x > 1$, $\zeta(x) = \displaystyle\sum_{k=1}^{\infty} \frac{1}{k^x}$ is a convergent p-series. If $x \leq 1$, $\zeta(x)$ is a divergent p-series.

61. $\zeta(4) = \displaystyle\sum_{k=1}^{\infty} \frac{1}{k^4} \approx \sum_{k=1}^{100} \frac{1}{k^4} \approx 1.0823$

63. $\zeta(8) = \displaystyle\sum_{k=1}^{\infty} \frac{1}{k^8} \approx \sum_{k=1}^{100} \frac{1}{k^8} \approx 1.0041$

65. First we'll compare $(\ln k)^{\ln k}$ to k^2. Let $L = \displaystyle\lim_{k \to \infty} \frac{(\ln k)^{\ln k}}{k^2}$ and calculate as follows:

$$L = \lim_{k \to \infty} \frac{(\ln k)^{\ln k}}{k^2}$$

$$\ln L = \lim_{k \to \infty} \ln \left[\frac{(\ln k)^{\ln k}}{k^2}\right]$$

$$= \lim_{k \to \infty} \left[\ln \left[(\ln k)^{\ln k}\right] - \ln(k^2)\right]$$

$$= \lim_{k \to \infty} \left[(\ln k)\ln(\ln k) - 2\ln k\right]$$

$$= \lim_{k \to \infty} (\ln k)\left[\ln(\ln k) - 2\right]$$

$$= \infty$$

Thus,

$$\lim_{k \to \infty} \frac{(\ln k)^{\ln k}}{k^2} = \infty.$$

This means that eventually $(\ln k)^{\ln k} > k^2$. In other words, there exists a natural number N such that for all $k \geq N$,

$$\frac{1}{(\ln k)^{\ln k}} < \frac{1}{k^2}$$

Thus, by the Comparison Test, $\displaystyle\sum_{k=2}^{\infty} \frac{1}{(\ln k)^{\ln k}}$ converges.

Now we'll use the Comparison Test to show that $\displaystyle\sum_{k=2}^{\infty} \frac{1}{(\ln k)^k}$ converges. We begin by noting that

$$\ln k < k$$
$$\Rightarrow (\ln k)^{\ln k} < (\ln k)^k$$
$$\Rightarrow \frac{1}{(\ln k)^k} < \frac{1}{(\ln k)^{\ln k}}$$

Therefore, by the Comparison Test,

$\displaystyle\sum_{k=2}^{\infty} \frac{1}{(\ln k)^k}$ converges.

67. $\displaystyle\int_1^{\infty} \frac{x}{2^x}\,dx = \lim_{R\to\infty} \int_1^R \frac{x}{2^x}\,dx$

$= \displaystyle\lim_{R\to\infty} \frac{1}{2^x}\left(-\frac{1}{(\ln 2)^2} - \frac{x}{\ln 2}\right)\Big|_2^R$

This integral converges since $\displaystyle\lim_{R\to\infty} \frac{R}{2^R} = 0$ by l'Hôpital's Rule.

$$\sum_{k=1}^{\infty} k\left(\frac{1}{2}\right)^k \approx \sum_{k=1}^{20} k\left(\frac{1}{2}\right)^k \approx 2.0$$

69.

$$\sum_{k=1}^{\infty} \frac{9k}{10^k} = 9 \sum_{k=1}^{\infty} k\frac{1}{10^k} =$$

$9\Bigg(\dfrac{1}{10}$

$+ \dfrac{1}{10^2} + \dfrac{1}{10^2}$

$+ \dfrac{1}{10^3} + \dfrac{1}{10^3} + \dfrac{1}{10^3}$

$+ \dfrac{1}{10^4} + \dfrac{1}{10^4} + \dfrac{1}{10^4} + \dfrac{1}{10^4}$

$+ \dfrac{1}{10^5} + \dfrac{1}{10^5} + \dfrac{1}{10^5} + \dfrac{1}{10^5} + \dfrac{1}{10^5} \cdots\Bigg)$

$= 9\left(\dfrac{1}{10} + \dfrac{1}{10^2} + \dfrac{1}{10^3} + \dfrac{1}{10^4} + \dfrac{1}{10^5} + \cdots\right.$

$+ \dfrac{1}{10^2} + \dfrac{1}{10^3} + \dfrac{1}{10^4} + \dfrac{1}{10^5} + \cdots$

$+ \dfrac{1}{10^3} + \dfrac{1}{10^4} + \dfrac{1}{10^5} \cdots\Big)$

$= 9\left(\displaystyle\sum_{k=1}^{\infty} \frac{1}{10^k} + \sum_{k=2}^{\infty} \frac{1}{10^k} + \sum_{k=3}^{\infty} \frac{1}{10^k} + \cdots\right)$

Each of these sums is a geometric series with $r = 1/10$, so we get

$9\left(\dfrac{1/10}{1-\frac{1}{10}} + \dfrac{1/10^2}{1-\frac{1}{10}} + \dfrac{1/10^3}{1-\frac{1}{10}} + \cdots\right)$

$= \dfrac{9}{9/10} \displaystyle\sum_{k=1}^{\infty} \frac{1}{10^k} = \left(\dfrac{9}{9/10}\right)\left(\dfrac{1/10}{1-\frac{1}{10}}\right)$

$= \left(\dfrac{9}{9/10}\right)\left(\dfrac{1/10}{9/10}\right) = \dfrac{10}{9}.$

71. $1 + \dfrac{10}{9} + \dfrac{10}{8} + \cdots + \dfrac{10}{1} \approx 29.29$

73. For n cards in the set, you will need an average of

$$\frac{n}{n} + \frac{n}{n-1} + \frac{n}{n-2} + \cdots \frac{n}{1}$$

attempts. The ratio is $\displaystyle\sum_{k=1}^{n} \frac{1}{k}$.

9.4 ALTERNATING SERIES

1. $\displaystyle\lim_{k\to\infty} a_k = \lim_{k\to\infty} \frac{3}{k} = 0$, and

$$0 < a_{k+1} = \frac{3}{k+1} \leq \frac{3}{k} = a_k$$

for all $k \geq 1$, so by the Alternating Series Test, the original series converges.

3. $\displaystyle\lim_{k\to\infty} a_k = \lim_{k\to\infty} \frac{4}{\sqrt{k}} = 0$ and

$$0 < a_{k+1} = \frac{4}{\sqrt{k+1}} < \frac{4}{\sqrt{k}} = a_k$$

for all $k \geq 1$. Thus by the Alternating Series Test, the original series converges.

5.
$$\lim_{k \to \infty} a_k = \lim_{k \to \infty} \frac{k}{k^2 + 2} = 0$$

and

$$\frac{a_{k+1}}{a_k} = \frac{k+1}{(k+1)^2 + 2} \cdot \frac{k^2 + 2}{k}$$

$$= \frac{k^3 + k^2 + 2k + 2}{k^3 + 2k^2 + 3k} \leq 1$$

for all $k \geq 2$, so $a_{k+1} < a_k$ for all $k \geq 2$. Thus by the Alternating Series Test, the original series converges.

7. By l'Hôpital's Rule,

$$\lim_{k \to \infty} a_k = \lim_{k \to \infty} \frac{k}{2^k} = \lim_{k \to \infty} \frac{1}{2^k \ln 2} = 0$$

and

$$\frac{a_{k+1}}{a_k} = \frac{k+1}{2^{k+1}} \cdot \frac{2^k}{k} = \frac{k+1}{2k} < 1$$

for all $k \geq 5$, so $a_{k+1} \leq a_k$ for all $k \geq 5$.

Thus by the Alternating Series Test, the original series converges.

9.
$$\lim_{k \to \infty} a_k = \lim_{k \to \infty} \frac{4^k}{k^2} = \lim_{k \to \infty} \frac{4^k \ln 4}{2k}$$

$$= \lim_{k \to \infty} \frac{4^k (\ln 4)^2}{2} = \infty \text{ (by l'Hôpital's Rule)}$$

So by the kth-Term Test for Divergence,
$$\sum_{k=1}^{\infty} (-1)^k \frac{4^k}{k^2} \text{ diverges.}$$

11.
$$\lim_{k \to \infty} a_k = \lim_{k \to \infty} \frac{2k}{k+1} = 2,$$
so by the kth-Term Test for Divergence,
$$\sum_{k=1}^{\infty} \frac{2k}{k+1} \text{ diverges.}$$

13.
$$\lim_{k \to \infty} a_k = \lim_{k \to \infty} \frac{3}{\sqrt{k+1}} = 0$$
and

$$\frac{a_{k+1}}{a_k} = \frac{3}{\sqrt{k+2}} \cdot \frac{\sqrt{k+1}}{3} = \frac{\sqrt{k+1}}{\sqrt{k+2}} < 1$$

for all $k \geq 3$, so $a_{k+1} < a_k$ for all $k \geq 3$. Thus by the Alternating Series Test, $\sum_{k=3}^{\infty} (-1)^k \frac{3}{\sqrt{k+1}}$ converges.

15.
$$\lim_{k \to \infty} a_k = \lim_{k \to \infty} \frac{2}{k!} = 0 \text{ and}$$

$$\frac{a_{k+1}}{a_k} = \frac{2}{(k+1)!} \cdot \frac{k!}{2} = \frac{1}{k+1} < 1$$

for all $k \geq 1$, so $a_{k+1} < a_k$ for all $k \geq 1$.

Thus by the Alternating Series Test,
$$\sum_{k=1}^{\infty} (-1)^{k+1} \frac{2}{k!} \text{ converges.}$$

17.
$$a_k = \frac{k!}{2^k} = \frac{1 \cdot 2 \cdot 3 \dots k}{2 \cdot 2 \cdot 2 \dots 2} \geq \frac{1}{2} \cdot \frac{k}{2} = \frac{k}{4}$$
and

$$\lim_{k \to \infty} = k = \infty$$

so by the kth-Term Test for Divergence
$$\sum_{k=2}^{\infty} (-1)^k \frac{k!}{2^k} \text{ diverges.}$$

19.
$$\lim_{k \to \infty} a_k = \lim_{k \to \infty} 2e^{-k} = 0 \text{ and}$$

$$\frac{a_{k+1}}{a_k} = \frac{2e^{-(k+1)}}{2e^{-k}} = \frac{1}{e} < 1$$

for all $k \geq 0$, so $a_{k+1} < a_k$ for all $k \geq 0$. Thus by the Alternating Series Test, $\sum_{k=5}^{\infty} (-1)^{k+1} 2e^{-k}$ converges.

21. Note that

$$\lim_{k \to \infty} a_k = \lim_{k \to \infty} \ln k = \infty,$$

so by the kth-Term Test for Divergence,
$$\sum_{k=2}^{\infty} (-1)^k \ln k \text{ diverges.}$$

23.
$$\lim_{k \to \infty} a_k = \lim_{k \to \infty} \frac{1}{2^k} = 0$$
and

$$\frac{a_{k+1}}{a_k} = \frac{1}{2^{k+1}} \cdot \frac{2^k}{1} = \frac{1}{2} < 1$$

for all $k \geq 0$, so $a_{k+1} < a_k$ for all $k \geq 0$. Thus by the Alternating Series Test, $\displaystyle\sum_{k=0}^{\infty} (-1)^{k+1} \frac{1}{2^k}$ converges.

25. $\left| S - S_k \right| \leq a_{k+1} = \dfrac{4}{(k+1)^3} \leq 0.01$

and $a_7 = 4/8^3 < 0.01$, so
$S \approx S_7 \approx 3.61$.

27. $\left| S - S_k \right| \leq a_{k+1} = \dfrac{k+1}{2^{k+1}} \leq 0.01$

If $k = 9$, $a_{10} = \dfrac{10}{2^{10}} \approx 0.00977 < 0.01$
So $S \approx S_9 \approx -0.22$.

29. $\left| S - S_k \right| \leq a_{k+1} = \dfrac{3}{(k+1)!} \leq 0.01$

If $k = 5$, $a_6 = \dfrac{3}{6!} \approx 0.0042 < 0.01$
So $S \approx S_5 = 1.10$.

31. $\left| S - S_k \right| \leq a_{k+1} = \dfrac{4}{(k+1)^4} \leq 0.01$,

so $400 \leq (k+1)^4$, then $\sqrt[4]{400} \leq k + 1$,
so $k \geq \sqrt[4]{400} - 1 \approx 3.47$ so $k \geq 4$.
Thus $S \approx S_4 \approx -0.21$.

33. $\left| S - S_k \right| \leq a_{k+1} = \dfrac{2}{k+1} < 0.0001$,

so $k + 1 > 20{,}000$ and then $k > 19{,}999$. Thus $k = 20{,}000$.

35. $\left| S - S_k \right| \leq a_{k+1} = \dfrac{10^{k+1}}{(k+1)!}$

Since $a_{34} = \dfrac{10^{34}}{34!} \approx 0.00003387$ is the first term such that $a_k < 0.0001$, 33 terms are needed.

37. If the derivative of a function $f(k) = a_k$ is negative, it means that the function is decreasing, i.e., each successive term is smaller than the one before. If

$f(k) = a_k = \dfrac{k}{k^2 + 2}$, then

$f'(k) = \dfrac{-k^2 + 2}{(k^2 + 2)^2} < 0$ for all $k \geq 2$

so a_k is decreasing.

39. The sum of the odd terms is

$1 + \dfrac{1}{3} + \dfrac{1}{5} + \dfrac{1}{7} + \cdots,$

which diverges to ∞. The sum of the even terms is $-\displaystyle\sum_{k=1}^{\infty} \dfrac{1}{(2k)^2}$, which is a convergent p-series. So the series diverges to ∞.

41. $\dfrac{3}{4} - \dfrac{3}{4}\left(\dfrac{3}{4}\right) + \ldots = \displaystyle\sum_{k=0}^{\infty} \left(\dfrac{3}{4}\right)\left(-\dfrac{3}{4}\right)^k$

which is a geometric series with $a = \dfrac{3}{4}$ and $|r| = \dfrac{3}{4} < 1$. So

$\displaystyle\sum_{k=0}^{\infty} \left(\dfrac{3}{4}\right)\left(-\dfrac{3}{4}\right)^k = \dfrac{3/4}{1 + 3/4} = \dfrac{3}{7}.$

The person ends up $\dfrac{3}{7}$ of the distance from home.

43. $S_{2n} = \displaystyle\sum_{i=1}^{2n} \dfrac{(-1)^{k+1}}{k} = 1 - \dfrac{1}{2} + \dfrac{1}{3} - \dfrac{1}{4} + \cdots - \dfrac{1}{2n}$

$= 1 + \left(\dfrac{1}{2} - 1\right) + \dfrac{1}{3} + \left(\dfrac{1}{4} - \dfrac{1}{2}\right)$

$+ \cdots + \left(\dfrac{1}{2n} - \dfrac{1}{n}\right)$

$= \left(1 + \dfrac{1}{2} + \dfrac{1}{3} + \dfrac{1}{4} + \cdots + \dfrac{1}{2n}\right)$

$- \left(1 + \dfrac{1}{2} + \cdots + \dfrac{1}{n}\right)$

$= \displaystyle\sum_{k=1}^{2n} \dfrac{1}{k} - \sum_{k=1}^{n} \dfrac{1}{k} = \sum_{k=n+1}^{2n} \dfrac{1}{k}$

Let $k = r + n$, so that $k = n + 1 \Leftrightarrow r = 1$ and $k = 2n \Leftrightarrow r = n$. Thus,

$$S_{2n} = \sum_{r=1}^{n} \frac{1}{r+n}.$$

Now relabel r as k to get

$$S_{2n} = \sum_{k=1}^{n} \frac{1}{k+n} = \frac{1}{n} \sum_{k=1}^{n} \frac{1}{1+\frac{k}{n}}.$$

The previous line is a Riemann sum for $\int_{1}^{2} \frac{1}{x}\, dx$. Thus,

$$\lim_{n\to\infty} S_{2n} = \int_{1}^{2} \frac{1}{x}\, dx = [\ln x]_{1}^{2}$$
$$= \ln 2 - \ln 1 = \ln 2.$$

45. An example is

$$a_k = b_k = \frac{(-1)^k}{\sqrt{k}}.$$

Then $\sum_{k=1}^{\infty} a_k$ and $\sum_{k=1}^{\infty} b_k$ converge by the Alternating Series Test. However,

$$\sum_{k=1}^{\infty} a_k b_k = \sum_{k=1}^{\infty} \frac{(-1)^{2k}}{k} = \sum_{k=1}^{\infty} \frac{1}{k}$$

which diverges.

9.5 ABSOLUTE CONVERGENCE AND THE RATIO TEST

1. By the Ratio Test,

$$\lim_{k\to\infty} \left| \frac{a_{k+1}}{a_k} \right| = \lim_{k\to\infty} \frac{3}{(k+1)!} \cdot \frac{k!}{3}$$
$$= \lim_{k\to\infty} \frac{1}{k+1} = 0 < 1$$

so $\sum_{k=0}^{\infty} (-1)^k \frac{3}{k!}$ converges absolutely.

3. By the Ratio Test,

$$\lim_{k\to\infty} \left| \frac{a_{k+1}}{a_k} \right| = \lim_{k\to\infty} \frac{2^{k+1}}{2^k}$$
$$= \lim_{k\to\infty} 2 = 2 > 1,$$

so $\sum_{k=0}^{\infty} (-1)^k 2^k$ diverges. (Or use kth-Term Test.)

5. By the Alternating Series Test,

$$\lim_{k\to\infty} a_k = \lim_{k\to\infty} \frac{k}{k^2+1} = 0$$

and $\frac{a_{k+1}}{a_k} = \frac{k+1}{(k+1)^2+1} \cdot \frac{k^2+1}{k}$

$$= \frac{k^3 + k^2 + k + 1}{k^3 + 2k^2 + 2k} < 1$$

for all $k \geq 1$, so $a_{k+1} < a_k$ for all $k \geq 1$, so the series converges.

But by the Limit Comparison Test, letting

$$a_k = \frac{k}{k^2+1} \text{ and } b_k = \frac{1}{k},$$

$$\lim_{k\to\infty} \frac{a_k}{b_k} = \lim_{k\to\infty} \frac{k^2}{k^2+1} = 1 > 0$$

and $\sum_{k=1}^{\infty} \frac{1}{k}$ is the divergent harmonic series.

Therefore $\sum_{k=1}^{\infty} |a_k| = \sum_{k=1}^{\infty} \frac{k}{k^2+1}$ diverges.

Thus $\sum_{k=1}^{\infty} (-1)^{k+1} \frac{k}{k^2+1}$ converges conditionally.

7. By the Ratio Test,

$$\lim_{k\to\infty} \left| \frac{a_{k+1}}{a_k} \right| = \lim_{k\to\infty} \frac{3^{k+1}}{(k+1)!} \cdot \frac{k!}{3^k}$$
$$= \lim_{k\to\infty} \frac{3}{k+1} = 0 < 1,$$

so $\sum_{k=3}^{\infty} (-1)^k \frac{3^k}{k!}$ converges absolutely.

9. $\lim_{k\to\infty} |a_k| = \lim_{k\to\infty} \frac{k}{2k+1} = \frac{1}{2}$

so by the kth-Term Test for Divergence,
$$\sum_{k=2}^{\infty} (-1)^{k+1} \frac{k}{2k+1} \text{ diverges.}$$

11. Using the Ratio Test,

$$\lim_{k \to \infty} \left| \frac{a_{k+1}}{a_k} \right| = \lim_{k \to \infty} \frac{(k+1)2^{k+1}}{3^{k+1}} \cdot \frac{3^k}{k2^k}$$

$$= \lim_{k \to \infty} \frac{2(k+1)}{3k} = \frac{2}{3} < 1,$$

so $\displaystyle\sum_{k=6}^{\infty} (-1)^k \frac{k2^k}{3^k}$ converges absolutely.

13. Using the Root Test,

$$\lim_{k \to \infty} \sqrt[k]{|a_k|} = \lim_{k \to \infty} \sqrt[k]{\left(\frac{4k}{5k+1} \right)^k}$$

$$= \lim_{k \to \infty} \frac{4k}{5k+1} = \frac{4}{5} < 1$$

so $\displaystyle\sum_{k=1}^{\infty} \left(\frac{4k}{5k+1} \right)^k$ converges absolutely.

15. $\displaystyle\sum_{k=1}^{\infty} \frac{1}{k}$ is the divergent harmonic series; so

$$\sum_{k=1}^{\infty} \frac{-2}{k} \text{ diverges.}$$

17. Using the Alternating Series Test, $\displaystyle\lim_{k \to \infty} a_k =$

$$\lim_{k \to \infty} \frac{\sqrt{k}}{k+1} = 0 \text{ and } \frac{a_{k+1}}{a_k} = \frac{\sqrt{k+1}}{k+2} \cdot \frac{k+1}{\sqrt{k}}$$

$$= \frac{(k+1)^{3/2}}{k^{3/2} + 2k^{1/2}} < 1 \text{ for all } k \geq 1,$$

so $a_{k+1} < a_k$ for all $k \geq 1$ so the series converges.

But by the Limit Comparison Test, letting

$$a_k = \frac{\sqrt{k}}{k+1} \text{ and } b_k = \frac{1}{k^{1/2}},$$

$$\lim_{k \to \infty} \frac{a_k}{b_k} = \lim_{k \to \infty} \frac{\sqrt{k}}{k+1} \cdot \frac{k^{1/2}}{1}$$

$$= \lim_{k \to \infty} \frac{k}{k+1} = 1 > 0,$$

and $\displaystyle\sum_{k=0}^{\infty} \frac{1}{k^{1/2}}$ is a divergent p-series

$$\left(p = \frac{1}{2} < 1 \right).$$

Therefore,

$$\sum_{k=0}^{\infty} |a_k| = \sum_{k=0}^{\infty} \frac{\sqrt{k}}{k+1} \text{ diverges.}$$

So $\displaystyle\sum_{k=0}^{\infty} (-1)^{k+1} \frac{\sqrt{k}}{k+1}$ converges conditionally.

19. Using the Ratio Test,

$$\lim_{k \to \infty} \left| \frac{a_{k+1}}{a_k} \right| = \lim_{k \to \infty} \frac{(k+1)^2}{e^{k+1}} \cdot \frac{e^k}{k^2}$$

$$= \lim_{k \to \infty} \frac{(k+1)^2}{ek^2} = \frac{1}{e} < 1,$$

so $\displaystyle\sum_{k=7}^{\infty} \frac{k^2}{e^k}$ converges absolutely.

21. Using the Root Test,

$$\lim_{k \to \infty} \sqrt[k]{|a_k|} = \lim_{k \to \infty} \sqrt[k]{\left(\frac{e^3}{k^3} \right)^k}$$

$$= \lim_{k \to \infty} \frac{e^3}{k^3} = 0 < 1,$$

so $\displaystyle\sum_{k=2}^{\infty} \frac{e^{3k}}{k^{3k}}$ converges absolutely.

23. Since $|\sin k| \leq 1$ for all k,

$$\left| \frac{\sin k}{k^2} \right| = \frac{|\sin k|}{k^2} \leq \frac{1}{k^2},$$

and $\displaystyle\sum_{k=1}^{\infty} \frac{1}{k^2}$ is a convergent p-series ($p = 2 >$

1) so by the Comparison Test, $\displaystyle\sum_{k=1}^{\infty} \left| \frac{\sin k}{k^2} \right|$

converges, and by Theorem 5.1, $\displaystyle\sum_{k=1}^{\infty} \frac{\sin k}{k^2}$

converges absolutely.

25. Since $\cos k\pi = (-1)^k$ for all k,

$$\left| \frac{\cos k\pi}{k} \right| = \left| \frac{(-1)^k}{k} \right| = \frac{1}{k}$$

and $\displaystyle\sum_{k=1}^{\infty} \frac{1}{k}$ is the divergent harmonic series,

so $\displaystyle\sum_{k=1}^{\infty} \left| \frac{\cos k\pi}{k} \right|$ diverges by the Comparison Test.

But, using the Alternating Series Test,

$$\lim_{k\to\infty} a_k = \lim_{k\to\infty} \frac{1}{k} = 0$$

and

$$\frac{a_{k+1}}{a_k} = \frac{1}{k+1} \cdot \frac{k}{1} = \frac{k}{k+1} < 1$$

for all $k \geq 1$, so $a_{k+1} < a_k$ for all $k \geq 1$, so $\displaystyle\sum_{k=1}^{\infty} \frac{(-1)^k}{k}$ converges.

Thus $\displaystyle\sum_{k=1}^{\infty} \frac{(-1)^k}{k} = \sum_{k=1}^{\infty} \frac{\cos k\pi}{k}$ converges conditionally.

27. Using the Alternating Series Test,

$$\lim_{k\to\infty} a_k = \lim_{k\to\infty} \frac{1}{\ln k} = 0$$

and

$$\frac{a_{k+1}}{a_k} = \frac{1}{\ln(k+1)} \cdot \frac{\ln k}{1} = \frac{\ln k}{\ln(k+1)} < 1$$

for all $k \geq 2$, so $a_{k+1} < a_k$ for all $k \geq 2$. So $\displaystyle\sum_{k=2}^{\infty} \frac{(-1)^k}{\ln k}$ converges.

But by the Comparison Test, because $\ln k < k$ for all $k \geq 2$, $\dfrac{1}{\ln k} > \dfrac{1}{k}$ for all $k \geq 2$ and $\displaystyle\sum_{k=2}^{\infty} \frac{1}{k}$ is the divergent harmonic series. Therefore, $\displaystyle\sum_{k=2}^{\infty} |a_k| = \sum_{k=2}^{\infty} \frac{1}{\ln k}$ diverges.

Thus $\displaystyle\sum_{k=2}^{\infty} \frac{(-1)^k}{\ln k}$ converges conditionally.

29. $\displaystyle\sum_{k=1}^{\infty} |a_k| = \sum_{k=1}^{\infty} \frac{1}{k^{3/2}}$, which is a convergent

p-series $\left(p = \dfrac{3}{2} > 1 \right)$, so by Theorem 5.1,

$\displaystyle\sum_{k=1}^{\infty} \frac{(-1)^k}{k\sqrt{k}}$ converges absolutely.

31. Consider $\displaystyle\sum_{k=1}^{\infty} \frac{1}{k^k}$. Using the Root Test,

$$\lim_{k\to\infty} \sqrt[k]{|a_k|} = \lim_{k\to\infty} \sqrt[k]{\left(\frac{1}{k}\right)^k}$$

$$= \lim_{k\to\infty} \frac{1}{k} = 0 < 1,$$

so $\displaystyle\sum_{k=1}^{\infty} \frac{1}{k^k}$ converges absolutely, thus

$3\displaystyle\sum_{k=3}^{\infty} \frac{1}{k^k} = \sum_{k=3}^{\infty} \frac{3}{k^k}$ converges absolutely.

33. Using the Ratio Test,

$$\lim_{k\to\infty} \left| \frac{a_{k+1}}{a_k} \right| = \lim_{k\to\infty} \frac{(k+1)!}{4^{k+1}} \cdot \frac{4^k}{k!}$$

$$= \lim_{k\to\infty} \frac{k+1}{4} = \infty > 1,$$

so $\displaystyle\sum_{k=6}^{\infty} (-1)^{k+1} \frac{k!}{4^k}$ diverges.

35. Using the Ratio Test,

$$\lim_{k\to\infty} \left| \frac{a_{k+1}}{a_k} \right| = \lim_{k\to\infty} \frac{(k+1)^{10}}{(2k+2)!} \cdot \frac{2k!}{k^{10}}$$

$$= \lim_{k\to\infty} \frac{(k+1)^{10}}{k^{10}(2k+1)(2k+2)}$$

$$= \lim_{k\to\infty} \frac{(k+1)^{10}}{4k^{12} + 6k^{11} + 2k^{10}}$$

$$= 0 < 1$$

so $\displaystyle\sum_{k=1}^{\infty} (-1)^{k+1} \frac{k^{10}}{(2k)!}$ converges absolutely.

37. Using the Ratio Test,

$$\lim_{k\to\infty}\left|\frac{a_{k+1}}{a_k}\right|$$

$$= \lim_{k\to\infty}\left|\frac{(-2)^{k+1}(k+2)}{5^{k+1}}\cdot\frac{5^k}{(-2)^k(k+1)}\right|$$

$$= \lim_{k\to\infty}\frac{2(k+2)}{5(k+1)} = \frac{2}{5} < 1$$

so $\displaystyle\sum_{k=0}^{\infty}\frac{(-2)^k(k+1)}{5^k}$ converges absolutely.

39. $\displaystyle\frac{\sqrt{8}}{9801}\sum_{k=0}^{0}\frac{(4k)!(1103+26{,}390k)}{(k!)^4 396^{4k}}$

$$= \frac{\sqrt{8}}{9801}(1103) \approx 0.318309878 \approx \frac{1}{\pi}$$

so $\pi \approx 3.14159273$

$$\frac{\sqrt{8}}{9801}\sum_{k=0}^{1}\frac{(4k)!(1103+26{,}390k)}{(k!)^4 396^{4k}}$$

$$= \frac{\sqrt{8}}{9801}(1103) + \frac{\sqrt{8}}{9801}\cdot\frac{4!(27{,}493)}{396^4}$$

$$\approx 0.318309886183791 \approx \frac{1}{\pi}$$

For comparison, the value of $1/\pi$ to 15 places is 0.318309886183791, so two terms of the series give this value correct to 15 places.

41. Using the Ratio Test,

$$\lim_{k\to\infty}\left|\frac{a_{k+1}}{a^k}\right| = \lim_{k\to\infty}\frac{(k+1)!}{(k+1)^{k+1}}\cdot\frac{k^k}{k!}$$

$$= \lim_{k\to\infty}\frac{k^k}{(k+1)^k} = \lim_{k\to\infty}\left(\frac{k}{1+k}\right)^k$$

$$= \frac{1}{e} < 1,$$

so $\displaystyle\sum_{k=1}^{\infty}\frac{k!}{k^k}$ converges.

43. Use the Ratio Test,

$$\lim_{k\to\infty}\left|\frac{a_{k+1}}{a_k}\right| = \lim_{k\to\infty}\left|\frac{p^{k+1}/(k+1)}{p^k/k}\right|$$

$$= \lim_{k\to\infty}\left|p\frac{k}{k+1}\right|$$

$$= |p|\lim_{k\to\infty}\left|\frac{k}{k+1}\right| = |p|.$$

Thus, the series converges if $|p| < 1$; that is, if $-1 < p < 1$. If $p = 1$, we have the harmonic series, which diverges. If $p = -1$, we have the alternating harmonic series, which converges conditionally by the Alternating Series Test.

Thus, the series converges if $-1 \le p < 1$.

9.6 POWER SERIES

1. $\displaystyle f(x) = \frac{2}{1-x} = 2\sum_{k=0}^{\infty}x^k = \sum_{k=0}^{\infty}2x^k$

This is a geometric series that converges when $|x| < 1$, so the interval of convergence is $(-1, 1)$ and $r = 1$.

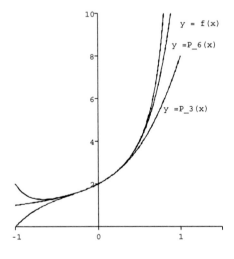

3. $\displaystyle f(x) = \frac{3}{1+x^2} = 3\left[\frac{1}{1-(-x^2)}\right]$

$$= 3\sum_{k=0}^{\infty}(-x^2)^k = \sum_{k=0}^{\infty}3(-1)^k x^{2k}$$

This is a geometric series that converges when $|x^2| < 1$, so the interval of convergence is $(-1, 1)$ and $r = 1$.

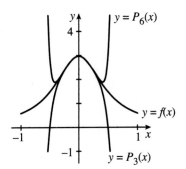

5.
$$f(x) = \frac{2x}{1 - x^3} = 2x \sum_{k=0}^{\infty} (x^3)^k$$
$$= \sum_{k=0}^{\infty} 2x^{3k+1}$$

This is a geometric series that converges when $|x^3| < 1$, so the interval of convergence is $(-1, 1)$ and $r = 1$.

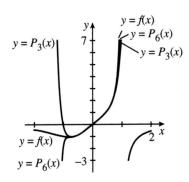

7.
$$f(x) = \frac{2}{4 + x} = \frac{1}{2} \left[\frac{1}{1 - (-x/4)} \right]$$
$$= \frac{1}{2} \sum_{k=0}^{\infty} \left(-\frac{x}{4} \right)^k$$

This is a geometric series that converges when $|x/4| < 1$, so the interval of convergence is $(-4, 4)$ and $r = 4$.

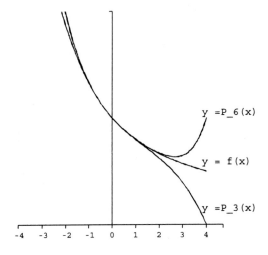

9. This is a geometric series
$$\sum_{k=0}^{\infty} (x + 2)^k = \frac{1}{1 - (x + 2)} = \frac{-1}{1 + x}$$
which converges for $|x + 2| < 1$, so $-1 < x + 2 < 1$ or $-3 < x < -1$. Thus, the interval of convergence is $(-3, -1)$ and $r = 1$.

11. This is a geometric series
$$\sum_{k=0}^{\infty} (2x - 1)^k = \frac{1}{1 - (2x - 1)}$$
$$= \frac{1}{2 - 2x}$$

which converges for $|2x - 1| < 1$, so $-1 < 2x - 1 < 1$ or $0 < x < 1$. Thus, the interval of convergence is $(0, 1)$ and $r = \dfrac{1}{2}$.

13. This is a geometric series
$$\sum_{k=0}^{\infty} (-1)^k \left(\frac{x}{2} \right)^k = \sum_{k=0}^{\infty} \left(\frac{-x}{2} \right)^k$$
$$= \frac{1}{1 + x/2} = \frac{2}{2 + x}$$

which converges for $\left| \dfrac{-x}{2} \right| < 1$, so $-1 < \dfrac{-x}{2} < 1$ or $2 > x > -2$. Thus, the interval of convergence is $(-2, 2)$ and $r = 2$.

15. Using the Ratio Test,

$$\lim_{k \to \infty} \left| \frac{a_{k+1}}{a_k} \right| = \lim_{k \to \infty} \left| \frac{2^{k+1} k! (x-2)^{k+1}}{2^k (k+1)! (x-2)^k} \right|$$

$$= 2|x-2| \lim_{k \to \infty} \frac{1}{k+1}$$

$= 0$ and $0 < 1$ for all x, so the series converges absolutely for $x \in (-\infty, \infty)$. The interval of convergence is $(-\infty, \infty)$ and $r = \infty$.

17. Using the Ratio Test,

$$\lim_{k \to \infty} \left| \frac{a_{k+1}}{a_k} \right| = \lim_{k \to \infty} \left| \frac{(k+1)4^k x^{k+1}}{k 4^{k+1} x^k} \right|$$

$$= \frac{|x|}{4} \lim_{k \to \infty} \frac{k+1}{k} = \frac{|x|}{4}$$

and $\dfrac{|x|}{4} < 1$ when $|x| < 4$ or $-4 < x < 4$. So the series converges absolutely for $x \in (-4, 4)$.

When $x = 4$, $\displaystyle\sum_{k=0}^{\infty} \frac{k}{4^k} 4^k = \sum_{k=0}^{\infty} k$ and $\lim_{k \to \infty} k = \infty$, so the series diverges by the kth-Term Test for Divergence.

When $x = -4$,

$$\sum_{k=0}^{\infty} \frac{k}{4^k} (-4)^k = \sum_{k=0}^{\infty} (-1)^k k, \text{ which diverges}$$

by the kth-Term Test for Divergence.

Thus the interval of convergence is $(-4, 4)$ and $r = 4$.

19. Using the Ratio Test,

$$\lim_{k \to \infty} \left| \frac{a_{k+1}}{a_k} \right|$$

$$= \lim_{k \to \infty} \left| \frac{(-1)^{k+1} k 3^k (x-1)^{k+1}}{(-1)^k (k+1) 3^{k+1} (x-1)^k} \right|$$

$$= \frac{|x-1|}{3} \lim_{k \to \infty} \frac{k}{k+1}$$

$$= \frac{|x-1|}{3}$$

and $\dfrac{|x-1|}{3} < 1$ when $-3 < x - 1 < 3$ or $-2 < x < 4$, so the series converges absolutely for $x \in (-2, 4)$.

When $x = -2$,

$$\sum_{k=0}^{\infty} \frac{(-1)^k}{k 3^k} (-3)^k = \sum_{k=0}^{\infty} \frac{1}{k}$$

which is the divergent harmonic series.

When $x = 4$,

$$\sum_{k=0}^{\infty} \frac{(-1)^k}{k 3^k} 3^k = \sum_{k=0}^{\infty} \frac{(-1)^k}{k}$$

which converges by the Alternating Series Test.

So the interval of convergence is $(-2, 4]$ and $r = 3$.

21. Using the Ratio Test,

$$\lim_{k \to \infty} \left| \frac{a_{k+1}}{a_k} \right| = \lim_{k \to \infty} \left| \frac{(k+1)! (x+1)^{k+1}}{k! (x+1)^k} \right|$$

$$= \lim_{k \to \infty} (k+1) |x+1||$$

$$= \begin{cases} 0 & \text{if } x = -1 \\ \infty & \text{if } x \neq -1 \end{cases}$$

so this series converges absolutely for $x = -1$ and $r = 0$.

23. Using the Ratio Test,

$$\lim_{k \to \infty} \left| \frac{a_{k+1}}{a_k} \right| = \lim_{k \to \infty} \left| \frac{(k+1)^2 (x-3)^{k+1}}{k^2 (x-3)^k} \right|$$

$$= |x-3| \lim_{k \to \infty} \frac{(k+1)^2}{k^2}$$

$$= |x-3| < 1$$

when $-1 < x - 3 < 1$ or $2 < x < 4$, so the series converges absolutely for $x \in (2, 4)$.

If $x = 2$, $\displaystyle\sum_{k=2}^{\infty} k^2 (-1)^k$ and $\lim_{k \to \infty} k^2 = \infty$, so the series diverges by the kth-Term Test for Divergence.

If $x = 4$, $\displaystyle\sum_{k=2}^{\infty} k^2$ and $\lim_{k \to \infty} k^2 = \infty$, so the series diverges by the kth-Term Test for Divergence.

Thus, the interval of convergence is $(2, 4)$ and $r = 1$.

25. Using the Ratio Test,

$$\lim_{k \to \infty} \left| \frac{a_{k+1}}{a_k} \right| = \lim_{k \to \infty} \left| \frac{(k + 1)!(2k)!x^{k+1}}{k!(2k + 2)!x^k} \right|$$

$$= |x| \lim_{k \to \infty} \frac{k + 1}{(2k + 2)(2k + 1)} = 0 < 1$$

for all x, so the series converges for all $x \in (-\infty, \infty)$.

Thus the interval of convergence is $(-\infty, \infty)$ and $r = \infty$.

27. Using the Ratio Test,

$$\lim_{k \to \infty} \left| \frac{a_{k+1}}{a_k} \right| = \lim_{k \to \infty} \left| \frac{2^{k+1}k^2(x + 2)^{k+1}}{2^k(k + 1)^2(x + 2)^k} \right|$$

$$= 2 |x + 2| \lim_{k \to \infty} \frac{k^2}{(k + 1)^2}$$

$$= 2 |x + 2| \text{ and } 2 |x + 2| < 1$$

when $-\dfrac{1}{2} < x + 2 < \dfrac{1}{2}$

or $-\dfrac{5}{2} < x < -\dfrac{3}{2}$,

so the series converges absolutely for $x \in \left(-\dfrac{5}{2}, -\dfrac{3}{2} \right)$.

If $x = -\dfrac{5}{2}$, then

$$\sum_{k=0}^{\infty} \frac{2^k}{k^2} \left(-\frac{1}{2} \right)^k = \sum_{k=0}^{\infty} \frac{(-1)^k}{k^2}$$

and

$$\lim_{k \to \infty} \frac{1}{k^2} = 0$$

and

$$\frac{a_{k+1}}{a_k} = \frac{k^2}{(k + 1)^2} < 1$$

for all $k \geq 0$, so $a_{k+1} < a_k$ for all $k \geq 0$, so the series converges by the Alternating Series Test.

If $x = -\dfrac{3}{2}$, then

$$\sum_{k=0}^{\infty} \frac{2^k}{k^2} \left(\frac{1}{2} \right)^k = \sum_{k=0}^{\infty} \frac{1}{k^2}$$

which is a convergent p-series ($p = 2 > 1$).

So the interval of convergence is

$$\left[-\frac{5}{2}, -\frac{3}{2} \right] \text{ and } r = \frac{1}{2}.$$

29. Using the Ratio Test,

$$\lim_{k \to \infty} \left| \frac{a_{k+1}}{a_k} \right| = \lim_{k \to \infty} \left| \frac{4^{k+1}k^{1/2}x^{k+1}}{4^k(k + 1)^{1/2}x^k} \right|$$

$$= 4 |x| \lim_{k \to \infty} \frac{k^{1/2}}{(k + 1)^{1/2}} = 4 |x|$$

and $4 |x| \leq 1$ when

$-\dfrac{1}{4} < x < \dfrac{1}{4}$, so the series converges absolutely for $x \in \left(-\dfrac{1}{4}, \dfrac{1}{4} \right)$.

If $x = -\dfrac{1}{4}$, then

$$\sum_{k=0}^{\infty} \frac{4^k}{\sqrt{k}} \left(-\frac{1}{4} \right)^k = \sum_{k=0}^{\infty} \frac{(-1)^k}{\sqrt{k}}$$

and

$$\lim_{k \to \infty} \frac{1}{\sqrt{k}} = 0$$

and

$$\frac{a_{k+1}}{a_k} = \frac{\sqrt{k}}{\sqrt{k + 1}} < 1$$

for all $k \geq 0$, so $a_{k+1} < a_k$ for all $k \geq 0$. Therefore, the series converges by the Alternating Series Test.

If $x = \dfrac{1}{4}$,

$$\sum_{k=0}^{\infty} \frac{4^k}{\sqrt{k}} \left(\frac{1}{4} \right)^k = \sum_{k=0}^{\infty} \frac{1}{\sqrt{k}}$$

which is a divergent p-series $\left(p = \dfrac{1}{2} < 1 \right)$.

Thus the interval of convergence is $\left[-\dfrac{1}{4}, \dfrac{1}{4}\right)$

and $r = \dfrac{1}{4}$.

31. We have seen that

$$\frac{3}{1+x^2} = \sum_{k=0}^{\infty} (-1)^k 3x^{2k}$$

with $r = 1$.

Integrating both sides gives

$$\int \frac{3}{1+x^2}\, dx = \sum_{k=0}^{\infty} 3(-1)^k \int x^{2k}\, dx$$

$$3\tan^{-1} x = \sum_{k=0}^{\infty} \frac{3(-1)^k x^{2k+1}}{2k+1} + c$$

Taking $x = 0$,

$$3\tan^{-1} 0 = \sum_{k=0}^{\infty} \frac{3(-1)^k (0)^{2k+1}}{2k+1} + c = c$$

so that

$$c = 3\tan^{-1}(0) = 0.$$

Thus

$$3\tan^{-1} x = \sum_{k=0}^{\infty} \frac{3(-1)^k x^{2k+1}}{2k+1}$$

with $r = 1$.

33. We have seen that

$$\frac{2}{1-x^2} = \sum_{k=0}^{\infty} 2x^{2k}$$

with $r = 1$. Taking the derivative of both sides gives

$$\frac{d}{dx}\left(\frac{2}{1-x^2}\right) = \sum_{k=0}^{\infty} 2\frac{d}{dx} x^{2k}$$

$$\frac{4x}{(1-x^2)^2} = \sum_{k=1}^{\infty} 2\cdot 2k x^{2k-1}$$

$$\frac{2x}{(1-x^2)^2} = \sum_{k=1}^{\infty} 2k x^{2k-1} \text{ with } r = 1.$$

35. We have seen that

$$\frac{3x}{1+x^2} = \sum_{k=0}^{\infty} (-1)^k 3x^{2k+1}$$

with $r = 1$. Integrating both sides gives

$$\int \frac{3x}{1+x^2}\, dx = \sum_{k=0}^{\infty} (-1)^k 3 \int x^{2k+1}\, dx$$

$$\frac{3}{2}\ln(1+x^2) = \sum_{k=0}^{\infty} \frac{(-1)^k 3x^{2k+2}}{2k+2} + c$$

$$\ln(1+x^2) = \sum_{k=0}^{\infty} \frac{(-1)^k x^{2k+2}}{k+1} + c$$

Taking $x = 0$,

$$\ln(1) = \sum_{k=0}^{\infty} \frac{(-1)^k (0)^{2k+2}}{k+1} + c = c$$

so that $c = \ln(1) = 0$.

Thus,

$$\ln(1+x^2) = \sum_{k=0}^{\infty} \frac{(-1)^k x^{2k+2}}{k+1}$$

with $r = 1$.

37. Since $\left|\cos(k^3 x)\right| \le 1$ for all x,

$$\left|\frac{\cos(k^3 x)}{k^2}\right| \le \frac{1}{k^2}$$

for all x, and $\displaystyle\sum_{k=0}^{\infty} \frac{1}{k^2}$ is a convergent p-series, so by the Comparison Test, $\displaystyle\sum_{k=0}^{\infty} \frac{\cos(k^3 x)}{k^2}$ converges absolutely for all x. So the interval of convergence is $(-\infty, \infty)$ and $r = \infty$.

The series of derivatives is

$$\sum_{k=0}^{\infty} \frac{d}{dx}\left[\frac{\cos(k^3 x)}{k^2}\right] = \sum_{k=0}^{\infty} (-k)\sin(k^3 x)$$

and $\displaystyle\lim_{k\to\infty} (-k)\sin(k^3 x)$ diverges if $x \ne 0$, while

$$\sum_{k=0}^{\infty} (-k)\sin(k^3(0)) = \sum_{k=0}^{\infty} 0 = 0,$$

thus the series converges absolutely if $\sin(k^3 x) = 0$.

39. Using the Ratio Test,

$$\lim_{k \to \infty} \left| \frac{a_{k+1}}{a_k} \right| = \lim_{k \to \infty} \left| \frac{e^{(k+1)x}}{e^{kx}} \right| = e^x$$

and $e^x < 1$ when $x < 0$, so the series converges absolutely for all $x \in (-\infty, 0)$.

When $x = 0$,

$$\sum_{x=0}^{\infty} e^0 = \sum_{x=0}^{\infty} 1$$

which diverges by the kth-Term Test for Divergence because $\lim_{k \to \infty} 1 = 1$. So the interval of convergence is $(-\infty, 0)$.

The series of derivatives is

$$\sum_{k=0}^{\infty} \frac{d}{dx} e^{kx} = \sum_{k=0}^{\infty} k e^{kx}.$$

Using the Ratio Test,

$$\lim_{k \to \infty} \left| \frac{a_{k+1}}{a_k} \right| = \lim_{k \to \infty} \left| \frac{(k+1) e^{(k+1)x}}{k e^{kx}} \right|$$

$$= e^x \lim_{k \to \infty} \frac{k+1}{k}$$

$$= e^x$$

and $e^x < 1$ when $x < 0$, so the series converges absolutely for all $x \in (-\infty, 0)$.

When $x = 0$,

$$\sum_{k=0}^{\infty} k e^0 = \sum_{k=0}^{\infty} k$$

which diverges by the kth-Term Test for Divergence because $\lim_{k \to \infty} k = \infty$. So the interval of convergence is $(-\infty, 0)$.

41. Using the Ratio Test,

$$\lim_{k \to \infty} \left| \frac{a_{k+1}}{a_k} \right| = \lim_{k \to \infty} \left| \frac{(x-a)^{(k+1)} b^k}{b^{k+1}(x-a)^k} \right|$$

$$= \frac{|x-a|}{b} \lim_{k \to \infty} 1 = \frac{|x-a|}{b}$$

and $\dfrac{|x - a|}{b} < 1$ when $-b < x - a < b$ or $a - b < x < a + b$. So the series converges absolutely for $x \in (a - b, a + b)$.

If $x = a - b$, $\displaystyle\sum_{k=0}^{\infty} \frac{(a - b - a)^k}{b^k}$

$$= \sum_{k=0}^{\infty} \frac{(-1)^k b^k}{b^k} = \sum_{k=0}^{\infty} (-1)^k,$$ which diverges by the kth-Term Test for Divergence because

$$\lim_{k \to \infty} 1 = 1.$$

If $x = a + b$,

$$\sum_{k=0}^{\infty} \frac{(a + b - a)^k}{b^k} = \sum_{k=0}^{\infty} \frac{b^k}{b^k} = \sum_{k=0}^{\infty} 1$$

which diverges by the kth-Term Test for Divergence because $\lim_{k \to \infty} 1 = 1$.

So the interval of convergence is $(a - b, a + b)$ and $r = b$.

43. If the radius of convergence of $\displaystyle\sum_{k=0}^{\infty} a_k x^k$ is r, then $-r < x < r$. For any constant c, $-r - c < x - c < r - c$.

Thus, the radius of convergence of $\displaystyle\sum_{k=0}^{\infty} a_k (x - c)^k$ is

$$\frac{(r - c) - (-r - c)}{2} = r.$$

45. $\dfrac{1}{1-x} = \displaystyle\sum_{k=0}^{\infty} x^k$

$$\frac{d}{dx} \left(\frac{1}{1-x} \right) = \frac{1}{(1-x)^2}$$

$$= \sum_{k=0}^{\infty} \frac{d}{dx} x^k = \sum_{k=0}^{\infty} k x^{k-1}$$

$$\frac{x+1}{(1-x)^2} = \frac{x}{(1-x)^2} + \frac{1}{(1-x)^2}$$

$$= \sum_{k=0}^{\infty} kx^k + \sum_{k=0}^{\infty} kx^{k-1}$$

$$= \sum_{k=0}^{\infty} kx^k + \sum_{k=0}^{\infty} (k+1)x^k$$

$$= \sum_{k=0}^{\infty} (2k+1)x^k$$

$$= 1 + 3x + 5x^2 + 7x^3 + \cdots$$

Since $\sum_{k=0}^{\infty} x^k$ converges for $|x| < 1$,

$$\frac{d}{dx}\left(\sum_{k=0}^{\infty} x^k\right) \text{ and } \frac{d}{dx}\left(\sum_{k=0}^{\infty} k^x\right) \text{ converge}$$

for $|x| < 1$. Hence $\dfrac{x+1}{(1-x)^2}$ converges for $|x| < 1$, so $r = 1$.

If $x = \dfrac{1}{1000}$, then $\dfrac{1,001,000}{998,001}$

$$= 1 + \frac{3}{1000} + \frac{5}{(1000)^2} + \frac{7}{(1000)^3} + \cdots$$

$$= 1.003005007\ldots.$$

47. If $x = 1$,

$$\ldots + 1 + 1 + 1 + 1 + 1 + \cdots \neq 0.$$

$\dfrac{1}{1 - \frac{1}{x}} = \displaystyle\sum_{k=0}^{\infty} \left(\frac{1}{x}\right)^k$ is a geometric series that

converges for $\dfrac{1}{|x|} < 1$ or $|x| > 1$.

$\dfrac{x}{1-x} = \displaystyle\sum_{k=0}^{\infty} x^{k+1}$ is a geometric series that

converges for $|x| < 1$. Euler's mistake was that there are no x's for which both series converge.

49. Note that for $-1 < x < 1$,

$$\frac{1}{1+x} = \sum_{k=0}^{\infty} (-1)^k x^k.$$

We can differentiate both sides and get

$$\frac{-1}{(1+x)^2} = \sum_{k=0}^{\infty} (-1)^k kx^{k-1}.$$

Similarly, for $-1 < x < 1$, we have

$$\frac{1}{1-x} = \sum_{k=0}^{\infty} x^k.$$

Therefore,

$$\frac{1}{(1-x)^2} = \sum_{k=0}^{\infty} kx^{k-1}.$$

Putting these together, we get

$$E(x) = \frac{kq}{(x-1)^2} - \frac{kq}{(x+1)^2}$$

$$= kq \sum_{k=0}^{\infty} kx^{k-1} + kq \sum_{k=0}^{\infty} (-1)^k kx^{k-1}$$

$$= \sum_{k=0}^{\infty} [1 + (-1)^k] k^2 q x^{k-1}$$

$$= \sum_{k \text{ even}} 2k^2 q x^{k-1}$$

for $-1 < x < 1$.

9.7 TAYLOR SERIES

1. $f(x) = \cos x,\ f'(x) = -\sin x$

$f''(x) = -\cos x,\ f'''(x) = \sin x$

$f(0) = 1,\ f'(0) = 0$

$f''(0) = -1,\ f'''(0) = 0$

Therefore,

$$\cos x = 1 - \frac{1}{2}x^2 + \cdots = \sum_{k=0}^{\infty} \frac{(-1)^k}{(2k)!} x^{2k}.$$

From the Ratio Test,

$$\lim_{n\to\infty} \left|\frac{a_{n+1}}{a_n}\right| = \lim_{n\to\infty} \frac{(2n)!x^{2n+2}}{(2n+2)!x^{2n}}$$

$$= x^2 \lim_{n\to\infty} \frac{1}{(2n+1)(2n+2)} = 0$$

we see that the interval of convergence is $(-\infty, \infty)$.

3. $f(x) = e^{2x}$, $f'(x) = 2e^{2x}$,

$f''(x) = 4e^{2x}$, $f'''(x) = 8e^{2x}$

$f(0) = 1$, $f'(0) = 2$

$f''(0) = 4$, $f'''(0) = 8$

Therefore,

$$e^{2x} = 1 + 2x + \cdots = \sum_{k=0}^{\infty} \frac{2^k}{k!} x^k.$$

From the Ratio Test,

$$\lim_{n\to\infty} \left| \frac{a_{n+1}}{a_n} \right| = x \lim_{n\to\infty} \frac{2}{n+1} = 0$$

we see that the interval of convergence is $(-\infty, \infty)$.

5. $f(x) = \ln(1+x)$, $f'(x) = \dfrac{1}{1+x}$,

$f''(x) = \dfrac{-1}{(1+x)^2}$, $f'''(x) = \dfrac{2}{(1+x)^3}$,

$f(0) = 0$, $f'(0) = 1$

$f''(0) = -1$, $f'''(0) = 2$

Therefore,

$$\ln(1+x) = x - \frac{1}{2}x^2 + \cdots = \sum_{k=0}^{\infty} \frac{(-1)^k x^{k+1}}{k+1}.$$

From the Ratio Test,

$$\lim_{n\to\infty} \left| \frac{a_{n+1}}{a_n} \right|$$

$$= \lim_{n\to\infty} \left| \frac{(-1)^{n+1}(n+1)x^{n+2}}{(-1)^n(n+2)x^{n+1}} \right|$$

$$= |x| \lim_{n\to\infty} \frac{n+1}{n+2} = |x|$$

and $|x| < 1$ when $-1 < x < 1$. So the series converges absolutely for all $x \in (-1, 1)$.

When $x = 1$,

$$\sum_{k=0}^{\infty} \frac{(-1)^k(1)^{k+1}}{k+1} = \sum_{k=0}^{\infty} \frac{(-1)^k}{k+1}$$

and $\lim_{k=0} \dfrac{1}{k+1} = 0$ and

$$\frac{a_{k+1}}{a_k} = \frac{k}{k+1} < 1$$

for all $k \geq 0$ so $a_{k+1} < a_k$, for all $k \geq 0$, so the series converges by the Alternating Series Test.

When $x = -1$,

$$\sum_{k=0}^{\infty} \frac{(-1)^k(-1)^{k+1}}{k+1} = \sum_{k=0}^{\infty} \frac{-1}{k+1}$$

which is the negative of the harmonic series, so it diverges.

Hence the interval of convergence is $(-1, 1]$.

7. $f^{(k)}(x) = (-1)^k(k+1)!(1+x)^{-2-k}$. Therefore,

$$\frac{1}{(1+x)^2} = \sum_{k=0}^{\infty}(-1)^k(k+1)x^k.$$

From the Ratio Test,

$$\lim_{k\to\infty} \left| \frac{a_{k+1}}{a_k} \right| = |x| \lim_{k\to\infty} \frac{k+2}{k+1} = |x|$$

and $|x| < 1$ when $-1 < x < 1$. So the series converges absolutely for $x \in (-1, 1)$.

When $x = -1$,

$$\sum_{k=0}^{\infty}(k+1) \text{ and } \lim_{k\to\infty}(k+1) = \infty, \text{ so the se-}$$

ries diverges by the kth-Term Test for Divergence.

When $x = 1$, $\sum_{k=0}^{\infty}(-1)(k+1)^k$ and $\lim_{k\to\infty} k + 1 = \infty$, so the series diverges by the kth-Term Test for Divergence.

So the interval of convergence is $(-1, 1)$.

9. $f^{(k)}(x) = e^{x-1}$, $f^{(k)}(1) = 1$

$$e^{x-1} = 1 + (x-1) + \cdots = \sum_{k=0}^{\infty} \frac{(x-1)^k}{k!}$$

Using the Ratio Test,

$$\lim_{k\to\infty} \left| \frac{a_{k+1}}{a_k} \right| = |x-1| \lim_{k\to\infty} \frac{1}{k+1} = 0.$$

So the interval of convergence is $(-\infty, \infty)$.

11. $f(x) = \ln x$ and for $k > 1$,

$$f^{(k)}(x) = (-1)^{k+1}(k-1)!\frac{1}{x^k}$$

$$f(e) = 1, \; f^{(k)}(e) = (-1)^{k+1}(k-1)!\frac{1}{e^k}$$

$$\ln x = 1 + \sum_{k=1}^{\infty} \frac{(-1)^{k+1}}{e^k k}(x-e)^k$$

By the Ratio Test,

$$\lim_{k \to \infty} \left| \frac{a_{k+1}}{a_k} \right| = |x-e| \lim_{k \to \infty} \frac{k}{e(k+1)}$$

$$= \frac{1}{e}|x-e|.$$

The interval of convergence is $(0, 2e]$.

13. $f^{(k)}(x) = (-1)^k k!\dfrac{1}{x^{k+1}}$,

$f^{(k)}(1) = (-1)^k k!$

Therefore,

$$\frac{1}{x} = \sum_{k=0}^{\infty} (-1)^k (x-1)^k$$

From the Ratio Test,

$$\lim_{k \to \infty} \left| \frac{a_{k+1}}{a_k} \right|$$

$$= |x-1| \lim_{k \to \infty} |-1| = |x-1|$$

we see that the interval of convergence is $(0, 2)$.

15.

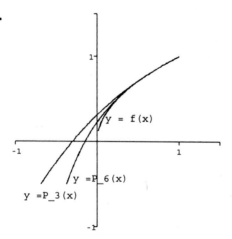

17.

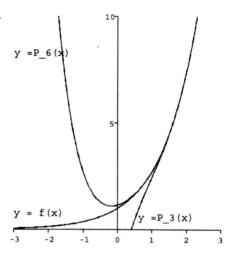

19.

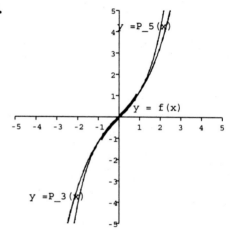

21. For any fixed x there exists a z between 0 and x such that

$$|R_n(x)| = \left| \frac{f^{(n+1)}(z)x^{n+1}}{(n+1)!} \right|$$

$$\leq \left| \frac{x^{n+1}}{(n+1)!} \right| \to 0 \text{ as } n \to \infty,$$

since $\left| f^{(n+1)}(z) \right| \leq 1$ for all n.

23. For any fixed x there exists a z between x and 1 such that

$$|R_n(x)| = \left| \frac{f^{(n+1)}(z)(x-1)^{n+1}}{(n+1)!} \right|.$$

Observe that for all n,

$$\left| f^{(n+1)}(z) \right| = \frac{n!}{|z|^{n+1}}, \text{ so}$$

$$|R_n(x)| = \left| \frac{(x-1)^{n+1}}{(n+1)|z|^{n+1}} \right|$$

$$= \left| \frac{x-1}{z} \right|^{n+1} \frac{1}{n+1} \to 0 \text{ as } n \to \infty.$$

25. (a) Expand $f(x) = \ln x$ into a Taylor series about $c = 1$. Recall

$$\ln(x) = \sum_{k=1}^{\infty} \frac{(-1)^{k+1}}{k} (x-1)^k.$$

So $P_4(x) = \sum_{k=1}^{4} \frac{(-1)^{k+1}}{k!} (x-1)^k$

$$= (x-1) - \frac{1}{2}(x-1)^2 + \frac{1}{3}(x-1)^3$$

$$- \frac{1}{4}(x-1)^4.$$

Letting $x = 1.05$ gives

$$\ln(1.05) \approx P_4(1.05) \approx 0.04879.$$

(b) $|\text{Error}| = |\ln(1.05) - P_4(1.05)|$

$$= |R_n(1.05)|$$

$$= \left| \frac{f^{4+1}(z)}{(4+1)!} (1.05-1)^{4+1} \right|$$

$$= \frac{4! |z|^{-5} (0.05)^5}{5!},$$

so because $1 < z < 1.05$,

$$\frac{1}{z} < \frac{1}{1} = 1. \text{ Thus, we have}$$

$$|\text{Error}| = \frac{(0.05)^5}{5z^5} < \frac{(0.05)^5}{5(1)^5} = \frac{(0.05)^5}{5}.$$

(c) From part (b) we have for $1 < z < 1.05$,

$$|R_n(1.05)|$$

$$= \left| \frac{f^{n+1}(z)}{(n+1)!} (1.05-1)^{n+1} \right|$$

$$= \frac{n!(0.05)^{n+1}}{(n+1)! z^{n+1}} = \frac{(0.05)^{n+1}}{(n+1) z^{n+1}}$$

$$< \frac{(0.05)^{n+1}}{n+1}$$

$$< 10^{-10} \text{ if } n = 7.$$

27. (a) Expand $f(x) = \sqrt{x}$ into a Taylor series about $c = 1$.

$$P_4(x) = 1 + \frac{(x-1)}{2} - \frac{(x-1)^2}{8} + \frac{(x-1)^3}{16}$$

$$- \frac{5(x-1)^4}{128}$$

Letting $x = 1.1$ gives

$$\sqrt{1.1} \approx P_4(1.1) = 1.0488.$$

(b) $|\text{Error}| = |\sqrt{1.1} - P_4(1.1)|$

$$= |R_n(1.1)|$$

$$= \left| \frac{f^{4+1}(z)}{(4+1)!} (1.1-1)^{4+1} \right|$$

$$= \frac{\frac{105}{32} |z|^{-9/2} (0.1)^5}{5!},$$

so because $1 < z < 1.1$,

$$|\text{Error}| = \frac{7(0.1)^5}{256 z^{9/2}} < \frac{7(0.1)^5}{256}.$$

(c) From part (b) we have for $1 < z < 1.1$,

$$|R_n(1.1)|$$

$$= \left| \frac{f^{n+1}(z)}{(n+1)!} (1.1-1)^{n+1} \right|$$

$$= \frac{(2n-1)(2n-3) \cdots 3 \cdot 1 (0.1)^{n+1}}{2^{n+1} (n+1)! z^{(2n+1)/2}}$$

$$< \frac{(2n-1)(2n-3) \cdots 3 \cdot 1 (0.1)^{n+1}}{2^{n+1} (n+1)!}$$

$$< 10^{-10} \text{ if } n = 9.$$

29. $\displaystyle\sum_{k=0}^{\infty} \frac{1}{k!} x^k = e^x$

Replacing x with 2, we have

$\displaystyle\sum_{k=0}^{\infty} \frac{2^k}{k!} = e^2.$

31. $\displaystyle\sum_{k=0}^{\infty} \frac{(-1)^k}{2k+1} x^{2k+1} = \tan^{-1} x$

Replacing x with 1, we have

$\displaystyle\sum_{k=0}^{\infty} \frac{(-1)^k}{2k+1} = \tan^{-1}(1) = \frac{\pi}{4}.$

33. $\displaystyle e^x = \sum_{k=0}^{\infty} \frac{x^k}{k!}$

with $r = \infty$. Replacing x with $-3x$, we have

$\displaystyle e^{-3x} = \sum_{k=0}^{\infty} \frac{(-1)^k (3x)^k}{k!}$

with $r = \infty$.

35. $\displaystyle e^x = \sum_{k=0}^{\infty} \frac{x^k}{k!}$

with $r = \infty$. By replacing x with $-x^2$ and multiplying by x we have

$\displaystyle xe^{-x^2} = \sum_{k=0}^{\infty} \frac{(-1)^k x^{2k+1}}{k!}$

with $r = \infty$.

37. $\displaystyle \sin x = \sum_{k=0}^{\infty} \frac{(-1)^k x^{2k+1}}{(2k+1)!}$

with $r = \infty$. Replacing x with $2x$ and multiplying by x, we have

$\displaystyle x \sin 2x = x \left(\sum_{k=0}^{\infty} \frac{(-1)^k (2x)^{2k+1}}{(2k+1)!} \right)$

$\displaystyle = \sum_{k=0}^{\infty} \frac{(-1)^k 2^{2k+1} x^{2k+2}}{(2k+1)!}$ with $r = \infty$.

39. Let us calculate $f'(0)$ and $f''(0)$ using the definition.

$\displaystyle f'(0) = \lim_{h \to 0} \frac{f(h) - f(0)}{h}$

$\displaystyle = \lim_{h \to 0} \frac{e^{-1/h^2} - 0}{h} = 0$

When $x \neq 0$, $f'(x) = \dfrac{2}{x^3} e^{-1/x^2}$ and then,

$\displaystyle f''(0) = \lim_{h \to 0} \frac{f'(h) - f'(0)}{h}$

$\displaystyle = \lim_{h \to 0} \frac{2/h^3 e^{-1/h^2} - 0}{h}$

$\displaystyle = 2 \lim_{h \to 0} \frac{e^{-1/h^2}}{h^4} = 0.$

41. $f(x) = |x|$

$f'(x) = \begin{cases} 1 & \text{if} \quad x > 0 \\ -1 & \text{if} \quad x < 0 \end{cases}$

$f(1) = 1$, $f'(1) = 1$,

$f^{(n)}(1) = 0$ for all $n \geq 2$

Thus, the Taylor series for f about $c = 1$ is $1 + (x - 1) = x$. Since $f(x) \neq x$ for $x < 0$, the Taylor series does not converge to $f(x)$ for $x < 0$. The Taylor series converges to f for all $x \geq 0$.

43. Since $\displaystyle \lim_{x \to a} \frac{f(x) - g(x)}{(x-a)^2} = 0$ it follows that

$\displaystyle \lim_{x \to a} \left[f(x) - g(x) \right] = 0$

$\Rightarrow f(a) - g(a) = 0$

$\Rightarrow f(a) = g(a).$

Using l'Hôpital's Rule on the given limit, we get $\displaystyle \lim_{x \to a} \frac{f'(x) - g'(x)}{2(x-a)} = 0$

$\Rightarrow \displaystyle \lim_{x \to a} \left[f'(x) - g'(x) \right] = 0$

$\Rightarrow f'(a) - g'(a) = 0$

$\Rightarrow f'(a) = g'(a).$

Similarly,

$$\lim_{x \to a} \frac{f''(x) - g''(x)}{2} = 0$$

$$\Rightarrow \lim_{x \to a} \left[f''(x) - g''(x) \right] = 0$$

$$\Rightarrow f''(a) - g''(a) = 0$$

$$\Rightarrow f''(a) = g''(a).$$

Thus, the first three terms of the Taylor series for f and g are identical.

45. The 5th term of the series is $\frac{1}{9!}$, which is less than 10^{-5}. So 4 terms will give the desired accuracy. In Simpson's Rule, the error is bounded by $\frac{1}{180n^4}$. Solving $\frac{1}{180n^4} \leq 10^{-5}$, we get $n > 4.9$, which means n must be 6, since n is an even integer. The conclusion is that the Taylor method is more efficient.

47. $f(x) = e^x \sin x, \; f(0) = 0$

$f'(x) = e^x (\sin x + \cos x), \; f'(0) = 1$

$f''(x) = 2e^x \cos x, \; f''(0) = 2$

$f'''(x) = -2e^x (\sin x - \cos x), \; f'''(0) = 2$

$f^{(4)}(x) = -4e^x \sin x, \; f^{(4)}(0) = 0$

$f^{(5)}(x) = -4e^x (\sin x + \cos x),$

$f^{(5)}(0) = -4$

$f^{(6)}(x) = -8e^x \cos x, \; f^{(6)}(0) = -8$

Thus, $e^x \sin x$

$$= x + \frac{2}{2!} x^2 + \frac{2}{3!} x^3 - \frac{4}{5!} x^5 - \frac{8}{6!} x^6 + \cdots$$

$$= x + x^2 + \frac{1}{3} x^3 - \frac{1}{30} x^5 - \frac{1}{90} x^6 + \cdots .$$

Now, the Taylor series for e^x and $\sin x$ are

$$e^x = 1 + x + \frac{x^2}{2} + \frac{x^3}{6} + \frac{x^4}{24} + \frac{x^5}{120} + \cdots$$

$$\sin x = x - \frac{x^3}{3!} + \frac{x^5}{5!} - \cdots$$

Multiplying together the series for e^x and $\sin x$ and collecting like terms gives

$$x - \frac{x^3}{3!} + \frac{x^5}{5!} + x^2 - \frac{x^4}{3!} + \frac{x^6}{5!} + \frac{x^3}{2}$$

$$-\frac{x^5}{12} + \frac{x^4}{6} - \frac{x^6}{36} + \frac{x^5}{4!} + \frac{x^6}{5!} + \cdots$$

$$= x + x^2 + x^3 \left[-\frac{1}{3!} + \frac{1}{2} \right]$$

$$+ x^5 \left[\frac{1}{120} - \frac{1}{12} + \frac{1}{24} \right]$$

$$+ x^6 \left[\frac{1}{5!} + \frac{1}{5!} - \frac{1}{36} \right]$$

$$= x + x^2 + \frac{x^3}{3} - \frac{x^5}{30} - \frac{x^6}{90}.$$

The results are the same in each case.

49. $f(x) = \begin{cases} \dfrac{\sin x}{x} & \text{if} \quad x \neq 0 \\ 1 & \text{if} \quad x = 0 \end{cases}$

First note that for $x \neq 0$,

$$f'(x) = \frac{x \cos x - \sin x}{x^2}$$

$$f''(x) = \frac{(2 - x^2) \sin x - 2x \cos x}{x^3}$$

$f'''(x)$

$$= \frac{(3x^2 - 6) \sin x + (6x - x^3) \cos x}{x^4}$$

$f^{(4)}(x)$

$$= \frac{(x^4 - 12x^2 + 24) \sin x + (4x^3 - 24x) \cos x}{x^5}$$

$f^{(5)}(x)$

$$= \frac{1}{x^6} \Big[(-5x^4 + 60x^2 - 120) \sin x$$

$$(x^5 - 20x^3 + 120x) \cos x \Big]$$

as can be verified with detailed calculations or with a CAS.

Using l'Hôpital's Rule when needed, we can determine the values of the derivatives of f evaluated at $x = 0$:

$$f'(0) = \lim_{x \to 0} \frac{f(x) - f(0)}{x - 0}$$

$$= \lim_{x \to 0} \left[\frac{\frac{\sin x}{x} - 1}{x} \right] = \lim_{x \to 0} \frac{\sin x - x}{x^2}$$

$$= \lim_{x \to 0} \frac{\cos x - 1}{2x} = \lim_{x \to 0} \frac{-\sin x}{2}$$

$$f'(0) = 0,$$

$$f''(0) = \lim_{x \to 0} \frac{f'(x) - f'(0)}{x - 0}$$

$$= \lim_{x \to 0} \frac{\frac{x \cos x - \sin x}{x^2} - 0}{x}$$

$$= \lim_{x \to 0} \frac{x \cos x - \sin x}{x^3}$$

$$= \lim_{x \to 0} \frac{-x \sin x}{3x^2}$$

$$= -\frac{1}{3} \lim_{x \to 0} \frac{\sin x}{x}$$

$$= -\frac{1}{3}$$

$$f'''(0) = \lim_{x \to 0} \frac{f''(x) - f''(0)}{x - 0}$$

$$= \lim_{x \to 0} \frac{\frac{(2 - x^2) \sin x - 2x \cos x}{x^3} - \left(-\frac{1}{3}\right)}{x}$$

$$= \lim_{x \to 0} \frac{(2 - x^2) \sin x - 2x \cos x + \frac{1}{3}x^3}{x^4}$$

$$= \lim_{x \to 0} \frac{1}{4x^3} \left[(2 - x^2) \cos x - 2x \sin x - 2(\cos x - x \sin x) + x^2 \right]$$

$$= \lim_{x \to 0} \frac{-x^2 \cos x + x^2}{4x^3}$$

$$= \lim_{x \to 0} \frac{1 - \cos x}{4x}$$

$$= \lim_{x \to 0} \frac{\sin x}{4}$$

$$f'''(0) = 0$$

$$f^{(4)}(0) = \lim_{x \to 0} \frac{f'''(x) - f'''(0)}{x - 0}$$

$$= \lim_{x \to 0} \frac{(3x^2 - 6) \sin x + (6x - x^3) \cos x}{x^5}$$

$$= \lim_{x \to 0} \frac{1}{5x^4} \left[(3x^2 - 6) \cos x + 6x \sin x + (6 - 3x^2) \cos x - (6x - x^3) \sin x \right]$$

$$= \lim_{x \to 0} \frac{x^3 \sin x}{5x^4}$$

$$= \frac{1}{5} \lim_{x \to 0} \frac{\sin x}{x}$$

$$f^{(4)}(0) = \frac{1}{5}$$

$$f^{(5)}(0) = \lim_{x \to 0} \frac{f^{(4)}(x) - f^{(4)}(0)}{x - 0}$$

$$= \lim_{x \to 0} \frac{1}{x^6} \left[(x^4 - 12x^2 + 24) \sin x + (4x^3 - 24x) \cos x - \frac{1}{5}x^5 \right]$$

$$= \lim_{x \to 0} \frac{1}{6x^5} \left[\cos x (x^4 - 12x^2 + 24 + 12x^2 - 24) + \sin x (4x^3 - 24x - 4x^3 + 24x) - x^4 \right]$$

$$= \frac{1}{6} \lim_{x \to 0} \frac{x^4 \cos x - x^4}{x^5}$$

$$= \frac{1}{6} \lim_{x \to 0} \frac{\cos x - 1}{x}$$

$$= \frac{1}{6} \lim_{x \to 0} [-\sin x]$$

$$f^{(5)}(0) = 0$$

$$f^{(6)}(0) = \lim_{x \to 0} \frac{f^{(5)}(x) - f^{(5)}(0)}{x - 0}$$

$$= \lim_{x \to 0} \frac{1}{x^7} \Big[(-5x^4 + 60x^2 - 120) \sin x +$$

$$(x^5 - 20x^3 + 120x) \cos x \Big]$$

$$= \lim_{x \to 0} \frac{1}{7x^6} \Big[\sin x(-20x^3 + 120x - x^5$$

$$+ 20x^3 - 120x) + \cos x(-5x^4$$

$$+ 60x^2 - 120 + 5x^4 - 60x^2 + 120) \Big]$$

$$= \frac{1}{7} \lim_{x \to 0} \frac{-x^5 \sin x}{x^6}$$

$$= -\frac{1}{7} \lim_{x \to 0} \frac{\sin x}{x}$$

$$f^{(6)}(0) = -\frac{1}{7}$$

Thus, the first 4 nonzero terms of the Maclaurin series for f are $1 - \frac{1}{3}\frac{x^2}{2!} + \frac{1}{5}\frac{x^4}{4!} - \frac{1}{7}\frac{x^6}{6!} + \cdots$

$$= 1 - \frac{x^2}{3!} + \frac{x^4}{5!} - \frac{x^6}{7!} + \cdots$$

The Maclaurin series for $\sin x$ is

$$x - \frac{x^3}{3!} + \frac{x^5}{5!} - \frac{x^7}{7!} + \cdots$$

which, when divided by x, is equal to the Maclaurin series for $f(x)$.

51.
$$f(t) = \sum_{k=0}^{3} \frac{f^{(0)}(0)}{k!} t^k = 10 + 10t + t^2 - \frac{t^3}{3!}$$

$$f(2) = 10 + 10(2) + (2)^2 - \frac{(2)^3}{6} = \frac{98}{3} \text{ miles}$$

53. $f^{(k)}(x) = e^x$, $f^{(k)}(c) = e^c$, so

$$e^x = \sum_{k=0}^{\infty} \frac{f^{(k)}(c)}{k!}(x - c)^k = \sum_{k=0}^{\infty} \frac{e^c}{k!}(x - c)^k$$

55. To generate the series, differentiate $(1 + x)^r$ repeatedly. The series terminates if r is a positive integer, since all derivatives after the rth derivative will be 0. Otherwise the series is infinite.

57. This is the case when $r = \frac{1}{2}$, and then, the Maclaurin series is

$$1 + \sum_{k=1}^{\infty} \frac{(1/2)\,(-1/2)\cdots(1/2 - k + 1)}{k!} x^k.$$

59.
$$\cosh x = 1 + \frac{1}{2}x^2 + \frac{1}{24}x^4 + \frac{1}{720}x^6 + \cdots$$

$$\sinh x = x + \frac{1}{6}x^3 + \frac{1}{120}x^5 + \frac{1}{5040}x^7 + \cdots$$

These match the cosine and sine series except that here all signs are positive.

9.8 APPLICATIONS OF TAYLOR SERIES

1. Using the first three terms of the sine series around the center $\pi/2$ we get $\sin 1.61 \approx 0.99923163442$.

3. Using the first five terms of the cosine series around the center 0 we get $\cos 0.34 \approx 0.94275466553$.

5. Using the first nine terms of the exponential series around the center 0, we get $e^{-0.2} \approx 0.81873075307$.

7. $\lim_{x \to 0} \frac{\cos x^2 - 1}{x^4}$

$$= \lim_{x \to 0} \frac{\left(1 - \frac{x^4}{2!} + \frac{x^8}{4!} - \frac{x^{12}}{6!} \cdots\right) - 1}{x^4}$$

$$= -\frac{1}{2}$$

9. Using

$$\ln x \approx (x - 1) - \frac{1}{2}(x - 1)^2 + \frac{1}{3}(x - 1)^3$$

we get

$$\lim_{x \to 1} \frac{\ln x - (x - 1)}{(x - 1)^2}$$

$$= \lim_{x \to 1} \frac{-\frac{1}{2}(x - 1)^2 + \frac{1}{3}(x - 1)^3}{(x - 1)^2}$$

$$= -\frac{1}{2}.$$

11. $\displaystyle\lim_{x \to 0} \frac{e^x - 1}{x}$

$$= \lim_{k \to \infty} \frac{\left(1 + x + \frac{x^2}{2} + \frac{x^3}{3!}x^3 + \cdots\right) - 1}{x}$$

$$= 1$$

13. $\displaystyle\int_{-1}^{1} \frac{\sin x}{x} \, dx$

$$\approx \int_{-1}^{1} \left(1 - \frac{x^2}{6} + \frac{x^4}{120}\right) dx$$

$$= x - \frac{x^3}{18} + \frac{x^5}{600}\Big|_{-1}^{1}$$

$$= \frac{1703}{900} \approx 1.8922$$

15. $\displaystyle\int_{-1}^{1} e^{-x^2} \, dx$

$$\approx \int_{-1}^{1} \left(1 - x^2 + \frac{x^4}{2} - \frac{x^6}{6} + \frac{x^8}{24}\right) dx$$

$$= x - \frac{x^3}{3} + \frac{x^5}{10} - \frac{x^7}{42} + \frac{x^9}{(24)(9)}\Big|_{-1}^{1}$$

$$= \frac{5651}{3780} \approx 1.495$$

17. $\displaystyle\int_{1}^{2} \ln x \, dx$

$$\approx \int_{1}^{2} \left[(x - 1) - \frac{1}{2}(x - 1)^2 \right.$$

$$\left. + \frac{1}{3}(x - 1)^3 - \frac{1}{4}(x - 1)^4 + \frac{1}{5}(x - 1)^5\right] dx$$

$$= \left(\frac{(x - 1)^2}{2} - \frac{(x - 1)^3}{6} + \right.$$

$$\left. \frac{(x - 1)^4}{12} - \frac{(x - 1)^5}{20} + \frac{(x - 1)^6}{30}\right)\Big|_{1}^{2}$$

$$= \frac{2}{5}$$

19. $\displaystyle\left|\frac{a_{n+1}}{a_n}\right| \leq \frac{x^2}{2^2(n + 1)(n + 2)}$

Since this ratio tends to 0, the radius of convergence for $J_1(x)$ is ∞.

21. We need the first neglected term to be less than 0.04. The kth term is bounded by

$$\frac{10^{2k+2}}{2^{2k+2}k!(k + 2)!},$$

which is equal to 0.0357 for $k = 12$. Thus we will need the terms up through $k = 11$, that is, the first 12 terms of the series.

23. Let $f(x) = \dfrac{1}{\sqrt{1 - x}}$. Then

$$f(x) \approx f(0) + f'(0)x = 1 + \frac{1}{2}x.$$

Now let $x = \dfrac{v^2}{c^2}$. Then

$$\frac{1}{\sqrt{1 - \dfrac{v^2}{c^2}}} \approx 1 + \frac{v^2}{2c^2}.$$

Thus

$$m(v) = m_0/\sqrt{1 - v^2/c^2} \approx m_0\left(1 + \frac{v^2}{2c^2}\right).$$

To increase mass by 10%, we want

$$1 + \frac{v^2}{2c^2} = 1.1.$$

Solving for v, we have $\frac{1}{2c^2}v^2 = 0.1$, so $v^2 = 0.2c^2$, thus

$$v = \sqrt{0.2}c \approx 83{,}000 \text{ miles per second.}$$

25.
$$w(x) = \frac{mgR^2}{(R+x)^2}, \; w(0) = mg$$

$$w'(x) = \frac{-2mgR^2}{(R+x)^3}, \; w'(0) = \frac{-2mg}{R}, \; \text{so}$$

$$w(x) \approx mg\left(1 - \frac{2x}{R}\right).$$

To reduce weight by 10%, we want $1 - \frac{2x}{R} = 0.9$. Solving for x, we have $\frac{-2x}{R} = -0.1$, so

$$x = \frac{R}{20} \approx 200 \text{ miles.}$$

27. No, since x is much smaller than R for high-altitude locations on Earth. You have to go out to an altitude of more than 100 miles before you weigh significantly less.

29. Using only the first term of the expansion,
$$v(t) \approx v_c\left(\frac{g}{v_c}t + \frac{1}{2}\right) = gt + \frac{1}{2}v_c. \text{ The first ne-}$$
glected term is negative, so this estimate is too large.

31. Use the first two terms of the series for tanh:

$$\tanh x \approx x - \frac{1}{3}x^3.$$

Substitute $\sqrt{\frac{g}{40m}}\,t$ for x, multiply by $\sqrt{40mg}$, and simplify. The result is

$$gt - \frac{g^2}{120m}t^3.$$

33. $\dfrac{1}{\sqrt{1-x}} = (1 + (-x))^{-1/2}$

$$= 1 - \frac{1}{2}(-x) + \frac{\left(-\frac{1}{2}\right)\left(-\frac{3}{2}\right)}{2!}(-x)^2$$

$$+ \frac{\left(-\frac{1}{2}\right)\left(-\frac{3}{2}\right)\left(-\frac{5}{2}\right)}{3!}(-x)^3$$

$$+ \frac{\left(-\frac{1}{2}\right)\left(-\frac{3}{2}\right)\left(-\frac{5}{2}\right)\left(-\frac{7}{2}\right)}{4!}(-x)^4$$

$$+ \cdots$$

$$= 1 + \frac{1}{2}x + \frac{3}{8}x^2 + \frac{5}{16}x^3 + \frac{35}{128}x^4 + \cdots$$

35. $\dfrac{6}{\sqrt[3]{1+3x}} = 6(1+(3x))^{-1/3}$

$$= 6\left[1 - \frac{1}{3}(3x) + \frac{\left(-\frac{1}{3}\right)\left(-\frac{4}{3}\right)}{2!}(3x)^2 + \right.$$

$$\frac{\left(-\frac{1}{3}\right)\left(-\frac{4}{3}\right)\left(-\frac{7}{3}\right)}{3!}(3x)^3 +$$

$$\left. \frac{\left(-\frac{1}{3}\right)\left(-\frac{4}{3}\right)\left(-\frac{7}{3}\right)\left(-\frac{10}{3}\right)}{4!}(3x)^4 + \cdots\right]$$

$$= 6\left[1 - x + 2x^2 - \frac{14}{3}x^3 + \frac{35}{3}x^4 - \cdots\right]$$

$$= 6 - 6x + 12x^2 - 28x^3 + 70x^4 - \cdots$$

37. (a)
$$\sqrt{26} = 5\sqrt{1 + \frac{1}{25}}$$

Since $\dfrac{1}{25}$ is in the interval of convergence $-1 < x < 1$, the binomial series can be used to get

$$\sqrt{1 + \frac{1}{25}}$$

$$= 1 + \frac{1}{2}\left(\frac{1}{25}\right) - \frac{1}{8}\left(\frac{1}{25}\right)^2 +$$

$$\frac{1}{16}\left(\frac{1}{25}\right)^3 - \frac{5}{128}\left(\frac{1}{25}\right)^4 + \cdots.$$

Using the first four terms of the series implies that the error will be less than

$$\frac{5}{128}\left(\frac{1}{25}\right)^4 = 10^{-7}.$$

Thus, $\sqrt{26}$

$$\approx 5\left[1 + \frac{1}{2}\left(\frac{1}{25}\right) - \frac{1}{8}\left(\frac{1}{25}\right)^2\right.$$
$$\left. + \frac{1}{16}\left(\frac{1}{25}\right)^3\right]$$

$$\approx 5.0990200.$$

(b) $\sqrt{24} = 5\sqrt{1 - \frac{1}{25}}$

$$\approx 5\left[1 - \frac{1}{2}\left(\frac{1}{25}\right) - \frac{1}{8}\left(\frac{1}{25}\right)^2\right.$$
$$\left. - \frac{1}{16}\left(\frac{1}{25}\right)^3\right]$$

$$\approx 4.8989800$$

39. $(4+x)^3 = \left(4\left[1 + \frac{x}{4}\right]\right)^3$

$$= 4^3\left[1 + \frac{x}{4}\right]^3$$

$$= 4^3\left[1 + 3\left(\frac{x}{4}\right) + 3\left(\frac{x}{4}\right)^2 + \left(\frac{x}{4}\right)^3\right]$$

$$= 4^3 + 3(4^2)x + 3(4)x^2 + x^3$$

$$= 64 + 48x + 12x^2 + x^3$$

$(1-2x)^4 = [1 + (-2x)]^4$

$$= 1 + 4(-2x) + \frac{(4)(3)}{2}(-2x)^2$$
$$+ \frac{(4)(3)(2)}{3!}(-2x)^3 + (-2x)^4$$

$$= 1 - 8x + 24x^2 - 32x^3 + 16x^4$$

For a positive integer n, there are $n+1$ non-zero terms in the binomial expansion.

41. $\frac{1}{\sqrt{1-x}} = 1 + \frac{1}{2}x + \frac{3}{8}x^2 + \frac{5}{16}x^3$

$$+ \frac{35}{128}x^4 + \cdots$$

Therefore,

$$\frac{1}{\sqrt{1-x^2}} = 1 + \frac{1}{2}x^2 + \frac{3}{8}x^4 + \frac{5}{16}x^6$$

$$+ \frac{35}{128}x^8 + \cdots$$

and so $\sin^{-1} x = \int \frac{1}{\sqrt{1-x^2}}\, dx$

$$= x + \frac{1}{6}x^3 + \frac{3}{40}x^5 + \frac{5}{112}x^7$$

$$+ \frac{35}{1152}x^9 + \cdots + \text{constant}.$$

Since $\sin^{-1}(0) = 0$, it follows that the constant in the previous equation is 0. Thus, the Maclaurin series for the inverse sine function is

$$\sin^{-1} x = x + \frac{x^3}{6} + \frac{3x^5}{40} + \frac{5x^7}{112} + \frac{35x^9}{1152} + \cdots.$$

43. $\sin x = x - \frac{x^3}{3!} + \frac{x^5}{5!} - \frac{x^7}{7!}$

$$+ \frac{x^9}{9!} - \frac{x^{11}}{11!} + \frac{x^{13}}{13!} - \cdots \qquad \frac{\sin x}{x}$$

$$= 1 - \frac{x^2}{3!} + \frac{x^4}{5!} - \frac{x^6}{7!}$$

$$+ \frac{x^8}{9!} - \frac{x^{10}}{11!} + \frac{x^{12}}{13!} - \cdots$$

$$\int_0^\pi \frac{\sin x}{x}\, dx$$

$$= \int_0^\pi \left[1 - \frac{x^2}{3!} + \frac{x^4}{5!} - \frac{x^6}{7!}\right.$$

$$\left. + \frac{x^8}{9!} - \frac{x^{10}}{11!} + \frac{x^{12}}{13!} - \cdots\right] dx$$

$$\approx 1.851937$$

45. Using $e^x = \sum_{n=0}^{\infty} \frac{x^n}{n!}$ and $\frac{hc}{k} \approx 0.014$, we get

$$g(\lambda) = \sum_{n=1}^{\infty} \frac{\lambda^5}{n!} \left(\frac{hc}{k\lambda T} \right)^n \approx \sum_{n=1}^{\infty} \frac{\lambda^5 \, 0.014^n}{(\lambda T)^n n!}.$$

Therefore,

$$\frac{dg}{d\lambda} \approx \sum_{n=1}^{6} (5-n) \frac{\lambda^4 \, 0.014^n}{(\lambda T)^n n!}$$

and the above expression equals 0 when

$$\lambda = \frac{0.0016}{T}.$$

This doesn't agree with Wien's Law, but the value of λ might be too large as the radius of convergence for the Maclaurin series is infinite.

47. Using the series for $(1+x)^{3/2}$ around the center $x = 0$, we get $S(d)$

$$\approx \frac{8\pi c^2}{3} \left(\frac{3}{32} \cdot \frac{d^2}{c^2} + \frac{3}{2048} \cdot \frac{d^4}{c^4} - \frac{1}{65{,}536} \cdot \frac{d^6}{c^6} \right)$$

If we ignore the d^4 and d^6 terms, this simplifies to $\dfrac{\pi d^2}{4}$.

9.9 FOURIER SERIES

1.
$$a_0 = \frac{1}{\pi} \int_{-\pi}^{\pi} f(x)\, dx = \frac{1}{\pi} \int_{-\pi}^{\pi} x\, dx$$

$$= \frac{1}{\pi} \frac{x^2}{2} \Big|_{-\pi}^{\pi} = 0$$

$$a_k = \frac{1}{\pi} \int_{-\pi}^{\pi} x \cos(kx)\, dx$$

$$= \frac{1}{\pi} \frac{x}{k} \sin(kx) + \frac{1}{k^2} \cos(kx) \Big|_{-\pi}^{\pi}$$

$$= 0$$

$$b_k = \frac{1}{\pi} \int_{-\pi}^{\pi} x \sin(kx)\, dx$$

$$= \frac{1}{\pi} \frac{-x}{k} \cos kx + \frac{1}{k^2} \sin kx \Big|_{-\pi}^{\pi}$$

$$= \frac{-2}{k} (-1)^k$$

Hence,

$$f(x) = \sum_{k=1}^{\infty} (-1)^{k+1} \frac{2}{k} \sin(kx)$$

for $-\pi < x < \pi$.

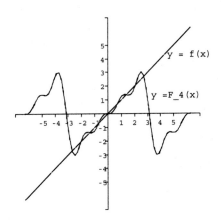

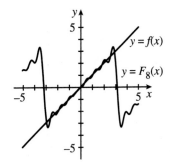

3.
$$a_0 = \frac{1}{\pi} \int_{-\pi}^{\pi} f(x)\, dx$$

$$= \frac{1}{\pi} \int_{-\pi}^{0} (-2x)\, dx + \frac{1}{\pi} \int_{0}^{\pi} 2x\, dx$$

$$= \frac{1}{\pi} (-x^2) \Big|_{-\pi}^{0} + \frac{1}{\pi} x^2 \Big|_{0}^{\pi} = 2\pi$$

$$a_k = \frac{1}{\pi} \int_{-\pi}^{\pi} f(x) \cos(kx)\, dx$$

$$= \frac{1}{\pi} \int_{-\pi}^{0} -2x \cos(kx)\, dx$$

$$+ \frac{1}{\pi} \int_{0}^{\pi} 3x \cos(kx)\, dx$$

$$= \frac{1}{\pi} \left[-\frac{2x}{k} \sin(kx) - \frac{2}{k^2} \cos(kx) \right] \Big|_{-\pi}^{0}$$

$$+ \frac{1}{\pi}\left[\frac{2x}{k}\sin(kx) + \frac{2}{k^2}\cos(kx)\right]\Big|_0^\pi$$

$$= \frac{4}{k^2\pi}[(-1)^k - 1]$$

So $a_{2k} = 0$ and $a_{2k-1} = \dfrac{-8}{(2k-1)^2\pi}$.

$$b_k = \frac{1}{\pi}\int_{-\pi}^{\pi} f(x)\sin(kx)\,dx$$

$$= \frac{1}{\pi}\int_{-\pi}^{0}(-2x\sin(kx))\,dx$$

$$+ \frac{1}{\pi}\int_0^\pi 2x\sin(kx)\,dx$$

$$= \frac{1}{\pi}\frac{2x}{k}\cos(kx) - \frac{2}{k^2}\sin(kx)\Big|_{-\pi}^0$$

$$+ \frac{1}{\pi}\frac{-2x}{k}\cos(kx) + \frac{2}{k^2}\sin(kx)\Big|_0^\pi$$

$$= \frac{2}{k}\cos(k\pi) - \frac{2}{k}\cos(k\pi)$$

$$= 0$$

So

$$f(x) = \frac{2\pi}{2} + \sum_{k=1}^{\infty}\frac{-8}{(2k-1)^2\pi}\cos[(2k-1)x]$$

$$f(x) = \pi - \sum_{k=1}^{\infty}\frac{8}{(2k-1)^2\pi}\cos[(2k-1)x]$$

for $-\pi < x < \pi$.

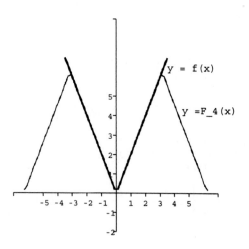

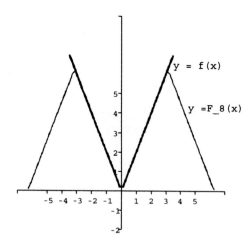

5.
$$a_0 = \frac{1}{\pi}\int_{-\pi}^0 1\,dx + \frac{1}{\pi}\int_0^\pi (-1)\,dx = 0$$

$$a_k = \frac{1}{\pi}\int_{-\pi}^0 \cos(kx)\,dx$$

$$+ \frac{1}{\pi}\int_0^\pi[-\cos(kx)]\,dx$$

$$= \frac{1}{\pi}\frac{1}{k}\sin kx\Big|_{-\pi}^0 - \frac{1}{\pi}\frac{1}{k}\sin kx\Big|_0^\pi$$

$$= 0$$

$$b_k = \frac{1}{\pi}\int_\pi^0 \sin(kx)\,dx$$

$$+ \frac{1}{\pi}\int_0^\pi[-\sin(kx)]\,dx$$

$$= \frac{1}{\pi}\left(-\frac{1}{k}\cos kx\right)\Big|_{-\pi}^0 - \frac{1}{\pi}\left(-\frac{1}{k}\cos kx\right)\Big|_0^\pi$$

So $b_{2k} = 0$ and $b_{2k-1} = \dfrac{-4}{(2k-1)\pi}$. Thus

$$f(x) = \sum_{k=1}^{\infty}\frac{-4}{(2k-1)\pi}\sin[(2k-1)x]$$

for $-\pi < x < \pi$.

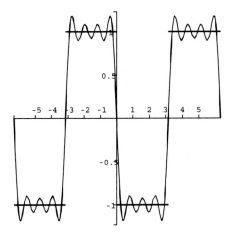

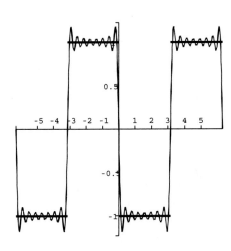

7. $f(x) = 3\sin 2x$ is already periodic on $[-2\pi, 2\pi]$.

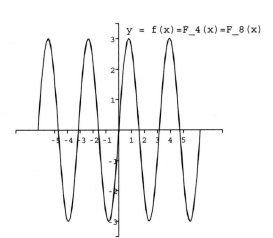

9.

$$a_0 = \int_{-1}^{1} (-x)\,dx = \left.\frac{-x^2}{2}\right|_{-1}^{1} = 0$$

$$a_k = \int_{-1}^{1} (-x)\cos(k\pi x)\,dx$$

$$= \left.\frac{-x}{k\pi}\sin k\pi x - \frac{1}{(k\pi)^2}\cos k\pi x\right|_{-1}^{1}$$

$$= \frac{-1}{k\pi}\sin k\pi - \frac{1}{(k\pi)^2}\cos k\pi$$

$$\quad - \left[\frac{1}{k\pi}\sin(-k\pi) - \frac{1}{(k\pi)^2}\cos(-k\pi)\right]$$

$$= 0$$

$$b_k = \int_{-1}^{1} (-x)\sin(k\pi x)\,dx$$

$$= \left.\frac{x}{k\pi}\cos k\pi x - \frac{1}{(k\pi)^2}\sin k\pi x\right|_{-1}^{1}$$

$$= \frac{1}{k\pi}\cos k\pi - \frac{1}{(k\pi)^2}\sin k\pi$$

$$\quad - \left[\frac{-1}{k\pi}\cos(-k\pi) - \frac{1}{(k\pi)^2}\sin(-k\pi)\right]$$

$$= \frac{2}{k\pi}\cos k\pi = \frac{2}{k\pi}(-1)^k$$

So

$$f(x) = \sum_{k=1}^{\infty} (-1)^k \frac{2}{k\pi}\sin k\pi x.$$

11.

$$a_0 = \frac{1}{1}\int_{-1}^{1} x^2\,dx = \left.\frac{x^3}{3}\right|_{-1}^{1} = \frac{2}{3}$$

$$a_k = \frac{1}{1}\int_{-1}^{1} x^2\cos k\pi x\,dx$$

$$= \frac{x^2}{k\pi}\sin k\pi x + \frac{2x}{(k\pi)^2}\cos k\pi x$$

$$\quad - \left.\frac{2}{(k\pi)^3}\sin k\pi x\right|_{-1}^{1}$$

$$= \frac{4}{(k\pi)^2}(-1)^k$$

$$b_k = \frac{1}{1} \int_{-1}^{1} x^2 \sin k\pi x \, dx$$

$$= -\frac{x^2}{k\pi} \cos k\pi x + \frac{2x}{(k\pi)^2} \sin k\pi x$$

$$- \frac{2}{(k\pi)^3} \cos k\pi x \Big|_{-1}^{1} = 0$$

Hence,

$$f(x) = \frac{1}{3} + \sum_{k=1}^{\infty} \frac{(-1)^k 4}{(k\pi)^2} \cos k\pi x$$

for $-1 < x < 1$.

15.

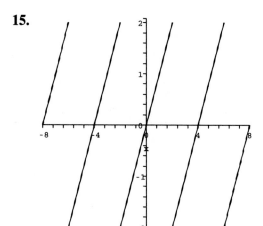

13.

$$a_0 = \int_{-1}^{1} f(x) \, dx = \int_{-1}^{0} 0 \, dx + \int_{0}^{1} x \, dx$$

$$= \frac{x^2}{2} \Big|_{0}^{1} = \frac{1}{2}$$

$$a_k = \int_{-1}^{0} 0 \, dx + \int_{0}^{1} x \cos(k\pi x) \, dx$$

$$= \frac{x}{k\pi} \sin(k\pi x) + \frac{1}{(k\pi)^2} \cos(k\pi x) \Big|_{0}^{1}$$

So $a_{2k} = 0$ and $a_{2k-1} = \dfrac{-2}{(2k-1)^2 \pi^2}$.

$$b_k = \int_{-1}^{0} 0 \, dx + \int_{0}^{1} x \cos(k\pi x) \, dx$$

$$= \frac{-x}{k\pi} \cos(k\pi x) + \frac{1}{(k\pi)^2} \sin(k\pi x) \Big|_{0}^{1}$$

$$= \frac{-1}{k\pi} (-1)^k$$

Hence,

$$f(x) = \frac{1}{4}$$

$$+ \sum_{k=1}^{\infty} \left[\frac{-2}{(2k-1)^2 \pi^2} \cos \left[(2k-1)\pi x \right] \right.$$

$$\left. + \frac{(-1)^{k+1}}{k\pi} \sin k\pi x \right]$$

for $-1 < x < 1$.

17.

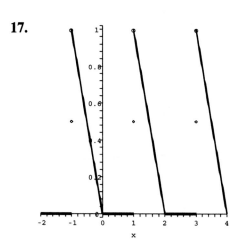

19.

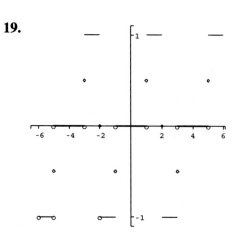

21.
$$f(x) = x^2 = \frac{1}{3} + \sum_{k=1}^{\infty} (-1)^k \frac{4}{k^2\pi^2} \cos k\pi x$$

$$f(1) = 1 = \frac{1}{3} + \sum_{k=1}^{\infty} (-1)^k \frac{4}{k^2\pi^2} (-1)^k$$

$$\frac{2}{3} = \sum_{k=1}^{\infty} \frac{4}{k^2\pi^2}, \text{ so}$$

$$\frac{2}{3} = \frac{4}{\pi^2} \sum_{k=1}^{\infty} \frac{1}{k^2}, \text{ so}$$

$$\frac{\pi^2}{6} = \sum_{k=1}^{\infty} \frac{1}{k^2}.$$

23.
$$f(x) = \frac{\pi}{2} - \sum_{k=1}^{\infty} \frac{4}{(2k-1)^2\pi} \cos[(2k-1)x]$$

$$f(0) = 0 = \frac{\pi}{2} - \sum_{k=1}^{\infty} \frac{4}{(2k-1)^2\pi} \cos 0$$

$$\frac{\pi}{2} = \sum_{k=1}^{\infty} \frac{4}{(2k-1)^2\pi} = \frac{4}{\pi} \sum_{k=1}^{\infty} \frac{1}{(2k-1)^2}$$

$$\frac{\pi^2}{8} = \sum_{k=1}^{\infty} \frac{1}{(2k-1)^2}.$$

25. Note that f is continuous except at $x = k\pi$, where $k \in \mathbb{Z}$. Also $f'(x) = 0$ except at $x = k\pi$. Thus, f' is continuous for $-\pi < x < \pi$, except at $x = 0$. By the Fourier Convergence Theorem (Theorem 9.1), the Fourier series for f converges to f for all x, except at $x = k\pi$, where the Fourier series converges to $\frac{1}{2}(-1+1) = 0$.

27. Note that f is continuous for all x except at $x = 2k + 1$, where $k \in \mathbb{Z}$. Also note that $f'(x) = -1$ except at $x = 2k + 1$, so that f' is continuous except at $x = 2k + 1$. By the Fourier Convergence Theorem (Theorem 9.1), the Fourier series for f converges to f for all x except at $x = 2k + 1$, where the Fourier series converges to $\frac{1}{2}(-1+1) = 0$.

29. If $f(x) = \cos x$, then $f(-x) = \cos(-x) = \cos x = f(x)$ so $\cos x$ is even.

If $f(x) = \sin x$, then $f(-x) = \sin(-x) = -\sin x = -f(x)$ so $\sin x$ is odd.

If $f(x) = \cos x + \sin x$, then $f(-x) = \cos(-x) + \sin(-x) = \cos x - \sin x \neq f(x)$ and $\cos x - \sin x \neq -f(x)$ so $\cos x + \sin x$ is neither even nor odd.

31. If $f(x)$ is odd and $g(x) = f(x) \cos x$, then
$$g(-x) = f(-x) \cos(-x)$$
$$= -f(x) \cos x = -g(x) \text{ is odd.}$$

Also if $h(x) = f(x) \sin x$, then
$$h(-x) = f(-x) \sin(-x)$$
$$= -f(x)(-\sin x)$$
$$= f(x) \sin x = h(x),$$
so $h(x)$ is even.

If $f(x)$ and $g(x)$ are even, then $fg(-x) = f(-x)g(-x) = f(x)g(x) = fg(x)$, so fg is even.

33. We have the Fourier series expansion
$$f(x) = \frac{a_0}{2}$$

$$+ \sum_{k=1}^{\infty} \left[a_k \cos\left(\frac{k\pi x}{l}\right) + b_k \sin\left(\frac{k\pi x}{l}\right) \right].$$

Multiply both sides of this equation by $\cos\left(\frac{n\pi x}{l}\right)$ and integrate with respect to x on the interval $[-l, l]$ to get

$$\int_{-l}^{l} f(x) \cos\left(\frac{n\pi x}{l}\right) dx$$

$$= \int_{-\ell}^{\ell} \frac{a_0}{2} \cos\left(\frac{n\pi x}{l}\right) dx$$

$$+ \sum_{k=1}^{\infty} \left[a_k \int_{-l}^{l} \cos\left(\frac{k\pi x}{l}\right) \cos\left(\frac{n\pi x}{l}\right) dx \right.$$

$$\left. + b_k \int_{-l}^{l} \sin\left(\frac{k\pi x}{l}\right) \cos\left(\frac{n\pi x}{l}\right) dx \right].$$

Then $\int_{-l}^{l} \cos\left(\frac{n\pi x}{l}\right) dx = 0$ for all n.

And,

$$\int_{-l}^{l} \sin\left(\frac{k\pi x}{l}\right) \cos\left(\frac{n\pi x}{l}\right) dx = 0.$$

$$\int_{-l}^{l} \cos\left(\frac{k\pi x}{l}\right) \cos\left(\frac{n\pi x}{l}\right)$$

$$= \begin{cases} 0 & \text{if } n \neq k \\ l & \text{if } n = k. \end{cases}$$

So when $n = k$, we have

$$\int_{-l}^{l} f(x) \cos\left(\frac{k\pi x}{l}\right) dx = a_k l, \text{ so}$$

$$a_k = \frac{1}{l} \int_{-l}^{l} f(x) \cos\left(\frac{k\pi x}{l}\right) dx.$$

Now multiply both sides of the original equation by $\sin\left(\frac{n\pi x}{l}\right)$ and integrate on $[-l, l]$, we have

$$\int_{-l}^{l} f(x) \sin\left(\frac{n\pi x}{l}\right) dx$$

$$= \int_{-l}^{l} \frac{a_0}{2} \sin\left(\frac{n\pi x}{l}\right) dx$$

$$+ \sum_{k=1}^{\infty} \left[a_k \int_{-l}^{l} \cos\left(\frac{k\pi x}{l}\right) \sin\left(\frac{n\pi x}{l}\right) dx \right.$$

$$\left. + b_k \int_{-l}^{l} \sin\left(\frac{k\pi x}{l}\right) \sin\left(\frac{n\pi x}{l}\right) dx \right].$$

Then $\int_{-l}^{l} \sin\left(\frac{n\pi x}{l}\right) dx = 0$. And

$$\int_{-l}^{l} \cos\left(\frac{k\pi x}{l}\right) \sin\left(\frac{n\pi x}{l}\right) dx = 0.$$

Also $\int_{-l}^{l} \sin\left(\frac{k\pi x}{l}\right) \sin\left(\frac{n\pi x}{l}\right) dx$

$$= \begin{cases} 0 & \text{if } n \neq k \\ l & \text{if } n = k \end{cases}.$$

So when $n = k$ we have

$$\int_{-l}^{l} f(x) \sin\left(\frac{k\pi x}{l}\right) dx = b_k l. \text{ So}$$

$$b_k = \frac{1}{l} \int_{-l}^{l} f(x) \sin\left(\frac{k\pi x}{l}\right) dx.$$

35. Since $f(x) = x^3$ is odd, the Fourier series will contain only sine.

37. Since $f(x) = e^x$ is neither even nor odd, the Fourier series will contain both.

39. Because $g(x)$ is an odd function, its series contains only sine terms, so $f(x)$ contains sine terms and the constant 1.

41. The graph of the limiting function for the function in exercise 9, with $k = 1000$:

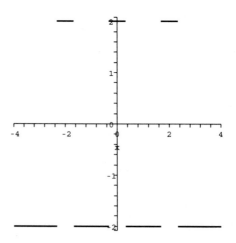

The graph of a pulse wave of width 1/3:

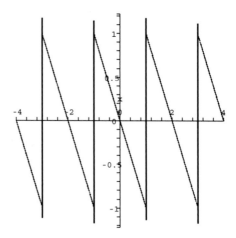

The graph of a pulse wave of width 1/4:

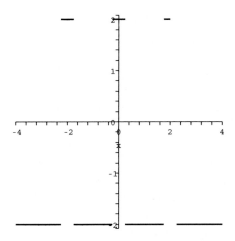

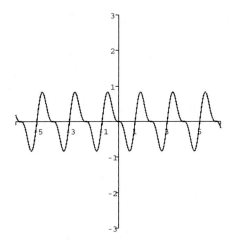

$$F_2(x) = \frac{8}{\pi}\left[\sin \pi x + \frac{1}{3}\sin 3\pi x\right]$$

43. The cutoff frequency, n, corresponds to the partial sum $F_n(x)$ because in both, everything after the nth term is zero.

$f(x) = -x$ on $[-1, 1]$, so $f(x)$

$$= \sum_{k=1}^{\infty} \frac{(-1)^k 2}{k\pi}\sin k\pi x, \text{ thus}$$

$$F_2(x) = \frac{2}{\pi}\left[-\sin \pi x + \frac{1}{2}\sin 2\pi x\right].$$

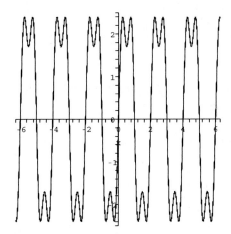

$$F_4(x) = \frac{8}{\pi}\left[\sin \pi x + \frac{1}{3}\sin 3\pi x\right.$$
$$\left. +\frac{1}{5}\sin 5\pi x + \frac{1}{7}\sin 7\pi x\right]$$

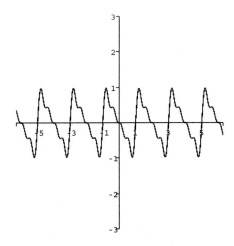

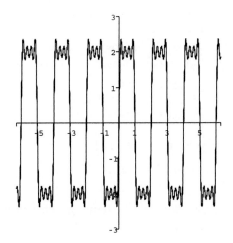

$$F_4(x) = \frac{2}{\pi}\left[-\sin \pi x + \frac{1}{2}\sin 2\pi\right.$$
$$\left. -\frac{1}{3}\sin 3\pi x + \frac{1}{4}\sin 4\pi x\right]$$

Because the wave is smoother.

45. The amplitude varies slowly because the frequency of $2\cos(0.2t)$ is small compared to the frequency of $\sin(8.1t)$. The variation of the amplitude explains why the volume varies, since the volume is proportional to the amplitude.

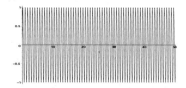

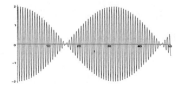

47. $f(x) = \dfrac{a_0}{2} + \displaystyle\sum_{k=1}^{\infty} \left[a_k \cos kx + b_k \sin kx \right]$

$[f(x)]^2 = \dfrac{a_0^2}{4}$

$\qquad + a_0 \displaystyle\sum_{k=1}^{\infty} \left[a_k \cos kx + b_k \sin kx \right]$

$\qquad + \left[\displaystyle\sum_{k=1}^{\infty} \left(a_k \cos kx + b_k \sin kx \right) \right]^2$

Now, note that

$\dfrac{1}{\pi} \displaystyle\int_{-\pi}^{\pi} \dfrac{a_0^2}{4}\, dx = \dfrac{a_0^2}{4\pi}(2\pi) = \dfrac{a_0^2}{2}$ and

$\dfrac{1}{\pi} \displaystyle\int_{-\pi}^{\pi} a_0 \sum_{k=1}^{\infty} [a_k \cos kx + b_k \sin kx]\, dx$

$\qquad = \dfrac{a_0}{\pi} \displaystyle\sum_{k=1}^{\infty} \left[a_k \int_{-\pi}^{\pi} \cos kx\, dx \right.$

$\qquad\qquad\qquad \left. + b_k \int_{-\pi}^{\pi} \sin kx\, dx \right]$

$= 0$

Now consider the last term, which can be expanded as

$\left[\displaystyle\sum_{k=1}^{\infty} \left(a_k \cos kx + b_k \sin kx \right) \right]^2$

$= \left[\displaystyle\sum_{k=1}^{\infty} \left(a_k \cos kx + b_k \sin kx \right) \right]$

$\quad \cdot \left[\displaystyle\sum_{m=1}^{\infty} \left(a_m \cos mx + b_m \sin mx \right) \right]$

$= \displaystyle\sum_{k=1}^{\infty} \sum_{m=1}^{\infty} \left[a_k a_m \cos kx \cos mx \right.$

$\qquad + a_k b_m \cos kx \sin mx$

$\qquad + a_m b_k \cos mx \sin kx$

$\qquad \left. + b_k b_m \sin kx \sin mx \right].$

Since $\displaystyle\int_{-\pi}^{\pi} \cos kx \sin mx\, dx = 0$ and

$\displaystyle\int_{-\pi}^{\pi} \cos kx \cos mx\, dx$

$= \displaystyle\int_{-\pi}^{\pi} \sin kx \sin mx\, dx$

$= \begin{cases} 1 & \text{if } k = m \\ 0 & \text{otherwise,} \end{cases}$

it follows that

$\dfrac{1}{\pi} \displaystyle\int_{-\pi}^{\pi} \sum_{k=1}^{\infty} [f(x)]^2\, dx$

$= \dfrac{a_0^2}{2}$

$\quad + \dfrac{1}{\pi} \displaystyle\sum_{k=1}^{\infty} \sum_{m=1}^{\infty} \left[a_k a_m \int_{-\pi}^{\pi} \cos kx \cos mx\, dx \right.$

$\qquad\qquad \left. + b_k b_m \int_{-\pi}^{\pi} \sin kx \sin mx\, dx \right]$

$= \dfrac{a_0^2}{2} + \dfrac{1}{\pi} \displaystyle\sum_{k=1}^{\infty} \left[a_k^2 \pi + b_k^2 \pi \right]$

$= \dfrac{a_0^2}{2} + \displaystyle\sum_{k=1}^{\infty} \left[a_k^2 + b_k^2 \right]$

$= \dfrac{a_0^2}{2} + \displaystyle\sum_{k=1}^{\infty} A_k^2.$

49. The plots from a CAS indeed show that the modified Fourier series

$$\frac{1}{2} + \sum_{k=1}^{n} \frac{2n}{[(2k-1)\pi]^2} \sin \frac{(2k-1)\pi}{n}$$

$$\sin [(2k-1)x]$$

reduces the Gibbs phenomenon in this case.

The case $n = 4$

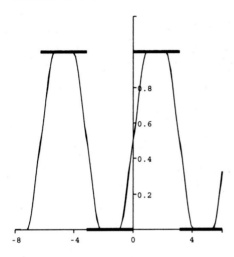

The case $n = 8$

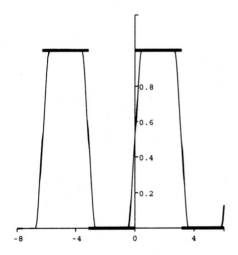

Chapter 9 Review Exercises

1. $\lim_{n\to\infty} \frac{4}{3+n} = 0$

The sequence converges to 0.

3. $\lim_{n\to\infty} \frac{n}{n^2+4} = \lim_{n\to\infty} \frac{1/n}{1+4/n^2} = 0$

The sequence converges to 0.

5. $0 < \frac{4 \cdot 4 \cdot 4 \cdot \ldots \cdot 4}{1 \cdot 2 \cdot 3 \ldots n} < 4 \cdot \frac{4}{2} \cdot \frac{4}{3} \cdot \frac{4}{4} \cdot \frac{4}{n} = \frac{128}{3n}$

and

$$\lim_{n\to\infty} \frac{128}{3n} = 0,$$

so by the Squeeze Theorem, the original series converges to 0.

7. $\{\cos \pi n\}_{n=1}^{\infty} = \{-1, 1, -1, \ldots\}$ diverges.

9. diverges

11. can't tell

13. diverges

15. converges

17. converges

19. $\sum_{k=0}^{\infty} 4\left(\frac{1}{2}\right)^k$ is a geometric series with $a = 4$

and $|r| = \frac{1}{2} < 1$ so the series converges to

$$\frac{a}{1-r} = \frac{4}{1-1/2} = 8.$$

21. $\sum_{k=0}^{\infty} \left(\frac{1}{4}\right)^k$ is a geometric series with $a = 1$

and $|r| = \frac{1}{4} < 1$, so the series converges to

$$\frac{a}{1-r} = \frac{1}{1-1/4} = \frac{4}{3}.$$

23. $|S - S_4| \le a_5 \le .01$ so $S \approx S_4 \approx -0.41$.

25. $\lim_{k\to\infty} \frac{2k}{k+3} = \lim_{k\to\infty} \frac{2}{1+\frac{3}{k}} = 2 \ne 0$ so by the

kth-Term Test for Divergence, the original series diverges.

27. $\lim\limits_{k\to\infty} |a_k| = \lim\limits_{k\to\infty} \dfrac{4}{(k+1)^{1/2}} = 0$ and $\dfrac{a_{k+1}}{a_k} =$

$\dfrac{4}{(k+2)^{1/2}} \cdot \dfrac{(k+1)^{1/2}}{4} = \dfrac{(k+1)^{1/2}}{(k+2)^{1/2}} < 1$ for
all $k \geq 0$ so $a_{k+1} < a_k$ for all $k \geq 0$, so by
the Alternating Series Test, the original series
converges.

29. diverges, by the p-test: $p = \dfrac{7}{8} < 1$

31. Using the Limit Comparison Test, let $a_k = \dfrac{\sqrt{k}}{k^3 + 1}$ and $b_k = \dfrac{1}{k^{5/2}}$. Then

$$\lim\limits_{k\to\infty} \dfrac{a_k}{b_k} = \lim\limits_{k\to\infty} \dfrac{\sqrt{k}}{k^3 + 1} \cdot \dfrac{k^{5/2}}{1} = 1 > 0$$

so because $\sum\limits_{k=1}^{\infty} \dfrac{1}{k^{5/2}}$ is a convergent p-series

$\left(p = \dfrac{5}{2} > 1\right)$, the original series converges.

33. Using the Alternating Series Test, $\lim\limits_{k\to\infty} \dfrac{4^k}{k!} = 0$
and

$$\dfrac{a_{k+1}}{a_k} = \dfrac{4^{k+1}}{(k+1)!} \cdot \dfrac{k!}{4} \leq 1$$

for $k \geq 3$, so $a_{k+1} \leq a_k$ for $k \geq 3$, thus the
original series converges.

35. converges, using the Alternating Series Test,
since

$$\lim\limits_{k\to\infty} |a_k| = \lim\limits_{k\to\infty} \ln\left(1 + \dfrac{1}{k}\right) = \ln 1 = 0$$

and

$$a_{k+1} = \ln\left(\dfrac{k+2}{k+1}\right) < \ln\left(\dfrac{k+1}{k}\right) = a_k$$

37. Using the Limit Comparison Test, let $a_k = \dfrac{2}{(k+3)^2}$ and $b_k = \dfrac{1}{k^2}$, so

$$\lim\limits_{k\to\infty} \dfrac{a_k}{b_k} = \lim\limits_{k\to\infty} \dfrac{2k^2}{(k+3)^2} = 2 > 0$$

so because $\sum\limits_{k=1}^{\infty} \dfrac{1}{k^2}$ is a convergent p-series
$(p = 2 > 1)$, the original series converges.

39. $\lim\limits_{k\to\infty} \dfrac{k!}{3^k} = \infty$

diverges, by the kth-Term Test

41. converges, by the Comparison Test, ($e^{1/k} < e$), and the p-series test $(p = 2 > 1)$

43. Converges, by the Ratio Test:

$$\lim\limits_{k\to\infty} \left|\dfrac{a_{k+1}}{a_k}\right| = \lim\limits_{k\to\infty} \left|\dfrac{4^{k+1}k!^2}{4^k(k+1)!^2}\right|$$

$$= \lim\limits_{k\to\infty} \dfrac{4}{(k+1)^2} = 0 < 1.$$

45. The original series converges, by the Alternat-
ing Series Test. However, $\sum\limits_{k=1}^{\infty} \dfrac{k}{k^2 + 1}$ diverges,
which can be shown using the Limit Compar-
ison Test with the harmonic series. The series
converges conditionally.

47. $\left|\dfrac{\sin k}{k^{3/2}}\right| = \dfrac{|\sin k|}{k^{3/2}} \leq \dfrac{1}{k^{3/2}}$

because $|\sin k| \leq 1$ for all k. And $\sum\limits_{k=1}^{\infty} \dfrac{1}{k^{3/2}}$ is a

convergent p-series $\left(p = \dfrac{3}{2} > 1\right)$. So by the

Comparison Test, $\sum\limits_{k=1}^{\infty} \left|\dfrac{\sin k}{k^{3/2}}\right|$ converges, so
the original series converges absolutely.

49. Using the Limit Comparison Test, $a_k = \dfrac{2}{(3+k)^p}$ and $b_k = \dfrac{1}{k^p}$, so

$$\lim\limits_{k\to\infty} \dfrac{a_k}{b_k} = \lim\limits_{k\to\infty} \dfrac{2k^p}{(3+k)^p} = 2 > 0$$

so the series $\displaystyle\sum_{k=1}^{\infty} \frac{2}{(3+k)^p}$ converges if and only if the p-series $\displaystyle\sum_{k=1}^{\infty} \frac{1}{k^p}$ converges, which happens when $p > 1$.

51. $|S - S_k| \le a_{k+1} = \dfrac{3}{(k+1)^2} \le 10^{-6}$,
so $1732.05 \le k+1$, so $1731.05 \le k$. Hence $k = 1732$ terms.

53. $f(x) = \dfrac{1}{4+x} = \displaystyle\sum_{k=0}^{\infty} \frac{(-1)^k x^k}{4^{k+1}}$, which is a geometric series that converges when $\left|-\dfrac{x}{4}\right| < 1$, or $-4 < x < 4$. Thus, $r = 4$.

55. $f(x) = \dfrac{3}{3+x^2} = \displaystyle\sum_{k=0}^{\infty} \frac{(-x^2)^k}{3^k}$ is a geometric series that converges for $\left|-\dfrac{x^2}{3}\right| < 1$ so $x^2 < 3$ or $|x| < \sqrt{3}$, so $-\sqrt{3} < x < \sqrt{3}$. Thus, $r = \sqrt{3}$.

57. $\dfrac{1}{4+x} = \displaystyle\sum_{k=0}^{\infty} \frac{(-1)^k x^k}{4^{k+1}}$ with $r = 4$. By integrating both sides, we get

$$\int \frac{1}{4+x}\, dx = \sum_{k=0}^{\infty} \frac{(-1)^k}{4^{k+1}} \int x^k\, dx$$

so

$$\ln(4+x) = \sum_{k=0}^{\infty} \frac{(-1)^k x^{k+1}}{(k+1)4^{k+1}} + \ln 4$$

with $r = 4$.

59. The Ratio Test gives $|x| < 1$, so $|r| = 1$. Since the series diverges for both $x = -1$ and $x = 1$, the interval of convergence is $(-1, 1)$.

61. The Ratio Test gives $|x| < 1$, so the radius of convergence is $r = 1$. The series diverges for $x = -1$ and converges for $x = 1$. Thus the interval of convergence is $(-1, 1]$.

63. The Ratio Test gives that the series converges absolutely for all $x \in (-\infty, \infty)$. Thus the interval of convergence is $(-\infty, \infty)$.

65. The Ratio Test gives $3|x - 2| < 1$, and this is when $-\dfrac{1}{3} < x - 2 < \dfrac{1}{3}$ or $\dfrac{5}{3} < x < \dfrac{7}{3}$. The series diverges for both $x = \dfrac{5}{3}$ and $x = \dfrac{7}{3}$. So the interval of convergence is $\left(\dfrac{5}{3}, \dfrac{7}{3}\right)$.

67. $\sin x = \displaystyle\sum_{k=0}^{\infty} \frac{(-1)^k x^{2k+2}}{(2k+1)!}$

69. $P_4(x) = (x-1) - \dfrac{1}{2}(x-1)^2 + \dfrac{1}{3}(x-1)^3 - \dfrac{1}{4}(x-1)^4$

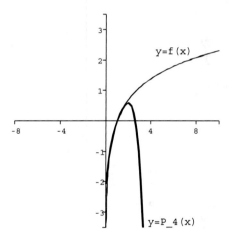

71. $|R_n(1.2)| = \left|\dfrac{f^{(n+1)}(z)(1.2-1)^{n+1}}{(n+1)!}\right|$ for some $z \in (1, 1.2)$.
Hence

$$|R_n(1.2)| = \frac{n!|z|^{-n-1}}{(n+1)!}(0.2)^{n+1} < \frac{(0.2)^{n+1}}{n+1}$$

and
$R_{10}(1.2) < 10^{-8}$. Consequently, $\ln(1.2) \approx P_{10}(1.2) \approx 0.182321498$.

73. Since $e^x = \displaystyle\sum_{k=0}^{\infty} \frac{x^k}{k!}$ for all x, we have

$$e^{-3x^2} = \sum_{k=0}^{\infty} \frac{(-3x^2)^k}{k!}$$

for all x, and so the radius of convergence is ∞.

75. $\displaystyle\int_0^1 \tan^{-1} x \, dx$

$$= \int_0^1 \left(x - \frac{1}{3}x^3 + \frac{1}{5}x^5 - \frac{1}{7}x^7 + \frac{1}{9}x^9 \right) dx$$

$$= \frac{x^2}{2} - \frac{x^4}{12} + \frac{x^6}{30} - \frac{x^8}{56} + \frac{x^{10}}{90} \Big|_0^1$$

$$= \frac{1117}{2520} \approx 0.4432539683$$

Compare this estimation with the actual value

$$\int_0^1 \tan^{-1} x \, dx \approx 0.4388245732.$$

77. $\displaystyle a_0 = \frac{1}{2} \int_{-2}^{2} x \, dx = \frac{1}{2} \frac{x^2}{2} \Big|_{-2}^{2} = 0$

$$a_k = \frac{1}{2} \int_{-2}^{2} x \cos\left(\frac{k\pi x}{2}\right) dx = 0$$

$$b_k = \frac{1}{2} \int_{-2}^{2} x \sin\left(\frac{k\pi x}{2}\right) dx$$

$$= -\frac{4}{k\pi}(-1)^k$$

So

$$f(x) = \sum_{k=1}^{\infty} \frac{(-1)^{k+1}4}{k\pi} \sin\left(\frac{k\pi x}{2}\right)$$

for $-2 < x < 2$.

79.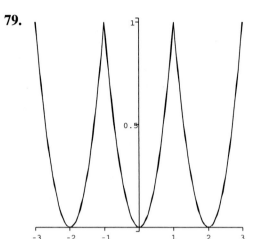

81.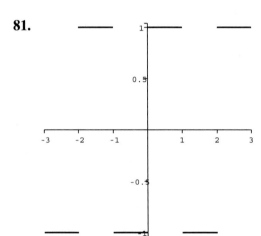

83. $\displaystyle \frac{1}{2} + \frac{1}{8} + \frac{1}{32} = \cdots = \sum_{k=0}^{\infty} \frac{1}{2}\left(\frac{1}{4}\right)^4$

is a geometric series with $a = \dfrac{1}{2}$ and $|r| = \dfrac{1}{4} < 1$, so it converges to

$$\left(\frac{\frac{1}{2}}{1 - \frac{1}{4}} \right) = \left(\frac{\frac{1}{2}}{\frac{3}{4}} \right) = \frac{2}{3}.$$

85. $a_{n+1} = a_n + a_{n-1}$,

$$\frac{a_{n+1}}{a_n} = \frac{a_n}{a_n} + \frac{a_{n-1}}{a_n} = 1 + \frac{a_{n-1}}{a_n}$$

Let $r = \displaystyle\lim_{n \to \infty} \frac{a_{n+1}}{a_n}$, then taking the limit as $n \to \infty$ on both sides of the equation above, we get

$$\lim_{n\to\infty} \frac{a_{n+1}}{a_n} = \lim_{n\to\infty} 1 + \lim_{n\to\infty} \frac{a_{n-1}}{a_n}.$$

$$r = 1 + \frac{1}{r}, \; r^2 - r - 1 = 0$$

$$r = \frac{1 \pm \sqrt{5}}{2}$$

But r is the limit of the ratios of positive integers, so r cannot be negative, and this gives

$$r = \lim_{n\to\infty} \frac{a_{n+1}}{a_n} = \frac{1 + \sqrt{5}}{2}.$$

87. Let $a_1 = 1 + \dfrac{1^2}{2}$ and so on.

n	a_n
1	1.5
2	1.153846154
3	1.381578947
4	1.197718631
5	1.344065167
6	1.218123800
7	1.325788792
8	1.229873907
9	1.314993721
10	1.237502842

Parametric Equations and Polar Coordinates

10.1 PLANE CURVES AND PARAMETRIC EQUATIONS

1. Equation: $\left(\dfrac{x}{2}\right)^2 + \left(\dfrac{y}{3}\right)^2 = 1$

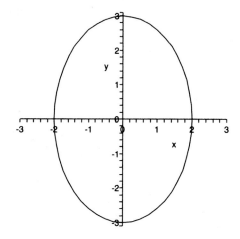

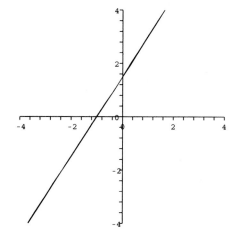

3. $t = \dfrac{y}{3}$

$x = -1 + 2\left(\dfrac{y}{3}\right)$

$\dfrac{2}{3}y = x + 1$

$y = \dfrac{3}{2}x + \dfrac{3}{2}$

This is a line, slope 3/2, y-intercept 3/2.

5. $t = x - 1$
$y = (x - 1)^2 + 2$
$y = x^2 - 2x + 3$

This is a parabola, with vertex (1, 2) opening up.

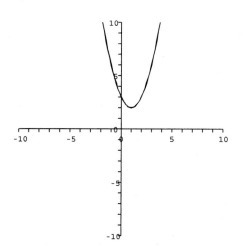

7. $t = \dfrac{y}{2}$

$x = \left(\dfrac{y}{2}\right)^2 - 1 = \dfrac{1}{4}y^2 - 1$

This is a parabola, vertex $(-1, 0)$ opening to the right.

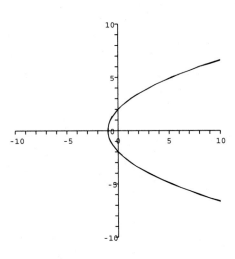

9. Equation: $y = 3x - 1, \; -1 \le x \le 1$

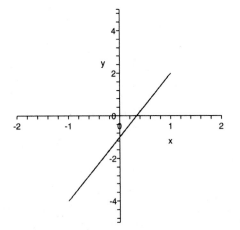

11. Graph is:

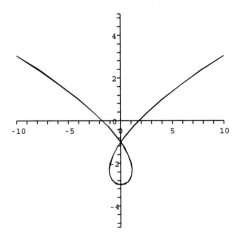

13. Graph is:

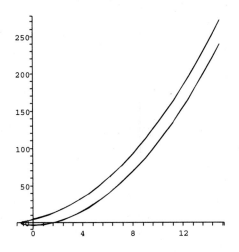

15. Graph is:

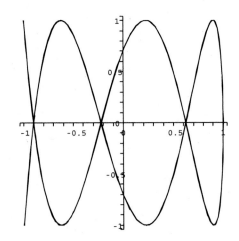

17. Graph is:

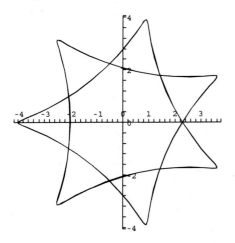

Graph for $k = 3$:

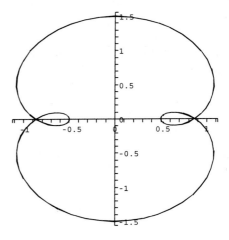

19. Graph is:

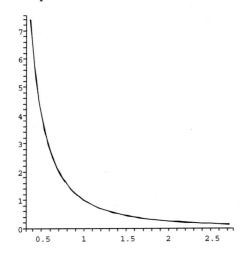

Graph for $k = 4$:

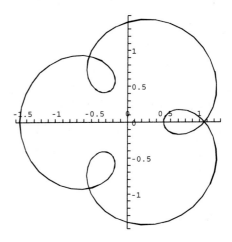

21. Integer values for k lead to closed curves, but irrational values for k do not.

Graph for $k = 5$:

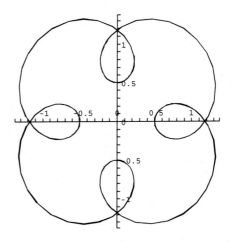

23. Graph for $k = 2$:

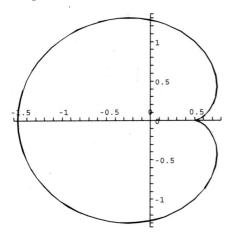

For $k > 2$, the graph has $k - 1$ "inner loops."

25. Since $(x + 1)^2 = y$, with $x \geq -1$, this is the right half of an upward-opening parabola. It has to be C.

27. x is bounded below by -1. y is bounded below by -1 and above by 1. From

$$y = \sin(t) = \sin(\pm\sqrt{x + 1})$$
$$= \pm \sin(\sqrt{x + 1}),$$

it has some of the features of a double sine curve, but the length of the cycles gets longer as x increases. On the basis of this alone, it could be B or E, but the first y-intercept after $x = -1$ will be when

$$\sqrt{x + 1} = \pi$$
$$x = \pi^2 - 1 \approx 8.9$$

and this is B.

29. x and y both oscillate between -1 and 1, but with different periods. This has to be A.

31. Use the model $x = a + tb$, $y = c + dt$, $(0 \leq t \leq 1)$, with $t = 0$ corresponding to $(0, 1)$ and $t = 1$ corresponding to $(3, 4)$.

$t = 0$: $0 = x = a$, $1 = y = c$
$t = 1$: $3 = x = 0 + b$, $4 = y = 1 + d$
So $b = 3$, $d = 3$, and the equations are
$x = 3t$, $y = 1 + 3t$.

33. Use the model $x = a + tb$, $y = c + dt$, $(0 \leq t \leq 1)$, with $t = 0$ corresponding to $(-2, 4)$ and $t = 1$ corresponding to $(6, 1)$.
$t = 0$: $-2 = x = a$, $4 = y = c$
$t = 1$: $6 = x = -2 + b$, $1 = y = 4 + d$
So $b = 8$, $d = -3$, and the equations are
$x = -2 + 8t$, $y = 4 - 3t$.

35. $x = t$ and $y = t^2 + 1$ from $t = 1$ to $t = 2$.

37. We set $x = -t + 2$ from $t = 0$ to $t = 2$ so that x travels from 2 to 0. So now we have $t = 2 - x$. Squaring this and taking the negative (to get the $-x^2$ in the formula for y) gives
$-t^2 = -(2 - x)^2 = -4 + 4x - x^2$.
There is no x term in the formula for y, so we will have to eliminate the x term above by

adding $4t$. We now have

$$-t^2 + 4t = -(2 - x)^2 + 4(2 - x)$$
$$= -4 + 4x - x^2 + 8 - 4x$$
$$= -x^2 + 4$$

So now we see that we need to subtract 2 to get y. Our final equations are:
$x = -t + 2$, $y = -t^2 + 4t - 2$ from $t = 0$ to $t = 2$.

39. Use the model
$x = a + b \cos t$, $y = c + b \sin t$
from $t = 0$ to $t = 2\pi$.
The center is at $(2, 1)$, therefore $a = 2$ and $c = 1$. The radius is 3, therefore $b = 3$, and the equations are

$$x = 2 + 3 \cos t, \quad y = 1 + 3 \sin t.$$

41. We set the x-values equal to each other, and the y-values equal to each other, giving us a system of two equations with two unknowns:
$t = 1 + s$
$t^2 - 1 = 4 - s$
The first equation is already solved for t, so we plug this expression for t into the second equation and then solve for s:
$(1 + s)^2 - 1 = 4 - s$
$1 + 2s + s^2 - 1 = 4 - s$
$s^2 + 3s - 4 = 0$
$(s + 4)(s - 1) = 0$
So the two possible solutions for s are $s = -4$ and $s = 1$. Since $t = 1 + s$, the corresponding t-values are $t = -3$ and $t = 2$, respectively. The points are $(-3, 8)$ and $(2, 3)$.

43. We must solve the system of equations:
$t + 3 = 1 + s$
$t^2 = 2 - s$
We plug $s = 2 - t^2$ into the first equation and solve for t:
$t + 3 = 1 + 2 - t^2$
$t^2 + t = 0$
$t(t + 1) = 0$
So $t = 0$ or $t = -1$. Since $s = 2 - t^2$, the corresponding s-values are $s = 2$ and $s = 1$, respectively. The points are $(3, 0)$ and $(2, 1)$.

45. The missile from Example 1.9 follows the path

$$\begin{cases} x = 100t \\ y = 80t - 16t^2 \end{cases}$$

for $0 \le t \le 5$. Set the x-values equal:
$100t = 500 - 500(t - 2) = 1500 - 500t$

$600t = 1500$

$t = 5/2$
Now check the y-values at $t = 5/2$:

$$80(\frac{5}{2}) - 16(\frac{5}{2})^2 = 200 - 100 = 100$$

$$208(\frac{5}{2} - 2) - 16(\frac{5}{2} - 2)^2 = 104 - 4 = 100$$

Therefore the interceptor missile will hit its target at time $t = 5/2$.

47. The initial missile is launched at time $t = 0$ and the interceptor missile is launched 2 minutes later. Let T be the time variable for the interceptor missile. Then when $T = 0$, $t = 2$ and from then on $T = t - 2$.

Let k be the length of the time delay. The flight path of the interceptor missile is then
$x = 500 - 200(t - k)$,
$y = 80(t - k) - 16(t - k)^2$, for $k \le t \le k + 5$.
Set the x-values equal:
$100t = 500 - 200(t - k)$
$300t = 500 + 200k$

$$t = \frac{5}{3} + \frac{2}{3}k$$

Set the y-values equal with this value of t and solve for k.

$$80\left(\frac{5}{3} + \frac{2}{3}k\right) - 16\left(\frac{5}{3} + \frac{2}{3}k\right)^2 = 80\left(\frac{5}{3} + \frac{2}{3}k - k\right)$$

$$- 16\left(\frac{5}{3} + \frac{2}{3}k - k\right)^2$$

$$\frac{64}{9}k^2 + \frac{160}{9}k + \frac{800}{9} = -\frac{16}{9}k^2 - \frac{80}{9}k + \frac{800}{9}$$

$$\frac{16}{3}k^2 - \frac{80}{3}k = 0$$

$$\frac{16}{3}k(k - 5) = 0$$

Either there is no time delay, $k = 0$, or the time delay is 5 seconds. However, since the missile lands at your location when $t = 5$, a 5-second delay isn't effective. Thus, there is no time delay that will work.

49. At time t, the position of the sound wave, as described by these parametric equations, is a circle of radius t centered at the origin. This makes sense as the sound wave would propagate equally in all directions.

51. The graph will look as follows, where we have marked the position of the jet with a diamond:

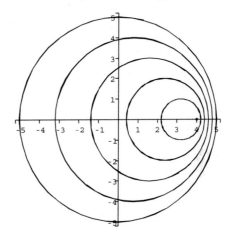

53. The graph will look as follows, where we have marked the position of the jet with a diamond:

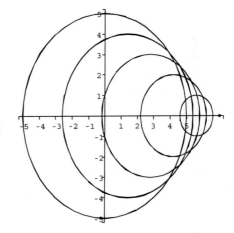

55. The shock waves are the lines formed by connecting the tops of the circles. In three dimensions, these lines will radiate out equally in all directions, i.e., will have circular cross sections. The three-dimensional figure formed by revolving the shock wave lines about the x-axis is a cone, which has circular cross sections, as expected.

57. Since distance = rate · time, the distance from the point $(0, D)$ to the position of the object at (x, y) is $v \cdot t$. We then find:

$$\begin{cases} x = (v \sin \theta)t \\ y = D - (v \cos \theta)t \end{cases}$$

59.
$$h(t) = \lim_{dt \to 0} \frac{x(t + dt) - x(t)}{\Delta T}$$

$$= \lim_{dt \to 0} \frac{v \sin \theta (t + dt) - v \sin \theta t}{\Delta T}$$

$$= \lim_{dt \to 0} \frac{v \sin \theta \, dt}{\Delta T}$$

$$= \lim_{dt \to 0} \frac{v \sin \theta}{\Delta T / dt}$$

$$= \frac{v \sin \theta}{\lim_{dt \to 0} \frac{\Delta T}{dt}}$$

$$= \frac{v \sin \theta}{T'(t)}$$

Since $T = t + L(t)$, then $T'(t) = 1 + L'(t)$. Thus,

$$h(t) = \frac{v \sin \theta}{1 + L'(t)}.$$

61. From exercise 60,

$$h(0) = \frac{cv \sin \theta}{c - v \cos \theta}.$$

The maximum value of $h(0)$ occurs when $(d/d\theta)h(0) = 0$:

$$\frac{1}{cv} \frac{d}{d\theta}[h(0)] =$$

$$= \frac{(c - v \cos \theta) \cos \theta - \sin \theta (v \sin \theta)}{(c - v \cos \theta)^2}$$

$$= \frac{c \cos \theta - v \cos^2 \theta - v \sin^2 \theta}{(c - v \cos \theta)^2}$$

$$= \frac{c \cos \theta - v(\cos^2 \theta + \sin^2 \theta)}{(c - v \cos \theta)^2}$$

$$= \frac{c \cos \theta - v}{(c - v \cos \theta)^2}.$$

Thus,

$$\frac{d}{d\theta}[h(0)] = 0 \Leftrightarrow 0 = c \cos \theta - v$$

$$\Leftrightarrow \cos \theta = \frac{v}{c}.$$

Thus, the maximum value of $h(0)$ occurs when $\cos \theta = v/c$. When $\cos \theta = v/c$,

$$\sin \theta = \sqrt{1 - \cos^2 \theta}$$

$$= \sqrt{1 - \frac{v^2}{c^2}}$$

$$= \gamma^{-1}$$

The maximum value of $h(0)$ is

$$h(0)_{max} = \frac{cv\gamma^{-1}}{c - v(v/c)}$$

$$= \frac{v\gamma^{-1}}{1 - \frac{v^2}{c^2}}$$

$$= \frac{v\gamma^{-1}}{\gamma^{-2}}$$

$$= v\gamma.$$

63. $\begin{cases} x = \cos 2t \\ y = \sin t \end{cases}$

We have the identity $\cos 2t = 1 - 2 \sin^2 t$, so we have the x-y equation $x = 1 - 2y^2$.

This shows that the graph is part of a parabola, with vertex $(1, 0)$ and opening to the left. However, the moving point never goes left of $x = -1$, forever going back and forth on this parabola, making an about-face every time x reaches -1.

$$\begin{cases} x = \cos t \\ y = \sin 2t \end{cases}$$

We have the identity
$$\sin 2t = 2\cos t \sin t$$ so
$$\sin^2 2t = 4\cos^2 t \sin^2 t$$
$$= 4\cos^2 t (1 - \cos^2 t)$$

Therefore $y^2 = 4x^2(1 - x^2)$ or
$y = \pm 2x\sqrt{1 - x^2}$.

This is a considerably more complicated graph, a figure-eight in which the moving point cycles smoothly around, starting out (when $t = 0$) by moving upward from $(1, 0)$ and completing the figure every time t passes an integral multiple of 2π. The moving point passes the origin twice during each cycle (first when $t = \pi/2$ and later when $t = 3\pi/2$).

65. The plot (including both pieces) is:

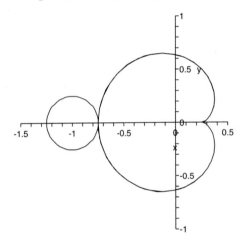

10.2 CALCULUS AND PARAMETRIC EQUATIONS

1. $x'(t) = 2t$
$y'(t) = 3t^2 - 1$

(a) $\dfrac{3(-1)^2 - 1}{2(-1)} = -1$

(b) $\dfrac{3(1)^2 - 1}{2(1)} = 1$

(c) We need to solve the system of equations:
$$t^2 - 2 = -2t^3 - t = 0$$

The only solution to the first equation is $t = 0$. As this is also a solution to the second equation, this solution will work (and is the only solution). The slope at $(-2, 0)$ (i.e., the slope at $t = 0$) is
$$\frac{3(0)^2 - 1}{2(0)} = \frac{-1}{0}.$$

So we see that there is a vertical tangent line at this point.

3. $x'(t) = -2\sin t$
$y'(t) = 3\cos t$

(a) $\dfrac{3\cos(\pi/4)}{-2\sin(\pi/4)} = \dfrac{-3}{2}$

(b) $\dfrac{3\cos(\pi/2)}{-2\sin(\pi/2)} = \dfrac{0}{-2} = 0$

(c) We need to solve the system of equations:
$$2\cos t = 0$$
$$3\sin t = 3$$
The solutions to the first equation are $t = \pi/2 + k\pi$ for any integer k. The solutions to the second equation are $t = \pi/2 + 2k\pi$ for any integer k. So the values of t that solve both are $t = \pi/2 + 2k\pi$. From (b), we see that these values all give slope 0.

5. $x'(t) = -2\sin 2t$
$y'(t) = 4\cos 4t$

(a) $\dfrac{4\cos(\pi)}{-2\sin(\pi/2)} = \dfrac{-4}{-2} = 2$

(b) $\dfrac{4\cos(2\pi)}{-2\sin(\pi)} = \dfrac{4}{0}$
Thus there is a vertical tangent line at this point.

(c) We need to solve the system of equations:
$$\cos 2t = \sqrt{2}/2$$
$$\sin 4t = 1$$
The first equation requires that $2t = \pm\pi/4 + 2\pi k$ or $t = \pm\pi/8 + \pi k$ for any integer k. The second equation requires that $4t = \pi/2 + 2\pi k$ or $t = \pi/8 + \pi k/2$ for any integer k. Thus we have

$t = \pi/8 + \pi k$. At these values of t, we have slope:

$$\frac{4\cos(4(\pi/8 + \pi k))}{-2\sin(2(\pi/8 + \pi k))}$$

$$= \frac{4\cos(\pi/2)}{-2\sin(\pi/4)} = \frac{0}{-\sqrt{2}} = 0$$

7. $x'(t) = 2t$
$y'(t) = 3t^2 - 1$
We must solve the system of equations:
$t^2 - 2 = -1$
$t^3 - t = 0$
The solutions to the first equation are $t = \pm 1$. The solutions to the second are $t = 0$ and $t = \pm 1$, so $t = \pm 1$ are the solutions to the system of equations. These are the values at which we must find slopes.

$$t = 1: \quad \frac{3-1}{2(1)} = 1$$

$$t = -1: \quad \frac{3-1}{2(-1)} = -1$$

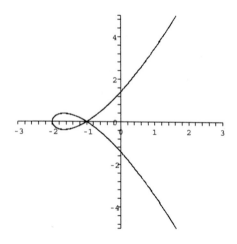

9. $x'(t) = -2\sin 2t$
$y'(t) = 4\cos 4t$

(a) We need $\cos 4t = 0$, so $4t = \pi/2 + k\pi$ or $t = \pi/8 + k\pi/4$ for any integer k. We need to check that the denominator is not 0 at these values.
$\sin(2(\pi/8 + k\pi/4))$
$= \sin(\pi/4 + k\pi/2)$
This is nonzero for all k, so all of these

t values give horizontal tangent lines. We plug these t values into the expressions for x and y to find the corresponding points. Depending on the value of k, we get the points $(\sqrt{2}/2, 1)$, $(-\sqrt{2}/2, -1)$, $(-\sqrt{2}/2, 1)$, and $(\sqrt{2}/2, -1)$.

(b) We need $\sin 2t = 0$, so $2t = k\pi$ or $t = k\pi/2$ for any integer k. We need to check that the numerator is not 0 at these values.

$$\cos(4(k\pi/2)) = \cos(2k\pi) \neq 0$$

The corresponding points are $(1, 0)$ and $(-1, 0)$.

11. $x'(t) = 2t$
$y'(t) = 4t^3 - 4$

(a) We want $4(t^3 - 1) = 0$ so the only solution is $t = 1$, which works since the denominator $(2t)$ is not zero for this t. This t-value corresponds to the x-y point $(0, -3)$.

(b) We want $2t = 0$ so $t = 0$, which does indeed give an undefined slope as the numerator will be -4. The corresponding point is $(-1, 0)$.

13. $x'(t) = -2\sin t + 2\cos 2t$
$y'(t) = 2\cos t - 2\sin 2t$

(a) We are looking for values of t for which $y'(t) = 0$, i.e., we want $\cos t = \sin 2t$. Since $\sin 2t = 2\sin t \cos t$, we need $\cos t = 2\sin t \cos t$. This happens when $\cos t = 0$ (when $t = \pi/2 + k\pi$ for any integer k) or when $\sin t = 1/2$. This occurs when $t = \pi/6 + 2k\pi$ or $t = 5\pi/6 + 2k\pi$ for any integer k.

When $t = \dfrac{\pi}{2} + k\pi$,

$$x'(t) = -2\sin\left(\frac{\pi}{2} + k\pi\right) + 2\cos\left(2\left(\frac{\pi}{2} + k\pi\right)\right)$$

$$= 2\sin\left(\frac{\pi}{2} + k\pi\right) + 2\cos(\pi + k\pi)$$

$$= -2\sin\left(\frac{\pi}{2} + k\pi\right) - 2$$

which is 0 when k is odd. Thus the points are when $t = \dfrac{\pi}{2} + k\pi$. When $\sin t = \dfrac{1}{2}$, $x'(t) = 0$ (see part (b)), so these points do not have horizontal tangent lines.

$t = \dfrac{\pi}{2} + 2k\pi$ gives the point $(0, 1)$.

(b) We are looking for values of t for which $x'(t) = 0$, i.e., we want $\sin t = \cos 2t$. Since $\cos 2t = 1 - 2 \sin^2 t$, we need to solve $2 \sin^2 t + \sin t - 1 = 0$.

Using the quadratic formula (with $\sin t$ replacing the usual x), we see that $\sin t = 1/2$ or $\sin t = -1$. From part (a), $\sin t = 1/2$ gives $y'(t) = 0$. $\sin t = -1$ gives $t = 3\pi/2 + 2k\pi$ for any integer k, which yields the point $(0, -3)$.

15. $x' = -2 \sin t,\ y' = 2 \cos t$

    (a) At $t = 0$, $x' = -2 \sin 0 = 0$,

      $y' = 2 \cos 0 = 2$.
      Speed $= \sqrt{0^2 + 2^2} = 2$.
      Motion is up.

    (b) At $t = \pi/2$, $x' = -2 \sin(\pi/2) = -2$,

      $y' = 2 \cos(\pi/2) = 0$.
      Speed $= \sqrt{(-2)^2 + 0^2} = 2$.
      Motion is to the left.

17. $x' = 20,\ y' = -2 - 32t$

    (a) At $t = 0$, $x' = 20$, $y' = -2$.
      Speed $= \sqrt{20^2 + (-2)^2} = 2\sqrt{101}$.
      Motion is to the right and slightly down.

    (b) At $t = 2$, $x' = 20$, $y' = -2 - 64 = -66$.
      Speed $= \sqrt{20^2 + (-66)^2}$
          $= 2\sqrt{1189}$.
      Motion is to the right and down.

19. $x' = -4 \sin 2t + 5 \cos 5t$,
    $y' = 4 \cos 2t - 5 \sin 5t$

    (a) At $t = 0$,
      $x' = -4 \sin 0 + 5 \cos 0 = 5$,
      $y' = 4 \cos 0 - 5 \sin 0 = 4$.
      Speed $= \sqrt{5^2 + 4^2} = \sqrt{41}$.
      Motion is to the right and up.

(b) At $t = \pi/2$,
    $x' = -4 \sin \pi + 5 \cos(5\pi/2) = 0$,
    $y' = 4 \cos \pi - 5 \sin(5\pi/2) = -9$.
    Speed $= \sqrt{0^2 + (-9)^2} = 9$.
    Motion is down.

21. $x'(t) = -3 \sin t$ and this curve is traced out counterclockwise, so

$$A = -\int_0^{2\pi} 2 \sin t (-3 \sin t)\, dt$$

$$= 6 \int_0^{2\pi} \sin^2 t\, dt$$

$$= 6 \left(\frac{1}{2}t - \frac{1}{2} \sin t \cos t \right) \Big|_0^{2\pi}$$

$$= 6\pi.$$

23. $x'(t) = -\frac{1}{2} \sin t + \frac{1}{2} \sin 2t$ and this curve is traced out counterclockwise, so

$$A = -\int_0^{2\pi} y(t) x'(t)\, dt$$

$$\approx 1.178$$

where $y(t) = \frac{1}{2} \sin t - \frac{1}{4} \sin 2t$.

25. $x'(t) = -\sin t$ and this curve is traced out clockwise, so

$$A = \int_{\pi/2}^{3\pi/2} \sin 2t (-\sin t)\, dt$$

$$= -2 \int_0^{2\pi} \sin^2 t \cos t\, dt$$

$$= -\frac{2}{3} \sin^3 t \Big|_{\pi/2}^{3\pi/2}$$

$$= -\frac{2}{3}((-1)^3 - 1^3) = \frac{4}{3}.$$

27. $x'(t) = 3t^2 - 4$ and this curve is traced out clockwise, so

$$A = \int_{-2}^{2} (t^2 - 3)(3t^2 - 4)\, dt$$

$$= \int_{-2}^{2} (3t^4 - 13t^2 + 12)\, dt$$

$$= \left[\frac{3t^5}{5} - \frac{13t^3}{3} + 12t \right] \Big|_{-2}^{2}$$

$$= \frac{256}{15}.$$

29. Crossing $x - $ axis $\Rightarrow y = 0 \Rightarrow t = n\pi$
$x' = -4\cos t \sin t - 2\sin t,$
$y' = 2\sin^2 t + 2(1 - \cos t)\cos t$

At $t = n\pi$, n even, $(x, y) = (3, 0)$ and $x' = 0$, $y' = 0$, speed $= 0$. At $t = n\pi$, n odd, $(x, y) = (-1, 0)$ and $x' = 0$, $y' = -4$, speed $= 4$.

31. The $3t$ and $5t$ indicate a ratio of 5-to-3.

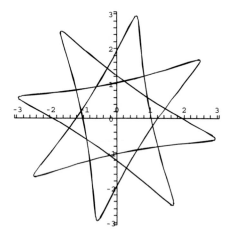

33. Use $x = 2\cos t + \sin 3t$,
$y = 2\sin t + \cos 3t$.

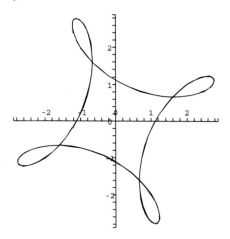

$x' = -2\sin t + 3\cos 3t$
$y' = 2\cos t - 3\sin 3t$
$s(t) = \sqrt{(x'(t))^2 + (y'(t))^2}$
$\quad = \sqrt{4 + 9 - 12(\sin t \cos 3t + \cos t \sin 3t)}$
$\quad = \sqrt{13 - 12\sin 4t}$
Minimum speed $= \sqrt{13 - 12} = 1$;
maximum speed $= \sqrt{13 + 12} = 5$.

35. $x' = 4\cos 4t$, $y' = 4\sin 4t$

$s(t) = \sqrt{(4\cos 4t)^2 + (4\sin 4t)^2}$

$\quad = \sqrt{16(\sin^2 4t + \cos^2 4t)}$

$\quad = 4$

The slope of the tangent line is $y'/x' = \tan 4t$, and the slope of the origin-to-object line is $y/x = -\cot 4t$, and the product of the two slopes is -1.

37. $\dfrac{y(t)}{x(t)} = \tan\theta$, where θ is the angle the object makes with the observer at the origin and the positive x-axis. The derivative

$$\left[\frac{y(t)}{x(t)} \right]' = \sec^2\theta \frac{d\theta}{dt}$$

will be positive if $\dfrac{d\theta}{dt}$ is positive, and the object will be moving counterclockwise. If the derivative is negative, then $\dfrac{d\theta}{dt}$ is negative, and the object will be moving clockwise.

39. We decide for convenience that we will have the circle rolling out on the positive x-axis with the center starting at $(0, r)$. The center moves out in proportion to time, i.e., $x_c = vt$, $y_c = r$ (constant).

If the circle were rotating in place, the motion of the point would be given by

$$x = r\cos\left(-\frac{r}{v}t \right), \quad y = r\sin\left(-\frac{r}{v}t \right). \text{ The}$$

negative signs come from the point moving in a clockwise direction as the circle rolls forward. The factor of $\dfrac{r}{v}$ comes from *rate · time =*

distance, which gives that at velocity v the time for one revolution is $t = \dfrac{2\pi r}{v}$.

In summary, the path (x_p, y_p) of the selected point on the rim satisfies

$$
\begin{cases}
x_p - x_c = r \cos\left(-\dfrac{r}{v}t\right) \\
\qquad = r \cos\left(\dfrac{r}{v}t\right), \\
y_p - y_c = r \sin\left(-\dfrac{r}{v}t\right) \\
\qquad = -r \sin\left(\dfrac{r}{v}t\right).
\end{cases}
$$

The final instructions are to *add* these numbers to the coordinates of the center. We can see why. It produces

$$
\begin{cases}
x_p = x_c + (x_p - x_c) \\
\qquad = vt + r \cos\left(\dfrac{r}{v}t\right), \\
y_p = y_c + (y_p - y_c) \\
\qquad = r - r \sin\left(\dfrac{r}{v}t\right).
\end{cases}
$$

To find the speed, we'll need

$$x'(t) = v - \frac{r^2}{v} \sin\left(\frac{r}{v}t\right) \text{ and}$$

$$y'(t) = -\frac{r^2}{v} \cos\left(\frac{r}{v}t\right).$$

Plugging these in and simplifying we find that the speed is given by

$$
s(t) = \sqrt{(x'(t))^2 + (y'(t))^2}
$$
$$
= \sqrt{v^2 - 2r^2 \sin\left(\frac{r}{v}t\right) + \frac{r^4}{v^2}}.
$$

To find the extrema, we take the derivative:

$$
s'(t) = \frac{-\dfrac{r^3}{v} \cos\left(\dfrac{r}{v}t\right)}{\sqrt{v^2 - 2r^2 \sin\left(\dfrac{r}{v}t\right) + \dfrac{r^4}{v^2}}}
$$

This is 0 when $\cos\left(\dfrac{r}{v}t\right) = 0$, i.e., when $\dfrac{r}{v}t = \dfrac{\pi}{2} + k\pi$ or $t = \dfrac{v\pi}{2r} + \dfrac{kv\pi}{r} = \dfrac{v\pi(1 + 2k)}{2r}$.

$1 + 2k$ is always odd, so this gives odd multiples of $\dfrac{v\pi}{2r}$. For these values of t, $\sin\left(\dfrac{r}{v}t\right) = \pm 1$ and $s(t) = \sqrt{v^2 \pm 2r^2 + \dfrac{r^4}{v^2}} = \left| v \pm \dfrac{r^2}{v} \right|$.

The maximum speed is $v + \dfrac{r^2}{v}$ at the top $(y = 2r)$. The minimum speed is $v - \dfrac{r^2}{v}$ at the bottom $(y = 0)$.

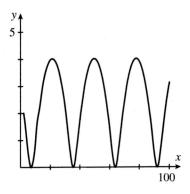

41. There are an infinite number of solutions, depending on the initial positions of both circles and the direction of motion. Assume that the center of the small circle begins at $(a - b, 0)$ and moves counterclockwise. Then the path of the center is $((a - b) \cos t, (a - b) \sin t)$. The path of a point about the center of the small circle is $(b \cos \theta, -b \sin \theta)$. To determine the relationship between θ and t, observe that when the small circle rotates through an angle θ, the distance $b\theta$ that its circumference moves matches up with an arc of length at on the large circle. Thus, $b\theta = at$, and so $\theta = \frac{a}{b}t$.

Thus, the path traced out by a point on the small circle is

$$
\begin{cases}
x = (a - b) \cos t + b \cos\left(\frac{a}{b}t\right) \\
y = (a - b) \sin t - b \sin\left(\frac{a}{b}t\right).
\end{cases}
$$

If $a = 2b$, then the path of a point on the small circle is

$(b \cos t + b \cos 2t, \, b \sin t - b \sin 2t)$.

The path now passes through the origin.

The graph for $a = 5$ and $b = 3$:

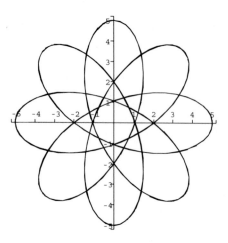

The equation for the slope of the tangent line at t is

$$\left.\frac{dy}{dx}\right|_t = \frac{\frac{dy}{dt}(t)}{\frac{dx}{dt}(t)}$$

$$= \frac{(a - b) \cos t - a \cos(\frac{a}{b}t)}{-(a - b) \sin t - a \sin(\frac{a}{b}t)}.$$

One point at which the tangent line is vertical is when $t = 0$, i.e., at the point $(a, 0)$.

43. Let $x = 2 \cos t$ and $y = 2 \sin t$. Then,

$$\frac{dx}{dt} = -2 \sin t$$

$$\frac{d^2x}{dt^2} = -2 \cos t$$

$$\frac{d^2x}{dt^2}(\pi/6) = -2\frac{\sqrt{3}}{2} = -\sqrt{3}$$

$$\frac{dy}{dt} = 2 \cos t$$

$$\frac{d^2y}{dt^2} = -2 \sin t$$

$$\frac{d^2y}{dt^2}(\pi/6) = -2\frac{1}{2} = -1$$

$$\frac{\frac{d^2y}{dt^2}(\pi/6)}{\frac{d^2x}{dt^2}(\pi/6)} = \frac{-1}{-\sqrt{3}}$$

$$= \frac{\sqrt{3}}{3}$$

Now, eliminating the parameter t to obtain an equation involving only the variables x and y, we get

$$\left(\frac{x}{2}\right)^2 + \left(\frac{y}{2}\right)^2 = \cos^2 t + \sin^2 t$$

$$\frac{x^2}{4} + \frac{y^2}{4} = 1$$

$$x^2 + y^2 = 4$$

Differentiating implicitly with respect to x:

$$2x + 2yy' = 0$$

$$x + yy' = 0 \quad (*)$$

$$1 + y'y' + yy'' = 0 \quad (**)$$

Solving equation $(*)$ for y', we get $y' = -x/y$. Substituting the result into equation $(**)$, we get

$$1 + \left(\frac{-x}{y}\right)\left(\frac{-x}{y}\right) + yy'' = 0$$

$$1 + \frac{x^2}{y^2} + yy'' = 0$$

$$y^2 + x^2 + y^3 y'' = 0$$

$$4 + y^3 y'' = 0$$

$$y'' = -\frac{4}{y^3}$$

Thus,

$$\frac{d^2y}{dx^2}\left(\sqrt{3}, 1\right) = -\frac{4}{1^3} = -4.$$

10.3 ARC LENGTH AND SURFACE AREA IN PARAMETRIC EQUATIONS

1. $x'(t) = -2 \sin t \; y'(t) = 4 \cos t$

$$s = \int_0^{2\pi} \sqrt{(-2 \sin t)^2 + (4 \cos t)^2} \, dt$$

$$= \int_0^{2\pi} \sqrt{4 \sin^2 t + 16 \cos^2 t} \, dt$$

$$\approx 19.3769$$

3. $x'(t) = 3t^2 - 4 \; y'(t) = 2t$

$$s = \int_{-2}^{2} \sqrt{(3t^2 - 4)^2 + (2t)^2} \, dt$$

$$\approx 15.6940$$

5. $x'(t) = -4 \sin 4t$
$y'(t) = 4 \cos 4t$

$$s = \int_0^{\pi/2} \sqrt{(-4 \sin 4t)^2 + (4 \cos 4t)^2} \, dt$$

$$= \int_0^{\pi/2} \sqrt{16 \sin^2 4t + 16 \cos^2 4t} \, dt$$

$$= \int_0^{\pi/2} \sqrt{16} \, dt$$

$$= \int_0^{\pi/2} 4 \, dt = 4 \left(\frac{\pi}{2} \right) = 2\pi$$

7. $x'(t) = \cos t - t \sin t$
$y'(t) = t \cos t + \sin t$

$$s = \int_{-1}^{1} \sqrt{(y'(t))^2 + (x'(t))^2} \, dt$$

$$= \int_{-1}^{1} \sqrt{t^2 + 1} \, dt$$

$$= 2 \int_0^1 \sqrt{t^2 + 1} \, dt$$

$$= \sqrt{2} + \ln(1 + \sqrt{2}) \approx 2.29559$$

9. $x'(t) = 2 \cos 2t \cos t - \sin 2t \sin t$
$y'(t) = 2 \cos 2t \sin t + \sin 2t \cos t$

Expanding and then combining like terms gives

$$(x'(t))^2 + (y'(t))^2 = 4 \cos^2 2t + \sin^2 2t \text{ so}$$

$$s = \int_0^{\pi/2} ((x'(t))^2 + (y'(t))^2)^{1/2} \, dt$$

$$= \int_0^{\pi/2} (4 \cos^2 2t + \sin^2 2t)^{1/2} \, dt$$

$$\approx 2.422.$$

11. $x'(t) = \cos t$
$y'(t) = \pi \cos \pi t$

$$s = \int_0^{\pi} (\cos^2 t + \pi^2 \cos^2 \pi t)^{1/2} \, dt$$

$$\approx 6.914$$

13. $x(0) = 0, \; y(0) = 0, \; x(1) = \pi, \; y(1) = 2$
$x'(t) = \pi$
$y'(t) = t^{-1/2}$

$$T = \int_0^1 k \sqrt{\frac{\pi^2 + 1/t}{2\sqrt{t}}} \, dt$$

$$= \int_0^1 k \left(\frac{\pi^2}{2t^{1/2}} + \frac{t^{-3/2}}{2} \right)^{1/2} \, dt$$

$$\approx k \cdot 4.4859$$

15. $x(0) = -\frac{1}{2}\pi(\cos 0 - 1) = 0,$
$y(0) = 0 + \frac{7}{10} \sin 0 = 0,$
$x(1) = -\frac{1}{2}\pi(\cos \pi - 1) = \pi,$
$y(1) = 2 + \frac{7}{10} \sin \pi = 2$

$x'(t) = \frac{\pi^2}{2} \sin \pi t$
$y'(t) = 2 + \frac{7}{10}\pi \cos \pi t$

$$T = \int_0^1 k \sqrt{\frac{(x'(t))^2 + (y'(t))^2}{y(t)}} \, dt$$

Here,

$$\frac{(x'(t))^2 + (y'(t))^2}{y(t)} =$$

$$\frac{\frac{\pi^4}{4}\sin^2 \pi t + 4 + \frac{14}{5}\pi \cos \pi t}{2t + \frac{7}{10}\sin \pi t}$$

$$+ \frac{\frac{49}{100}\pi^2 \cos^2 \pi t}{2t + \frac{7}{10}\sin \pi t}$$

so $T \approx k \cdot 4.4569$.

17. Slope is $\dfrac{t^{-1/2}}{\pi}$; so at $t = 0$ the slope is undefined.

$$s = \int_0^1 \left(\pi^2 + \frac{1}{t}\right)^{1/2} dt \approx 3.8897$$

The time for this path was approximately $k \cdot 4.4859$, which is slightly slower than the time for the cycloid path, which was approximately $k \cdot 4.443$.

19. Slope is

$$\frac{2 + \frac{7}{10}\pi \cos \pi t}{\frac{1}{2}\pi^2 \sin \pi t};$$

at $t = 0$ the slope is undefined, so there is a vertical tangent line. We have

$$s = \int_0^1 \sqrt{f(t)}\, dt$$

where

$$f(t) = \frac{1}{4}\pi^4 \sin^2 \pi t + \left(2 + \frac{7}{10}\pi \cos \pi t\right)^2$$

so $s \approx 4.066$. The time for this path was approximately $k \cdot 4.4569$, which is just slightly slower than the time for the cycloid path, which was approximately $k \cdot 4.443$.

21. $x'(t) = 2t$
$y'(t) = 3t^2 - 4$

$$s = \int_{-2}^0 2\pi|t^3 - 4t|\sqrt{(2t)^2 + (3t^2 - 4)^2}\, dt$$

$$= \int_{-2}^0 2\pi|t^3 - 4t|\sqrt{9t^4 - 20t^2 + 16}\, dt$$

$$\approx 85.823$$

23. $x'(t) = 2t$
$y'(t) = 3t^2 - 4$

$$s = \int_{-1}^1 2\pi|t^2 - 1|\sqrt{(2t)^2 + (3t^2 - 4)^2}\, dt$$

$$\approx 29.696$$

25. $x'(t) = 3t^2 - 4$
$y'(t) = 2t$

$$s = \int_0^2 2\pi|t^3 - 4t|\sqrt{(3t^2 - 4)^2 + (2t)^2}\, dt$$

$$\approx 85.823$$

27. Parametric equations for the midpoint of the ladder are given by

$$\begin{cases} x = 4\sin\theta \\ y = 4\cos\theta \end{cases}$$

for $0 \le \theta \le \pi/2$.

The distance, given by the formula for arc length, is

$$s = \int_0^{\pi/2} \sqrt{16\sin^2\theta + 16\cos^2\theta}\, d\theta$$

$$= \int_0^{\pi/2} 4\, d\theta$$

$$= 4\theta\Big|_0^{\pi/2}$$

$$= 2\pi.$$

29. $x'(t) = \cos(\pi t^2)$
$y'(t) = \sin(\pi t^2)$

(a) The arc length for $-2\pi \le t \le 2\pi$ is

$$\int_{-2\pi}^{2\pi} \sqrt{\left(\frac{dx}{dt}\right)^2 + \left(\frac{dy}{dt}\right)^2}\, dt$$

$$= \int_{-2\pi}^{2\pi} \sqrt{\cos^2(\pi t^2) + \sin^2(\pi t^2)}\, dt$$

$$= \int_{-2\pi}^{2\pi} dt$$

$$= [t]_{-2\pi}^{2\pi}$$

$$= 4\pi$$

(b) The arc length for $a \leq t \leq b$ is

$$\int_a^b dt$$
$$= [t]_a^b$$
$$= b - a.$$

Thus, the arc length of the curve is equal to the time interval; in other words, the parametrization being used is one for which the corresponding speed of a particle is 1. This means that as the size of the spiral decreases the spiralling rate increases.

10.4 POLAR COORDINATES

1. $r = 2, \theta = 0$
$x = 2 \cos 0 = 2$,
$y = 2 \sin 0 = 0$
Rectangular representation: $(2, 0)$

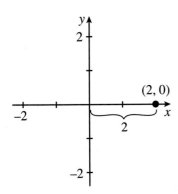

3. $r = -2, \theta = \pi$
$x = -2 \cos \pi = 2$,
$y = -2 \sin \pi = 0$
Rectangular representation: $(2, 0)$

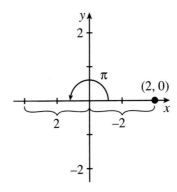

5. $r = 3, \theta = -\pi$
$x = 3 \cos(-\pi) = -3$,
$y = 3 \sin(-\pi) = 0$
Rectangular representation: $(-3, 0)$

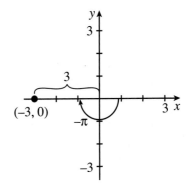

7. $r = \sqrt{2^2 + (-2)^2} = \pm\sqrt{8} = \pm 2\sqrt{2}$
$\tan \theta = \frac{-2}{2} = -1$
The point $(x, y) = (2, -2)$ is in Quadrant IV, so θ could be $-\pi/4$.

All polar representations: $\left(2\sqrt{2}, -\frac{\pi}{4} + 2\pi n\right)$
or
$\left(-2\sqrt{2}, \frac{3\pi}{4} + 2\pi n\right)$
where n is any integer.

9. $r = \sqrt{0^2 + 3^2} = \pm 3$
Since the point $(0, 3)$ is on the positive y-axis, we can take $\theta = \pi/2$.

All polar representations:
$\left(3, \frac{\pi}{2} + 2\pi n\right)$
or
$\left(-3, \frac{3\pi}{2} + 2\pi n\right)$
where n is any integer.

11. $r = \sqrt{3^2 + 4^2} = \pm 5$

$\tan \theta = \frac{4}{3}$

The point $(x, y) = (3, 4)$ is in Quadrant I;

$\theta = \tan^{-1}(\frac{4}{3}) \approx 0.9273$.

All polar representations:

$\left(5, \tan^{-1}(\frac{4}{3}) + 2\pi n \right)$

or

$\left(-5, \tan^{-1}(\frac{4}{3}) + \pi + 2\pi n \right)$

where n is any integer.

13. $r = 2, \theta = -\pi/3$

$x = 2 \cos \left(-\frac{\pi}{3} \right) = 1$

$y = 2 \sin \left(-\frac{\pi}{3} \right) = -\sqrt{3}$

Rectangular representation: $\left(1, -\sqrt{3} \right)$

15. $r = 0, \theta = 3$

$x = 0 \cos (3) = 0$

$y = 0 \sin (3) = 0$

Rectangular representation: $(0, 0)$

17. $r = 4, \theta = \pi/10$

$x = 4 \cos \left(\frac{\pi}{10} \right) \approx 3.8042$

$y = 4 \sin \left(\frac{\pi}{10} \right) \approx 1.2361$

Rectangular representation:
$(3.8042, 1.2361)$

19.

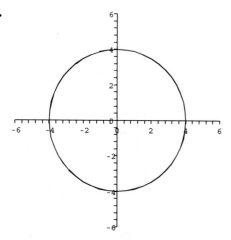

This is a circle centered at the origin with radius 4. The equation is

$x^2 + y^2 = 4^2 = 16$.

21.

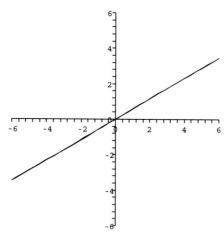

This is a line through the origin making an angle $\pi/6$ to the x-axis. We have

$\frac{y}{x} = \tan \theta = \tan \frac{\pi}{6} = \frac{1}{\sqrt{3}}$, so

$y = \frac{1}{\sqrt{3}} x$.

23.

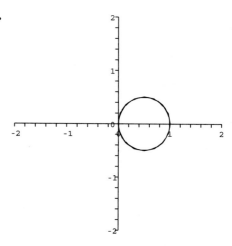

One knows from studying the examples that the figure is a circle centered at $(1/2, 0)$ with radius $1/2$. One could simply write down the equation as

$$\left(x - \frac{1}{2} \right)^2 + y^2 = \left(\frac{1}{2} \right)^2.$$

Failing the recognition, write

$r = \cos \theta$

$r^2 = r \cos \theta$

$x^2 + y^2 = x$

The two equations are the same.

25.

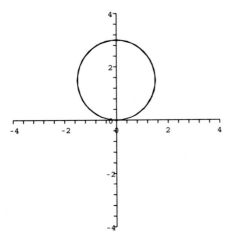

$$r = 3 \sin \theta$$
$$r^2 = 3r \sin \theta$$
$$x^2 + y^2 = 3y$$

27.

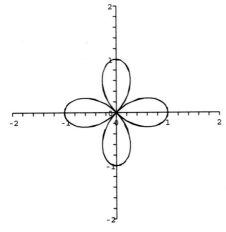

$r = \cos 2\theta = 0$ when $2\theta = \pi/2 + k\pi$ for any integer k, i.e., when $\theta = \pi/4 + k\pi/2$ for any integer k.

$0 \le \theta \le 2\pi$ produces one copy of the graph.

29. $r = \sin 3\theta = 0$ when $3\theta = k\pi$, i.e., $\theta = k\pi/3$ for any integer k.

$0 \le \theta \le \pi$ produces one copy of the graph.

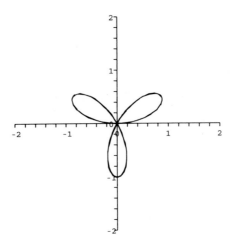

31. $r = 0$ when $3 + 2 \sin \theta = 0$ or $\sin \theta = -3/2$. This never happens so r is never 0.

$0 \le \theta \le 2\pi$ produces one copy of the graph.

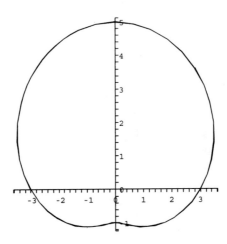

33. $r = 0$ when $2 - 4 \sin(\theta) = 0$, hence $\sin \theta = 1/2$, so $\theta = \pi/6 + 2k\pi$ or $\theta = 5\pi/6 + 2k\pi$ for integers k.

$0 \le \theta \le 2\pi$ produces one copy of the graph.

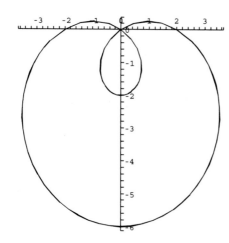

35. $r = 0$ when $\theta = 3\pi/2 + 2k\pi$ for integers k.

$0 \le \theta \le 2\pi$ produces one copy of the graph.

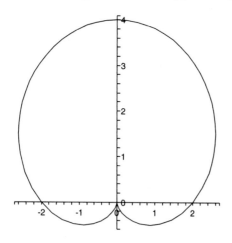

37. $r = \frac{1}{4}\theta = 0$ only when $\theta = 0$.

This graph does not repeat itself, i.e., for completion, one would have to graph for all real numbers θ.

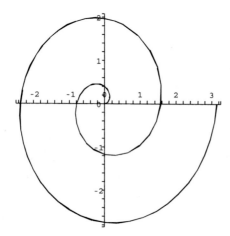

39. $\theta - \pi/4 = \pi/2 + k\pi$

$r = 2\cos(\theta - \pi/4) = 0$ when

$\theta - \pi/4 = \pi/2 + k\pi$, i.e.,

$\theta = 3\pi/4 + k\pi$ for integers k.

$0 \le \theta \le \pi$ produces one copy of the graph.

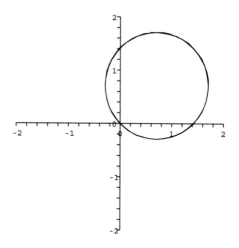

41. $r = \cos\theta + \sin\theta = 0$ when

$\theta = 3\pi/4 + 2k\pi$ for integers k.

$0 \le \theta \le \pi$ produces one copy of the graph.

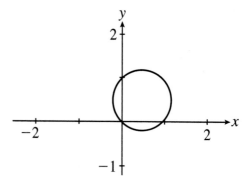

43. $r = \tan^{-1} 2\theta = 0$ only when $\theta = 0$.

This graph does not repeat itself, i.e., for completion, one would have to graph for all real numbers θ.

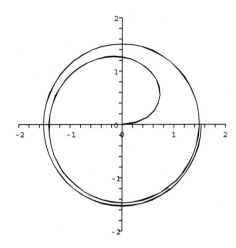

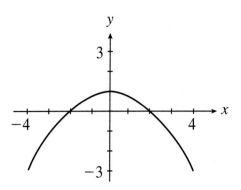

49. $r = \dfrac{2}{1 + \cos\theta} \neq 0$ for any θ. It is undefined when $\cos\theta = -1$, i.e., at $\theta = (2k + 1)\pi$ for integers k.

$\pi < \theta < 3\pi$ produces one copy of the graph.

45. $r = 2 + 4\cos 3\theta = 0$ when $\cos 3\theta = -1/2$, i.e., when $\theta = 2\pi/9 + 2k\pi/3$ or $\theta = 4\pi/9 + 2k\pi/3$ for any integer k.

$0 \leq \theta \leq 2\pi$ produces one copy of the graph.

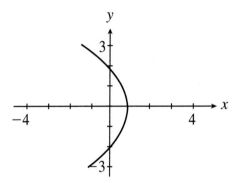

51.

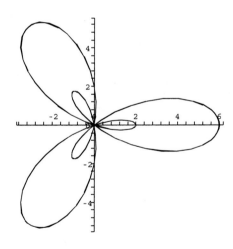

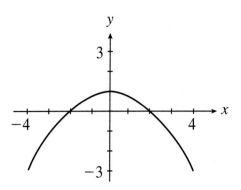

There is no graph to the left of the y-axis because the value of the x-coordinate of every point is positive, which we see as follows:

$x = r\cos\theta$

$\quad = 4\cos\theta\sin^2\theta\cos\theta$

$\quad = 4\cos^2\theta\sin^2\theta.$

47. $r = \dfrac{2}{1 + \sin\theta} \neq 0$ for any θ. It is undefined when $\sin\theta = -1$, i.e., at $\theta = -\pi/2 + 2k\pi$ for integers k.

$-\pi/2 < \theta < 3\pi/2$ produces one copy of the graph.

53. One guesses based on a number of examples in the text and the problems, that the figure is a circle of center $(a/2, 0)$ and radius $a/2$. To verify it:

$$r = a \cos \theta, \quad r^2 = ar \cos \theta$$

$$x^2 + y^2 = ax, \quad x^2 - ax + y^2 = 0$$

$$x^2 - ax + \frac{a^2}{4} + y^2 = \frac{a^2}{4}$$

$$\left(x - \frac{a}{2}\right)^2 + y^2 = \left(\frac{a}{2}\right)^2$$

55. $\cos n\theta$ is a rose with n petals if n is odd, and $2n$ petals if n is even.

57. $y^2 - x^2 = 4$,

$r^2 \sin^2 \theta - r^2 \cos^2 \theta = 4$,

$r^2(\cos^2 \theta - \sin^2 \theta) = -4$,

$$r^2 = \frac{-4}{\cos(2\theta)}$$

This is an acceptable answer. Before going any farther, one should note that this will require $\cos(2\theta) < 0$, which, as a subset of $[0, 2\pi)$, puts θ in one of the two open intervals

$$\left(\frac{\pi}{4}, \frac{3\pi}{4}\right) \cup \left(\frac{5\pi}{4}, \frac{7\pi}{4}\right).$$

Subject to that quantification, one could go

$$r = \frac{2}{\sqrt{-\cos(2\theta)}} = 2\sqrt{-\sec(2\theta)},$$

eschewing the possible minus sign since the use of the minus sign merely duplicates points already "present and accounted for."

All this confirms what we might already know about the curve: it is a hyperbola, opening up and down, with asymptotes formed by the two lines $y = x$ and $y = -x$ (which correspond to the endpoints of the stated domain-intervals for θ).

59. $r = 4$

61. $r \sin \theta = 3$

$$r = \frac{3}{\sin \theta} = 3 \csc \theta$$

63. Graph for $0 \le \theta \le \pi$:

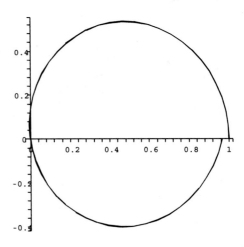

Graph for $0 \le \theta \le 2\pi$:

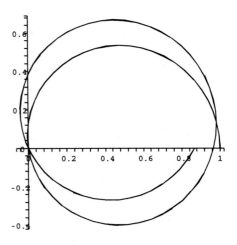

Graph for $0 \le \theta \le 3\pi$:

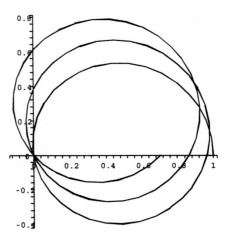

Graph for $0 \leq \theta \leq 6\pi$:

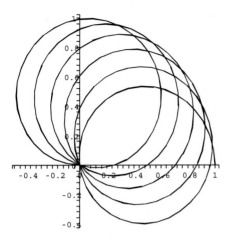

Graph for $0 \leq \theta \leq 24\pi$:

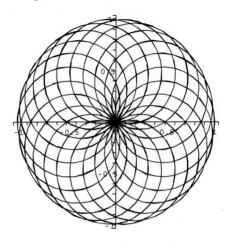

After 24π, the graph will repeat itself.

Graph for $0 \leq \theta \leq 12\pi$:

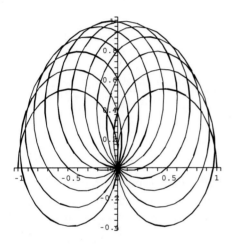

Graph for $0 \leq \theta \leq 23\pi$:

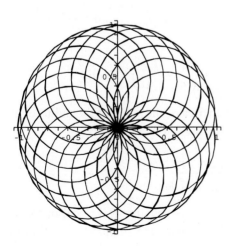

65. It is clear enough that there is a critical angle formed by the two lines from the ball, tangent to the hole. In one possible setup, the ball is at the origin O, the hole is centered at $D = (d, 0)$ (on the x-axis, at distance d), and the critical angle A_0 is being measured from the tangent line to the center line of the hole (the x-axis). If the upper point of tangency is denoted by T, one must recall that in the triangle TOD, the right angle is at T rather than at D. This makes OD the hypotenuse,

$$\sin A_0 = \frac{\text{opposite}}{\text{hypotenuse}} = \frac{h}{d},$$

and $A_0 = \sin^{-1}(h/d)$. Any acceptable angle A has to satisfy $-A_0 < A < A_0$.

67. There are no more calculations to be done at this point. According to the conclusions of exercise 65, it must be the case that $A_1 = -A_0$ while $A_2 = A_0$. From exercise 66 one finds

$$r_1(A) = d \cos(A) - \sqrt{d^2 \cos^2 A - (d^2 - h^2)}$$

and given in this problem is

$$r_2(A) = d + b \left(1 - \left[\frac{A}{\sin^{-1}(h/d)} \right]^2 \right)$$

$$= d + b \left(1 - \left[\frac{A}{A_0} \right]^2 \right).$$

One can observe, if the ball is to just drop into the side of the hole ($A \approx A_0$), that r_1 is just about $\sqrt{d^2 - h^2}$ and r_2 is just about d. There is very little margin for error on the speed.

10.5 CALCULUS AND POLAR COORDINATES

1. $f(\theta) = \sin 3\theta$
$f'(\theta) = 3\cos 3\theta$

$$\frac{dy}{dx}\Big|_{\theta=\frac{\pi}{3}} = \frac{3\cos\pi\sin\frac{\pi}{3} + \sin\pi\cos\frac{\pi}{3}}{3\cos\pi\cos\frac{\pi}{3} - \sin\pi\sin\frac{\pi}{3}}$$

$$= \frac{3(\frac{\sqrt{3}}{2})}{3(\frac{1}{2})} = \sqrt{3}$$

3. $f(\theta) = \cos 2\theta$
$f'(\theta) = -2\sin 2\theta$

$$\frac{dy}{dx}\Big|_{\theta=0} = \frac{-2\sin 0\sin 0 + \cos 0\cos 0}{-2\sin 0\cos 0 - \cos 0\sin 0}$$

$$= \frac{1}{0} \text{ undefined}$$

5. $f(\theta) = 3\sin\theta$
$f'(\theta) = 3\cos\theta$

$$\frac{dy}{dx}\Big|_{\theta=0} = \frac{3\cos 0\sin 0 + 3\sin 0\cos 0}{3\cos 0\cos 0 - 3\sin 0\sin 0}$$

$$= \frac{0}{3} = 0$$

7. $f(\theta) = \sin 4\theta$
$f'(\theta) = 4\cos 4\theta$

$$\frac{dy}{dx}\Big|_{\theta=\frac{\pi}{4}} = \frac{4\cos\pi\sin\frac{\pi}{4} + \sin\pi\cos\frac{\pi}{4}}{4\cos\pi\cos\frac{\pi}{4} - \sin\pi\sin\frac{\pi}{4}}$$

$$= \frac{-4(\frac{\sqrt{2}}{2})}{-4(\frac{\sqrt{2}}{2})} = 1$$

9. $f(\theta) = \cos 3\theta$
$f'(\theta) = -3\sin 3\theta$

$$\frac{dy}{dx}\Big|_{\theta=\frac{\pi}{6}} = \frac{-3\sin\frac{\pi}{2}\sin\frac{\pi}{6} + \cos\frac{\pi}{2}\cos\frac{\pi}{6}}{-3\sin\frac{\pi}{2}\cos\frac{\pi}{6} - \cos\frac{\pi}{2}\sin\frac{\pi}{6}}$$

$$= \frac{-3(\frac{1}{2})}{-3(\frac{\sqrt{3}}{2})} = \frac{1}{\sqrt{3}}$$

11. $|r|$ is a maximum when $\sin 3\theta = \pm 1$. This occurs when $3\theta = \pi/2 + k\pi$ or $\theta = \pi/6 + k\pi/3$ for any integer k. For $\theta = \frac{\pi}{6} + \frac{k\pi}{3}$, the points $(r(\theta)\cos\theta, r(\theta)\sin\theta)$ are $\left(\frac{\sqrt{3}}{2}, \frac{1}{2}\right)$, $(0, -1)$, and $\left(-\frac{\sqrt{3}}{2}, \frac{1}{2}\right)$.

We have $f(\theta) = \sin 3\theta$ so $f'(\theta) = 3\cos 3\theta$. Thus the slope of the tangent line is:

$$\frac{dy}{dx}\Big|_{\theta=\frac{\pi}{6}+\frac{k\pi}{3}}$$

$$= \frac{0 + \sin(\frac{\pi}{2} + k\pi)\cos(\frac{\pi}{6} + \frac{k\pi}{3})}{0 - \sin(\frac{\pi}{2} + k\pi)\sin(\frac{\pi}{6} + \frac{k\pi}{3})}$$

$$= \frac{\sin(\frac{\pi}{2} + k\pi)\cos(\frac{\pi}{6} + \frac{k\pi}{3})}{-\sin(\frac{\pi}{2} + k\pi)\sin(\frac{\pi}{6} + \frac{k\pi}{3})}$$

$$= -\frac{\cos(\frac{\pi}{6} + \frac{k\pi}{3})}{\sin(\frac{\pi}{6} + \frac{k\pi}{3})}$$

Here, the first terms of the numerator and denominator are always 0 since both have a factor of $\cos(\frac{\pi}{2} + k\pi)$. At $\theta = \frac{\pi}{6} + \frac{k\pi}{3}$, $r = \sin(\frac{\pi}{2} + k\pi)$ so the point in question is either

$$\left(\cos\left(\frac{\pi}{6} + \frac{k\pi}{3}\right), \sin\left(\frac{\pi}{6} + \frac{k\pi}{3}\right)\right)$$

or

$$\left(-\cos\left(\frac{\pi}{6} + \frac{k\pi}{3}\right), -\sin\left(\frac{\pi}{6} + \frac{k\pi}{3}\right)\right).$$

In either case, the slope of the radius connecting the point to the origin is

$$\frac{\sin(\frac{\pi}{6} + \frac{k\pi}{3})}{\cos(\frac{\pi}{6} + \frac{k\pi}{3})}$$

which is the negative reciprocal of the slope of the tangent line found above. Therefore the tangent line is perpendicular to the radius connecting the point to the origin.

13. $|r|$ is a maximum when $\sin 2\theta = -1$. This occurs when $2\theta = 3\pi/2 + 2k\pi$ or $\theta = 3\pi/4 + k\pi$ for any integer k. For $\theta = \frac{3\pi}{4} + k\pi$, the points $(r(\theta)\cos\theta, r(\theta)\sin\theta)$ are $(-3\sqrt{2}, 3\sqrt{2})$ and $(3\sqrt{2}, -3\sqrt{2})$. We have $f(\theta) = 2 - 4\sin(2\theta)$, so $f'(\theta) = -8\cos(2\theta)$. Thus the slope of the tangent line is

$$\frac{dy}{dx}\bigg|_{\theta=\frac{3\pi}{4}+k\pi}$$

$$= \frac{0 + \left(2 - 4\sin\left(\frac{3\pi}{2} + 2k\pi\right)\right)\cos\left(\frac{3\pi}{4} + k\pi\right)}{0 - \left(2 - 4\sin\left(\frac{3\pi}{2} + 2k\pi\right)\right)\sin\left(\frac{3\pi}{4} + k\pi\right)}$$

$$= \frac{\left(2 - 4\sin\left(\frac{3\pi}{2} + 2k\pi\right)\right)\cos\left(\frac{3\pi}{4} + k\pi\right)}{-\left(2 - 4\sin\left(\frac{3\pi}{2} + 2k\pi\right)\right)\sin\left(\frac{3\pi}{4} + k\pi\right)}$$

$$= -\frac{\cos\left(\frac{3\pi}{4} + k\pi\right)}{\sin\left(\frac{3\pi}{4} + k\pi\right)}.$$

Here, the first terms of the numerator and denominator are always 0 since both have a factor of $\cos\left(\frac{3\pi}{2} + 2k\pi\right)$. At $\theta = \frac{3\pi}{4} + k\pi$,

$$r = 2 - 4\sin\left(\frac{3\pi}{2} + 2k\pi\right)$$

so the point in question is

$$\left(6\cos\left(\frac{3\pi}{4} + k\pi\right), 6\sin\left(\frac{3\pi}{4} + k\pi\right)\right)$$

and the slope of the radius connecting the point to the origin is $\dfrac{\sin\left(\frac{3\pi}{4} + k\pi\right)}{\cos\left(\frac{3\pi}{4} + k\pi\right)}$, which is the negative reciprocal of the slope of the tangent line found above. Therefore the tangent line is perpendicular to the radius connecting the point to the origin.

15. One leaf is traced out over the range $\pi/6 \le \theta \le \pi/2$, so the area A is

$$A = \frac{1}{2}\int_{\pi/6}^{\pi/2}\cos^2 3\theta\, d\theta$$

$$= \frac{1}{6}\left(\frac{1}{2}\cdot 3\theta + \frac{1}{2}\sin 3\theta \cos 3\theta\right)\bigg|_{\pi/6}^{\pi/2}$$

$$= \frac{1}{6}\left(\frac{3\pi}{4} - \frac{\pi}{4}\right) = \frac{\pi}{12}.$$

17. Endpoints for the inner loop are given by $\theta = \sin^{-1}(3/4) \approx 0.848$ and $\theta = \pi - \sin^{-1}(3/4) \approx 2.294$, so the area A is

$$A = \frac{1}{2}\int_{0.848}^{2.294}(3 - 4\sin\theta)^2\, d\theta$$

$$= \frac{1}{2}\int_{0.848}^{2.294}(9 - 24\sin\theta + 16\sin^2\theta)\, d\theta$$

$$= \frac{1}{2}\int_{0.848}^{2.294}(17 - 24\sin\theta - 8\cos 2\theta)\, d\theta$$

$$= \frac{1}{2}(17\theta + 24\cos\theta - 4\sin 2\theta)\bigg|_{0.848}^{2.294}$$

$$\approx 0.3806.$$

19. The curve is traced out over the range $0 \le \theta \le \pi$, so the area A is

$$A = \frac{1}{2}\int_0^{\pi} 4\cos^2\theta\, d\theta$$

$$= 2\int_0^{\pi}\cos^2\theta\, d\theta$$

$$= 2\left(\frac{1}{2}\theta + \frac{1}{2}\sin\theta\cos\theta\right)\bigg|_0^{\pi}$$

$$= 2\left(\frac{\pi}{2}\right) = \pi.$$

21. A small loop is traced out over the range $7\pi/12 \le \theta \le 11\pi/12$, so the area A is

$$A = \frac{1}{2} \int_{7\pi/12}^{11\pi/12} (1 + 2 \sin 2\theta)^2 \, d\theta$$

$$= \frac{1}{2} \int_{7\pi/12}^{11\pi/12} (1 + 4 \sin 2\theta + 4 \sin^2 2\theta) \, d\theta$$

$$= \frac{1}{2} \int_{7\pi/12}^{11\pi/12} (1 + 4 \sin 2\theta$$

$$+ \, 2(1 - \cos 4\theta)) \, d\theta$$

$$= \frac{1}{2} \left(3\theta - 2 \cos 2\theta - \frac{1}{2} \sin 4\theta \right) \Big|_{7\pi/12}^{11\pi/12}$$

$$= \frac{\pi}{2} - \frac{3\sqrt{3}}{4} \approx 0.2718.$$

23. Endpoints for the inner loop are given by

$$\theta = \frac{\pi - \sin^{-1}(-2/3)}{3} \approx 1.2904$$

and

$$\theta = \frac{2\pi + \sin^{-1}(-2/3)}{3} \approx 1.8512,$$

so the area A is

$$A = \frac{1}{2} \int_{1.2904}^{1.8512} (2 + 3 \sin 3\theta)^2 \, d\theta$$

$$= \frac{1}{2} \int_{1.2904}^{1.8512} (4 + 12 \sin 3\theta + 9 \sin^2 3\theta) \, d\theta$$

$$\approx 0.14696.$$

25. This region is traced out over $-\pi/6 \le \theta \le 7\pi/6$, so the area A is given by

$$A = A_1 - A_2$$

where

$$A_1 = \frac{1}{2} \int_{-\pi/6}^{7\pi/6} (3 + 2 \sin \theta)^2 \, d\theta$$

and

$$A_2 = \frac{1}{2} \int_{-\pi/6}^{7\pi/6} 2^2 \, d\theta.$$

This works out to

$$A = \frac{11\sqrt{3}}{2} + \frac{14\pi}{3} \approx 24.187.$$

27. The curves intersect at $\theta = \frac{5\pi}{6}$ and $\theta = \frac{13\pi}{6}$. However, over this interval $r = 1 + 2 \sin \theta$ traces out an inner loop between $\theta = \frac{7\pi}{6}$ and $\theta = \frac{11\pi}{6}$. Thus the area inside $r = 2$ and outside both loops of $r = 1 + 2 \sin \theta$ is given by $A = A_1 - A_2$ where

$$A_1 = \frac{1}{2} \int_{5\pi/6}^{13\pi/6} 2^2 \, d\theta$$

and

$$A_2 = \frac{1}{2} \int_{5\pi/6}^{7\pi/6} (1 + 2 \sin \theta)^2 \, d\theta$$

$$+ \, \frac{1}{2} \int_{11\pi/6}^{13\pi/6} (1 + 2 \sin \theta)^2 \, d\theta.$$

This works out to $A = \frac{5\pi}{3} + \sqrt{3} \approx 6.9680.$

29. Since the graph is symmetric with respect to the x-axis, we can take the portions of the area we need above the x-axis and multiply these by 2. We then need the area of $r = 1$ on the region $0 \le \theta \le \pi/2$ (multiplied by 2) plus the area of $r = 1 + \cos \theta$ on the region $\pi/2 \le \theta \le \pi$ (multiplied by 2). We find that the area A is

$$A = \frac{1}{2} \cdot 2 \int_0^{\pi/2} 1^2 \, d\theta$$

$$+ \, \frac{1}{2} \cdot 2 \int_{\pi/2}^{\pi} (1 + \cos \theta)^2 \, d\theta$$

$$= \int_0^{\pi/2} 1 \, d\theta + \int_{\pi/2}^{\pi} (1 + \cos \theta)^2 \, d\theta$$

$$= \frac{\pi}{2} + \frac{3\pi}{4} - 2$$

$$= \frac{5\pi}{4} - 2 \approx 1.92699.$$

31. To find the points of intersection, we graph the function

$$y = 1 - 2 \sin x - 2 \cos x$$

and look for the roots. We find $x \approx 1.9948$ and $x \approx 5.8592$. These are the θ-values where the curves intersect. The coordinates of the intersection points are $(0.3386, -0.7500)$ and $(1.6614, -0.7499)$. Note that both graphs pass through the origin, so $(0, 0)$ is also a point of intersection.

33. To find the points of intersection in this case, we just set the two expressions for r equal to each other and solve for θ:

$$1 + \sin \theta = 1 + \cos \theta$$
$$\sin \theta = \cos \theta$$

This occurs when $\theta = \pi/4$ or $\theta = 5\pi/4$.

The coordinates of the intersection points are $(1.2071, 1.2071)$ and $(-0.2071, -0.2071)$. Note that both graphs pass through the origin, so $(0, 0)$ is also a point of intersection.

35. $f(\theta) = 2 - 2 \sin \theta$
$f'(\theta) = -2 \cos \theta$
The arc length is

$$s = \int_0^{2\pi} \sqrt{4 \cos^2 \theta + (2 - 2 \sin \theta)^2} \, d\theta$$

$$= \int_0^{2\pi} \sqrt{4 \cos^2 \theta + 4 - 8 \sin \theta + 4 \sin^2 \theta} \, d\theta$$

$$= \int_0^{2\pi} \sqrt{8(1 - \sin \theta)} \, d\theta$$

$$= 16.$$

37. $f(\theta) = \sin 3\theta$
$f'(\theta) = 3 \cos 3\theta$
The arc length is

$$s = \int_0^{\pi} \sqrt{\sin^2 3\theta + 9 \cos^2 3\theta} \, d\theta$$

$$\approx 6.6824.$$

39. $f(\theta) = 1 + 2 \sin 2\theta$
$f'(\theta) = 4 \cos 2\theta$
The arc length is

$$s = \int_0^{2\pi} \sqrt{16 \cos^2 2\theta + (1 + 2 \sin 2\theta)^2} \, d\theta$$

$$\approx 20.016.$$

41. We use the same setup as in Example 5.7, except here we use the line $y = -0.6$, which corresponds to $r = -0.6 \csc \theta$. To find the limits of integration, we must solve the equation

$$2 = -0.6 \csc \theta.$$

We find $\theta_1 \approx 3.4463$ and $\theta_2 \approx 5.9785$. With these limits of integration, we find the area A of the filled region:

$$A = \int_{\theta_1}^{\theta_2} 2 \, d\theta - \int_{\theta_1}^{\theta_2} \frac{1}{2}(-0.6 \csc \theta)^2 \, d\theta$$

$$= (2\theta + 0.18 \cot \theta) \Big|_{\theta_1}^{\theta_2}$$

$$\approx 3.9197.$$

The fraction of oil remaining in the tank is then approximately $\dfrac{3.9197}{4\pi} \approx 0.3119$ or about 31.2% of the total capacity of the tank.

43. We use the same setup as in Example 5.7, except here we use the line $y = 0.4$, which corresponds to $r = 0.4 \csc \theta$. To find the limits of integration, we must solve the equation

$$2 = 0.4 \csc \theta.$$

We find $\theta_1 \approx 0.2014$ and $\theta_2 \approx 2.9402$. With these limits of integration, we find the area A of the filled region:

$$A = \int_{\theta_1}^{\theta_2} 2 \, d\theta - \int_{\theta_1}^{\theta_2} \frac{1}{2}(0.4 \csc \theta)^2 \, d\theta$$

$$= (2\theta + 0.08 \cot \theta) \Big|_{\theta_1}^{\theta_2}$$

$$\approx 4.694.$$

The fraction of oil remaining in the tank is then approximately $\dfrac{4.694}{4\pi} \approx 0.3735$ or about 37.4% of the total capacity of the tank.

45. To show that the curve passes through the origin at each of these three values of θ, we simply show that $r = 0$ for each.

$\theta = 0: r = \sin 0 = 0$
$\theta = \pi/3: r = \sin \pi = 0$
$\theta = 2\pi/3: r = \sin 2\pi = 0$

We now find the slope at each of the three angles:

$$\frac{dy}{dx}\bigg|_{\theta=0} = \frac{3\sin 0 + 0}{3 - 0} = 0$$

$$\frac{dy}{dx}\bigg|_{\theta=\pi/3} = \frac{-3(\frac{\sqrt{3}}{2}) + 0}{-3(\frac{1}{2}) - 0} = \sqrt{3}$$

$$\frac{dy}{dx}\bigg|_{\theta=\pi/3} = \frac{3(\frac{\sqrt{3}}{2}) + 0}{3(-\frac{1}{2}) - 0} = -\sqrt{3}$$

As the graph passes through the origin at each of these three angles, it does so at a different slope.

47. If $r = cf(\theta)$ then $r' = cf'(\theta)$. The arc length is

$$s = \int_a^b \sqrt{(cf(\theta))^2 + (cf'(\theta))^2}\, d\theta$$

$$= \int_a^b \sqrt{c^2((f(\theta))^2 + (f'(\theta))^2)}\, d\theta$$

$$= |c| \int_a^b \sqrt{(f(\theta))^2 + (f'(\theta))^2}\, d\theta$$

$$= |c|L.$$

49. Let s represent the arc length from $\theta = d$ to $\theta = c$. Also note that since $f(\theta) = ae^{b\theta}$, it follows that $f'(\theta) = abe^{b\theta}$. Then,

$$s = \int_d^c \sqrt{[f'(\theta)]^2 + [f(\theta)]^2}\, d\theta$$

$$= \int_d^c \sqrt{a^2b^2e^{2b\theta} + a^2e^{2b\theta}}\, d\theta$$

$$= \int_d^c ae^{b\theta}\sqrt{b^2 + 1}\, d\theta$$

$$= a\sqrt{b^2 + 1} \int_d^c e^{b\theta}\, d\theta$$

$$= a\sqrt{b^2 + 1}\left(\frac{1}{b}\right)\left[e^{b\theta}\right]_d^c$$

$$= \frac{a\sqrt{b^2 + 1}}{b}\left[e^{bc} - e^{bd}\right].$$

Therefore, the total arc length is

$$\lim_{d\to-\infty} s = \lim_{d\to-\infty} \frac{a\sqrt{b^2 + 1}}{b}\left[e^{bc} - e^{bd}\right]$$

$$= \frac{a\sqrt{b^2 + 1}}{b}\left[e^{bc} - \lim_{d\to-\infty} e^{bd}\right]$$

$$= \frac{a\sqrt{b^2 + 1}}{b}e^{bc}$$

$$= \frac{\sqrt{b^2 + 1}}{b}R$$

since the distance from the origin to the starting point is $R = ae^{bc}$.

10.6 CONIC SECTIONS

1. Since the focus is $(0, -1)$ and the directrix is $y = 1$, we see that the vertex must be $(0, 0)$. Thus, $b = 0$, $c = 0$, and $a = -\frac{1}{4}$. The equation for the parabola is

$$y = -\frac{1}{4}x^2.$$

3. Since the focus is $(3, 0)$ and the directrix is $x = 1$, we see that the vertex must be $(2, 0)$. Thus, $b = 0$, $c = 2$, and $a = \frac{1}{4}$. The equation for the parabola is

$$x = \frac{1}{4}y^2 + 2.$$

5. From the given foci and vertices we find that $x_0 = 0$, $y_0 = 3$, $c = 2$, $a = 4$, and $b = \sqrt{12}$. The equation for the ellipse is

$$\frac{x^2}{12} + \frac{(y - 3)^2}{16} = 1.$$

7. From the given foci and vertices we find that $y_0 = 1$, $x_0 = 4$, $c = 2$, $a = 4$, and $b = \sqrt{12}$.

The equation for the ellipse is

$$\frac{(x-4)^2}{16} + \frac{(y-1)^2}{12} = 1.$$

9. From the given foci and vertices we find that $y_0 = 0$, $x_0 = 2$, $c = 2$, $a = 1$, and $b = \sqrt{3}$. The equation for the hyperbola is

$$\frac{(x-2)^2}{1} - \frac{y^2}{3} = 1.$$

11. From the given foci and vertices we find that $x_0 = 2$, $y_0 = 4$, $c = 2$, $a = 1$, and $b = \sqrt{3}$. The equation for the hyperbola is

$$\frac{(y-4)^2}{1} - \frac{(x-2)^2}{3} = 1.$$

13. This is a parabola with $a = 2$, $b = -1$, and $c = -1$.

vertex: $(-1, -1)$
focus: $(-1, -1 + \frac{1}{8})$
directrix: $y = -1 - \frac{1}{8}$

15. This is an ellipse with $x_0 = 1$, $y_0 = 2$, $a = 3$, $b = 2$, and $c = \sqrt{5}$.

foci: $(1, 2 - \sqrt{5})$, $(1, 2 + \sqrt{5})$
vertices: $(1, 5)$, $(1, -1)$

17. This is a hyperbola with $x_0 = 1$, $y_0 = 0$, $a = 3$, $b = 2$, and $c = \sqrt{13}$.

foci: $(1 - \sqrt{13}, 0)$, $(1 + \sqrt{13}, 0)$
vertices: $(4, 0)$, $(-2, 0)$

19. This is a hyperbola with $y_0 = -1$, $x_0 = -2$, $a = 4$, $b = 2$, and $c = \sqrt{20}$.

foci: $(-2, -1 + \sqrt{20})$, $(-2, -1 - \sqrt{20})$
vertices: $(-2, 3)$, $(-2, -5)$

21. Dividing both sides by 9 gives

$$\frac{(x-2)^2}{9} + y^2 = 1.$$

This is an ellipse with $x_0 = 2$, $y_0 = 0$, $a = 3$, $b = 1$, and $c = \sqrt{8}$.

foci: $(2 - \sqrt{8}, 0)$, $(2 + \sqrt{8}, 0)$
vertices: $(5, 0)$, $(-1, 0)$

23. Solving for y gives

$$y = \frac{1}{4}(x+1)^2 - 2.$$

This is a parabola with $a = 1/4$, $b = -1$,, and $c = -2$.

vertex: $(-1, -2)$
focus: $(-1, -1)$
directrix: $y = -3$

25. This is a parabola with focus $(2, 1)$ and directrix $y = -3$. Thus the vertex is $(2, -1)$ and $b = 2$, $c = -1$, and $a = \frac{1}{8}$. The equation is

$$y = \frac{1}{8}(x-2)^2 - 1.$$

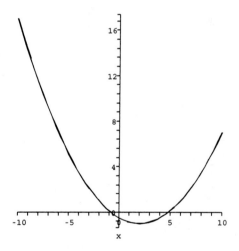

27. This is an ellipse with foci $(0, 2)$ and $(4, 2)$ and center $(2, 2)$. Since the sum of the distances to the foci must be 8, the vertices are $(-2, 2)$ and $(6, 2)$. Thus $y_0 = 2$, $x_0 = 2$, $c = 2$, $a = 4$, and $b = \sqrt{12}$. The equation is

$$\frac{(x-2)^2}{16} + \frac{(y-2)^2}{12} = 1.$$

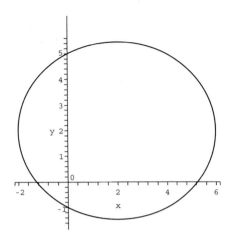

29. This is a hyperbola with foci $(0, 4)$ and $(0, -2)$ and center $(0, 1)$. Since the difference of the distances to the foci must be 4, the vertices are $(0, 3)$ and $(0, -1)$. Thus $x_0 = 0$, $y_0 = 1$, $c = 3$, $a = 2$, and $b = \sqrt{5}$. The equation is

$$\frac{(y - 1)^2}{4} - \frac{x^2}{5} = 1.$$

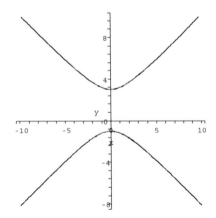

31. Since $x = 4y^2$, we have $a = 4$, $b = 0$, and $c = 0$. The lightbulb should be placed at the focus, i.e., $(\frac{1}{16}, 0)$.

33. Since $y = 2x^2$, we have $a = 2$, $b = 0$, and $c = 0$. The microphone should be placed at the focus, i.e., $(0, \frac{1}{8})$.

35. We have $c = \sqrt{a^2 - b^2} = \sqrt{124 - 24} = 10$, so that the foci are 20 inches apart. The transducer should be located 20 inches away from the kidney stone and aligned so that the line segment from the kidney stone to the trans-

ducer lies along the major axis of the elliptical reflector.

37. From the equation, we have $y_0 = -4$, $x_0 = 0$, $a = 1$, $b = \sqrt{15}$, and $c = 4$. Thus the foci are $(0, -8)$ and $(0, 0)$. Light rays following the path $y = cx$ would go through the focus $(0, 0)$, so they are reflected toward the focus at $(0, -8)$.

39. From the equation, we have $x_0 = 0$, $y_0 = 0$, $a = \sqrt{3}$, $b = 1$ and $c = 2$. Thus the foci are $(-2, 0)$ and $(2, 0)$. Light rays following the path $y = c(x - 2)$ would go through the focus $(2, 0)$, so they are reflected toward the focus at $(-2, 0)$.

41. We have $x_0 = 0$, $y_0 = 0$, and $c = \sqrt{400 - 100} = 10\sqrt{3}$. The foci are $(\pm 10\sqrt{3}, 0)$ so the desks should be placed along the major axis of the ellipse $10\sqrt{3}$ units from the center (at $(0, 0)$) i.e., at the foci.

43. Let $f(t)$ be the object's distance from the spectator at time t. We know this is a quadratic function, so it must be of the form $f(t) = at^2 + bt + c$ for some constants a, b and c. At time $t = 0$, the spectator has the object, so the distance from the spectator is 0. Thus $f(0) = a(0)^2 + b(0) + c = c = 0$. So now our function is of the form $f(t) = at^2 + bt$. Plugging in $t = 2$ and $t = 4$ (and their respective values of $f(t)$) gives the following two equations:

$$4a + 2b = 28$$
$$16a + 4b = 48$$

Multiplying the first equation by 2 and then subtracting the result from the second equation gives $8a = -8$ or $a = -1$. Plugging $a = -1$ into either of the above equations gives $b = 16$. Thus the equation for the distance of the object from the spectator is $f(t) = -t^2 + 16t$. Graphing this function gives:

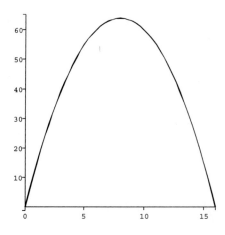

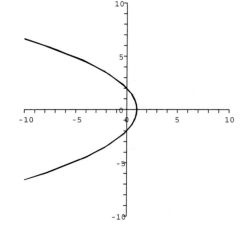

Since this object returns to the spectator (at time $t = 16$ seconds the object is again at 0 meters from the spectator) we guess that the object is a boomerang.

10.7 CONIC SECTIONS IN POLAR COORDINATES

1.   From Theorem 7.2 part (i) we have

$$r = \frac{0.6(2)}{0.6\cos\theta + 1} = \frac{1.2}{0.6\cos\theta + 1}.$$

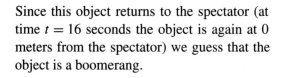

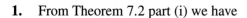

3.   From Theorem 7.2 part (i) we have

$$r = \frac{2}{\cos\theta + 1}.$$

5.   From Theorem 7.2 part (iii) we have

$$r = \frac{1.2}{0.6\sin\theta + 1}.$$

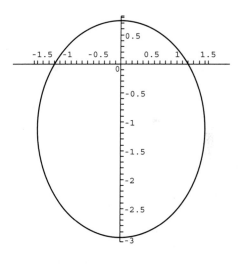

7.   From Theorem 7.2 part (iii) we have

$$r = \frac{2}{\sin\theta + 1}.$$

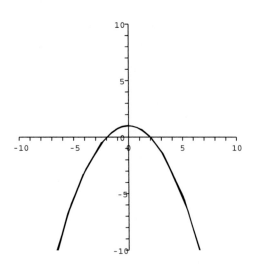

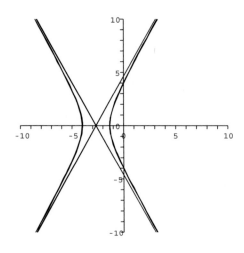

9. From Theorem 7.2 part (ii) we have

$$r = \frac{-0.8}{0.4 \cos \theta - 1}.$$

13. From Theorem 7.2 part (iv) we have

$$r = \frac{-0.8}{0.4 \sin \theta - 1}.$$

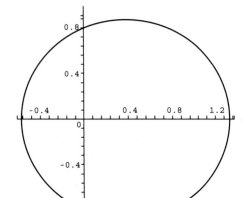

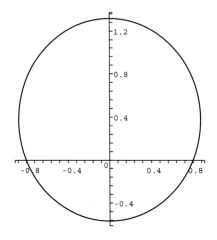

11. From Theorem 7.2 part (ii) we have

$$r = \frac{-4}{2 \cos \theta - 1}.$$

15. From Theorem 7.2 part (iv) we have

$$r = \frac{-2}{\sin \theta - 1}.$$

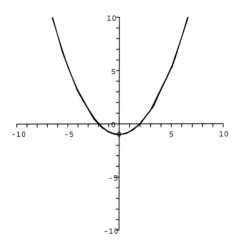

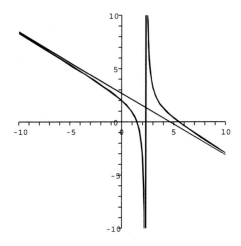

17. This is a hyperbola since $e = 2$:

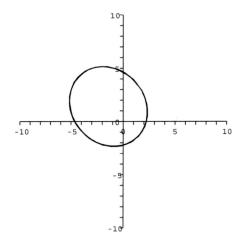

19. Divide the numerator and denominator by 2 to get $r = \dfrac{-3}{\frac{1}{2}\sin\left(\theta - \frac{\pi}{4}\right) - 1}$. This is an ellipse since $e = \frac{1}{2}$:

21. Divide the numerator and denominator by 2 to get $r = \dfrac{-\frac{3}{2}}{\cos\left(\theta + \frac{\pi}{4}\right) - 1}$. This is a parabola since $e = 1$:

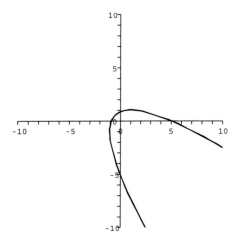

23. This is an ellipse with center $(-1, 1)$ and major axis parallel to the x-axis. The parametric equations (for $0 \le t \le 2\pi$) are:
$$\begin{cases} x = 3\cos t - 1 \\ y = 2\sin t + 1. \end{cases}$$

25. The right half of this hyperbola is given by
$$\begin{cases} x = 4\cosh t - 1 \\ y = 3\sinh t \end{cases}$$
while the left half is given by
$$\begin{cases} x = -4\cosh t - 1 \\ y = 3\sinh t. \end{cases}$$

27. We solve for y to get
$$y = -\frac{x^2}{4} + 1.$$

The parametric equations are then given by
$$\begin{cases} x = t \\ y = -t^2/4 + 1. \end{cases}$$

29. We have
$$A_1 = \frac{1}{2} \int_0^{\pi/2} \left(\frac{2}{\sin\theta + 2}\right)^2 d\theta$$
$$\approx 0.473$$

$$A_2 = \frac{1}{2} \int_{3\pi/2}^{4.953} \left(\frac{2}{\sin\theta + 2} \right)^2 d\theta$$

$$\approx 0.472$$

and

$$s_1 = \int_0^{\pi/2} g(\theta)\, d\theta \approx 1.266$$

$$s_2 = \int_{3\pi/2}^{4.953} g(\theta)\, d\theta \approx 0.481$$

where

$$g(\theta) = \sqrt{\frac{4\cos^2\theta}{(\sin\theta + 2)^4} + \frac{4}{(\sin\theta + 2)^2}}.$$

Therefore the two areas are approximately the same, while the average speed on the portion of the orbit from $\theta = 0$ to $\theta = \pi/2$ is a little more than two-and-a-half times the average speed on the portion of the orbit from $\theta = 3\pi/2$ to $\theta = 4.953$.

31. For any point (x, y) on the curve, the distance to the focus is $\sqrt{x^2 + y^2}$ and the distance to the directrix is $x - d$. We then have

$$\sqrt{x^2 + y^2} = e(x - d)$$

$$r = e(r\cos\theta - d)$$

$$r - er\cos\theta = -ed$$

$$r(1 - e\cos\theta) = -ed$$

$$r = \frac{-ed}{1 - e\cos\theta}$$

$$r = \frac{ed}{e\cos\theta - 1}.$$

33. For any point (x, y) on the curve, the distance to the focus is $\sqrt{x^2 + y^2}$ and the distance to the directrix is $y - d$. We then have

$$\sqrt{x^2 + y^2} = e(y - d)$$

$$r = e(r\sin\theta - d)$$

$$r(1 - e\sin\theta) = -ed$$

$$r = \frac{ed}{e\sin\theta - 1}.$$

Chapter 10 Review Exercises

1. The given parametric equations for x and y satisfy the following x-y equation:

$$\left(\frac{x+1}{3} \right)^2 + \left(\frac{y-2}{9} \right)^2 = 1$$

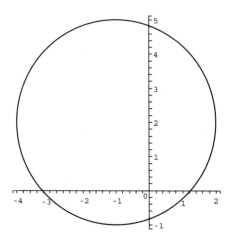

3. From the first equation we have $t^2 = x - 1$. We plug this into the second equation to get

$$y = (t^2)^2 = (x - 1)^2$$

which is valid for $x \geq 1$.

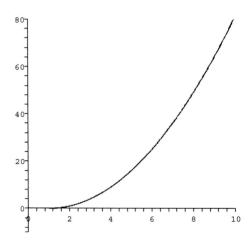

5.

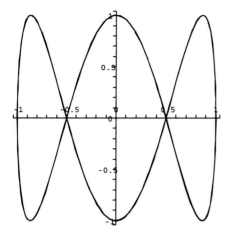

7.

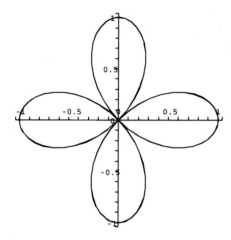

9. Solving the first equation for t gives $t = \sqrt{x+1}$, so $x \geq -1$. Plugging into the second equation gives $y = (x+1)^{3/2}$. This matches Figure C.

11. Sketching the graph, we find that this corresponds to Figure B.

13. We want equations of the form
$$\begin{cases} x = at + b \\ y = ct + d \end{cases}$$
on the interval $0 \leq t \leq 1$. At $t = 0$, we need $x = b = 2$ and $y = d = 1$. At $t = 1$, we then have $x = a + 2 = 4$ so $a = 2$ and $y = c + 1 = 7$ so $c = 6$. Therefore our parametric equations for $0 \leq t \leq 1$ are
$$\begin{cases} x = 2t + 2 \\ y = 6t + 1. \end{cases}$$

15. $x'(t) = 3t^2 - 3$
$y'(t) = 2t - 1$

(a) $\left. \dfrac{dy}{dx} \right|_{t=0} = \dfrac{y'(0)}{x'(0)} = \dfrac{-1}{-3} = \dfrac{1}{3}$

(b) $\left. \dfrac{dy}{dx} \right|_{t=1} = \dfrac{y'(1)}{x'(1)} = \dfrac{1}{0}$ undefined

(c) We need to find a value for t such that $x = t^3 - 3t = 2$ and $y = t^2 - t + 1 = 3$. The latter gives $t^2 - t - 2 = 0$. Solving for t gives $t = 2$ or $t = -1$. Both solve the former equation, so we check the slope at both values of t:

$$\left. \dfrac{dy}{dx} \right|_{t=-1} = \dfrac{y'(-1)}{x'(-1)} = \dfrac{-3}{0} \text{ undefined}$$

and

$$\left. \dfrac{dy}{dx} \right|_{t=2} = \dfrac{y'(2)}{x'(2)} = \dfrac{3}{9} = \dfrac{1}{3}.$$

17. $x'(t) = 3t^2 - 3$
$y'(t) = 2t + 2$
$x'(0) = -3$; $y'(0) = 2$, so the motion is up and to the left.
$\text{speed} = \sqrt{(-3)^2 + 2^2} = \sqrt{13}$

19. $x'(t) = 3 \cos t$
With $0 \leq t \leq 2\pi$, the curve is traced out clockwise, so the area A is

$$A = \int_0^{2\pi} 2 \cos t \cdot 3 \cos t \, dt$$

$$= 6 \int_0^{2\pi} \cos^2 t \, dt$$

$$= 6 \left(\frac{1}{2} t + \frac{1}{2} \sin t \cos t \right) \Big|_0^{2\pi}$$

$$= 6\pi.$$

21. From $t = -1$ to $t = 1$, the curve is traced counterclockwise.
$x'(t) = -2 \sin 2t$

$$A = -\int_{-1}^{1} \sin(\pi t)(-2\sin(2t))\, dt$$

$$= 2\int_{-1}^{1} \sin(\pi t)\sin(2t)\, dt$$

$$\approx 1.947$$

23. $x'(t) = -2\sin 2t$
$y'(t) = \pi\cos\pi t$

$$s = \int_{-1}^{1} \sqrt{4\sin^2 2t + \pi^2\cos^2\pi t}\, dt$$

$$\approx 5.2495$$

25. $x'(t) = -4\sin 4t$
$y'(t) = 5\cos 5t$

$$s = \int_{0}^{2\pi} \sqrt{16\sin^2 4t + 25\cos^2 5t}\, dt$$

$$\approx 27.185$$

27. $x'(t) = 3t^2 - 4$
$y'(t) = 4t^3 - 4$
Let $f(t) = (3t^2 - 4)^2 + (4t^3 - 4)^2$. Then

$$SA = \int_{-1}^{1} 2\pi |t^4 - 4t| \sqrt{f(t)}\, dt$$

$$\approx 128.075.$$

29. Multiplying both sides by r gives
$r^2 = 3r\cos\theta$. Substitution then gives
$x^2 + y^2 = 3x$.
We then complete the square (as follows) to
obtain an equation:

$$x^2 - 3x + y^2 = 0$$

$$\left(x^2 - 3x + \frac{9}{4}\right) + y^2 - \frac{9}{4} = 0$$

$$\left(x - \frac{3}{2}\right)^2 + y^2 = \left(\frac{3}{2}\right)^2$$

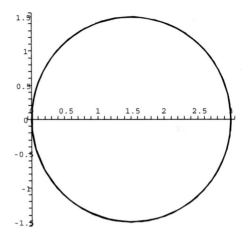

31. $2\sin\theta = 0$ when $\theta = k\pi$ for any integer k.
One copy of the graph will be produced by the
range $0 \le \theta \le \pi$.

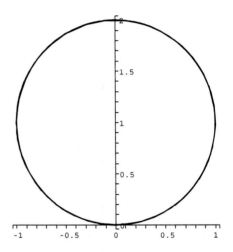

33. $2 - 3\sin\theta = 0$ when $\sin\theta = 2/3$ or $\theta = \sin^{-1}(2/3)$. One copy of the graph will be
produced by the range $0 \le \theta \le 2\pi$.

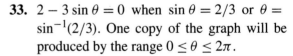
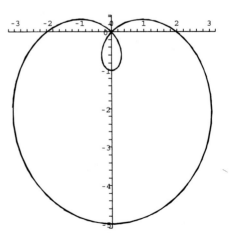

35. $r = 0$ when $\sin 2\theta = 0$, i.e., when $2\theta = k\pi$ or $\theta = k\pi/2$ for any integer k. One copy of the graph will be produced by the range $0 \le \theta \le 2\pi$.

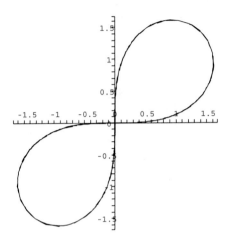

37. $\dfrac{2}{1 + 2\sin\theta}$ is not equal to 0 for any value of θ. One copy of the graph will be produced by the range $0 \le \theta \le 2\pi$.

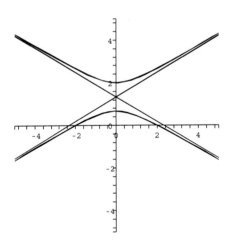

39. This is a circle of radius 3 centered at the origin, so a polar equation is $r = 3$ for $0 \le \theta \le 2\pi$.

41. If $f(\theta) = \cos 3\theta$ then $f'(\theta) = -3\sin 3\theta$. The slope of the tangent line is

$$\frac{dy}{dx}\bigg|_{\theta=\frac{\pi}{6}} = \frac{-3\sin\frac{\pi}{2}\sin\frac{\pi}{6} + \cos\frac{\pi}{2}\cos\frac{\pi}{6}}{-3\sin\frac{\pi}{2}\cos\frac{\pi}{6} - \cos\frac{\pi}{2}\sin\frac{\pi}{6}}$$

$$= \frac{-3\sin\frac{\pi}{2}\sin\frac{\pi}{6}}{-3\sin\frac{\pi}{2}\cos\frac{\pi}{6}}$$

$$= \frac{\sin\frac{\pi}{6}}{\cos\frac{\pi}{6}}$$

$$= \frac{\frac{1}{2}}{\frac{\sqrt{3}}{2}} = \frac{1}{\sqrt{3}}.$$

43. One leaf of $r = \sin 5\theta$ is traced out by $0 \le \theta \le \pi/5$ so the area A is

$$A = \int_0^{\pi/5} \frac{1}{2}(\sin^2 5\theta)\, d\theta$$

$$= \frac{1}{2}\int_0^{\pi/5}\frac{1}{2}(1 - \cos 10\theta)\, d\theta$$

$$= \frac{1}{4}\left(\theta - \frac{\sin 10\theta}{10}\right)\bigg|_0^{\pi/5}$$

$$= \frac{1}{4}\left(\frac{\pi}{5} - 0\right) = \frac{\pi}{20} \approx 0.157.$$

45. Endpoints for the inner loop are given by $\theta = \pi/6$ and $\theta = 5\pi/6$ so the area A is

$$A = \frac{1}{2}\int_{\pi/6}^{5\pi/6}(1 - 2\sin\theta)^2\, d\theta$$

$$= \frac{1}{2}\left(-3\sqrt{3} + 2\pi\right)$$

$$= \frac{-3\sqrt{3} + 2\pi}{2}$$

$$\approx 0.5435.$$

47. $1 + \sin\theta = 1 + \cos\theta$ when $\theta = \pi/4$ and $\theta = 5\pi/4$. We want the region from $\pi/4$ to $5\pi/4$, so we find that the area A is

$$A = \frac{1}{2} \int_{\pi/4}^{5\pi/4} (1 + \sin\theta)^2 \, d\theta$$

$$- \frac{1}{2} \int_{\pi/4}^{5\pi/4} (1 + \cos\theta)^2 \, d\theta$$

$$= \frac{1}{2} \left(\frac{3\pi}{2} + 2\sqrt{2} \right)$$

$$- \frac{1}{2} \left(\frac{3\pi}{2} - 2\sqrt{2} \right)$$

$$= 2\sqrt{2}.$$

49. $r = f(\theta) = 3 - 4\sin\theta$

$f'(\theta) = -4\cos\theta$

$$s = \int_0^{2\pi} \sqrt{f'(\theta)^2 + f(\theta)^2} \, d\theta$$

$$= \int_0^{2\pi} \sqrt{16 + 9 - 24\sin\theta} \, d\theta$$

$$= \int_0^{2\pi} \sqrt{25 - 24\sin\theta} \, d\theta$$

$$\approx 28.814$$

51. Since the focus is $(1, 2)$ and the directrix is $y = 0$, the vertex must be $(1, 1)$. Then $b = 1$, $c = 1$, and $a = 1/4$. The equation is

$$y = \frac{1}{4}(x - 1)^2 + 1.$$

53. Since the foci are $(2, 0)$ and $(2, 4)$ and the vertices are $(2, 1)$ and $(2, 3)$, we see that the center is $(2, 2)$. We then have $x_0 = 2$, $y_0 = 2$, $c = 2$, $a = 1$, and $b = \sqrt{3}$. The equation is

$$\frac{(y-2)^2}{1} - \frac{(x-2)^2}{3} = 1.$$

55. This is an ellipse with $x_0 = -1$, $y_0 = 3$, $a = 5$, $b = 3$, and $c = 4$.

foci: $(-1, 7)$ and $(-1, -1)$
vertices: $(-1, 8)$ and $(-1, -2)$

57. Solving for y gives $y = (x - 1)^2 - 4$.
This is a parabola with $a = 1$, $b = 1$, and $c = -4$.

vertex: $(1, -4)$
focus: $(1, -4 + \frac{1}{4})$
directrix: $y = -4 - \frac{1}{4}$

59. This is a parabola with vertex $(0, 0)$ and $a = \frac{1}{2}$. The microphone should be placed at the focus,

i.e., at $\left(0, 0 + \dfrac{1}{4\left(\frac{1}{2}\right)} \right) = \left(0, \dfrac{1}{2} \right)$.

61. Theorem 7.2 part (i) gives

$$r = \frac{(0.8)(3)}{0.8\cos\theta + 1} = \frac{2.4}{0.8\cos\theta + 1}.$$

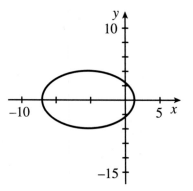

63. Theorem 7.2 part (iii) gives

$$r = \frac{2.8}{1.4\sin\theta + 1}.$$

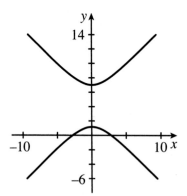

65. This is an ellipse with center $(-1, 3)$. Parametric equations are

$$\begin{cases} x = 3\cos t - 1 \\ y = 5\sin t + 3 \end{cases}$$

with $0 \le t \le 2\pi$.

Vectors and the Geometry of Space

11.1 VECTORS IN THE PLANE

1. Sketch of vectors:

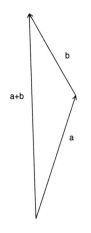

3. $\mathbf{a} + \mathbf{b} = \langle 2 + 3, 4 - 1 \rangle = \langle 5, 3 \rangle$

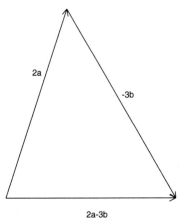

$$\mathbf{a} - 2\mathbf{b} = \langle 2, 4 \rangle - 2\langle 3, -1 \rangle$$
$$= \langle 2 - 6, 4 + 2 \rangle$$
$$= \langle -4, 6 \rangle$$

$$3\mathbf{a} = 3\langle 2, 4 \rangle = \langle 6, 12 \rangle$$

$$5\mathbf{b} - 2\mathbf{a} = 5\langle 3, -1 \rangle - 2\langle 2, 4 \rangle$$
$$= \langle 15 - 4, -5 - 8 \rangle = \langle 11, -13 \rangle$$

$$\|5\mathbf{b} - 2\mathbf{a}\| = \sqrt{11^2 + (-13)^2} = \sqrt{290}$$

5. $\mathbf{a} + \mathbf{b} = (\mathbf{i} + 2\mathbf{j}) + (3\mathbf{i} - \mathbf{j}) = 4\mathbf{i} + \mathbf{j}$

$\mathbf{a} - 2\mathbf{b} = (\mathbf{i} + 2\mathbf{j}) - 2(3\mathbf{i} - \mathbf{j}) = -5\mathbf{i} + 4\mathbf{j}$

$3\mathbf{a} = 3\mathbf{i} + 6\mathbf{j}$

$5\mathbf{b} - 2\mathbf{a} = 5(3\mathbf{i} - \mathbf{j}) - 2(\mathbf{i} + 2\mathbf{j}) = 13\mathbf{i} - 9\mathbf{j}$

$$\|5\mathbf{b} - 2\mathbf{a}\| = \sqrt{13^2 + (-9)^2}$$
$$= \sqrt{250} = 5\sqrt{10}$$

7. Sketch of vectors for exercise 3:

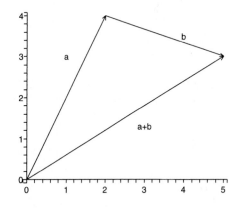

Sketch of vectors for exercise 4:

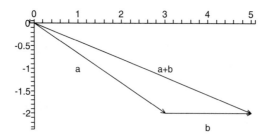

9. $\mathbf{b} = -2\mathbf{a}$, so these are parallel.

11. Suppose $\langle -2, 3 \rangle = c\langle 4, 6 \rangle$

x component: $-2 = 4c \Rightarrow c = -\frac{1}{2}$

y component: $3 = 6c \Rightarrow c = \frac{1}{2}$

Not parallel

13. $\mathbf{b} = 3\mathbf{a}$, so these are parallel.

15. The vector $\mathbf{b} - \mathbf{a}$ has initial point $\mathbf{a}$ and terminal point $\mathbf{b}$.
$\mathbf{b} - \mathbf{a} = \langle 5 - 2, 4 - 3 \rangle = \langle 3, 1 \rangle$

17. The vector $\mathbf{b} - \mathbf{a}$ has initial point $\mathbf{a}$ and terminal point $\mathbf{b}$.
$\mathbf{b} - \mathbf{a} = \langle 1 - (-1), -1 - 2 \rangle = \langle 2, -3 \rangle$

19. $\|\langle 4, -3 \rangle\| = \sqrt{4^2 + (-3)^2} = 5$, so a unit vector in the same direction is $\langle \frac{4}{5}, -\frac{3}{5} \rangle$, and the vector in polar form is $5\langle \frac{4}{5}, -\frac{3}{5} \rangle$.

21. $\|2\mathbf{i} - 4\mathbf{j}\| = \sqrt{2^2 + (-4)^2} = 2\sqrt{5}$, so a unit vector in the same direction is $\frac{1}{\sqrt{5}}\mathbf{i} - \frac{2}{\sqrt{5}}\mathbf{j}$, and the vector in polar form is $2\sqrt{5}\langle \frac{1}{\sqrt{5}}, -\frac{2}{\sqrt{5}} \rangle$.

23. $\|\langle 5 - 2, 2 - 1 \rangle\| = \sqrt{3^2 + 1^2} = \sqrt{10}$, so a unit vector in the same direction is $\langle \frac{3}{\sqrt{10}}, \frac{1}{\sqrt{10}} \rangle$, and the vector in polar form is $\sqrt{10}\langle \frac{3}{\sqrt{10}}, \frac{1}{\sqrt{10}} \rangle$.

25. A unit vector in the direction of $\mathbf{v}$ is $\mathbf{u} = \frac{1}{\|\mathbf{v}\|}\langle 3, 4 \rangle = \langle \frac{3}{5}, \frac{4}{5} \rangle$. Vector in this direction with magnitude 3 is $3\langle \frac{3}{5}, \frac{4}{5} \rangle = \langle \frac{9}{5}, \frac{12}{5} \rangle$.

27. A unit vector in the direction of $\mathbf{v}$ is $\mathbf{u} = \frac{1}{\|\mathbf{v}\|}\langle 2, 5 \rangle = \langle \frac{2}{\sqrt{29}}, \frac{5}{\sqrt{29}} \rangle$. Vector in this direction with magnitude 29 is $29\langle \frac{2}{\sqrt{29}}, \frac{5}{\sqrt{29}} \rangle = \langle 2\sqrt{29}, 5\sqrt{29} \rangle$.

29. A unit vector in the direction of $\mathbf{v}$ is $\mathbf{u} = \frac{1}{\|\mathbf{v}\|}\langle 3, 0 \rangle = \langle 1, 0 \rangle$. Vector in this direction with magnitude 4 is $4\langle 1, 0 \rangle = \langle 4, 0 \rangle$.

31. The force due to gravity is $\mathbf{g} = \langle 0, -150 \rangle$ and the force due to air resistance is $\mathbf{w} = \langle 20, 140 \rangle$, so the net force is $\mathbf{g} + \mathbf{w} = \langle 20, -10 \rangle$.

33. The force due to gravity is $\mathbf{g} = \langle 0, -200 \rangle$ and the net force is $\mathbf{g} + \mathbf{w} = \langle 30, -10 \rangle$, so the force due to air resistance is $\mathbf{w} = \langle 30, -10 \rangle - \langle 0, -200 \rangle = \langle 30, 190 \rangle$.

35. The force due to rope A is $\mathbf{a} = \langle -164, 115 \rangle$, the force due to rope B is $\mathbf{b} = \langle 177, 177 \rangle$, and the force due to gravity is $\mathbf{g} = \langle 0, -275 \rangle$. The net force is $\mathbf{a} + \mathbf{b} + \mathbf{g} = \langle 13, 17 \rangle$, and the crate will move up and to the right.

37. Let $\mathbf{v} = \langle x, y \rangle$ be the direction of the plane and $\mathbf{w} = \langle 30, -20 \rangle$ represent the wind velocity. We want $\mathbf{v} + \mathbf{w} = \langle c, 0 \rangle$, where $c < 0$ so the plane is traveling due west. We have $y + (-20) = 0$, so $y = 20$. We are also given $\|\mathbf{v}\| = 300$, so $\sqrt{x^2 + y^2} = 300$.

$x^2 + 400 = 90{,}000$
$x^2 = 89{,}600$
$x = -\sqrt{89{,}600} = -80\sqrt{14}$

(We take the negative square root to have the plane moving west.) This points up and left at an angle of $\tan^{-1}\frac{20}{80\sqrt{14}} \approx 3.8°$ north of west.

39. Let $\mathbf{v} = \langle x, y \rangle$ be the direction of the plane and $\mathbf{w} = \langle -20, 30 \rangle$ represent the wind velocity. We want $\mathbf{v} + \mathbf{w} = \langle 0, c \rangle$, where $c > 0$ so the plane is traveling due north. We have $x + (-20) = 0$, so $x = 20$. We are also given $\|\mathbf{v}\| = 400$, so $\sqrt{x^2 + y^2} = 400$.

$400 + y^2 = 160{,}000$
$y^2 = 159{,}600$
$y = \sqrt{159{,}600} = 20\sqrt{399}$

(We take the positive square root to have the plane moving north.) This points up and right at an angle of $\tan^{-1} \frac{20}{20\sqrt{399}} \approx 2.9°$ east of north.

41. The paper will travel with velocity $\langle -50, 10 \rangle$ feet per second. The paper will take 1 second to travel 50 feet to the left, so it will travel 10 feet up the road in that time. He should release the paper 10 feet up the road.

43. The weight of the hose is $\langle 0, -20 \rangle$, and the force of the water is $\langle -200, 0 \rangle$. The force required to hold the hose horizontal is $\langle 200, 20 \rangle$, or $\sqrt{200^2 + 20^2} = 20\sqrt{101}$ pounds at angle $\tan^{-1} \frac{20}{200} \approx 5.7°$ above horizontal.

45. The speed of the current is $\langle 1, 0 \rangle$ (parallel to shore), and speed due to paddling is $\langle 0, 4 \rangle$ (toward the shore). The net velocity is then $\langle 1, 4 \rangle$. The speed is $\sqrt{1^2 + 4^2} = \sqrt{17}$ and the angle between the kayak's direction and the shore is $\tan^{-1} \frac{4}{1} \approx 76°$ (downstream of perpendicular).

47. The largest magnitude of $\mathbf{a} + \mathbf{b}$ is 7 (if the vectors point in the same direction). The smallest magnitude is 1 (if the vectors point in opposite directions). If the vectors are perpendicular, then $\mathbf{a} + \mathbf{b}$ can be viewed as the hypotenuse of a right triangle with sides $\mathbf{a}$ and $\mathbf{b}$, so it has length $\sqrt{3^2 + 4^2} = 5$.

49. Let $\mathbf{a} = \langle a_1, a_2 \rangle$, $\mathbf{b} = \langle b_1, b_2 \rangle$, and $\mathbf{c} = \langle c_1, c_2 \rangle$.

$$\mathbf{a} + (\mathbf{b} + \mathbf{c}) = \langle a_1, a_2 \rangle + (\langle b_1, b_2 \rangle + \langle c_1, c_2 \rangle)$$
$$= \langle a_1, a_2 \rangle + \langle b_1 + c_1, b_2 + c_2 \rangle$$
$$= \langle a_1 + b_1 + c_1, a_2 + b_2 + c_2 \rangle$$
$$= \langle a_1 + b_1, a_2 + b_2 \rangle + \langle c_1, c_2 \rangle$$
$$= (\langle a_1, a_2 \rangle + \langle b_1, b_2 \rangle) + \langle c_1, c_2 \rangle$$
$$= (\mathbf{a} + \mathbf{b}) + \mathbf{c}$$

51. $\|\mathbf{a}\| = \sqrt{2^2 + 3^2} = \sqrt{13}$,
$\|\mathbf{b}\| = \sqrt{1^2 + 4^2} = \sqrt{17}$, and
$\|\mathbf{a} + \mathbf{b}\| = \sqrt{3^2 + 7^2} = \sqrt{58}$
$\|\mathbf{a}\| + \|\mathbf{b}\| \approx 7.73$
$\|\mathbf{a} + \mathbf{b}\| \approx 7.62$
For every choice of vectors $\|\mathbf{a} + \mathbf{b}\| \leq \|\mathbf{a}\| + \|\mathbf{b}\|$. In Figure 11.6 we see that $\mathbf{a} + \mathbf{b}$ can be viewed as one side of a triangle with $\mathbf{a}$ and $\mathbf{b}$ as the other two sides. The third side of a triangle cannot be longer than the sum of the other two sides.

53. $\|\mathbf{a} + \mathbf{b}\| = \|\mathbf{a}\| + \|\mathbf{b}\|$ if and only if $\mathbf{a}$ and $\mathbf{b}$ are in the same direction. In other words, we have equality if $\mathbf{a} = c\mathbf{b}$ for some constant $c \geq 0$.
$\|\mathbf{a} + \mathbf{b}\|^2 = \|\mathbf{a}\|^2 + \|\mathbf{b}\|^2$ when $\mathbf{a}$ and $\mathbf{b}$ are perpendicular so that $\mathbf{a} + \mathbf{b}$ forms the hypotenuse of a right triangle.

In general, $\|\mathbf{a}\|^2 + \|\mathbf{b}\|^2$ is larger than $\|\mathbf{a} + \mathbf{b}\|^2$. Again this comes from the geometry of triangles, and in particular the Law of Cosines:

$$c^2 = a^2 + b^2 - 2ab \cos \theta.$$

11.2 VECTORS IN SPACE

1. (a)

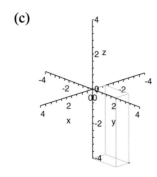

(b)

(c)

3.   Sketch of z-axis:

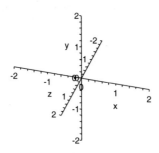

5. $d = \sqrt{(x_2 - x_1)^2 + (y_2 - y_1)^2 + (z_2 - z_1)^2}$
$d = \sqrt{(2-5)^2 + (1-5)^2 + (2-2)^2} = 5$

7. $d = \sqrt{(x_2 - x_1)^2 + (y_2 - y_1)^2 + (z_2 - z_1)^2}$
$d = \sqrt{(1-(-1))^2 + (2-0)^2 + (3-2)^2} = 3$

9. $\mathbf{a} + \mathbf{b} = \langle 2+1, 1+3, -2+0 \rangle = \langle 3, 4, -2 \rangle$
$\mathbf{a} - 3\mathbf{b} = \langle 2, 1, -2 \rangle - 3\langle 1, 3, 0 \rangle$
$= \langle -1, -8, -2 \rangle$
$\|4\mathbf{a} + 2\mathbf{b}\| = \|4\langle 2, 1, -2 \rangle + 2\langle 1, 3, 0 \rangle\|$
$= \|\langle 10, 10, -8 \rangle\| = \sqrt{264} = 2\sqrt{66}$

11. $\mathbf{a} + \mathbf{b} = (3\mathbf{i} - \mathbf{j} + 4\mathbf{k}) + (5\mathbf{i} + \mathbf{j})$
$= (3+5)\mathbf{i} + (-1+1)\mathbf{j} + (4+0)\mathbf{k}$
$= 8\mathbf{i} + 4\mathbf{k} \; \mathbf{a} - 3\mathbf{b} = (3\mathbf{i} - \mathbf{j} + 4\mathbf{k}) - 3(5\mathbf{i} + \mathbf{j})$
$= -12\mathbf{i} - 4\mathbf{j} + 4\mathbf{k}$
$\|4\mathbf{a} + 2\mathbf{b}\| = \|22\mathbf{i} - 2\mathbf{j} + 16\mathbf{k}\|$
$= \sqrt{22^2 + (-2)^2 + 16^2} = 2\sqrt{186}$

13.   **(a)** $\|\langle 3, 1, 2 \rangle\| = \sqrt{14}$, so the two unit vectors are $\pm\left\langle \frac{3}{\sqrt{14}}, \frac{1}{\sqrt{14}}, \frac{2}{\sqrt{14}} \right\rangle$.

(b) $\langle 3, 1, 2 \rangle = \sqrt{14}\left\langle \frac{3}{\sqrt{14}}, \frac{1}{\sqrt{14}}, \frac{2}{\sqrt{14}} \right\rangle$

15.   **(a)** $\|2\mathbf{i} - \mathbf{j} + 2\mathbf{k}\| = 3$, so the two unit vectors are $\pm\frac{1}{3}(2\mathbf{i} - \mathbf{j} + 2\mathbf{k})$.

(b) $2\mathbf{i} - \mathbf{j} + 2\mathbf{k} = 3(\frac{2}{3}\mathbf{i} - \frac{1}{3}\mathbf{j} + \frac{2}{3}\mathbf{k})$

17.   **(a)** $\|\langle 3-1, 2-2, 1-3 \rangle\| = 2\sqrt{2}$, so the two unit vectors are $\pm\langle \frac{1}{\sqrt{2}}, 0, -\frac{1}{\sqrt{2}} \rangle$.

(b) $\langle 2, 0, -2 \rangle = 2\sqrt{2}\langle \frac{1}{\sqrt{2}}, 0, -\frac{1}{\sqrt{2}} \rangle$

19. $\|\mathbf{v}\| = 3$, so a unit vector in the direction of $\mathbf{v}$ is $\langle \frac{2}{3}, \frac{2}{3}, -\frac{1}{3} \rangle$, and the desired vector is $6\langle \frac{2}{3}, \frac{2}{3}, -\frac{1}{3} \rangle = \langle 4, 4, -2 \rangle$.

21. $\|\mathbf{v}\| = \sqrt{14}$, so a unit vector in the direction of $\mathbf{v}$ is $\frac{1}{\sqrt{14}}(2\mathbf{i} - \mathbf{j} + 3\mathbf{k})$, and the desired vector is $4(\frac{1}{\sqrt{14}})(2\mathbf{i} - \mathbf{j} + 3\mathbf{k})$.

23. $(x-3)^2 + (y-1)^2 + (z-4)^2 = 4$

25. $(x-\pi)^2 + (y-1)^2 + (z+3)^2 = 5$

27.   A sphere of radius 2 and center $(1, 0, -2)$.

29.   Complete the squares to get
$(x^2 - 2x + 1) + y^2 + (z^2 - 4z + 4) = 1 + 4$
$(x-1)^2 + y^2 + (z-2)^2 = 5$
A sphere of radius $\sqrt{5}$ and center $(1, 0, 2)$.

31.   Parallel to xz-plane:

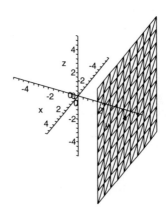

33.   Parallel to xy-plane:

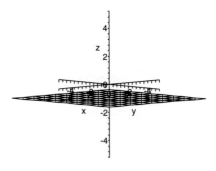

35. $y = 0$

37. $x = 0$

39.   Let $\mathbf{a} = \langle a_1, a_2, a_3 \rangle$ and $\mathbf{b} = \langle b_1, b_2, b_3 \rangle$. Then,

$$\mathbf{a} + \mathbf{b} = \langle a_1, a_2, a_3 \rangle + \langle b_1, b_2, b_3 \rangle$$
$$= \langle a_1 + b_1, a_2 + b_2, a_3 + b_3 \rangle$$
$$= \langle b_1 + a_1, b_2 + a_2, b_3 + a_3 \rangle$$
$$= \langle b_1, b_2, b_3 \rangle + \langle a_1, a_2, a_3 \rangle$$
$$= \mathbf{b} + \mathbf{a}.$$

41. Let $\mathbf{a} = \langle a_1, a_2, a_3 \rangle$, $\mathbf{b} = \langle b_1, b_2, b_3 \rangle$, and d and e be constants. Then,

$$d(\mathbf{a} + \mathbf{b}) = d(\langle a_1, a_2, a_3 \rangle + \langle b_1, b_2, b_3 \rangle)$$
$$= d\langle a_1 + b_1, a_2 + b_2, a_3 + b_3 \rangle$$
$$= \langle d(a_1 + b_1), d(a_2 + b_2), d(a_3 + b_3) \rangle$$
$$= \langle da_1 + db_1, da_2 + db_2, da_3 + db_3 \rangle$$
$$= \langle da_1, da_2, da_3 \rangle + \langle db_1, db_2, db_3 \rangle$$
$$= d\langle a_1, a_2, a_3 \rangle + d\langle b_1, b_2, b_3 \rangle$$
$$= d\mathbf{a} + d\mathbf{b}$$

$$(d + e)\mathbf{a} = (d + e)\langle a_1, a_2, a_3 \rangle$$
$$= \langle (d + e)a_1, (d + e)a_2, (d + e)a_3 \rangle$$
$$= \langle da_1 + ea_1, da_2 + ea_2, da_3 + ea_3 \rangle$$
$$= \langle da_1, da_2, da_3 \rangle + \langle ea_1, ea_2, ea_3 \rangle$$
$$= d\langle a_1, a_2, a_3 \rangle + e\langle a_1, a_2, a_3 \rangle$$
$$= d\mathbf{a} + e\mathbf{a}$$

43. $\mathbf{PQ} = \langle 4 - 2, 2 - 3, 2 - 1 \rangle = \langle 2, -1, 1 \rangle$
$\mathbf{QR} = \langle 8 - 4, 0 - 2, 4 - 2 \rangle = \langle 4, -2, 2 \rangle$
$\mathbf{QR} = 2\mathbf{PQ}$, so the vectors are parallel and the points are colinear.

45. $P = (0, 1, 1)$, $Q = (2, 4, 2)$, $R = (3, 1, 4)$
$\mathbf{PQ} = \langle 2, 3, 1 \rangle$, so $\|\mathbf{PQ}\| = \sqrt{14}$
$\mathbf{QR} = \langle 1, -3, 2 \rangle$, so $\|\mathbf{QR}\| = \sqrt{14}$
$\mathbf{PR} = \langle 3, 0, 3 \rangle$, so $\|\mathbf{RP}\| = \sqrt{18}$
No, the points do not form an equilateral triangle.

47. $P = (3, 1, -2)$, $Q = (1, 0, 1)$, $R = (4, 2, -1)$
$\mathbf{PQ} = \langle -2, -1, 3 \rangle$, so $\|\mathbf{PQ}\| = \sqrt{14}$
$\mathbf{QR} = \langle 3, 2, -2 \rangle$, so $\|\mathbf{QR}\| = \sqrt{17}$
$\mathbf{RP} = \langle -1, -1, -1 \rangle$, so $\|\mathbf{RP}\| = \sqrt{3}$
$\sqrt{3}^2 + \sqrt{14}^2 = \sqrt{17}^2$, so yes, these points form a right triangle.

49. Let $P = (2, 1, 0)$, $Q = (5, -1, 2)$, $R = (0, 3, 3)$, and $S = (3, 1, 5)$. There are six

pairs, four will correspond to sides (side length s if square) and two will be diagonals (length $s\sqrt{2}$ if square).
$\mathbf{PQ} = \langle 3, -2, 2 \rangle$, so $\|\mathbf{PQ}\| = \sqrt{17}$
$\mathbf{QR} = \langle -5, 4, 1 \rangle$, so $\|\mathbf{QR}\| = \sqrt{42}$
Since these are not equal, and do not differ by a factor of $\sqrt{2}$, these points cannot form a square.

51. Let the force due to rope A be
$\mathbf{a} = \langle 10, -130, 200 \rangle$
and the force due to rope B be
$\mathbf{b} = \langle -20, 180, 160 \rangle$, and write the force due to gravity as $\mathbf{w} = \langle 0, 0, -500 \rangle$. The net force is $\mathbf{a} + \mathbf{b} + \mathbf{w} = \langle -10, 50, -140 \rangle$. In order to compensate, rope C must exert a force of $\langle 10, -50, 140 \rangle$, or 149 pounds in direction $\langle 1, -5, 14 \rangle$.

53. The velocity vector for the plane in still air is $600\langle \frac{2}{3}, \frac{2}{3}, \frac{1}{3} \rangle = \langle 400, 400, 200 \rangle$. The net velocity with wind speed $\langle 10, -20, 0 \rangle$ will be $\langle 410, 380, 200 \rangle$. The speed is $\sqrt{410^2 + 380^2 + 200^2} \approx 593.72$ mph.

55. $\langle 2, 3, 1, 5 \rangle + 2\langle 1, -2, 3, 1 \rangle$
$= \langle 2 + 2, 3 - 4, 1 + 6, 5 + 2 \rangle$
$= \langle 4, -1, 7, 7 \rangle$

57. $\langle 3, -2, 4, 1, 0, 2 \rangle - 3\langle 1, 2, -2, 0, 3, 1 \rangle$
$= \langle 3 - 3, -2 - 6, 4 + 6, 1 - 0, 0 - 9, 2 - 3 \rangle$
$= \langle 0, -8, 10, 1, -9, -1 \rangle$

59. $\| \langle 3, 1, -2, 4, 1 \rangle \|$
$= \sqrt{3^2 + 1^2 + (-2)^2 + 4^2 + 1^2} = \sqrt{31}$

61. $\| \langle 1, -2, 4, 1 \rangle + \langle -1, 4, 2, -4 \rangle \|$
$= \| \langle 0, 2, 6, -3 \rangle \|$
$= \sqrt{0^2 + 2^2 + 6^2 + (-3)^2} = 7$

63. Call the point of intersection between the inscribed circle and the unit circle in the first quadrant P. Symmetry forces P to lie on the line $y = x$, which passes through the center of the inscribed circle, $(0, 0)$, and the center of the unit circle in the first quadrant, $(1, 1)$. The distance from $(0, 0)$ to $(1, 1)$ is $\sqrt{2}$ and the distance from P to $(1, 1)$ is 1 (since this is the radius of a unit circle). Therefore, the radius of the inscribed circle is the distance from $(0, 0)$ to P, or $\sqrt{2} - 1$.

65. Let P be the point of intersection of the inscribed hypersphere (centered at the origin) and the unit hypersphere centered at $\langle 1, 1, 1, \ldots, 1 \rangle$. The distance from the origin to $\langle 1, 1, 1, \ldots, 1 \rangle$ is $\sqrt{n}$. The distance from P to $\langle 1, 1, 1, \ldots, 1 \rangle$ is 1 because this is the radius of a unit hypersphere. Therefore the radius of the inscribed hypersphere is $\sqrt{n} - 1$. For $n \geq 10$ this radius will be larger than 2, so the inscribed hypersphere extends outside of the box along the coordinate axes.

11.3 THE DOT PRODUCT

1. $\langle 3, 1 \rangle \cdot \langle 2, 4 \rangle = 6 + 4 = 10$

3. $\langle 2, -1, 3 \rangle \cdot \langle 0, 2, 4 \rangle = 0 - 2 + 12 = 10$

5. $(2\mathbf{i} - \mathbf{k}) \cdot (4\mathbf{j} - \mathbf{k}) = 0 + 0 + 1 = 1$

7. $\cos \theta = \dfrac{\mathbf{a} \cdot \mathbf{b}}{\|\mathbf{a}\| \|\mathbf{b}\|} = \dfrac{3 - 2}{\sqrt{13}\sqrt{2}}$, so the angle between the vectors is $\cos^{-1} \dfrac{1}{\sqrt{26}} \approx 1.37$.

9. $\cos \theta = \dfrac{\mathbf{a} \cdot \mathbf{b}}{\|\mathbf{a}\| \|\mathbf{b}\|} = \dfrac{-6 + 2 - 4}{\sqrt{26}\sqrt{9}}$, so the angle between the vectors is $\cos^{-1} \dfrac{-8}{3\sqrt{26}} \approx 2.12$.

11. $\mathbf{a} \cdot \mathbf{b} = 4 - 4 = 0$. Orthogonal.

13. $\mathbf{a} \cdot \mathbf{b} = -6 + 6 = 0$. Orthogonal.

15. $\langle 1, 2 \rangle$ or any multiple.

17. $\mathbf{i} - 3\mathbf{j}$ is one possible answer. Find any vector so that the dot product is zero.

19. $\text{comp}_\mathbf{b} \mathbf{a} = \dfrac{\mathbf{a} \cdot \mathbf{b}}{\|\mathbf{b}\|} = \dfrac{6 + 4}{\sqrt{9 + 16}} = 2$

$\text{proj}_\mathbf{b} \mathbf{a} = \dfrac{\mathbf{a} \cdot \mathbf{b}}{\|\mathbf{b}\|} \dfrac{\mathbf{b}}{\|\mathbf{b}\|} = \dfrac{6 + 4}{5} \dfrac{\langle 3, 4 \rangle}{5}$

$= \langle \dfrac{6}{5}, \dfrac{8}{5} \rangle$

21. $\text{comp}_\mathbf{b} \mathbf{a} = \dfrac{\mathbf{a} \cdot \mathbf{b}}{\|\mathbf{b}\|} = \dfrac{2 - 2 + 6}{\sqrt{1 + 4 + 4}} = \dfrac{6}{3} = 2$

$\text{proj}_\mathbf{b} \mathbf{a} = \dfrac{\mathbf{a} \cdot \mathbf{b}}{\|\mathbf{b}\|} \dfrac{\mathbf{b}}{\|\mathbf{b}\|} = \dfrac{6}{3} \dfrac{\langle 1, 2, 2 \rangle}{3}$

$= \langle \dfrac{2}{3}, \dfrac{4}{3}, \dfrac{4}{3} \rangle$

23. $\text{comp}_\mathbf{b} \mathbf{a} = \dfrac{\mathbf{a} \cdot \mathbf{b}}{\|\mathbf{b}\|} = \dfrac{0 + 0 - 8}{\sqrt{0 + 9 + 16}} = \dfrac{-8}{5}$

$\text{proj}_\mathbf{b} \mathbf{a} = \dfrac{\mathbf{a} \cdot \mathbf{b}}{\|\mathbf{b}\|} \dfrac{\mathbf{b}}{\|\mathbf{b}\|} = \dfrac{-8}{5} \dfrac{\langle 0, -3, 4 \rangle}{5}$

$= \langle 0, \dfrac{24}{25}, \dfrac{-32}{25} \rangle$

25. $\mathbf{F} = 40 \langle \cos \frac{\pi}{3}, \sin \frac{\pi}{3} \rangle = 40 \langle \frac{1}{2}, \frac{\sqrt{3}}{2} \rangle$. The horizontal component of the force is 20, so the work done is $W = Fd = 20 \cdot 5280 = 105{,}600$ foot-pounds of work.

27. The force exerted in exercises 25 and 26 is the same, but in exercise 26 the force is exerted in a direction that is closer to the actual direction of motion. This means that the contribution the force makes to the motion is greater, and more work is done.

29. Let $\mathbf{F} = \langle 30, 20 \rangle$, and $\mathbf{d} = \langle 24, 10 \rangle$. The component of force in the direction of motion is $\text{comp}_\mathbf{d} \mathbf{F} = \dfrac{\mathbf{F} \cdot \mathbf{d}}{\|\mathbf{d}\|} = \dfrac{720 + 200}{\sqrt{24^2 + 10^2}} = \dfrac{920}{26}$. The distance traveled is $\|\mathbf{d}\| = 26$, and the work done is $W = \dfrac{920}{26} \cdot 26 = 920$ foot-pounds.

31. (a) False; counterexample: $\mathbf{a} = \langle 1, 0 \rangle$ $\mathbf{b} = \langle 0, 1 \rangle$, and $\mathbf{c} = \langle 0, 2 \rangle$

(b) True; if $\mathbf{b} = \mathbf{c}$, then the computations will be identical.

(c) True; $\|\mathbf{a}\|^2 = \sqrt{a_1^2 + a_2^2 + a_3^2}^2 = a_1^2 + a_2^2 + a_3^2 = \mathbf{a} \cdot \mathbf{a}$.

(d) False; counterexample: $\mathbf{a} = \langle 100, 1 \rangle$, $\mathbf{b} = \langle 1, 2 \rangle$, and $\mathbf{c} = \langle 0, 1 \rangle$.

(e) False; counterexample: $\mathbf{a} = \langle 1, 0 \rangle$ and $\mathbf{b} = \langle \frac{\sqrt{2}}{2}, \frac{\sqrt{2}}{2} \rangle$.

33. We have equality in the Cauchy-Schwartz Inequality if the cosine of the angle between the vectors is ± 1. This happens exactly when the vectors point in the same or opposite directions. In other words, when $\mathbf{a} = c\mathbf{b}$ for some constant c.

35. $\|\mathbf{a}\| = \|(\mathbf{a} - \mathbf{b}) + \mathbf{b}\| \le \|\mathbf{a} - \mathbf{b}\| + \|\mathbf{b}\|$ by the Triangle Inequality, so $\|\mathbf{a}\| - \|\mathbf{b}\| \le \|\mathbf{a} - \mathbf{b}\|$.

37. The maximum would occur when $\mathbf{a} \cdot \mathbf{b} = |\mathbf{a} \cdot \mathbf{b}| = \|\mathbf{a}\| \|\mathbf{b}\| = 15$.

39. In n dimensions, apply the Cauchy-Schwartz Inequality to the vectors $\langle |a_1|, |a_2|, \ldots, |a_n| \rangle$ and $\langle |b_1|, |b_2|, \ldots, |b_n| \rangle$. We get

$$|a_1 b_1| + |a_2 b_2| + \ldots + |a_n b_n|$$
$$\le \sqrt{a_1^2 + a_2^2 + \ldots + a_n^2} \sqrt{b_1^2 + b_2^2 + \ldots + b_n^2}.$$

Squaring both sides gives

$$\left(\sum_{k=1}^n |a_k b_k| \right)^2 \le \left(\sum_{k=1}^n a_k^2 \right) \left(\sum_{k=1}^n b_k^2 \right).$$

If $\sum_{k=1}^\infty a_k^2$ and $\sum_{k=1}^\infty b_k^2$ both converge, then we conclude that $\sum_{k=1}^\infty |a_k b_k|$ must also converge. If $a_k = \frac{1}{k}$ and $b_k = \frac{1}{k^2}$, then since we know

$$\sum_{k=1}^\infty \frac{1}{k^2} = \frac{\pi^2}{6} \text{ and } \sum_{k=1}^\infty \frac{1}{k^4} = \frac{\pi^4}{90}, \text{ we also know}$$

$$\sum_{k=1}^n \frac{1}{k^3} \le \frac{\pi^3}{\sqrt{540}} \approx 1.3343.$$

41. Apply the Cauchy-Schwartz Inequality to the vectors

$\mathbf{a}_1 = \langle |a_1|^{1/3}, |a_2|^{1/3}, \ldots, |a_n|^{1/3} \rangle$ and
$\mathbf{a}_2 = \langle |a_1|^{2/3}, |a_2|^{2/3}, \ldots, |a_n|^{2/3} \rangle$.

We get $\mathbf{a}_1 \cdot \mathbf{a}_2 = \sum_{k=1}^n |a_k| \le \|\mathbf{a}_1\| \|\mathbf{a}_2\|$

$$= \left(\sum_{k=1}^n |a_k|^{2/3} \right)^{1/2} \left(\sum_{k=1}^n |a_k|^{4/3} \right)^{1/2}.$$

43. Use the inequality from exercise 42.

$$\sum_{k=1}^n p_k \le \sqrt{n} \left(\sum_{k=1}^n p_k^2 \right)^{1/2}. \text{ Now since } \sum_{k=1}^n p_k =$$

1, we have $\frac{1}{\sqrt{n}} \le \left(\sum_{k=1}^n p_k^2 \right)^{1/2}$ and squaring both sides gives $\frac{1}{n} \le \sum_{k=1}^n p_k^2$.

45. Consider the product $\left(\sum_{k=1}^n a_k^2 \right) \left(\sum_{k=1}^n b_k^2 \right)$.

This can be expanded as $a_1^2 \left(\sum_{k=1}^n b_k^2 \right) + a_2^2 \left(\sum_{k=1}^n b_k^2 \right) + \ldots + a_n^2 \left(\sum_{k=1}^n b_k^2 \right)$. We can see

that every term in $\sum_{k=1}^n a_k^2 b_k^2 = a_1^2 b_1^2 + a_2^2 b_2^2 + \ldots + a_n^2 b_n^2$ occurs once in the expansion, but that there are many other nonnegative terms in the product. Therefore

$$\sum_{k=1}^n a_k^2 b_k^2 \le \left(\sum_{k=1}^n a_k^2 \right) \left(\sum_{k=1}^n b_k^2 \right).$$

To see the second inequality, first apply the Cauchy-Schwartz Inequality to the vectors $\mathbf{v}_1 = \langle a_1 b_1, a_2 b_2, \ldots, a_n b_n \rangle$ and $\mathbf{v}_2 = \langle c_1, c_2, \ldots c_n \rangle$ to see that

$\left| \sum_{k=1}^n (a_k b_k) c_k \right| \le \|\mathbf{v}_1\| \|\mathbf{v}_2\|$. Squaring both sides gives

$$\left(\sum_{k=1}^n a_k b_k c_k \right)^2 \le \left(\sum_{k=1}^n a_k^2 b_k^2 \right) \left(\sum_{k=1}^n c_k^2 \right).$$

Now apply the first inequality to $\left(\sum_{k=1}^n a_k^2 b_k^2 \right)$ to see that

$$\left(\sum_{k=1}^n a_k b_k c_k \right)^2$$
$$\le \left(\sum_{k=1}^n a_k^2 \right) \left(\sum_{k=1}^n b_k^2 \right) \left(\sum_{k=1}^n c_k^2 \right).$$

47. The vectors from the carbon atom to the four hydrogen atoms are

$\mathbf{a} = \langle 0 - \frac{1}{2}, 0 - \frac{1}{2}, 0 - \frac{1}{2} \rangle = \langle -\frac{1}{2}, -\frac{1}{2}, -\frac{1}{2} \rangle$,

$\mathbf{b} = \langle 1 - \frac{1}{2}, 1 - \frac{1}{2}, 0 - \frac{1}{2} \rangle = \langle \frac{1}{2}, \frac{1}{2}, -\frac{1}{2} \rangle$,

$\mathbf{c} = \langle 1 - \frac{1}{2}, 0 - \frac{1}{2}, 1 - \frac{1}{2} \rangle = \langle \frac{1}{2}, -\frac{1}{2}, \frac{1}{2} \rangle$, and

$\mathbf{d} = \langle 0 - \frac{1}{2}, 1 - \frac{1}{2}, 1 - \frac{1}{2} \rangle = \langle -\frac{1}{2}, \frac{1}{2}, \frac{1}{2} \rangle$.

$\|\mathbf{a}\| = \|\mathbf{b}\| = \|\mathbf{c}\| = \|\mathbf{d}\| = \frac{\sqrt{3}}{2}$

$\mathbf{a}\cdot\mathbf{b} = \mathbf{a}\cdot\mathbf{c} = \mathbf{a}\cdot\mathbf{d} = \mathbf{b}\cdot\mathbf{c} = \mathbf{b}\cdot\mathbf{d} = \mathbf{c}\cdot\mathbf{d} = -\frac{1}{4}$

$\cos\theta = \dfrac{-\frac{1}{4}}{\left(\frac{\sqrt{3}}{2}\right)\left(\frac{\sqrt{3}}{2}\right)} = -\dfrac{1}{3}$

$\theta = \cos^{-1}\left(\frac{-1}{3}\right) \approx 109.5°$

49. $\operatorname{comp}_\mathbf{c}(\mathbf{a}+\mathbf{b}) = \dfrac{(\mathbf{a}+\mathbf{b})\cdot\mathbf{c}}{\|\mathbf{c}\|} = \dfrac{\mathbf{a}\cdot\mathbf{c}+\mathbf{b}\cdot\mathbf{c}}{\|\mathbf{c}\|}$

$= \dfrac{\mathbf{a}\cdot\mathbf{c}}{\|\mathbf{c}\|} + \dfrac{\mathbf{b}\cdot\mathbf{c}}{\|\mathbf{c}\|} = \operatorname{comp}_\mathbf{c}\mathbf{a} + \operatorname{comp}_\mathbf{c}\mathbf{b}$

51. The component of the given force along the beam is $\operatorname{comp}_\mathbf{b}\mathbf{F} = \dfrac{\mathbf{b}\cdot\mathbf{F}}{\|\mathbf{b}\|}$

$= \dfrac{0(10)-200(1)+0(5)}{\sqrt{126}} \approx -17.8$ N.

53. The vector $\mathbf{b} = \langle\cos 10°, \sin 10°\rangle$ is a unit vector that makes a 10° angle with horizontal, so it is parallel to the road. The weight of the car is $\mathbf{w} = \langle 0, -2000\rangle$. The component of the weight in the direction of the bank is $\operatorname{comp}_\mathbf{b}\mathbf{w} = \dfrac{\mathbf{b}\cdot\mathbf{w}}{\|\mathbf{b}\|} = \sin 10°(-2000) \approx -347.3$ lbs toward the inside of the curve.

55. If we assume the track is circular and the circumference of the track is 0.533 mile, then the radius will be $\frac{0.533}{2\pi} \approx 0.085$ mile. The gravitational constant $g = 32\frac{\text{ft}}{\text{sec}^2}$ is equal to $32\frac{\text{ft}}{\text{sec}^2}\left(\frac{3600^2\text{sec}^2}{\text{hr}^2}\right)\left(\frac{1\text{mile}}{5280\text{ft}}\right) \approx 78{,}545\frac{\text{miles}}{\text{hour}^2}$. Now the force equation is $\frac{v^2}{0.085} = 78{,}545\sin 36°$, which leads to a speed of only $v = 62.6$ miles per hour.

The extra speed could be supported in part by friction and by the aerodynamics of the car.

57. $\mathbf{a}\cdot\mathbf{b}$ represents a sum of products of the form (number sold) × (price), and this gives the total revenue.

59. If we have $a = b$, then the position vectors are $\langle a\cos t, a\sin t\rangle$ and $\langle a\cos(t+\frac{\pi}{2}), a\sin(t+\frac{\pi}{2})\rangle$.

The dot product of these vectors is $a^2(\cos t\cos(t+\frac{\pi}{2}) + \sin t\sin(t+\frac{\pi}{2}))$ $= a^2\cos(t-(t+\frac{\pi}{2})) = a^2\cos\frac{-\pi}{2} = 0$.

If we have $a \neq b$, then the position vectors are $\langle a\cos t, b\sin t\rangle$ and $\langle a\cos(t+\frac{\pi}{2}), b\sin(t+\frac{\pi}{2})\rangle$.

The dot product of these vectors is $a^2\cos t\cos(t+\frac{\pi}{2}) + b^2\sin t\sin(t+\frac{\pi}{2})$, and this is not zero if $a \neq b$.

61. $\mathbf{v}\cdot\mathbf{n} = \langle\cos\theta, \sin\theta\rangle\cdot\langle-\sin\theta, \cos\theta\rangle$ $= -\cos\theta\sin\theta + \sin\theta\cos\theta = 0$ So, $\mathbf{v}$ and $\mathbf{n}$ are perpendicular. The component of $\mathbf{w}$ along $\mathbf{v}$ is $-w\sin\theta$. The component of $\mathbf{w}$ along $\mathbf{n}$ is $-w\cos\theta$.

63. The angle is $\cos^{-1}\left(\dfrac{\mathbf{a}\cdot\mathbf{b}}{\|\mathbf{a}\|\|\mathbf{b}\|}\right)$

$= \cos^{-1}\left(\dfrac{50{,}000}{\sqrt{50{,}000}\sqrt{100{,}000}}\right)$

$= \cos^{-1}\dfrac{1}{\sqrt{2}} = 45°$.

65. $\mathbf{s}\cdot\mathbf{p} = (3000)(\$20) + (2000)(\$15) + (4000)(\$25) = \$190{,}000$. This is the monthly revenue.

67. Apply the Cauchy-Schwartz Inequality to the vectors

$\mathbf{a} = \langle\sqrt{a_1}, \sqrt{a_2}, \ldots, \sqrt{a_n}\rangle$

$\mathbf{b} = \langle\frac{1}{1^p}, \frac{1}{2^p}, \frac{1}{3^p}, \ldots, \frac{1}{n^p}\rangle$.

$\mathbf{a}\cdot\mathbf{b} = \displaystyle\sum_{k=1}^n \frac{\sqrt{a_k}}{k^p}$

$\|\mathbf{a}\| = \sqrt{\displaystyle\sum_{k=1}^n \sqrt{a_n}^2} = \sqrt{\displaystyle\sum_{k=1}^n a_n}$

$\|\mathbf{b}\| = \sqrt{\displaystyle\sum_{k=1}^n \left(\frac{1}{k^p}\right)^2} = \sqrt{\displaystyle\sum_{k=1}^n \frac{1}{k^{2p}}}$

The Cauchy-Schwartz Inequality says that

$\displaystyle\sum_{k=1}^n \frac{\sqrt{a_k}}{k^p} \leq \sqrt{\displaystyle\sum_{k=1}^n a_n}\sqrt{\displaystyle\sum_{k=1}^n \frac{1}{k^{2p}}}$.

69. $|x^2 - c| \geq |x|^2 - |c|$, so it is enough to show $|x|^2 - |c| > |x|$, or $|x|^2 - |x| - c > 0$. This will be true as long as $|x|$ is not between the two roots of this quadratic. The roots are given by the quadratic formula to be $\frac{1 \pm \sqrt{1+4c}}{2} = \frac{1}{2} \pm \sqrt{\frac{1}{4} + c}$. If $|x|$ is greater than the larger of these two roots, namely, $\frac{1}{2} + \sqrt{\frac{1}{4} + c}$, we have $|x|^2 - |x| - c > 0$ and $|x^2 - c| \geq |x|^2 - |c| > |x|$.

11.4 THE CROSS PRODUCT

1. $\begin{vmatrix} 2 & 0 & -1 \\ 1 & 1 & 0 \\ -2 & -1 & 1 \end{vmatrix}$

$= 2 \begin{vmatrix} 1 & 0 \\ -1 & 1 \end{vmatrix} - 0 \begin{vmatrix} 1 & 0 \\ -2 & 1 \end{vmatrix} + (-1) \begin{vmatrix} 1 & 1 \\ -2 & -1 \end{vmatrix}$

$= 2(1 - 0) - 0 - 1(-1 + 2) = 1$

3. $\begin{vmatrix} 2 & 3 & -1 \\ 0 & 1 & 0 \\ -2 & -1 & 3 \end{vmatrix}$

$= 2 \begin{vmatrix} 1 & 0 \\ -1 & 3 \end{vmatrix} - 3 \begin{vmatrix} 0 & 0 \\ -2 & 3 \end{vmatrix} + (-1) \begin{vmatrix} 0 & 1 \\ -2 & -1 \end{vmatrix}$

$= 2(3 - 0) - 3(0 - 0) - 1(0 + 2) = 4$

5. $\mathbf{a} \times \mathbf{b} = \begin{vmatrix} \mathbf{i} & \mathbf{j} & \mathbf{k} \\ 1 & 2 & -1 \\ 1 & 0 & 2 \end{vmatrix}$

$= \mathbf{i} \begin{vmatrix} 2 & -1 \\ 0 & 2 \end{vmatrix} - \mathbf{j} \begin{vmatrix} 1 & -1 \\ 1 & 2 \end{vmatrix} + \mathbf{k} \begin{vmatrix} 1 & 2 \\ 1 & 0 \end{vmatrix}$

$= (4 - 0)\mathbf{i} - (2 + 1)\mathbf{j} + (0 - 2)\mathbf{k} = \langle 4, -3, -2 \rangle$

7. $\mathbf{a} \times \mathbf{b} = \begin{vmatrix} \mathbf{i} & \mathbf{j} & \mathbf{k} \\ 0 & 1 & 4 \\ -1 & 2 & -1 \end{vmatrix}$

$= (-1 - 8)\mathbf{i} - (0 + 4)\mathbf{j} + (0 + 1)\mathbf{k} = \langle -9, -4, 1 \rangle$

9. $\mathbf{a} \times \mathbf{b} = \begin{vmatrix} \mathbf{i} & \mathbf{j} & \mathbf{k} \\ 2 & 0 & -1 \\ 0 & 4 & 1 \end{vmatrix}$

$= (0 + 4)\mathbf{i} - (2 - 0)\mathbf{j} + (8 + 0)\mathbf{k} = \langle 4, -2, 8 \rangle$

11. $\mathbf{a} \times \mathbf{b} = \begin{vmatrix} \mathbf{i} & \mathbf{j} & \mathbf{k} \\ 1 & 0 & 4 \\ 1 & -4 & 2 \end{vmatrix}$

$= (0 + 16)\mathbf{i} - (2 - 4)\mathbf{j} + (-4 - 0)\mathbf{k} = \langle 16, 2, -4 \rangle$

Two orthogonal unit vectors to $\mathbf{a}$ and $\mathbf{b}$ are

$\pm \frac{\mathbf{a} \times \mathbf{b}}{\|\mathbf{a} \times \mathbf{b}\|} = \pm \frac{\langle 16, 2, -4 \rangle}{\sqrt{276}} = \pm \frac{1}{\sqrt{69}} \langle 8, 1, -2 \rangle$.

13. $\mathbf{a} \times \mathbf{b} = \begin{vmatrix} \mathbf{i} & \mathbf{j} & \mathbf{k} \\ 2 & -1 & 0 \\ 1 & 0 & 3 \end{vmatrix}$

$= (-3 - 0)\mathbf{i} - (6 - 0)\mathbf{j} + (0 + 1)\mathbf{k}$

$= \langle -3, -6, 1 \rangle$

Two orthogonal unit vectors to $\mathbf{a}$ and $\mathbf{b}$ are

$\pm \frac{\mathbf{a} \times \mathbf{b}}{\|\mathbf{a} \times \mathbf{b}\|} = \pm \frac{\langle -3, -6, 1 \rangle}{\sqrt{46}}$.

15. $\mathbf{a} \times \mathbf{b} = \begin{vmatrix} \mathbf{i} & \mathbf{j} & \mathbf{k} \\ 3 & -1 & 0 \\ 0 & 4 & 1 \end{vmatrix}$

$= (-1 - 0)\mathbf{i} - (3 - 0)\mathbf{j} + (12 - 0)\mathbf{k}$

$= \langle -1, -3, 12 \rangle$

Two orthogonal unit vectors to $\mathbf{a}$ and $\mathbf{b}$ are

$\pm \frac{\mathbf{a} \times \mathbf{b}}{\|\mathbf{a} \times \mathbf{b}\|} = \pm \frac{\langle -1, -3, 12 \rangle}{\sqrt{154}}$.

17. $\|\mathbf{a}\| = \sqrt{17}$, $\|\mathbf{b}\| = \sqrt{5}$

$\mathbf{a} \times \mathbf{b} = \begin{vmatrix} \mathbf{i} & \mathbf{j} & \mathbf{k} \\ 1 & 0 & 4 \\ 2 & 0 & 1 \end{vmatrix}$

$= (0 - 0)\mathbf{i} - (1 - 8)\mathbf{j} + (0 - 0)\mathbf{k} = \langle 0, 7, 0 \rangle$

$\|\mathbf{a} \times \mathbf{b}\| = 7$

$\|\mathbf{a} \times \mathbf{b}\| = \|\mathbf{a}\| \|\mathbf{b}\| \sin \theta$, so

$\theta = \sin^{-1} \frac{7}{\sqrt{17}\sqrt{5}} \approx 0.862$.

19. $\|\mathbf{a}\| = \sqrt{10}$, $\|\mathbf{b}\| = \sqrt{17}$

$\mathbf{a} \times \mathbf{b} = \begin{vmatrix} \mathbf{i} & \mathbf{j} & \mathbf{k} \\ 3 & 0 & 1 \\ 0 & 4 & 1 \end{vmatrix}$

$= (0 - 4)\mathbf{i} - (3 - 0)\mathbf{j} + (12 - 0)\mathbf{k}$

$= \langle -4, -3, 12 \rangle$

$\|\mathbf{a} \times \mathbf{b}\| = 13$

$\|\mathbf{a} \times \mathbf{b}\| = \|\mathbf{a}\|\|\mathbf{b}\| \sin \theta$, so

$\theta = \sin^{-1} \dfrac{13}{\sqrt{10}\sqrt{17}} \approx 1.49.$

21. The distance is $\dfrac{\|\mathbf{PQ} \times \mathbf{PR}\|}{\|\mathbf{PR}\|}$.

$\mathbf{PQ} = \langle 1 - 0, 2 - 1, 0 - 2 \rangle$ and
$\mathbf{PR} = \langle 3 - 0, 1 - 1, 1 - 2 \rangle$

$\mathbf{PQ} \times \mathbf{PR} = \begin{vmatrix} \mathbf{i} & \mathbf{j} & \mathbf{k} \\ 1 & 1 & -2 \\ 3 & 0 & -1 \end{vmatrix}$

$= (-1 - 0)\mathbf{i} - (-1 + 6)\mathbf{j} + (0 - 3)\mathbf{k}$
$= \langle -1, -5, -3 \rangle$, so
$\|\mathbf{PQ} \times \mathbf{PR}\| = \sqrt{35}.$
$\|\mathbf{PR}\| = \sqrt{10}.$

The distance from Q to the line is $\sqrt{\dfrac{7}{2}} \approx 1.87.$

23. The distance is $\dfrac{\|\mathbf{PQ} \times \mathbf{PR}\|}{\|\mathbf{PR}\|}$.

$\mathbf{PQ} = \langle 3 - 2, -2 - 1, 1 + 1 \rangle$ and
$\mathbf{PR} = \langle 1 - 2, 1 - 1, 1 + 1 \rangle$

$\mathbf{PQ} \times \mathbf{PR} = \begin{vmatrix} \mathbf{i} & \mathbf{j} & \mathbf{k} \\ 1 & -3 & 2 \\ -1 & 0 & 2 \end{vmatrix}$

$= (-6 - 0)\mathbf{i} - (2 + 2)\mathbf{j} + (0 - 3)\mathbf{k}$
$= \langle -6, -4, -3 \rangle$, so
$\|\mathbf{PQ} \times \mathbf{PR}\| = \sqrt{61}.$
$\|\mathbf{PR}\| = \sqrt{5}.$

The distance from Q to the line is $\sqrt{\dfrac{61}{5}} \approx 3.49.$

25.
$\|\tau\| = \|\mathbf{r}\|\|\mathbf{F}\| \sin \theta = \dfrac{8}{12}(20) \sin \dfrac{\pi}{4}$

$= \dfrac{40}{3}\left(\dfrac{\sqrt{2}}{2}\right) = \dfrac{20\sqrt{2}}{3} \approx 9.43 \text{ft-lbs}$

27.
$\|\tau\| = \|\mathbf{r}\|\|\mathbf{F}\| \sin \theta = \dfrac{8}{12}(30) \sin \dfrac{\pi}{6}$

$= 30\left(\dfrac{2}{3}\right)\left(\dfrac{1}{2}\right) = 10 \text{ft-lbs}$

29. (a) Direction of $\mathbf{s} \times \mathbf{v}$ is up.

(b) Direction of $\mathbf{s} \times \mathbf{v}$ is up and right.

31. (a) Direction of $\mathbf{s} \times \mathbf{v}$ is down and left.

(b) Direction of $\mathbf{s} \times \mathbf{v}$ is up and left

33. Rising fastball is hard to hit.

35. Topspin shot will dive.

37. Spiral pass should fly true.

39. Ball will rise.

41. False. Example: $\mathbf{a} = \langle 1, 0, 0 \rangle$, $\mathbf{b} = \langle 1, 1, 1 \rangle$, $\mathbf{c} = \langle 2, 1, 1 \rangle$.

43. False. $\mathbf{a} \times \mathbf{a}$ is a vector, not a number.

45. True. Torque is the cross product of direction and force. $\mathbf{r} \times (2\mathbf{F}) = 2(\mathbf{r} \times \mathbf{F}) = 2\tau$.

47. $\mathbf{a} \times \mathbf{b} = \begin{vmatrix} \mathbf{i} & \mathbf{j} & \mathbf{k} \\ 2 & 3 & 0 \\ 1 & 4 & 0 \end{vmatrix}$

$= (0 - 0)\mathbf{i} - (0 - 0)\mathbf{j} + (8 - 3)\mathbf{k} = 5\mathbf{k}.$
$\|\mathbf{a} \times \mathbf{b}\| = 5.$

49. $\mathbf{a} \times \mathbf{b} = \begin{vmatrix} \mathbf{i} & \mathbf{j} & \mathbf{k} \\ 2 & 3 & -1 \\ 3 & -1 & 4 \end{vmatrix}$

$= (12 - 1)\mathbf{i} - (8 + 3)\mathbf{j} + (-2 - 9)$
$\mathbf{k} = \langle 11, -11, -11 \rangle.$

Area of triangle is $\dfrac{1}{2}\|\mathbf{a} \times \mathbf{b}\| = \dfrac{11\sqrt{3}}{2}.$

51. $\begin{vmatrix} 2 & 1 & 0 \\ -1 & 2 & 0 \\ 1 & 1 & 2 \end{vmatrix} = 2(4 - 0) - (-2 - 0) + 0$
$= 10$

53. $(\mathbf{j} \times \mathbf{k})$ is $\mathbf{i}$ and $\mathbf{i} \times \mathbf{i}$ is $\langle 0, 0, 0 \rangle$.

55. $\mathbf{j} \times (\mathbf{j} \times \mathbf{i}) = \mathbf{j} \times -\mathbf{k} = -\mathbf{i}$

57. $\mathbf{i} \times (3\mathbf{k}) = 3(\mathbf{i} \times \mathbf{k}) = -3\mathbf{j}$

59. $\begin{vmatrix} 2 & 3 & 1 \\ 1 & 0 & 2 \\ 0 & 3 & -3 \end{vmatrix}$
$= 2(0 - 6) - 3(-3 - 0) + 1(3 - 0)$
$= -12 + 9 + 3 = 0.$ Since the volume of the parallelepiped is 0, the vectors are coplanar.

61. $\begin{vmatrix} 1 & 0 & -2 \\ 3 & 0 & 1 \\ 2 & 1 & 0 \end{vmatrix}$
$= 1(0 - 1) - 0 - 2(3 - 0)$
$= -1 - 0 - 6 = -7.$ Since the volume of the

parallelepiped is not 0, the vectors are not coplanar.

63. $\|\mathbf{a}\|^2\|\mathbf{b}\|^2 = \|\mathbf{a}\|^2\|\mathbf{b}\|^2(\sin^2\theta + \cos^2\theta)$
$= \|\mathbf{a}\|^2\|\mathbf{b}\|^2\sin^2\theta + \|\mathbf{a}\|^2\|\mathbf{b}\|^2\cos^2\theta$
$= \|\mathbf{a}\times\mathbf{b}\|^2 + (\mathbf{a}\cdot\mathbf{b})^2$, so $\|\mathbf{a}\times\mathbf{b}\|^2$
$= \|\mathbf{a}\|^2\|\mathbf{b}\|^2 - (\mathbf{a}\cdot\mathbf{b})^2$.

65. $\|\mathbf{a}\times\mathbf{b}\|$ is larger in Figure A. This follows from the formula $\|\mathbf{a}\times\mathbf{b}\| = \|\mathbf{a}\|\|\mathbf{b}\|\sin\theta$. As the angle θ increases from 0 to $\frac{\pi}{2}$, the sine of θ increases from 0 to 1. The maximum possible value for $\|\mathbf{a}\times\mathbf{b}\|$ is 12, and this occurs when the vectors are perpendicular.

67. $\begin{vmatrix} \mathbf{a}\cdot\mathbf{c} & \mathbf{b}\cdot\mathbf{c} \\ \mathbf{a}\cdot\mathbf{d} & \mathbf{b}\cdot\mathbf{d} \end{vmatrix} = (\mathbf{a}\cdot\mathbf{c})(\mathbf{b}\cdot\mathbf{d}) - (\mathbf{b}\cdot\mathbf{c})(\mathbf{a}\cdot\mathbf{d})$
$= (a_1c_1 + a_2c_2 + a_3c_3)(b_1d_1 + b_2d_2 + b_3d_3)$
$-(b_1c_1 + b_2c_2 + b_3c_3)(a_1d_1 + a_2d_2 + a_3d_3)$
$= (a_2b_3 - a_3b_2)(c_2d_3 - c_3d_2)$
$+(a_3b_1 - a_1b_3)(c_3d_1 - c_1d_3)$
$+(a_1b_2 - a_2b_1)(c_1d_2 - c_2d_1)$
$= \left[(a_2b_3 - a_3b_2)\mathbf{i} + (a_3b_1 - a_1b_3)\mathbf{j}\right.$
$+ (a_1b_2 - a_2b_1)\mathbf{k}\left]\cdot\right[(c_2d_3 - c_3d_2)\mathbf{i}$
$+ (c_3d_1 - c_1d_3)\mathbf{j} + (c_1d_2 - c_2d_1)\mathbf{k}]$
$= (\mathbf{a}\times\mathbf{b})\cdot(\mathbf{c}\times\mathbf{d})$

11.5 LINES AND PLANES IN SPACE

1. (a) $) x = 1 + 2t, y = 2 - t, z = -3 + 4t$

(b) $\dfrac{x-1}{2} = \dfrac{y-2}{-1} = \dfrac{z+3}{4}$

3. The direction is $\langle 4 - 2, 0 - 1, 4 - 3\rangle$.

(a) $x = 2 + 2t, y = 1 - t, z = 3 + t$

(b) $\dfrac{x-2}{2} = \dfrac{y-1}{-1} = \dfrac{z-3}{1}$

5. (a) $x = 1 - 3t, y = 4 + 0t, z = 1 + t$

(b) $\dfrac{x-1}{-3} = \dfrac{z-1}{1}, y = 4$

7. $\begin{vmatrix} \mathbf{i} & \mathbf{j} & \mathbf{k} \\ 1 & 0 & 2 \\ 0 & 2 & 1 \end{vmatrix} = \langle -4, -1, 2\rangle$ is in the direction perpendicular to both vectors.

(a) $x = 2 - 4t, y = 0 - t, z = 1 + 2t$

(b) $\dfrac{x-2}{-4} = \dfrac{y}{-1} = \dfrac{z-1}{2}$

9. $\langle 2, -1, 3\rangle$ is normal to the plane.

(a) $x = 1 + 2t, y = 2 - t, z = -1 + 3t$

(b) $\dfrac{x-1}{2} = \dfrac{y-2}{-1} = \dfrac{z+1}{3}$

11. Parallel to line 1: $\mathbf{v_1} = \langle -3, 4, 1\rangle$
Parallel to line 2: $\mathbf{v_2} = \langle 2, -2, 1\rangle$
$\cos\theta = \dfrac{\mathbf{v_1}\cdot\mathbf{v_2}}{\|\mathbf{v_1}\|\|\mathbf{v_2}\|} = \dfrac{-6-8+1}{\sqrt{26}\sqrt{9}} = \dfrac{-13}{\sqrt{234}}$
$\theta = \cos^{-1}\dfrac{-13}{\sqrt{234}} \approx 2.59$ radians

13. Parallel to line 1: $\mathbf{v_1} = \langle 2, 0, 1\rangle$
Parallel to line 2: $\mathbf{v_2} = \langle -1, 5, 2\rangle$
$\cos\theta = \dfrac{\mathbf{v_1}\cdot\mathbf{v_2}}{\|\mathbf{v_1}\|\|\mathbf{v_2}\|} = \dfrac{-2+0+2}{\sqrt{5}\sqrt{30}} = 0$
These vectors are perpendicular.

15. Parallel to line 1: $\mathbf{v_1} = \langle 2, 4, -6\rangle$
Parallel to line 2: $\mathbf{v_2} = \langle -1, -2, 3\rangle$
Since $\mathbf{v_1} = -2\mathbf{v_2}$, the lines are parallel.

17. Parallel to line 1: $\mathbf{v_1} = \langle 1, 0, 2\rangle$
Parallel to line 2: $\mathbf{v_2} = \langle 2, 2, 4\rangle$
Since $\mathbf{v_1} \neq c\mathbf{v_2}$, the lines are not parallel.
$x \Rightarrow 4 + t = 2 + 2s \Rightarrow t = -2 + 2s$
$y \Rightarrow 2 = 2s \Rightarrow s = 1$ and $t = 0$
$z \Rightarrow 3 + 2t = -1 + 4s \Rightarrow 3 = 3$
The lines intersect at the point $(4, 2, 3)$.

19. Parallel to line 1: $\mathbf{v_1} = \langle 2, 0, -4\rangle$
Parallel to line 2: $\mathbf{v_2} = \langle -1, 0, 2\rangle$
Since $\mathbf{v_1} = -2\mathbf{v_2}$, the lines are parallel.

21. $2(x - 1) - 1(y - 3) + 5(z - 2) = 0$

23. $P = (2, 0, 3)$, $Q = (1, 1, 0)$, $R = (3, 2, -1)$
$\mathbf{PQ} = (-1, 1, -3)$, $\mathbf{PR} = (1, 2, -4)$
$$\mathbf{PQ} \times \mathbf{PR} = \begin{vmatrix} \mathbf{i} & \mathbf{j} & \mathbf{k} \\ -1 & 1 & -3 \\ 1 & 2 & -4 \end{vmatrix} = \langle 2, -7, -3 \rangle$$
is normal to the plane.
$2(x - 2) - 7y - 3(z - 3) = 0$

25. $P = (-2, 2, 0)$, $Q = (-2, 3, 2)$, $R = (1, 2, 2)$
$\mathbf{PQ} = (0, 1, 2)$, $\mathbf{PR} = (3, 0, 2)$
$$\mathbf{PQ} \times \mathbf{PR} = \begin{vmatrix} \mathbf{i} & \mathbf{j} & \mathbf{k} \\ 0 & 1 & 2 \\ 3 & 0 & 2 \end{vmatrix} = \langle 2, 6, -3 \rangle$$
is normal to the plane.
$2(x + 2) + 6(y - 2) - 3z = 0$

27. Normal vector is $\langle -2, 4, 0 \rangle$, so
$-2x + 4(y + 2) = 0$.

29. Normal vector must be perpendicular to the normal vectors of both planes.
$$\mathbf{n}_1 \times \mathbf{n}_2 = \begin{vmatrix} \mathbf{i} & \mathbf{j} & \mathbf{k} \\ 1 & 1 & 0 \\ 2 & 1 & -1 \end{vmatrix} = \langle -1, 1, -1 \rangle, \text{ so}$$
$-1(x - 1) + 1(y - 2) - 1(z - 1) = 0$.

31. $x + y + z = 4$

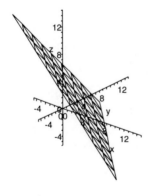

33. $3x + 6y - z = 6$

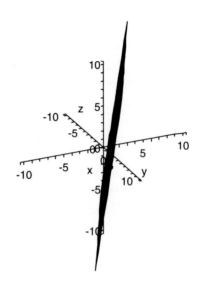

35. $x = 4$

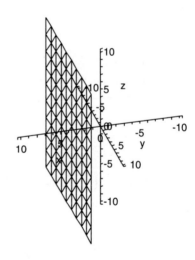

37. $z = 2$

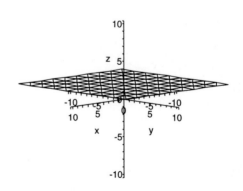

39. $2x - z = 2$

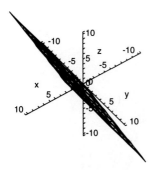

41. Solve the equations for z and equate:
$2x - y - 4 = 2y - 3x \Rightarrow y = \frac{5x-4}{3}$.
Substitute this into the first equation and solve
for z: $2x - \frac{5x-4}{3} - z = 4 \Rightarrow z = \frac{x-8}{3}$.
Using $x = t$ as a parameter, we get the line
$x = t$, $y = \frac{5t-4}{3}$, $z = \frac{t-8}{3}$.

43. Solve the equations for x and equate:
$\frac{-4y+1}{3} = -y + z + 3 \Rightarrow y = -3z - 8$.
Substitute this into the first equation and solve
for x:
$3x + 4(-3z - 8) = 1 \Rightarrow x = 4z + 11$.
Using $z = t$ as a parameter, we get the line
$x = 4t + 11$, $y = -3t - 8$, $z = t$.

45. $d = \dfrac{|2(2) - 0 + 2(1) - 4|}{\sqrt{4 + 1 + 4}} = \dfrac{2}{3}$.

47. $d = \dfrac{|2 - (-1) + (-1) - 4|}{\sqrt{1 + 1 + 1}} = \dfrac{2}{\sqrt{3}}$

49. $(2, 0, 0)$ is a point on the plane $2x - y - z = 4$.
$d = \dfrac{|2(2) - 1(0) - 1(0) - 1|}{\sqrt{4 + 1 + 1}} = \dfrac{3}{\sqrt{6}}$

51. If (x_0, y_0, z_0) is a point on the plane $ax + by + cz = d_2$, then the formula for the distance from this point to the plane $ax + by + cz = d_1$ is
$\dfrac{|ax_0 + by_0 + cz_0 - d_1|}{\sqrt{a^2 + b^2 + c^2}} = \dfrac{|d_2 - d_1|}{\sqrt{a^2 + b^2 + c^2}}$.

53. Parallel to line 1: $\mathbf{v_1} = \langle 1, 0, 2 \rangle$
Parallel to line 2: $\mathbf{v_2} = \langle 2, 2, 4 \rangle$
$\mathbf{v_1} \times \mathbf{v_2} = \begin{vmatrix} \mathbf{i} & \mathbf{j} & \mathbf{k} \\ 1 & 0 & 2 \\ 2 & 2 & 4 \end{vmatrix} = \langle -4, 0, 2 \rangle$ is normal
to the plane. Plane with this normal vector

through point $(4, 2, 3)$ is $-4(x - 4) + 2(z - 3) = 0$.

55. True. Can be both if the planes coincide.

57. False. Can be a point, a line, or a plane, or can be empty.

59. False. (True if we take all lines perpendicular to a given line through a given point.)

61. True.

63. Direction of line 1: $\mathbf{v_1} = \langle -2, 3, 1 \rangle$
Direction of line 2: $\mathbf{v_2} = \langle 4, -6, -2 \rangle$
Since $\mathbf{v_2} = -2\mathbf{v_1}$, the lines are parallel. The point $(1, 3, -1)$ lies on the second line, using $t = 0$. This point also lies on the first line, using $t = 1$. The lines are the same.

65. Simplify $2(x - 1) - (y + 2) + (z - 3) = 0$ to get $2x - y + z = 7$. Multiply by 2 to get $4x - 2y + 2z = 14$. This is parallel to $4x - 2y + 2z = 2$, but not the same plane.

67. The flight paths are lines. Set the x coordinates equal to see that if the paths intersect, $s = 1$. If $s = 1$ and the paths intersect, then the y coordinates force $6 - 2t = 3 + 1 \Rightarrow t = 1$. The z coordinates also agree when $s = t = 1$. The airplanes collide at $(3, 4, 4)$ at time $s = t = 1$.

11.6 SURFACES IN SPACE

1. Cylinder

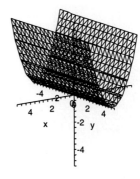

3. Ellipsoid

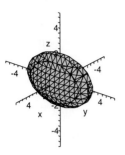

5. Circular paraboloid

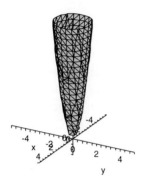

7. Elliptic cone

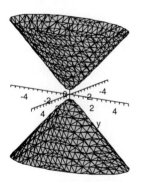

9. Hyperbolic paraboloid

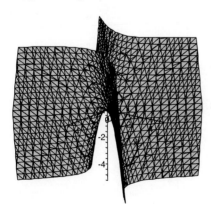

11. Hyperboloid of one sheet

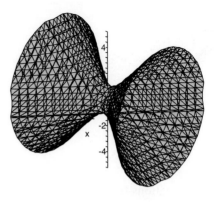

13. Hyperboloid of two sheets

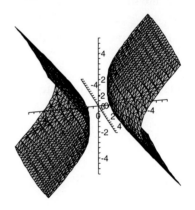

15. Cylinder

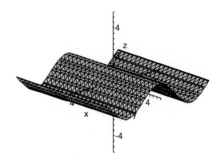

17. Circular Paraboloid

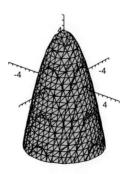

19. Cylinder

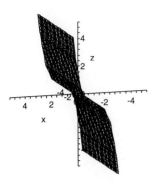

21. Top half of circular cone

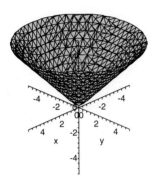

23. Cylinder

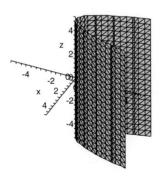

25. Circular paraboloid

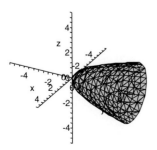

27. Ellipsoid

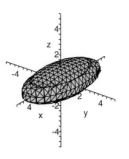

29. Hyperbolic paraboloid

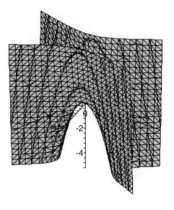

31. Hyperboloid of one sheet

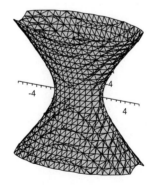

33. Hyperboloid of two sheets

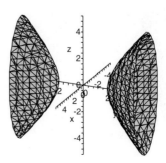

35. Plane

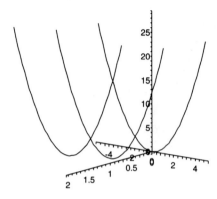

37. Circular cylinder

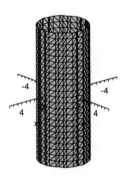

39. Circular paraboloid

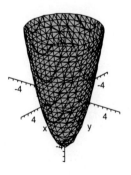

41. $x = 0, 1, 2$

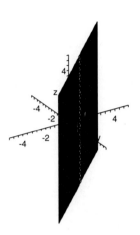

43. $x = 0, 1, 2$

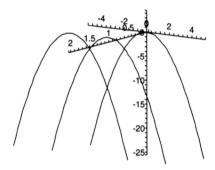

45. For constant $y = c$, the equation is $z = x^2 - c^2$. $z' = 2x = 0$ when $x = 0$. $z'' = 2$ at $x = 0$ so this is a minimum.

Similarly, at $x = 0$ the equation is $z = -y^2$, which is a parabola with maximum at $y = 0$. Water will run toward $x = 0$ in the y direction and away from $y = 0$ in the x direction. The two primary runoff points will be $(0, 1, -1)$ and $(0, -1, -1)$.

47. $x = a \sin s \cos t, \; y = b \sin s \sin t, \; z = c \cos s$

$$\frac{x^2}{a^2} + \frac{y^2}{b^2} + \frac{z^2}{c^2} =$$

$$\frac{a^2 \sin^2 s \cos^2 t}{a^2} + \frac{b^2 \sin^2 s \sin^2 t}{b^2} + \frac{c^2 \cos^2 s}{c^2}$$

$$= \sin^2 s \cos^2 t + \sin^2 s \sin^2 t + \cos^2 s$$
$$= \sin^2 s (\cos^2 t + \sin^2 t) + \cos^2 s$$
$$= \sin^2 s + \cos^2 s = 1$$

So, (x, y, z) lies on $\dfrac{x^2}{a^2} + \dfrac{y^2}{b^2} + \dfrac{z^2}{c^2} = 1.$

49. $x = as\cos t,\ y = bs\sin t,\ z = s$

$$\frac{x^2}{a^2} + \frac{y^2}{b^2} = \frac{(as\cos t)^2}{a^2} + \frac{(bs\sin t)^2}{b^2}$$
$$= (s\cos t)^2 + (s\sin t)^2$$
$$= s^2(\cos^2 t + \sin^2 t) = s^2 = z^2$$

So, (x, y, z) lies on $\dfrac{x^2}{a^2} + \dfrac{y^2}{b^2} = z^2$.

51. $x = a\cosh s,\ y = b\sinh s\cos t,$
$z = c\sinh s\sin t$

$$\frac{x^2}{a^2} - \frac{y^2}{b^2} - \frac{z^2}{c^2}$$
$$= \frac{a^2\cosh^2 s}{a^2} - \frac{b^2\sinh^2 s\cos^2 t}{b^2}$$
$$- \frac{c^2\sinh^2 s\sin^2 t}{c^2}$$
$$= \cosh^2 s - \sinh^2 s\cos^2 t - \sinh^2 s\sin^2 t$$
$$= \cosh^2 s - \sinh^2 s(\cos^2 t + \sin^2 t)$$
$$= \cosh^2 s - \sinh^2 s = 1$$

So, (x, y, z) lies on $\dfrac{x^2}{a^2} - \dfrac{y^2}{b^2} - \dfrac{z^2}{c^2} = 1$. Since $a > 0$, we have $x = a\cosh s > 0$, and this point is on the right half.

53. For the surface in exercise 3: $x = \sin s\cos t$,
$y = 3\sin s\sin t,\ z = 2\cos s$.

For the surface in exercise 5: $x = \dfrac{s}{2}\cos t$,

$y = \dfrac{s}{2}\sin t,\ z = s^2$.

For the surface in exercise 7: $x = \dfrac{s}{2}\cos t$,

$y = s\sin t,\ z = s$.

55. For the surface in exercise 17: $x = s\cos t$.
$y = s\sin t,\ z = -s^2 + 4$.

57.

Limiting the z range does avoid the peaks and make the graph appear as in Figure 11.57b.

Chapter 11 Review Exercises

1. $\mathbf{a} + \mathbf{b} = \langle -2 + 1, 3 + 0\rangle = \langle -1, 3\rangle$ $4\mathbf{b} = 4\langle 1, 0\rangle = \langle 4, 0\rangle$ $2\mathbf{b} - \mathbf{a} = 2\langle 1, 0\rangle - \langle -2, 3\rangle = \langle 4, -3\rangle$ $\|2\mathbf{b} - \mathbf{a}\| = \sqrt{16 + 9} = 5$

3. $\mathbf{a} + \mathbf{b} = (10\mathbf{i} + 2\mathbf{j} - 2\mathbf{k}) + (-4\mathbf{i} + 3\mathbf{j} + 2\mathbf{k}) = 6\mathbf{i} + 5\mathbf{j}$
$4\mathbf{b} = 4(-4\mathbf{i} + 3\mathbf{j} + 2\mathbf{k}) = -16\mathbf{i} + 12\mathbf{j} + 8\mathbf{k}$
$2\mathbf{b} - \mathbf{a} = 2(-4\mathbf{i} + 3\mathbf{j} + 2\mathbf{k}) - (10\mathbf{i} + 2\mathbf{j} - 2\mathbf{k}) = -18\mathbf{i} + 4\mathbf{j} + 6\mathbf{k}$
$\|2\mathbf{b} - \mathbf{a}\| = \sqrt{324 + 16 + 36} = \sqrt{376}$
$= 2\sqrt{94}$

5. $\mathbf{a} \cdot \mathbf{b} = 2(4) + 3(5) = 23 \neq 0$. The vectors are not orthogonal.
$\mathbf{a} \neq c\mathbf{b}$. The vectors are not parallel.

7. $\mathbf{a} \cdot \mathbf{b} = -2(4) + 3(-6) + 1(-2) = -28 \neq 0$.
The vectors are not orthogonal.
$\mathbf{b} = -2\mathbf{c}$. The vectors are parallel.

9. $\mathbf{PQ} = \langle 2 - 3, -1 - 1, 1 + 2\rangle = \langle -1, -2, 3\rangle$

11. $\dfrac{1}{\|\langle 3, 6\rangle\|}\langle 3, 6\rangle = \dfrac{1}{\sqrt{9 + 36}}\langle 3, 6\rangle$

$\quad = \langle \dfrac{1}{\sqrt{5}}, \dfrac{2}{\sqrt{5}}\rangle$

13. $\dfrac{1}{\|10\mathbf{i} + 2\mathbf{j} - 2\mathbf{k}\|}(10\mathbf{i} + 2\mathbf{j} - 2\mathbf{k})$

$\quad = \dfrac{1}{\sqrt{100 + 4 + 4}}(10\mathbf{i} + 2\mathbf{j} - 2\mathbf{k})$

$\quad = \dfrac{10}{\sqrt{108}}\mathbf{i} + \dfrac{2}{\sqrt{108}}\mathbf{j} - \dfrac{2}{\sqrt{108}}\mathbf{k}$

$\quad = \dfrac{5}{3\sqrt{3}}\mathbf{i} + \dfrac{1}{3\sqrt{3}}\mathbf{j} - \dfrac{1}{3\sqrt{3}}\mathbf{k}$

15. Displacement vector is $\langle 1 - 4, 1 - 1, 6 - 2\rangle$.
Unit vector in this direction is

$$\frac{1}{\|\langle -3, 0, 4\rangle\|}\langle -3, 0, 4\rangle = \frac{1}{\sqrt{9 + 16}}\langle -3, 0, 4\rangle$$

$$= \langle \frac{-3}{5}, 0, \frac{4}{5} \rangle.$$

17. $d = \sqrt{(3-0)^2 + (4+2)^2 + (1-2)^2} = \sqrt{46}$

19. $2\frac{\mathbf{v}}{\|\mathbf{v}\|} = \frac{2}{\sqrt{12}}(2\mathbf{i} - 2\mathbf{j} + 2\mathbf{k})$

$$= \frac{2}{\sqrt{3}}(\mathbf{i} - \mathbf{j} + \mathbf{k})$$

21. Suppose that the velocity of the plane is $\mathbf{v} = \langle x, y \rangle$, and the wind velocity is $\mathbf{w} = \langle 20, -80 \rangle$. Then $\|\mathbf{v}\| = 500$, and we want $\mathbf{v} + \mathbf{w} = \langle c, 0 \rangle$ for some positive constant c. This forces $y = 80$, and then we solve for x in $500^2 = x^2 + 80^2$ to get $x = 20\sqrt{609}$. The vector $\mathbf{v} = \langle 20\sqrt{609}, 80 \rangle$ and the direction is $\sin^{-1}\frac{80}{500} \approx 9.2°$ north of east.

23. $x^2 + (y+2)^2 + z^2 = 36$

25. $\mathbf{a} \cdot \mathbf{b} = 2(2) + (-1)(4) = 0$

27. $\mathbf{a} \cdot \mathbf{b} = 3(-2) + 1(2) + (-4)(1) = -8$

29. $\cos\theta = \frac{\mathbf{a} \cdot \mathbf{b}}{\|\mathbf{a}\|\|\mathbf{b}\|} = \frac{-3+2+2}{\sqrt{14}\sqrt{6}} = \frac{1}{\sqrt{84}}$

$$\theta = \cos^{-1}\left(\frac{1}{\sqrt{84}}\right) \approx 1.46$$

31. $\text{comp}_{\mathbf{b}}\mathbf{a} = \frac{3(1) + 1(2) + (-4)(1)}{\sqrt{1+4+1}} = \frac{1}{\sqrt{6}}$

$$\text{proj}_{\mathbf{b}}\mathbf{a} = \frac{1}{\sqrt{6}}\left(\frac{\mathbf{i} + 2\mathbf{j} + \mathbf{k}}{\sqrt{1+4+1}}\right) = \frac{1}{6}\mathbf{i} + \frac{1}{3}\mathbf{j} + \frac{1}{6}\mathbf{k}$$

33. $\mathbf{a} \times \mathbf{b} = \begin{vmatrix} \mathbf{i} & \mathbf{j} & \mathbf{k} \\ 1 & -2 & 1 \\ 2 & 0 & 1 \end{vmatrix}$

$$= \langle -2 - 0, -(1-2), 0+4 \rangle = \langle -2, 1, 4 \rangle$$

35. $\mathbf{a} \times \mathbf{b} = \begin{vmatrix} \mathbf{i} & \mathbf{j} & \mathbf{k} \\ 0 & 2 & 1 \\ 4 & 2 & -1 \end{vmatrix}$

$$= (-2-2)\mathbf{i} - (0-4)\mathbf{j} + (0-8)\mathbf{k}$$
$$= -4\mathbf{i} + 4\mathbf{j} - 8\mathbf{k}$$

37. The two vectors will be $\pm\dfrac{\mathbf{a} \times \mathbf{b}}{\|\mathbf{a} \times \mathbf{b}\|}$.

$$\mathbf{a} \times \mathbf{b} = \begin{vmatrix} \mathbf{i} & \mathbf{j} & \mathbf{k} \\ 2 & 0 & 1 \\ -1 & 2 & -1 \end{vmatrix}$$
$$= (0-2)\mathbf{i} - (-2+1)\mathbf{j} + (4-0)\mathbf{k}$$
$$= -2\mathbf{i} + \mathbf{j} + 4\mathbf{k}$$
$$\|\mathbf{a} \times \mathbf{b}\| = \sqrt{4+1+16} = \sqrt{21}$$

The unit vectors orthogonal to both given vectors are $\pm\frac{1}{\sqrt{21}}(-2\mathbf{i} + \mathbf{j} + 4\mathbf{k})$.

39. Displacement is $\mathbf{d} = \langle 60 - 1, 22 - 0 \rangle$.

$$W = \mathbf{F} \cdot \mathbf{d} = 40 \cdot 59 + (-30) \cdot 22$$
$$= 1700 \text{ foot-pounds}$$

41. Two points on the line: $t = 0$ gives $P = (1, -1, 3)$ and $t = 1$ gives $R = (2, 1, 3)$.
$\mathbf{PQ} = \langle 0, 0, -3 \rangle$, $\mathbf{PR} = \langle 1, 2, 0 \rangle$
$$\text{proj}_{\mathbf{PR}}\mathbf{PQ} = \frac{\mathbf{PQ} \cdot \mathbf{PR}}{\|\mathbf{PR}\|^2}\mathbf{PR} = \mathbf{0}$$

The distance from point Q to the line through P and R is $\|\mathbf{PQ} - \text{proj}_{\mathbf{PR}}\mathbf{PQ}\|$
$$= \|\langle 0, 0, -3 \rangle\| = 3.$$

43. $\begin{vmatrix} \mathbf{i} & \mathbf{j} & \mathbf{k} \\ 2 & 0 & 1 \\ 0 & 1 & -3 \end{vmatrix} = \langle 0 - 1, -(-6 - 0), 2 - 0 \rangle$

$$\|\langle -1, 6, 2 \rangle\| = \sqrt{41}$$

45. $\|\tau\| = \|\mathbf{r}\|\|\mathbf{F}\|\sin\theta = \frac{6}{12}(50)\sin\frac{\pi}{6} = 25(\frac{1}{2}) = 12.5$ foot-pounds

47. The direction is
$\langle 0 - 2, 2 - (-1), -3 - (-3) \rangle$.

(a) $x = 2 - 2t, y = -1 + 3t, z = -3$

(b) $\dfrac{x - 2}{-2} = \dfrac{y + 1}{3}, z = -3$

49. $\langle 2, \frac{1}{2}, -3 \rangle$ is parallel to the line.

(a) $x = 2 + 2t, y = -1 + \frac{1}{2}t, z = 1 - 3t$

(b) $\dfrac{x - 2}{2} = 2(y + 1) = \dfrac{z - 1}{-3}$

51. $\mathbf{v}_1 = \langle 1, 0, 2 \rangle$ is parallel to the first line.
$\mathbf{v}_2 = \langle 2, 2, 4 \rangle$ is parallel to the second line.

$$\cos \theta = \frac{\mathbf{v}_1 \cdot \mathbf{v}_2}{\|\mathbf{v}_1\| \|\mathbf{v}_2\|} = \frac{10}{\sqrt{5}\sqrt{24}}$$

$$\theta = \cos^{-1} \frac{10}{\sqrt{120}} = \cos^{-1} \frac{5}{\sqrt{30}} \approx 0.42$$

53. $\mathbf{v}_1 = \langle 2, 1, 4 \rangle$ is parallel to the first line.
$\mathbf{v}_2 = \langle 0, 1, 1 \rangle$ is parallel to the second line.
$\mathbf{v}_1 \neq c\mathbf{v}_2$, so the vectors are not parallel. If they intersect, then equating x-coordinates shows that $2t = 4$, so $t = 2$. This makes the point on the first line $(4, 5, 7)$. Now equating y-coordinates shows $5 = 4 + s$, so $s = 1$. This makes the point on the second line $(4, 5, 4)$. Since these points are not the same, the lines are skew.

55. $4(x + 5) + y - 2(z - 1) = 0$

57. $P = (2, 1, 3)$, $Q = (2, -1, 2)$, $R = (3, 3, 2)$
$\mathbf{PQ} = \langle 0, -2, -1 \rangle$, $\mathbf{PR} = \langle 1, 2, -1 \rangle$
$\mathbf{PQ} \times \mathbf{PR} = \langle 4, -1, 2 \rangle$ is normal to the plane.
$4(x - 2) - (y - 1) + 2(z - 3) = 0$

59. Elliptic paraboloid

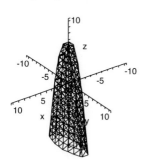

61. Cylinder

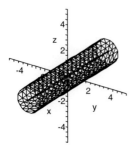

63. Sphere

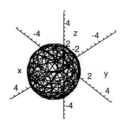

65. Plane

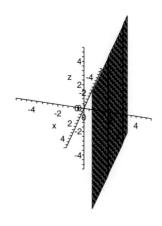

67. Plane

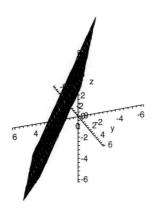

69. Hyperboloid of one sheet

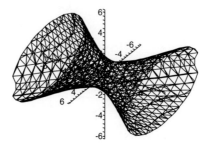

71. Hyperboloid of two sheets

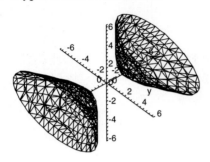

Vector-Valued Functions

12.1 VECTOR-VALUED FUNCTIONS

1. $t = 0 :< \langle 0, 0, -1 \rangle$,
$t = 1 : \langle 3, 1, 1 \rangle$,
$t = 2 : \langle 6, 4, 3 \rangle$

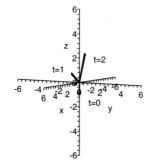

3. $t = \frac{-\pi}{2} : \langle 0, 2, -1 \rangle$,
$t = 0 : \langle 1, 2, -1 \rangle$,
$t = \frac{\pi}{2} : \langle 0, 2, -1 \rangle$

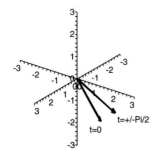

5.

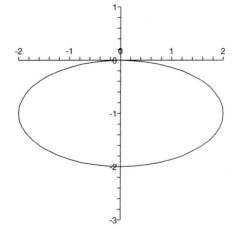

7.

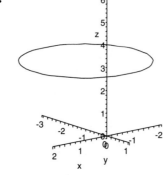

9.

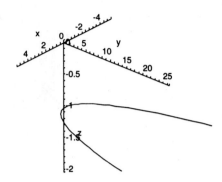

11.

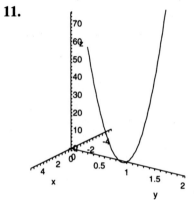

13.

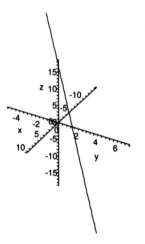

15.

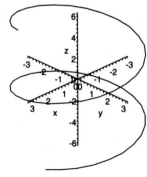

17.

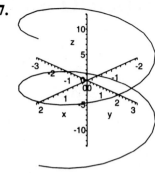

19.

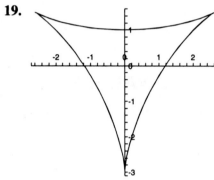

21.

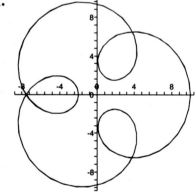

23.

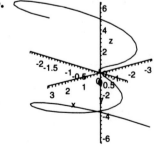

25.

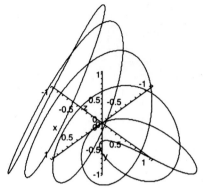

27.

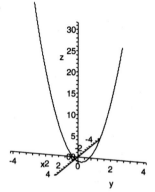

29.

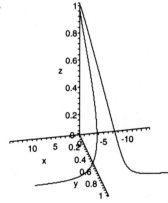

31. (a) Graph F. x is between -1 and 1, and graph follows plane $y = z$.

(b) Graph C. Only graph with all coordinates between -1 and 1.

(c) Graph E. x and y between -1 and 1. Coils spread out as $\sqrt{t}$ grows slower than t.

(d) Graph A. x and y between -1 and 1. Coils get closer as t^2 grows faster than t.

(e) Graph B. Parabola in plane $x = y$.

(f) Graph D. None of the coordinates are bounded.

33.

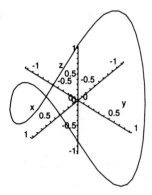

$\mathbf{r}'(t) = \langle -\sin t, \cos t, -2 \sin 2t \rangle$
$s = \int_0^{2\pi} \sqrt{1 + 4 \sin^2 2t}\, dt$
≈ 10.54

35.

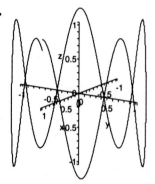

$\mathbf{r}'(t) = \langle -\pi \sin \pi t, \pi \cos \pi t, -16 \sin 16t \rangle$
$s = \int_0^2 \sqrt{\pi^2 + (-16 \sin 16t)^2}\, dt$
≈ 21.56

37.

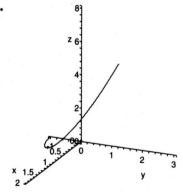

$\mathbf{r}'(t) = \langle 1, 2t, 3t^2 \rangle$
$s = \int_0^2 \sqrt{1 + 4t^2 + 9t^4}\, dt$
≈ 9.57

39. Substituting $\mathbf{r}(t)$ into $z = x^2 - y^2$ gives the trigonometric identity
$\cos 2t = \cos^2 t - \sin^2 t$.

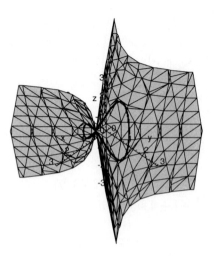

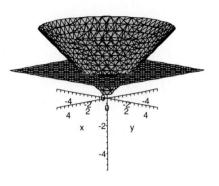

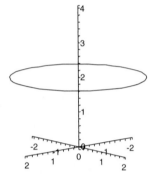

41. The curve

$$\mathbf{r}(t) = \langle t, t^2 - 1, t^3 \rangle, 0 \le t \le 2,$$

can be parametrized as

$$\mathbf{r}(s) = \langle 2s, 4s^2 - 1, 8s^3 \rangle, 0 \le s \le 1,$$

by substituting $s = \frac{t}{2}$. Since both equations represent exactly the same curve, they will give us the same arc length.

43. The curve $\mathbf{g}(t) = \langle \cos t, \cos^2 t, \cos^2 t \rangle$ covers the part of $\mathbf{r}(t) = \langle t, t^2, t^2 \rangle$ with coordinates between -1 and 1. $\mathbf{h}(t) = \langle \sqrt{t}, t, t \rangle$ covers the part of $\mathbf{r}(t)$ with positive x-coordinate.

45. Substitute $z = 2$ into $z = \sqrt{x^2 + y^2}$ to see that the intersection is the circle $x^2 + y^2 = 4$ in the plane $z = 2$. This curve can be parameterized as
$x = 2 \cos t, y = 2 \sin t,$
$z = t, 0 \le t \le 2\pi.$

47. Solve for x and z in terms of y, and use $y = t$ as a parameter to get
$x = \pm\sqrt{9 - t^2}, y = t$, and $z = 2 - t$.

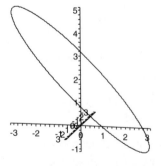

49. Possible parametric equations for the helix are $x = 3 \cos t$, $y = 3 \sin t$, and $z = \dfrac{10t}{4\pi}$ for $0 \le t \le 4\pi$. The length is then

$$s = \int_0^{4\pi} \sqrt{9 \sin^2 t + 9 \cos^2 t + \frac{100}{16\pi^2}} \, dt$$

$$\approx 39.00 \text{ feet.}$$

51.

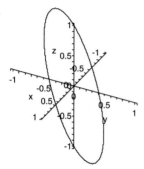

The components all have period 2π, so for larger ranges the graph retraces this. As you use large and larger domains, the graphing utility uses the same number of subdivisions, and so it "connects-the-dots" of points farther and farther apart. This causes the graph to look like a tangle of line segments.

53. If we unroll the helix to lay it flat, the arc length will be the hypotenuse of a right triangle with height 2π and base 2π, and so it will have length $2\pi\sqrt{2}$.

12.2 THE CALCULUS OF VECTOR-VALUED FUNCTIONS

1. The component functions are all continuous at $t = 0$. Therefore,
$$\lim_{t \to 0} \langle t^2 - 1, e^{2t}, \sin t \rangle = \langle -1, 1, 0 \rangle.$$

3. The limits of the component functions all exist. Therefore,
$$\lim_{t \to 0} \langle \frac{\sin t}{t}, \cos t, \frac{t+1}{t-1} \rangle = \langle 1, 1, -1 \rangle.$$

5. $\lim_{t \to 0} \ln t$ does not exist. Therefore,
$$\lim_{t \to 0} \langle \ln t, \sqrt{t^2 + 1}, t - 3 \rangle \text{ does not exist.}$$

7. $\dfrac{t+1}{t-1}$ is not continuous at $t = 1$. The component functions are all continuous at $t \neq 1$. Therefore, $\mathbf{r}(t)$ is continuous for all $t \neq 1$.

9. $\sin t^2$, $\cos t$, are continuous everywhere. However, $\tan t$ is continuous everywhere except $t = \frac{n\pi}{2}$ for all n odd. Therefore, $\mathbf{r}(t)$ is continuous for all $t \neq \frac{n\pi}{2}$ where n is odd.

11. $\sqrt{t}$ is only defined for $t \geq 0$. The component functions are all continuous for $t \geq 0$. Therefore $\mathbf{r}(t)$ is continuous for all $t \geq 0$.

13. $\mathbf{r}'(t) = \langle 4t^3, \dfrac{1}{2\sqrt{t+1}}, \dfrac{-6}{t^3} \rangle$

15. $\mathbf{r}'(t) = \langle \cos t, 2t \cos t^2, -\sin t \rangle$

17. $\mathbf{r}'(t) = \langle 2t e^{t^2}, 2t, 2 \sec 2t \tan 2t \rangle$

19.

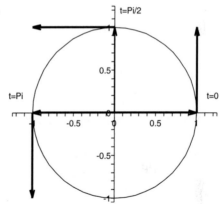

21.

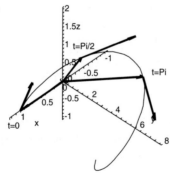

23. $\langle \dfrac{3}{2}t^2 - t, \dfrac{2}{3}t^{3/2} \rangle + \mathbf{c}$ where $\mathbf{c}$ is an arbitrary constant vector.

25. $\langle \frac{1}{3} \sin 3t, -\cos t, \frac{1}{4}e^{4t} \rangle + \mathbf{c}$ where $\mathbf{c}$ is an arbitrary constant vector.

27. Using $u = t^2$ in the first component, integration by parts with $u = t$ and $dv = \sin t$ in the second component, and $u = t^2 + 1$ in the third component of the vector gives
$\langle \frac{1}{2}e^{t^2}, 3(\sin t - t \cos t), \frac{3}{2} \ln(t^2 + 1) \rangle + \mathbf{c}$
where $\mathbf{c}$ is an arbitrary constant vector.

29. $\langle \frac{-2}{3}, \frac{3}{2} \rangle$

31. Using integration by parts where $u = t$ and $dv = e^t$ to integrate the third component of the vector yields $\langle 4 \ln 3, 1 - e^{-2}, e^2 + 1 \rangle$.

33. $\mathbf{r}(t) = \langle \cos t, \sin t \rangle$ and $\mathbf{r}'(t) = \langle -\sin t, \cos t \rangle$ are perpendicular for all t.

35. $\mathbf{r}(t) = \langle t, t, t^2 - 1 \rangle$ and $\mathbf{r}'(t) = \langle 1, 1, 2t \rangle$. $\mathbf{r}(t) \cdot \mathbf{r}'(t) = 2t^3 = 0$ only when $t = 0$. Therefore $\mathbf{r}(t)$ and $\mathbf{r}'(t)$ are perpendicular only when $t = 0$.

37. In exercise 33 the vectors are perpendicular for all t so they cannot be parallel.

In exercise 34 if there existed a t_0 and $k \neq 0$ such that
$\langle 2 \cos t_0, \sin t_0 \rangle = k \langle -2 \sin t_0, \cos t_0 \rangle$, then we would have $\sin t_0 = k \cos t_0$ and $\cos t_0 = -k \sin t_0$. This would mean that $\sin t_0 = -k^2 \sin t_0$, and this cannot happen for real k unless $\sin t_0 = 0$. But, if $\sin t_0 = 0$, then $\cos t_0 \neq 0$, so this cannot occur.

39. $\mathbf{r}'(t)$ is parallel to the xy-plane when the third component is 0. $\frac{d}{dt}(t^3 - 3) = 3t^2 = 0$ when $t = 0$.

41. $\mathbf{r}'(t)$ is parallel to the xy-plane when the third component is 0. $\frac{d}{dt} \sin 2t = 2 \cos 2t = 0$ when $t = \frac{n\pi}{4}$ for any odd integer n.

43. For $\mathbf{r}(t) = \langle f(t), g(t), h(t) \rangle$,
$$\frac{d}{dt} \langle cf(t), cg(t), ch(t) \rangle$$

$$= \langle c\frac{d}{dt}f(t), \frac{d}{dt}g(t), \frac{d}{dt}h(t) \rangle = c\mathbf{r}'(t).$$

45. (iii) For scalar $f(t)$ and $\mathbf{r}(t) = \langle u(t), v(t), w(t) \rangle$, we have
$$\frac{d}{dt}[f(t)\mathbf{r}(t)]$$
$$= \frac{d}{dt} \langle f(t)u(t), f(t)v(t), f(t)w(t) \rangle$$
$$= \langle f'(t)u(t) + f(t)u'(t),$$
$$f'(t)v(t) + f(t)v'(t),$$
$$f'(t)w(t) + f(t)w'(t) \rangle$$
$$= \langle f'(t)u(t), f'(t)v(t), f'(t)w(t) \rangle$$
$$+ \langle f(t)u'(t), f(t)v'(t), f(t)w'(t) \rangle$$
$$= f'(t)\mathbf{r}(t) + f(t)\mathbf{r}'(t)$$

(iv) For $\mathbf{r}(t) = \langle f_1(t), g_1(t), h_1(t) \rangle$, and $\mathbf{s}(t) = \langle f_2(t), g_2(t), h_2(t) \rangle$, we have
$$\frac{d}{dt}[\mathbf{r}(t) \cdot \mathbf{s}(t)]$$
$$= \frac{d}{dt} \langle f_1(t)f_2(t), g_1(t)g_2(t), h_1(t)h_2(t) \rangle$$
$$= \langle f_1'(t)f_2(t) + f_1(t)f_2'(t),$$
$$g_1'(t)g_2(t) + g_1(t)g_2'(t),$$
$$h_1'(t)h_2(t) + h_1(t)h_2'(t) \rangle$$
$$= \langle f_1'(t)f_2(t), g_1'(t)g_2(t), h_1'(t)h_2(t) \rangle$$
$$+ \langle f_1(t)f_2'(t), g_1(t)g_2'(t), h_1(t)h_2'(t) \rangle$$
$$= \mathbf{r}'(t) \cdot \mathbf{s}(t) + \mathbf{r}(t) \cdot \mathbf{s}'(t)$$

47. False. For any function $\mathbf{r}(t)$, $\mathbf{u}(t)$ is a unit vector for all t, and $\|\mathbf{u}(t)\| = 1$ is constant. So by Theorem 2.4, $\mathbf{u}(t) \cdot \mathbf{u}'(t)$ is 0 for any function $\mathbf{r}(t)$, but there are clearly functions $\mathbf{r}(t)$ with $\|\mathbf{r}(t)\|$ not constant, so for these functions $\mathbf{r}(t) \cdot \mathbf{r}'(t)$ will not equal 0.

49. Recall from part (i) of Theorem 2.4 that $\mathbf{r}(t) \cdot \mathbf{r}(t) = \|\mathbf{r}(t)\|^2$ and that $\frac{d}{dt}[\mathbf{r}(t) \cdot \mathbf{r}(t)] = 2\mathbf{r}(t) \cdot \mathbf{r}'(t)$. If $\mathbf{r}(t) \cdot \mathbf{r}'(t) = 0$ for all t, then $\|\mathbf{r}(t)\|^2$ is constant, so $\|\mathbf{r}(t)\|$ must be constant.

51. We have

$$QQ' = \langle a(\cos(t - \tfrac{\pi}{2}) - \cos(t + \tfrac{\pi}{2})),$$

$$b(\sin(t - \tfrac{\pi}{2}) - \sin(t + \tfrac{\pi}{2}))\rangle$$

$$= \langle 2a \sin t \sin \tfrac{\pi}{2}, -2b \cos t \sin \tfrac{\pi}{2}\rangle$$

$$= -2 \sin \tfrac{\pi}{2} \langle -a \sin t, b \cos t\rangle.$$

The tangent vector at P is $\langle -a \sin t, b \cos t\rangle$, so we see that QQ' is a multiple of the tangent vector at P.

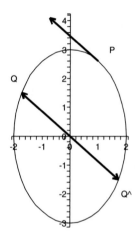

53. Using parts (iv) and (v) of Theorem 2.3 we see that

$$\frac{d}{dt}[\mathbf{f}(t) \cdot (\mathbf{g}(t) \times \mathbf{h}(t))]$$

$$= \mathbf{f}'(t) \cdot [\mathbf{g}(t) \times \mathbf{h}(t)]$$
$$\quad + \mathbf{f}(t) \cdot \frac{d}{dt}[\mathbf{g}(t) \times \mathbf{h}(t)]$$

$$= \mathbf{f}'(t) \cdot [\mathbf{g}(t) \times \mathbf{h}(t)]$$
$$\quad + \mathbf{f}(t) \cdot [\mathbf{g}'(t) \times \mathbf{h}(t) + \mathbf{g}(t) \times \mathbf{h}'(t)].$$

12.3 MOTION IN SPACE

1. $\mathbf{v}(t) = \mathbf{r}'(t) = \langle -10 \sin 2t, 10 \cos 2t\rangle$
$\mathbf{a}(t) = \mathbf{r}''(t) = \langle -20 \cos 2t, -20 \sin 2t\rangle$

3. $\mathbf{v}(t) = \mathbf{r}'(t) = \langle 25, -32t + 15\rangle$
$\mathbf{a}(t) = \mathbf{r}''(t) = \langle 0, -32\rangle$

5. $\mathbf{v}(t) = \mathbf{r}'(t) = \langle 4e^{-2t} - 8te^{-2t}, -4e^{-2t}, -32t\rangle$
$\mathbf{a}(t) = \mathbf{r}''(t)$
$\quad = \langle -16e^{-2t} + 16te^{-2t}, 8e^{-2t}, -32\rangle$

7. $\mathbf{r}(t) = \int \mathbf{v}(t) \, dt = \langle 10t + c_1, -16t^2 + 4t + c_2\rangle$
$\mathbf{r}(0) = \langle 3, 8\rangle \Rightarrow c_1 = 3$ and $c_2 = 8$
$\mathbf{r}(t) = \langle 10t + 3, -16t^2 + 4t + 8\rangle$

9. $\mathbf{v}(t) = \int \mathbf{a}(t) \, dt = \langle b_1, -32t + b_2\rangle$
$\mathbf{v}(0) = \langle 5, 0\rangle \Rightarrow b_1 = 5$ and $b_2 = 0$
$\mathbf{v}(t) = \langle 5, -32t\rangle$
$\mathbf{r}(t) = \int \mathbf{v}(t) \, dt = \langle 5t + c_1, -16t^2 + c_2\rangle$
$\mathbf{r}(0) = \langle 0, 16\rangle \Rightarrow c_1 = 0$ and $c_2 = 16$
$\mathbf{r}(t) = \langle 5t, -16t^2 + 16\rangle$

11. $\mathbf{r}(t) = \int \mathbf{v}(t) \, dt$
$$\quad = \langle 10t + c_1, -3e^{-t} + c_2, -16t^2 + 4t + c_3\rangle$$
$\mathbf{r}(0) = \langle 0, -6, 20\rangle \Rightarrow$
$\quad c_1 = 0, c_2 = -3$ and $c_3 = 20$
$\mathbf{r}(t) = \langle 10t, -3e^{-t} - 3, -16t^2 + 4t + 20\rangle$

13. $\mathbf{v}(t) = \int \mathbf{a}(t) \, dt$
$$\quad = \langle \tfrac{1}{2}t^2 + b_1, b_2, -16t + b_3\rangle$$
$\mathbf{v}(0) = \langle 12, -4, 0\rangle \Rightarrow$
$\quad b_1 = 12, b_2 = -4$ and $b_3 = 0$
$\mathbf{v}(t) = \langle \tfrac{1}{2}t^2 + 12, -4, -16t\rangle$
$\mathbf{r}(t) = \int \mathbf{v}(t) \, dt$
$$\quad = \langle \tfrac{1}{6}t^3 + 12t + c_1, -4t + c_2, -8t^2 + c_3\rangle$$
$\mathbf{r}(0) = \langle 5, 0, 2\rangle \Rightarrow$
$\quad c_1 = 5, c_2 = 0$ and $c_3 = 2$
$\mathbf{r}(t) = \langle \tfrac{t^3}{6} + 12t + 5, -4t, -8t^2 + 2\rangle$

15. Since $\mathbf{F}(t) = m\mathbf{a}(t) = m\mathbf{r}''(t)$,
$\mathbf{F}(t) = -160\langle \cos 2t, \sin 2t \rangle$.

17. Since $\mathbf{F}(t) = m\mathbf{a}(t) = m\mathbf{r}''(t)$,
$\mathbf{F}(t) = -960\langle \cos 4t, \sin 4t \rangle$.

19. Since $\mathbf{F}(t) = m\mathbf{a}(t) = m\mathbf{r}''(t)$,
$\mathbf{F}(t) = \langle -120 \cos 2t, -200 \sin 2t \rangle$.

21. Since $\mathbf{F}(t) = m\mathbf{a}(t) = m\mathbf{r}''(t)$,
$\mathbf{F}(t) = \langle 60, 0 \rangle$.

23. $\mathbf{a}(t) = -32\mathbf{j}$

$\mathbf{v}(0) = 100\langle \cos \frac{\pi}{3}, \sin \frac{\pi}{3} \rangle = \langle 50, 50\sqrt{3} \rangle$

$\mathbf{v}(t) = 50\mathbf{i} + (50\sqrt{3} - 32t)\mathbf{j}$

$\mathbf{r}(0) = \langle 0, 0 \rangle$

$\mathbf{r}(t) = 50t\mathbf{i} + (50\sqrt{3}t - 16t^2)\mathbf{j}$

Maximum altitude occurs when vertical component of velocity is 0, which occurs at $t = \frac{50\sqrt{3}}{32}$. $(50\sqrt{3}t - 16t^2)|_{\frac{50\sqrt{3}}{32}} \approx 117.2$ ft is the maximum altitude.
Impact occurs when the height is 0, which occurs at $t = \frac{50\sqrt{3}}{16}$.
The horizontal range is $50t|_{\frac{50\sqrt{3}}{16}} \approx 270.6$.
The speed at impact is $\|\mathbf{v}(\frac{50\sqrt{3}}{16})\| = 100\frac{\text{ft}}{\text{s}}$.

25. $\mathbf{a}(t) = -32\mathbf{j}$

$\mathbf{v}(0) = 160\langle \cos \frac{\pi}{4}, \sin \frac{\pi}{4} \rangle = \langle 80\sqrt{2}, 80\sqrt{2} \rangle$

$\mathbf{v}(t) = 80\sqrt{2}\mathbf{i} + (80\sqrt{2} - 32t)\mathbf{j}$

$\mathbf{r}(0) = \langle 0, 10 \rangle$

$\mathbf{r}(t) = 80\sqrt{2}t\mathbf{i} + (80\sqrt{2}t - 16t^2 + 10)\mathbf{j}$

$80\sqrt{2}t - 16t^2 + 10|_{\frac{80\sqrt{2}}{32}} = 210$ ft is the maximum altitude.
$80\sqrt{2}t|_{\frac{80\sqrt{2}+8\sqrt{210}}{32}} \approx 809.9$ ft is the horizontal range.
$\|\mathbf{v}(\frac{80\sqrt{2}+8\sqrt{210}}{32})\| \approx 162\frac{\text{ft}}{\text{s}}$ is the speed at impact.

27. $\mathbf{a}(t) = -32\mathbf{j}$

$\mathbf{v}(0) = 320\langle \cos \frac{\pi}{4}, \sin \frac{\pi}{4} \rangle = \langle 160\sqrt{2}, 160\sqrt{2} \rangle$

$\mathbf{v}(t) = 160\sqrt{2}\mathbf{i} + (160\sqrt{2} - 32t)\mathbf{j}$

$\mathbf{r}(0) = \langle 0, 10 \rangle$

$\mathbf{r}(t) = 160\sqrt{2}t\mathbf{i} + (160\sqrt{2}t - 16t^2 + 10)\mathbf{j}$

$160\sqrt{2}t - 16t^2 + 10|_{\frac{160\sqrt{2}}{32}} = 810$ ft is the maximum altitude.
$160\sqrt{2}t|_{\frac{160\sqrt{2}+72\sqrt{10}}{32}} \approx 3210$ ft is the horizontal range.
$\|\mathbf{v}(\frac{160\sqrt{2}+72\sqrt{10}}{32})\| \approx 321\frac{\text{ft}}{\text{s}}$ is the speed at impact.

29. Doubling the initial speed approximately quadruples the horizontal range.

31. $\mathbf{a}(t) = -g\mathbf{j}$

$\mathbf{v}(0) = v_0\langle \cos \theta, \sin \theta \rangle$

$\mathbf{v}(t) = v_0 \cos \theta\mathbf{i} + (v_0 \sin \theta - gt)\mathbf{j}$

$\mathbf{r}(0) = \langle 0, h \rangle$

$\mathbf{r}(t) = v_0 \cos \theta\mathbf{i} + (v_0 \sin \theta t - \frac{g}{2}t^2 + h)\mathbf{j}$

33. $\mathbf{a}(t) = -32\mathbf{j}$

$\mathbf{v}(0) = 120\langle \cos(30°), \sin(30°) \rangle$

$\mathbf{v}(t) = 60\sqrt{3}\mathbf{i} + (60 - 32t)\mathbf{j}$

$\mathbf{r}(0) = \langle 0, 3 \rangle$

$\mathbf{r}(t) = 60\sqrt{3}t\mathbf{i} + (3 + 60t - 16t^2)\mathbf{j}$

$3 + 60t - 16t^2 = 6$ when $t \approx 0.051, 3.699$.
Since $60\sqrt{3}t|_{3.699} \approx 384.4$ ft, we see that this is just short of a home run.

Alternatively, $60\sqrt{3}t = 385$ when $t = \frac{77\sqrt{3}}{36}$.
Since $(3 + 60t - 16t^2)|_{\frac{77\sqrt{3}}{36}} \approx 5.7$, the ball hits the wall 5.7 feet from the ground.

35. $\mathbf{v}(0) = \langle 130, 0 \rangle$ and $\mathbf{a}(t) = -32\mathbf{j}$

$\mathbf{v}(t) = \langle 130, -32t \rangle$ and $\mathbf{r}(0) = \langle 0, 6 \rangle$

$\mathbf{r}(t) = \langle 130t, 6 - 16t^2 \rangle$.

The ball reaches the plate when $130t = 60$, so that $t = \frac{6}{13}$. The height of the ball then is $6 - 16(\frac{6}{13})^2 \approx 2.59$ ft.

37. $\mathbf{a}(t) = -32\mathbf{j}$ and $\mathbf{v}(0) = \langle 120, 0 \rangle$
$\mathbf{v}(t) = \langle 120, -32t \rangle$ and $\mathbf{r}(0) = \langle 0, 8 \rangle$
$\mathbf{r}(t) = \langle 120t, 8 - 16t^2 \rangle$

The ball reaches the net when $120t = 39$, so that $t = \frac{39}{120}$. The ball is at height $8 - 16(\frac{39}{120})^2 \approx 6.31$ ft at this time, so it easily clears the net. The ball lands when $8 - 16t^2 = 0$ or when $t = \frac{1}{\sqrt{2}}$ and goes a horizontal distance of $120\frac{1}{\sqrt{2}} \approx 84.9$ ft. The serve is long.

39. Since 55 mph $= \frac{242}{3}\frac{\text{ft}}{\text{s}}$, we have that $\mathbf{r}(t) = \langle \frac{242}{3}\cos(50°)t, \frac{242}{3}\sin(50°)t - 16t^2 \rangle$. The punt hits the ground when $\frac{242}{3}\sin(50°)t - 16t^2 = 0$, so the hang time is $t = \frac{242\sin(50°)}{48} \approx 3.86$ sec.

41. Since the projectile has mass 1 slug and the force of the wind is 8 pounds north, the acceleration is $\mathbf{a}(t) = \langle 0, 8, -32 \rangle$.
The velocity and position then become
$\mathbf{v}(t) = \langle 50, 8t, 50\sqrt{3} - 32t \rangle$ and
$\mathbf{r}(t) = \langle 50t, 4t^2, 50\sqrt{3}t - 16t^2 \rangle$. The object lands when $t = \frac{50\sqrt{3}}{16}$ at
$\mathbf{r}(\frac{50\sqrt{3}}{16}) \approx \langle 270.6, 117.2, 0 \rangle$.

43. We have $\mathbf{a} = \mathbf{g} + \mathbf{w} + \mathbf{e} = \langle 2t, 1, -8 \rangle$ for $0 \leq t \leq 1$, so that $\mathbf{v}(t) = \int \mathbf{a}(t)\,dt = \langle t^2 + c_1, t + c_2, -8t + c_3 \rangle$.
Since $\mathbf{v}(0) = \langle 100, 0, 10 \rangle$, we have that $\mathbf{v}(t) = \langle t^2 + 100, t, -8t + 10 \rangle$ for $0 \leq t \leq 1$.
For $t > 1$, we have $\mathbf{a} = \langle 2t, 2, -8 \rangle$, so that $\mathbf{v}(t) = \int \mathbf{a}(t)\,dt = \langle t^2 + a, 2t + b, -8t + c \rangle$. The velocity should be continuous, so that the movement of the airplane is relatively smooth, i.e., $\lim_{t \to 1^+} \mathbf{v}(t) = \mathbf{v}(1) = \langle 101, 1, 2 \rangle$. This gives $\mathbf{v}(t) = \langle t^2 + 100, 2t - 1, -8t + 10 \rangle$.

The velocity function is not differentiable at $t = 1$, since the acceleration (and force) function is discontinuous at that point.

45. Given $\mathbf{r}(t) = \langle 100\cos \omega t, 100\sin \omega t \rangle$, we need to find the value of ω that produces $\|\mathbf{a}(t)\| = 32\frac{\text{ft}}{\text{sec}}$, and then find $\|\mathbf{v}(t)\|$.
$\|\mathbf{a}(t)\| = 100\omega^2 = 32 \Rightarrow \omega = \sqrt{\frac{32}{100}}$, then $\|\mathbf{v}(t)\| = 100\omega \approx 56.57$ ft/s.

47. For circular motion $\|\mathbf{a}\| = r\omega^2$ and $\|\mathbf{v}\| = r\omega$. Solving the equations $r\omega^2 = 5g = 5(9.8\frac{\text{m}}{\text{sec}^2}) = 5(127{,}008\frac{\text{km}}{\text{hr}^2})$ and $r\omega = 900\frac{\text{km}}{\text{hr}}$ gives $r = 1.2755$ km $= 1275.5$ m.

49. The torque has magnitude $\tau = (20)(5) = 100$ foot $=$ pounds. Since $\tau = I\alpha$, we get that $\alpha = 10$ for $0 \leq t \leq 0.5$. The change in angular velocity is given by $\int_0^{0.5} \alpha\,dt = 5\frac{\text{rad}}{\text{sec}}$.

51. Since $\theta(t) = \frac{\alpha t^2}{2}$, we know that $\theta = 0$ and $\theta = \pi$ correspond to $t = 0$ and $t = \sqrt{\frac{2\pi}{\alpha}}$. The change in angular velocity is then $\int_0^{\sqrt{\frac{2\pi}{\alpha}}} \alpha\,dt = 15$, and $\alpha\sqrt{\frac{2\pi}{\alpha}} = 15$, so that $\alpha = \frac{225}{2\pi}\frac{\text{rad}}{s^2}$.

53. $\theta(t) = \frac{\alpha t^2}{2} = 2\pi$ gives $t = \sqrt{\frac{4\pi}{\alpha}}$
$\theta(t) = 4\pi$ gives $t = \sqrt{\frac{8\pi}{\alpha}}$.
The angular speed for a rotation of 2π is then $\int_0^{\sqrt{\frac{4\pi}{\alpha}}} \alpha\,dt = \alpha\sqrt{\frac{4\pi}{\alpha}}$.
The angular speed for a rotation of 4π is then $\int_0^{\sqrt{\frac{8\pi}{\alpha}}} \alpha\,dt = \alpha\sqrt{\frac{8\pi}{\alpha}}$, a factor of $\sqrt{2}$ more.

55. Since the derivative of angular momentum, $\mathbf{L}'(t) = \tau$, if the net torque $\tau = 0$, the angular momentum, $\mathbf{L}(t)$, must be a constant.

57. Torque is defined to be $\tau = \mathbf{r} \times \mathbf{F}$. Recall that the cross product of parallel vectors is $\mathbf{0}$, and that force is parallel to acceleration. If acceleration is parallel to position, then so is force, and the torque is **0**.

If acceleration is parallel to the position, then this will change the linear velocity, not the angular velocity. Since the angular velocity does not change, there will be no change in angular momentum.

59. For geosynchronous orbit,

$$\omega = \frac{2\pi \, \text{rad}}{\text{sidereal day}} = \frac{2\pi \, \text{rad}}{86,164 \text{sec}}.$$

Solving for b now gives

$$b = \sqrt[3]{\frac{39.87187 \times 10^{13}}{\left(\frac{2\pi}{86,164}\right)^2}} \approx 42,168.3 \text{ km}$$

61. If the projectile passes through point (x_1, y_1, z_1) at time t_1 and point (x_2, y_2, z_2) at time t_2, then we can find the initial position $(x_0, y_0, 0)$.

The acceleration, velocity and position are given by
$\mathbf{a}(t) = \langle 0, 0, -32 \rangle$,
$\mathbf{v}(t) = \langle c_1, c_2, c_3 - 32t \rangle$, and
$\mathbf{x}(t) = \langle x_0 + c_1 t, y_0 + c_2 t, c_3 t - 16t^2 \rangle$
(where $\langle c_1, c_2, c_3 \rangle$ is the initial velocity).

$(x_1, y_1, z_1) = \langle x_0 + c_1 t_1, y_0 + c_2 t_1, c_3 t_1 - 16t_1^2 \rangle$
$(x_2, y_2, z_2) = \langle x_0 + c_1 t_2, y_0 + c_2 t_2, c_3 t_2 - 16t_2^2 \rangle$
Use the x-coordinates to solve for $c_1 = \frac{x_2 - x_1}{t_2 - t_1}$
and the y-coordinates to write $c_2 = \frac{y_2 - y_1}{t_2 - t_1}$.
Substitute these into either known point and simplify to find that
$x_0 = \frac{x_1 t_2 - x_2 t_1}{t_2 - t_1}$ and $y_0 = \frac{y_1 t_2 - y_2 t_1}{t_2 - t_1}$.

63. A satellite 15,000 mi above the earth's surface travels at a velocity of $2.24 \frac{\text{mi}}{\text{s}}$. A satellite 20,000 mi above the earth's surface travels at a velocity of $2.00 \frac{\text{mi}}{\text{s}}$. The velocity must decrease by $0.24 \frac{\text{mi}}{\text{s}}$ for the height of the orbit to increase from 15,000 mi to 20,000 mi.

12.4 CURVATURE

1. A circle of radius 2 centered at the origin is parameterized by the equations
$x = 2 \cos t, y = 2 \sin t$ for $0 \le t \le 2\pi$.
$s = \int_0^t \sqrt{(-2 \sin u)^2 + (2 \cos u)^2} \, du = 2t$.
Therefore, $t = \frac{s}{2}$ so that the curve is parameterized by
$x = 2 \cos(\frac{s}{2})$, $y = 2 \sin(\frac{s}{2})$, for $0 \le s \le 4\pi$.

3. The line segment from $(0, 0)$ to $(3, 4)$ can be parameterized by
$x = 3t, y = 4t$ for $0 \le t \le 1$.
$s = \int_0^t \sqrt{3^2 + 4^2} \, du = 5t$. Therefore, $t = \frac{s}{5}$
so that the curve is parameterized by
$x = \frac{3}{5}s, y = \frac{4}{5}s$ for $0 \le s \le 5$.

5. $\mathbf{r}'(t) = \langle 3, 2t \rangle$ and $\|\mathbf{r}'(t)\| = \sqrt{9 + 4t^2}$, so that
$\mathbf{T}(t) = \langle \frac{3}{\sqrt{9+4t^2}}, \frac{2t}{\sqrt{9+4t^2}} \rangle$.
$\mathbf{T}(0) = \langle 1, 0 \rangle$, $\mathbf{T}(-1) = \langle \frac{3}{\sqrt{13}}, \frac{-2}{\sqrt{13}} \rangle$, and
$\mathbf{T}(1) = \langle \frac{3}{\sqrt{13}}, \frac{2}{\sqrt{13}} \rangle$.

7. $\mathbf{r}'(t) = \langle -3 \sin t, 2 \cos t \rangle$ and
$\|\mathbf{r}'(t)\| = \sqrt{9 \sin^2 t + 4 \cos^2 t}$, so that
$\mathbf{T}(t) = \langle \frac{-3 \sin t}{\sqrt{9 \sin^2 t + 4 \cos^2 t}}, \frac{2 \cos t}{\sqrt{9 \sin^2 t + 4 \cos^2 t}} \rangle$.
$\mathbf{T}(0) = \langle 0, 1 \rangle$, $\mathbf{T}(-\frac{\pi}{2}) = \langle 1, 0 \rangle$, and
$\mathbf{T}(\frac{\pi}{2}) = \langle -1, 0 \rangle$,

9. $\mathbf{r}'(t) = \langle 3, -2 \sin 2t, 2 \cos 2t \rangle$ and
$\|\mathbf{r}'(t)\| = \sqrt{9 + 4 \sin^2 2t + 4 \cos^2 2t} = \sqrt{13}$
so that $\mathbf{T}(t) = \langle \frac{3}{\sqrt{13}}, \frac{-2 \sin 2t}{\sqrt{13}}, \frac{2 \cos 2t}{\sqrt{13}} \rangle$.
$\mathbf{T}(0) = \mathbf{T}(-\pi) = \mathbf{T}(\pi) = \frac{1}{\sqrt{13}} \langle 3, 0, 2 \rangle$.

11.

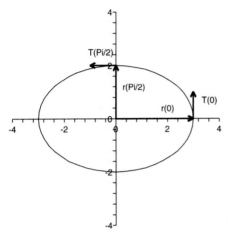

13.

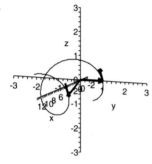

15. $\mathbf{r}'(t) = \langle -2e^{-2t}, 2, 0 \rangle$ and
$\mathbf{r}''(t) = \langle 4e^{-2t}, 0, 0 \rangle$, so $\mathbf{r}'(0) = \langle -2, 2, 0 \rangle$
and $\mathbf{r}''(0) = \langle 4, 0, 0 \rangle$.

$$\mathbf{r}' \times \mathbf{r}''|_{t=0} = \begin{vmatrix} \mathbf{i} & \mathbf{j} & \mathbf{k} \\ -2 & 2 & 0 \\ 4 & 0 & 0 \end{vmatrix} = \langle 0, 0, -8 \rangle$$

$$\kappa = \frac{\|\mathbf{r}' \times \mathbf{r}''\|}{\|\mathbf{r}'\|^3} = \frac{8}{8^{3/2}} = 2^{-3/2} \approx 0.3536$$

17. $\mathbf{r}'(t) = \langle 1, 2\cos 2t, 3 \rangle$ and
$\mathbf{r}''(t) = \langle 0, -4\sin 2t, 0 \rangle$, so
$\mathbf{r}'(0) = \langle 1, 2, 3 \rangle$ and $\mathbf{r}''(0) = \langle 0, 0, 0 \rangle$.

$$\mathbf{r}' \times \mathbf{r}''|_{t=0} = \begin{vmatrix} \mathbf{i} & \mathbf{j} & \mathbf{k} \\ 1 & 2 & 3 \\ 0 & 0 & 0 \end{vmatrix} = \langle 0, 0, 0 \rangle$$

$$\kappa = \frac{\|\mathbf{r}' \times \mathbf{r}''\|}{\|\mathbf{r}'\|^3} = 0$$

19. $f'(x) = 6x$ and $f''(x) = 6$.
$$\kappa = \frac{|f''(1)|}{(1+(f'(1))^2)^{3/2}} = \frac{6}{37^{3/2}} \approx 0.0267$$

21. $f'(x) = \cos x$ and $f''(x) = -\sin x$.

$$\kappa = \frac{|f''(\frac{\pi}{2})|}{(1+(f'(\frac{\pi}{2}))^2)^{3/2}} = 1$$

23. Referring to exercise 21, we see the curvature
at $x = \frac{3\pi}{2}$ is $\kappa = \frac{|-\sin\frac{3\pi}{2}|}{(1+\cos^2\frac{3\pi}{2})^{3/2}} = 1.$

This makes sense due to the symmetry of the
graph. The graph of $\sin x$ is almost straight at
$x = \pi$, so we predict the curvature will be less
there.

25.

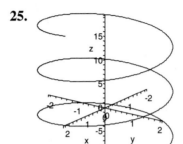

$\mathbf{r}'(t) = \langle -4\sin 2t, 4\cos 2t, 3 \rangle$ and
$\mathbf{r}''(t) = \langle -8\cos 2t, -8\sin 2t, 0 \rangle$.

$$\mathbf{r}' \times \mathbf{r}'' = \begin{vmatrix} \mathbf{i} & \mathbf{j} & \mathbf{k} \\ -4\sin 2t & 4\cos 2t & 3 \\ -8\cos 2t & -8\sin 2t & 0 \end{vmatrix}$$

$$= \langle 24\sin 2t, -24\cos 2t, 32 \rangle$$

$\|\mathbf{r}' \times \mathbf{r}''\| = \sqrt{24^2 + 32^2} = 40$ and
$\|\mathbf{r}'(t)\| = \sqrt{4^2 + 3^2} = 5$, so
$\kappa = \frac{40}{5^3} = \frac{8}{25}$ for all t.

27.

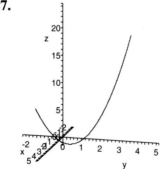

$\mathbf{r}'(t) = \langle 1, 1, 2t \rangle$ and $\mathbf{r}''(t) = \langle 0, 0, 2 \rangle$.

$$\mathbf{r}' \times \mathbf{r}'' = \begin{vmatrix} \mathbf{i} & \mathbf{j} & \mathbf{k} \\ 1 & 1 & 2t \\ 0 & 0 & 2 \end{vmatrix} = \langle 2, -2, 0 \rangle$$

$\|\mathbf{r}' \times \mathbf{r}''\| = \sqrt{8}$ and

$\|\mathbf{r}'(t)\| = \sqrt{2 + 4t^2}$, so

$\kappa(t) = \frac{\sqrt{8}}{(\sqrt{2+4t^2})^3}$. $\kappa(0) = 1$ and $\kappa(2) = \frac{1}{27}$.

29.

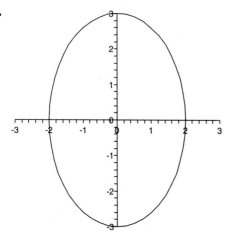

$\mathbf{r}'(t) = \langle -2 \sin t, 3 \cos t \rangle$ and

$\|\mathbf{r}'(t)\| = \sqrt{4 + 5 \cos^2 t}$, so that

$\mathbf{T}(t) = \langle \frac{-2 \sin t}{\sqrt{4+5 \cos^2 t}}, \frac{3 \cos t}{\sqrt{4+5 \cos^2 t}} \rangle$ and

$\mathbf{T}'(t) = \langle \frac{-18 \cos t}{(4+5 \cos^2 t)^{3/2}}, \frac{-12 \sin t}{(4+5 \cos^2 t)^{3/2}} \rangle$.

Therefore $\|\mathbf{T}'(t)\| = \frac{6}{4+5 \cos^2 t}$ and

$\kappa = \frac{\|\mathbf{T}'(t)\|}{\|\mathbf{r}'\|} = \frac{6}{(4+5 \cos^2 t)^{3/2}}$. We see that κ is maximized/minimized when $(4 + 5 \cos^2 t)$ is minimized/maximized, so κ is maximized when t is an odd multiple of $\frac{\pi}{2}$ and κ is minimized when t is an even multiple of $\frac{\pi}{2}$. That is, the curvature is maximum at $(0, \pm 3)$ and minimum at $(\pm 2, 0)$.

31. $f'(x) = 8x$ and $f''(x) = 8$, so we have

$\kappa(x) = \frac{|f''(x)|}{(1+(f'(x))^2)^{3/2}} = \frac{8}{(1+64x^2)^{3/2}}$.

$\kappa'(x) = \frac{-1536x}{(1+64x^2)^{5/2}} = 0$ only when $x = 0$, and $\kappa''(0) < 0$ so the curvature is maximum only at $x = 0$. There is no minimum curvature.

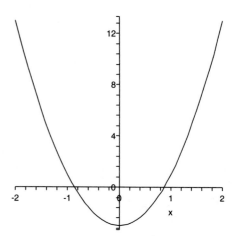

33. $f'(x) = 2e^{2x}$ and $f''(x) = 4e^{2x}$, so we have $\kappa = \frac{|f''(x)|}{(1+(f'(x))^2)^{3/2}} = \frac{4e^{2x}}{(1+4e^{4x})^{3/2}}$. As $x \to \infty$, the curvature tends to 0.

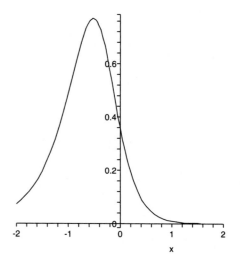

35. $f'(x) = 3x^2$ and $f''(x) = 6x$, so we have $\kappa = \frac{|f''(x)|}{(1+(f'(x))^2)^{3/2}} = \frac{|6x|}{(1+9x^4)^{3/2}}$. As $x \to \infty$, the curvature tends to 0.

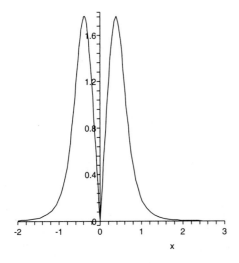

37. The graphs of the functions become straighter as $x \to \infty$, so the curvature goes to 0.

39. False. If the function has a maximum or a minimum at a point, we cannot be sure that the curve is turning more sharply there than at points nearby.

41. True. $\|\mathbf{r}'(t) \times \mathbf{r}''(t)\| = |f''(t)|$ and $\|\mathbf{r}'(t)\| = \sqrt{1 + f'(x)^2}$, so the curvature of $\mathbf{r}(t)$ is $\frac{|f''(t)|}{(1+(f'(t))^2)^{3/2}}$, the same as the curvature of $f(x)$.

43. We can write the polar curve $r = f(\theta)$ parametrically as $x = f(\theta) \cos \theta$, $y = f(\theta) \sin \theta$. Then $\dfrac{dy}{dx} = \dfrac{f'(\theta) \sin \theta + f(\theta) \cos \theta}{f'(\theta) \cos \theta - f(\theta) \sin \theta}$. The second derivative then simplifies to

$$\frac{d^2y}{dx^2} = \frac{\frac{d}{d\theta} \frac{dy}{dx}}{\frac{dx}{d\theta}}$$

$$= \frac{2f'(\theta)^2 - f(\theta)f''(\theta) + f(\theta)^2}{(f'(\theta) \cos \theta - f(\theta) \sin \theta)^2}.$$

We simplify $\kappa = \dfrac{\left|\frac{d^2y}{dx^2}\right|}{(1+(\frac{dy}{dx})^2)^{3/2}}$, and arrive at

$\kappa = \dfrac{|2f'(\theta)^2 - f(\theta)f''(\theta) + f(\theta)^2|}{(f'(\theta)^2 + f(\theta)^2)^{3/2}}$, as desired.

45. By exercise 43, with $f(\theta) = \sin 3\theta$, $f'(\theta) = 3 \cos 3\theta$, and $f''(\theta) = -9 \sin 3\theta$, we get $\kappa(\theta) = \dfrac{|18 \cos^2 3\theta + 10 \sin^2 3\theta|}{(9 \cos^2 3\theta + \sin^2 3\theta)^{3/2}}$.

$\kappa(0) = \frac{18}{9^{3/2}} = \frac{2}{3}$, and

$\kappa(\frac{\pi}{6}) = \frac{10}{1^{3/2}} = 10$.

47. By exercise 43, with $f(\theta) = 3e^{2\theta}$, $f'(\theta) = 6e^{2\theta}$, and $f''(\theta) = 12e^{2\theta}$, we have $\kappa(\theta) = \dfrac{45e^{4\theta}}{(45e^{4\theta})^{3/2}} = \dfrac{1}{\sqrt{45}} e^{-2\theta}$.

$\kappa(0) = \frac{1}{\sqrt{45}}$, and

$\kappa(1) = \frac{e^{-2}}{\sqrt{45}}$.

49. $\mathbf{r}'(t) = \langle 2 \cos t, -2 \sin t, \frac{2}{5} \rangle$ and $\mathbf{r}''(t) = \langle -2 \sin t, -2 \cos t, 0 \rangle$.

$$\mathbf{r}'(t) \times \mathbf{r}''(t) = \begin{vmatrix} \mathbf{i} & \mathbf{j} & \mathbf{k} \\ 2\cos t & -2\sin t & \frac{2}{5} \\ -2\sin t & -2\cos t & 0 \end{vmatrix}$$

$= \langle \frac{4}{5} \cos t, -\frac{4}{5} \sin t, -4 \rangle$.

$\|\mathbf{r}'(t) \times \mathbf{r}''(t)\| = \sqrt{\frac{416}{25}}$, and

$\|\mathbf{r}'(t)\| = \sqrt{\frac{104}{25}}$. So, $\kappa = \dfrac{\sqrt{\frac{416}{25}}}{\left(\sqrt{\frac{104}{25}}\right)^3} = \frac{25}{52} \approx$

0.48. This is larger than the curvature of the helix in example 4.5, which is stretched out more and therefore has less of a curve.

51. Given $x = t - \sin t$ and $y = 1 - \cos t$, we have $\dfrac{dy}{dx} = \dfrac{\sin t}{1 - \cos t}$, and

$$\frac{d^2y}{dx^2} = \frac{\frac{d}{dt} \frac{dy}{dx}}{\frac{dx}{dt}} = \frac{\cos t - \cos^2 t - \sin^2 t}{(1 - \cos t)^3}.$$

Simplifying $\kappa = \dfrac{\left|\frac{d^2y}{dx^2}\right|}{(1+(\frac{dy}{dx})^2)^{3/2}}$ yields

$$\kappa = \frac{\left|\frac{-1}{(1-\cos t)^2}\right|}{\left(1 + \left(\frac{\sin t}{1-\cos t}\right)^2\right)^{3/2}} = \frac{1 - \cos t}{(2(1 - \cos t))^{3/2}}$$

$$= \frac{1}{\sqrt{8(1 - \cos t)}} = \frac{1}{\sqrt{8y}}.$$

53. By exercise 43, with $f(\theta) = ae^{b\theta}$,
$f'(\theta) = abe^{b\theta}$, and $f''(\theta) = ab^2 e^{b\theta}$, we have

$$\kappa(\theta) = \frac{|2a^2 b^2 e^{2b\theta} - a^2 b^2 e^{2b\theta} + a^2 e^{2b\theta}|}{(a^2 b^2 e^{2b\theta} + a^2 e^{2b\theta})^{3/2}}$$

$$= \frac{a^2 e^{2b\theta}(1 + b^2)}{a^3 e^{3b\theta}(1 + b^2)^{3/2}} = \frac{e^{-b\theta}}{a\sqrt{1 + b^2}}.$$

As $b \to 0$, κ approaches the constant $\frac{1}{a}$, so the spiral approaches a circle of radius a.

12.5 TANGENT AND NORMAL VECTORS

1. $\mathbf{r}'(t) = \langle 1, 2t \rangle$ and $||\mathbf{r}'(t)|| = \sqrt{1 + 4t^2}$, so we have $\mathbf{T}(t) = \langle \frac{1}{\sqrt{1 + 4t^2}}, \frac{2t}{\sqrt{1 + 4t^2}} \rangle$. This yields $\mathbf{T}(0) = \langle 1, 0 \rangle$ and $\mathbf{T}(1) = \langle \frac{1}{\sqrt{5}}, \frac{2}{\sqrt{5}} \rangle$. Also, since

$$\mathbf{T}'(t) = \left\langle \frac{-4t}{(1 + 4t^2)^{3/2}}, \frac{2}{(1 + 4t^2)^{3/2}} \right\rangle \text{ and }$$

$$||\mathbf{T}'(t)|| = \frac{2}{1 + 4t^2}, \text{ we have}$$

$$\mathbf{N}(t) = \left\langle \frac{-2t}{\sqrt{1 + 4t^2}}, \frac{1}{\sqrt{1 + 4t^2}} \right\rangle. \text{ This yields}$$

$$\mathbf{N}(0) = \langle 0, 1 \rangle \text{ and } \mathbf{N}(1) = \left\langle \frac{-2}{\sqrt{5}}, \frac{1}{\sqrt{5}} \right\rangle.$$

3. $\mathbf{r}'(t) = \langle -2\sin 2t, 2\cos 2t \rangle$ and $||\mathbf{r}'(t)|| = 2$, so we have $\mathbf{T}(t) = \langle -\sin 2t, \cos 2t \rangle$. This yields $\mathbf{T}(0) = \langle 0, 1 \rangle$ and $\mathbf{T}(\frac{\pi}{4}) = \langle -1, 0 \rangle$.
Also, since $\mathbf{T}'(t) = \langle -2\cos 2t, -2\sin 2t \rangle$ and $||\mathbf{T}'(t)|| = 2$,
we have $\mathbf{N}(t) = \langle -\cos 2t, -\sin 2t \rangle$. This yields $\mathbf{N}(0) = \langle -1, 0 \rangle$ and $\mathbf{N}\left(\frac{\pi}{4}\right) = \langle 0, -1 \rangle$.

5. $\mathbf{r}'(t) = \langle -2\sin 2t, 1, 2\cos 2t \rangle$ and $||\mathbf{r}'(t)|| = \sqrt{5}$, so we have

$$\mathbf{T}(t) = \left\langle \frac{-2\sin 2t}{\sqrt{5}}, \frac{1}{\sqrt{5}}, \frac{2\cos 2t}{\sqrt{5}} \right\rangle.$$

This yields $\mathbf{T}(0) = \frac{1}{\sqrt{5}} \langle 0, 1, 2 \rangle$ and

$$\mathbf{T}(\frac{\pi}{2}) = \frac{1}{\sqrt{5}} \langle 0, 1, -2 \rangle.$$

Also, since $\mathbf{T}'(t) = \left\langle \frac{-4\cos 2t}{\sqrt{5}}, 0, \frac{-4\sin 2t}{\sqrt{5}} \right\rangle$ and

$$||\mathbf{T}'(t)|| = \frac{4}{\sqrt{5}}, \text{ we have}$$

$\mathbf{N}(t) = \langle -\cos 2t, 0, -\sin 2t \rangle$.
This yields $\mathbf{N}(0) = \langle -1, 0, 0 \rangle$ and

$$\mathbf{N}(\frac{\pi}{2}) = \langle 1, 0, 0 \rangle.$$

7. $\mathbf{r}'(t) = \langle 1, 2t, 1 \rangle$ and $||\mathbf{r}'(t)|| = \sqrt{2 + 4t^2}$, so we have $\mathbf{T}(t) = \left\langle \frac{1}{\sqrt{2 + 4t^2}}, \frac{2t}{\sqrt{2 + 4t^2}}, \frac{1}{\sqrt{2 + 4t^2}} \right\rangle$.

This yields $\mathbf{T}(0) = \left\langle \frac{1}{\sqrt{2}}, 0, \frac{1}{\sqrt{2}} \right\rangle$ and

$$\mathbf{T}(1) = \frac{1}{\sqrt{6}} \langle 1, 2, 1 \rangle. \text{ Also, since}$$

$$\mathbf{T}'(t) = \left\langle \frac{-4t}{(2 + 4t^2)^{3/2}}, \frac{4}{(2 + 4t^2)^{3/2}}, \frac{-4t}{(2 + 4t^2)^{3/2}} \right\rangle$$

and $||\mathbf{T}'(t)|| = \frac{4\sqrt{1 + 2t^2}}{(2 + 4t^2)^{3/2}}$, we have

$$\mathbf{N}(t) = \left\langle \frac{-t}{\sqrt{1 + 2t^2}}, \frac{1}{\sqrt{1 + 2t^2}}, \frac{-t}{\sqrt{1 + 2t^2}} \right\rangle.$$

This yields $\mathbf{N}(0) = \langle 0, 1, 0 \rangle$ and

$$\mathbf{N}(1) = \frac{1}{\sqrt{3}} \langle -1, 1, -1 \rangle.$$

9. From exercise 1, we have $||\mathbf{T}'(t)|| = \frac{2}{1 + 4t^2}$

and $||\mathbf{r}'(t)|| = \sqrt{1 + 4t^2}$, so that

$\kappa(t) = \frac{2}{(1 + 4t^2)^{3/2}}$ and $\kappa(0) = 2$. The osculating circle is a circle with radius

$\rho = \frac{1}{\kappa} = \frac{1}{2}$. The center is distance $\frac{1}{2}$ from the point $\mathbf{r}(0) = \langle 0, 0 \rangle$ in the direction of

$\mathbf{N}(0) = \langle 0, 1 \rangle$ (from exercise 1). The center is at $(0, \frac{1}{2})$, and the equation for the circle is $x^2 + (y - \frac{1}{2})^2 = \frac{1}{4}$.

11. From exercise 3, we see $\|\mathbf{T}'(t)\| = 2$ and $\|\mathbf{r}'(t)\| = 2$, so that $\kappa = 1$. The osculating circle is a circle with radius $\rho = \frac{1}{\kappa} = 1$. The center is distance 1 from the point $\mathbf{r}(\frac{\pi}{4}) = \langle 0, 1 \rangle$ in the direction of $\mathbf{N}(\frac{\pi}{4}) = \langle 0, -1 \rangle$ (from exercise 3). The center is the origin, and the equation for the circle is $x^2 + y^2 = 1$ (the same as the circle traced out by $\mathbf{r}(t)$).

13. $\mathbf{v}(t) = \langle 8, 16 - 32t \rangle$ and $\mathbf{a}(t) = \langle 0, -32 \rangle$.

$\|\mathbf{v}(t)\| = \frac{ds}{dt} = 8\sqrt{5 - 16t + 16t^2}$, so the tangential component is

$a_T = \frac{d^2s}{dt^2} = \frac{64(2t - 1)}{\sqrt{5 - 16t + 16t^2}}$.

The normal component is

$a_N = \sqrt{\|\mathbf{a}(t)\|^2 - a_T^2} = \frac{32}{\sqrt{5 - 16t + 16t^2}}$.

At $t = 0$, $a_T = \frac{-64}{\sqrt{5}}$ and $a_N = \frac{32}{\sqrt{5}}$. At $t = 1$, $a_T = \frac{64}{\sqrt{5}}$ and $a_N = \frac{32}{\sqrt{5}}$.

15. $\mathbf{v}(t) = \langle -2\sin 2t, 2t, 2\cos 2t \rangle$ and $\mathbf{a}(t) = \langle -4\cos 2t, 2, -4\sin 2t \rangle$.

$\|\mathbf{v}(t)\| = \frac{ds}{dt} = 2\sqrt{1 + t^2}$, so the tangential component is $a_T = \frac{d^2s}{dt^2} = \frac{2t}{\sqrt{1 + t^2}}$.

The normal component is

$a_N = \sqrt{\|\mathbf{a}(t)\|^2 - a_T^2} = 2\sqrt{\frac{5 + 4t^2}{1 + t^2}}$.

At $t = 0$, $a_T = 0$ and $a_N = 2\sqrt{5}$.

At $t = \frac{\pi}{4}$, $a_T = \frac{2\pi}{\sqrt{16 + \pi^2}}$ and

$a_N = 4\sqrt{\frac{20 + \pi^2}{16 + \pi^2}}$.

17. Since speed is simply $\|\mathbf{v}(t)\| = \frac{ds}{dt}$ and $a_T = \frac{d^2s}{dt^2}$ we may look at a_T to see if the speed is increasing or decreasing. When $t = 0$, the speed is neither increasing or decreasing since $a_T = 0$. When $t = \frac{\pi}{4}$, the speed is increasing since $a_T > 0$.

19. $\mathbf{v}(t) = \langle -a\sin t, a\cos t, b \rangle$ and $\mathbf{a}(t) = \langle -a\cos t, -a\sin t, 0 \rangle$.

$\|\mathbf{v}(t)\| = \frac{ds}{dt} = \sqrt{a^2 + b^2}$, so the tangential component is $a_T = \frac{d^2s}{dt^2} = 0$.

The normal component is

$a_N = \sqrt{\|\mathbf{a}\|^2 - a_T^2} = \sqrt{a^2 - 0^2} = a$.

21. $\mathbf{r}'(t) = \langle 1, 2, 2t \rangle$ and $\|\mathbf{r}'(t)\| = \sqrt{5 + 4t^2}$

$\mathbf{T}(t) = \left\langle \frac{1}{\sqrt{5 + 4t^2}}, \frac{2}{\sqrt{5 + 4t^2}}, \frac{2t}{\sqrt{5 + 4t^2}} \right\rangle$.

$\mathbf{T}(0) = \left\langle \frac{1}{\sqrt{5}}, \frac{2}{\sqrt{5}}, 0 \right\rangle$, and $\mathbf{T}(1) = \left\langle \frac{1}{3}, \frac{2}{3}, \frac{2}{3} \right\rangle$.

$\mathbf{T}'(t) = \left\langle \frac{-4t}{(5 + 4t^2)^{3/2}}, \frac{-8t}{(5 + 4t^2)^{3/2}}, \frac{10}{(5 + 4t^2)^{3/2}} \right\rangle$

$\|\mathbf{T}'(t)\| = \frac{2\sqrt{5}}{5 + 4t^2}$, so

$\mathbf{N}(t) = \left\langle \frac{-2t}{\sqrt{5}\sqrt{5 + 4t^2}}, \frac{-4t}{\sqrt{5}\sqrt{5 + 4t^2}}, \frac{\sqrt{5}}{\sqrt{5 + 4t^2}} \right\rangle$.

$\mathbf{N}(0) = \langle 0, 0, 1 \rangle$ and $\mathbf{N}(1) = \left\langle \frac{-2}{3\sqrt{5}}, \frac{-4}{3\sqrt{5}}, \frac{\sqrt{5}}{3} \right\rangle$.

$$B(0) = T(0) \times N(0) = \begin{vmatrix} i & j & k \\ \frac{1}{\sqrt{5}} & \frac{2}{\sqrt{5}} & 0 \\ 0 & 0 & 1 \end{vmatrix}$$

$$= \left\langle \frac{2}{\sqrt{5}}, \frac{-1}{\sqrt{5}}, 0 \right\rangle$$

$$B(1) = T(1) \times N(1) = \begin{vmatrix} i & j & k \\ \frac{1}{3} & \frac{2}{3} & \frac{2}{3} \\ \frac{-2}{3\sqrt{5}} & \frac{-4}{3\sqrt{5}} & \frac{\sqrt{5}}{3} \end{vmatrix}$$

$$= \left\langle \frac{2}{\sqrt{5}}, \frac{-1}{\sqrt{5}}, 0 \right\rangle$$

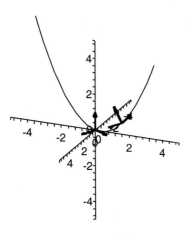

23. $r'(t) = \langle -4\pi \sin \pi t, 4\pi \cos \pi t, 1 \rangle$ and

$\|r'(t)\| = \sqrt{16\pi^2 + 1}$.

$$T(t) = \left\langle \frac{-4\pi \sin \pi t}{\sqrt{16\pi^2 + 1}}, \frac{4\pi \cos \pi t}{\sqrt{16\pi^2 + 1}}, \frac{1}{\sqrt{16\pi^2 + 1}} \right\rangle,$$

so

$$T(0) = \left\langle 0, \frac{4\pi}{\sqrt{16\pi^2 + 1}}, \frac{1}{\sqrt{16\pi^2 + 1}} \right\rangle, \text{ and }$$

$$T(1) = \left\langle 0, \frac{-4\pi}{\sqrt{16\pi^2 + 1}}, \frac{1}{\sqrt{16\pi^2 + 1}} \right\rangle.$$

$$T'(t) = \left\langle \frac{-4\pi^2 \cos \pi t}{\sqrt{16\pi^2 + 1}}, \frac{-4\pi^2 \sin \pi t}{\sqrt{16\pi^2 + 1}}, 0 \right\rangle \text{ and}$$

$$\|T'(t)\| = \frac{4\pi^2}{\sqrt{16\pi^2 + 1}}, \text{ so}$$

$N(t) = \langle -\cos \pi t, -\sin \pi t, 0 \rangle,$

$N(0) = \langle -1, 0, 0 \rangle$, and

$N(1) = \langle 1, 0, 0 \rangle$.

$$B(0) = T(0) \times N(0)$$

$$= \begin{vmatrix} i & j & k \\ 0 & \frac{4\pi}{\sqrt{16\pi^2+1}} & \frac{1}{\sqrt{16\pi^2+1}} \\ -1 & 0 & 0 \end{vmatrix}$$

$$= \left\langle 0, \frac{-1}{\sqrt{16\pi^2 + 1}}, \frac{4\pi}{\sqrt{16\pi^2 + 1}} \right\rangle$$

$$B(1) = T(1) \times N(1)$$

$$= \begin{vmatrix} i & j & k \\ 0 & \frac{-4\pi}{\sqrt{16\pi^2+1}} & \frac{1}{\sqrt{16\pi^2+1}} \\ 1 & 0 & 0 \end{vmatrix}$$

$$= \left\langle 0, \frac{1}{\sqrt{16\pi^2 + 1}}, \frac{4\pi}{\sqrt{16\pi^2 + 1}} \right\rangle$$

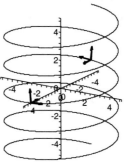

25. True. We know that $0 = T \cdot T' = T \cdot \dfrac{dT}{ds} \dfrac{ds}{dt}$,

but $\dfrac{ds}{dt} > 0$. This means $T \cdot \dfrac{dT}{ds} = 0$.

27. True. Since $T \cdot T = \|T\|^2$ and T has constant

length 1, $\dfrac{d}{dt}(T \cdot T) = 0$. But

$\dfrac{d}{dt}(T \cdot T) = \dfrac{d}{ds}(T \cdot T) \dfrac{ds}{dt}$, and $\dfrac{ds}{dt} > 0$.

Therefore, $\dfrac{d}{ds}(T \cdot T) = 0$.

29. Since $\|v(t)\| = 100\pi$, $\|a(t)\| = 100\pi^2$, and

$a_T = \dfrac{d}{dt}\|v(t)\| = 0$, we know

$a_N = \sqrt{\|a(t)\|^2 - a_T^2} = 100\pi^2$. Thus,

$F_s(t) = m a_N N(t)$

$= 10{,}000\pi^2 \langle -\cos \pi t, -\sin \pi t \rangle$.

31. Since $\|v(t)\| = 200\pi$, $\|a(t)\| = 400\pi^2$, and

$a_T = \dfrac{d}{dt}\|\mathbf{v}(t)\| = 0$, we know

$a_N = \sqrt{\|\mathbf{a}(t)\|^2 - a_T^2} = 400\pi^2$. Thus,

$\mathbf{F}_s(t) = m a_N \mathbf{N}(t)$

$= 40{,}000\pi^2\,\langle -\cos 2\pi t,\, -\sin 2\pi t\rangle.$

33. The required friction force doubles when the radius doubles.

35. $\kappa(x) = \dfrac{|f''(x)|}{(1 + (f'(x))^2)^{3/2}} = \dfrac{|-\cos x|}{(1 + \sin^2 x)^{3/2}}.$

$\kappa(0) = 1$, but $\kappa\left(\dfrac{\pi}{4}\right) = \dfrac{2}{3\sqrt{3}} \approx 0.3849$. The

radius of the osculating circle, $\rho = \dfrac{1}{\kappa}$, is larger

at $x = \dfrac{\pi}{4}$ where the curve is straighter. The

magnitude of the concavity

$|f''(x)| = |-\cos x|$ is 1 when $x = 0$ and $\dfrac{1}{\sqrt{2}}$

when $x = \dfrac{\pi}{4}$. The radius of the osculating

circle is larger where the magnitude of the concavity is smaller.

37. The curve $y = x^2$ can be parameterized by $\langle t, t^2\rangle$. The unit tangent is then

$\mathbf{T}(t) = \left\langle \dfrac{1}{\sqrt{1 + 4t^2}}, \dfrac{2t}{\sqrt{1 + 4t^2}}\right\rangle$ and the unit normal is

$\mathbf{N}(t) = \left\langle \dfrac{-2t}{\sqrt{1 + 4t^2}}, \dfrac{1}{\sqrt{1 + 4t^2}}\right\rangle.$

$\|\mathbf{T}'(t)\| = \dfrac{2}{1 + 4t^2}$ and $\|\mathbf{r}'(t)\| = \sqrt{1 + 4t^2},$

so $\kappa(t) = \dfrac{2}{(1 + 4t^2)^{3/2}}$ and the radius of the

osculating circle is $\dfrac{(1 + 4t^2)^{3/2}}{2}$. The center of

the osculating circle is at

$\langle t, t^2\rangle + \dfrac{(1 + 4t^2)^{3/2}}{2}\left\langle \dfrac{-2t}{\sqrt{1 + 4t^2}}, \dfrac{1}{\sqrt{1 + 4t^2}}\right\rangle$

$= \left\langle -4t^3, \dfrac{1}{2} + 3t^2\right\rangle.$

This is equivalent to $\left\langle 4t^3, \dfrac{1}{2} + 3t^2\right\rangle.$

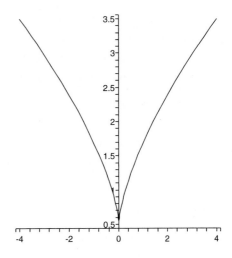

39. If $\mathbf{r} = \langle r\cos\theta,\, r\sin\theta,\, 0\rangle$,

then $\mathbf{v} = \left\langle -r\sin\theta\,\dfrac{d\theta}{dt},\, r\cos\theta\,\dfrac{d\theta}{dt},\, 0\right\rangle$, so

$\mathbf{r} \times \mathbf{v} = \begin{vmatrix} \mathbf{i} & \mathbf{j} & \mathbf{k} \\ r\cos\theta & r\sin\theta & 0 \\ -r\sin\theta\,\frac{d\theta}{dt} & r\cos\theta\,\frac{d\theta}{dt} & 0 \end{vmatrix}$

$= \left\langle 0, 0, r^2\dfrac{d\theta}{dt}\right\rangle = r^2\dfrac{d\theta}{dt}\mathbf{k}.$

Therefore, $\|\mathbf{r} \times \mathbf{v}\| = r^2\dfrac{d\theta}{dt}.$

$\dfrac{dA}{dt} = \dfrac{dA}{d\theta}\dfrac{d\theta}{dt} = \dfrac{1}{2}r^2\dfrac{d\theta}{dt}$, by the Fundamental

Theorem of Calculus. Since $\dfrac{dA}{dt} = \dfrac{1}{2}\|\mathbf{r} \times \mathbf{v}\|$

is a constant, equal areas are swept out in equal times.

41. **(a)** We have $\dfrac{d\mathbf{B}}{ds} = \dfrac{d}{ds}(\mathbf{T} \times \mathbf{N})$

$= \dfrac{d\mathbf{T}}{ds} \times \mathbf{N} + \mathbf{T} \times \dfrac{d\mathbf{N}}{ds}.$

Now, $\dfrac{d\mathbf{T}}{ds} = \dfrac{\mathbf{T}'}{\frac{ds}{dt}}$ is parallel to $\mathbf{N}$, so that

$\dfrac{d\mathbf{T}}{ds} \times \mathbf{N} = 0.$

This means that $\mathbf{T} \cdot \dfrac{d\mathbf{B}}{ds} = \mathbf{T} \cdot \left(\mathbf{T} \times \dfrac{d\mathbf{N}}{ds}\right) =$

$(\mathbf{T} \times \mathbf{T}) \cdot \dfrac{d\mathbf{N}}{ds} = 0$

and $\dfrac{d\mathbf{B}}{ds}$ is orthogonal to $\mathbf{T}$.

(b) We have $\dfrac{d\mathbf{B}}{ds} = \dfrac{d\mathbf{B}}{dt}\dfrac{dt}{ds} = \dfrac{\mathbf{B}'}{\frac{ds}{dt}}$. Since $\mathbf{B}$ is a vector of constant length, Theorem 2.4 tells us $\mathbf{B}$ is orthogonal to $\mathbf{B}'$, and is therefore orthogonal to $\dfrac{d\mathbf{B}}{ds}$.

43. If $\mathbf{r}(t) = \langle f(t),\, g(t),\, k \rangle$, then $\mathbf{r}'(t) = \langle f'(t),\, g'(t),\, 0 \rangle$. This forces the third component of $\mathbf{T}$ and $\mathbf{N}$ to be zero, and then the first two coordinates of $\mathbf{B}$ and $\dfrac{d\mathbf{B}}{ds}$ will also be zero. Since the first two coordinates of $\dfrac{d\mathbf{B}}{ds}$ are zero and the third coordinate of $\mathbf{N}$ is zero, $\tau = -\dfrac{d\mathbf{B}}{ds} \cdot \mathbf{N} = 0$.

45. (a) $\mathbf{r}'(t) = \mathbf{T}s'(t)$, so
$\mathbf{r}''(t) = \mathbf{T}'s'(t) + \mathbf{T}s''(t).$
Using the Frenet-Serret formula (a) from exercise 44, we get
$\mathbf{r}''(t) = \mathbf{T}s''(t) + \kappa[s'(t)]^2\mathbf{N}.$

(b) Combining $\mathbf{r}'(t) = \mathbf{T}s'(t)$ with part (a)
$\mathbf{r}'(t) \times \mathbf{r}''(t)$
$= s'(t)\mathbf{T} \times (s''(t)\mathbf{T} + \kappa[s'(t)]^2\mathbf{N})$
$= s'(t)s''(t)(\mathbf{T} \times \mathbf{T}) + \kappa[s'(t)]^3(\mathbf{T} \times \mathbf{N})$
$= \kappa[s'(t)]^3\mathbf{B}.$

(c) $\mathbf{r}''(t) = s''(t)\mathbf{T} + \kappa(t)[s'(t)]^2\mathbf{N}$, so
$\mathbf{r}'''(t) = s'''(t)\mathbf{T} + s''(t)\mathbf{T}' + \kappa'(t)[s'(t)]^2\mathbf{N}$
$+ 2\kappa(t)s'(t)s''(t)\mathbf{N} + \kappa(t)[s'(t)]^2\mathbf{N}'.$

We use the Frenet-Serret formulas from exercise 44 to write $\mathbf{T}' = \kappa(t)s'(t)\mathbf{N}$ and $\mathbf{N}' = -\kappa(t)s'(t)\mathbf{T} + \tau s'(t)\mathbf{B}$ and substitute to get

$\mathbf{r}'''(t) = s'''(t)\mathbf{T} + \kappa(t)s'(t)s''(t)\mathbf{N}$
$\quad + \kappa'(t)[s'(t)]^2\mathbf{N} + 2\kappa(t)s'(t)s''(t)\mathbf{N}$
$\quad - \kappa(t)^2[s'(t)]^3\mathbf{T} + \kappa(t)\tau[s'(t)]^3\mathbf{B}$
$\mathbf{r}'''(t) = \{s'''(t) - \kappa^2[s'(t)]^3\}\mathbf{T}$
$\quad + \{3\kappa s'(t)s''(t) + \kappa'(t)[s'(t)]^2\}\mathbf{N}$
$\quad + \kappa\tau[s'(t)]^3\mathbf{B}.$

(d) $\dfrac{(\mathbf{r}'(t) \times \mathbf{r}''(t)) \cdot \mathbf{r}'''(t)}{\|\mathbf{r}'(t) \times \mathbf{r}''(t)\|^2} = \dfrac{\kappa^2\tau[s'(t)]^6}{\kappa^2[s'(t)]^6} = \tau$

12.6 PARAMETRIC SURFACES

1. Paraboloid

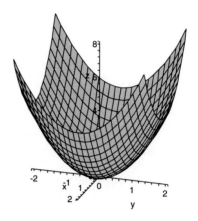

3. Paraboloid

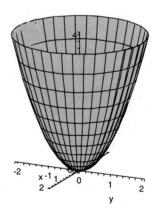

5.

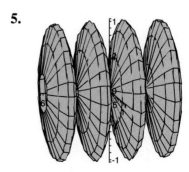

7. Sphere

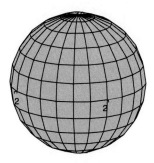

9. Half a hyperbolic paraboloid ($y \geq 0$)

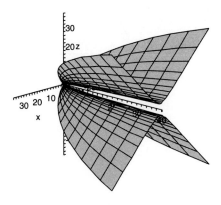

11. One sheet of a hyperboloid of two sheets

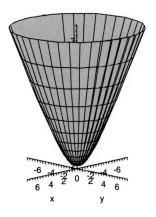

13. One possible parametric representation:
$x = u$, $y = v$, $z = 3u + 2v$.

15. One possible parametric representation of this hyperboloid is
$x = \cos u \cosh v$, $y = \sin u \cosh v$,
and $z = \sinh v$
for $0 \leq u \leq 2\pi$ and $-\infty < v < \infty$.

17. Circular cylinder from $z = 0$ to $z = 2$:
$x = 2 \cos \theta$, $y = 2 \sin \theta$, $z = t$
for $0 \leq \theta \leq 2\pi$ and $0 \leq t \leq 2$.

19. Downward opening paraboloid with positive z coordinate:
$x = r \cos \theta$, $y = r \sin \theta$, and $z = 4 - r^2$
for $0 \leq \theta \leq 2\pi$ and $0 \leq r \leq 2$.

21. Hyperboloid of two sheets:
$x = \pm \cosh u$, $y = \sinh u \cos v$,
and $z = \sinh u \sin v$
where $-\infty \leq u \leq \infty$ and $0 \leq v \leq 2\pi$.

23. (a) is Surface A, the only one unbounded in the z direction.

(b) is Surface C, since the position along the x-axis determines the direction of a line in the yz cross section.

(c) is Surface B, since the position along the x axis determines the radius of a circle in the yz cross section.

25. The trace of the sphere $x^2 + y^2 + z^2 = 4$ at height $z = k$ is a circle of radius $\sqrt{4 - k^2}$. If $2 \cos v = k$, then $4 \cos^2 v = k^2$ and $4(1 - \sin^2 v) = k^2$, so that $2 \sin v = \pm\sqrt{4 - k^2}$. If we restrict v to the range $[0, \pi]$, $z = 2 \cos v$ still takes on all values between -2 and 2, and we can take the positive square root. Then $x = \sqrt{4 - k^2} \cos u$ and $y = \sqrt{4 - k^2} \sin u$ describes a circle of radius $\sqrt{4 - k^2}$. Since the traces are the same, the surfaces must be the same.

27. The surface determined by $\rho = 3$ is a sphere centered at $(0, 0, 0)$ with radius 3.

29. This is the half of a circular cone with axis along the z-axis and $z \geq 0$.

31. This is the half-plane $x = y$ with the z-axis as the boundary ($x \geq 0$ and $y \geq 0$ since $0 \leq \phi \leq \pi$).

33. The surface is the top half of a circular cone if $0 \le \phi < \frac{\pi}{2}$, the xy-plane if $\phi = \frac{\pi}{2}$, and the bottom half of a circular cone if $\frac{\pi}{2} < \phi \le \pi$.

35. $x = 3\cos\theta\sin\phi$, $y = 3\sin\theta\sin\phi$, and $z = 3\cos\phi$ defines the top half-sphere when $0 \le \theta \le 2\pi$ and $0 \le \phi \le \frac{\pi}{2}$.

37. $x = \rho\cos\theta\sin\frac{\pi}{4}$, $y = \rho\sin\theta\sin\frac{\pi}{4}$, and $z = \rho\cos\frac{\pi}{4}$ defines the desired cone when $0 \le \theta \le 2\pi$ and $\rho \ge 0$.

39. The part of the cone inside the sphere of radius 2 is given by

$x = \rho\cos\theta\sin\frac{\pi}{4}$, $y = \rho\sin\theta\sin\frac{\pi}{4}$,

and $z = \rho\cos\frac{\pi}{4}$ for values $0 \le \theta \le 2\pi$ and $0 \le \rho \le 2$. The part of the sphere cut out by the cone is given by $x = 2\cos\theta\sin\phi$, $y = 2\sin\theta\sin\phi$, and $z = 2\cos\phi$ for values $0 \le \theta \le 2\pi$ and $0 \le \phi \le \frac{\pi}{4}$.

41. $u = v = 0$ gives $(2, -1, 3)$.
$u = 1$ and $v = 0$ gives $(3, 1, 0)$.
$u = 0$ and $v = 1$ gives $(4, -2, 5)$. The displacement vectors start at the first point and end at the other two points.

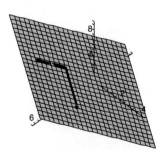

43. See the graph in exercise 41. $\mathbf{v}_1$ and $\mathbf{v}_2$ are the displacement vectors. $\mathbf{r}_0$ is the base of these vectors. A vector normal to the plane is given by $\mathbf{v}_1 \times \mathbf{v}_2$.

45. $\mathbf{r}_0 = \langle 3, 1, 1 \rangle$
$\mathbf{v}_1 = \langle 2, -1, 3 \rangle - \langle 3, 1, 1 \rangle = \langle -1, -2, 2 \rangle$
$\mathbf{v}_2 = \langle 4, 2, 1 \rangle - \langle 3, 1, 1 \rangle = \langle 1, 1, 0 \rangle$.
An equation for the plane is
$\mathbf{r} = \langle 3, 1, 1 \rangle + u\langle -1, -2, 2 \rangle + v\langle 1, 1, 0 \rangle$.

47. $r = \cos 2\theta$

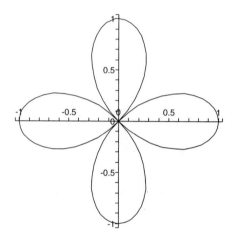

$x = r\cos\theta$, $y = r\sin\theta$, $z = z$

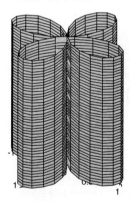

Substituting $r = \cos 2\theta$ and $z = v$ gives the desired parameterization. This is a cylinder of height 1 over the curve in the plane.

49. $x = r\cos\theta$, $y = r\sin\theta$, $z = z$ gives the desired region with $0 \le \theta \le \frac{\pi}{4}$, $0 \le r \le 2$ and $0 \le z \le 1$.

51. $f(t) = e^{-t^2}$ for $t \geq 0$.

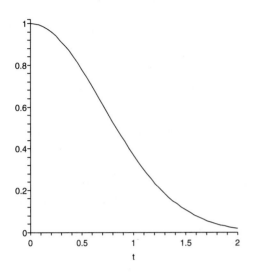

$z = e^{-x^2 - y^2}$

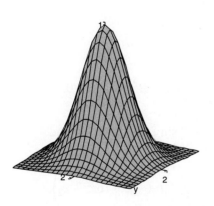

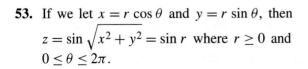

This is the graph of $f(t)$ revolved around the
z-axis. A parameterization of this surface is
$x = u \cos v$, $y = u \cos v$, and $z = e^{-u^2}$ for
$0 \leq v \leq 2\pi$ and $u \geq 0$.

53. If we let $x = r \cos \theta$ and $y = r \sin \theta$, then
$z = \sin \sqrt{x^2 + y^2} = \sin r$ where $r \geq 0$ and
$0 \leq \theta \leq 2\pi$.

Chapter 12 Review Exercises

1.

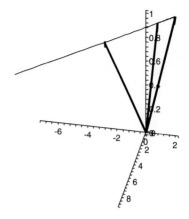

3.

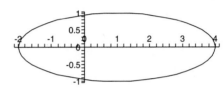

5.

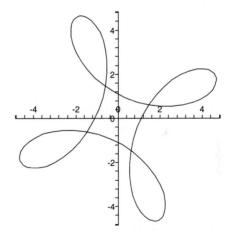

7.

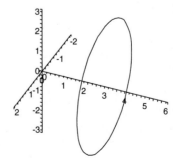

9.

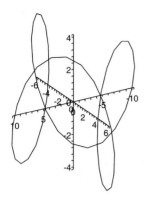

11.

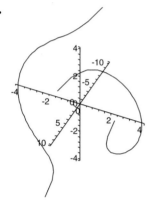

13. (a) Graph B: linear along y-axis, not circular.

(b) Graph C: linear along x-axis.

(c) Graph A: circular along y-axis.

(d) Graph F: coordinates bounded, not in one plane.

(e) Graph D: in plane $x = z$.

(f) Graph E: in plane $y = x + 1$.

15. Let $\mathbf{r}(t) = \langle f(t), g(t), h(t) \rangle$.
The arc length is equal to
$$\int_0^{2\pi} \sqrt{(f'(t))^2 + (g'(t))^2 + (h'(t))^2}\, dt$$
$$= \int_0^{2\pi} \sqrt{1 + 36}\, dt = 2\pi\sqrt{37}.$$

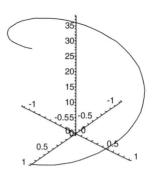

17. $\displaystyle\lim_{t \to 1} \left\langle t^2 - 1,\, e^{2t},\, \cos \pi t \right\rangle = \left\langle 0,\, e^2,\, -1 \right\rangle$

19. $\mathbf{r}(t)$ fails to be continuous where $\ln t^2$ fails to be continuous. $\mathbf{r}(t)$ is continuous for all $t \neq 0$.

21. $\mathbf{r}'(t) = \left\langle \dfrac{t}{\sqrt{t^2 + 1}},\, 4\cos 4t,\, \dfrac{1}{t} \right\rangle$

23. $\displaystyle\int \left\langle e^{-4t},\, \dfrac{2}{t^3},\, 4t - 1 \right\rangle dt$
$$= \left\langle \dfrac{-1}{4} e^{-4t},\, \dfrac{-1}{t^2},\, 2t^2 - t \right\rangle + \mathbf{c}$$

25. $\displaystyle\int_0^1 \langle \cos \pi t,\, 4t,\, 2 \rangle\, dt$
$$= \left\langle \dfrac{1}{\pi} \sin \pi t,\, 2t^2,\, 2t \right\rangle \Big|_0^1 = \langle 0,\, 2,\, 2 \rangle$$

27. $\mathbf{v}(t) = \langle -8\sin 2t,\, 8\cos 2t,\, 4 \rangle$ and
$\mathbf{a}(t) = \langle -16\cos 2t,\, -16\sin 2t,\, 0 \rangle$.

29. $\mathbf{r}(t) = \left\langle t^2 + 4t + c_1,\, -16t^2 + c_2 \right\rangle$.
To make $\mathbf{r}(0) = \langle 2, 1 \rangle$, we must have
$\mathbf{r}(t) = \left\langle t^2 + 4t + 2,\, -16t^2 + 1 \right\rangle$.

31. $\mathbf{v}(t) = \langle c_1,\, -32t + c_2 \rangle$ and $\mathbf{v}(0) = \langle 4, 3 \rangle$, so
$\mathbf{v}(t) = \langle 4,\, -32t + 3 \rangle$.
Then $\mathbf{r}(t) = \left\langle 4t + b_1,\, -16t^2 + 3t + b_2 \right\rangle$
and $\mathbf{r}(0) = \langle 2, 6 \rangle$, so
$\mathbf{r}(t) = \left\langle 4t + 2,\, -16t^2 + 3t + 6 \right\rangle$.

33. $\mathbf{F}(t) = m\mathbf{a}(t) = 4\langle 0, -32 \rangle = -128\langle 0, 1 \rangle$

35. $\mathbf{a}(t) = -32\mathbf{j}$

$\mathbf{v}(0) = 80 \left\langle \cos \dfrac{\pi}{12}, \sin \dfrac{\pi}{12} \right\rangle$

$\mathbf{v}(t) = 80 \cos \dfrac{\pi}{12}\mathbf{i} + (80 \sin \dfrac{\pi}{12} - 32t)\mathbf{j}$

$\mathbf{r}(0) = \langle 0, 0 \rangle$

$\mathbf{r}(t) = 80 \cos(\dfrac{\pi}{12})t\mathbf{i} + (80 \sin(\dfrac{\pi}{12})t - 16t^2)\mathbf{j}$

Maximum altitude occurs when vertical component of velocity is 0, which occurs at

$t = \dfrac{80 \sin \frac{\pi}{12}}{32}.$

$(80 \sin(\dfrac{\pi}{12})t - 16t^2)\Big|_{t = \frac{80 \sin \frac{\pi}{12}}{32}} = 25(2 - \sqrt{3})$

≈ 6.7 feet.

Impact occurs when the height is 0, which

occurs at $t = \dfrac{80 \sin \frac{\pi}{12}}{16}.$

The horizontal range is

$80 \cos(\dfrac{\pi}{12})t \Big|_{t = \frac{80 \sin \frac{\pi}{12}}{16}} = 100$ feet.

The speed at impact is

$\left\| \mathbf{v}\left(\dfrac{80 \sin \frac{\pi}{12}}{16} \right) \right\| = 80 \dfrac{\text{ft}}{\text{s}}.$

37. $\mathbf{r}'(t) = \left\langle -2e^{-2t}, 2, 0 \right\rangle$ and

$\|\mathbf{r}'(t)\| = \sqrt{4e^{-4t} + 4},$

so $\mathbf{T}(0) = \left\langle \dfrac{-2}{\sqrt{8}}, \dfrac{2}{\sqrt{8}}, 0 \right\rangle$

and $\mathbf{T}(1) = \dfrac{1}{2\sqrt{e^{-4} + 1}} \left\langle -2e^{-2}, 2, 0 \right\rangle.$

39. Since $\mathbf{r}'(t) = \langle -\sin t, \cos t, \cos t \rangle$, we have

$\|\mathbf{r}'(t)\| = \sqrt{1 + \cos^2 t}.$

$\mathbf{T}(t) = \left\langle \dfrac{-\sin t}{\sqrt{1 + \cos^2 t}}, \dfrac{\cos t}{\sqrt{1 + \cos^2 t}}, \dfrac{\cos t}{\sqrt{1 + \cos^2 t}} \right\rangle,$

$\mathbf{T}'(t) = \left\langle \dfrac{-2\cos t}{(1 + \cos^2 t)^{3/2}}, \dfrac{-\sin t}{(1 + \cos^2 t)^{3/2}}, \dfrac{-\sin t}{(1 + \cos^2 t)^{3/2}} \right\rangle,$

and $\|\mathbf{T}'(t)\| = \dfrac{\sqrt{2}}{1 + \cos^2 t}.$

$\kappa = \dfrac{\|\mathbf{T}'(t)\|}{\|\mathbf{r}'(t)\|}$, so $\kappa(0) = \dfrac{1}{2}$ and $\kappa(\dfrac{\pi}{4}) = \dfrac{4}{3\sqrt{3}}.$

41. $\mathbf{r}'(t) = \langle 0, 3 \rangle$ and $\|\mathbf{r}'(t)\| = 3.$ $\mathbf{T}(t) = \langle 0, 1 \rangle$ and $\mathbf{T}'(t) = \langle 0, 0 \rangle$, so $\|\mathbf{T}'(t)\| = 0$ for all t. This means that $\kappa = 0$ for all t.

43. From exercise 39, we have

$\mathbf{T}(t) = \left\langle \dfrac{-\sin t}{\sqrt{1 + \cos^2 t}}, \dfrac{\cos t}{\sqrt{1 + \cos^2 t}}, \dfrac{\cos t}{\sqrt{1 + \cos^2 t}} \right\rangle,$

$\mathbf{T}'(t) = \left\langle \dfrac{-2\cos t}{(1 + \cos^2 t)^{3/2}}, \dfrac{-\sin t}{(1 + \cos^2 t)^{3/2}}, \dfrac{-\sin t}{(1 + \cos^2 t)^{3/2}} \right\rangle,$

and $\|\mathbf{T}'(t)\| = \dfrac{\sqrt{2}}{1 + \cos^2 t}.$

Recall that $\mathbf{N}(t) = \dfrac{\mathbf{T}'(t)}{\|\mathbf{T}'(t)\|}.$

$\mathbf{T}(0) = \left\langle 0, \dfrac{1}{\sqrt{2}}, \dfrac{1}{\sqrt{2}} \right\rangle$ and $\mathbf{N}(0) = \langle -1, 0, 0 \rangle.$

45. Since $\mathbf{v}(t) = \langle 2, 2t, 0 \rangle$, we have

$\|\mathbf{v}(t)\| = 2\sqrt{t^2 + 1}.$ Also since $\dfrac{ds}{dt} = \|\mathbf{v}(t)\|,$

we have $a_T = \dfrac{d^2 s}{dt^2} = \dfrac{2t}{\sqrt{t^2 + 1}}.$ Furthermore, since $\mathbf{a}(t) = \langle 0, 2, 0 \rangle$ then

$a_N = \sqrt{\|\mathbf{a}\|^2 - a_T^2} = \sqrt{4 - \dfrac{4t^2}{t^2 + 1}} = \dfrac{2}{\sqrt{t^2 + 1}}.$

At $t = 0$, $a_T = 0$ and $a_N = 2$.

At $t = 1$, $a_T = \sqrt{2}$ and $a_N = \sqrt{2}$.

47. Since $\mathbf{r}(t)$ describes a circle we can see quickly that $\mathbf{N}(t) = \langle -\cos 6t, -\sin 6t \rangle$,

$\|\mathbf{v}(t)\| = 480$, and $\|\mathbf{a}(t)\| = 2880.$ Since

$\|\mathbf{v}(t)\| = \dfrac{ds}{dt}$ we see that $a_T = \dfrac{d^2 s}{dt^2} = 0$ and

$a_N = \sqrt{\|\mathbf{a}\|^2 - a_T^2} = 2880.$ Therefore,

$\mathbf{F}_s(t) = m a_N \mathbf{N} = -345{,}600 \langle \cos 6t, \sin 6t \rangle.$

49.

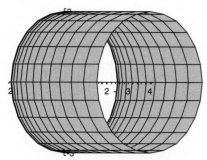

51.

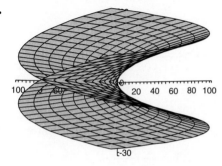

53. (a) Surface B: quadratic in x and z.

(b) Surface C: quadratic in the x direction.

(c) Surface A: quadratic in the z direction.

Functions of Several Variables and Partial Differentiation

13.1 FUNCTIONS OF SEVERAL VARIABLES

1. Domain $= \{(x, y) | y \neq -x\}$

3. Domain $= \{(x, y) | x + y + 2 > 0\}$

5. Domain $= \{(x, y, z) | x^2 + y^2 + z^2 < 4\}$

7. Range $= \{z | z > 0\}$

9. Range $= \{z | z \geq -1\}$

11. $f(1, 2) = 1^2 + 2 = 3$
    $f(0, 3) = 0^2 + 3 = 3$

13. $f(0, 1, 2, 3) = \cos 0 - \dfrac{2 \cdot 1 \cdot 3}{2 + 3} = \dfrac{-1}{5}$

    $f(\pi, 2, 0, -1) = \cos \pi - \dfrac{2 \cdot 2 \cdot -1}{0 - 1} = -5$

15. (a) $R(150, 1000) = 312$

    (b) $R(150, 2000) = 333$

    (c) $R(150, 3000) = 350$

    (d) The distance gained varies from 17 feet to 21 feet, an average of about 19 feet.

17.

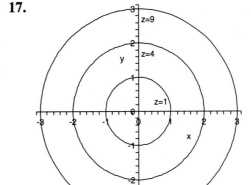

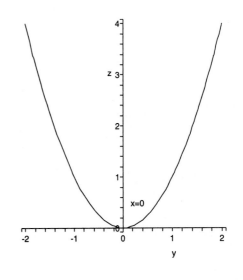

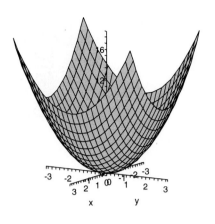

21.

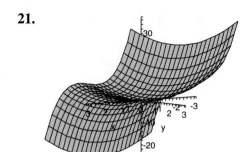

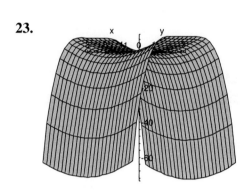

23.

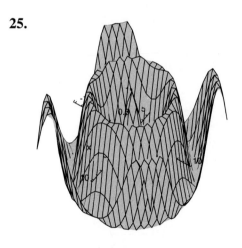

19.

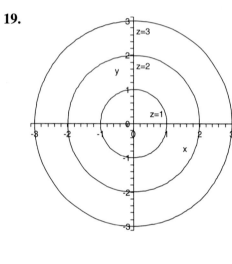

25.

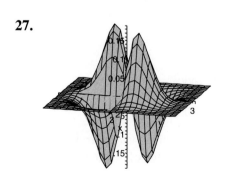

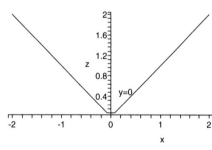

27.

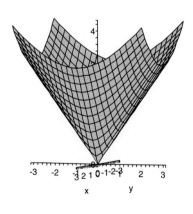

29.

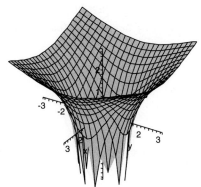

39.

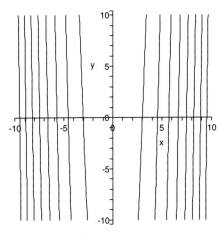

31. Point A is at height 480 and "straight up" is to the northeast. Point B is at height 470 and "straight up" is to the southwest. Point C is at height between 470 and 480 and "straight up" is to the northwest.

33. It appears that the highest levels of reflectivity are near the center, and that the tornado is traveling northwest.

41.

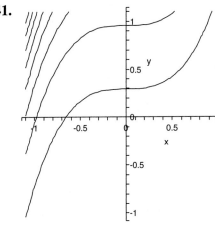

35. $H(80, 20) = 77.4$, $H(80, 40) = 80.4$, and $H(80, 60) = 82.8$. It appears that at $80°$, increasing the humidity by 20% increases the heat index by about $2.7°$.

37.

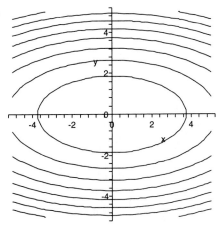

43.

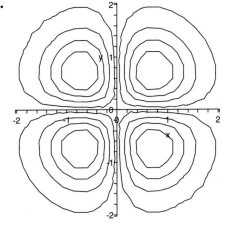

45.

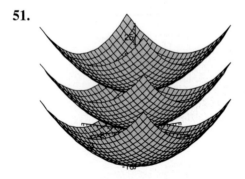

47. Surface a ⟶ Contour Plot A
Surface b ⟶ Contour Plot D
Surface c ⟶ Contour Plot C
Surface d ⟶ Contour Plot B

49. Function a ⟶ Surface B
Function b ⟶ Surface D
Function c ⟶ Surface A
Function d ⟶ Surface F
Function e ⟶ Surface C
Function f ⟶ Surface E

51.

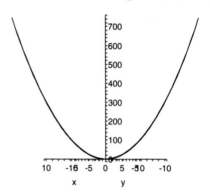

53. Viewed from positive x-axis: View B
Viewed from positive y-axis: View A

55. This is accurate since every (x, y) in the plane is in the domain of the function, but it is not helpful because no height is visible.

57. The graph is a sine wave traveling in the direction of the line $y = x$. Seen from a great distance along the line $y = x$, the graph will look rectangular, since all the points on the

lines $y + x = \pm\frac{\pi}{2}$ (perpendicular to $y = x$), will have the same height ± 1 (respectively).

59. As viewed from the point $(10, 10\sqrt{3}, 0)$.

61. The rock concert is in the upper left. The other level curves could be restaurants and roads.

63. The point of maximum power will be inside all the contours, slightly toward the handle from the center. This is maximum because power increases away from the rim of the racket.

65. It is not possible to have a PGA of 4.0. If a student earned a 4.0 grade point average in high school, and 1600 on the SAT's, their PGA would be 3.942.

It is possible to have a negative PGA, if the high school grade point average is close to 0, and the SAT score is the lowest possible.

It seems the high school grade point average is the most important. The maximum possible contribution from it is 2.832. The maximum possible contribution from SAT verbal is 1.44, and from SAT math is 0.80.

67. If you drive d miles at x mph, it will take you $\frac{d}{x}$ hours. Similarly, driving d miles at y miles per hour takes $\frac{d}{y}$ hours. The total distance traveled is $2d$, and the time taken is $\frac{d}{x} + \frac{d}{y} = \frac{d(x+y)}{xy}$. The average speed is total distance divided by total time, so $S(x, y) = \frac{2xy}{x+y}$.

If $x = 30$, then $S(30, y) = \frac{60y}{30+y} = 40$. We solve to get $60y = 1200 + 40y$, $20y = 1200$, and $y = 60$ mph.

If we replace 40 with 60 in the above solution, we see that there is no solution. It is not possible to average 60 mph in this situation.

69. Plot of $z = x^2 + y^2$:

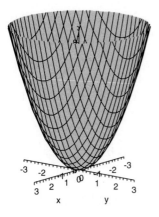

Plot of $x = r \cos t$, $y = r \sin t$, $z = r^2$:

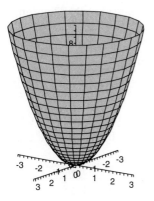

The graphs are the same surface. The grid is different.

71. Parametric equations are
$x = r \cos t$, $y = r \sin t$, $z = \cos r^2$.
The graphs are the same surface, but the parametric equations make the graph look much cleaner:

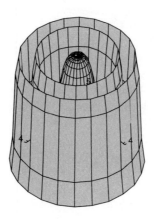

13.2 LIMITS AND CONTINUITY

1. $\displaystyle\lim_{(x,y)\to(1,3)} \frac{x^2 y}{4x^2 - y} = 3$

3. $\displaystyle\lim_{(x,y)\to(\pi,1)} \frac{\cos xy}{y^2 + 1} = \frac{-1}{2}$

5. $\displaystyle\lim_{(x,y,z)\to(1,0,2)} \frac{4xz}{y^2 + z^2} = 2$

7. Along the path $x = 0$,
$\displaystyle\lim_{(0,y)\to(0,0)} \frac{0}{y^2} = 0.$
Along the path $y = 0$,
$\displaystyle\lim_{(x,0)\to(0,0)} \frac{3x^2}{x^2} = 3.$
Since the limits along these two paths do not agree, the limit does not exist.

9. Along the path $x = 0$,
$\displaystyle\lim_{(0,y)\to(0,0)} \frac{0}{3y^2} = 0.$
Along the path $y = x$,
$\displaystyle\lim_{(x,x)\to(0,0)} \frac{4x^2}{2x^2} = 2.$
Since the limits along these two paths do not agree, the limit does not exist.

11. Along the path $x = 0$,
$\displaystyle\lim_{(0,y)\to(0,0)} \frac{0}{y^2} = 0.$

Along the path $y = x^2$,

$$\lim_{(x,x^2)\to(0,0)} \frac{2x^4}{x^4 + x^4} = 1.$$

Since the limits along these two paths do not agree, the limit does not exist.

13. Along the path $x = 0$,

$$\lim_{(0,y)\to(0,0)} \frac{0}{y^3} = 0.$$

Along the path $x = y^3$,

$$\lim_{(y^3,y)\to(0,0)} \frac{y^3}{2y^3} = \frac{1}{2}.$$

Since the limits along these two paths do not agree, the limit does not exist.

15. Along the path $x = 0$,

$$\lim_{(0,y)\to(0,0)} \frac{0}{y^2} = 0.$$

Along the path $y = x$,

$$\lim_{(x,x)\to(0,0)} \frac{x \sin x}{2x^2} = \frac{1}{2}.$$

Since the limits along these two paths do not agree, the limit does not exist.

17. Along the path $x = 1$,

$$\lim_{(1,y)\to(1,2)} \frac{0}{y^2 - 4y + 4} = 0.$$

Along the path $y = x + 1$,

$$\lim_{(x,x+1)\to(1,2)} \frac{x^2 - 2x + 1}{2x^2 - 4x + 2} = \frac{1}{2}.$$

Since the limits along these two paths do not agree, the limit does not exist.

19. Along the path $x = 0$, $y = 0$,

$$\lim_{(0,0,z)\to(0,0,0)} \frac{0}{z^2} = 0.$$

Along the path $x^2 = y^2 + z^2$,

$$\lim_{(y^2+z^2,y,z)\to(0,0,0)} \frac{3(y^2 + z^2)}{2(y^2 + z^2)} = \frac{3}{2}.$$

Since the limits along these two paths do not agree, the limit does not exist.

21. Along the path $x = 0$, $y = 0$,

$$\lim_{(0,0,z)\to(0,0,0)} \frac{0}{z^3} = 0.$$

Along the path $x = y = z$,

$$\lim_{(x,x,x)\to(0,0,0)} \frac{x^3}{3x^3} = \frac{1}{3}.$$

Since the limits along these two paths do not agree, the limit does not exist.

23. If the limit exists, it must be equal to 0 (to see this use the path $x = 0$). To show that $L = 0$,

$$|f(x, y) - L| = \left| \frac{xy^2}{x^2 + y^2} \right| \le \left| \frac{xy^2}{y^2} \right| = |x|.$$

Since $\displaystyle\lim_{(x,y)\to(0,0)} |x| = 0$, Theorem 2.1 gives us that

$$\lim_{(x,y)\to(0,0)} \frac{xy^2}{x^2 + y^2} = 0.$$

25. If the limit exists, it must be equal to 0 (to see this, use the path $x = 0$). To show that $L = 0$,

$$|f(x, y) - L| = \left| \frac{2x^2 \sin y}{2x^2 + y^2} \right| \le \left| \frac{2x^2 \sin y}{2x^2} \right|$$
$$= |\sin y|.$$

Since $\displaystyle\lim_{(x,y)\to(0,0)} |\sin y| = 0$, Theorem 2.1 gives us that

$$\lim_{(x,y)\to(0,0)} \frac{2x^2 \sin y}{2x^2 + y^2} = 0.$$

27. If the limit exists, it must be equal to 2 (to see this, use the path $x = 0$). To show that $L = 2$,

$$|f(x, y) - L| = \left| \frac{x^3 + 4x^2 + 2y^2}{2x^2 + y^2} - 2 \right|$$
$$= \left| \frac{x^3}{2x^2 + y^2} \right|$$
$$\le \left| \frac{x^3}{2x^2} \right|$$
$$= \left| \frac{x}{2} \right|.$$

Since $\displaystyle\lim_{(x,y)\to(0,0)} \left| \frac{x}{2} \right| = 0$, Theorem 2.1 gives us that

$$\lim_{(x,y)\to(0,0)} \frac{x^3 + 4x^2 + 2y^2}{2x^2 + y^2} = 2.$$

29. If the limit exists, it must be equal to 0 (to see this, use the path $x = 0$, $y = 0$). To show that $L = 0$,

$$|f(x, y, z) - L| = \left| \frac{3x^3}{x^2 + y^2 + z^2} \right|$$

$$\leq \left| \frac{3x^3}{x^2} \right| = |3x|.$$

Since $\lim\limits_{(x,y,z)\to(0,0,0)} |3x| = 0$, Theorem 2.1 gives us that

$$\lim_{(x,y,z)\to(0,0,0)} \frac{3x^3}{x^2 + y^2 + z^2} = 0.$$

31. The density plot shows sharp color changes near the origin.

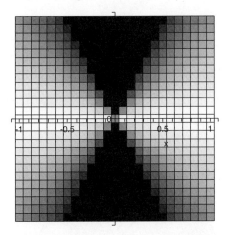

33. The density plot shows sharp color changes near the origin.

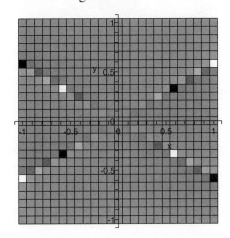

35. Since $\sqrt{t}$ is continuous for all $t \geq 0$ and because $9 - x^2 - y^2$ is a polynomial, f is continuous where $x^2 + y^2 \leq 9$.

37. Since $\ln t$ is continuous for all $t > 0$ and $3 - x^2 + y$ is a polynomial, f is continuous where $x^2 - y < 3$.

39. Since $\sqrt{t}$ is continuous for all $t \geq 0$ and because $x^2 + y^2 + z^2 - 4$ is a polynomial, f is continuous where $x^2 + y^2 + z^2 \geq 4$.

41. Along any line $y = y_0$, for $y_0 \neq 2$, the limit $\lim\limits_{(x,y_0)\to(0,y_0)} (y_0 - 2) \cos \dfrac{1}{x^2}$ does not exist. If $y_0 = 2$, then $\lim\limits_{(x,2)\to(0,2)} f(x, y) = 0 = f(0, 2)$. f is continuous for $x \neq 0$ and at the point $(0, 2)$.

43. Since $\lim\limits_{(x,y)\to(x_0,x_0)} \dfrac{x^2 - y^2}{x - y}$

$$= \lim_{(x,y)\to(x_0,x_0)} \frac{(x - y)(x + y)}{x - y}$$

$$= \lim_{(x,y)\to(x_0,x_0)} (x + y) = 2x_0 = f(x_0, x_0) \text{ for}$$

any x_0, the function $f(x, y)$ is continuous for all (x, y).

45.

x	y	$f(x, y)$
0.1	0.1	0.4545
−0.1	−0.1	0.5555
0.01	0.01	0.4950
−0.01	−0.01	0.5050
0.001	0.001	0.4995
−0.001	−0.001	0.5005

Thus, we estimate that the limit is 0.5.

47. True. If the limit is L, then the limit computed along the line $y = b$ must also be L.

49. False. The limit along two paths being L does not imply that the limit is L. The limit must be the same along **any** path.

51. The limit along the line $y = kx$:

$$\lim_{(x,kx)\to(0,0)} \frac{xy^2}{x^2+y^4} = \lim_{x\to0} \frac{k^2x^3}{x^2+k^4x^4}$$
$$= \lim_{x\to0} \frac{k^2x}{1+k^4x^2} = 0.$$

53. As shown in Example 2.5, the limit of f as $(x, y) \to (0, 0)$ does not exist, therefore the function cannot be continuous there.

55. Converting to polar coordinates,

$$\lim_{(x,y)\to(0,0)} \frac{\sqrt{x^2+y^2}}{\sin\sqrt{x^2+y^2}} = \lim_{r\to0} \frac{r}{\sin r} = 1$$

by l'Hôpital's Rule.

57. Converting to polar coordinates,

$$\lim_{(x,y)\to(0,0)} \frac{xy^2}{x^2+y^2} = \lim_{r\to0} \frac{r^3\cos\theta\sin^2\theta}{r^2}$$
$$= \lim_{r\to0} r\cos\theta\sin^2\theta = 0.$$

13.3 PARTIAL DERIVATIVES

1. $f_x = 3x^2 - 4y^2, \quad f_y = -8xy + 4y^3$

3. $f_x = 2x\sin xy + x^2y\cos xy,$
$f_y = x^3\cos xy - 9y^2$

5. $f_x = \dfrac{4}{y}e^{x/y} + \dfrac{y}{x^2}, \quad f_y = \dfrac{-4x}{y^2}e^{x/y} - \dfrac{1}{x}$

7. $f_x = 3\sin y + 12x^2y^2z$
$f_y = 3x\cos y + 8x^3yz$
$f_z = 4x^3y^2$

9. $\dfrac{\partial f}{\partial x} = 3x^2 - 4y^2,$

$\dfrac{\partial f}{\partial y} = -8xy + 3, \quad \dfrac{\partial^2 f}{\partial x^2} = 6x,$

$\dfrac{\partial^2 f}{\partial y^2} = -8x, \quad \dfrac{\partial^2 f}{\partial y\partial x} = -8y$

11. $f_x = 4x^3 - 6xy^3, f_{xx} = 12x^2 - 6y^3,$
$f_{xy} = -18xy^2, f_{xyy} = -36xy$

13. $f_x = 3x^2y^2, f_{xx} = 6xy^2$
$f_y = 2x^3y - z\cos yz,$
$f_{yz} = -\cos yz + yz\sin yz$
$f_{xy} = 6x^2y, f_{xyz} = 0$

15. $f_w = 2wxy - ze^{wz}, f_{ww} = 2xy - z^2e^{wz},$
$f_{wx} = 2wy, f_{wxy} = 2w,$
$f_{wwx} = 2y, f_{wwxy} = 2, f_{wwxyz} = 0$

17. We consider the $y = 1$ trace. $\dfrac{\partial f}{\partial x}(1, 1) = -2$ is the slope of this trace at $(1, 1, 2)$.

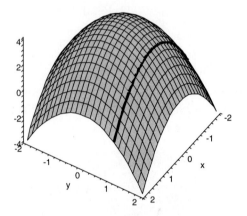

19. We consider the $x = 2$ trace. $\dfrac{\partial f}{\partial y}(2, 0) = 0$ is the slope of this trace at $(2, 0, 0)$.

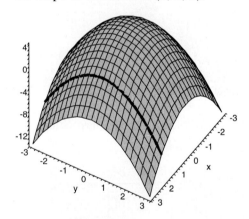

21. Take the partial derivative implicitly:
$$\left(P + \frac{n^2a}{V^2}\right)\frac{\partial V}{\partial T} + \left(\frac{-2n^2a}{V^3}\frac{\partial V}{\partial T}\right)(V - nb)$$
$= nR$. Multiply through by V^3 and solve for $\dfrac{\partial V}{\partial T}$ to get

$$(PV^3 - n^2aV + 2n^3ab)\frac{\partial V}{\partial T} = nRV^3$$

$$\frac{\partial V}{\partial T} = \frac{nRV^3}{PV^3 - n^2aV + 2n^3ab}$$

23. Find $\dfrac{\partial P}{\partial T}$ implicitly in

$$\left(P + \frac{14}{V^2}\right)(V - 0.004) = 12T,$$

$$\frac{\partial P}{\partial T}(V - 0.004) = 12, \text{ and } \frac{\partial P}{\partial T} = \frac{12}{V - 0.004}.$$

The increase in pressure due to an increase in one degree will be $\frac{12}{V}$ (assuming V is much larger than 0.004).

25. $S = \dfrac{cL^4}{wh^3} \quad \dfrac{\partial S}{\partial w} = -\dfrac{cL^4}{w^2h^3} = -\dfrac{1}{w}\dfrac{cL^4}{wh^3} = -\dfrac{1}{w}S$

27. The variable with the largest exponent has the largest proportional effect. In this case h has the greatest proportional effect.

29. $\dfrac{\partial f}{\partial x} = 2x, \quad \dfrac{\partial f}{\partial y} = 2y$

$$\frac{\partial f}{\partial x} = 0 \text{ at } x = 0.$$

$$\frac{\partial f}{\partial y} = 0 \text{ at } y = 0.$$

This means there are horizontal tangent lines to the trace in the $y = 0$ plane and the $x = 0$ plane at $(0, 0)$. This corresponds to the minimum value of the function.

31. $\dfrac{\partial f}{\partial x} = \cos x \sin y, \quad \dfrac{\partial f}{\partial y} = \sin x \cos y$

$$\frac{\partial f}{\partial x} = 0 \text{ when either } x = \frac{\pi}{2} + n\pi, \text{ or } y = m\pi.$$

$$\frac{\partial f}{\partial y} = 0 \text{ when either } x = n\pi, \text{ or } y = \frac{\pi}{2} + m\pi.$$

When $x = \frac{\pi}{2} + n\pi$ and $y = \frac{\pi}{2} + m\pi$, $f(x, y) = 1$ if m and n are both even and if m and n are both odd, and $f(x, y) = -1$ if one is odd and the other is even. These are maximum and minimum points. If $x = n\pi$ and $y = m\pi$, $f(x, y) = 0$ and these points are neither minima nor maxima.

33. $\dfrac{\partial f}{\partial x} \approx 4, \quad \dfrac{\partial f}{\partial y} \approx 2$

35. $\dfrac{\partial f}{\partial x} \approx 1, \quad \dfrac{\partial f}{\partial y} \approx -\dfrac{2}{3}$

37. $\dfrac{\partial C}{\partial t}(10, 10) \approx 1.4, \quad \dfrac{\partial C}{\partial s}(10, 10) \approx -2.4$

When the temperature is $10°$ and the wind speed is 10 mph, an increase in temperature of $1°$ will increase the wind chill by approximately 1.4 degrees, whereas an increase in wind speed of 1 mph will decrease the wind chill by about 2.4 degrees.

If $\dfrac{\partial C}{\partial t}(10, 10) = 1$, then a $1°$ increase in temperature would correspond to a $1°$ increase in the wind chill temperature. It is perhaps surprising that a $1°$ increase in temperature leads to a greater increase in the "felt" temperature (when the wind speed is 10 mph).

39. $\dfrac{\partial f}{\partial v}(170, 3000) \approx 2.2$ feet per ft/sec. An increase of 1 foot per second of velocity increases the range by approximately 2.2 feet.

$\dfrac{\partial f}{\partial w}(170, 3000) \approx 0.0195$ feet per rpm. An increase in 1 rpm increases the range by 0.0195 feet.

41. $\dfrac{\partial f}{\partial x}(x, y, z) = \lim\limits_{h \to 0} \dfrac{f(x + h, y, z) - f(x, y, z)}{h}$

$\dfrac{\partial f}{\partial y}(x, y, z) = \lim\limits_{h \to 0} \dfrac{f(x, y + h, z) - f(x, y, z)}{h}$

$\dfrac{\partial f}{\partial z}(x, y, z) = \lim\limits_{h \to 0} \dfrac{f(x, y, z + h) - f(x, y, z)}{h}$

43. $\dfrac{\partial f_n}{\partial x} = n\pi \cos n\pi x \cos n\pi ct$

$\dfrac{\partial^2 f_n}{\partial x^2} = -n^2\pi^2 \sin n\pi x \cos n\pi ct$

$\dfrac{\partial f_n}{\partial t} = -n\pi c \sin n\pi x \sin n\pi ct$

$\dfrac{\partial^2 f_n}{\partial t^2} = -n^2\pi^2 c^2 \sin n\pi x \cos n\pi ct$

So, $c^2 \dfrac{\partial^2 f_n}{\partial x^2} = \dfrac{\partial^2 f_n}{\partial t^2}$.

45. $\dfrac{\partial V}{\partial I} = -5000\left[\dfrac{(1+0.1(1-T))^5}{(1+I)^6}\right]$

$= \dfrac{-5}{1+I}V$

$\dfrac{\partial V}{\partial T} = (5)(-0.1)1000\left[\dfrac{(1+0.1(1-T))^4}{(1+I)^5}\right]$

$= \dfrac{-0.5}{1+0.1(1-T)}V$

The inflation rate has a greater influence on V.

47. $\dfrac{\partial p}{\partial x} = \cos x\cos t$. This describes the change in the position of the string at a fixed time as the distance along the string changes.

$\dfrac{\partial p}{\partial t} = -\sin x\sin t$. This describes the change in position of the string at a fixed distance from the end as time changes.

49. $G/T = \dfrac{H}{T} - S$, so $\dfrac{\partial(G/T)}{\partial T} = \dfrac{-H}{T^2}$.

51. $\dfrac{\partial R}{\partial R_1} = \dfrac{(R_1R_2 + R_1R_3 + R_2R_3)(R_2R_3)}{(R_1R_2 + R_1R_3 + R_2R_3)^2}$

$- \dfrac{(R_1R_2R_3)(R_2 + R_3)}{(R_1R_2 + R_1R_3 + R_2R_3)^2}$

$= \dfrac{R_2^2 R_3^2}{(R_1R_2 + R_1R_3 + R_2R_3)^2} = \left(\dfrac{R}{R_1}\right)^2$

Due to the symmetry we can easily write

$\dfrac{\partial R}{\partial R_2} = \left(\dfrac{R}{R_2}\right)^2$, and $\dfrac{\partial R}{\partial R_3} = \left(\dfrac{R}{R_3}\right)^2$.

53. $P(100, 60, 15) = \dfrac{(100)(60)}{15} = 400$. Since 100 animals were tagged, we estimate that we tagged $\frac{1}{4}$ of the population.

$\dfrac{\partial P}{\partial t} = -\dfrac{TS}{t^2}$, so $\dfrac{\partial P}{\partial t}(100, 60, 15) \approx -27$. If one more recaptured animal were tagged, our estimate of the total population would decrease by 27 animals.

55. $\dfrac{\partial P}{\partial L} = 0.75L^{-0.25}K^{0.25}$

$\dfrac{\partial P}{\partial K} = 0.25L^{0.75}K^{-0.75}$

57. $\dfrac{\partial D_1}{\partial p_2} = -5$ and $\dfrac{\partial D_2}{\partial p_1} = -6$. They are complementary because an increase in the price of one decreases the demand for the other.

59. $\dfrac{\partial P}{\partial L} = 10K^{1/3}L^{-1/2}$, so $\dfrac{\partial P}{\partial L}(125, 900) = \dfrac{5}{3}$. Increasing the workforce by 1 unit (a thousand workers) will increase the output by $\frac{5}{3}$.

61. $f_x(x, y) = \dfrac{y(x^4 + 4x^2y^2 - y^4)}{(x^2 + y^2)^2}$, and

$f_y(x, y) = \dfrac{x(x^4 - 4x^2y^2 - y^4)}{(x^2 + y^2)^2}$ for $(x, y) \neq (0, 0)$. For $(x, y) = (0, 0)$, the limit definition gives $f_x = f_y = 0$. We compute

$f_{xy}(0, 0) = \lim_{h\to 0}\dfrac{f_x(0, 0 + h) - f_x(0, 0)}{h}$

$\lim_{h\to 0}\dfrac{-h^5}{h^5} = -1$

$f_{yx}(0, 0) = \lim_{h\to 0}\dfrac{f_y(0 + h, 0) - f_y(0, 0)}{h}$

$\lim_{h\to 0}\dfrac{h^5}{h^5} = 1$

The mixed partial derivatives are not continuous on an open set containing $(0, 0)$.

63. $\dfrac{\partial f}{\partial x} = \dfrac{-1}{x^2}\sin xy^2 + \dfrac{y^2}{x}\cos xy^2$

$\dfrac{\partial^2 f}{\partial x\partial y} = \dfrac{-2xy}{x^2}\cos xy^2 + \dfrac{2y}{x}\cos xy^2$

$+ \dfrac{y^2}{x}(-2xy)\sin xy^2$

$= -2y^3\sin xy^2$

$\dfrac{\partial f}{\partial y} = \dfrac{2yx}{x}\cos xy^2 = 2y\cos xy^2$

$\dfrac{\partial^2 f}{\partial y\partial x} = -2y^3\sin xy^2$

$\dfrac{\partial^2 f}{\partial x\partial y} = \dfrac{\partial^2 f}{\partial y\partial x}$, but differentiating with respect to y first was much easier!

65. $\dfrac{\partial f}{\partial x}(x_0, y_0)$ is equal to the slope at x_0 of the curve obtained by intersecting the surface $z = f(x, y)$ with the plane $y = y_0$.

$\dfrac{\partial^2 f}{\partial x^2}(x_0, y_0)$ is the concavity of this curve at the point $x = x_0$.

13.4 TANGENT PLANES AND LINEAR APPROXIMATIONS

1. $f_x = 2x, \quad f_y = 2y$

(a) $f_x(2, 1) = 4, \quad f_y(2, 1) = 2$

The tangent plane at $(2, 1, 4)$ is

$4(x - 2) + 2(y - 1) - (z - 4) = 0$.

The normal line is

$x = 2 + 4t, \ y = 1 + 2t, \ z = 4 - t$.

(b) $f_x(0, 2) = 0, \quad f_y(0, 2) = 4$

The tangent plane at $(0, 2, 3)$ is

$0(x - 0) + 4(y - 2) - (z - 3) = 0$.

The normal line is

$x = 0, \ y = 2 + 4t, \ z = 3 - t$.

3. $f_x = \cos x \cos y, \quad f_y = -\sin x \sin y$

(a) $f_x(0, \pi) = -1, \quad f_y(0, \pi) = 0$

The tangent plane at $(0, \pi, 0)$ is

$-1(x - 0) + 0(y - \pi) - (z - 0) = 0$.

The normal line is

$x = -t, \ y = \pi, \ z = -t$.

(b) $f_x(\frac{\pi}{2}, \pi) = 0, \quad f_y(\frac{\pi}{2}, \pi) = 0$

The tangent plane at $(\frac{\pi}{2}, \pi, -1)$ is

$0(x - \frac{\pi}{2}) + 0(y - \pi) - (z + 1) = 0$.

The normal line is

$x = \frac{\pi}{2}, \ y = \pi, \ z = -1 - t$.

5. $f_x = \dfrac{x}{\sqrt{x^2 + y^2}}, \quad f_y = \dfrac{y}{\sqrt{x^2 + y^2}}$

(a) $f_x(-3, 4) = \frac{-3}{5}, \quad f_y(-3, 4) = \frac{4}{5}$

The tangent plane at $(-3, 4, 5)$ is

$\dfrac{-3}{5}(x + 3) + \dfrac{4}{5}(y - 4) - (z - 5) = 0$.

The normal line is

$x = -3 - \frac{3}{5}t, \ y = 4 + \frac{4}{5}t, \ z = 5 - t$.

(b) $f_x(8, -6) = \frac{4}{5}, \quad f_y(8, -6) = -\frac{3}{5}$.

The tangent plane at $(8, -6, 10)$ is

$\dfrac{4}{5}(x - 8) - \dfrac{3}{5}(y + 6) - (z - 10) = 0$.

The normal line is

$x = 8 + \frac{4}{5}t, \ y = -6 - \frac{3}{5}t, \ z = 10 - t$.

7. $f_x = \dfrac{x}{\sqrt{x^2 + y^2}}, \quad f_y = \dfrac{y}{\sqrt{x^2 + y^2}}$

(a) $f_x(3, 0) = 1, \quad f_y(3, 0) = 0$

$L(x, y) = 3 + 1(x - 3) + 0(y - 0) = x$

(b) $f_x(0, -3) = 0, \quad f_y(0, -3) = -1$

$L(x, y) = 3 + 0(x - 0) - 1(y + 3)$
$= -y$

9. $f_x = (1 + xy^2)e^{xy^2}, \quad f_y = 2x^2ye^{xy^2} + 6y$

(a) $f_x(0, 1) = 1, \quad f_y(0, 1) = 6$

$L(x, y) = 3 + 1(x - 0) + 6(y - 1)$
$= x + 6y - 3$

(b) $f_x(2, 0) = 1, \quad f_y(2, 0) = 0$

$L(x, y) = 2 + (x - 2) + 0(y - 0) = x$

11. $f_w = 2wxy - yze^{wyz}$,
$f_x = w^2y$,
$f_y = w^2x - wze^{wyz}$,
$f_z = -wye^{wyz}$

(a) $f_w(-2, 3, 1, 0) = -12$
$f_x(-2, 3, 1, 0) = 4$
$f_y(-2, 3, 1, 0) = 12$
$f_z(-2, 3, 1, 0) = 2$

$L(w, x, y, z) = -11 - 12(w + 2)$
$+ 4(x - 3) + 12(y - 1) + 2(z - 0)$
$= -12w + 4x + 12y + 2z - 37$

(b) $f_w(0, 1, -1, 2) = 2$
$f_x(0, 1, -1, 2) = 0$

$$f_y(0, 1, -1, 2) = 0$$
$$f_z(0, 1, -1, 2) = 0$$
$$L(w, x, y, z)$$
$$= -1 + 2(w - 0) = 2w - 1$$

13. $L(x, y) = x$

x	y	$L(x, y)$	$f(x, y)$
3	−0.1	3	3.00167
3.1	0	3.1	3.1
3.1	−0.1	3.1	3.10161

15. $L(x, y) = -x$

x	y	$L(x, y)$	$f(x, y)$
0	3	0	0
0.1	π	−0.1	−0.09983
0.1	3	−0.1	−0.09883

17. As in example 4.5,

$$\frac{\partial S}{\partial L}(36, 2, 6) = 0.1728$$

$$\frac{\partial S}{\partial h}(36, 2, 6) = -0.7776$$

$$\frac{\partial S}{\partial w}(36, 2, 6) = -0.7776$$

$$S(36, 2, 6) = 1.5552.$$

The maximum sag occurs if $(L - 36) = 0.5$, $(w - 2) = -0.2$ and $(h - 6) = -0.5$. The linear approximation predicts the change in sag will be $0.5(0.1728) + 0.2(0.7776) + 0.5(0.7776) = 0.6307$.

The range of sags will be 1.5552 ± 0.6307.

19. $g(9.9, 930) \approx 4 + 0.3(-0.1) - 0.004(30)$
$= 3.85$

21. The linear approximation will be
$$g(s, t) \approx 4 + 0.1(s - 10) - 0.001(t - 900)$$
$$g(10.2, 890) \approx 4.03.$$

23. $f_x = 2y, \quad f_y = 2x + 2y$

$$\Delta z = f(x + \Delta x, y + \Delta y) - f(x, y)$$
$$= 2(x + \Delta x)(y + \Delta y) + (y + \Delta y)^2$$
$$- (2xy + y^2)$$
$$= (2y)\Delta x + (2x + 2y)\Delta y$$
$$+ (2\Delta y)\Delta x + (\Delta y)\Delta y$$

25. $f_x = 2x, \quad f_y = 2y$

$$\Delta z = f(x + \Delta x, y + \Delta y) - f(x, y)$$
$$= (x + \Delta x)^2 + (y + \Delta y)^2 - (x^2 + y^2)$$
$$= (2x)\Delta x + (2y)\Delta y + (\Delta x)\Delta x + (\Delta y)\Delta y$$

27. $f_x = 2x + 3y, \quad f_y = 3x$

$$\Delta z = f(x + \Delta x, y + \Delta y) - f(x, y)$$
$$= (x + \Delta x)^2 + 3(x + \Delta x)(y + \Delta y)$$
$$- (x^2 + 3xy)$$
$$= (2x + 3y)\Delta x + (3x)\Delta y$$
$$+ (\Delta x)\Delta x + (3\Delta x)\Delta y$$
$$= f_x \Delta x + f_y \Delta y + \varepsilon_1 \Delta x + \varepsilon_2 \Delta y$$

where $\varepsilon_1 = \Delta x$ and $\varepsilon_2 = 3\Delta x$.
Since $\varepsilon_1 \to 0$ and $\varepsilon_2 \to 0$ as
$(\Delta x, \Delta y) \to (0, 0)$, f is differentiable.

29. $f_x = ye^x + \cos x, \quad f_y = e^x$
$$dz = (ye^x + \cos x)dx + e^x dy$$

31. $f_x(0, 0) = \lim\limits_{h \to 0} \dfrac{f(0 + h, 0) - f(0, 0)}{h} = 0$

$f_y(0, 0) = \lim\limits_{h \to 0} \dfrac{f(0, 0 + h) - f(0, 0)}{h} = 0$

Using Definition 4.1, at the origin we have

$$\Delta z = \frac{2(0 + \Delta x)(0 + \Delta y)}{(0 + \Delta x)^2 + (0 + \Delta y)^2}$$
$$= \frac{2\Delta x \Delta y}{\Delta x^2 + \Delta y^2}.$$

The function is differentiable if $\Delta z = f_x \Delta x + f_y \Delta y + \varepsilon_1 \Delta x + \varepsilon_2 \Delta y$ where ε_1 and ε_2 both go to zero as $(\Delta x, \Delta y) \to (0, 0)$. If the function is differentiable we must be able to write

$\dfrac{2\Delta x\Delta y}{\Delta x^2 + \Delta y^2} = \varepsilon_1\Delta x + \varepsilon_2\Delta y$, but the function on the left does not have a limit as $(\Delta x, \Delta y) \to (0, 0)$. (The limit is different along the lines $\Delta y = \Delta x$ and along $\Delta y = -\Delta x$.)

33. Use level curves for z-values between 0.9 and 1.1 with a graphing window of $-0.1 \le x \le 0.1$ and $-0.1 \le y \le 0.1$.

To move from the $z = 1.00$ level curve to the $z = 1.05$ level curve, you move 0.025 to the right, so $\frac{\partial f}{\partial x} \approx \frac{0.05}{0.025} = 2$.

To move from the $z = 1.00$ level curve to the $z = 1.05$ level curve you move 0.05 down, so $\frac{\partial f}{\partial y} \approx \frac{0.05}{-0.05} \approx -1$.

We also have $f(0, 0) = 1$. Therefore

$L(x, y) \approx 1 + 2(x - 0) - 1(y - 0)$.

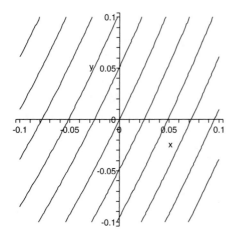

35. $f(0, 0) = 6$. We can get from the $z = 6$ level curve to the $z = 8$ level curve by moving 1 in the y direction, or by moving 0.5 in the x direction.

$\dfrac{\partial z}{\partial x} \approx \dfrac{2}{0.5} = 4$

$\dfrac{\partial z}{\partial y} \approx \dfrac{2}{1} = 2$

$L(x, y) = 6 + 4x + 2y$

37. $f(0, 0) = 3$

$\dfrac{\partial z}{\partial x}(0, 0) \approx \dfrac{1}{1} = 1$

$\dfrac{\partial z}{\partial y}(0, 0) \approx \dfrac{-1}{1.5} = -\dfrac{2}{3}$

$L(x, y) = 3 + x - \dfrac{2}{3}y$

39. (See exercise 37 from Section 13.3.)

$\dfrac{\partial w}{\partial t} \approx 1.4$

$\dfrac{\partial w}{\partial x} \approx -2.4$

$L(t, s) = -9 + 1.4(t - 10) - 2.4(s - 10)$

$L(12, 13) = -13.4$

41. With window $-1 \le x \le 1$ and $-1 \le y \le 1$, the contour plot is:

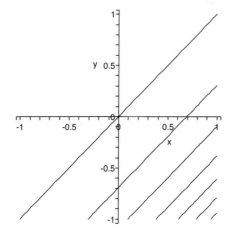

Moving from the $z = 1$ contour to the $z = 2$ contour moves 0.7 in the x direction or -0.7 in the y direction. This makes our approximation of $\dfrac{\partial z}{\partial x} \approx 1.43$ and $\dfrac{\partial z}{\partial y} \approx -1.43$.

The exact values are $\dfrac{\partial z}{\partial x} = 1$ and $\dfrac{\partial z}{\partial y} = -1$.

Zoomed in so the level curves are equally spaced , we get (with $-0.1 \le x \le 0.1$ and $-0.1 \le y \le 0.1$):

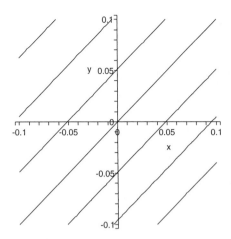

Moving from the $z = 1$ contour to the $z = 1.05$ contour moves 0.05 in the x direction or -0.05 in the y direction. This makes our approximation of $\dfrac{\partial z}{\partial x} \approx 1.0$ and $\dfrac{\partial z}{\partial y} \approx -1.0$.

In the first estimate, the function values were changing much more rapidly away from $(0, 0)$ than they were at $(0, 0)$. In the second estimate, the spacing between contours was even, so the function values were changing roughly the same amount throughout the window.

43. $\displaystyle\lim_{(x,y) \to (0,0)} f(x, y) = 0$ because

$$\left| \frac{x^2 y}{x^2 + y^2} - 0 \right| \le \left| \frac{x^2 y}{x^2} \right| = |y|,$$

and $\displaystyle\lim_{(x,y) \to (0,0)} |y| = 0$. Therefore $f(x, y)$ is continuous at $(0, 0)$.

That $f(x, y)$ is not differentiable at $(0, 0)$ is seen as follows.

$$f_x(0, 0) = \lim_{h \to 0} \frac{f(0 + h, 0) - f(0, 0)}{h} = 0$$

$$f_y(0, 0) = \lim_{h \to 0} \frac{f(0, 0 + h) - f(0, 0)}{h} = 0$$

Using Definition 4.1, at the origin we have

$$\Delta z = \frac{(0 + \Delta x)(0 + \Delta y)^2}{(0 + \Delta x)^2 + (0 + \Delta y)^2}$$

$$= \frac{\Delta x \, \Delta y^2}{\Delta x^2 + \Delta y^2}.$$

The function is differentiable if $\Delta z = f_x \Delta x + f_y \Delta y + \varepsilon_1 \Delta x + \varepsilon_2 \Delta y$ where ε_1 and ε_2 both go

to zero as $(\Delta x, \Delta y) \to (0, 0)$. If the function is differentiable, we must be able to write

$$\frac{\Delta x \, \Delta y^2}{\Delta x^2 + \Delta y^2} = \varepsilon_1 \Delta x + \varepsilon_2 \Delta y.$$

To see that this is impossible, assume that we have such an expression, solve for ε_1, and examine the limit:

$$\lim_{(\Delta x, \Delta y) \to (0,0)} \varepsilon_1 =$$

$$\lim_{(\Delta x, \Delta y) \to (0,0)} \left[\frac{\Delta y^2}{\Delta x^2 + \Delta y^2} - \varepsilon_2 \frac{\Delta y}{\Delta x} \right].$$

Along the line $\Delta y = \Delta x$, this gives $0 = \frac{1}{2} + 0$. Therefore the function f is not differentiable.

45. $\mathbf{r} = \langle 2u, v, 4uv \rangle$ and

$\mathbf{r}(1, 2) = \langle 2, 2, 8 \rangle$ and

$\mathbf{r}_u = \langle 2, 0, 4v \rangle$.

$\mathbf{r}_u(1, 2) = \langle 2, 0, 8 \rangle$

$\mathbf{r}_v = \langle 0, 1, 4u \rangle$

$\mathbf{r}_v(1, 2) = \langle 0, 1, 4 \rangle$

$\mathbf{r}_u(1, 2) \times \mathbf{r}_v(1, 2) = \langle -8, -8, 2 \rangle$

Therefore the tangent plane is

$-8(x - 2) - 8(y - 2) + 2(z - 8) = 0.$

47. $\mathbf{r} = \langle \cos u, \sin u, v \rangle$ for $0 \le u \le 2\pi$ and $0 \le v \le 2$. The point $(1,0,1)$ corresponds to $(u, v) = (0, 1)$.

$\mathbf{r}(0, 1) = \langle 1, 0, 1 \rangle$ and

$\mathbf{r}_u = \langle -\sin u, \cos u, 0 \rangle$.

$\mathbf{r}_u(0, 1) = \langle 0, 1, 0 \rangle$

$\mathbf{r}_v = \langle 0, 0, 1 \rangle$

$\mathbf{r}_v(0, 1) = \langle 0, 0, 1 \rangle$

$\mathbf{r}_u(0, 1) \times \mathbf{r}_v(0, 1) = \langle 1, 0, 0 \rangle$

Therefore the tangent plane is

$(x - 1) = 0.$

13.5 THE CHAIN RULE

1. $g(t) = (t^2 - 1)^2 e^{\sin t}$

$g'(t) = 2(t^2 - 1)(2t)e^{\sin t} + (t^2 - 1)^2 \cos t \, e^{\sin t}$

3. $g'(t) = \dfrac{\partial f}{\partial x}\dfrac{dx}{dt} + \dfrac{\partial f}{\partial y}\dfrac{dy}{dt}$

$\dfrac{\partial f}{\partial x} = 2xy, \qquad \dfrac{\partial f}{\partial y} = x^2 - \cos y$

$\dfrac{dx}{dt} = \dfrac{t}{\sqrt{t^2 + 1}}, \qquad \dfrac{dy}{dt} = e^t$

$g'(t) = 2xy\dfrac{t}{\sqrt{t^2 + 1}} + (x^2 - \cos y)e^t$

$= 2\sqrt{t^2 + 1}\,e^t\dfrac{t}{\sqrt{t^2 + 1}}$

$+ [(t^2 + 1) - \cos e^t]e^t$

$= (2t + t^2 + 1 - \cos e^t)e^t$

5. $\dfrac{\partial g}{\partial u} = \dfrac{\partial f}{\partial x}\dfrac{\partial x}{\partial u} + \dfrac{\partial f}{\partial y}\dfrac{\partial y}{\partial u}$

$= 8xy^3(3u^2 - v\cos u) + 12x^2y^2(8u)$

$= 8(u^3 - v\sin u)(4u^2)^3(3u^2 - v\cos u)$

$+ 12(u^3 - v\sin u)^2(4u^2)^2(8u)$

$\dfrac{\partial g}{\partial v} = \dfrac{\partial f}{\partial x}\dfrac{\partial x}{\partial v} + \dfrac{\partial f}{\partial y}\dfrac{\partial y}{\partial v}$

$= 8xy^3(-\sin u) + 12x^2y^2(0)$

$= 8(u^3 - v\sin u)(4u^2)^3(-\sin u)$

7. $g'(t) = \dfrac{\partial f}{\partial x}\dfrac{dx}{dt} + \dfrac{\partial f}{\partial y}\dfrac{dy}{dt} + \dfrac{\partial f}{\partial z}\dfrac{dz}{dt}$

9. $\dfrac{\partial g}{\partial u} = \dfrac{\partial f}{\partial x}\dfrac{\partial x}{\partial u} + \dfrac{\partial f}{\partial y}\dfrac{\partial y}{\partial u}$

$\dfrac{\partial g}{\partial v} = \dfrac{\partial f}{\partial x}\dfrac{\partial x}{\partial v} + \dfrac{\partial f}{\partial y}\dfrac{\partial y}{\partial v}$

$\dfrac{\partial g}{\partial w} = \dfrac{\partial f}{\partial x}\dfrac{\partial x}{\partial w} + \dfrac{\partial f}{\partial y}\dfrac{\partial y}{\partial w}$

11. $\dfrac{\partial P}{\partial k}(6, 4) \approx 3.6889$

$\dfrac{\partial P}{\partial l}(6, 4) \approx 16.6002$

$k'(t) = 0.1, \quad l'(t) = -0.06$

$g'(t) = \dfrac{\partial P}{\partial k}k'(t) + \dfrac{\partial P}{\partial l}l'(t)$

$\approx (3.6889)(0.1) + (16.6002)(-0.06)$

$= -0.6271$

13. $\dfrac{\partial P}{\partial k} = \dfrac{16}{3}k^{-2/3}l^{2/3}, \qquad \dfrac{\partial P}{\partial l} = \dfrac{32}{3}k^{1/3}l^{-1/3}$

$\dfrac{\partial P}{\partial k}(4, 3) \approx 4.4026, \qquad \dfrac{\partial P}{\partial l}(4, 3) \approx 11.7402$

$k'(t) = -0.2, \quad l'(t) = 0.08$

$g'(t) = \dfrac{\partial P}{\partial k}k'(t) + \dfrac{\partial P}{\partial l}l'(t)$

$= (4.4026)(-0.2) + (11.7402)(0.08)$

$= 0.0587$

15. $I(t) = q(t)p(t)$

$\dfrac{dq}{dt} = 0.05q(t), \qquad \dfrac{dp}{dt} = 0.03p(t)$

$\dfrac{dI}{dt} = \dfrac{\partial I}{\partial q}\dfrac{dq}{dt} + \dfrac{\partial I}{\partial p}\dfrac{dp}{dt}$

$= p(t)\dfrac{dq}{dt} + q(t)\dfrac{dp}{dt}$

$= p(t)[0.05q(t)] + q(t)[0.03p(t)]$

$= 0.08p(t)q(t)$

$= 0.08I(t)$

Income increases at a rate of 8% as claimed.

17. $g'(t) = f_x x'(t) + f_y y'(t)$

$g''(t) = f_x x''(t) + (f_{xx}x'(t) + f_{xy}y'(t))x'(t)$

$+ f_y y''(t) + (f_{yx}x'(t) + f_{yy}y'(t))y'(t)$

$= f_{xx}(x'(t))^2 + 2f_{xy}x'(t)y'(t)$

$+ f_{yy}(y'(t))^2 + f_x x''(t) + f_y y''(t)$

19. $\dfrac{\partial g}{\partial u} = \dfrac{\partial f}{\partial x}\dfrac{\partial x}{\partial u} + \dfrac{\partial f}{\partial y}\dfrac{\partial y}{\partial u}$

$\dfrac{\partial^2 g}{\partial u^2} = \dfrac{\partial f}{\partial x}\dfrac{\partial^2 x}{\partial u^2} + \left(\dfrac{\partial^2 f}{\partial x^2}\dfrac{\partial x}{\partial u} + \dfrac{\partial^2 f}{\partial x \partial y}\dfrac{\partial y}{\partial u}\right)\dfrac{\partial x}{\partial u}$

$+ \dfrac{\partial f}{\partial y}\dfrac{\partial^2 y}{\partial u^2} + \left(\dfrac{\partial^2 f}{\partial y \partial x}\dfrac{\partial x}{\partial u} + \dfrac{\partial^2 f}{\partial y^2}\dfrac{\partial y}{\partial u}\right)\dfrac{\partial y}{\partial u}$

$= \dfrac{\partial^2 f}{\partial x^2}\left(\dfrac{\partial x}{\partial u}\right)^2 + 2\dfrac{\partial^2 f}{\partial x \partial y}\dfrac{\partial x}{\partial u}\dfrac{\partial y}{\partial u}$

$+ \dfrac{\partial^2 f}{\partial y^2}\left(\dfrac{\partial y}{\partial u}\right)^2 + \dfrac{\partial f}{\partial x}\dfrac{\partial^2 x}{\partial u^2} + \dfrac{\partial f}{\partial y}\dfrac{\partial^2 y}{\partial u^2}$

21. $F(x, y, z) = 3x^2z + 2z^3 - 3yz$

$$F_x = 6xz$$

$$F_y = -3z$$

$$F_z = 3x^2 + 6z^2 - 3y$$

$$\frac{\partial z}{\partial x} = -\frac{F_x}{F_z} = \frac{-6xz}{3x^2 + 6z^2 - 3y}$$

$$\frac{\partial z}{\partial y} = -\frac{F_y}{F_z} = \frac{3z}{3x^2 + 6z^2 - 3y}$$

23. $F(x, y, z) = 3e^{xyz} - 4xz^2 + x \cos y - 2$

$$F_x = 3yze^{xyz} - 4z^2 + \cos y$$

$$F_y = 3xze^{xyz} - x \sin y$$

$$F_z = 3xye^{xyz} - 8xz$$

$$\frac{\partial z}{\partial x} = -\frac{F_x}{F_z} = \frac{-3yze^{xyz} + 4z^2 - \cos y}{3xye^{xyz} - 8xz}$$

$$\frac{\partial z}{\partial y} = -\frac{F_y}{F_z} = \frac{-3xze^{xyz} + x \sin y}{3xye^{xyz} - 8xz}$$

25. The chain rule gives $f_\theta = f_x \dfrac{\partial x}{\partial \theta} + f_y \dfrac{\partial y}{\partial \theta}$.

$$\frac{\partial x}{\partial \theta} = -r \sin \theta \text{ and } \frac{\partial y}{\partial \theta} = r \cos \theta, \text{ so}$$

$$f_\theta = -f_x r \sin \theta + f_y r \cos \theta.$$

27. From exercises 25 and 26, and Example 5.4, we have

$$f_r = f_x \cos \theta + f_y \sin \theta$$

$$f_{rr} = f_{xx} \cos^2 \theta + 2f_{xy} \cos \theta \sin \theta + f_{yy} \sin^2 \theta$$

$$f_\theta = -f_x r \sin \theta + f_y r \cos \theta$$

$$f_{\theta\theta} = f_{xx}r^2 \sin^2 \theta - 2f_{xy}r^2 \sin \theta \cos \theta$$

$$+ f_{yy}r^2 \cos^2 \theta - f_x r \cos \theta - f_y r \sin \theta$$

$$f_{rr} + \frac{1}{r}f_r + \frac{1}{r^2}f_{\theta\theta}$$

$$= (f_{xx} \cos^2 \theta + 2f_{xy} \cos \theta \sin \theta + f_{yy} \sin^2 \theta)$$

$$+ \frac{1}{r}(f_x \cos \theta + f_y \sin \theta)$$

$$+ \frac{1}{r^2}(f_{xx}r^2 \sin^2 \theta - 2f_{xy}r^2 \sin \theta \cos \theta$$

$$+ f_{yy}r^2 \cos^2 \theta - f_x r \cos \theta - r_y \sin \theta)$$

$$= f_{xx} + f_{yy}$$

29. Make the change of variables $X = \dfrac{x}{L}$ and $T = \dfrac{\alpha^2}{L^2}t$. Then,

$$u_t = u_X \frac{\partial X}{\partial t} + u_T \frac{\partial T}{\partial t}$$

$$= \frac{\alpha^2}{L^2}u_T$$

$$u_x = u_X \frac{\partial X}{\partial x} + u_T \frac{\partial T}{\partial x}$$

$$= \frac{1}{L}u_X$$

$$u_{xx} = \frac{1}{L}\left(u_{XX}\frac{\partial X}{\partial x} + u_{XT}\frac{\partial T}{\partial x}\right)$$

$$= \frac{1}{L^2}u_{XX}$$

The heat equation then becomes $\alpha^2 \dfrac{1}{L^2}u_{XX} = \dfrac{\alpha^2}{L^2}u_T$, or simply $u_{XX} = u_T$.

The dimensions of X are $\dfrac{\text{ft}}{\text{ft}} = 1$, and the dimensions of T are $\dfrac{\text{ft}^2/\text{sec}}{\text{ft}^2}\text{sec} = 1$. Both X and T are dimensionless.

31. Since $g(h) = f(x + hu_1, y + hu_2)$, it is clear that $g(0) = f(x, y)$.

$$g'(h) = f_x \frac{\partial(x + hu_1)}{\partial h} + f_y \frac{\partial(y + hu_2)}{\partial h}$$

$$= f_x(x + hu_1, y + hu_2)u_1$$

$$+ f_y(x + hu_1, y + hu_2)u_2$$

$$g'(0) = f_x(x, y)u_1 + f_y(x, y)u_2$$

$g''(h) = f_{xx}u_1^2 + f_{xy}u_1u_2 + f_{yx}u_2u_1 + f_{yy}u_2^2$
$= f_{xx}u_1^2 + 2f_{xy}u_1u_2 + f_{yy}u_2^2$ where each second partial of f is evaluated at $(x + hu_1, y + hu_2)$. Therefore,

$g''(0) = f_{xx}u_1^2 + 2f_{xy}u_1u_2 + f_{yy}u_2^2$ where each second partial of f is evaluated at (x, y).

Continuing in this vein, we see that

$g'''(0) = f_{xxx}u_1^3 + 3f_{xxy}u_1^2u_2$
$+ 3f_{xyy}u_1u_2^2 + f_{yyy}u_2^3$

$g^{(4)}(0) = f_{xxxx}u_1^4 + 4f_{xxxy}u_1^3u_2$
$+ 6f_{xxyy}u_1^2u_2^2$
$+ 4f_{xyyy}u_1u_2^3 + f_{yyyy}u_2^4$

The coefficients are from the binomial expansion (Pascal's triangle), the number of partial derivatives with respect to x match the powers of u_1, and the number of partial derivatives with respect to y match the powers of u_2.

33. $f(x, y) = \sin x \cos y$ $\qquad f(0, 0) = 0$
$f_x(x, y) = \cos x \cos y$ $\qquad f_x(0, 0) = 1$
$f_y(x, y) = -\sin x \sin y$ $\qquad f_y(0, 0) = 0$
$f_{xx}(x, y) = -\sin x \cos y$ $\qquad f_{xx}(0, 0) = 0$
$f_{xy}(x, y) = -\cos x \sin y$ $\qquad f_{xy}(0, 0) = 0$
$f_{yy}(x, y) = -\sin x \cos y$ $\qquad f_{yy}(0, 0) = 0$
$f_{xxx}(x, y) = -\cos x \cos y$ $\qquad f_{xxx}(0, 0) = -1$
$f_{xxy}(x, y) = \sin x \sin y$ $\qquad f_{xxy}(0, 0) = 0$
$f_{xyy}(x, y) = -\cos x \cos y$ $\qquad f_{xyy}(0, 0) = -1$
$f_{yyy}(x, y) = \sin x \sin y$ $\qquad f_{yyy}(0, 0) = 0$
$f(\Delta x, \Delta y) \approx 1\Delta x + \frac{1}{3!}[1(-1)\Delta x^3 +$
$3(-1)\Delta x \Delta y^2] = \Delta x - \frac{1}{6}\Delta x^3 - \frac{1}{2}\Delta x \Delta y^2$

35. $f(x, y) = \sin xy$ $\qquad f(0, 0) = 0$
$f_x(x, y) = y \cos xy$ $\qquad f_x(0, 0) = 0$
$f_y(x, y) = x \cos xy$ $\qquad f_y(0, 0) = 0$
$f_{xx}(x, y) = -y^2 \sin xy$ $\qquad f_{xx}(0, 0) = 0$
$f_{xy}(x, y) = \cos xy - xy \sin xy$
$f_{xy}(0, 0) = 1$
$f_{yy}(x, y) = -x^2 \sin xy$ $\qquad f_{yy}(0, 0) = 0$

$f_{xxx}(0, 0) = f_{xxy}(0, 0) = f_{xyy}(0, 0)$
$= f_{yyy}(0, 0) = 0$

$f(\Delta x, \Delta y) \approx \frac{1}{2!}2(1)\Delta x \Delta y = \Delta x \Delta y$

37. $f(x, y) = e^{2x+y}$ $\qquad f(0, 0) = 1$
$f_x(x, y) = 2e^{2x+y}$ $\qquad f_x(0, 0) = 2$
$f_y(x, y) = e^{2x+y}$ $\qquad f_y(0, 0) = 1$
$f_{xx}(x, y) = 4e^{2x+y}$ $\qquad f_{xx}(0, 0) = 4$
$f_{xy}(x, y) = 2e^{2x+y}$ $\qquad f_{xy}(0, 0) = 2$
$f_{yy}(x, y) = e^{2x+y}$ $\qquad f_{yy}(0, 0) = 1$
$f_{xxx}(x, y) = 8e^{2x+y}$ $\qquad f_{xxx}(0, 0) = 8$
$f_{xxy}(x, y) = 4e^{2x+y}$ $\qquad f_{xxy}(0, 0) = 4$
$f_{xyy}(x, y) = 2e^{2x+y}$ $\qquad f_{xyy}(0, 0) = 2$
$f_{yyy}(x, y) = e^{2x+y}$ $\qquad f_{yyy}(0, 0) = 1$
$f(\Delta x, \Delta y) \approx 1 + 2\Delta x + \Delta y$
$+ \frac{1}{2!}[4\Delta x^2 + 2(2)\Delta x \Delta y + \Delta y^2]$
$+ \frac{1}{3!}[8\Delta x^3 + 3(4)\Delta x^2 \Delta y$
$+ 3(2)\Delta x \Delta y^2 + \Delta y^3]$

$\quad = 1 + 2\Delta x + \Delta y + 2\Delta x^2 + 2\Delta x \Delta y$
$+ \frac{1}{2}\Delta y^2 + \frac{4}{3}\Delta x^3 + 2\Delta x^2 \Delta y$
$+ \Delta x \Delta y^2 + \frac{1}{6}\Delta y^3$

39. $R(c, h) = \dfrac{1}{\frac{0.55}{c} + \frac{0.45}{h}}$ $\quad R(1, 1) = 1$

$R_c(1, 1) = 0.55 \qquad R_h(1, 1) = 0.45$

$R(c, h) \approx 1 + 0.55\Delta c + 0.45\Delta h$

41. If $E = f(P, T)$ and $P = g(T, V)$, then substituting we get $E = f(g(T, V), T) = h(T, V)$.
Using the chain rule:

$$\left(\frac{\partial E}{\partial T}\right)_V = \frac{\partial f}{\partial T}\frac{\partial T}{\partial T} + \frac{\partial f}{\partial P}\frac{\partial g}{\partial T}$$

$$= \left(\frac{\partial E}{\partial T}\right)_P + \left(\frac{\partial E}{\partial P}\right)_T \left(\frac{\partial P}{\partial T}\right)_V .$$

43. $a' = \dfrac{bh' - hb'}{b^2} = \dfrac{b - h}{b^2}$

At $h = 50$ and $b = 200$, $a = 0.250$, and $a' = 0.00375 \approx 4$ points.

If $h = 100$ and $b = 400$, $a = 0.250$ still, and $a' = 0.01875 \approx 2$ points.

In general if the number of hits and at bats are doubled, then the rate of change of the average is halved.

45. Apply the natural log to write: $\ln g(t) = v(t) \ln u(t)$ and differentiate to get $\dfrac{1}{g(t)} g'(t) = v'(t) \ln u(t) + v(t) \dfrac{1}{u(t)} u'(t)$. Now solve for $g'(t)$ (using $g(t) = u(t)^{v(t)}$).

$g' = u^v (v' \ln u + \frac{v}{u} u')$

Applying this to $g(t) = (2t + 1)^{3t^2}$ yields

$g'(t) = (2t + 1)^{3t^2} (6t \ln(2t + 1) + \frac{3t^2}{2t+1}(2))$

$= (2t + 1)^{3t^2} (6t \ln(2t + 1) + \frac{6t^2}{2t+1}).$

13.6 THE GRADIENT AND DIRECTIONAL DERIVATIVES

1. $\nabla f = \left\langle 2x + 4y^2, 8xy - 5y^4 \right\rangle$

3. $\nabla f = \left\langle e^{xy^2} + xy^2 e^{xy^2}, 2x^2 y e^{xy^2} - 2y \sin y^2 \right\rangle$

5. $\nabla f = \left\langle \frac{8}{y} e^{4x/y} - 2, \frac{-8x}{y^2} e^{4x/y} \right\rangle$

$\nabla f(2, -1) = \left\langle -8e^{-8} - 2, -16e^{-8} \right\rangle$

7. $\nabla f = \left\langle 6xy + z \sin x, 3x^2, -\cos x \right\rangle$

$\nabla f(0, 2, -1) = \langle 0, 0, -1 \rangle$

9. $\nabla f = \langle 2w \cos x, -w^2 \sin x + 3yze^{xz},$
$\qquad 3e^{xz}, 3xye^{xz} \rangle$

$\nabla f(2, \pi, 1, 4) = \left\langle -4, 12e^{4\pi}, 3e^{4\pi}, 3\pi e^{4\pi} \right\rangle$

11. $\nabla f = \left\langle 2xy, x^2 + 8y \right\rangle$

$\nabla f(2, 1) = \langle 4, 12 \rangle$

$D_{\mathbf{u}} f(2, 1) = \langle 4, 12 \rangle \cdot \left\langle \frac{1}{2}, \frac{\sqrt{3}}{2} \right\rangle$

$\qquad = 2 + 6\sqrt{3}$

13. $\nabla f = \left\langle \dfrac{x}{\sqrt{x^2 + y^2}}, \dfrac{y}{\sqrt{x^2 + y^2}} \right\rangle$

$\nabla f(3, -4) = \left\langle \dfrac{3}{5}, \dfrac{-4}{5} \right\rangle$

$\mathbf{u} = \left\langle \dfrac{3}{\sqrt{13}}, \dfrac{-2}{\sqrt{13}} \right\rangle$

$D_{\mathbf{u}} f(3, -4) = \left\langle \dfrac{3}{5}, \dfrac{-4}{5} \right\rangle \cdot \left\langle \dfrac{3}{\sqrt{13}}, \dfrac{-2}{\sqrt{13}} \right\rangle$

$\qquad = \dfrac{17}{5\sqrt{13}}$

15. $\nabla f = \langle -2 \sin(2x - y), \sin(2x - y) \rangle$

$\nabla f(\pi, 0) = \langle 0, 0 \rangle$

$\mathbf{u} = \left\langle \dfrac{1}{\sqrt{2}}, \dfrac{1}{\sqrt{2}} \right\rangle$

$D_{\mathbf{u}} f(\pi, 0) = \langle 0, 0 \rangle \cdot \left\langle \dfrac{1}{\sqrt{2}}, \dfrac{1}{\sqrt{2}} \right\rangle$

$\qquad = 0$

17. $\nabla f = \langle 2x - 2y, -2x + 2y \rangle$

$\nabla f(-2, -1) = \langle -2, 2 \rangle$

$\mathbf{u} = \left\langle \dfrac{4}{\sqrt{20}}, \dfrac{-2}{\sqrt{20}} \right\rangle$

$= \left\langle \dfrac{2}{\sqrt{5}}, \dfrac{-1}{\sqrt{5}} \right\rangle$

$D_{\mathbf{u}} f(-2, -1) = \langle -2, 2 \rangle \cdot \left\langle \dfrac{2}{\sqrt{5}}, -\dfrac{1}{\sqrt{5}} \right\rangle$

$\qquad = -\dfrac{6}{\sqrt{5}}$

19. $\nabla f = \left\langle 3x^2 yz^2 - 4y, x^3 z^2 - 4x, 2x^3 yz \right\rangle$

$\nabla f(1, -1, 2) = \langle -8, 0, -4 \rangle$

$\mathbf{u} = \left\langle \dfrac{2}{\sqrt{5}}, 0, \dfrac{-1}{\sqrt{5}} \right\rangle$

$$D_{\mathbf{u}}f(1, -1, 2) = \langle -8, 0, -4 \rangle \cdot \left\langle \frac{2}{\sqrt{5}}, 0, \frac{-1}{\sqrt{5}} \right\rangle$$

$$= -\frac{12}{\sqrt{5}}$$

21. $\nabla f = \langle ye^{xy+z}, xe^{xy+z}, e^{xy+z} \rangle$

$\nabla f(1, -1, 1) = \langle -1, 1, 1 \rangle$

$\mathbf{u} = \left\langle \frac{4}{\sqrt{29}}, \frac{-2}{\sqrt{29}}, \frac{3}{\sqrt{29}} \right\rangle$

$D_{\mathbf{u}}f(1, -1, 1)$

$$= \langle -1, 1, 1 \rangle \cdot \left\langle \frac{4}{\sqrt{29}}, \frac{-2}{\sqrt{29}}, \frac{3}{\sqrt{29}} \right\rangle$$

$$= -\frac{3}{\sqrt{29}}$$

23. $\nabla f = \left\langle 2w\sqrt{x^2+1}, \frac{w^2 x}{\sqrt{x^2+1}} + 3z^2 e^{xz}, \right.$

$\left. 0, 3e^{xz} + 3xze^{xz} \right\rangle$

$\nabla f(2, 0, 1, 0) = \langle 4, 0, 0, 3 \rangle$

$\mathbf{u} = \left\langle \frac{1}{\sqrt{30}}, \frac{3}{\sqrt{30}}, \frac{4}{\sqrt{30}}, \frac{-2}{\sqrt{30}} \right\rangle$

$D_{\mathbf{u}}f(2, 0, 1, 0)$

$$= \langle 4, 0, 0, 3 \rangle \cdot \left\langle \frac{1}{\sqrt{30}}, \frac{3}{\sqrt{30}}, \frac{4}{\sqrt{30}}, \frac{-2}{\sqrt{30}} \right\rangle$$

$$= \frac{-2}{\sqrt{30}}$$

25. $\nabla f = \left\langle \frac{2x_1}{x_2}, \frac{-x_1^2}{x_2^2}, \frac{-2}{\sqrt{1-4x_3^2}}, \right.$

$\left. \frac{3}{2}\frac{\sqrt{x_5}}{\sqrt{x_4}}, \frac{3}{2}\frac{\sqrt{x_4}}{\sqrt{x_5}} \right\rangle$

$\nabla f(2, 1, 0, 1, 4) = \left\langle 4, -4, -2, 3, \frac{3}{4} \right\rangle$

$\mathbf{u} = \left\langle \frac{1}{5}, 0, \frac{-2}{5}, \frac{4}{5}, \frac{-2}{5} \right\rangle$

$D_{\mathbf{u}}f(2, 1, 0, 1, 4)$

$$= \left\langle 4, -4, -2, 3, \frac{3}{4} \right\rangle$$

$$\cdot \left\langle \frac{1}{5}, 0, \frac{-2}{5}, \frac{4}{5}, \frac{-2}{5} \right\rangle$$

$$= \frac{37}{10}$$

27. $\nabla f = \langle 2x, -3y^2 \rangle$

$\nabla f(2, 1) = \langle 4, -3 \rangle$

$\|\nabla f(2, 1)\| = 5$

The maximum change is 5; in the direction $\langle 4, -3 \rangle$.

The minimum change is -5; in the direction $\langle -4, 3 \rangle$.

29. $\nabla f = \langle 4y^2 e^{4x}, 2ye^{4x} \rangle$

$\nabla f(0, -2) = \langle 16, -4 \rangle$.

$\|\nabla f(0, -2)\| = \sqrt{272}$

The maximum change is $\sqrt{272}$; in the direction $\langle 16, -4 \rangle$.

The minimum change is $-\sqrt{272}$; in the direction $\langle -16, 4 \rangle$.

31. $\nabla f = \langle \cos 3y, -3x \sin 3y \rangle$

$\nabla f(2, 0) = \langle 1, 0 \rangle$

$\|\nabla f(2, 0)\| = 1$

The maximum change is 1; in the direction $\langle 1, 0 \rangle$.

The minimum change is -1; in the direction $\langle -1, 0 \rangle$.

33. $\nabla f = \left\langle \frac{2x}{\sqrt{2x^2-y}}, \frac{-1}{2\sqrt{2x^2-y}} \right\rangle$

$\nabla f(3, 2) = \left\langle \frac{3}{2}, -\frac{1}{8} \right\rangle$

$\|\nabla f(3, 2)\| = \frac{\sqrt{145}}{8}$

The maximum change is $\frac{\sqrt{145}}{8}$; in the direction $\left\langle \frac{3}{2}, -\frac{1}{8} \right\rangle$.

The minimum change is $-\dfrac{\sqrt{145}}{8}$; in the direction $\left\langle -\dfrac{3}{2}, \dfrac{1}{8} \right\rangle$.

35. $\nabla f = \langle 8xyz^3, 4x^2z^3, 12x^2yz^2 \rangle$
$\nabla f(1, 2, 1) = \langle 16, 4, 24 \rangle$
$\| \nabla f(1, 2, -2) \| = \sqrt{848}$

The maximum change is $\sqrt{848}$; in the direction $\langle 16, 4, 24 \rangle$.

The minimum change is $-\sqrt{848}$; in the direction $\langle -16, -4, -24 \rangle$.

37. The level curves (surfaces) are circles (spheres) centered at the origin. The gradients will be orthogonal to these, and therefore be parallel to the position vectors.

39. $\nabla \sin(x + y) = \langle \cos(x + y), \cos(x + y) \rangle$
$= \cos(x + y)\langle 1, 1 \rangle$

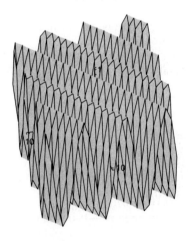

The gradient points in the direction of maximum change, and this will be the direction the wave is traveling.

$\nabla \sin(2x - y) = \langle 2\cos(2x - y), -\cos(2x - y) \rangle = \cos(2x - y)\langle 2, -1 \rangle$
This wave is traveling in the direction $\langle 2, -1 \rangle$.

41. $f(x, y, z) = x^2 + y^3 - z$
$\nabla f = \langle 2x, 3y^2, -1 \rangle$
$\nabla f(1, -1, 0) = \langle 2, 3, -1 \rangle$

The tangent plane is
$2(x - 1) + 3(y + 1) - z = 0.$

The normal line has parametric equations
$x = 1 + 2t, \; y = -1 + 3t, \; z = -t.$

43. $f(x, y, z) = x^2 + y^2 + z^2 - 6$
$\nabla f = \langle 2x, 2y, 2z \rangle$
$\nabla f(-1, 2, 1) = \langle -2, 4, 2 \rangle$

The tangent plane is
$-2(x + 1) + 4(y - 2) + 2(z - 1) = 0.$

The normal line has parametric equations
$x = -1 - 2t$
$y = 2 + 4t$
$z = 1 + 2t$

45. $\nabla f = \langle 4x - 4y, -4x + 4y^3 \rangle$

$\nabla f = \langle 0, 0 \rangle$ when $x = y$ and $-x + x^3 = 0$.
$x^3 - x$ factors as $x(x - 1)(x + 1)$, so the places where the tangent plane is parallel to the xy-plane are at $(0, 0, 0)$, $(1, 1, -1)$, and $(-1, -1, -1)$.

These are possible local extrema.

47.

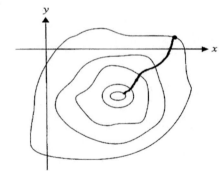

49.

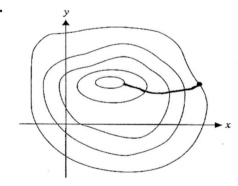

51. $\nabla f(0, 0) \approx \left\langle -\dfrac{4}{3}, -2 \right\rangle$

53. $f_x(0, 0) \approx \dfrac{2.2 - 1.8}{0.1 - (-0.1)} = 2$

$f_y(0, 0) \approx \dfrac{1.6 - 2.4}{0.2 - (-0.2)} = -2$

$\nabla f(0, 0) \approx \langle 2, -2 \rangle$

55. The gradient is
$\nabla f = \langle -\tan 10°, \tan 6° \rangle \approx \langle -0.176, 0.105 \rangle$,
and this gives the direction of maximum ascent. The rate of change in this direction is the magnitude of the gradient, $\|\nabla f\| \approx 0.205$, and the rise in degrees is $\tan^{-1} 0.205 \approx 11.6°$.

57. $\nabla f = \langle -8x, -2y \rangle$
$\nabla f(1, 2) = \langle -8, -4 \rangle$
The rain will run in direction $\langle 8, 4 \rangle$.

59. Heat flows in the direction of $-\nabla T$ because this is the direction of maximum *decrease* of the function T, i.e., the direction toward the colder region.

It is reasonable to expect the heat flow to be proportional to $\|\nabla T\|$ because the heat will flow in the direction of $-\nabla T$, and the directional derivative in this direction is equal to $\|\nabla T\|$.

61. The y-coordinate of position needs to decrease in order to decrease the error. The x-coordinate of velocity needs to increase in order to decrease the error.

63. If the shark moves toward a higher electrical charge, it should move in direction $\langle 12, -20, 5 \rangle$.

65. (a) True.
$\nabla(f + g) = \left\langle \frac{\partial f}{\partial x} + \frac{\partial g}{\partial x}, \frac{\partial f}{\partial y} + \frac{\partial g}{\partial y} \right\rangle$
$= \left\langle \frac{\partial f}{\partial x}, \frac{\partial f}{\partial y} \right\rangle + \left\langle \frac{\partial g}{\partial x}, \frac{\partial g}{\partial y} \right\rangle$
$= \nabla f + \nabla g$

(b) True.
$\nabla(fg) = \left\langle \frac{\partial f}{\partial x}g + \frac{\partial g}{\partial x}f, \frac{\partial f}{\partial y}g + \frac{\partial g}{\partial y}f \right\rangle$
$= \left\langle \frac{\partial f}{\partial x}, \frac{\partial f}{\partial y} \right\rangle g + \left\langle \frac{\partial g}{\partial x}, \frac{\partial g}{\partial y} \right\rangle f$

$= (\nabla f)g + f(\nabla g)$

67. We have the function $g(w, s, t)$, with $g(4, 10, 900) = 4$. Then $\frac{\partial g}{\partial w}(4, 10, 900) = \frac{0.04}{0.05} = 0.8$, $\frac{\partial g}{\partial s}(4, 10, 900) = \frac{0.06}{0.2} = 0.3$, and $\frac{\partial g}{\partial t}(4, 10, 900) = \frac{-0.04}{10} = -0.004$. Then $\nabla g(4, 10, 900) = \langle 0.8, 0.3, -0.004 \rangle$ and this gives the direction of maximum increase of gauge.

13.7 EXTREMA OF FUNCTIONS OF SEVERAL VARIABLES

1. $f_x = -2xe^{-x^2}(y^2 + 1)$
$f_y = 2ye^{-x^2}$
$f_{xx} = (4x^2 - 2)e^{-x^2}(y^2 + 1)$
$f_{yy} = 2e^{-x^2}$
$f_{xy} = -4xye^{-x^2}$
Solving $\nabla f = \langle 0, 0 \rangle$ gives critical point $(x, y) = (0, 0)$. $D(0, 0) = (-2)2 - 0^2 = -4$, so f has a saddle point at $(0, 0)$.

3. $f_x = 3x^2 - 3y$
$f_y = -3x + 3y^2$
$f_{xx} = 6x$
$f_{yy} = 6y$
$f_{xy} = -3$
$D = 36xy - 9$

Solving $\nabla f = \langle 0, 0 \rangle$ gives equations $y = x^2$ and $x = y^2$.

Substituting gives
$x = (x^2)^2 = x^4$
$x^4 - x = 0$
$x(x^3 - 1) = 0$
so $x = 0$ or $x = 1$, and the critical points are $(0, 0)$ and $(1,1)$.

$D(0, 0) < 0$ so $(0, 0)$ is a saddle point.

$D(1, 1) > 0$ and $f_{xx} > 0$, so $(1, 1)$ is a local minimum.

5. $f_x = 2xy + 2x$
$f_y = 2y + x^2 - 2$
$f_{xx} = 2y + 2$
$f_{yy} = 2$
$f_{xy} = 2x$
$D = 2y + 4 - 4x^2$

Solving $\nabla f = \langle 0, 0 \rangle$ gives equations
$2x(y + 1) = 0$ and $x^2 + 2y - 2 = 0$.
Either $x = 0$ or $y = -1$.

This gives critical points at
$(0, 1)$ and $(\pm 2, -1)$.

$D(0, 1) = 6 > 0$ and $f_{xx} = 4 > 0$ so $(0, 1)$ is a
local minimum.

$D(\pm 2, -1) = -14 < 0$, so $(\pm 2, -1)$ are both
saddle points.

7. $f_x = -2xe^{-x^2-y^2}$
$f_y = -2ye^{-x^2-y^2}$
$f_{xx} = (4x^2 - 2)e^{-x^2-y^2}$
$f_{yy} = (4y^2 - 2)e^{-x^2-y^2}$
$f_{xy} = 4xye^{-x^2-y^2}$
$D = (4x^2 - 2)(4y^2 - 2)e^{-2x^2-2y^2}$
$\quad - 16x^2y^2e^{-2x^2-2y^2}$

Solving $\nabla f = \langle 0, 0 \rangle$ gives critical point $(0, 0)$.
$D(0, 0) = 4$, $f_{xx}(0, 0) = -2$
and so there is a local maximum at $(0, 0)$.

9. $f_x = 2x - \dfrac{4y}{y^2 + 1}$
$f_y = 4x \dfrac{y^2 - 1}{(y^2 + 1)^2}$

Solving $f_y = 0$ gives $x = 0$ or $y = \pm 1$. Now
$f_x = 0$ gives critical points $(0, 0)$, $(1, 1)$, and
$(-1, -1)$.

Using a CAS, we find that:

$D(0, 0) < 0$ so $(0, 0)$ is a saddle point.

$D(1, 1) > 0$ and $f_{xx}(1, 1) > 0$, so $(1, 1)$ is a
local minimum.

$D(-1, -1) > 0$ and $f_{xx}(-1, -1) > 0$, so
$(-1, -1)$ is a local minimum.

11. $f_x = (1 - 2x^2)e^{-x^2-y^2}$ $f_y = -2xye^{-x^2-y^2}$

Solving $f_x = 0$ gives $x = \pm \dfrac{1}{\sqrt{2}}$. Now $f_y = 0$
gives critical points $(\pm \dfrac{1}{\sqrt{2}}, 0)$.

Using a CAS, we find that:

$D(\dfrac{1}{\sqrt{2}}, 0) > 0$ and $f_{xx}(\dfrac{1}{\sqrt{2}}, 0) < 0$, so $(\dfrac{1}{\sqrt{2}}, 0)$
is a local maximum.

$D(-\dfrac{1}{\sqrt{2}}, 0) > 0$ and $f_{xx}(-\dfrac{1}{\sqrt{2}}, 0) > 0$, so
$(-\dfrac{1}{\sqrt{2}}, 0)$ is a local minimum.

13. $f_x = y(1 - 2x^2)e^{-x^2-y^2}$
$f_y = x(1 - 2y^2)e^{-x^2-y^2}$

Solving $f_x = 0$ gives $y = 0$ or $x = \pm \dfrac{1}{\sqrt{2}}$.
Then $f_y = 0$ gives critical points $(0, 0)$ and
$(\pm \dfrac{1}{\sqrt{2}}, \pm \dfrac{1}{\sqrt{2}})$.

$D(0, 0) < 0$, so $(0, 0)$ is a saddle point.

$D(\pm \dfrac{1}{\sqrt{2}}, \pm \dfrac{1}{\sqrt{2}}) > 0$, so all four of these
points are extrema. Using f_{xx} we see that
$\pm \left(\dfrac{1}{\sqrt{2}}, \dfrac{1}{\sqrt{2}} \right)$ are both local maxima, and
$\pm(\dfrac{1}{\sqrt{2}}, -\dfrac{1}{\sqrt{2}})$ are both local minima.

15. $f_x = y^2 - 2x + \frac{1}{4}x^3$
$f_y = 2xy - 1$

Equation $f_y = 0$ gives $y = \frac{1}{2x}$. Substituting this into $f_x = 0$ and clearing denominators yields $x^5 - 8x^3 + 1 = 0$. Numerically, this gives solutions at approximately
$(0.5054, 0.9892)$, $(2.8205, 0.1773)$, and
$(-2.8362, -0.1763)$.

$D(0.5054, 0.9892) \approx -5.7420$, so this is a
saddle point.

$D(2.8205, 0.1773) \approx 22.2488$ and
$f_{xx}(2.8205, 0.1773) \approx 3.966$ so this is a relative minimum.

$D(-2.8362, -0.1763) \approx -23.0012$, so this is
a saddle point.

17. $f_x = 2x(1 - x^2 + y^3)e^{-x^2-y^2}$

$f_y = y(2y^3 - 2x^2 - 3y)e^{-x^2-y^2}$

Using a CAS to find and analyze the critical points, we get:

$(0, 0)$ is a saddle point.

$(\pm 1, 0)$ are local maxima.

$(0, \sqrt{\frac{3}{2}})$ is a local minimum.

$(0, -\sqrt{\frac{3}{2}})$ is a local maximum.

$(\pm \frac{\sqrt{57}}{9}, \frac{-2}{3})$ are both saddle points.

19. The sum of the squares of the residuals is given by

$$f(a, b) = \sum_{k=1}^{n}(ax_k + b - y_k)^2.$$

The partial with respect to b is

$$f_b = \sum_{k=1}^{n} 2(ax_k + b - y_k)$$

and the partial with respect to a is

$$f_a = \sum_{k=1}^{n} 2x_k(ax_k + b - y_k).$$

Setting each of these equal to 0, dividing by 2, and distributing sums, we get equations

$$a\left(\sum_{k=1}^{n} x_k\right) + b\left(\sum_{k=1}^{n} 1\right) - \left(\sum_{k=1}^{n} y_k\right) = 0$$

and

$$a\left(\sum_{k=1}^{n} x_k^2\right) + b\left(\sum_{k=1}^{n} x_k\right) - \left(\sum_{k=1}^{n} x_k y_k\right) = 0$$

as desired.

21. The sum of the squares of the residuals is given by

$$f(a, b) = (a + b - 0.6)^2 + (4a + b - 2.0)^2 + (9a + b - 6.4)^2 + (16a + b - 21)^2.$$

$f_a = 708a + 60b - 804.4$

$f_b = 60a + 8b - 60$

Solving $f_a = f_b = 0$ yields

$a \approx 1.37$, and $b \approx -2.80$,

so the linear model is $y = 1.37x - 2.80$.

23. The sum of the squares of the residuals is given by

$$f(a, b) = (0a + b - 8910)^2 + (2a + b -$$

$$8800)^2 + (4a + b - 9040)^2$$
$$+ (6a + b - 9040)^2 + (8a + b - 9050)^2.$$

$f_a = 240a + 40b - 360,800$

$f_b = 40a + 10b - 89,680$

Solving $f_a = f_b = 0$ yields $a = 26$ and $b = 8864$, so the linear model is $y = 26x + 8864$.

This model predicts the average on day 12 will be 9176.

A linear model is not appropriate because the average fluctuates wildly, and is influenced by many external factors.

25. The sum of the squares of the residuals is given by

$$f(a, b) = (68a + b - 160)^2 + (70a + b - 172)^2 + (70a + b - 184)^2 + (71a + b - 180)^2.$$

$f_a = 38,930a + 558b - 97,160$

$f_b = 558a + 8b - 1392$

Solving $f_a = f_b = 0$ gives us

$a \approx 7.16$, $b \approx -325.26$,

so the linear model is $y = 7.16x - 325.26$.

This model predicts the weight of a $6'8''$ person will be $y(80) = 248$ pounds, and the weight of a $5'0''$ person will be $y(60) = 104$ pounds.

There are many other factors besides height that influence a person's weight.

27. The sum of the squares of the residuals is given by

$$f(a, b) = (15a + b - 4.57)^2 + (35a + b - 3.17)^2 + (55a + b - 1.54)^2 + (75a + b - 0.24)^2 + (95a + b + 1.25)^2.$$

Solving $f_a = f_b = 0$ gives us

$a \approx -0.07285$, $b \approx 5.66$,

so the linear model is $y = -0.07285x + 5.66$.

This model predicts the average number of points scored starting from the 60-yard line will be $y(60) = 1.29$, and the average number of points scored starting from the 40-yard line will be $y(40) = 2.75$.

29. $(x_0, y_0) = (0, -1)$

$\nabla f = \left\langle 2y - 4x, 2x + 3y^2\right\rangle$

$\nabla f(0, -1) = \langle -2, 3 \rangle$

$g(h) = f(0 - 2h, -1 + 3h)$

$g'(h) = -2 f_x(-2h, -1 + 3h)$

$\qquad + 3 f_y(-2h, -1 + 3h)$

$\qquad = -2\,[2(-1 + 3h) - 4(-2h)]$

$\qquad + 3[2(-2h) + 3(-1 + 3h)^2]$

$\qquad = 13 - 94h + 81h^2$

The smallest positive solution to $g'(h) = 0$ is $h \approx 0.16049$. This leads us to

$(x_1, y_1) = (0, -1) + 0.16049(-2, 3)$

$= (-0.32098, -0.51853)$

$\nabla f(-0.32098, -0.51853)$

$= \langle 0.24686, 0.16466 \rangle$

$g(h) = f(-0.32098 + 0.24686h,$

$-0.51853 + 0.16466h)$

$g'(h) = 0.24686 \cdot f_x(-0.32098 + 0.24686h,$

$-0.51853 + 0.16466h)$

$+0.16466 \cdot f_y(-0.32098 + 0.24686h,$

$-0.51853 + 0.16466h)$

$= 0.08805 - 0.16552h + 0.01339h^2$

Solving $g'(h) = 0$ gives $h \approx 11.80144, 0.55709$. Using the first positive value means we arrive at

$(x_2, y_2) = (-0.32098, -0.51853)$

$+0.55709\,\langle 0.24686, 0.16466 \rangle$

$\approx (-0.18346, -0.42680).$

31. $\nabla f = \left\langle 1 - 2xy^4, -4x^2 y^3 + 2y \right\rangle$

$\nabla f(1, 1) = \langle -1, -2 \rangle$

$g(h) = f(1 - h, 1 - 2h)$

$g'(h) = -1 \cdot f_x(1 - h, 1 - 2h)$

$\qquad - 2 \cdot f_y(1 - h, 1 - 2h)$

$\qquad = 5 - 74h + 264h^2 - 416h^3 + 320h^4$

$\qquad - 96h^5$

The smallest positive solution to $g'(h) = 0$ is approximately $h = 0.09563$, and we arrive at

$(x_1, y_1) = (1, 1) + 0.09563\,\langle -1, -2 \rangle$

$= (0.90437, 0.80874)$

$\nabla f(0.90437, 0.80874)$

$= \langle 0.22623, -0.11305 \rangle$

$g(h) = f(0.90437 + 0.22623h,$

$0.80874 - 0.11305h)$

$g'(h) = 0.22623 \cdot f_x(0.90437 + 0.22623h,$

$0.80874 - 0.11305h)$

$- 0.11305 \cdot f_y(0.90437 + 0.22623h,$

$0.80874 - 0.11305h)$

$= 0.06396 + 0.09549h - 0.01337h^2$

$- 0.00315h^3 + 0.00086h^4$

$- 0.00005h^5$

The smallest positive solution to $g'(h) = 0$ is approximately $h = 10.56164$, and we arrive at

$(x_2, y_2) = (0.90437, 0.80874)$

$+ 10.56164\,\langle 0.22623, -0.11305 \rangle$

$\approx (3.29373, -0.38525).$

33. Refer to exercise 29: $\nabla f(0, 0) = \langle 0, 0 \rangle$

$g(h) = f(0 + 0h, 0 + 0h) = 0$, $g'(h) = 0$, and there is no smallest positive solution to $g'(h) = 0$. Graphically, we started at a point where the tangent plane was horizontal (a saddle point in this case), so the gradient didn't tell us which direction to move!

35. In the interior:

$f_x = 2x - 3y \qquad f_y = 3 - 3x$

Solving $f_y = 0$ gives $x = 1$. Then $f_x = 0$ gives $y = \dfrac{2}{3}$, and we have one critical point $\left(1, \dfrac{2}{3}\right)$.

Along $y = x$: $f(x, x) = g(x) = x^2 + 3x - 3x^2 = 3x - 2x^2$

$g'(x) = 3 - 4x = 0$ at $x = \dfrac{4}{3}$

This gives a local maximum at $\left(\dfrac{4}{3}, \dfrac{4}{3}\right)$.

Along $y = 0$:

$f(x, 0) = x^2$, which has a minimum at $(0, 0)$.

Along $x = 2$:

$f(2, y) = h(y) = 4 + 3y - 6y = 4 - 3y$

This will give a minimum and maximum value at the intersection points along the boundary.

The intersection points of the boundaries are $(0, 0)$, $(2, 2)$, and $(2, 0)$.

The function values at the points of interest are

$f(0, 0) = 0$

$f(2, 0) = 4$

$f(2, 2) = -2$

$f\left(1, \dfrac{2}{3}\right) = 1$

$f\left(\dfrac{4}{3}, \dfrac{4}{3}\right) = \dfrac{4}{9}$

The absolute maximum is 4 and the absolute minimum is -2.

37. In the interior:

$f_x = 2x \qquad f_y = 2y$

Solving $f_x = 0$, $f_y = 0$ gives critical point $(0, 0)$.

On the circle $(x - 1)^2 + y^2 = 4$, we substitute $y^2 = 4 - (x - 1)^2$ to get $f(x, y) = g(x) = x^2 + 4 - (x - 1)^2 = 3 + 2x$. This has no critical points, but is maximized for the largest value of x and minimized for the smallest value of x. The point with the largest value of x on the circle is $(3, 0)$. The point with the smallest value of x on the circle is $(-1, 0)$.

Finding functional values at all these points:

$f(0, 0) = 0$

$f(3, 0) = 9$

$f(-1, 0) = 1$

Therefore the absolute maximum is 9 and the absolute minimum is 0.

39. Let x, y, z be the dimensions. In this case, the amount of material used would be $2(xy + xz + yz) = 96$.

Solving for z, we get

$z = \dfrac{48 - xy}{x + y}$.

This gives volume $V(x, y) = xyz$

$= \dfrac{(48 - xy)xy}{x + y} = \dfrac{48xy - x^2y^2}{x + y}$

$V_x = \dfrac{(48y - 2xy^2)(x + y) - (48xy - x^2y^2)}{(x + y)^2}$

$= \dfrac{y^2(48 - 2xy - x^2)}{(x + y)^2}$

$V_y = \dfrac{(48x - 2x^2y)(x + y) - (48xy - x^2y^2)}{(x + y)^2}$

$= \dfrac{x^2(48 - 2xy - y^2)}{(x + y)^2}$

Solving $V_x = 0$ and $V_y = 0$ gives equations

$y^2(48 - 2xy - x^2) = 0$

$x^2(48 - 2xy - y^2) = 0$

Since maximum volume can not occur when x or y is 0, we assume both x and y are nonzero. Solving $48 - 2xy - x^2 = 0$ for y gives $y = \dfrac{48 - x^2}{2x}$. Substituting this into $48 - y^2 - 2xy = 0$ gives

$0 = 48 - \left(\dfrac{48 - x^2}{2x}\right)^2 - 2x\left(\dfrac{48 - x^2}{2x}\right)$

$= \dfrac{3x^4 + 96x^2 - 2304}{4x^2}$.

The only positive solution to this equation is $x = 4$.

Thus, our critical point is $(4, 4, 4)$.

A quick check of the discriminant assures that this gives the maximum volume of $V(4, 4, 4) = 64$.

41. We first simplify the calculations by noting that we may maximize $B = A^2$ instead of A (since x^2 is an increasing function for positive x).

We solve $s = \frac{1}{2}(a + b + c)$ for c, and substitute into B to find:

$c = 2s - a - b$ and

$B = s(s - a)(s - b)(a + b - s)$.

Treating s as a constant, we see that

$B_a = -s(s - b)(a + b - s)$

$+ s(s - a)(s - b)$

$B_b = -s(s - a)(a + b - s)$

$+ s(s - a)(s - b)$

Subtracting $B_b = 0$ from $B_a = 0$, we get $s(s - a)(a + b - s) - s(s - b)(a + b - s) = 0$. Note that the semi-perimeter $s \neq 0$, and that if $a + b - s = 0$ then $c = 0$, giving the triangle 0 area. Therefore $s - a = s - b$, so $a = b$.

Substituting $b = a$ into B_a yields $-s(s-a)(2a-s) + s(s-a)^2$ and this is 0 when $a = s$ or $a = \frac{2}{3}s$. If $a = b = s$, then $c = 0$ and the area is 0. If $a = b = \frac{2}{3}s$, then $c = \frac{2}{3}s$, as well, and we see that an equilateral triangle gives the maximum area for a fixed perimeter. (That this is indeed a maximum can be verified using Theorem 7.2.)

43. For the function $f(x, y) = x^2y^2$,
$f_x = 2xy^2$ and $f_y = 2x^2y$.
$f_x = f_y = 0$ whenever $x = 0$ or $y = 0$, so we have critical points at $(x, 0)$ and $(0, y)$ for all x and y.

$f_{xx} = 2y^2$
$f_{xy} = 4xy$
$f_{yy} = 2x^2$
$D(x, y) = 4x^2y^2 - 16x^2y^2$
$D(0, y) = D(x, 0) = 0$ for all critical points, so Theorem 7.2 fails to identify them.

$f(x, 0) = f(0, y) = 0$ for all x and y. If x and y are both not zero, then $f(x, y) > 0$. This means that all the critical points are minima.

For the function $f(x, y) = x^{2/3}y^2$,
$f_x = \frac{2}{3}x^{-1/3}y^2$ and $f_y = 2x^{2/3}y$
$f_x = f_y = 0$ whenever $y = 0$, so we have critical points at $(x, 0)$ for all x. When $x = 0$, f_x is undefined and $f_y = 0$, so we have critical points at $(0, y)$ for all y.

$f_{xx} = \frac{-2}{9}x^{-4/3}y^2$
$f_{xy} = \frac{4}{3}x^{-1/3}y$
$f_{yy} = 2x^{2/3}$

$D(x, y) = \frac{-4}{9}x^{-2/3}y^2 - \frac{16}{9}x^{-2/3}y^2$
$D(x, 0) = 0$ for all critical points $(x, 0)$ with $x \neq 0$, so Theorem 7.2 fails to identify them. Theorem 7.2 also fails to identify critical points $(0, y)$, since the second partial derivatives are not continuous there.

$f(x, 0) = f(0, y) = 0$ for all x and y. If x and y are both not zero, then $f(x, y) > 0$. This means that all the critical points are minima.

45. We substitute $y = kx$ into $z = x^3 - 3xy + y^3$ and get $z = x^3 - 3x(kx) + (kx)^3$
$= (1 + k^3)x^3 - 3kx^2$.
If we set $f(x) = (1 + k^3)x^3 - 3kx^2$, we find $f'(x) = 3(1 + k^3)x^2 - 6kx = 0$ when $x = 0$, so this is a critical point. Then $f''(x) = 6(1 + k^3)x - 6k$, so $f''(0) = -6k$. The Second Derivative Test then shows $f(x)$ has a local maximum if $k > 0$ and a local minimum if $k < 0$.

47. We substitute $y = kx$ into $z = x^3 - 2y^2 - 2y^4 + 3x^2y$ to get $f(x) = x^3 - 2(kx)^2 - 2(kx)^4 + 3x^2(kx)$
$= (1 + 3k)x^3 - 2k^2x^2 - 2k^4x^4$ $f'(x) = 3(1 + 3k)x^2 - 4k^2x - 8k^4x^3$, so $f(x)$ has a critical point at $x = 0$.
$f''(x) = 6(1 + 3k)x - 4k^2 - 24k^4x^2$, so $f''(0) = -4k^2$. The Second Derivative Test shows that $f(x)$ has a local maximum for all $k \neq 0$. When $k = 0$ the graph looks like x^3, which has an inflection point at $x = 0$.

49. $f_x = z - 1$, $f_y = 3y^2 - 3$, $f_z = x$
These all equal zero at $(0, 1, 1)$, so this is a critical point.

$f(\Delta x, 1 + \Delta y, 1 + \Delta z)$
$= \Delta x(1 + \Delta z) - \Delta x + (1 + \Delta y)^3$
$ - 3(1 + \Delta y)$
$= \Delta x \Delta z + 3\Delta y^2 + \Delta y^3 - 2$
$= \Delta x \Delta z + 3\Delta y^2 + \Delta y^3 + f(0, 1, 1)$

So with $\Delta y = 0$, as we move away from the critical point with $\Delta x \Delta z > 0$, f increases, while with $\Delta x \Delta z < 0$, f decreases.

The critical point is therefore neither a local maximum or local minimum.

51. False. The partial derivatives could be 0 or undefined.

53. False. There do not have to be any local minima.

55. The extrema occur in the centers of the four circles. The saddle points occur in the nine crosses between the circles.

57. Extrema at approximately $(-0.8, -0.8)$ and $(0.8, 0.8)$.

Saddle at approximately $(0, 0)$.

59. Extrema at approximately $(\pm 0.1, 0.1)$.

Saddle at approximately $(0, 0)$.

61. The distance from a point $(x, y, 4 - x^2 - y^2)$ to the point $(3, -2, 1)$ is
$$d(x, y)$$
$$= \sqrt{(x - 3)^2 + (y + 2)^2 + (3 - x^2 - y^2)^2}.$$

To minimize this it is useful to note that we can minimize $g(x, y) = d(x, y)^2$ instead.

$$g_x = 2(x - 3) - 4x(3 - x^2 - y^2)$$
$$g_y = 2(y + 2) - 4y(3 - x^2 - y^2)$$
$$g_{xx} = -10 + 12x^2 + 4y^2$$
$$g_{yy} = -10 + 12y^2 + 4x^2$$
$$g_{xy} = 8xy$$

Solving $g_x = g_y = 0$ numerically yields $(1.55, -1.03)$.

$D(1.55, -1.03) \approx 121.6$ and $g_{xx}(1.55, -1.03) \approx 23.1$, therefore this point is a minimum.

The closest point on the paraboloid to the point $(3, -2, 1)$ is approximately $(1.55, -1.03, 0.54)$.

63. The closest point on the sphere will be below the xy plane. The distance from a point $(x, y, -\sqrt{9 - x^2 - y^2})$ to the point $(2, 1, -3)$ squared is $g(x, y)$
$$= (x - 2)^2 + (y - 1)^2$$
$$+ (-\sqrt{9 - x^2 - y^2} + 3)^2$$
$$= -4x - 2y - 6\sqrt{9 - x^2 - y^2} + 23$$

$$g_x = -4 + \frac{6x}{\sqrt{9 - x^2 - y^2}}$$

$$g_y = -2 + \frac{6y}{\sqrt{9 - x^2 - y^2}}$$

$$g_{xx} = \frac{54 - 6y^2}{(9 - x^2 - y^2)^{3/2}}$$

$$g_{yy} = \frac{54 - 6x^2}{(9 - x^2 - y^2)^{3/2}}$$

$$g_{xy} = \frac{6xy}{(9 - x^2 - y^2)^{3/2}}$$

Solving $g_x = g_y = 0$ numerically yields $(1.6, 0.8)$.

$D(1.6, 0.8) \approx 9.4$ and $g_{xx}(1.6, 0.8) \approx 3.6$, therefore this point is a minimum.

The closest point on the sphere to the point $(2, 1, -3)$ is approximately $(1.6, 0.8, -2.4)$.

65. $f_x = 5e^y - 5x^4, \qquad f_y = 5xe^y - 5e^{5y}$
$f_{xx} = -20x^3, \quad f_{yy} = 5xe^y - 25e^{5y}$
$f_{xy} = 5e^y$

Solving $f_x = 0$ gives $e^y = x^4$. Substitute this into $f_y = 0$ to see that $5x^5 - 5x^{20} = 0$. The solution $x = 0$ is extraneous, leaving us with solution $x = 1$ and $y = 0$.

$D(1, 0) = 375 > 0$
$f_{xx}(1, 0) = -20 < 0$

Therefore $f(1, 0) = 3$ is a local maximum. Since $f(-5, 0) = 3099$, (for example) it is clear that $(1, 0)$ is not a global maximum.

67. Suppose that $f(x)$ is a differentiable function with local minima at $x = c_1$ and $x = c_2$. Assume $c_1 < c_2$. At c_1 the derivative changes from negative to positive. Then at c_2 this derivative changes from negative to positive. At some point in between, the derivative must change from positive to negative, giving a local maximum.

13.8 CONSTRAINED OPTIMIZATION AND LAGRANGE MULTIPLIERS

1. $f(x, y) = x^2 + y^2$
$g(x, y) = 3x - 4 - y$
$\nabla f = \langle 2x, 2y \rangle$
$\nabla g = \langle 3, -1 \rangle$

$\nabla f = \lambda \nabla g$

$2x = 3\lambda, \qquad 2y = -\lambda$

Eliminating λ we get $y = -\frac{x}{3}$.

Substituting this into the constraint

$y = 3x - 4$,

$-\frac{x}{3} = 3x - 4$

$x = \frac{6}{5}, \qquad y = -\frac{2}{5}$

3. $f(x, y) = (x - 4)^2 + y^2$

$g(x, y) = 2x + y - 3$

$\nabla f = \langle 2(x - 4), 2y \rangle$

$\nabla g = \langle 2, 1 \rangle$

$\nabla f = \lambda \nabla g$

$2(x - 4) = 2\lambda, \qquad 2y = \lambda$

Eliminating λ we get $y = \frac{x}{2} - 2$.

Substituting this into the constraint

$y = 3 - 2x$

$\frac{x}{2} - 2 = 3 - 2x$

$x = 2, \qquad y = -1$

5. $f(x, y) = (x - 3)^2 + y^2$

$g(x, y) = x^2 - y$

$\nabla f = \langle 2(x - 3), 2y \rangle$

$\nabla g = \langle 2x, -1 \rangle$

$\nabla f = \lambda \nabla g$

$2(x - 3) = 2x\lambda, \qquad 2y = -\lambda$

Eliminating λ gives $x - 3 = -2xy$.

Applying the constraint

$y = x^2$

$2x^3 + x - 3 = 0$

$x = 1, \qquad y = 1$

7. $f(x, y) = (x - 2)^2 + (y - \frac{1}{2})^2$

$g(x, y) = x^2 - y$

$\nabla f = \langle 2(x - 2), 2(y - \frac{1}{2}) \rangle$

$\nabla g = \langle 2x, -1 \rangle$

$\nabla f = \lambda \nabla g$

$2(x - 2) = 2x\lambda, \qquad 2(y - \frac{1}{2}) = -\lambda$

Eliminating λ we get $y = \frac{1}{x}$.

Substituting this into the constraint

$y = x^2$

$\frac{1}{x} = x^2$

$x^3 = 1$

$x = 1, \qquad y = 1$

9. $g(x, y) = x^2 + y^2 - 8 = 0$

$\nabla f = \langle 4y, 4x \rangle$

$\nabla g = \langle 2x, 2y \rangle$

$\nabla f = \lambda \nabla g$

$4y = 2x\lambda$

$4x = 2y\lambda$

Eliminating λ we get $y = \pm x$.

Substituting this into the constraint

$x^2 + y^2 - 8 = 0$

$x^2 + x^2 - 8 = 0$

$x^2 = 4$

$x = \pm 2$

This gives the points

$(2, 2), (2, -2), (-2, 2), (-2, -2)$.

$f(2, 2) = 16$, maximum

$f(-2, -2) = 16$, maximum

$f(2, -2) = -16$, minimum

$f(-2, 2) = -16$, minimum

11. $g(x, y) = x^2 + y^2 - 3 = 0$

$\nabla f = \langle 8xy, 4x^2 \rangle$

$\nabla g = \langle 2x, 2y \rangle$

$\nabla f = \lambda \nabla g$

$8xy = 2x\lambda$

$4x^2 = 2y\lambda$

Eliminating λ we get $y = \pm \frac{1}{\sqrt{2}}x$.

Substituting this into the constraint

$x^2 + y^2 - 3 = 0$

$x^2 + \frac{x^2}{2} - 3 = 0$

$x = \pm\sqrt{2}$

This gives the points

$(\sqrt{2}, 1), (-\sqrt{2}, 1), (\sqrt{2}, -1), (-\sqrt{2}, -1)$.

$f(\sqrt{2}, 1) = 8$, maximum

$f(-\sqrt{2}, -1) = -8$, minimum

$f(-\sqrt{2}, 1) = 8$, maximum

$f(\sqrt{2}, -1) = -8$, minimum

13. $g(x, y) = x^2 + y^2 - 2 = 0$

$\nabla f = \langle e^y, xe^y \rangle$

$\nabla g = \langle 2x, 2y \rangle$

$\nabla f = \lambda \nabla g$

$e^y = 2x\lambda$
$xe^y = 2y\lambda$
Eliminating λ we get $y = x^2$.

Substituting this into the constraint
$x^2 + y^2 - 2 = 0$
$x^2 + x^4 - 2 = 0$
$x = \pm 1$

This gives the points
$(1, 1), (-1, 1)$.
$f(1, 1) = e$, maximum
$f(-1, 1) = -e$, minimum

15. $g(x, y) = x^2 + y^2 - 3 = 0$
$\nabla f = \langle 2xe^y, x^2 e^y \rangle$
$\nabla g = \langle 2x, 2y \rangle$
$\nabla f = \lambda \nabla g$
$2xe^y = 2x\lambda$
$x^2 e^y = 2y\lambda$
Eliminating λ we get either $x = 0$ or $y = \frac{1}{2}x^2$.

If $x = 0$, the constraint gives $y = \pm\sqrt{3}$.
If $y = \frac{1}{2}x^2$, the constraint gives:
$x^2 + y^2 - 3 = 0$
$x^2 + \frac{x^4}{4} - 3 = 0$
$(x^2 + 6)(x^2 - 2) = 0$
$x = \pm\sqrt{2}$

This gives the points
$(\pm\sqrt{2}, 1), (\pm\sqrt{2}, -1)$ and $(0, \pm\sqrt{3})$.
$f(\pm\sqrt{2}, 1) = 2e$, maximum
$f(\pm\sqrt{2}, -1) = \frac{2}{e}$, neither
$f(0, \pm\sqrt{3}) = 0$, minimum

17. On the boundary, $x^2 + y^2 = 8$.
$g(x, y) = x^2 + y^2 - 8 = 0$
$\nabla f = \langle 4y, 4x \rangle$
$\nabla g = \langle 2x, 2y \rangle$
$\nabla f = \lambda \nabla g$
$4y = 2x\lambda$
$4x = 2y\lambda$
Eliminating λ we get $y = \pm x$.

Substituting this into the constraint
$x^2 + y^2 - 8 = 0$
$x^2 + x^2 - 8 = 0$
$x = \pm 2$

This gives the points
$(2, 2), (2, -2), (-2, 2), (-2, -2)$.

In the interior, solving $\nabla f = \langle 0, 0 \rangle$ gives the critical point $(0, 0)$.

$f(2, 2) = f(-2, -2) = 16$, maxima
$f(-2, 2) = f(2, -2) = -16$, minima
$f(0, 0) = 0$

19. On the boundary, $x^2 + y^2 = 3$
$g(x, y) = x^2 + y^2 - 3 = 0$
$\nabla f = \langle 8xy, 4x^2 \rangle$
$\nabla g = \langle 2x, 2y \rangle$
$\nabla f = \lambda \nabla g$
$8xy = 2x\lambda$
$4x^2 = 2y\lambda$
Eliminating λ we get $y = \pm\frac{1}{\sqrt{2}}x$.

Substituting this into the constraint
$x^2 + y^2 - 3 = 0$
$x^2 + \frac{1}{2}x^2 - 3 = 0$
$x = \pm\sqrt{2}$

This gives the points
$(\pm\sqrt{2}, 1), (\pm\sqrt{2}, -1)$

In the interior, solving $\nabla f = \langle 0, 0 \rangle$ gives the critical points $(0, y)$.

$f(\pm\sqrt{2}, 1) = 8$, maxima
$f(\pm\sqrt{2}, -1) = -8$, minima
$f(0, y) = 0$

21. We use the constraint
$g(t, u) = u^2 t - 11{,}000 = 0$, so
$\nabla g = \langle u^2, 2ut \rangle$ remains the same as in Example 8.2.

As in the example, $u = \frac{128}{3}$. Applying the constraint gives
$t = \frac{11{,}000}{(128/3)^2} = 6.04$.

$z = \frac{1}{2}\left(\frac{128}{3} - 32\right)(6.04)^2 \approx 195$ feet

23. Substituting $t = \frac{10{,}000}{u^2}$ into $f(t, u)$ gives
$$f(u) = \frac{1}{2}(u - 32)\left(\frac{10{,}000}{u^2}\right)^2$$

$$= 50{,}000{,}000 \left(\frac{1}{u^3} - \frac{32}{u^4} \right)$$

$$f'(u) = 50{,}000{,}000 \left(\frac{-3}{u^4} + \frac{128}{u^5} \right)$$

$$f''(u) = 50{,}000{,}000 \left(\frac{12}{u^5} - \frac{640}{u^6} \right)$$

$f'(u)$ is undefined when $u = 0$, but this clearly does not lead to a maximum.
$f'(u) = 0$ when $u = \frac{128}{3}$. This is a maximum because $f''(\frac{128}{3}) = -1.06 < 0$.

25. $\nabla P = \langle 3, 6, 6 \rangle$, so there are no critical points in the interior.

On the boundary, $2x^2 + y^2 + 4z^2 = 8800$.
$g(x, y) = 2x^2 + y^2 + 4z^2 - 8800 = 0$
$\nabla g = \langle 4x, 2y, 8z \rangle$
$\nabla P = \lambda \nabla g$
$3 = 4x\lambda$
$6 = 2y\lambda$
$6 = 8z\lambda$

Solving these three equations gives
$$x = \frac{3}{4\lambda}, \; y = \frac{6}{2\lambda}, \; z = \frac{6}{8\lambda}.$$
Substituting into the constraint gives
$2x^2 + y^2 + 4z^2 - 8800 = 0$
$$2 \left(\frac{3}{4\lambda} \right)^2 + \left(\frac{6}{2\lambda} \right)^2 + 4 \left(\frac{6}{8\lambda} \right)^2 = 8800$$
$$\lambda^2 = \frac{9}{6400}$$
Using $\lambda = \frac{3}{80}$, we get $(20, 80, 20)$.
Using $\lambda = -\frac{3}{80}$, we get $(-20, -80, -20)$. Of course, the production levels cannot be negative (this would give a minimum of the profit function), so the maximum profit is $P(20, 80, 20) = 660$.

27. In exercise 25, the value of λ is $\lambda = \frac{3}{80}$.

Following the work in exercise 25, we see that the constraint equation gives us
$\lambda^2 = \frac{792}{64k}$, and $\lambda = \sqrt{\frac{792}{64k}}$.

We use the positive square root so that x, y, and z are all positive, and write the profit function as a function of k.

$$P(x, y, z) = P \left(\frac{3}{4\lambda}, \frac{6}{2\lambda}, \frac{6}{8\lambda} \right)$$

$$= \frac{198\sqrt{k}}{\sqrt{792}}$$

Differentiating this function of k yields
$$P'(k) = \frac{99}{\sqrt{792k}},$$
and $P'(8800) = \frac{3}{80} = \lambda$.

29. A rectangle with sides x and y has perimeter $P(x, y) = 2x + 2y$ and area xy. If we are given area c, we get constraint $g(x, y) = xy - c = 0$.
$\nabla P = \langle 2, 2 \rangle$
$\nabla g = \langle y, x \rangle$
$\nabla P = \lambda \nabla g$ gives equations
$2 = y\lambda$
$2 = x\lambda$
Eliminating λ gives $y = x$. This gives the minimum perimeter.

For a given area, the rectangle with the smallest perimeter is a square.

31. Maximize the function $f(x, y) = y - x$ subject to the constraint $g(x, y) = x^2 + y^2 - 1 = 0$.
$\nabla f = \langle -1, 1 \rangle$
$\nabla g = \langle 2x, 2y \rangle$
$\nabla f = \lambda \nabla g$ gives equations $-1 = 2x\lambda$
$1 = 2y\lambda$
Eliminating λ yields $y = -x$. Substituting this into the constraint gives
$x^2 + (-x)^2 = 1$, so that $x = \pm \frac{\sqrt{2}}{2}$.
$f(-\frac{\sqrt{2}}{2}, \frac{\sqrt{2}}{2}) = \sqrt{2}$ is a maximum.

33. The angles α, β, and θ sum to the angle between due east and due north, so $\alpha + \beta + \theta = \frac{\pi}{2}$.

We maximize $f(\alpha, \beta, \theta) = \sin \alpha \sin \beta \sin \theta$ subject to the constraint $g(\alpha, \beta, \theta) = \alpha + \beta + \theta - \frac{\pi}{2} = 0$.

$\nabla f = \langle \cos \alpha \sin \beta \sin \theta, \sin \alpha \cos \beta \sin \theta,$
$\sin \alpha \sin \beta \cos \theta \rangle$
$\nabla g = \langle 1, 1, 1 \rangle$
$\nabla f = \lambda \nabla g$ gives equations
$\cos \alpha \sin \beta \sin \theta = \lambda$
$\sin \alpha \cos \beta \sin \theta = \lambda$
$\sin \alpha \sin \beta \cos \theta = \lambda$

Using these equations in pairs, we get $\tan \alpha = \tan \beta = \tan \theta$.
Since these are angles between 0 and $\frac{\pi}{2}$, they must all be equal.
$\alpha = \beta = \theta = \frac{\pi}{6}$
and the maximum northward component of force is
$f(\frac{\pi}{6}, \frac{\pi}{6}, \frac{\pi}{6}) = \frac{1}{8}.$

35. Find the extreme values of $f(x, y) = xy^2$ subject to the constraint $g(x, y) = x + y = 0$.

$\nabla f = \langle y^2, 2xy \rangle$
$\nabla g = \langle 1, 1 \rangle$
$\nabla f = \lambda \nabla g$ gives equations
$y^2 = \lambda$
$2xy = \lambda$

Eliminating λ yields $y(y - 2x) = 0$. Substituting either $y = 0$ or $y = 2x$ into the constraint yields $x = y = 0$, so $(0, 0)$ is identified as a critical point.

Graphically, this is seen to be a saddle point.

37. $\nabla P = \left\langle \dfrac{400 K^{1/3}}{3 L^{1/3}}, \dfrac{200 L^{2/3}}{3 K^{2/3}} \right\rangle$

On the interior, the critical points occur where K or L is zero, but this gives production zero (not a maximum).

On the boundary, $g(L, K) = 2L + 5K - 150 = 0$.
$\nabla g = \langle 2, 5 \rangle$

$\nabla P = \lambda \nabla g$ gives equations
$\dfrac{400 K^{1/3}}{3 L^{1/3}} = 2\lambda$
$\dfrac{200 L^{2/3}}{3 K^{2/3}} = 5\lambda$

Eliminating λ yields $L = 5K$. Substituting this into the constraint gives
$2(5K) + 5K = 150$, so $K = 10$ and $L = 50$.

Production is maximized when $K = 10$ and $L = 50$.

39. $\nabla C = \langle 25, 100 \rangle$
On the interior, there are no critical points.

On the boundary,
$g(L, K) = 60 L^{2/3} K^{1/3} - 1920 = 0$.
$\nabla g = \left\langle \dfrac{40 K^{1/3}}{L^{1/3}}, \dfrac{20 L^{1/3}}{K^{2/3}} \right\rangle$

$\nabla C = \lambda \nabla g$ gives equations
$25 = \dfrac{40 K^{1/3}}{L^{1/3}} \lambda$
$100 = \dfrac{20 L^{2/3}}{K^{2/3}} \lambda$

Eliminating λ yields $L = 8K$. Substituting this into the constraint gives $60(8K)^{2/3} K^{1/3} = 1920$, so $K = 8$ and $L = 64$ gives minimum cost.

The minimum cost is $C(64, 8) = 2400$.

41. $f(c, d) = 10 c^{0.4} d^{0.6}$
$g(c, d) = 10c + 15d - 300 = 0$
$\nabla f = \left\langle \dfrac{4 d^{0.6}}{c^{0.6}}, \dfrac{6 c^{0.4}}{d^{0.4}} \right\rangle$
$\nabla g = \langle 10, 15 \rangle$

$\nabla f = \lambda \nabla g$ gives the equations:
$\dfrac{4 d^{0.6}}{c^{0.6}} = 10\lambda$
$\dfrac{6 c^{0.4}}{d^{0.4}} = 15\lambda$

Eliminating λ gives $c = d$. Using the constraint, we find that $10c + 15c = 300$, so that $c = d = 12$ maximizes the utility function.

43. $f(x, y, z) = x^2 + y^2 + z^2$
$g(x, y, z) = x + 2y + 3z - 6 = 0$
$h(x, y, z) = y + z = 0$
$\nabla f = \langle 2x, 2y, 2z \rangle$
$\nabla g = \langle 1, 2, 3 \rangle$
$\nabla h = \langle 0, 1, 1 \rangle$

Setting $\nabla f = \lambda \nabla g + \mu \nabla h$ gives the equations:

$2x = \lambda$

$2y = 2\lambda + \mu$

$2z = 3\lambda + \mu$

The first and second equations give $\lambda = 2x$ and $\mu = 2y - 4x$. Then the third equation yields $z = x + y$.

Substituting this into $h(x, y, z) = 0$ gives $x = -2y$, and using these relationships in $g(x, y, z) = 0$ then shows $y = -2$, $z = 2$ and $x = 4$.

The minimum value of $f(x, y, z)$ is 24.

45. $f(x, y, z) = xyz$

$g(x, y, z) = x + y + z - 4 = 0$

$h(x, y, z) = x + y - z = 0$

$\nabla f = \langle yz, xz, xy \rangle$

$\nabla g = \langle 1, 1, 1 \rangle$

$\nabla h = \langle 1, 1, -1 \rangle$

Setting $\nabla f = \lambda \nabla g + \mu \nabla h$ gives the equations:

$yz = \lambda + \mu$

$xz = \lambda + \mu$

$xy = \lambda - \mu$

Subtracting h from g shows that $z = 2$ and $y = 2 - x$. The above equations become

$4 - 2x = \lambda + \mu$

$2x = \lambda + \mu$

$2x - x^2 = \lambda - \mu$

Equating the two expressions for $\lambda + \mu$ gives critical points $x = 1$, $y = 1$ and $z = 2$.

The maximum value of $f(x, y, z) = 2$.

47. We find the extreme values of

$f(x, y, z) = x^2 + y^2 + z^2$

subject to the constraints

$g(x, y, z) = x^2 + y^2 - 1 = 0$ and

$h(x, y, z) = x^2 + z^2 - 1 = 0$.

$\nabla f = \langle 2x, 2y, 2z \rangle$

$\nabla g = \langle 2x, 2y, 0 \rangle$

$\nabla h = \langle 2x, 0, 2z \rangle$

$\nabla f = \lambda \nabla g + \mu \nabla h$ gives the equations:

$2x = 2x\lambda + 2x\mu$

$2y = 2y\lambda$

$2z = 2z\mu$

If y and z are not equal to zero we are led to solution $\lambda = \mu = 1$ and $x = 0$. The constraints then show that $y = \pm 1$ and $z = \pm 1$.

We also have a solution when $y = 0$. The constraints then give $x = \pm 1$ and $z = 0$. (We get the same solutions if we start with $z = 0$.)

$f(\pm 1, 0, 0) = 1$ are minima.

$f(0, \pm 1, \pm 1) = 2$ are maxima.

49. We minimize the square of the distance from (x, y) to $(0, 1)$, $f(x, y) = x^2 + (y - 1)^2$, subject to the constraint $g(x, y) = x^n - y = 0$.

$\nabla f = \langle 2x, 2(y - 1) \rangle$

$\nabla g = \langle nx^{n-1}, -1 \rangle$

$\nabla f = \lambda \nabla g$ leads to equations:

$2x = \lambda n x^{n-1}$

$2y - 2 = -\lambda$

Eliminating λ yields:

$2x + 2n x^{2n-1} - 2n x^{n-1} = 0$.

We always have solution $(0, 0)$.

In order to tell whether this is a minimum or a maximum, we notice that by substituting $y = x^n$ into $f(x, y)$, we get

$f(x, x^n) = x^2 + (x^n - 1)^2$, and

$f'(x) = 2x + 2n x^{2n-1} - 2n x^{n-1}$

(Not surprisingly, $f'(x) = 0$ is the relation we were led to by the method of Lagrange multipliers.)

$f''(x) = 2 + 2n(2n - 1)x^{2n-2}$

$- 2n(n - 1)x^{n-2}$

$f''(0) = 2$ if $n > 2$ and this is a local minimum.

$f''(0) = -2$ if $n = 2$ and this is a local maximum.

The last part of this question is best explored with a CAS. The function $f(x)$ has absolute minimum at its largest critical value. As n increases, this critical value approaches $x = 1$. At $x = 1$ the distance to the point $(0, 1)$ is one, the same as the distance at $x = 0$. Therefore the difference between the absolute minimum value and the local minimum at $x = 0$ goes to 0.

51. We minimize the square of the distance,
$$f(x, y) = (x - 1)^2 + y^2 + z^2$$
subject to the constraint
$$g(x, y, z) = x^2 + y^2 - z = 0.$$
$\nabla f = \langle 2(x - 1), 2y, 2z \rangle$
$\nabla g = \langle 2x, 2y, -1 \rangle$
$\nabla f = \lambda \nabla g$ leads to equations:
$2x - 2 = 2x\lambda$
$2y = 2y\lambda$
$2z = -\lambda$

If $y \neq 0$ then $\lambda = 1$ and the first equation is inconsistent.

If $y = 0$, the constraint gives $z = x^2$, and combining the first and third equation above yields
$$4x^3 + 2x - 2 = 0$$
which we numerically solve to find
$x = 0.5898$, $y = 0$, $z = 0.3478$.

Chapter 13 Review Exercises

1.

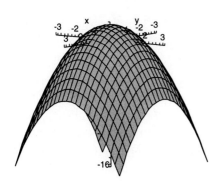

3.

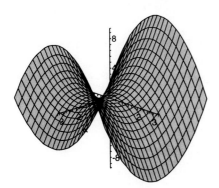

5.

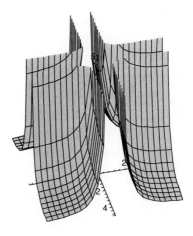

7.

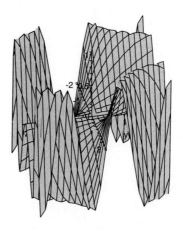

9.

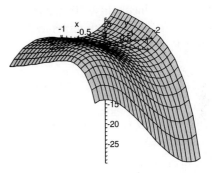

11. **(a)** Surface D

(b) Surface C

(c) Surface A

(d) Surface F

(e) Surface E

13. **(a)** Contour C

(b) Contour A

(c) Contour D

(d) Contour B

15. Along the line $x = 0$, we have
$$\lim_{(0,y)\to(0,0)} \frac{0}{0 + y^2} = 0.$$

Along the curve $y = x^2$, we have
$$\lim_{(x,x^2)\to(0,0)} \frac{3x^4}{x^4 + x^4} = \frac{3}{2}.$$

Since these limits are different, the limit does not exist.

17. Along the line $x = 0$, we have
$$\lim_{(0,y)\to(0,0)} \frac{y^2}{y^2} = 1.$$

Along the curve $y = x$, we have
$$\lim_{(x,x)\to(0,0)} \frac{x^2 + x^2}{x^2 + x^2 + x^2} = \frac{2}{3}.$$

Since these limits are different, the limit does not exist.

19. We use Theorem 2.1.
$$\left|\frac{x^3 + xy^2}{x^2 + y^2}\right| \le \left|\frac{x^3}{x^2 + y^2}\right| + \left|\frac{xy^2}{x^2 + y^2}\right|$$

by the triangle inequality. We make the denominators smaller in both terms to get
$$\le \left|\frac{x^3}{x^2}\right| + \left|\frac{xy^2}{y^2}\right|$$
$$= |x| + |x| = 2|x|$$
$$\lim_{(x,y)\to(0,0)} 2|x| = 0, \text{ therefore}$$
$$\lim_{(x,y)\to(0,0)} \frac{x^3 + xy^2}{x^2 + y^2} = 0.$$

21. $f(x, y)$ is continuous unless $x = 0$.

23. $f_x = \dfrac{4}{y} + xye^{xy} + e^{xy}$

$f_y = \dfrac{-4x}{y^2} + x^2 e^{xy}$

25. $f_x = 6xy \cos y - \dfrac{1}{2\sqrt{x}}$

$f_y = 3x^2 \cos y - 3x^2 y \sin y$

27. $f_x = f_{xx} = e^x \sin y$

$f_y = e^x \cos y \qquad f_{yy} = -e^x \sin y$

Therefore,

$f_{xx} + f_{yy} = e^x \sin y - e^x \sin y = 0.$

29. $\dfrac{\partial f}{\partial x}(0, 0) \approx \dfrac{1.6 - 2.4}{10 - (-10)} = -0.04$

$\dfrac{\partial f}{\partial y}(0, 0) \approx \dfrac{2.6 - 1.4}{10 - (-10)} = 0.06$

31. $f_x = \dfrac{3xy}{\sqrt{x^2 + 5}}, \qquad f_y = 3\sqrt{x^2 + 5}$

$f(-2, 5) = 45$

$f_x(-2, 5) = -10, \quad f_y(-2, 5) = 9$

$L(x, y) = 45 - 10(x + 2) + 9(y - 5)$

33. $f_x = \sec^2(x + 2y), \qquad f_y = 2\sec^2(x + 2y)$

$f\left(\pi, \dfrac{\pi}{2}\right) = 0$

$f_x\left(\pi, \dfrac{\pi}{2}\right) = 1, \quad f_y\left(\pi, \dfrac{\pi}{2}\right) = 2$

$L(x, y) = (x - \pi) + 2\left(y - \dfrac{\pi}{2}\right)$

35. $f_x = 8x^3 y + 6xy^2$

$f_y = 2x^4 + 6x^2 y$

$f_{xx} = 24x^2 y + 6y^2$

$f_{yy} = 6x^2$

$f_{xy} = 8x^3 + 12xy$

37. $f_x = 2xy + 2, \qquad f_y = x^2 - 2y$

$f(1, -1) = 0$

$f_x(1, -1) = 0, \quad f_y(1, -1) = 3$

$3(y + 1) - z = 0$

39. $f(x, y, z) = x^2 + 2xy + y^2 + z^2 = 5$

$\nabla f = \langle 2x + 2y, 2x + 2y, 2z \rangle$

$\nabla f(0, 2, 1) = \langle 4, 4, 2 \rangle$

$4(x - 0) + 4(y - 2) + 2(z - 1) = 0$

41. $g'(t) = f_x(x(t), y(t))x'(t)$

$\qquad + f_y(x(t), y(t))y'(t)$

$f_x = 2xy \qquad f_y = x^2 + 2y$

$x'(t) = 4e^{4t} \qquad y'(t) = \cos t$

$g'(t) = 2e^{4t} \sin t (4e^{4t}) + (e^{8t} + 2\sin t) \cos t$

$\qquad = 8e^{8t} \sin t + (e^{8t} + 2\sin t) \cos t$

43. $g'(t) = f_x(x(t), y(t), z(t), w(t))x'(t)$
$+ f_y(x(t), y(t), z(t), w(t))y'(t)$
$+ f_z(x(t), y(t), z(t), w(t))z'(t)$
$+ f_w(x(t), y(t), z(t), w(t))w'(t)$

45. $F(x, y, z) = x^2 + 2xy + y^2 + z^2 - 1$
$F_x = 2x + 2y \qquad F_y = 2x + 2y$
$F_z = 2z$
$$\frac{\partial z}{\partial x} = -\frac{F_x}{F_z} = -\frac{x+y}{z}$$
$$\frac{\partial z}{\partial y} = -\frac{F_y}{F_z} = -\frac{x+y}{z}$$

47. $\nabla f = \langle 3\sin 4y - \frac{\sqrt{y}}{2\sqrt{x}}, 12x\cos 4y - \frac{\sqrt{x}}{2\sqrt{y}} \rangle$

$\nabla f(\pi, \pi) = \langle -\frac{1}{2}, 12\pi - \frac{1}{2} \rangle$

49. $\nabla f = \langle 3x^2 y, x^3 - 8y \rangle$
$\nabla f(-2, 3) = \langle 36, -32 \rangle$
$D_{\mathbf{u}} f(-2, 3) = \langle 36, -32 \rangle \cdot \langle \frac{3}{5}, \frac{4}{5} \rangle$
$= \frac{-20}{5} = -4$

51. $\nabla f = \langle 3ye^{3xy}, 3xe^{3xy} - 2y \rangle$
$\nabla f(0, -1) = \langle -3, 2 \rangle$
$\mathbf{u} = \langle \frac{1}{\sqrt{5}}, \frac{-2}{\sqrt{5}} \rangle$
$D_{\mathbf{u}} f(0, -1) = \langle -3, 2 \rangle \cdot \langle \frac{1}{\sqrt{5}}, \frac{-2}{\sqrt{5}} \rangle$
$= \frac{-7}{\sqrt{5}}$

53. $\nabla f = \langle 3x^2 y, x^3 - 8y \rangle$
$\nabla f(-2, 3) = \langle 36, -32 \rangle$
$\|\nabla f(-2, 3)\| = 4\sqrt{145}$

The direction of maximum change is
$\langle 36, -32 \rangle$.
The maximum rate of change is $4\sqrt{145}$.

The direction of minimum change is
$\langle -36, 32 \rangle$.
The minimum rate of change is $-4\sqrt{145}$.

55. $\nabla f = \langle \frac{2x^3}{\sqrt{x^4 + y^4}}, \frac{2y^3}{\sqrt{x^4 + y^4}} \rangle$

$\nabla f(2, 0) = \langle 4, 0 \rangle$
$\|\nabla f(2, 0)\| = 4$
The direction of maximum change is $\langle 4, 0 \rangle$.
The maximum rate of change is 4.

The direction of minimum change is $\langle -4, 0 \rangle$.
The minimum rate of change is -4.

57. $\nabla f = \langle -8x, -2 \rangle$
$\nabla f(2, 1) = \langle -16, -2 \rangle$

The rain will run in the direction $\langle 16, 2 \rangle$.

59. $\nabla f = \langle 8x^3 - y^2, -2xy + 4y \rangle$
$f_{xx} = 24x^2, \qquad f_{yy} = -2x + 4, \qquad f_{xy} = -2y$
Solving $\nabla f = \langle 0, 0 \rangle$ gives equations
$8x^3 = y^2$
$2y(2 - x) = 0$
The second equation gives us $y = 0$ or $x = 2$.

If $y = 0$ then the first equation gives us $x = 0$
and we have the critical point $(0, 0)$.

If $x = 2$ and $8x^3 = y^2$, then $y = \pm 8$ and we get
the critical points $(2, \pm 8)$.

$D(0, 0) = 0$, so Theorem 7.2 provides no
information. But, along every trace $y = cx$,
$f(x, cx) = 2x^4 - c^2 x^3 + 2c^2 x^2$, and the second derivative test shows this to be a minimum.

$D(2, \pm 8) = -256 < 0$ so these are both saddle
points.

61. $\nabla f = \langle 4y - 3x^2, 4x - 4y \rangle$
$f_{xx} = -6x, \qquad f_{yy} = -4, \qquad f_{xy} = 4$
Solving $\nabla f = \langle 0, 0 \rangle$ gives equations
$4y - 3x^2 = 0$
$4x = 4y$

The second equation gives us $x = y$.

The first equation then becomes
$x(4 - 3x) = 0$, so that $x = 0$ or $x = \frac{4}{3}$.

We get the critical points $\left(\frac{4}{3}, \frac{4}{3}\right)$, $(0, 0)$.

$D(0, 0) = -16 < 0$, so this is a saddle point.

$$D\left(\frac{4}{3}, \frac{4}{3}\right) = 16 > 0$$

$$f_{xx}\left(\frac{4}{3}, \frac{4}{3}\right) = -8 < 0,$$

so $f\left(\frac{4}{3}, \frac{4}{3}\right) = \frac{32}{27}$ is a local maximum.

63. The residuals are

$64a + b - 140, \qquad 66a + b - 156$
$70a + b - 184, \qquad 71a + b - 190$

$g(a, b) = (64a + b - 140)^2 + (66a + b - 156)^2$
$+ (70a + b - 184)^2 + (71a + b - 190)^2$

$$\frac{\partial g}{\partial a} = 36{,}786a + 542b - 91{,}252$$

$$\frac{\partial g}{\partial b} = 542a + 8b - 1340$$

Solving $\dfrac{\partial g}{\partial a} = \dfrac{\partial g}{\partial b} = 0$ we get

$a = \dfrac{934}{131} \approx 7.130$

$b = -\dfrac{41{,}336}{131} \approx -315.542$

$y = 7.130x - 315.542$
$y(74) \approx 212$
$y(60) \approx 112$

65. $\nabla f = \langle 8x^3 - y^2, -2xy + 4y \rangle$
$f_{xx} = 24x^2, \quad f_{yy} = -2x + 4, \quad f_{xy} = -2y$
Solving $\nabla f = \langle 0, 0 \rangle$ gives equations
$y^2 = 8x^3$
$2y(2 - x) = 0$

The second equation gives us $x = 2$ or $y = 0$.

If $y = 0$ then the first equation gives us $x = 0$ and we have the critical point $(0, 0)$.

If $x = 2$ and $y^2 = 8x^3$, then $y = \pm 8$ and we get the critical points $(2, \pm 8)$, neither of which are in the region.

Along $y = 0$, $f(x, 0) = 2x^4$ which has a critical point at $x = 0$ that gives us the critical point $(0, 0)$ (we already had this point).

Along $y = 2$, $f(x, 2) = 2x^4 - 4x + 8$, which has a critical point at $x = \frac{1}{\sqrt[3]{2}}$ and the only critical point in the region is $\left(\frac{1}{\sqrt[3]{2}}, 2\right)$.

Along $x = 0$, $f(0, y) = 2y^2$, which has a critical point at $y = 0$ and we get the same critical point of $(0, 0)$.

Along $x = 4$, $f(4, y) = 512 - 2y^2$, which has a critical point at $y = 0$ and we get the same critical point of $(0, 0)$.

In addition, the intersection points of our boundaries are $(0, 0)$, $(4, 0)$, $(0, 2)$, $(4, 2)$.

$f(0, 0) = 0$, minimum
$f\left(\frac{1}{\sqrt[3]{2}}, 2\right) \approx 5.619$
$f(0, 2) = 8$
$f(4, 0) = 512$, maximum
$f(4, 2) = 504$

67. $g(x, y) = x^2 + y^2 - 5 = 0$
$\nabla f = \langle 1, 2 \rangle$
$\nabla g = \langle 2x, 2y \rangle$
$\nabla f = \lambda \nabla g$
$1 = 2x\lambda$
$2 = 2y\lambda$
Eliminating λ gives $y = 2x$.

Substituting this into the constraint
$x^2 + y^2 - 5 = 0$
$x^2 + 4x^2 - 5 = 0$
$x = \pm 1$

This gives the points $(1, 2)$, $(1, -2)$, $(-1, -2)$, $(-1, 2)$.
Maximum: $f(1, 2) = 5$
Minimum: $f(-1, -2) = -5$

69. $g(x, y) = x^2 + y^2 - 1 = 0$
$\nabla f = \langle y, x \rangle$
$\nabla g = \langle 2x, 2y \rangle$
$\nabla f = \lambda \nabla g$
$y = 2x\lambda$
$x = 2y\lambda$
Eliminating λ we see that $y = \pm x$. Substituting this into the constraint yields $x^2 + x^2 = 1$, so that $x = \pm\frac{1}{\sqrt{2}}$.

Therefore our critical points are
$(\frac{1}{\sqrt{2}}, \frac{1}{\sqrt{2}}), (-\frac{1}{\sqrt{2}}, \frac{1}{\sqrt{2}}),$
$(\frac{1}{\sqrt{2}}, -\frac{1}{\sqrt{2}}), (-\frac{1}{\sqrt{2}}, -\frac{1}{\sqrt{2}}).$
$f(\frac{1}{\sqrt{2}}, \frac{1}{\sqrt{2}}) = f(-\frac{1}{\sqrt{2}}, -\frac{1}{\sqrt{2}}) = \frac{1}{2}$, maximum
$f(\frac{1}{\sqrt{2}}, -\frac{1}{\sqrt{2}}) = f(-\frac{1}{\sqrt{2}}, \frac{1}{\sqrt{2}}) = \frac{-1}{2}$, minimum

71. We want to minimize
$f(x, y) = (x - 4)^2 + y^2$
subject to the constraint $y = x^3$.

$g(x, y) = x^3 - y = 0$
$\nabla f = \langle 2(x - 4), 2y \rangle$
$\nabla g = \langle 3x^2, -1 \rangle$

$\nabla f = \lambda \nabla g 2(x - 4) = 3x^2 \lambda 2y = -\lambda$
Eliminating λ gives
$y = \dfrac{4 - x}{3x^2}.$

Substituting this into our constraint gives
$\dfrac{4 - x}{3x^2} = x^3$
or
$3x^5 + x - 4 = 0.$

With the aid of a CAS, we find $x = 1$ is the only real solution. This gives closest point $(1, 1)$.

Multiple Integrals

14.1 DOUBLE INTEGRALS

1. $f(x, y) = x + 2y^2, n = 4$
$0 \leq x \leq 2, -1 \leq y \leq 1$

The centers of the four rectangles are
$\left(\frac{1}{2}, -\frac{1}{2}\right), \left(\frac{1}{2}, \frac{1}{2}\right), \left(\frac{3}{2}, -\frac{1}{2}\right), \left(\frac{3}{2}, \frac{1}{2}\right)$.
Since the rectangles are the same size,
$\Delta A_i = 1$.

$$V \approx \sum_{i=1}^{4} f(u_i, v_i) \Delta A_i$$

$$= f\left(\frac{1}{2}, -\frac{1}{2}\right)(1) + f\left(\frac{1}{2}, \frac{1}{2}\right)(1)$$

$$+ f\left(\frac{3}{2}, -\frac{1}{2}\right)(1) + f\left(\frac{3}{2}, \frac{1}{2}\right)$$

$$= 1 + 1 + 2 + 2 = 6$$

3. $f(x, y) = x + 2y^2, n = 16$
$0 \leq x \leq 2, -1 \leq y \leq 1$

The centers of the 16 rectangles are
$\left(\frac{1}{4}, \frac{1}{4}\right), \left(\frac{3}{4}, \frac{1}{4}\right), \left(\frac{5}{4}, \frac{1}{4}\right), \left(\frac{7}{4}, \frac{1}{4}\right),$

$\left(\frac{1}{4}, \frac{3}{4}\right), \left(\frac{3}{4}, \frac{3}{4}\right), \left(\frac{5}{4}, \frac{3}{4}\right), \left(\frac{7}{4}, \frac{3}{4}\right),$

$\left(\frac{1}{4}, -\frac{1}{4}\right), \left(\frac{3}{4}, -\frac{1}{4}\right), \left(\frac{5}{4}, -\frac{1}{4}\right), \left(\frac{7}{4}, -\frac{1}{4}\right),$

$\left(\frac{1}{4}, -\frac{3}{4}\right), \left(\frac{3}{4}, -\frac{3}{4}\right), \left(\frac{5}{4}, -\frac{3}{4}\right), \left(\frac{7}{4}, -\frac{3}{4}\right)$.
Since the rectangles are the same size,
$\Delta A_i = \frac{1}{4}$.

$$V \approx \sum_{i=1}^{16} f(u_i, v_i) \Delta A_i$$

$$= f\left(\frac{1}{4}, \frac{1}{4}\right)\left(\frac{1}{4}\right) + f\left(\frac{3}{4}, \frac{1}{4}\right)\left(\frac{1}{4}\right)$$

$$+ f\left(\frac{5}{4}, \frac{1}{4}\right)\left(\frac{1}{4}\right) + f\left(\frac{7}{4}, \frac{1}{4}\right)\left(\frac{1}{4}\right)$$

$$+ f\left(\frac{1}{4}, \frac{3}{4}\right)\left(\frac{1}{4}\right) + f\left(\frac{3}{4}, \frac{3}{4}\right)\left(\frac{1}{4}\right)$$

$$+ f\left(\frac{5}{4}, \frac{3}{4}\right)\left(\frac{1}{4}\right) + f\left(\frac{7}{4}, \frac{3}{4}\right)\left(\frac{1}{4}\right)$$

$$+ f\left(\frac{1}{4}, -\frac{1}{4}\right)\left(\frac{1}{4}\right) + f\left(\frac{3}{4}, -\frac{1}{4}\right)\left(\frac{1}{4}\right)$$

$$+ f\left(\frac{5}{4}, -\frac{1}{4}\right)\left(\frac{1}{4}\right) + f\left(\frac{7}{4}, -\frac{1}{4}\right)\left(\frac{1}{4}\right)$$

$$+ f\left(\frac{1}{4}, -\frac{3}{4}\right)\left(\frac{1}{4}\right) + f\left(\frac{3}{4}, -\frac{3}{4}\right)\left(\frac{1}{4}\right)$$

$$+ f\left(\frac{5}{4}, -\frac{3}{4}\right)\left(\frac{1}{4}\right) + f\left(\frac{7}{4}, -\frac{3}{4}\right)\left(\frac{1}{4}\right)$$

$$= \left(\frac{1}{4}\right)\left(\frac{3}{8} + \frac{7}{8} + \frac{11}{8} + \frac{15}{8} + \frac{11}{8} + \frac{15}{8}\right.$$

$$+ \frac{19}{8} + \frac{23}{8} + \frac{3}{8} + \frac{7}{8} + \frac{11}{8}$$

$$\left. + \frac{15}{8} + \frac{11}{8} + \frac{15}{8} + \frac{19}{8} + \frac{23}{8}\right)$$

$$= \frac{13}{2}$$

5. $f(x, y) = 3x - y$
$0 \le x \le 4, 0 \le y \le 2$
The centers of the rectangles are
$\left(\frac{1}{2}, \frac{1}{2}\right), \left(\frac{1}{2}, \frac{3}{2}\right), \left(\frac{5}{2}, \frac{1}{2}\right), \left(\frac{5}{2}, \frac{3}{2}\right).$
The areas are
$\Delta A_1 = \Delta A_2 = 1, \Delta A_3 = \Delta A_4 = 3.$

$$V \approx \sum_{i=1}^{4} f(u_i, v_i) \Delta A_i$$

$$= f\left(\frac{1}{2}, \frac{1}{2}\right)(1) + f\left(\frac{1}{2}, \frac{3}{2}\right)(1)$$

$$+ f\left(\frac{5}{2}, \frac{1}{2}\right)(3) + f\left(\frac{5}{2}, \frac{3}{2}\right)(3)$$

$$= 1 \cdot 1 + 0 \cdot 1 + 7 \cdot 3 + 6 \cdot 3$$

$$= 40$$

7. $\iint_R (x^2 - 2y)\, dA$

$$= \int_{-1}^{1} \int_0^2 (x^2 - 2y)\, dx\, dy$$

$$= \int_{-1}^{1} \left[\frac{x^3}{3} - 2xy\right]_{x=0}^{x=2} dy$$

$$= \int_{-1}^{1} \left(\frac{8}{3} - 4y\right) dy$$

$$= \left[\frac{8y}{3} - 2y^2\right]_{-1}^{1} = \frac{16}{3}$$

9. $\iint_R (1 - ye^{xy})\, dA$

$$= \int_0^3 \int_0^2 (1 - ye^{xy})\, dx\, dy$$

$$= \int_0^3 \left[x - e^{xy}\right]_{x=0}^{x=2} dy$$

$$= \int_0^3 \left(3 - e^{2y}\right) dy$$

$$= \left[3y - \frac{e^{2y}}{2}\right]_0^3 = \frac{19 - e^6}{2}$$

11.

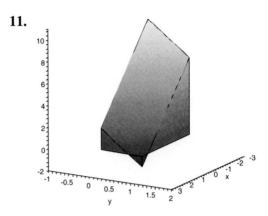

13.

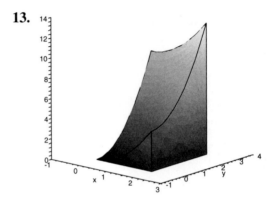

15. $\int_0^1 \int_0^{2x} (x + 2y)\, dy\, dx$

$$= \int_0^1 \left[xy + y^2\right]_{y=0}^{y=2x} dx$$

$$= \int_0^1 6x^2\, dx = \left[2x^3\right]_0^1 = 2$$

17. $\int_0^1 \int_0^{2y} (4x\sqrt{y} + y)\, dx\, dy$

$$= \int_0^1 \left[2x^2\sqrt{y} + xy\right]_{x=0}^{x=2y} dy$$

$$= \int_0^1 \left(8y^{5/2} + 2y^2\right) dy$$

$$= \left[\frac{16}{7}y^{7/2} + \frac{2}{3}y^3\right]_0^1 = \frac{62}{21}$$

19. $\displaystyle\int_0^2 \int_0^{2y} e^{y^2}\, dx\, dy$

$$= \int_0^2 \left[x e^{y^2} \right]_{x=0}^{x=2y} dy$$

$$= \int_0^2 \left(2y e^{y^2} \right) dy$$

$$= \left[e^{y^2} \right]_0^2 = e^4 - 1$$

21. $\displaystyle\int_1^4 \int_0^{1/x} \cos xy\, dy\, dx$

$$= \int_1^4 \left[\frac{\sin xy}{x} \right]_{y=0}^{y=1/x} dx$$

$$= \int_1^4 \left(\frac{\sin 1}{x} \right) dx$$

$$= [(\sin 1)(\ln x)]_1^4 = 2(\ln 2)(\sin 1)$$

23. $\displaystyle\int_0^1 \int_0^{2x} x^2\, dy\, dx$

$$= \int_0^1 \left[x^2 y \right]_{y=0}^{y=2x} dx$$

$$= \int_0^1 \left(2x^3 \right) dx$$

$$= \left[\frac{x^4}{2} \right]_0^1 = \frac{1}{2}$$

$$\int_0^2 \int_0^{y/2} x^2\, dx\, dy$$

$$= \int_0^2 \left[\frac{x^3}{3} \right]_{x=0}^{x=y/2} dy$$

$$= \int_0^2 \left(\frac{y^3}{24} \right) dy$$

$$= \left[\frac{y^4}{96} \right]_0^2 = \frac{1}{6}$$

25. The region of integration is a rectangle in the xy-plane.

$$V = \int_1^4 \int_0^3 (x^2 + y^2)\, dx\, dy$$

$$= \int_1^4 \left[\frac{x^3}{3} + xy^2 \right]_{x=0}^{x=3} dy$$

$$= \int_1^4 \left(9 + 3y^2 \right) dy$$

$$= \left[9y + y^3 \right]_1^4 = 90$$

27. The solid lies under the surface $z = x^2 + y^2$ and over the region R in the xy-plane pictured below.

Region of Integration

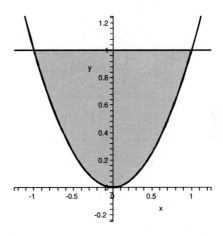

$$V = \int_{-1}^1 \int_{x^2}^1 (x^2 + y^2)\, dy\, dx$$

$$= \int_{-1}^1 \left[x^2 y + \frac{y^3}{3} \right]_{y=x^2}^{y=1} dx$$

$$= \int_{-1}^1 \left(\frac{1}{3} + x^2 - x^4 - \frac{x^6}{3} \right) dy$$

$$= 2 \int_0^1 \left(\frac{1}{3} + x^2 - x^4 - \frac{x^6}{3} \right) dy$$

$$= 2 \left[\frac{x}{3} + \frac{x^3}{3} - \frac{x^5}{5} - \frac{x^7}{21} \right]_{-1}^1$$

$$= \frac{88}{105}$$

where the fourth line is by symmetry.

29. The solid in question lies under the plane $z = 6 - x - y$ and over the region in the xy-plane pictured below.

Region of Integration

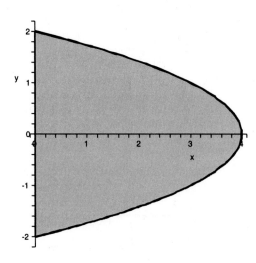

$$V = \int_{-2}^{2} \int_{0}^{4-y^2} (6 - x - y)\, dx\, dy$$

$$= \int_{-2}^{2} \left[6x - \frac{x^2}{2} - xy \right]_{x=0}^{x=4-y^2} dy$$

$$= \int_{-2}^{2} \left(16 - 4y - 2y^2 - \frac{y^4}{2} + y^3 \right) dy$$

Using symmetry we disregard the odd powers.

$$V = 2 \int_{0}^{2} \left(16 - 2y^2 - \frac{y^4}{2} \right) dy$$

$$= 2 \left[16y - \frac{2y^3}{3} - \frac{y^5}{10} \right]_{0}^{2}$$

$$= \frac{704}{15}$$

31. The solid lies under the surface $z = y^2$ and over the region in the xy-plane pictured below.

Region of Integration

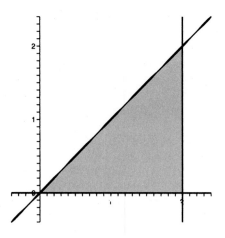

$$V = \int_{0}^{2} \int_{0}^{x} y^2\, dy\, dx = \int_{0}^{2} \left[\frac{y^3}{3} \right]_{y=0}^{y=x} dx$$

$$= \int_{0}^{2} \left(\frac{x^3}{3} \right) dx = \left[\frac{x^4}{12} \right]_{0}^{2} = \frac{4}{3}$$

33. From the graph, y ranges from $\sin x$ to $1 - x^2$. For the outer limits of integration, we solve the equation (one method is to use a graphing calculator)
$$\sin x = 1 - x^2$$
and we obtain
$$x \approx -1.4096,\ 0.6367.$$

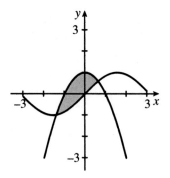

$$\int_{-1.4096}^{0.6367} \int_{\sin x}^{1-x^2} (2x - y) \, dy \, dx$$

$$= \int_{-1.4096}^{0.6367} \left[2xy - \frac{y^2}{2} \right]_{y=\sin x}^{y=1-x^2} dx$$

$$= \int_{-1.4096}^{0.6367} \left[-\frac{1}{2} + 2x + x^2 - 2x^3 \right.$$

$$\left. -\frac{x^4}{2} - 2x \sin x + \frac{\sin^2 x}{2} \right] dx$$

$$\approx -1.59454$$

The integral can be evaluated using integration techniques discussed in earlier chapters or by using a table of integrals.

35. The limits of integration can be found from the graph of the region of integration.

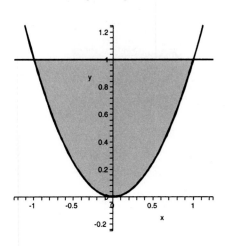

Region of Integration

$$\int_{-1}^{1} \int_{x^2}^{1} e^{x^2} \, dy \, dx = \int_{-1}^{1} e^{x^2} \left[\int_{x^2}^{1} dy \right] dx$$

$$= \int_{-1}^{1} e^{x^2} \left[y \right]_{y=x^2}^{y=1} dx$$

$$= \int_{-1}^{1} e^{x^2} (1 - x^2) \, dx$$

$$\approx 1.6697$$

The final integral must be computed using technology or a numerical method.

37.

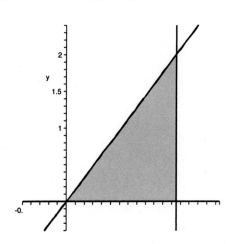

Region of Integration

$$\int_{0}^{1} \int_{0}^{2x} f(x, y) \, dy \, dx$$

$$= \int_{0}^{2} \int_{y/2}^{1} f(x, y) \, dx \, dy$$

39.

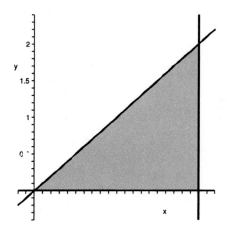

Region of Integration

$$\int_{0}^{2} \int_{2y}^{4} f(x, y) \, dx \, dy$$

$$= \int_{0}^{4} \int_{0}^{x/2} f(x, y) \, dy \, dx$$

41.

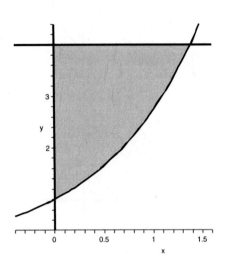

Region of Integration

$$\int_0^2 \int_x^2 2e^{y^2} \, dy \, dx$$

$$= \int_0^2 \int_0^y 2e^{y^2} \, dx \, dy$$

$$= \int_0^2 \left[2xe^{y^2} \right]_{x=0}^{x=y} \, dy$$

$$= \int_0^2 2ye^{y^2} \, dy$$

$$= \left[e^{y^2} \right]_0^2 = e^4 - 1$$

$$\int_0^{\ln 4} \int_{e^x}^4 f(x, y) \, dy \, dx$$

$$= \int_1^4 \int_0^{\ln y} f(x, y) \, dx \, dy$$

43.

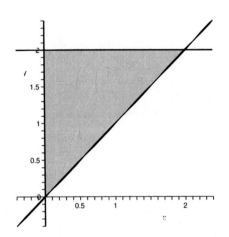

Region of Integration

45.

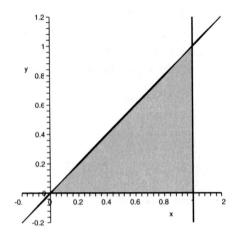

Region of Integration

$$\int_0^1 \int_y^1 3xe^{x^3} \, dx \, dy$$

$$= \int_0^1 \int_0^x 3xe^{x^3} \, dy \, dx$$

$$= \int_0^1 \left[3xye^{x^3} \right]_{y=0}^{y=x} \, dx$$

$$= \int_0^1 3x^2 e^{x^3} \, dx$$

$$= \left[e^{x^3} \right]_0^1 = e - 1$$

47. The CAS we are using (Maple) will not evaluate these integrals. Of course, we can change the order of integration in the second integral as in exercise 43.

49.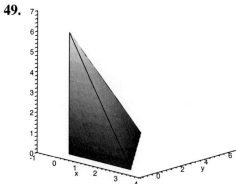

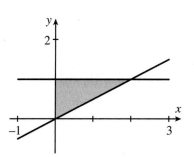

51.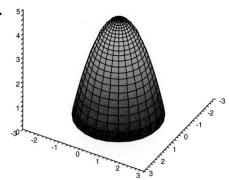

53. As you can see from the graphs of the regions, the regions are different regions in the xy-plane.

First 'Region 'of 'Integration

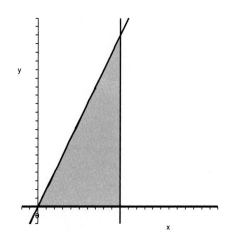

55. The figure is a hemisphere of radius 1. Therefore

$$\int_{-1}^{1} \int_{-\sqrt{1-x^2}}^{\sqrt{1-x^2}} \sqrt{1 - x^2 - y^2} \, dy \, dx$$

$$= \text{Volume of hemisphere}$$

$$= \frac{1}{2} \cdot \frac{4}{3} \pi (1)^3 = \frac{2}{3} \pi.$$

57.
$$\int_{a}^{b} \int_{c}^{d} f(x)g(y) \, dy \, dx$$

$$= \int_{a}^{b} \left(\int_{c}^{d} f(x)g(y) \, dy \right) dx$$

$$= \int_{a}^{b} \left(f(x) \int_{c}^{d} g(y) \, dy \right) dx$$

$$= \int_{a}^{b} f(x) \left(\int_{c}^{d} g(y) \, dy \right) dx$$

$$= \left(\int_{a}^{b} f(x) \, dx \right) \left(\int_{c}^{d} g(y) \, dy \right)$$

59. The upper-left corners are
(0.0, 0.25), (0.25, 0.25), (0.5, 0.25),
(0.75, 0.25), (0.0, 0.5), (0.25, 0.5),
(0.5, 0.5), (0.75, 0.5), (0.0, 0.75),
(0.25, 0.75), (0.5, 0.75), (0.75, 0.75),
(0.0, 1.0), (0.25, 1.0), (0.5, 1.0), (0.75, 1.0).

Since the rectangles are the same size,
$\Delta A_i = 0.0625$.

$$\int_0^1 \int_0^1 f(x, y) \, dy \, dx \approx \sum_{i=1}^{16} f(u_i, v_i) \Delta A_i$$

$$= f(0, 0.25)(0.0625) + f(0.25, 0.25)(0.0625)$$

$$+ \cdots + f(0.75, 1)(0.0625)$$

$$= (0.0625)(2.3 + 2.1 + 1.8 + 1.6$$
$$+ 2.5 + 2.3 + 2.0 + 1.8$$
$$+ 2.8 + 2.6 + 2.3 + 2.2$$
$$+ 3.2 + 3.0 + 2.8 + 2.7)$$

$$= 2.375$$

61. The upper-left corners are

(0.0, 0.25), (0.25, 0.25), (0.5, 0.25), (0.75, 0.25), (0.0, 0.5), (0.25, 0.5), (0.5, 0.5), (0.75, 0.5),.

Since the rectangles are the same size, $\Delta A_i = 0.0625$.

$$\int_0^1 \int_0^{0.5} f(x, y) \, dy \, dx$$

$$\approx \sum_{i=1}^{8} f(u_i, v_i) \Delta A_i$$

$$= f(0, 0.25)(0.0625)$$
$$+ f(0.25, 0.25)(0.0625)$$
$$+ \cdots + f(0.75, 0.5)(0.0625)$$

$$= (0.0625)(2.3 + 2.1 + 1.8 + 1.6$$
$$+ 2.5 + 2.3 + 2.0 + 1.8)$$

$$= 1.025$$

63. The lower-right corners we are interested in are
(0.25, 0.0), (0.5, 0.0), (0.75, 0.0),
(0.25, 0.25), (0.5, 0.25), (0.75, 0.25),
(0.25, 0.5), (0.5, 0.5).
Since the rectangles are the same size, $\Delta A_i = 0.0625$.

$$\int_0^1 \int_0^{\sqrt{1-x^2}} f(x, y) \, dy \, dx \approx \sum_{i=1}^{8} f(u_i, v_i) \Delta A_i$$

$$= f(0.25, 0.0)(0.0625)$$
$$+ f(0.5, 0.0)(0.0625)$$
$$+ \cdots + f(0.5, 0.5)(0.0625)$$

$$= (0.0625)(2.0 + 1.7 + 1.4 + 2.1$$
$$+ 1.8 + 1.6 + 2.3 + 2.0)$$
$$= 0.93125$$

65. We approximate the value of the integral on each subrectangle using the average of the function values at the 4 corners of each rectangle.

There are 16 rectangles, all the same size, with $\Delta A_i = 0.0625$.

Lower-left corner	f_{ave}
(0.0, 0.0)	2.15
(0.25, 0.0)	1.9
(0.5, 0.0)	1.625
(0.75, 0.0)	1.275
(0.0, 0.25)	2.3
(0.25, 0.25)	2.05
(0.5, 0.25)	1.8
(0.75, 0.25)	1.475
(0.0, 0.5)	2.55
(0.25, 0.5)	2.3
(0.5, 0.5)	2.075
(0.75, 0.5)	1.8
(0.0, 0.75)	2.9
(0.25, 0.75)	2.675
(0.5, 0.75)	2.5
(0.75, 0.75)	2.3

$$\int_0^1 \int_0^1 f(x, y) \, dy \, dx \approx \sum_{i=1}^{16} f_{ave}(u_i, v_i) \Delta A_i$$

$$= (0.0625)(2.15 + 1.9 + 1.625$$
$$+ 1.275 + 2.3 + 2.05 + 1.8$$
$$+ 1.475 + 2.55 + 2.3 + 2.075$$
$$+ 1.8 + 2.9 + 2.675 + 2.5 + 2.3)$$

$$= 2.1046875$$

67. We can take advantage of the symmetry in the problem:

$$\int_{-1}^1 \int_0^1 f(x, y) \, dy \, dx = 2 \int_0^1 \int_0^1 f(x, y) \, dy \, dx.$$

The contour lines divide the right half into five pieces. We estimate the area of these. The first (which is roughly half an ellipse) is approximately 0.1. The next three pieces are roughly vertical strips with areas approximately equal to 0.2, 0.3, 0.2 and 0.2.

Using the contours for the value of the function (just using a midpoint) and summing:

$$\int_{-1}^{1} \int_{0}^{1} f(x, y) \, dy \, dx = 2 \int_{0}^{1} \int_{0}^{1} f(x, y) \, dy \, dx$$

$$\approx 2 \sum_{i=1}^{5} \text{value} \cdot \text{area}$$

$$= 2\,[4.5(0.1) + 3.5(0.2) + 2.5(0.3)$$
$$+ 1.5(0.2) + 0.5(0.2)]$$
$$= 4.6$$

This gives the best estimate as (c), 4.

69.

$$\int_{c}^{d} \int_{a}^{b} f_{xy}(x, y) \, dx \, dy$$

$$= \int_{c}^{d} \int_{a}^{b} \frac{\partial}{\partial x} f_y(x, y) \, dx \, dy$$

$$= \int_{c}^{d} \left[f_y(x, y) \right]_{x=a}^{x=b} dy$$

$$= \int_{c}^{d} \left[f_y(b, y) - f_y(a, y) \right] dy$$

$$= \int_{c}^{d} \frac{\partial}{\partial y} \left[f(b, y) - f(a, y) \right] dy$$

$$= \left[f(b, y) - f(a, y) \right]_{c}^{d}$$

$$= \left[f(b, d) - f(a, d) \right] - \left[f(b, c) - f(a, c) \right]$$

$$= f(a, c) - f(a, d) - f(b, c) + f(b, d)$$

For the integral given, we have $a = 0$, $b = 1$, $c = 0$ and $d = 1$ and we just use the formula,

$$\int_{0}^{1} \int_{0}^{1} 24xy^2 \, dx \, dy$$

$$= f(0, 0) - f(0, 1) - f(1, 0) + f(1, 1)$$
$$= 0 - 1 - 3 + 8 = 4.$$

71.

$$\int_{0}^{2} \left[\tan^{-1}(4 - x) - \tan^{-1} x \right] dx$$

$$= \int_{0}^{2} \int_{x}^{4-x} \frac{1}{y^2 + 1} \, dy \, dx$$

$$= \int_{0}^{2} \int_{0}^{y} \frac{1}{y^2 + 1} \, dx \, dy$$

$$+ \int_{2}^{4} \int_{0}^{4-y} \frac{1}{y^2 + 1} \, dx \, dy$$

$$= \int_{0}^{2} \frac{y}{y^2 + 1} \, dy$$

$$+ \int_{2}^{4} \frac{4}{y^2 + 1} - \frac{y}{y^2 + 1} \, dy$$

$$= \left[\frac{1}{2} \ln(y^2 + 1) \right]_{0}^{2}$$

$$+ \left[4 \tan^{-1} y - \frac{1}{2} \ln(y^2 + 1) \right]_{2}^{4}$$

$$= \frac{1}{2} \ln 5 + 4 \tan^{-1} 4 - 4 \tan^{-1} 2$$

$$- \frac{1}{2} \ln 17 + \frac{1}{2} \ln 5$$

$$= \ln 5 - \frac{1}{2} \ln 17 + 4 \tan^{-1} 4 - 4 \tan^{-1} 2$$

73. We split the region of integration into two regions:

$$R_1 = \{(x, y) \in R : y \geq 2x\}$$
$$R_2 = \{(x, y) \in R : y < 2x\}$$

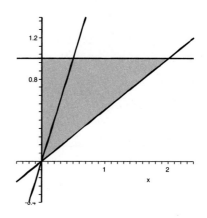

Region of Integration

R_1 is the region to the left of the line $y = 2x$ and R_2 is the region to the right of this line. In the region R_1 we have $f(x, y) = 2x$ and in the region R_2 we have $f(x, y) = y$.

$$\iint_R f(x, y)\, dA$$

$$= \iint_{R_1} f(x, y)\, dA + \iint_{R_2} f(x, y)\, dA$$

$$= \iint_{R_1} 2x\, dA + \iint_{R_2} y\, dA$$

$$= \int_0^2 \int_0^{y/2} 2x\, dx\, dy + \int_0^2 \int_{y/2}^{2y} y\, dx\, dy$$

$$= \int_0^2 \frac{y^2}{4}\, dy + \int_0^2 \frac{3y^2}{2}\, dy$$

$$= \frac{2}{3} + 4 = \frac{14}{3}$$

14.2 AREA, VOLUME, AND CENTER OF MASS

1. To find the limits of integration:

$$x^2 = 8 - x^2$$
$$x^2 = 4$$
$$x = \pm 2$$

Region of Integration

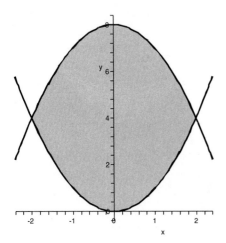

$$A = \iint_R dA$$

$$= \int_{-2}^{2} \int_{x^2}^{8-x^2} dy\, dx$$

$$= \int_{-2}^{2} [y]_{y=x^2}^{y=8-x^2}\, dx$$

$$= \int_{-2}^{2} (8 - 2x^2)\, dx$$

$$= \left[8x - \frac{2x^3}{3} \right]_{-2}^{2} = \frac{64}{3}$$

3. The region is a triangle, as pictured below.

Region of Integration

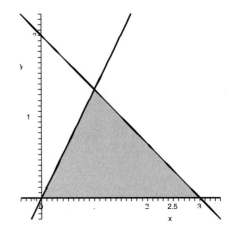

$$A = \iint_R dA$$

$$= \int_0^2 \int_{y/2}^{3-y} dx\, dy$$

$$= \int_0^2 [x]_{x=y/2}^{x=3-y}\, dy$$

$$= \int_0^2 \left(3 - \frac{3y}{2} \right) dy$$

$$= \left[3y - \frac{3y^2}{4} \right]_0^2 = 3$$

5. The region is pictured below.

Region 'of 'Integration

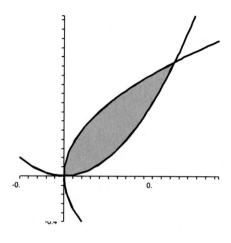

$$A = \iint_R dA$$

$$= \int_0^1 \int_{x^2}^{\sqrt{x}} dy\, dx$$

$$= \int_0^1 [y]_{y=x^2}^{y=\sqrt{x}} dx$$

$$= \int_0^1 \left(\sqrt{x} - x^2\right) dx$$

$$= \left[\frac{2}{3}x^{3/2} - \frac{x^3}{3}\right]_0^1 = \frac{1}{3}$$

7. The plane $2x + 3y + z = 6$ intersects the three coordinate axes at $(3, 0, 0)$, $(0, 2, 0)$, and $(0, 0, 6)$. Thus, we are to integrate the function $z = 6 - 2x - 3y$ over the region below.

Region 'of 'Integration

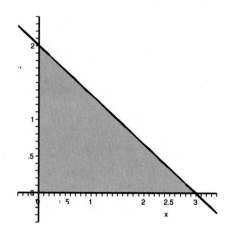

$$V = \iint_R f(x, y)\, dA$$

$$= \int_0^3 \int_0^{(6-2x)/3} (6 - 2x - 3y)\, dy\, dx$$

$$= \int_0^3 \left[6y - 2xy - \frac{3y^2}{2}\right]_{y=0}^{y=(6-2x)/3} dx$$

$$= \int_0^3 \left(6 - 4x + \frac{2x^2}{3}\right) dx$$

$$= \left[6x - 2x^2 + \frac{2x^3}{9}\right]_0^3 = 6$$

9.
$$V = \iint_R f(x, y)\, dA$$

$$= \int_{-1}^1 \int_{-1}^1 (4 - x^2 - y^2)\, dy\, dx$$

$$= \int_{-1}^1 \left[4y - x^2 y - \frac{y^3}{3}\right]_{y=-1}^{y=1} dx$$

$$= \int_{-1}^1 \left(\frac{22}{3} - 2x^2\right) dx$$

$$= \left[\frac{22x}{3} - \frac{2x^3}{3}\right]_{-1}^1 = \frac{40}{3}$$

11. The region lies above the square $1 \le x \le 2$ and $0 \le y \le 1$ in the xy-plane.

$$V = \iint_R f(x, y)\, dA$$

$$= \int_1^2 \int_0^1 (1 - y)\, dy\, dx$$

$$= \int_1^2 \left[y - \frac{y^2}{2}\right]_{y=0}^{y=1} dx$$

$$= \int_1^2 \frac{1}{2}\, dx = \frac{1}{2}$$

13. The solid is bounded above by $z = 1 - y^2$ and lies above the triangle in the xy-plane:

Region 'of 'Integration

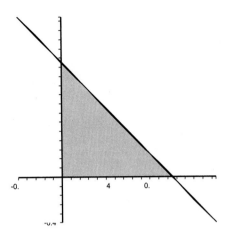

$$V = \iint_R f(x, y)\, dA$$

$$= \int_0^1 \int_0^{1-x} (1 - y^2)\, dy\, dx$$

$$= \int_0^1 \left[y - \frac{y^3}{3} \right]_{y=0}^{y=1-x} dx$$

$$= \int_0^1 \left(\frac{2}{3} - x^2 + \frac{x^3}{3} \right) dx$$

$$= \left[\frac{2x}{3} - \frac{x^3}{3} + \frac{x^4}{12} \right]_0^1 = \frac{5}{12}$$

15. The solid lies above the region in the xy-plane pictured below. This is our region of integration.

Region 'of 'Integration

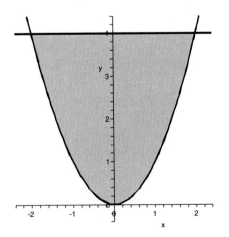

The solid lies above the plane $z = 1$ and below the surface $z = x^2 + y^2 + 3$. Therefore our integrand is $(x^2 + y^2 + 3) - 1$.

$$V = \iint_R f(x, y)\, dA$$

$$= \int_{-2}^2 \int_{x^2}^4 (x^2 + y^2 + 2)\, dy\, dx$$

$$= \int_{-2}^2 \left[yx^2 + \frac{y^3}{3} + 2y \right]_{y=x^2}^{y=4} dx$$

$$= \int_{-2}^2 \left(\frac{88}{3} + 2x^2 - x^4 - \frac{x^6}{3} \right) dx$$

$$= \left[\frac{88x}{3} - \frac{2x^3}{3} - \frac{x^5}{5} - \frac{x^7}{21} \right]_{-2}^2$$

$$= \frac{10,816}{105}$$

17. The solid lies above $z = y - 2$ and below $z = x + 2$. Our region of integration is the region in the xy-plane bounded by $y = x$ and $x = y^2 - 2$ pictured below.

Region 'of 'Integration

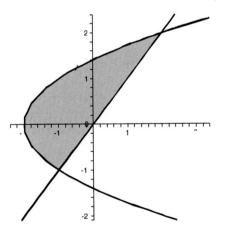

$$V = \iint_R dA$$

$$= \int_{-1}^{2} \int_{y^2-2}^{y} \left[(x+2) - (y-2) \right] dx \, dy$$

$$= \int_{-1}^{2} \left[-yx + 4x + \frac{x^2}{2} \right]_{x=y^2-2}^{x=y} dy$$

$$= \int_{-1}^{2} \left(-\frac{1}{2}y^4 + y^3 - \frac{5}{2}y^2 + 2y + 6 \right) dy$$

$$= \left[-\frac{1}{10}y^5 + \frac{1}{4}y^4 - \frac{5}{6}y^3 + y^2 + 6y \right]_{-1}^{2}$$

$$= \frac{279}{20}$$

19. This solid lies below the cone $z = \sqrt{x^2 + y^2}$ and above the region in the xy-plane pictured below:

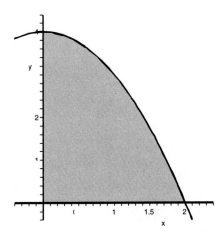

Region of Integration

The first integral can be computed using a trigonometric substitution or using the tables in the text. The difficulty in actually computing the volume is the second integral.

$$V = \iint_R \sqrt{x^2 + y^2} \, dA$$

$$= \int_0^2 \int_0^{4-x^2} \sqrt{x^2 + y^2} \, dy \, dx$$

$$= \int_0^2 \frac{1}{2} \left[x^2 \ln \left(y + \sqrt{x^2 + y^2} \right) \right.$$

$$+ \left. y\sqrt{x^2 + y^2} \right]_{y=0}^{y=4-x^2} dx$$

$$= \frac{1}{2} \int_0^2 \left[x^2 \ln \left(\frac{4 - x^2 + \sqrt{x^2 + (4 - x^2)^2}}{x} \right) \right.$$

$$+ \left. (4 - x^2)\sqrt{x^2 + (4 - x^2)^2} \right] dx$$

To approximate this integral, we can use Simpson's Rule. One problem is that that integrand is not defined at $x = 0$, but we can ignore this as the limit of the integrand at $x \to 0$ is 8, which is the functional value we will use. Using this and Simpson's Rule for $n = 10$ gives us an approximation of $V \approx 10.275$.

21. The solid is bounded above by $z = e^{xy}$ and below by the triangle in the xy-plane shown below:

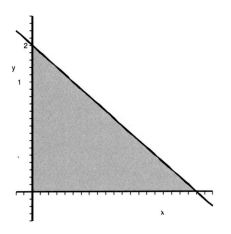

Region of Integration

$$V = \iint_R e^{xy}\, dA$$

$$= \int_0^2 \int_0^{4-2y} e^{xy}\, dx\, dy$$

$$= \int_0^2 \left[\frac{e^{xy}}{y}\right]_{x=0}^{x=4-2y} dy$$

$$= \int_0^2 \left(\frac{e^{y(4-2y)}}{y} - \frac{1}{y}\right) dy$$

This integral we estimate using Simpson's Rule. There is a problem because the integrand is not defined for $y = 0$, but the limit of the integrand as $y \to 0$ is 4, which we can use as the value of the integrand at $y = 0$.

With $n = 10$ Simpson's Rule gives $V \approx 9.003$.

23.

$$m = \int_0^1 \int_{x^3}^{x^2} 4\, dy\, dx$$

$$= \int_0^1 [4y]_{y=x^3}^{y=x^2}\, dx$$

$$= \int_0^1 \left(4x^2 - 4x^3\right) dx$$

$$= \left[\frac{4x^3}{3} - x^4\right]_0^1 = \frac{1}{3}$$

$$M_y = \int_0^1 \int_{x^3}^{x^2} 4x\, dy\, dx$$

$$= \int_0^1 [4xy]_{y=x^3}^{y=x^2}\, dx$$

$$= \int_0^1 \left(4x^3 - 4x^4\right) dx$$

$$= \left[x^4 - \frac{4x^5}{5}\right]_0^1 = \frac{1}{5}$$

$$\bar{x} = \frac{M_y}{m} = \frac{3}{5}$$

$$M_x = \int_0^1 \int_{x^3}^{x^2} 4y\, dy\, dx$$

$$= \int_0^1 \left[2y^2\right]_{y=x^3}^{y=x^2}\, dx$$

$$= \int_0^1 \left(2x^4 - 2x^6\right) dx$$

$$= \left[\frac{2x^5}{5} - \frac{2x^7}{7}\right]_0^1 = \frac{4}{35}$$

$$\bar{y} = \frac{M_x}{m} = \frac{12}{35}$$

25.

$$m = \int_{-1}^1 \int_{y^2}^1 (y^2 + x + 1)\, dx\, dy$$

$$= \int_{-1}^1 \left[y^2 x + \frac{x^2}{2} + x\right]_{x=y^2}^{x=1} dy$$

$$= \int_{-1}^1 \left(-\frac{3}{2}y^4 + \frac{3}{2}\right) dy$$

$$= \left[-\frac{3}{10}y^5 + \frac{3}{2}y\right]_{-1}^1 = \frac{12}{5}$$

$$M_y = \int_{-1}^1 \int_{y^2}^1 x(y^2 + x + 1)\, dx\, dy$$

$$= \int_{-1}^1 \left[\frac{x^3}{3} + \frac{x^2}{2}(y^2 + 1)\right]_{x=y^2}^{x=1} dy$$

$$= \int_{-1}^1 \left(-\frac{5y^6}{6} - \frac{y^4}{2} + \frac{y^2}{2} + \frac{5}{6}\right) dy$$

$$= \left[-\frac{5y^7}{42} - \frac{y^5}{10} + \frac{y^3}{6} + \frac{5y}{6}\right]_{-1}^1 = \frac{164}{105}$$

$$\bar{x} = \frac{M_y}{m} = \frac{41}{63}$$

$$M_x = \int_{-1}^1 \int_{y^2}^1 y(y^2 + x + 1)\, dx\, dy$$

$$= \int_{-1}^1 \left[y^3 x + \frac{x^2 y}{2} + xy\right]_{x=y^2}^{x=1} dy$$

$$= \int_{-1}^{1} \left(\frac{3y}{2} - \frac{3y^5}{2} \right) dy$$

$$= \left[\frac{3y^2}{4} - \frac{y^6}{4} \right]_{-1}^{1} = 0$$

$$\bar{y} = \frac{M_x}{m} = 0$$

We could have seen the fact that $\bar{y} = 0$ by the symmetry of the problem. The lamina and the density function are symmetric with respect to the x-axis. (They must both be symmetric to use symmetry in these problems!)

27. In this case we have $\rho = x$.

$$m = \int_0^2 \int_{x^2}^4 x \, dy \, dx$$

$$= \int_0^2 [xy]_{y=x^2}^{y=4} \, dx$$

$$= \int_0^2 \left(4x - x^3 \right) dx$$

$$= \left[2x^2 - \frac{x^4}{4} \right]_0^2 = 4$$

$$M_y = \int_0^2 \int_{x^2}^4 x^2 \, dy \, dx$$

$$= \int_0^2 [x^2 y]_{y=x^2}^{y=4} \, dx$$

$$= \int_0^2 \left(4x^2 - x^4 \right) dx$$

$$= \left[\frac{4x^3}{3} - \frac{x^5}{5} \right]_0^2 = \frac{64}{15}$$

$$\bar{x} = \frac{M_y}{m} = \frac{16}{15}$$

$$M_x = \int_0^2 \int_{x^2}^4 xy \, dy \, dx$$

$$= \int_0^2 \left[\frac{xy^2}{2} \right]_{y=x^2}^{y=4} dx$$

$$= \int_0^2 \left(8x - \frac{x^5}{2} \right) dx$$

$$= \left[4x^2 - \frac{x^6}{12} \right]_0^2 = \frac{32}{3}$$

$$\bar{y} = \frac{M_x}{m} = \frac{8}{3}$$

29. In exercise 25, both the lamina and the density function are symmetric. In exercise 26, the density function is not symmetric about the x-axis—the lamina is lighter per unit area in the lower half (where $y < 0$).

31. If the density function is symmetric about the y-axis ($\rho(x, y) = \rho(-x, y)$), then the center of mass will be on the y-axis.

33.

$$P = \int_0^1 \int_1^{y+1} 15{,}000 x e^{-x^2-y^2} \, dx \, dy$$

$$= 15{,}000 \int_0^1 e^{-y^2} \left[\int_1^{y+1} x e^{-x^2} \, dx \right] dy$$

$$= 15{,}000 \int_0^1 e^{-y^2} \left[-\frac{e^{-x^2}}{2} \right]_1^{y+1} dy$$

$$= 15{,}000 \int_0^1 e^{-y^2} \left(\frac{-e^{-(y+1)^2} + e^{-1}}{2} \right) dy$$

$$= \frac{15{,}000}{2e} \int_0^1 \left(-e^{-2y^2-2y} + e^{-y^2} \right) dy$$

This final integral can be approximated with Simpson's Rule. With $n = 10$, Simpson's Rule gives $P \approx 1164$.

35.

$$\int_0^{10} \int_0^4 20 e^{-t/6} \, dt \, dx$$

$$= \int_0^{10} 120 \left(1 - e^{-2/3} \right) dx$$

$$= 1200 \left(1 - e^{-2/3} \right) \approx 583.90$$

The units in the integrand are dollars per barrel per year. After the t-integration, we are left with dollars per barrel. After the x-integration, we are left with dollars. The integral represents the total revenue, in billions of dollars, generated by selling 10 billion barrels of oil between 2000 and 2004.

37.
$$m = \int_{-2}^{2} \int_{x^2}^{4} 1 \, dy \, dx = \frac{32}{3}$$

$$I_y = \int_{-2}^{2} \int_{x^2}^{4} x^2 \, dy \, dx = \frac{128}{15}$$

$$I_x = \int_{-2}^{2} \int_{x^2}^{4} y^2 \, dy \, dx = \frac{512}{7}$$

39. For the skater with extended arms, we add together the moments contributed by the central rectangles and the two rectangles representing the arms:

I_y(extended)

$$= \int_{-1}^{1} \int_{0}^{8} 1 \cdot x^2 \, dy \, dx + 2 \int_{1}^{3} \int_{6}^{7} x^2 \, dy \, dx$$

$$= \frac{16}{3} + \frac{52}{3} = \frac{68}{3}$$

I_y(extended)

$$= \int_{-1/2}^{1/2} \int_{0}^{10} 2 \cdot x^2 \, dy \, dx = \frac{5}{3}$$

Therefore the ratio of spin rates is
$\dfrac{68/3}{5/3} = 13.6$, the skater with the raised arms spins 13.6 times faster.

41. To save a bit of work, we compute the moments of the ellipse E bounded by $x^2 + 4y^2 = a^2$ (and we are interested in $a = 4$ and $a = 6$).

$$I_y(E) = \iint_{E} x^2 \rho \, dA$$

$$= \int_{-a}^{a} \int_{-\sqrt{a^2-x^2}/2}^{\sqrt{a^2-x^2}/2} x^2 \, dy \, dx$$

$$= \int_{-a}^{a} x^2 \sqrt{a^2 - x^2} \, dy \, dx$$

$$= \frac{\pi a^4}{8}$$

The last integral can be done either by using a trigonometric substitution or can be compute using a table of integral (or even a CAS).

With $a = 4$ (the smaller ellipse), the moment is 32π. With $a = 6$ (the larger ellipse), the moment is 162π. The ratio of these moments is $\frac{81}{16} \approx 5$.

43.
$$a = \int_{-2}^{2} \int_{x^2}^{4} dy \, dx$$

$$= \int_{-2}^{2} [y]_{y=x^2}^{y=4} \, dx$$

$$= \int_{-2}^{2} \left(4 - x^2\right) \, dx$$

$$= \left[4x - \frac{x^3}{3}\right]_{-2}^{2} = \frac{32}{3}$$

$$\iint_{R} f(x, y) \, dA = \int_{-2}^{2} \int_{x^2}^{4} y \, dy \, dx$$

$$= \int_{-2}^{2} \left[\frac{y^2}{2}\right]_{y=x^2}^{y=4} \, dx$$

$$= \int_{-2}^{2} \left(8 - \frac{x^4}{2}\right) \, dx$$

$$= \left[8x - \frac{x^5}{10}\right]_{-2}^{2} = \frac{128}{5}$$

Ave Value $= \dfrac{1}{a} \iint_{R} f(x, y) = \dfrac{12}{5}$

45. The average value and the y-coordinate of the center of mass are the same.

If the lamina had constant density, ρ, the total mass would be $m = \rho a = \dfrac{32\rho}{3}$.

The moment would be

$$M_x = \iint_{R} y\rho \, dA = \rho \iint_{R} y \, dA = \frac{128\rho}{5}.$$

In the center of mass, the factor of ρ will cancel:

$$\overline{y} = \frac{M_x}{m} = \frac{12}{5}$$

47. To find the limits of integration, we graph the region and determine where the curves cross.

Region of Integration

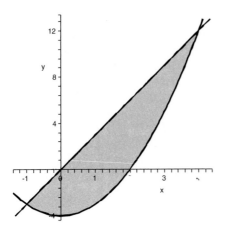

$$x^2 - 4 = 3x$$

$$x^2 - 3x - 4 = 0$$

$$(x - 4)(x + 1) = 0$$

So $x = -1, 4$ are the solutions.

$$a = \int_{-1}^{4} \int_{x^2-4}^{3x} dy \, dx$$

$$= \int_{-1}^{4} [y]_{y=x^2-4}^{y=3x} \, dx$$

$$= \int_{-1}^{4} \left(-x^2 + 3x + 4 \right) \, dx$$

$$= \left[-\frac{x^3}{3} + \frac{3x^2}{2} + 4x \right]_{-1}^{4} = \frac{125}{6}$$

$$\iint_R f(x, y) \, dA = \int_{-1}^{4} \int_{x^2-4}^{3x} \sqrt{x^2 + y^2} \, dy \, dx$$

$$= \int_{-1}^{4} \frac{1}{2} \left[y\sqrt{x^2 + y^2} \right.$$

$$\left. + x^2 \ln \left(y + \sqrt{x^2 + y^2} \right) \right]_{y=x^2-4}^{y=3x} \, dx$$

$$\approx 78.9937$$

Using a table of integrals gives the first integral above but then it gets messy. Possibilities include using Simpson's Rule or using a CAS to compute. In any case, one should arrive at something close to what we found above.

$$\text{Ave Value} = \frac{1}{a} \iint_R f(x, y)$$

$$\approx \frac{1}{125/6} (78.9937) \approx 3.7917$$

49. To find the limits of integration, we graph the region and determine where the curves cross.

Region of Integration

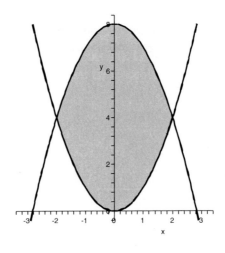

$$x^2 = 8 - x^2$$

$$2x^2 - 8 = 0$$

$$2(x - 2)(x + 2) = 0$$

So $x = -2, 2$.

$$a = \int_{-2}^{2} \int_{x^2}^{8-x^2} dy \, dx$$

$$= \int_{-2}^{2} [y]_{y=x^2}^{y=8-x^2} \, dx$$

$$= \int_{-2}^{2} \left(8 - 2x^2 \right) \, dx$$

$$= \left[8x - \frac{2x^3}{3} \right]_{-2}^{2} = \frac{64}{3}$$

$$\iint_R f(x, y)\, dA$$

$$= \int_{-2}^{2} \int_{x^2}^{8-x^2} \left[50 + \cos(2x + y) \right]\, dy\, dx$$

$$= \int_{-2}^{2} \left[50y + \sin(2x + y) \right]_{y=x^2}^{y=8-x^2}\, dx$$

$$= \int_{-2}^{2} \left[400 - 100x^2 + \sin(2x + 8 - x^2) \right.$$

$$\left. - \sin(2x + x^2) \right]\, dx$$

$$\approx 1069.084$$

This last integral can be approximated using Simpson's Rule or using a computer.

$$\text{Ave Value} = \frac{1}{a} \iint_R f(x, y)$$

$$\approx \frac{1069.084}{64/3} \approx 50.113$$

51. (a) $\displaystyle\iint_R f(x, y)\, dA$ represents the total rainfall in the region R.

(b) $\dfrac{\iint_R f(x, y)\, dA}{\iint_R 1\, dA}$ represents the average rainfall per unit area in the region R.

53. The area of the triangle is $a = \frac{c}{2}$. The equation of the hypotenuse is $y = -\frac{1}{c}x + 1$. The y-coordinate of the center of mass is

$$\overline{y} = \frac{1}{a} \iint_R y\, dA$$

$$= \frac{2}{c} \int_0^c \int_0^{-x/c+1} y\, dy\, dx$$

$$= \frac{2}{c} \int_0^c \left[\frac{y^2}{2} \right]_{y=0}^{y=-x/c+1}\, dx$$

$$= \frac{2}{c} \int_0^c \frac{(x - c)^2}{2c^2}\, dx$$

$$= \frac{2}{c} \left[\frac{(x - c)^3}{6c^2} \right]_0^c$$

$$= \frac{2}{c} \left(\frac{c}{6} \right) = \frac{1}{3}.$$

55. The volume of B is just abc, therefore we are to show that the volume of T is $abc/6$. T is bounded above by the plane
$\frac{x}{a} + \frac{y}{b} + \frac{z}{c} = 1$ or $z = c\left(1 - \frac{x}{a} - \frac{y}{b}\right)$.
The tetrahedron lies above the triangle in the xy-plane with vertices at $(0, 0, 0)$, $(a, 0, 0)$ and $(0, b, 0)$. The equation of the hypotenuse of this triangle is $y = b\left(1 - \frac{x}{a}\right)$.

$$V(T) = \int_0^a \int_0^{b(1-x/a)} c\left(1 - \frac{x}{a} - \frac{y}{b}\right) dy\, dx$$

$$= c \int_0^a \left[y - \frac{xy}{a} - \frac{y^2}{2b} \right]_{y=0}^{y=b(1-x/a)} dx$$

$$= c \int_0^a \frac{b(a - x)^2}{2a^2}\, dx$$

$$= bc \left[-\frac{(a - x)^3}{6a^2} \right]_0^a$$

$$= \frac{abc}{6}$$

57. Clearly, $f(x, y) \geq 0$ for all (x, y). The improper integrals can be computed individually or else save a little bit of work:

$$\iint f(x, y)\, dA$$

$$= \int_0^{\infty} \int_0^{\infty} e^{-x} e^{-y}\, dx\, dy$$

$$= \int_0^{\infty} e^{-y} \left(\int_0^{\infty} e^{-x}\, dx \right) dy$$

$$= \left(\int_0^{\infty} e^{-x}\, dx \right) \left(\int_0^{\infty} e^{-y}\, dy \right)$$

$$= \left(\int_0^{\infty} e^{-x}\, dx \right)^2$$

$$= \left(\lim_{t \to \infty} \int_0^t e^{-x}\, dx \right)^2$$

$$= \left(\lim_{t \to \infty} \left[1 - e^{-t} \right] \right)^2$$

$$= (1)^2 = 1$$

59. We need to solve for c:

$$1 = \int_0^2 \int_0^{3x} c(x + 2y)\, dy\, dx$$

$$= c \int_0^2 \left[xy + y^2 \right]_0^{3x} dx$$

$$= c \int_0^2 12x^2\, dx$$

$$= c \left[4x^3 \right]_0^2 = 32c$$

and therefore $c = \frac{1}{32}$.

61. The region in question is shown below:

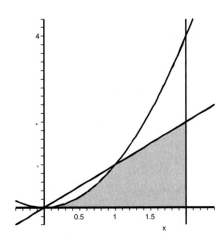

Region of Integration

The probability we are interested in is the region below the line $y = x$. To integrate over this region, we must use two integrals: one from $x = 0$ to $x = 1$ and then from $x = 1$ to $x = 2$.

$$P(y < x) = \int_0^1 \int_0^{x^2} f(x, y)\, dy\, dx$$

$$+ \int_1^2 \int_0^x f(x, y)\, dy\, dx$$

Another approach is to use the fact that the integral over the entire region is 1. Therefore we can integrate over the other region and subtract from 1:

$$P(y < x) = 1 - P(y \geq x)$$

$$= 1 - \int_1^2 \int_x^{x^2} f(x, y)\, dy\, dx$$

63. To find c we solve

$$1 = \iint_R c\, dA$$

$$= \int_0^2 \int_0^{4-x^2} c\, dy\, dx$$

$$= \frac{16c}{3}$$

and therefore $c = \frac{3}{16}$.

To find $P(y > x)$ we must find the intersection point of $y = x$ and $y = 4 - x^2$.

$$x = 4 - x^2$$

$$x^2 + x - 4 = 0$$

$$x = \frac{-1 \pm \sqrt{17}}{2}$$

Rather than writing this repeatedly, let $b = \frac{-1+\sqrt{17}}{2}$, and we will make the numerical substitution at the end. Then,

$$P(y > x) = \int_0^b \int_x^{4-x^2} \frac{3}{16}\, dy\, dx$$

$$= \int_0^b \left[\frac{3y}{16} \right]_{y=x}^{y=4-x^2} dx$$

$$= \int_0^b \left[\frac{3}{4} - \frac{3x}{16} - \frac{3x^2}{16} \right] dx$$

$$= \left[\frac{3x}{4} - \frac{3x^2}{32} - \frac{x^3}{16} \right]_0^b$$

$$= \frac{3b}{4} - \frac{3b^2}{32} - \frac{b^3}{16}$$

$$= \frac{-25 + 17\sqrt{17}}{64}$$

$$\approx 0.704575.$$

65. It should be clear from symmetry that $\bar{x} = 0$.

We compute the y-coordinate of the center of mass of the region.

$$m = \iint_R dA$$

$$= \text{Area of region}$$

$$= \frac{9\pi}{2} - \frac{8\pi}{2} = \frac{\pi}{2}$$

To find M_x we actually have to integrate. We will take advantage of the symmetry, but we will still need to split the integral up.

$$M_x = \iint_R y \, dA$$

$$= \int_{-3}^{-\sqrt{8}} \int_0^{\sqrt{9-x^2}} y \, dy \, dx$$

$$+ \int_{-\sqrt{8}}^{\sqrt{8}} \int_{\sqrt{8-x^2}}^{\sqrt{9-x^2}} y \, dy \, dx$$

$$+ \int_{\sqrt{8}}^{3} \int_0^{\sqrt{9-x^2}} y \, dy \, dx$$

$$= 2 \int_0^{\sqrt{8}} \int_{\sqrt{8-x^2}}^{\sqrt{9-x^2}} y \, dy \, dx$$

$$+ 2 \int_{\sqrt{8}}^{3} \int_0^{\sqrt{9-x^2}} y \, dy \, dx$$

$$= \left(2\sqrt{2}\right) + 2 \left(9 - \frac{19\sqrt{2}}{3}\right)$$

$$= 18 - \frac{32\sqrt{2}}{3}$$

$$\bar{y} = \frac{M_x}{m} = \frac{2}{\pi} \left(18 - \frac{32\sqrt{2}}{3}\right)$$

$$\approx 1.85578$$

Thus, the center of mass is below the bar, but since $(0, 2)$ is not in the region, the body doesn't touch the bar.

14.3 DOUBLE INTEGRALS IN POLAR COORDINATES

1.

$$A = \int_0^{2\pi} \int_0^{3+2\sin\theta} r \, dr \, d\theta$$

$$= \int_0^{2\pi} \left[\frac{r^2}{2}\right]_{r=0}^{r=3+2\sin\theta} d\theta$$

$$= \int_0^{2\pi} \left(\frac{13}{2} + 6\sin\theta - 2\cos^2\theta\right) d\theta$$

$$= 11\pi$$

3. One leaf is while θ is between 0 and $\frac{\pi}{3}$ (when r first returns to 0).

$$A = \int_0^{\pi/3} \int_0^{\sin 3\theta} r \, dr \, d\theta$$

$$= \int_0^{\pi/3} \left[\frac{r^2}{2}\right]_{r=0}^{r=\sin 3\theta} d\theta$$

$$= \frac{1}{2} \int_0^{\pi/3} \sin^2 3\theta \, d\theta$$

$$= \frac{1}{2} \left(\frac{\pi}{6}\right) = \frac{\pi}{12}$$

5. We will consider the first leaf, $\theta \in [0, \pi/3]$. Then $r = 2\sin 3\theta$ and $r = 1$ meet when

$$2\sin 3\theta = 1$$

$$\sin 3\theta = \frac{1}{2}$$

$$3\theta = \frac{\pi}{6}, \frac{5\pi}{6}, \ldots$$

$$\theta = \frac{\pi}{18}, \frac{5\pi}{18}, \ldots$$

$$A = \int_{\pi/18}^{5\pi/18} \int_1^{2\sin 3\theta} r \, dr \, d\theta$$

$$= \int_{\pi/18}^{5\pi/18} \left[\frac{r^2}{2}\right]_{r=1}^{r=2\sin 3\theta} d\theta$$

$$= \int_{\pi/18}^{5\pi/18} \left(\frac{3}{2} - 2\cos^2 3\theta\right) d\theta$$

$$= \frac{\pi}{9} + \frac{\sqrt{3}}{6}$$

7.
$$\iint_R \sqrt{x^2 + y^2} \, dA = \iint_R r \, dA$$
$$= \int_0^{2\pi} \int_0^3 r \cdot r \, dr \, d\theta$$
$$= \int_0^{2\pi} \int_0^3 r^2 \, dr \, d\theta$$
$$= \int_0^{2\pi} 9 \, d\theta = 18\pi$$

9.
$$\iint_R e^{-x^2 - y^2} \, dA = \iint_R e^{-r^2} \, dA$$
$$= \int_0^{2\pi} \int_0^2 e^{-r^2} r \, dr \, d\theta$$
$$= \int_0^{2\pi} \left[-\frac{e^{-r^2}}{2} \right]_0^2 \, d\theta$$
$$= \int_0^{2\pi} \left(\frac{1 - e^{-4}}{2} \right) d\theta$$
$$= 2\pi \left(\frac{1 - e^{-4}}{2} \right) = \pi(1 - e^{-4})$$

11.
$$\iint_R y \, dA = \int_0^{2\pi} \int_0^{2 - \cos\theta} r^2 \sin\theta \, dr \, d\theta$$
$$= \int_0^{2\pi} \frac{1}{3}(2 - \cos\theta)^3 \sin\theta \, d\theta$$
$$= \left[\frac{1}{12}(2 - \cos\theta)^4 \right]_0^{2\pi} = 0$$

This can also be seen from the symmetry of the problem without actually computing the integral.

13.
$$\iint_R (x^2 + y^2) \, dA = \iint_R r^2 \, dA$$
$$= \int_0^{2\pi} \int_0^3 r^3 \, dr \, d\theta$$
$$= \int_0^{2\pi} \frac{81}{4} \, d\theta$$
$$= (2\pi)\frac{81}{4} = \frac{81\pi}{2}$$

15. Either polar coordinates or Cartesian coordinates work in this problem (although the integrals are a bit easier using Cartesian coordinates).

In polar coordinates, $x = 2$ converts to $r \cos\theta = 2$ or $r = 2 \sec\theta$, $\theta \in [0, \pi/4]$. We use a table of integrals to evaluate the final integral.

$$\iint_R (x^2 + y^2) \, dA = \iint_R r^2 \, dA$$
$$= \int_0^{\pi/4} \int_0^{2 \sec\theta} r^3 \, dr \, d\theta$$
$$= \int_0^{\pi/4} 4 \sec^4\theta \, d\theta = 4 \int_0^{\pi/4} \sec^4\theta \, d\theta$$
$$= 4 \left(\frac{\sec^2\theta \tan\theta}{3} \Big|_0^{\pi/4} + \frac{2}{3} \int_0^{\pi/4} \sec^2\theta \, d\theta \right)$$
$$= 4 \left(\left[\frac{\sec^2\theta \tan\theta}{3} \right]_0^{\pi/4} + \left[\frac{2}{3} \tan\theta \right]_0^{\pi/4} \right)$$
$$= 4 \left(\frac{2}{3} + \frac{2}{3} \right) = \frac{16}{3}$$

Using Cartesian coordinates,
$$\iint_R (x^2 + y^2) \, dA = \int_0^2 \int_0^x (x^2 + y^2) \, dy \, dx$$
$$= \int_0^2 \frac{4x^3}{3} \, dx = \frac{16}{3}.$$

17. This is the same integral as exercise 13.
$$V = \iint_R (x^2 + y^2) \, dA = \iint_R r^2 \, dA$$
$$= \int_0^{2\pi} \int_0^3 r^3 \, dr \, d\theta$$
$$= \int_0^{2\pi} \frac{81}{4} \, d\theta$$
$$= (2\pi)\frac{81}{4} = \frac{81\pi}{2}$$

19. The top of the solid is the cone and the bottom is the disk in the xy-plane. We integrate over the disk of radius 2:

$$V = \iint_R \sqrt{x^2 + y^2}\, dA = \iint_R r\, dA$$

$$= \int_0^{2\pi} \int_0^2 r^2\, dr\, d\theta$$

$$= \int_0^{2\pi} \frac{8}{3}\, d\theta = \frac{16\pi}{3}$$

21. The solid is bounded on the top by $z = \sqrt{4 - x^2 - y^2}$ and the bottom by $z = 1$. The region of integration is the disk of radius $\frac{1}{2}$ in the xy-plane.

$$V = \iint_R \left(\sqrt{4 - x^2 - y^2} - 1 \right) dA$$

$$= \int_0^{2\pi} \int_0^{1/2} \left(\sqrt{4 - r^2} - 1 \right) r\, dr\, d\theta$$

$$= \int_0^{2\pi} \left[-\frac{1}{3}(4 - r^2)^{3/2} - \frac{r^2}{2} \right]_{r=0}^{r=1/2} d\theta$$

$$= \int_0^{2\pi} \left(\frac{61}{24} - \frac{5\sqrt{15}}{8} \right) d\theta$$

$$= 2\pi \left(\frac{61}{24} - \frac{5\sqrt{15}}{8} \right)$$

$$= \pi \left(\frac{61}{12} - \frac{5\sqrt{15}}{4} \right)$$

23. This is easiest done in rectangular coordinates. The region of integration is the triangle in the xy-plane with vertices $(0, 0)$, $(0, 6)$, $(6, 0)$.

$$V = \iint_R (6 - x - y)\, dA$$

$$= \int_0^6 \int_0^{6-x} (6 - x - y)\, dy\, dx$$

$$= \int_0^6 \left(18 - 6x + \frac{x^2}{2} \right) dx$$

$$= 36$$

25. Our region of integration will be the sector between $y = 0$ and $y = x$ which tells us that θ will range from 0 to $\frac{\pi}{4}$. To determine the range for r, we solve

$$4 - x^2 - y^2 = x^2 + y^2$$
$$x^2 + y^2 = 2$$

$$V = \iint_R \left[(4 - x^2 - y^2) - (x^2 - y^2) \right] dA$$

$$= \int_0^{\pi/4} \int_0^{\sqrt{2}} (4 - 2r^2) r\, dr\, d\theta$$

$$= \int_0^{\pi/4} 2\, d\theta = \frac{\pi}{2}$$

27.
$$\int_{-2}^2 \int_{-\sqrt{4-x^2}}^{\sqrt{4-x^2}} \sqrt{x^2 + y^2}\, dy\, dx$$

$$= \int_0^{2\pi} \int_0^2 r^2\, dr\, d\theta$$

$$= \int_0^{2\pi} \frac{8}{3}\, d\theta = \frac{16\pi}{3}$$

29.
$$\int_0^2 \int_{-\sqrt{4-x^2}}^{\sqrt{4-x^2}} e^{-x^2 - y^2}\, dy\, dx$$

$$= \int_{-\pi/2}^{\pi/2} \int_0^2 r e^{-r^2}\, dr\, d\theta$$

$$= \int_{-\pi/2}^{\pi/2} \left(\frac{1 - e^{-4}}{2} \right) d\theta$$

$$= \frac{\pi(1 - e^{-4})}{2}$$

31. As with all of these, drawing the region of integration is important:

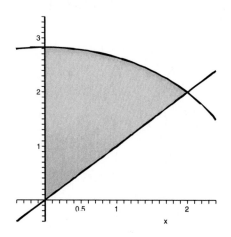

Region of Integration

$$\int_0^2 \int_x^{\sqrt{8-x^2}} (x^2 + y^2)^{3/2} \, dy \, dx$$

$$= \iint_R r^3 \, dA$$

$$= \int_{\pi/4}^{\pi/2} \int_0^{\sqrt{8}} r^4 \, dr \, d\theta$$

$$= \int_{\pi/4}^{\pi/2} \frac{128\sqrt{2}}{5} \, d\theta = \frac{32\pi\sqrt{2}}{5}$$

33. $P = \dfrac{1}{\pi} \displaystyle\iint_R e^{-x^2-y^2} \, dA$

$$= \frac{1}{\pi} \int_0^{2\pi} \int_0^{1/4} re^{-r^2} \, dr \, d\theta$$

$$= \frac{1}{2\pi} \int_0^{2\pi} \left[-e^{-r^2} \right]_{r=0}^{r=1/4} \, d\theta$$

$$= \frac{1}{2\pi} \int_0^{2\pi} \left(1 - e^{-1/16} \right) \, d\theta$$

$$= 1 - e^{-1/16} \approx 0.06059$$

35. $P = \dfrac{1}{\pi} \displaystyle\iint_R e^{-x^2-y^2} \, dA$

$$= \frac{1}{\pi} \int_{9\pi/20}^{11\pi/20} \int_{15/4}^{4} re^{-r^2} \, dr \, d\theta$$

$$= \frac{1}{2\pi} \int_{9\pi/20}^{11\pi/20} \left[-e^{-r^2} \right]_{r=15/4}^{r=4} \, d\theta$$

$$= \frac{1}{2\pi} \int_{9\pi/20}^{11\pi/20} \left(e^{-225/16} - e^{-16} \right) \, d\theta$$

$$= \frac{e^{-225/16} - e^{-16}}{20} \approx 3.34 \times 10^{-8}$$

37. $A = \displaystyle\iint_R dA$

$$= \int_{9\pi/20}^{11\pi/20} \int_{15/4}^{4} r \, dr \, d\theta$$

$$= \int_{9\pi/20}^{11\pi/20} \frac{31}{32} \, d\theta = \frac{31\pi}{320}$$

39. In polar coordinates, the equation
$x^2 + (y - 1)^2 = 1$
transforms to $r = 2 \sin \theta$. Notice as well that the θ limits are from
$\theta = 0$ to $\theta = \pi$.

$$m = \iint_R \rho \, dA = \iint_R \frac{1}{r} \, dA$$

$$= \int_0^{\pi} \int_0^{2\sin\theta} \left(\frac{1}{r} \right) r \, dr \, d\theta$$

$$= \int_0^{\pi} \int_0^{2\sin\theta} dr \, d\theta$$

$$= \int_0^{\pi} 2 \sin \theta \, d\theta = 4$$

$$M_y = \iint_R x\rho \, dA$$

$$= \int_0^{\pi} \int_0^{2\sin\theta} \left(\frac{r \cos \theta}{r} \right) r \, dr \, d\theta$$

$$= \int_0^{\pi} \int_0^{2\sin\theta} r \cos \theta \, dr \, d\theta$$

$$= \int_0^{\pi} 2 \cos \theta \sin^2 \theta \, d\theta = 0$$

$$\bar{x} = \frac{M_y}{m} = 0$$

$$M_x = \iint_R y\rho \, dA$$

$$= \int_0^{\pi} \int_0^{2\sin\theta} \left(\frac{r \sin \theta}{r} \right) r \, dr \, d\theta$$

$$= \int_0^{\pi} \int_0^{2\sin\theta} r \sin \theta \, dr \, d\theta$$

$$= \int_0^{\pi} 2 \sin^3 \theta \, d\theta$$

$$= 2 \int_0^{\pi} (1 - \cos^2 \theta) \sin \theta \, d\theta$$

$$= 2 \left[-\cos \theta + \frac{\cos^3 \theta}{3} \right]_0^{\pi} = \frac{8}{3}$$

$$\bar{y} = \frac{M_x}{m} = \frac{2}{3}$$

41. The population is found by integrating the density function over the region. We convert to polar coordinates.

$$P = \iint_R f(x, y)\, dA = \iint_R 20{,}000 e^{-r^2}\, dA$$

$$= 20{,}000 \int_0^{2\pi} \int_0^1 e^{-r^2} r\, dr\, d\theta$$

$$= 10{,}000 \int_0^{2\pi} \left[-e^{-r^2} \right]_0^1 d\theta$$

$$= 10{,}000 \int_0^{2\pi} \left(1 - e^{-1} \right) d\theta$$

$$= 20{,}000\pi \left(1 - e^{-1} \right) \approx 39{,}717$$

43.
$$I_y = \iint_R x^2 \rho\, dA = \iint_R r^2 \cos^2 \theta\, dA$$

$$= \int_0^{2\pi} \int_0^R r^3 \cos^2 \theta\, dr\, d\theta$$

$$= \left(\int_0^{2\pi} \cos^2 \theta\, d\theta \right) \left(\int_0^R r^3\, dr \right)$$

$$= \pi \left(\frac{R^4}{4} \right) = \frac{\pi R^4}{4}$$

Therefore if the radius, R, is doubled then the moment of inertia is multiplied by 16.

45. We will compute the volume in the first octant. In this case, the solid is bounded on the top by $z = \sqrt{a^2 - x^2 - y^2} = \sqrt{a^2 - r^2}$.
We integrate over the quarter disk in the xy-plane.

$$V = 8 \iint_R \sqrt{a^2 - r^2}\, dA$$

$$= 8 \int_0^{\pi/2} \int_0^a r\sqrt{a^2 - r^2}\, dr\, d\theta$$

$$= 8 \int_0^{\pi/2} \left[-\frac{1}{3}(a^2 - r^2)^{3/2} \right]_{r=0}^{r=a} d\theta$$

$$= 8 \int_0^{\pi/2} \frac{1}{3} a^3\, d\theta = \frac{4\pi a^3}{3}$$

47. The cylinder, in polar coordinates, is $r = 2\cos\theta$ with $\theta \in [-\pi/2, \pi/2]$.

The top of the solid is given by
$$z = \sqrt{9 - x^2 - y^2} = \sqrt{9 - r^2}.$$

Using symmetry we integrate only over $z \geq 0$ and we will multiply by two to get the volume (since the bottom of the cylinder is $z = -\sqrt{9 - x^2 - y^2}$).

$$V = 2 \iint_R \sqrt{9 - r^2}\, dA$$

$$= 2 \int_{-\pi/2}^{\pi/2} \int_0^{2\cos\theta} r\sqrt{9 - r^2}\, dr\, d\theta$$

$$= 2 \int_{-\pi/2}^{\pi/2} \left[-\frac{1}{3}(9 - r^2)^{3/2} \right]_0^{2\cos\theta} d\theta$$

$$= \frac{2}{3} \int_{-\pi/2}^{\pi/2} \left[27 - (9 - 4\cos^2\theta)^{3/2} \right] d\theta$$

$$\approx 17.1639$$

The last integral was computing using a CAS. It could have also been approximated using a Simpson's Rule (for example, with $n = 4$, Simpson's Rule gives 17.3621).

49. The top of the solid is given by
$$z = \sqrt{4 - x^2 - y^2} = \sqrt{4 - r^2}.$$
The bottom of the solid is given by
$$z = 2 - y = 2 - r\sin\theta.$$

What is not so clear is the region of integration, R—the projection of the solid to the xy-plane. To find R, we eliminate z from the system $x^2 + y^2 + z^2 = 4$ and $z = 2 - y$.

This gives
$$x^2 + y^2 + (2 - y)^2 = 4$$
$$x^2 + 2y^2 = 4y \quad \text{(an ellipse)}$$
$$r^2 \cos^2\theta + 2r^2 \sin^2\theta = 4r\sin\theta$$
$$r\left(\cos^2\theta + 2\sin^2\theta \right) = 4\sin\theta$$
$$r = \frac{4\sin\theta}{\cos^2\theta + 2\sin^2\theta} \qquad \theta \in [0, \pi]$$

$$V = \int_0^\pi \int_0^{\frac{4\sin\theta}{\cos^2\theta + 2\sin^2\theta}}$$

$$\left[\sqrt{4 - r^2} - (2 - r\sin\theta) \right] r\, dr\, d\theta$$

51. The cone $z = k - r$ lies above the disk $r \leq k$, this is our region of integration.

$$V(\text{Cone}) = \int_0^{2\pi} \int_0^k (k - r) r \, dr \, d\theta$$

$$= \int_0^{2\pi} \frac{1}{6} k^3 \, d\theta = \frac{1}{3} \pi k^3$$

The paraboloid $z = k - r^2$ lies above the disk $r \leq \sqrt{k}$, this is our region of integration.

$$V(\text{Paraboloid}) = \int_0^{2\pi} \int_0^{\sqrt{k}} (k - r^2) r \, dr \, d\theta$$

$$= \int_0^{2\pi} \frac{1}{4} k^2 \, d\theta = \frac{1}{2} \pi k^2$$

Notice that although the cone and the paraboloid have the same height, the area at the base are not the same. The area of the base of the cone is πk^2 whereas the area of the base of the paraboloid is πk. This explains the powers of k in the volumes.

53. The region described is pictured below.

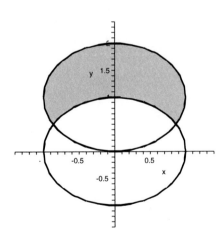

Region of Integration

We must find where the two circles intersect. In rectangular coordinates, these circles are $x^2 + y^2 = 1$ and $x^2 + (y - 1)^2 = 1$.

If we subtract these equations we get $2y - 1 = 0$ or $y = \frac{1}{2}$. Thus the circles intersect at $\left(-\frac{\sqrt{3}}{2}, \frac{1}{2} \right) \left(\frac{\sqrt{3}}{2}, \frac{1}{2} \right)$ or when $\theta = \frac{\pi}{6}, \frac{5\pi}{6}$.

$$\iint_R \frac{2}{1 + x^2 + y^2} \, dA$$

$$= \int_{\pi/6}^{5\pi/6} \int_1^{2 \sin \theta} \frac{2r}{1 + r^2} \, dr \, d\theta$$

$$= \int_{\pi/6}^{5\pi/6} \left[\ln(1 + r^2) \right]_1^{2 \sin \theta} d\theta$$

$$= \int_{\pi/6}^{5\pi/6} \left[\ln(1 + 4 \sin^2 \theta) - \ln 2 \right] d\theta$$

$$= \int_{\pi/6}^{5\pi/6} \ln \left(\frac{1 + 4 \sin^2 \theta}{2} \right) d\theta$$

$$\approx 1.2860$$

The last integral was approximated using Simpson's Rule with $n = 10$.

14.4 SURFACE AREA

1.
$$S = \int_0^4 \int_0^x \sqrt{(2x)^2 + (2)^2 + 1} \, dy \, dx$$

$$= \int_0^4 \int_0^x \sqrt{5 + 4x^2} \, dy \, dx$$

$$= \int_0^4 x \sqrt{5 + 4x^2} \, dx$$

$$= \left[\frac{1}{12} \left(5 + 4x^2 \right)^{3/2} \right]_0^4$$

$$= \frac{1}{12} \left(69^{3/2} - 5^{3/2} \right) \approx 46.8314$$

3. We change to polar coordinates.

$$S = \iint_R \sqrt{(-2x)^2 + (-2y)^2 + 1} \, dy \, dx$$

$$= \int_0^{2\pi} \int_0^2 r \sqrt{4r^2 + 1} \, dr \, d\theta$$

$$= \int_0^{2\pi} \left[\frac{1}{12} (4r^2 + 1)^{3/2} \right]_{r=0}^{r=2} d\theta$$

$$= \int_0^{2\pi} \frac{1}{12} \left(17^{3/2} - 1 \right) d\theta$$

$$= \frac{\pi}{6} \left(17^{3/2} - 1 \right) \approx 36.177$$

5. We will change to polar coordinates. Notice that since $z = r$, we have

$z^2 = r^2 = x^2 + y^2$

$2z \dfrac{\partial z}{\partial x} = 2x$

$\dfrac{\partial z}{\partial x} = \dfrac{x}{z} = \dfrac{r \cos \theta}{r} = \cos \theta$

Similarly, $\dfrac{\partial z}{\partial y} = \sin \theta$.

The region of integration is a disk of radius 2 in the xy-plane.

$$S = \iint_R \sqrt{(\cos \theta)^2 + (\sin \theta)^2 + 1} \, dA$$

$$= \iint_R \sqrt{2} \, dA$$

$$= \sqrt{2} \, (\text{Area of } R)$$

$$= \sqrt{2}(4\pi) \approx 17.7715$$

7. The surface is $z = 6 - x - 3y$ and lies above the triangle in the xy-plane with vertices $(0, 0, 0)$, $(6, 0, 0)$ and $(0, 2, 0)$. Notice that this triangle has area 6.

$$S = \iint_R \sqrt{(-1)^2 + (-3)^2 + 1} \, dA$$

$$= \iint_R \sqrt{11} \, dA$$

$$= \sqrt{11} \, (\text{Area of Triangle}) = 6\sqrt{11}$$

9. The surface is $z = \frac{x-y-4}{2}$ and lies below the triangle in the xy-plane with vertices $(0, 0, 0)$, $(4, 0, 0)$ and $(0, -4, 0)$. Notice that this triangle has area 8.

$$S = \iint_R \sqrt{(1/2)^2 + (-1/2)^2 + 1} \, dA$$

$$= \iint_R \sqrt{\frac{3}{2}} \, dA$$

$$= \sqrt{3/2} \, (\text{Area of Triangle})$$

$$= 8\sqrt{3/2} = 4\sqrt{6}$$

11. We convert to polar coordinates. The region of integration is the disk in the xy-plane $r \le 2$. The equation of the surface is

$z = \sqrt{4 - x^2 - y^2}$

$z^2 = 4 - x^2 - y^2 = 4 - r^2$

$2z \dfrac{\partial z}{\partial x} = -2x$

$\dfrac{\partial z}{\partial x} = -\dfrac{x}{z}$

$\dfrac{\partial z}{\partial x} = -\dfrac{r \cos \theta}{\sqrt{4 - r^2}}$

Similarly, $\dfrac{\partial z}{\partial y} = -\dfrac{r \sin \theta}{\sqrt{4-r^2}}$.

The integrand of the integral will be:

$\sqrt{(f_x)^2 + (f_y)^2 + 1}$

$$= \sqrt{\left(-\frac{r \cos \theta}{\sqrt{4 - r^2}}\right)^2 + \left(-\frac{r \sin \theta}{\sqrt{4 - r^2}}\right)^2 + 1}$$

$$= \sqrt{\frac{r^2}{4 - r^2}\left(\cos^2 \theta + \sin^2 \theta\right) + 1}$$

$$= \frac{2}{\sqrt{4 - r^2}}.$$

The region of integration is the disk $r \le 2$:

$$S = \iint_R \frac{2}{\sqrt{4 - r^2}} \, dA$$

$$= \int_0^{2\pi} \int_0^2 \frac{2r}{\sqrt{4 - r^2}} \, dr \, d\theta$$

$$= \int_0^{2\pi} \left[-2(4 - r^2)^{1/2}\right]_0^2 \, d\theta$$

$$= \int_0^{2\pi} 4 \, d\theta = 8\pi.$$

13. We will convert to polar coordinates. First,

$\dfrac{\partial z}{\partial x} = 2x e^{x^2 + y^2} = 2r e^{r^2} \cos \theta$

$\dfrac{\partial z}{\partial y} = 2r e^{r^2} \sin \theta$

The integrand will be

$\sqrt{(f_x)^2 + (f_y)^2 + 1}$

$$= \sqrt{\left(2r e^{r^2} \cos \theta\right)^2 + \left(2r e^{r^2} \sin \theta\right)^2 + 1}$$

$$= \sqrt{4r^2 e^{2r^2} + 1}.$$

$$S = \iint_R \sqrt{4r^2 e^{2r^2} + 1} \, dA$$

$$= \int_0^2 \int_0^{2\pi} r\sqrt{4r^2 e^{2r^2} + 1} \, d\theta \, dr$$

$$= 2\pi \int_0^2 r\sqrt{4r^2 e^{2r^2} + 1} \, dr$$

$$\approx 583.7692$$

Using Simpson's Rule with $n = 10$ gives this approximately as 586.8553.

15. We will convert to polar coordinates. The integrand will be

$$\sqrt{(f_x)^2 + (f_y)^2 + 1}$$

$$= \sqrt{(2x)^2 + (2y)^2 + 1}$$

$$= \sqrt{4r^2 + 1}.$$

$$S = \iint_R \sqrt{4r^2 + 1} \, dA$$

$$= \int_{\sqrt{5}}^{\sqrt{7}} \int_0^{2\pi} r\sqrt{4r^2 + 1} \, d\theta \, dr$$

$$= 2\pi \int_{\sqrt{5}}^{\sqrt{7}} r\sqrt{4r^2 + 1} \, dr$$

$$= 2\pi \left[\frac{1}{12}(1 + 4r^2)^{3/2} \right]_{\sqrt{5}}^{\sqrt{7}}$$

$$= \frac{\pi}{6}\left(29\sqrt{29} - 21\sqrt{21} \right)$$

$$\approx 31.3823$$

17.
$$S = \iint_R \sqrt{(0)^2 + (2y)^2 + 1} \, dA$$

$$= \int_{-2}^2 \int_{-2}^2 \sqrt{4y^2 + 1} \, dx \, dy$$

$$= 4 \int_{-2}^2 \sqrt{4y^2 + 1} \, dy$$

$$= 8\sqrt{17} + 2\ln(\sqrt{17} + 4)$$

$$\approx 37.1743$$

The last integral can be computed using trigonometric substitution or using tables. Of course, one could just use Simpson's Rule to approximate the integral as well.

19.
$$S = \iint_R \Big[(\cos x \cos y)^2$$

$$+ (-\sin x \sin y)^2 + 1 \Big]^{1/2} \, dA$$

$$= \iint_R \Big[\cos^2 x \cos^2 y$$

$$+ \sin^2 x \sin^2 y + 1 \Big]^{1/2} \, dA$$

To numerically approximate this integral, we do a Riemann sum—we divide the rectangle into a grid of n^2 subrectangles (n horizontal and n vertical subdivisions) and we use the upper-right corner for the value of the function. In this case, the area of these subrectangles is $\frac{\pi^2}{n^2}$. In general, the surface area will then be:

$$S \approx \sum_{i=1}^n \sum_{j=1}^n f\left(\frac{\pi i}{n}, \frac{\pi j}{n} \right) \frac{\pi^2}{n^2}$$

$$= \frac{\pi^2}{n^2} \sum_{i=1}^n \sum_{j=1}^n$$

$$\left[\cos^2\left(\frac{\pi i}{n} \right) \cos^2\left(\frac{\pi j}{n} \right) \right.$$

$$\left. + \sin^2\left(\frac{\pi i}{n} \right) \sin^2\left(\frac{\pi j}{n} \right) + 1 \right]^{1/2}.$$

Using only $n = 2$ or $n = 3$ gives a good approximation (and, in fact, can be done fine without even writing out the sum in full generality as above). Here are the values of the approximations for various values of n:

n	Approx S
1	13.95772840
2	11.91366640
3	12.04680517
4	12.04422713
5	12.04492766

21. In exercises 5 and 6, it is shown that the surface area is equal to $\sqrt{2}(\text{Area } R)$.

This computation is valid for any region R in the xy-plane.

23. If we have a plane $ax + by + cz = d$, the surface area integrand is a constant and equal to

$$\sqrt{(f_x)^2 + (f_y)^2 + 1}$$

$$= \sqrt{\left(-\frac{a}{c}\right)^2 + \left(-\frac{b}{c}\right)^2 + 1}$$

$$= \frac{\sqrt{a^2 + b^2 + c^2}}{|c|}.$$

The angle, θ, between the given plane and the xy-plane is really the angle between the normals of the planes. The normals are $\mathbf{n} = (a, b, c)$ and $\mathbf{k} = (0, 0, 1)$. Thus,

$$\cos \theta = \frac{\mathbf{n} \cdot \mathbf{k}}{\|\mathbf{n}\|\|\mathbf{k}\|} = \frac{c}{\sqrt{a^2 + b^2 + c^2}}.$$

Therefore

$$S = \iint_R \frac{\sqrt{a^2 + b^2 + c^2}}{|c|} \, dA$$

$$= \iint_R |\sec \theta| \, dA = |\sec \theta| (\text{Area of } R).$$

25. Given a set B in the xz-plane and an interval $[c, d]$ on the y-line, the cylinder is the region $C = \{(x, y, z) : (x, z) \in B, \ y \in [c, d]\}$. If B is a curve, then C is a surface.

Suppose we have such a cylindrical surface, given by a curve $z = f(x)$, $x \in [a, b]$. Then

$$\frac{\partial z}{\partial x} = f'(x)$$

$$\frac{\partial z}{\partial y} = 0$$

and the surface area integrand is

$$\sqrt{(f_x)^2 + (f_y)^2 + 1}$$

$$= \sqrt{(f'(x))^2 + (0)^2 + 1}$$

$$= \sqrt{1 + (f'(x))^2}$$

which is also the integrand for arc length. In any case, the surface area of the cylindrical

surface is

$$S = \iint_R \sqrt{1 + (f'(x))^2} \, dA$$

$$= \int_a^b \int_c^d \sqrt{1 + (f'(x))^2} \, dy \, dx$$

$$= (d - c) \int_a^b \sqrt{1 + (f'(x))^2} \, dx$$

$$= (d - c) \left(\text{Length of curve} \right)$$

$$= (d - c)L.$$

So, for exercises 17 and 18, the surface area is $4L$.

27. $k = 3$ will not work because there is more surface area between $z = 3$ and $z = 5$ than there is between $z = 1$ and $z = 3$ (this should be clear by looking at the pictures in the text).

To compute the surface area of the surface between $z = 1$ and $z = k$ we integrate. Notice that when we convert to polar coordinates the surface between $z = 1$ and $z = k$ corresponds to the region of integration $r \leq \sqrt{k - 1}$.

$$S_k = \int_0^{2\pi} \int_0^{\sqrt{k-1}} r\sqrt{4r^2 + 1} \, dr \, d\theta$$

$$= 2\pi \int_0^{\sqrt{k-1}} r\sqrt{4r^2 + 1} \, dr$$

$$= 2\pi \left[\frac{1}{12}(4r^2 + 1)^{3/2} \right]_0^{\sqrt{k-1}}$$

$$= \frac{\pi}{6} \left[(4k - 3)^{3/2} - 1 \right]$$

We want this to be half the total surface area, or

$$S_k = \frac{S}{2} = \frac{\pi}{12} \left(17^{3/2} - 1 \right)$$

which gives the equation

$$\frac{\pi}{6} \left[(4k - 3)^{3/2} - 1 \right] = \frac{\pi}{12} \left(17^{3/2} - 1 \right)$$

$$(4k - 3)^{3/2} = \frac{1}{2} \left(17^{3/2} + 1 \right)$$

$$k = \frac{\left(\frac{17^{3/2}+1}{2} \right)^{2/3} + 3}{4} \approx 3.4527$$

29. To condense the notation, we define $\langle a, b, c \rangle$ as

$$\langle a, b, c \rangle = \mathbf{r}_u \times \mathbf{r}_v$$
$$= \langle x_u, y_u, z_u \rangle \times \langle x_v, y_v, z_v \rangle$$
$$= \langle y_u z_v - y_v z_u,$$
$$x_v z_u - x_u z_v,$$
$$y_v x_u - y_u x_v \rangle.$$

According to the chain rule we have

$$z_u = \frac{\partial z}{\partial x} x_u + \frac{\partial z}{\partial y} y_u$$

$$z_v = \frac{\partial z}{\partial x} x_v + \frac{\partial z}{\partial y} y_v$$

This is a system of equations that we can solve for $\frac{\partial z}{\partial x}$ and $\frac{\partial z}{\partial y}$. (Cramer's Rule, substitution, etc., will all work.) This gives

$$\frac{\partial z}{\partial x} = \frac{y_v z_u - y_u z_v}{y_v x_u - y_u x_v} = \frac{-a}{c} = -\frac{a}{c}$$

$$\frac{\partial z}{\partial y} = \frac{x_v z_u - x_u z_v}{x_u y_v - x_v y_u} = \frac{b}{c}$$

Suppose V is a region in the uv-plane, S is the area of the surface image $\mathbf{r}(V)$, and R is the corresponding region in the xy-plane that is "under" the surface. Then,

$$S = \iint_R \sqrt{\left(\frac{\partial z}{\partial x}\right)^2 + \left(\frac{\partial z}{\partial y}\right)^2 + 1}\, dA_{xy}$$

$$= \iint_R \sqrt{\left(-\frac{a}{c}\right)^2 + \left(\frac{b}{c}\right)^2 + 1}\, dA_{xy}$$

where dA_{xy} means dA, but in the xy-plane (we will use dA_{uv} to signify when we move to the uv-plane).

Now, if a, b, c are constants, then S will be equal to

$$S = \frac{\sqrt{a^2 + b^2 + c^2}}{|c|}\,(\text{Area of } R).$$

This is important since we will approximate the surface integral by this constant approximation.

Next, if we assume that $\mathbf{r}$ is linear, i.e., that $\mathbf{r}(u, v) = \mathbf{A}u + \mathbf{B}v$ where $\mathbf{A}$ and $\mathbf{B}$ are vectors. If we take V to be the unit square $V = \{(u, v) : 0 \le u \le 1, 0 \le v \le 1\}$, then $\mathbf{r}(V)$ is

the parallelogram spanned by the vectors $\mathbf{A}$ and $\mathbf{B}$. The area will then be

$$\text{Area}(\mathbf{r}(V)) = \|\mathbf{A} \times \mathbf{B}\|$$
$$= \|\mathbf{A} \times \mathbf{B}\|(\text{Area of } V).$$

But, by looking at proportions, this relation will hold for any V and in the case where a, b are constant we arrive at

$$S = \frac{\sqrt{a^2 + b^2 + c^2}}{|c|}\,(\text{Area of } R)$$

$$S = \|\mathbf{r}_u \times \mathbf{r}_v\|\,(\text{Area of } V)$$

Of course in general $a, b,$ and c are not constants, but if we look at a very small piece of V, then $a, b,$ and c are approximately constant. Thus, we subdivide V into sufficiently small subrectangles:

$$V = V_1 \cup V_2 \cup \cdots \cup V_n$$

We will also let S_i be the surface area $\mathbf{r}(V_i)$ and we let R_i be the portion of the xy-plane corresponding to V_i. Then we will have

$$S_i \approx \frac{\sqrt{a^2 + b^2 + c^2}}{|c|}\,(\text{Area of } R_i)$$

$$S_i \approx \|\mathbf{r}_u \times \mathbf{r}_v\|\,(\text{Area of } V_i)$$

Summing these gives $\sum S_i = S$ on the left and a Riemann sum for an integral on the right, from which we conclude that

$$S = \iint_R \sqrt{\left(\frac{\partial z}{\partial x}\right)^2 + \left(\frac{\partial z}{\partial y}\right)^2 + 1}\, dA_{xy}$$

$$= \iint_V \|\mathbf{r}_u \times \mathbf{r}_v\|\, dA_{uv}.$$

Of course, there is no need to assume that V is a rectangle.

31.
$$\mathbf{r} = \langle u, v \cos u, v \sin u \rangle$$
$$\mathbf{r}_u = \langle 1, -v \sin u, v \cos u \rangle$$
$$\mathbf{r}_v = \langle 0, \cos u, \sin u \rangle$$
$$\mathbf{r}_u \times \mathbf{r}_v = \langle -v, -\sin u, \cos u \rangle$$
$$\|\mathbf{r}_u \times \mathbf{r}_v\| = \sqrt{v^2 + 1}$$

$$S = \int_0^1 \int_0^{2\pi} \sqrt{1+v^2} \, du \, dv$$

$$= 2\pi \int_0^1 \sqrt{1+v^2} \, dv$$

$$= 2\pi \left[\frac{1}{2} v \sqrt{1+v^2} \right.$$

$$\left. + \frac{1}{2} \ln \left(v + \sqrt{1+v^2} \right) \right]_0^1$$

$$= \pi \left[\sqrt{2} + \ln \left(1 + \sqrt{2} \right) \right]$$

$$\approx 7.2118$$

14.5 TRIPLE INTEGRALS

1.
$$\iiint_Q (2x + y - z) \, dV$$

$$= \int_0^2 \int_{-2}^2 \int_0^2 (2x + y - z) \, dz \, dy \, dx$$

$$= \int_0^2 \int_{-2}^2 \left[2xz + yz - \frac{z^2}{2} \right]_0^2 dy \, dx$$

$$= \int_0^2 \int_{-2}^2 (4x + 2y - 2) \, dy \, dx$$

$$= \int_0^2 \left[4xy + y^2 - 2y \right]_{-2}^2 dx$$

$$= \int_0^2 (16x - 8) \, dx$$

$$= \left[8x^2 - 8x \right]_0^2 = 16$$

3.
$$\iiint_Q \left(\sqrt{y} - 3z^2 \right) dV$$

$$= \int_2^3 \int_0^1 \int_{-1}^1 (\sqrt{y} - 3z^2) \, dz \, dy \, dx$$

$$= \int_2^3 \int_0^1 (2\sqrt{y} - 2) \, dy \, dx$$

$$= \int_2^3 \left(-\frac{2}{3} \right) dx = -\frac{2}{3}$$

5. In this case, because the integrand is independent of x, we integrate in the x-direction first.

$$\iiint_Q 4yz \, dV$$

$$= \int_0^1 \int_0^{2-2y} \int_0^{2-2y-z} 4yz \, dx \, dz \, dy$$

$$= \int_0^1 \int_0^{2-2y} 4yz(2 - 2y - z) \, dz \, dy$$

$$= \int_0^1 \frac{16}{3} \left(y - 3y^2 + 3y^3 - y^4 \right) dy$$

$$= \frac{4}{15}$$

7. In this case, notice that the region lies in below the xy-plane. Because the integrand is independent of x, we integrate x first.

$$\iiint_Q (3y^2 - 2z) \, dV$$

$$= \int_0^3 \int_{-6+2y}^0 \int_0^{(6-2y+z)/3} (3y^2 - 2z) \, dx \, dz \, dy$$

$$= \int_0^3 \int_{-6+2y}^0 (3y^2 - 2z) \frac{(6 - 2y + z)}{3} \, dz \, dy$$

$$= \int_0^3 \int_{-6+2y}^0 \left(-2y^3 + 6y^2 + y^2 z + \frac{4}{3} yz \right.$$

$$\left. - \frac{2}{3} z^2 - 4z \right) dz \, dy$$

$$= \int_0^3 \left(24 - 24y + 26y^2 - \frac{116}{9} y^3 + 2y^4 \right) dy$$

$$= \frac{171}{5}$$

9. For this problem, we first integrate in the z-direction and then use polar coordinates in the xy-plane.

$$\iiint_Q 2xy\,dV$$

$$= \iint_R \left\{ \int_0^{1-x^2-y^2} 2xy\,dz \right\} dA$$

$$= \iint_R (2xy)(1 - x^2 - y^2)\,dA$$

$$= \int_0^1 \int_0^{2\pi} 2r^3(1 - r^2)\cos\theta\sin\theta\,d\theta\,dr$$

$$= \int_0^1 \left[2r^3(1 - r^2)\frac{\sin^2\theta}{2} \right]_0^{2\pi} dr$$

$$= \int_0^1 0\,dr = 0$$

11. It is important to draw the region of integration. In this case the top of the region is bounded by the two planes $x + z = 2$ and $z - x = 2$. The bottom of the region is the plane $z = 1$.

$$\iiint_Q 2yz\,dV$$

$$= \int_1^2 \int_{z-2}^{2-z} \int_{-2}^2 2yz\,dy\,dx\,dz$$

$$= \int_1^2 \int_{z-2}^{2-z} \left[y^2 z \right]_{-2}^2 dx\,dz$$

$$= \int_1^2 \int_{z-2}^{2-z} 0\,dx\,dz = 0$$

13. $$\iiint_Q 15\,dV$$

$$= 15 \int_{-1}^1 \int_0^{1-y^2} \int_0^{4-2x-y} dz\,dx\,dy$$

$$= 15 \int_{-1}^1 \int_0^{1-y^2} (4 - 2x - y)\,dx\,dy$$

$$= 15 \int_{-1}^1 \left(3 - y - 2y^2 + y^3 - y^4 \right) dy$$

$$= 15 \left(\frac{64}{15} \right) = 64$$

15.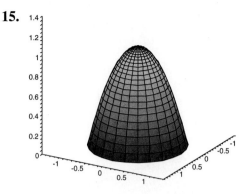

The integral in exercise 9 was zero because of the symmetry: the positive and negative parts of the integrand $2xy$ canceled each other. Note that for this to occur the region had to be symmetric as well.

If the integrand had been $2x^2y$ then the integral would still have been zero due to the symmetry in y.

If the integral had been $2x^2y^2$ then we lose the symmetry that caused the integral to be zero. In this case, you can work out the integral to be

$$\iiint_Q 2x^2y^2\,dV = \frac{\pi}{48}.$$

17. This is a cylinder in the y-direction. The base can be described as
$\{(x, z) : x^2 \le z \le 1, -1 \le x \le 1\}$.

$$V = \iiint_Q dV$$

$$= \int_{-1}^1 \int_{x^2}^1 \int_0^2 dy\,dz\,dx$$

$$= \int_{-1}^1 \int_{x^2}^1 2\,dz\,dx$$

$$= \int_{-1}^1 2(1 - x^2)\,dx = \frac{8}{3}$$

19. In this case the description actually describes two solids. What we have is a "tunnel" in the x-direction ($z = 1 - y^2$). This "tunnel" is cut by two planes: $x = 4$ and $2x + z = 4$ (which meet at $z = -4$). These two planes cut a wedge from the tunnel. Finally, the plane $z = 0$ cuts this wedge in two. We will find both these volumes.

Volume 1

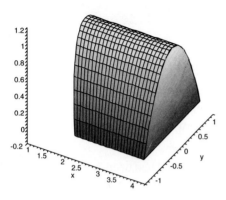

Volume 2

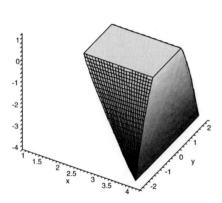

First the lower solid ($z \leq 0$):

$$V_1 = \iiint_Q dV$$

$$= \int_{-4}^{0} \int_{-\sqrt{1-z}}^{\sqrt{1-z}} \int_{(4-z)/2}^{4} dx\, dy\, dz$$

$$= \int_{-4}^{0} \int_{-\sqrt{1-z}}^{\sqrt{1-z}} \frac{z+4}{2} dy\, dz$$

$$= \int_{-4}^{0} (4+z)\sqrt{1-z}\, dz$$

$$= \left[-\frac{2(3z+22)}{15}(1-z)^{3/2} \right]_{-4}^{0}$$

$$= \frac{20\sqrt{5}}{3} - \frac{44}{15}$$

Now the upper solid ($z \geq 0$). Note that the integral is the same except for the z-limits.

$$V_2 = \iiint_Q dV$$

$$= \int_{0}^{1} \int_{-\sqrt{1-z}}^{\sqrt{1-z}} \int_{(4-z)/2}^{4} dx\, dy\, dz$$

$$= \left[-\frac{2(3z+22)}{15}(1-z)^{3/2} \right]_{0}^{1}$$

$$= \frac{44}{15}$$

21. $$V = \iiint_Q dV$$

$$= \int_{-\sqrt{10}}^{\sqrt{10}} \int_{-6}^{4-x^2} \int_{0}^{y+6} dz\, dy\, dx$$

$$= \int_{-\sqrt{10}}^{\sqrt{10}} \int_{-6}^{4-x^2} (y+6)\, dy\, dx$$

$$= \int_{-\sqrt{10}}^{\sqrt{10}} \left(\frac{x^4}{2} - 10x^2 + 50 \right) dx$$

$$= \frac{160\sqrt{10}}{3}$$

23. The equations $z = x^2$ and $z = 1$ give a cylindrical boundary (in the y-direction), with a base $\{(x, z) : x^2 \leq z \leq 1, -1 \leq x \leq 1\}$
y goes from 0 to $3 - x$.

$$V = \iiint_Q dV$$

$$= \int_{-1}^{1} \int_{x^2}^{1} \int_{0}^{3-x} dy\, dz\, dx$$

$$= \int_{-1}^{1} \int_{x^2}^{1} (3-x) \, dz \, dx$$

$$= \int_{-1}^{1} (3-x)(1-x^2) \, dx$$

$$= \int_{-1}^{1} (3 - x - 3x^2 + x^3) \, dx = 4$$

25. $$V = \iiint_{Q} dV$$

$$= \int_{0}^{1} \int_{z-1}^{1-z} \int_{z-1}^{1-z} dy \, dx \, dz$$

$$= \int_{0}^{1} \int_{z-1}^{1-z} 2(1-z) \, dx \, dz$$

$$= \int_{0}^{1} 4(1-z)^2 \, dz = \frac{4}{3}$$

27. This is a good problem for polar coordinates in the xy-plane.

$$V = \iiint_{Q} dV$$

$$= \int_{0}^{2\pi} \int_{0}^{2} \int_{0}^{4-r^2} r \, dz \, dr \, d\theta$$

$$= \int_{0}^{2\pi} \int_{0}^{2} r(4-r^2) \, dr \, d\theta$$

$$= \int_{0}^{2\pi} 4 \, d\theta = 8\pi$$

29. We convert to polar coordinates in the xy-plane.

$$m = \iiint_{Q} \rho(x, y, z) \, dV$$

$$= \int_{-2}^{2} \int_{-\sqrt{4-x^2}}^{\sqrt{4-x^2}} \int_{x^2+y^2}^{4} 4 \, dz \, dy \, dx$$

$$= \int_{-2}^{2} \int_{-\sqrt{4-x^2}}^{\sqrt{4-x^2}} 4(4 - x^2 - y^2) \, dy \, dx$$

$$= \int_{0}^{2\pi} \int_{0}^{2} 4r(4 - r^2) \, dr \, d\theta$$

$$= \int_{0}^{2\pi} 16 \, d\theta = 32\pi$$

$$M_{yz} = \iiint_{Q} x\rho(x, y, z) \, dV$$

$$= \int_{-2}^{2} \int_{-\sqrt{4-x^2}}^{\sqrt{4-x^2}} \int_{x^2+y^2}^{4} 4x \, dz \, dy \, dx$$

$$= \int_{-2}^{2} \int_{-\sqrt{4-x^2}}^{\sqrt{4-x^2}} 4x(4 - x^2 - y^2) \, dy \, dx$$

$$= \int_{0}^{2\pi} \int_{0}^{2} 4r^2(4 - r^2) \cos\theta \, dr \, d\theta$$

$$= \int_{0}^{2\pi} \frac{256}{15} \cos\theta \, d\theta = 0$$

$$\overline{x} = \frac{M_{yz}}{m} = 0$$

$$M_{xz} = \iiint_{Q} y\rho(x, y, z) \, dV$$

$$= \int_{-2}^{2} \int_{-\sqrt{4-x^2}}^{\sqrt{4-x^2}} \int_{x^2+y^2}^{4} 4y \, dz \, dy \, dx$$

$$= \int_{-2}^{2} \int_{-\sqrt{4-x^2}}^{\sqrt{4-x^2}} 4y(4 - x^2 - y^2) \, dy \, dx$$

$$= \int_{0}^{2\pi} \int_{0}^{2} 4r^2(4 - r^2) \sin\theta \, dr \, d\theta$$

$$= \int_{0}^{2\pi} \frac{256}{16} \sin\theta \, d\theta = 0$$

$$\overline{y} = \frac{M_{xz}}{m} = 0$$

$$M_{xy} = \iiint_{Q} z\rho(x, y, z) \, dV$$

$$= \int_{-2}^{2} \int_{-\sqrt{4-x^2}}^{\sqrt{4-x^2}} \int_{x^2+y^2}^{4} 4z \, dz \, dy \, dx$$

$$= \int_{-2}^{2} \int_{-\sqrt{4-x^2}}^{\sqrt{4-x^2}} 2\left[16 - (x^2 + y^2)^2\right] dy \, dx$$

$$= \int_0^{2\pi} \int_0^2 2r(16 - r^4)\, dr\, d\theta$$

$$= \int_0^{2\pi} \frac{128}{3}\, d\theta = \frac{256\pi}{3}$$

$$\bar{z} = \frac{M_{xy}}{m} = \frac{8}{3}$$

31. $\quad m = \iiint_Q \rho(x, y, z)\, dV$

$$= \int_0^6 \int_0^{-x/3+2} \int_0^{6-x-3y} (10 + x)\, dz\, dy\, dx$$

$$= \int_0^6 \int_0^{-x/3+2} (6 - x - 3y)(10 + x)\, dy\, dx$$

$$= \int_0^6 \left(60 - 14x - \frac{1}{3}x^2 + \frac{1}{6}x^3\right) dx$$

$$= 138$$

$$M_{yz} = \iiint_Q x\rho(x, y, z)\, dV$$

$$= \int_0^6 \int_0^{-x/3+2} \int_0^{6-x-3y} x(10 + x)\, dz\, dy\, dx$$

$$= \int_0^6 \int_0^{-x/3+2} (6 - x - 3y)x(10 + x)\, dy\, dx$$

$$= \int_0^6 \left(60x - 14x^2 - \frac{1}{3}x^3 + \frac{1}{6}x^4\right) dx$$

$$= \frac{1116}{5}$$

$$\bar{x} = \frac{M_{yz}}{m} = \frac{186}{115}$$

$$M_{xz} = \iiint_Q y\rho(x, y, z)\, dV$$

$$= \int_0^6 \int_0^{-x/3+2} \int_0^{6-x-3y} y(10 + x)\, dz\, dy\, dx$$

$$= \int_0^6 \int_0^{-x/3+2} (6 - x - 3y)y(10 + x)\, dy\, dx$$

$$= \int_0^6 \left(40 - 16x + \frac{4}{3}x^2 + \frac{4}{27}x^3 - \frac{1}{54}x^4\right) dx$$

$$= \frac{336}{5}$$

$$\bar{y} = \frac{M_{xz}}{m} = \frac{56}{115}$$

$$M_{xy} = \iiint_Q z\rho(x, y, z)\, dV$$

$$= \int_0^6 \int_0^{-x/3+2} \int_0^{6-x-3y} z(10 + x)\, dz\, dy\, dx$$

$$= \int_0^6 \int_0^{-x/3+2} \frac{1}{2}(6 - x - 3y)^2(10 + x)\, dy\, dx$$

$$= \int_0^6 \left(120 - 48x + 4x^2 + \frac{4}{9}x^3 - \frac{1}{18}x^4\right) dy\, dx$$

$$= \frac{1008}{5}$$

$$\bar{z} = \frac{M_{xy}}{m} = \frac{168}{115}$$

33. In exercise 29, the figure is completely symmetrical about the yz-plane. In addition, density is also symmetrical about the yz-plane (the density is constant).

In exercise 30, the figure is symmetrical but the density function is not. The figure is heavier on the positive x side of the figure, which pulls the center of gravity towards the positive x-axis.

35. $\quad \iiint_Q 4yz\, dV$

$$= \int_0^1 \int_0^{2-2y} \int_0^{2-2y-z} 4yz\, dx\, dz\, dy$$

$$= \int_0^2 \int_0^{1-x/2} \int_0^{2-x-2y} 4yz\, dz\, dy\, dx$$

$$= \int_0^2 \int_0^{2-x} \int_0^{1-x/2-z/2} 4yz\, dy\, dz\, dx$$

$$= \frac{4}{15}$$

37. $\displaystyle\int_0^2 \int_0^{4-2y} \int_0^{4-2y-z} dx\, dz\, dy$

$$= \int_0^2 \int_0^{4-2y} \int_0^{4-x-2y} dz\, dx\, dy$$

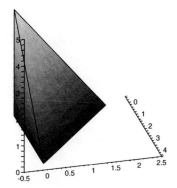

39. $\displaystyle\int_0^1 \int_0^{\sqrt{1-x^2}} \int_0^{\sqrt{1-x^2-y^2}} dz\, dy\, dx$

$$= \int_0^1 \int_0^{\sqrt{1-x^2}} \int_0^{\sqrt{1-x^2-z^2}} dy\, dz\, dx$$

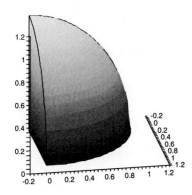

41. $\displaystyle\int_0^2 \int_0^{\sqrt{4-z^2}} \int_{x^2+z^2}^4 dy\, dx\, dz$

$$= \int_0^2 \int_{z^2}^4 \int_0^{\sqrt{y-z^2}} dx\, dy\, dz$$

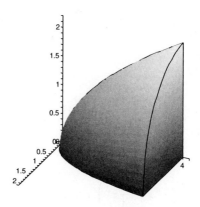

43. The total pollutant in the room is given by

$$\iiint_Q f(x, y, z)\, dV$$

$$= \int_0^{12} \int_0^{12} \int_0^8 f(x, y, z)\, dz\, dy\, dx$$

$$= \int_0^{12} \int_0^{12} \int_0^8 xyz e^{-x^2-2y^2-4z^2}\, dz\, dy\, dx$$

$$= \int_0^{12} \int_0^{12} \int_0^8 \left(xe^{-x^2}\right)\left(ye^{-2y^2}\right)\left(ze^{-4z^2}\right)\, dz\, dy\, dx$$

$$= \left[\int_0^{12} xe^{-x^2}\, dx\right] \cdot \left[\int_0^{12} ye^{-2y^2}\, dy\right]$$

$$\cdot \left[\int_0^8 ze^{-4z^2}\, dz\right]$$

$$= \left[\frac{1-e^{-144}}{2}\right]\left[\frac{1-e^{-288}}{4}\right]\left[\frac{1-e^{-256}}{8}\right]$$

$$= \frac{(1-e^{-144})(1-e^{-288})(1-e^{-256})}{64}$$

$$\approx 0.015625$$

The average density of pollutant in the room is the total divided by the volume of the room, which is $V = (12)(12)(8) = 1152.$

$$\text{Ave} = \frac{\text{Total Pollutant}}{\text{Volume}}$$

$$= \frac{\left[\frac{(1-e^{-144})(1-e^{-288})(1-e^{-256})}{64}\right]}{1152}$$

$$\approx 1.356 \times 10^{-5} \text{ grams per cubic foot}$$

45. We need to solve for c:

$$1 = \iiint_Q c\, dV$$

$$= c \int_0^2 \int_0^{1-x/2} \int_0^{2-x-2y} dz\, dy\, dx$$

$$= c \int_0^2 \int_0^{1-x/2} (2-x-2y)\, dy\, dx$$

$$= c \int_0^2 \left(1 - x - \frac{1}{4}x^2\right) dx$$

$$= \frac{2c}{3}.$$

Therefore $c = \frac{3}{2}$.

47. To make this as easy as possible, we will set up the integral so that z is the outer variable. We want to solve the following for k:

$$\frac{1}{2} = P(z < k)$$

$$= \int_0^k \int_0^{1-z/2} \int_0^{2-2y-z} \frac{3}{2}\, dx\, dy\, dz$$

$$= \int_0^k \int_0^{1-z/2} \left(3 - 3y - \frac{3}{2}z\right) dy\, dz$$

$$= \int_0^k \left(\frac{3}{2} - \frac{3}{2}z + \frac{3}{8}z^2\right) dz$$

$$= \frac{3}{2}k - \frac{3}{4}k^2 + \frac{1}{8}k^3.$$

This gives us the equation to solve:

$$\frac{3}{2}k - \frac{3}{4}k^2 + \frac{1}{8}k^3 = \frac{1}{2}$$
$$k^3 - 6k^2 + 12k - 4 = 0$$

A numerical solution of this polynomial can be found by graphing or using a polynomial root finder.

A simpler solution is to use the fact that if you cut the top off the tetrahedron at $z = k$, then the removed piece is proportional to the original, and the factor of proportionality is $(2-k)/2$ (as it is in the z-direction). Therefore, if we want to remove exactly half the volume then we must have

$$\left(\frac{2-k}{2}\right)^3 = \frac{1}{2}$$

Solving for k gives
$$k = 2 - 2^{2/3} \approx 0.4126.$$

49.
$$\int_a^b \int_c^d \int_r^s f(x)g(y)h(z)\, dz\, dy\, dx$$

$$= \int_a^b \int_c^d f(x)g(y)\left[\int_r^s h(z)\, dz\right] dy\, dx$$

$$= \left[\int_r^s h(z)\, dz\right]\left\{\int_a^b f(x)\left[\int_c^d g(y)\, dy\right] dx\right\}$$

$$= \left[\int_r^s h(z)\, dz\right]\left[\int_c^d g(y)\, dy\right]\left[\int_a^b f(x)\, dx\right]$$

In general this does not work—this can only be done if all the limits of integration are constants.

51. The volume of the tetrahedron is

$$V = \int_0^a \int_0^{b(1-x/a)} \int_0^{c(1-x/a-y/b)} dz\, dy\, dx$$

$$= \int_0^a \int_0^{b(1-x/a)} c\left(1 - \frac{x}{a} - \frac{y}{b}\right) dy\, dx$$

$$= \int_0^a \frac{bc}{2}\left(1 - \frac{x}{a}\right)^2 dx = \frac{abc}{6}$$

which is $\frac{1}{6}$ the volume of the parallelepiped, abc.

14.6 CYLINDRICAL COORDINATES

1. $x^2 + y^2 = 16$
$r^2 = 16$
$r = 4$

3. $(x-2)^2 + y^2 = 4$
$x^2 - 4x + 4 + y^2 = 4$
$x^2 - 4x + y^2 = 0$
$x^2 + y^2 - 4x = 0$
$r^2 - 4r\cos\theta = 0$

$$r(r - 4\cos\theta) = 0$$
$$r = 4\cos\theta$$

5. $z = x^2 + y^2$
$z = r^2$

7. $y = 2x$
$r\sin\theta = 2r\cos\theta$
$\frac{\sin\theta}{\cos\theta} = 2$
$\tan\theta = 2$

9. The lower boundary surface is
$z = \sqrt{x^2 + y^2} = r$.
The upper boundary surface is
$z = \sqrt{8 - x^2 - y^2} = \sqrt{8 - r^2}$.

These meet when $r = 2$. Therefore

$$\iiint_Q f(x, y, z)\, dV$$

$$= \int_0^{2\pi} \int_0^2 \int_r^{\sqrt{8-r^2}} f(r\cos\theta, r\sin\theta, z)$$
$$\cdot r\, dz\, dr\, d\theta$$

11. The upper boundary surface is $9 - r^2$.

The lower boundary surface is $z = 0$. These meet when $r = 3$.

$$\iiint_Q f(x, y, z)\, dV$$

$$= \int_0^{2\pi} \int_0^3 \int_0^{9-r^2} f(r\cos\theta, r\sin\theta, z)$$
$$\cdot r\, dz\, dr\, d\theta$$

13. $$\iiint_Q f(x, y, z)\, dV$$

$$= \int_0^{2\pi} \int_{\sqrt{3}}^{\sqrt{8}} \int_{r^2-1}^8 f(r\cos\theta, r\sin\theta, z)$$
$$\cdot r\, dz\, dr\, d\theta$$

15. Here, we use polar coordinates in the xz-plane. This means that we will have $x = r\cos\theta$, $y = y$, and $z = r\sin\theta$. In this case the integral will be

$$\iiint_Q f(x, y, z)\, dV$$

$$= \int_0^{2\pi} \int_0^2 \int_0^{4-r^2} f(r\cos\theta, y, r\sin\theta)$$
$$\cdot r\, dy\, dr\, d\theta$$

17. We use polar coordinates in the yz-plane.

$$\iiint_Q f(x, y, z)\, dV$$

$$= \int_0^{2\pi} \int_0^1 \int_{r^2}^{2-r^2} f(x, r\cos\theta, r\sin\theta)$$
$$\cdot r\, dx\, dr\, d\theta$$

19. We split the integral up:

$$\iiint_Q f(x, y, z)\, dV$$

$$= \int_0^{2\pi} \int_0^2 \int_2^3 f(r\cos\theta, r\sin\theta, z)$$
$$\cdot r\, dz\, dr\, d\theta$$
$$+ \int_0^{2\pi} \int_2^3 \int_r^3 f(r\cos\theta, r\sin\theta, z)$$
$$\cdot r\, dz\, dr\, d\theta$$

Of course, splitting the integral can be avoided like this:

$$\iiint_Q f(x, y, z)\, dV$$

$$= \int_0^{2\pi} \int_2^3 \int_0^z f(r\cos\theta, r\sin\theta, z)$$
$$\cdot r\, dr\, dz\, d\theta$$

21. $$\iiint_Q e^{x^2+y^2}\, dV$$

$$= \int_0^{2\pi} \int_0^2 \int_1^2 re^{r^2}\, dz\, dr\, d\theta$$

$$= \int_0^{2\pi} \int_0^2 re^{r^2}\, dr\, d\theta$$

$$= \int_0^{2\pi} \left(\frac{e^4 - 1}{2}\right) d\theta$$

$$= \pi(e^4 - 1)$$

23.
$$\iiint_Q (x + z)\, dV$$

$$= \int_0^6 \int_0^{(6-x)/2} \int_0^{(6-x-2y)/3} (x + z)\, dz\, dy\, dx$$

$$= \int_0^6 \int_0^{(6-x)/2} \left(\frac{4}{3}x - \frac{5}{18}x^2 - \frac{4}{9}xy \right.$$

$$\left. + 2 - \frac{4}{3}y + \frac{2}{9}y^2 \right) dy\, dx$$

$$= \int_0^6 \left(2 + 2x - \frac{5}{6}x^2 + \frac{2}{27}x^3 \right) dx$$

$$= 12$$

25.
$$\iiint_Q z\, dV$$

$$= \int_0^{2\pi} \int_0^{\sqrt 2} \int_r^{\sqrt{4-r^2}} zr\, dz\, dr\, d\theta$$

$$= 2\pi \int_0^{\sqrt 2} \int_r^{\sqrt{4-r^2}} zr\, dz\, dr$$

$$= \pi \int_0^{\sqrt 2} (4 - 2r^2)r\, dr = 2\pi$$

27.
$$\iiint_Q (x + y)\, dV$$

$$= \int_0^2 \int_0^{4-2y} \int_0^{4-x-2y} (x + y)\, dz\, dx\, dy$$

$$= \int_0^2 \int_0^{4-2y} (x + y)(4 - x - 2y)\, dx\, dy$$

$$= \int_0^2 \left(\frac{32}{3} - 8y + \frac{2}{3}y^3 \right) dy = 8$$

29.
$$\iiint_Q e^z\, dV$$

$$= \int_0^{2\pi} \int_{\sqrt 3}^2 \int_{-\sqrt{4-r^2}}^0 re^z\, dz\, dr\, d\theta$$

$$= 2\pi \int_{\sqrt 3}^2 \int_{-\sqrt{4-r^2}}^0 re^z\, dz\, dr$$

$$= 2\pi \int_{\sqrt 3}^2 r \left(1 - e^{-\sqrt{4-r^2}} \right) dr$$

This integral can be handled using substitution and then integration by parts. Another approach that gives a slightly easier integral is to change the order of integration from $dz\, dr$ to $dr\, dz$:

$$2\pi \int_{\sqrt 3}^2 \int_{-\sqrt{4-r^2}}^0 re^z\, dz\, dr$$

$$= 2\pi \int_{-1}^0 \int_{\sqrt 3}^{\sqrt{4-z^2}} re^z\, dr\, dz$$

$$= \pi \int_{-1}^0 (1 - z^2)e^z\, dz$$

$$= \pi \left[-(1 - 2z + z^2)e^z \right]_{-1}^0$$

$$= \pi(4e^{-1} - 1)$$

31. In polar coordinates the circle $x^2 + (y - 1)^2 = 1$ is given by $r = 2\sin\theta,\ 0 \le \theta \le \pi$.

$$\iiint_Q 2x\, dV$$

$$= \int_0^\pi \int_0^{2\sin\theta} \int_0^r 2r^2 \cos\theta\, dz\, dr\, d\theta$$

$$= 2 \int_0^\pi \int_0^{2\sin\theta} r^3 \cos\theta\, dr\, d\theta$$

$$= 2 \int_0^\pi 4\sin^4\theta r^3 \cos\theta\, dr\, d\theta$$

$$= 0$$

33.
$$\int_{-1}^1 \int_{-\sqrt{1-x^2}}^{\sqrt{1-x^2}} \int_0^{\sqrt{x^2+y^2}} 3z^2\, dz\, dy\, dx$$

$$= \int_0^{2\pi} \int_0^1 \int_0^r 3z^2 r\, dz\, dr\, d\theta$$

$$= 6\pi \int_0^1 \int_0^r z^2 r\, dz\, dr$$

$$= 6\pi \int_0^1 \frac{r^4}{3}\, dr = \frac{2\pi}{5}$$

35.
$$\int_0^2 \int_{-\sqrt{4-y^2}}^{\sqrt{4-y^2}} \int_{\sqrt{x^2+y^2}}^{\sqrt{8-x^2-y^2}} 2 \, dz \, dx \, dy$$

$$= 2 \int_0^\pi \int_0^2 \int_r^{\sqrt{8-r^2}} r \, dz \, dr \, d\theta$$

$$= 2\pi \int_0^2 \int_r^{\sqrt{8-r^2}} r \, dz \, dr$$

$$= 2\pi \int_0^2 \left(\sqrt{8-r^2} - r \right) r \, dr$$

$$= 2\pi \left[-\frac{(8-r^2)^{3/2}}{3} - \frac{r^3}{3} \right]_0^2$$

$$= \frac{32\pi}{3}(\sqrt{2} - 1)$$

37. We use cylindrical coordinates in the xz-plane.

$$\int_{-3}^3 \int_{-\sqrt{9-x^2}}^0 \int_0^{x^2+z^2} (x^2 + z^2) \, dy \, dz \, dx$$

$$= \int_\pi^{2\pi} \int_0^3 \int_0^{r^2} r^3 \, dy \, dr \, d\theta$$

$$= \int_\pi^{2\pi} \int_0^3 r^5 \, dr \, d\theta$$

$$= \pi \int_0^3 r^5 \, dr = \frac{243\pi}{2}$$

39.

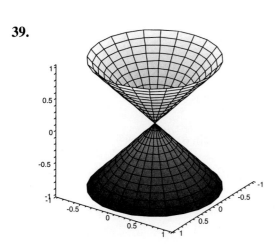

41.

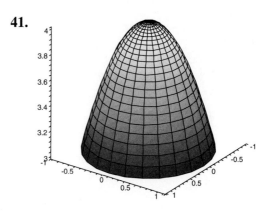

43. This is the plane $x = 2$.

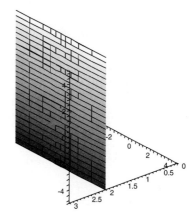

45. This is the plane $y = x$.

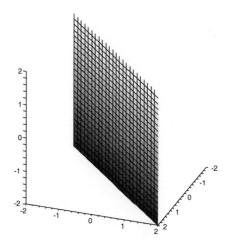

47.

$$m = \iiint_Q \rho(x, y, z)\, dV$$

$$= \int_0^{2\pi} \int_0^4 \int_r^4 r^2\, dz\, dr\, d\theta$$

$$= 2\pi \int_0^4 \int_r^4 r^2\, dz\, dr$$

$$= 2\pi \int_0^4 r^2(4 - r)\, dr = \frac{128\pi}{3}$$

Because of the radial symmetry, we must have $\bar{x} = \bar{y} = 0$.

$$M_{xy} = \iiint_Q z\rho(x, y, z)\, dV$$

$$= \int_0^{2\pi} \int_0^4 \int_r^4 zr^2\, dz\, dr\, d\theta$$

$$= 2\pi \int_0^4 \int_r^4 zr^2\, dz\, dr$$

$$= \pi \int_0^4 r^2(16 - r^2)\, dr$$

$$= \frac{2048\pi}{15}$$

$$\bar{z} = \frac{M_{xy}}{m} = \frac{16}{5}$$

49. This is similar to exercise 31.

$$m = \iiint_Q \rho(x, y, z)\, dV$$

$$= \int_0^\pi \int_0^{2\sin\theta} \int_{r^2}^4 4r\, dz\, dr\, d\theta$$

$$= \int_0^\pi \int_0^{2\sin\theta} 4r(4 - r^2)\, dr\, d\theta$$

$$= \int_0^\pi \left(32\sin^2\theta - 16\sin^4\theta\right) d\theta$$

$$= 10\pi$$

It should be clear from symmetry that $\bar{x} = 0$.

$$M_{xz} = \iiint_Q y\rho(x, y, z)\, dV$$

$$= \int_0^\pi \int_0^{2\sin\theta} \int_{r^2}^4 4r^2 \sin\theta\, dz\, dr\, d\theta$$

$$= \int_0^\pi \int_0^{2\sin\theta} (4 - r^2)4r^2 \sin\theta\, dr\, d\theta$$

$$= \int_0^\pi \left(\frac{32}{3}\sin^3\theta - \frac{32}{5}\sin^5\theta\right) 4\sin\theta\, d\theta$$

$$= 8\pi$$

$$\bar{y} = \frac{M_{xz}}{m} = \frac{4}{5}$$

$$M_{xy} = \iiint_Q z\rho(x, y, z)\, dV$$

$$= \int_0^\pi \int_0^{2\sin\theta} \int_{r^2}^4 4rz\, dz\, dr\, d\theta$$

$$= \int_0^\pi \int_0^{2\sin\theta} 2r(16 - r^4)\, dr\, d\theta$$

$$= \int_0^\pi \left(64\sin^2\theta - \frac{64}{3}\sin^6\theta\right) d\theta$$

$$= \frac{76\pi}{3}$$

$$\bar{z} = \frac{M_{xy}}{m} = \frac{38}{15}$$

51.

$$\hat{\mathbf{r}} = \frac{\langle r\cos\theta,\, r\sin\theta,\, 0\rangle}{r}$$

$$= \langle \cos\theta,\, \sin\theta,\, 0\rangle$$

53. By the chain rule we have

$$\frac{d\hat{\mathbf{r}}}{dt} = \frac{d}{dt}\langle \cos\theta,\, \sin\theta,\, 0\rangle$$

$$= \left\langle (-\sin\theta)\frac{d\theta}{dt},\, (\cos\theta)\frac{d\theta}{dt},\, 0\right\rangle$$

$$= \frac{d\theta}{dt}\hat{\boldsymbol{\theta}}$$

$$\frac{d\widehat{\boldsymbol{\theta}}}{dt} = \frac{d}{dt} \langle -\sin\theta, \cos\theta, 0\rangle$$

$$= \left\langle (-\cos\theta)\frac{d\theta}{dt}, (-\sin\theta)\frac{d\theta}{dt}, 0\right\rangle$$

$$= -\frac{d\theta}{dt}\widehat{\mathbf{r}}$$

55. If $\mathbf{v} = (3, 3, 0) - (1, 1, 0) = \langle 2, 2, 0\rangle$, then
$r = 2\sqrt{2}$ and $\theta = \frac{\pi}{4}$. $\mathbf{v} = \langle 2\sqrt{2}\cos\theta,$
$2\sqrt{2}\sin\theta, 0\rangle = 2\sqrt{2}\widehat{\mathbf{r}}$. Here, $c = 2\sqrt{2} = r$.

57. If $\mathbf{v} = (1, 1, 0) - (0, 2, 0) = \langle 0, 2, 0\rangle$, then
$\widehat{\mathbf{r}} = \langle 0, 1, 0\rangle$

$$\int_{-\pi/4}^{\pi/4} \widehat{\boldsymbol{\theta}}\, d\theta$$

$$= \left\langle \int_{-\pi/4}^{\pi/4} -\sin\theta\, d\theta, \int_{-\pi/4}^{\pi/4}\cos\theta\, d\theta, 0\right\rangle$$

$$= \left\langle 0, \sqrt{2}, 0\right\rangle = \sqrt{2}\widehat{\mathbf{r}}$$

Thus, $c = \sqrt{2}$.

59. Any vector $\mathbf{v}$ with third component zero has
$\widehat{\mathbf{r}} = \frac{\mathbf{v}}{\|\mathbf{v}\|}$
so that $\mathbf{v} = \|\mathbf{v}\|\widehat{\mathbf{r}}$.

$\widehat{\mathbf{r}}$ is a unit vector in the same direction as
$\mathbf{v}$. In addition, corresponding to $\mathbf{v}$ we have
the unit vector $\widehat{\boldsymbol{\theta}}$, which is rotated by $\pi/4$
counterclockwise from $\mathbf{v}$.

$\widehat{\boldsymbol{\theta}}$ does not normally appear in a representation
of $\mathbf{v}$. But, in this case we can write

$$(1, 1, 0) - (-1, -1, 0)$$

$$= \langle 2, 2, 0\rangle = 2\sqrt{2}\widehat{\mathbf{r}} = 2\sqrt{2}\int_{-3\pi/4}^{\pi/4}\widehat{\boldsymbol{\theta}}\, d\theta.$$

If $\mathbf{v}$ is a function of time, then $\widehat{\boldsymbol{\theta}}$ will appear in
the vector derivative $d\mathbf{v}/dt$. Using the product
rule on $\mathbf{v} = r\widehat{\mathbf{r}}$ and exercise 53:

$$\frac{d\mathbf{v}}{dt} = \frac{dr}{dt}\widehat{\mathbf{r}} + r\frac{d\widehat{\mathbf{r}}}{dt}$$

$$= \frac{dr}{dt}\widehat{\mathbf{r}} + r\frac{d\theta}{dt}\widehat{\boldsymbol{\theta}}$$

$$d\mathbf{v} = \widehat{\mathbf{r}}dr + r\widehat{\boldsymbol{\theta}}d\theta$$

14.7 SPHERICAL COORDINATES

1. $x = 4\sin 0\cos\pi = 0$
$y = 4\sin 0\sin\pi = 0$
$z = 4\cos 0 = 4$
$\quad (x, y, z) = (0, 0, 4)$

3. $x = 4\sin\frac{\pi}{2}\cos 0 = 4$
$y = 4\sin\frac{\pi}{2}\sin 0 = 0$
$z = 4\cos\frac{\pi}{2} = 0$
$\quad (x, y, z) = (4, 0, 0)$

5. $x = 2\sin\frac{\pi}{4}\cos 0 = \sqrt{2}$
$y = 2\sin\frac{\pi}{4}\sin 0 = 0$
$z = 2\cos\frac{\pi}{4} = \sqrt{2}$
$\quad (x, y, z) = (\sqrt{2}, 0, \sqrt{2})$

7. $x = \sqrt{2}\sin\frac{\pi}{6}\cos\frac{\pi}{3} = \frac{\sqrt{2}}{4}$
$y = \sqrt{2}\sin\frac{\pi}{6}\sin\frac{\pi}{3} = \frac{\sqrt{6}}{4}$
$z = \sqrt{2}\cos\frac{\pi}{6} = \frac{\sqrt{6}}{2}$
$\quad (x, y, z) = \left(\frac{\sqrt{2}}{4}, \frac{\sqrt{6}}{4}, \frac{\sqrt{6}}{2}\right)$

9. $x^2 + y^2 + z^2 = 9$
$\rho^2 = 9$
$\rho = 3$

11. $y = x$

$\rho \sin \phi \sin \theta = \rho \sin \phi \cos \theta$

$\rho \sin \phi \sin \theta - \rho \sin \phi \cos \theta = 0$

$\rho \sin \phi \, (\sin \theta - \cos \theta) = 0$

This means either $\rho = 0$ (the origin),
$\sin \phi = 0$ (the z-axis) or $\sin \theta = \cos \theta$. If
$\sin \theta = \cos \theta$ then $\tan \theta = 1$ or $\theta = \frac{\pi}{4}, \frac{5\pi}{4}$.

The case $\theta = \frac{\pi}{4}, \frac{5\pi}{4}$ includes $\rho = 0$ and
$\sin \phi = 0$, so these suffice for an answer.

13.
$$z = 2$$
$$\rho \cos \phi = 2$$
$$\rho = 2 \sec \phi$$

15.
$$z = \sqrt{3(x^2 + y^2)}$$
$$z^2 = 3(x^2 + y^2)$$
$$(\rho \cos \phi)^2 = 3(\rho \sin \phi)^2$$
$$\tan^2 \phi = \frac{1}{3}$$
$$\phi = \frac{\pi}{6} \quad \text{or} \quad \phi = \frac{5\pi}{6}$$

17. $x^2 + y^2 + z^2 = 4$

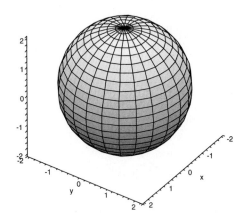

19. $z^2 = x^2 + y^2$
Here, both the upper and lower cones are shown. With $\rho \geq 0$, only the upper cone is generated.

21. This is the xz-plane, $y = 0$. If ρ is restricted to $\rho \geq 0$ then this would put the restriction on the plane that $x \geq 0$.

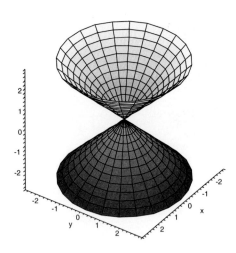

23. $z^2 = \frac{1}{3}(x^2 + y^2)$
Here, both the upper and lower cones are shown. With $\rho \geq 0$, only the upper cone is generated.

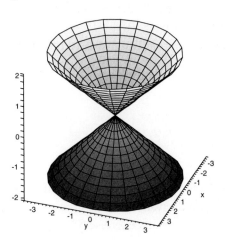

25.

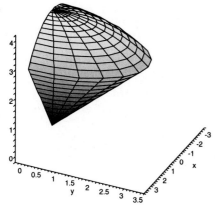

27.

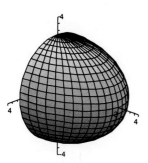

29.

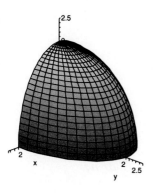

31. In this case, the picture does not show that this ball is not solid throughout. This is a ball of radius 3 with the ball of radius 2 "drilled" out of it.

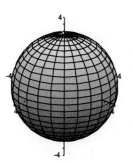

33. $\iiint_Q e^{(x^2+y^2+z^2)^{3/2}} \, dV$

$$= \int_0^{2\pi} \int_0^{\pi/2} \int_0^2 e^{\rho^3} \rho^2 \sin\phi \, d\rho \, d\phi \, d\theta$$

$$= 2\pi \int_0^{\pi/2} \int_0^2 e^{\rho^3} \rho^2 \sin\phi \, d\rho \, d\phi$$

$$= 2\pi \int_0^{\pi/2} \frac{1}{3}(e^8 - 1) \sin\phi \, d\rho \, d\phi$$

$$= (2\pi)\frac{1}{3}(e^8 - 1) = \frac{2\pi}{3}(e^8 - 1)$$

35. The condition that the solid is outside $x^2 + y^2 = 1$ means that $\rho \sin\phi \geq 1$, or $\rho \geq \csc\phi$. In addition, the cylinder intersects the sphere when $\phi = \pi/4, 3\pi/4$.

We will also take advantage of symmetry and only integrate in the upper half space: $\pi/4 \leq \phi \leq \pi/2$.

$$\iiint_Q (x^2 + y^2 + z^2)^{5/2} \, dV$$

$$= \int_0^{2\pi} \int_{\pi/4}^{3\pi/4} \int_{\csc\phi}^{\sqrt{2}} \rho^7 \sin\phi \, d\rho \, d\phi \, d\theta$$

$$= 2 \int_0^{2\pi} \int_{\pi/4}^{\pi/2} \int_{\csc\phi}^{\sqrt{2}} \rho^7 \sin\phi \, d\rho \, d\phi \, d\theta$$

$$= 4\pi \int_{\pi/4}^{\pi/2} \int_{\csc\phi}^{\sqrt{2}} \rho^7 \sin\phi \, d\rho \, d\phi$$

$$= \frac{\pi}{2} \int_{\pi/4}^{\pi/2} (16 - \csc^8\phi) \sin\phi \, d\phi$$

$$= \frac{\pi}{2} \int_{\pi/4}^{\pi/2} (16 \sin\phi - \csc^7\phi) \, d\phi$$

$$= \frac{\pi}{2} \left[8\sqrt{2} - \int_{\pi/4}^{\pi/2} \csc^7\phi \, d\phi \right]$$

$$\approx 14.238$$

37. This is a rectangular problem.

$$\iiint_Q dV$$

$$= \int_0^1 \int_1^2 \int_3^4 (x^2 + y^2 + z^2) \, dz \, dy \, dx$$

$$= \int_0^1 \int_1^2 \left(x^2 + y^2 + \frac{37}{3} \right) dy \, dx$$

$$= \int_0^1 \left(x^2 + \frac{44}{3} \right) dx = 15$$

39. Here we use cylindrical coordinates.

$$\iiint_Q (x^2 + y^2) \, dV$$

$$= \int_0^{2\pi} \int_0^2 \int_0^{4-r^2} r^3 \, dz \, dr \, d\theta$$

$$= 2\pi \int_0^2 \int_0^{4-r^2} r^3 \, dz \, dr$$

$$= 2\pi \int_0^2 (4 - r^2) r^3 \, dr = \frac{32\pi}{3}$$

41. The solid is the portion of the ball with $0 \le \phi \le \pi/4$.

$$\iiint_Q \sqrt{x^2 + y^2 + z^2} \, dV$$

$$= \int_0^{2\pi} \int_0^{\pi/4} \int_0^{\sqrt{2}} \rho^3 \sin \phi \, d\rho \, d\phi \, d\theta$$

$$= 2\pi \int_0^{\pi/4} \int_0^{\sqrt{2}} \rho^3 \sin \phi \, d\rho \, d\phi$$

$$= 2\pi \int_0^{\pi/4} \sin \phi \, d\phi$$

$$= 2\pi \left(1 - \frac{1}{\sqrt{2}} \right) = (2 - \sqrt{2})\pi$$

43. Note that the equation $x^2 + y^2 + z^2 = 4z$ transforms to $\rho^2 = 4\rho \cos \phi$ or $\rho = 4 \cos \phi$.

$$V = \iiint_Q dV$$

$$= \int_0^{2\pi} \int_0^{\pi/4} \int_0^{4 \cos \phi} \rho^2 \sin \phi \, d\rho \, d\phi \, d\theta$$

$$= 2\pi \int_0^{\pi/4} \int_0^{4 \cos \phi} \rho^2 \sin \phi \, d\rho \, d\phi$$

$$= 2\pi \int_0^{\pi/4} \frac{64 \cos^3 \phi \sin \phi}{3} \, d\rho \, d\phi$$

$$= 8\pi$$

45. We use cylindrical coordinates. To avoid splitting the integral up, we integrate r first instead of z:

$$V = \iiint_Q dV$$

$$= \int_0^{2\pi} \int_2^4 \int_0^{z/\sqrt{2}} r \, dr \, dz \, d\theta$$

$$= 2\pi \int_2^4 \int_0^{z/\sqrt{2}} r \, dr \, dz$$

$$= 2\pi \int_2^4 \frac{z^2}{4} \, dz = \frac{28\pi}{3}.$$

47. The z-limits would indicate cylindrical or spherical coordinates, but then the x- and y-limits cause complications.

$$V = \iiint_Q dV$$

$$= \int_{-1}^1 \int_{-1}^1 \int_0^{\sqrt{x^2+y^2}} dz \, dy \, dx$$

$$= \int_{-1}^1 \int_{-1}^1 \sqrt{x^2 + y^2} \, dy \, dx$$

$$\approx 3.06078$$

49. We use spherical coordinates.

$$V = \iiint_Q dV$$

$$= \int_0^{\pi/2} \int_0^{\pi/4} \int_0^2 \rho^2 \sin \phi \, d\rho \, d\phi \, d\theta$$

$$= \frac{\pi}{2} \int_0^{\pi/4} \int_0^2 \rho^2 \sin\phi \, d\rho \, d\phi$$

$$= \frac{\pi}{2} \int_0^{\pi/4} \frac{8}{3} \sin\phi \, d\phi$$

$$= \frac{4\pi}{3} \left(1 - \frac{1}{\sqrt{2}}\right) = \frac{2\pi}{3} \left(2 - \sqrt{2}\right)$$

51. We use cylindrical coordinates.

$$V = \iiint_Q dV$$

$$= \int_0^{2\pi} \int_0^2 \int_0^r r \, dz \, dr \, d\theta$$

$$= \int_0^{2\pi} \int_0^2 r^2 \, dr \, d\theta$$

$$= \int_0^{2\pi} \frac{8}{3} \, d\theta = \frac{16\pi}{3}$$

53. The region in this case is the interior of the unit ball, but only for $x \geq 0$. We naturally convert to spherical coordinates.

$$\int_0^1 \int_{-\sqrt{1-x^2}}^{\sqrt{1-x^2}} \int_{-\sqrt{1-x^2-y^2}}^{\sqrt{1-x^2-y^2}}$$

$$\sqrt{x^2 + y^2 + z^2} \, dz \, dy \, dx$$

$$= \int_{-\pi/2}^{\pi/2} \int_0^\pi \int_0^1 \rho^3 \sin\phi \, d\rho \, d\phi \, d\theta$$

$$= \int_{-\pi/2}^{\pi/2} \int_0^\pi \frac{1}{4} \sin\phi \, d\phi \, d\theta$$

$$= \int_{-\pi/2}^{\pi/2} \frac{1}{2} \, d\theta = \frac{\pi}{2}$$

55. The region of integration is inside the vertical cylinder $r = 4$, with upper boundary surface $\rho = \sqrt{8}$ and lower boundary $z = r$. Cylindrical coordinates will work but spherical coordinates are easier.

$$\int_{-2}^2 \int_0^{\sqrt{4-x^2}} \int_{\sqrt{x^2+y^2}}^{\sqrt{8-x^2-y^2}}$$

$$(x^2 + y^2 + z^2)^{3/2} \, dz \, dy \, dx$$

$$= \int_0^\pi \int_0^{\pi/4} \int_0^{\sqrt{8}} \rho^5 \sin\phi \, d\rho \, d\phi \, d\theta$$

$$= \int_0^\pi \int_0^{\pi/4} \frac{256}{3} \sin\phi \, d\phi \, d\theta$$

$$= \int_0^\pi \frac{256}{3} \left(1 - \frac{1}{\sqrt{2}}\right) \, d\theta$$

$$= \frac{256\pi}{3} \left(1 - \frac{1}{\sqrt{2}}\right)$$

$$= \frac{128\pi}{3} \left(2 - \sqrt{2}\right)$$

57. There is enough rotational symmetry to know that $\bar{x} = \bar{y} = 0$. Given that the density is constant, we can assume that the density is actually equal to 1.

$$m = \iint_Q \rho(x, y, z) \, dV$$

$$= \int_0^{2\pi} \int_0^{\pi/4} \int_0^2 \rho^2 \sin\phi \, d\rho \, d\phi \, d\theta$$

$$= \int_0^{2\pi} \int_0^{\pi/4} \frac{8}{3} \sin\phi \, d\phi \, d\theta$$

$$= \int_0^{2\pi} \frac{8}{3} \left(1 - \frac{1}{\sqrt{2}}\right) \, d\theta$$

$$= \frac{16\pi}{3} \left(1 - \frac{1}{\sqrt{2}}\right)$$

$$= \frac{8\pi}{3} \left(2 - \sqrt{2}\right)$$

$$M_{xy} = \iint_Q z\rho(x, y, z) \, dV$$

$$= \int_0^{2\pi} \int_0^{\pi/4} \int_0^2 \rho^3 \sin\phi \cos\phi \, d\rho \, d\phi \, d\theta$$

$$= \int_0^{2\pi} \int_0^{\pi/4} 4 \sin \phi \cos \phi \, d\phi \, d\theta$$

$$= \int_0^{2\pi} 1 \, d\theta$$

$$= 2\pi$$

$$\bar{z} = \frac{M_{xy}}{m} = \frac{3}{4(2 - \sqrt{2})}$$

59. In this case $\|\mathbf{r}\| = \rho$, so

$$\hat{\boldsymbol{\rho}} = \frac{\mathbf{r}}{\|\mathbf{r}\|}$$

$$= \langle \cos \theta \sin \phi, \sin \theta \sin \phi, \cos \phi \rangle$$

61. It is always true that $\mathbf{v} = \|\mathbf{v}\| \hat{\boldsymbol{\rho}}$. Therefore in this case,

$$\mathbf{v} = \langle 1, 1, \sqrt{2} \rangle,$$

and

$$\|\mathbf{v}\| = \sqrt{1 + 1 + 2} = 2$$

and therefore $c = 2$.

63. $\mathbf{v} = \langle \sqrt{2} - 1, \sqrt{2} - 1, -\sqrt{2} \rangle$. Then we have $\theta = \frac{\pi}{4}$ and

$$\int_{\pi/4}^{\pi/2} \hat{\boldsymbol{\phi}} \, d\phi$$

$$= \int_{\pi/4}^{\pi/2} \langle \cos \theta \cos \phi, \sin \theta \cos \phi, -\sin \phi \rangle \, d\phi$$

$$= \int_{\pi/4}^{\pi/2} \left\langle \frac{\cos \phi}{\sqrt{2}}, \frac{\cos \phi}{\sqrt{2}}, -\sin \phi \right\rangle \, d\phi$$

$$= \left\langle \int_{\pi/4}^{\pi/2} \frac{\cos \phi}{\sqrt{2}} \, d\phi, \int_{\pi/4}^{\pi/2} \frac{\cos \phi}{\sqrt{2}} \, d\phi, \right.$$

$$\left. \int_{\pi/4}^{\pi/2} -\sin \phi \, d\phi \right\rangle$$

$$= \left\langle \frac{\sqrt{2} - 1}{2}, \frac{\sqrt{2} - 1}{2}, \frac{-\sqrt{2}}{2} \right\rangle$$

$$= \frac{\mathbf{v}}{2}$$

and therefore $c = 2$.

65. Here is the cardioid in the xy-plane:

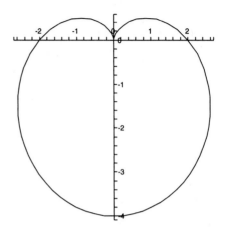

Cardioid r=2-2*sin(theta)

In the yz-plane, it is clear that ϕ and $\tilde{\theta}$ are complementary angles—their sum is $\frac{\pi}{2}$. This directly implies that
$\tilde{\theta} = \frac{\pi}{2} - \phi$ and
$\cos \phi = \sin \tilde{\theta}$.

Then, $2 - 2 \cos \phi = 2 - 2 \sin \tilde{\theta}$.

Because ϕ only ranges from 0 to π, the picture in the yz-plane will be the graph we drew in the xy-plane but with θ restricted to be between $-\pi/2$ and $\pi/2$. To get the entire graph in three-space, we rotate about the z-axis:

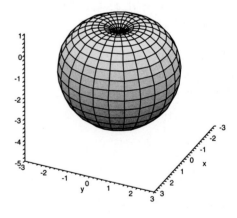

67. Here is the graph of $r = \cos^2(\theta)$ in the xy-plane:

r=[cos(theta)]^2

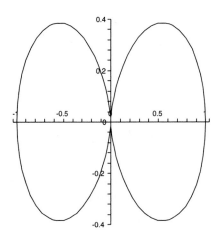

Using exercise 65, $\sin^2(\phi) = \cos^2(\tilde{\theta})$

Because ϕ only ranges from 0 to π, the picture in the yz-plane will be the graph we drew in the xy-plane but with θ restricted to be between $-\pi/2$ and $\pi/2$. To get the entire graph in three-space, we rotate about the z-axis:

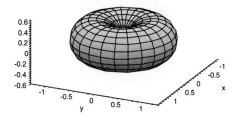

14.8 CHANGE OF VARIABLES IN MULTIPLE INTEGRALS

1.

$$\begin{array}{ll} y = 4x + 2 & \rightarrow \quad y - 4x = 2 \\ y = 4x + 5 & \rightarrow \quad y - 4x = 5 \end{array}$$

$$\underbrace{}$$
$$u = y - 4x$$

$$\begin{array}{ll} y = 3 - 2x & \rightarrow \quad y + 2x = 3 \\ y = 1 - 2x & \rightarrow \quad y + 2x = 1 \end{array}$$

$$\underbrace{}$$
$$v = y + 2x$$

$$\left. \begin{array}{l} u = y - 4x \\ v = y + 2x \end{array} \right\} \rightarrow \left\{ \begin{array}{l} x = \frac{-u+v}{6} \\ y = \frac{u+2v}{3} \end{array} \right.$$

$$2 \leq u \leq 5$$
$$1 \leq v \leq 3$$

3.

$$\begin{array}{ll} y = 1 - 3x & \rightarrow \quad y + 3x = 1 \\ y = 3 - 3x & \rightarrow \quad y + 3x = 3 \end{array}$$

$$\underbrace{}$$
$$u = y + 3x$$

$$\begin{array}{ll} y = x - 1 & \rightarrow \quad x - y = 1 \\ y = x - 3 & \rightarrow \quad x - y = 3 \end{array}$$

$$\underbrace{}$$
$$v = x - y$$

$$\left. \begin{array}{l} u = y + 3x \\ v = x - y \end{array} \right\} \rightarrow \left\{ \begin{array}{l} x = \frac{u+v}{4} \\ y = \frac{u-3v}{4} \end{array} \right.$$

$$1 \leq u \leq 3$$
$$1 \leq v \leq 3$$

5. This is perfect for polar coordinates.
$$x = r \cos \theta, \quad y = r \sin \theta$$
$$1 \leq r \leq 2$$
$$0 \leq \theta \leq \frac{\pi}{2}$$

7. Use polar coordinates:

$$x = r \cos \theta, \qquad y = r \sin \theta$$

$$2 \leq r \leq 3$$
$$\frac{\pi}{4} \leq \theta \leq \frac{3\pi}{4}$$

9.

$$\begin{array}{ll} y = x^2 & \rightarrow \quad y - x^2 = 0 \\ y = x^2 + 2 & \rightarrow \quad y - x^2 = 2 \end{array}$$

$$\underbrace{}$$
$$u = y - x^2$$

$$\begin{array}{ll} y = 4 - x^2 & \rightarrow \quad y + x^2 = 4 \\ y = 2 - x^2 & \rightarrow \quad y + x^2 = 2 \end{array}$$

$$\underbrace{}$$
$$v = y + x^2$$

$$\left. \begin{array}{l} u = y - x^2 \\ v = y + x^2 \end{array} \right\} \rightarrow \left\{ \begin{array}{l} x = \sqrt{\frac{v-u}{2}} \\ y = \frac{u+v}{2} \end{array} \right.$$

$$0 \leq u \leq 2$$
$$2 \leq v \leq 4$$

11.

$$\begin{array}{ll} y = e^x & \rightarrow \quad y - e^x = 0 \\ y = e^x + 1 & \rightarrow \quad y - e^x = 1 \end{array}$$

$$\underbrace{}$$
$$u = y - e^x$$

$$\begin{array}{ll} y = 3 - e^x & \rightarrow \quad y + e^x = 3 \\ y = 5 - e^x & \rightarrow \quad y + e^x = 5 \end{array}$$

$$\underbrace{}$$
$$v = y + e^x$$

$$\left.\begin{array}{c} u = y - e^x \\ v = y + e^x \end{array}\right\} \rightarrow \left\{\begin{array}{c} x = \ln\left(\frac{v-u}{2}\right) \\ y = \frac{v+u}{2} \end{array}\right.$$

$$0 \le u \le 1$$
$$3 \le v \le 5$$

13. $\dfrac{\partial(x, y)}{\partial(u, v)} = \begin{vmatrix} \frac{\partial x}{\partial u} & \frac{\partial x}{\partial v} \\ \frac{\partial y}{\partial u} & \frac{\partial y}{\partial v} \end{vmatrix} = \begin{vmatrix} -\frac{1}{6} & \frac{1}{6} \\ \frac{1}{3} & \frac{2}{3} \end{vmatrix} = -\dfrac{1}{6}$

$$\iint_R (y - 4x)\, dA$$

$$= \int_1^3 \int_2^5 \left[\frac{1}{3}(u + 2v) - 4\left(\frac{1}{6}\right)(v - u)\right]$$

$$\cdot \left|-\frac{1}{6}\right|\, du\, dv$$

$$= \frac{1}{6} \int_1^3 \int_2^5 u\, du\, dv$$

$$= \frac{1}{6} \int_1^3 \frac{21}{2}\, dv = \frac{7}{2}$$

15. $\dfrac{\partial(x, y)}{\partial(u, v)} = \begin{vmatrix} \frac{\partial x}{\partial u} & \frac{\partial x}{\partial v} \\ \frac{\partial y}{\partial u} & \frac{\partial y}{\partial v} \end{vmatrix} = \begin{vmatrix} \frac{1}{4} & \frac{1}{4} \\ \frac{1}{4} & -\frac{3}{4} \end{vmatrix} = -\dfrac{1}{4}$

$$\iint_R (y + 3x)^2\, dA$$

$$= \int_1^3 \int_1^3 \left[\frac{1}{4}(u - 3v) + 3\left(\frac{1}{4}\right)(u + v)\right]^2$$

$$\cdot \left|-\frac{1}{4}\right|\, du\, dv$$

$$= \frac{1}{4} \int_1^3 \int_1^3 u^2\, du\, dv$$

$$= \frac{1}{4} \int_1^3 \frac{26}{3}\, dv = \frac{13}{3}$$

17. $\dfrac{\partial(x, y)}{\partial(r, \theta)} = \begin{vmatrix} \frac{\partial x}{\partial r} & \frac{\partial x}{\partial \theta} \\ \frac{\partial y}{\partial r} & \frac{\partial y}{\partial \theta} \end{vmatrix}$

$$= \begin{vmatrix} \cos\theta & -r\sin\theta \\ \sin\theta & r\cos\theta \end{vmatrix} = r$$

$$\iint_R x\, dA$$

$$= \int_0^{\pi/2} \int_1^2 [r\cos\theta]\,|r|\, dr\, d\theta$$

$$= \int_0^{\pi/2} \int_1^2 r^2 \cos\theta\, dr\, d\theta$$

$$= \int_0^{\pi/2} \frac{7}{3} \cos\theta\, d\theta = \frac{7}{3}$$

19. From exercise 13, the Jacobian is equal to $-\frac{1}{6}$. Note that $u = y - 4x$ and $v = y + 2x$.

$$\iint_R \frac{e^{y-4x}}{y + 2x}\, dA$$

$$= \int_1^3 \int_2^5 \left[\frac{e^{(u+2v)/3 - 4(v-u)/6}}{(u + 2v)/3 + 2(v - u)/6}\right]$$

$$\cdot \left|-\frac{1}{6}\right|\, du\, dv$$

$$= \frac{1}{6} \int_1^3 \int_2^5 \frac{e^u}{v}\, du\, dv$$

$$= \frac{1}{6} \int_1^3 \frac{e^5 - e^2}{v}\, dv = \frac{(e^5 - e^2)\ln(3)}{6}$$

21. From exercise 13, the Jacobian is equal to $-\frac{1}{6}$.

$$\iint_R (x + y)\, dA$$

$$= \int_1^3 \int_2^5 \left[\frac{u + 2v}{3} + \frac{v - u}{6}\right]$$

$$\cdot \left|-\frac{1}{6}\right|\, du\, dv$$

$$= \frac{1}{6} \int_1^3 \int_2^5 \left(\frac{u}{6} + \frac{5v}{6}\right)\, du\, dv$$

$$= \frac{1}{6} \int_1^3 \left(\frac{7}{4} + \frac{5}{2}v\right)\, dv = \frac{9}{4}$$

23. $\dfrac{\partial(x, y)}{\partial(u, v)} = \begin{vmatrix} \frac{\partial x}{\partial u} & \frac{\partial x}{\partial v} \\ \frac{\partial y}{\partial u} & \frac{\partial y}{\partial v} \end{vmatrix}$

$$= \begin{vmatrix} e^v & ue^v \\ e^{-v} & -ue^{-v} \end{vmatrix}$$

$$= -2u$$

25.
$$\frac{\partial(x, y)}{\partial(u, v)} = \begin{vmatrix} \frac{\partial x}{\partial u} & \frac{\partial x}{\partial v} \\ \frac{\partial y}{\partial u} & \frac{\partial y}{\partial v} \end{vmatrix}$$

$$= \begin{vmatrix} \frac{1}{v} & -\frac{u}{v^2} \\ 0 & 2v \end{vmatrix}$$

$$= 2$$

27.
$$\left. \begin{array}{l} x + y + z = 1 \\ x + y + z = 2 \end{array} \right\} \quad u = x + y + z$$

$$\left. \begin{array}{l} x + 2y = 0 \\ x + 2y = 1 \end{array} \right\} \quad v = x + 2y$$

$$\left. \begin{array}{l} y + z = 2 \\ y + z = 4 \end{array} \right\} \quad w = y + z$$

$$\left. \begin{array}{l} u = x + y + z \\ v = x + 2y \\ w = y + z \end{array} \right\} \rightarrow \left\{ \begin{array}{l} x = u - w \\ y = \frac{-u + v + w}{2} \\ z = \frac{u - v + w}{2} \end{array} \right.$$

$$1 \le u \le 2$$
$$0 \le v \le 1$$
$$2 \le w \le 4$$

29.
$$\frac{\partial(x, y, z)}{\partial(u, v, w)} = \begin{vmatrix} \frac{\partial x}{\partial u} & \frac{\partial x}{\partial v} & \frac{\partial x}{\partial w} \\ \frac{\partial y}{\partial u} & \frac{\partial y}{\partial v} & \frac{\partial y}{\partial w} \\ \frac{\partial z}{\partial u} & \frac{\partial z}{\partial v} & \frac{\partial z}{\partial w} \end{vmatrix}$$

$$= \begin{vmatrix} 1 & 0 & -1 \\ -\frac{1}{2} & \frac{1}{2} & \frac{1}{2} \\ \frac{1}{2} & -\frac{1}{2} & \frac{1}{2} \end{vmatrix} = \frac{1}{2}$$

$$\iiint_Q dV$$

$$= \int_2^4 \int_0^1 \int_1^2 \frac{1}{2} \, du \, dv \, dw$$

$$= \frac{1}{2}(1)(1)(2) = 1$$

31.
$$\frac{\partial(x, y)}{\partial(u, v)} = \begin{vmatrix} \frac{\partial x}{\partial u} & \frac{\partial x}{\partial v} \\ \frac{\partial y}{\partial u} & \frac{\partial y}{\partial v} \end{vmatrix} = \begin{vmatrix} 1 & -1 \\ -2 & 2 \end{vmatrix} = 0$$

The equations combine to $y = -2x$ and the transformation maps the entire uv-plane to the line in the xy-plane.

33.
$$\frac{\partial(x, y)}{\partial(u, v)} = \begin{vmatrix} \frac{\partial x}{\partial u} & \frac{\partial x}{\partial v} \\ \frac{\partial y}{\partial u} & \frac{\partial y}{\partial v} \end{vmatrix}$$

$$= \begin{vmatrix} \dfrac{\cos u}{\cos v} & \dfrac{\sin u \sin v}{\cos^2 v} \\ \dfrac{\sin v \sin u}{\cos^2 u} & \dfrac{\cos v}{\cos u} \end{vmatrix}$$

$$= 1 - \tan^2 u \tan^2 v = 1 - x^2 y^2$$

Let's accept for the moment that the given transformation maps the triangle
$T = \{(u, v) : u \ge 0, v \ge 0, u + v < \pi/2\}$
onto the square
$S = \{(x, y) : 0 \le x < 1, 0 \le y < 1\}$
(this is exercise 34). Let's also accept for the moment that the Jacobian is positive in T. Then,

$$\iint_S \frac{1}{1 - x^2 y^2} \, dA_{xy}$$

$$= \iint_T \left(\frac{1}{1 - x^2 y^2} \right) \left| 1 - \tan^2 u \tan^2 v \right| dA_{uv}$$

$$= \iint_T dA_{uv} = \text{Area}(T) = \frac{\pi^2}{8}$$

The easiest way to see that the Jacobian is positive is to notice that the Jacobian is equal to $1 - x^2 y^2$ on the square, which is always positive (and the square does not include the point $(1, 1)$).

Chapter 14 Review Exercises

1. $f(x, y) = 5x - 2y, n = 4$
$1 \le x \le 3, 0 \le y \le 1$
The centers of the four rectangles are
$\left(\frac{3}{2}, \frac{1}{4} \right), \left(\frac{5}{2}, \frac{1}{4} \right), \left(\frac{3}{2}, \frac{3}{4} \right), \left(\frac{5}{2}, \frac{3}{4} \right).$
Since the rectangles are the same size,
$\Delta A_i = \frac{1}{2}.$

$$\sum_{i=1}^{4} f(u_i, v_i)\Delta A_i$$

$$= f\left(\frac{3}{2}, \frac{1}{4}\right)\left(\frac{1}{2}\right) + f\left(\frac{5}{2}, \frac{1}{4}\right)\left(\frac{1}{2}\right)$$

$$+ f\left(\frac{3}{2}, \frac{3}{4}\right)\left(\frac{1}{2}\right) + f\left(\frac{5}{2}, \frac{3}{4}\right)\left(\frac{1}{2}\right)$$

$$= \frac{1}{2}(7 + 12 + 6 + 11) = 18$$

3.
$$\iint_R (4x + 9x^2y^2)\,dA$$

$$= \int_0^3 \int_1^2 (4x + 9x^2y^2)\,dy\,dx$$

$$= \int_0^3 (4x + 21x^2)\,dx = 207$$

5. We convert to polar coordinates.

$$\iint_R e^{-x^2-y^2}\,dA$$

$$= \int_0^{2\pi} \int_1^2 re^{-r^2}\,dr\,d\theta$$

$$= \int_0^{2\pi} \frac{(e^{-1} - e^{-4})}{2}\,d\theta$$

$$= \pi(e^{-1} - e^{-4})$$

7.
$$\int_{-1}^1 \int_{x^2}^{2x} (2xy - 1)\,dy\,dx$$

$$= \int_{-1}^1 \left(4x^3 - 2x - x^5 + x^2\right)\,dx$$

$$= \frac{2}{3}$$

9. Notice that the curve $r = 2\cos\theta$ lies entirely in the right half plane ($x \geq 0$).

$$\iint_R xy\,dA$$

$$= \int_{-\pi/2}^{\pi/2} \int_0^{2\cos\theta} r^3 \cos\theta \sin\theta\,dr\,d\theta$$

$$= \int_{-\pi/2}^{\pi/2} 4\cos^5\theta \sin\theta\,d\theta$$

$$= \left[-\frac{4\cos^6\theta}{6}\right]_{-\pi/2}^{\pi/2} = 0$$

11. To find the limits of integration we solve
$$x^2 - 4 = \ln x.$$
Using a CAS we obtain
$$x \approx 0.0183218,\ 2.18689.$$

$$\iint_R 4xy\,dA$$

$$\approx 4\int_{0.183}^{2.187} \int_{x^2-4}^{\ln x} xy\,dy\,dx$$

$$= 2\int_{0.183}^{2.187} x\left[(\ln x)^2 - (x^2 - 4)^2\right]\,dx$$

$$\approx -19.9173$$

The last integral was computed using a CAS. It is possible to find an antiderivative or to use Simpson's Rule.

13.
$$V = \iiint_Q dV$$

$$= \int_0^1 \int_{-1}^1 \int_0^{1-x^2} dz\,dx\,dy$$

$$= \int_0^1 \int_{-1}^1 (1 - x^2)\,dx\,dy$$

$$= \int_0^1 \frac{4}{3}\,dy = \frac{4}{3}$$

15. We convert to cylindrical coordinates. The two surfaces ($z = r^2$ and $z = 8 - r^2$) meet at $r = 2$.

$$V = \iiint_Q dV$$

$$= \int_0^{2\pi} \int_0^2 \int_{r^2}^{8-r^2} r \, dz \, dr \, d\theta$$

$$= \int_0^{2\pi} \int_0^2 r(8 - 2r^2) \, dr \, d\theta$$

$$= \int_0^{2\pi} 8 \, d\theta = 16\pi$$

17.
$$V = \iiint_Q dV$$

$$= \int_0^4 \int_0^{8-2y} \int_0^{8-x-2y} dz \, dx \, dy$$

$$= \int_0^4 \int_0^{8-2y} (8 - x - 2y) \, dx \, dy$$

$$= \int_0^4 \left(32 - 16y + 2y^2\right) dy = \frac{128}{3}$$

19. We use cylindrical coordinates.

$$V = \iiint_Q dV$$

$$= \int_0^{2\pi} \int_0^4 \int_r^4 r \, dz \, dr \, d\theta$$

$$= \int_0^{2\pi} \int_0^4 r(4 - r) \, dr \, d\theta$$

$$= \int_0^{2\pi} \frac{32}{3} \, d\theta = \frac{64\pi}{3}$$

21. We use cylindrical coordinates (although spherical would probably be even easier).

$$V = \iiint_Q dV$$

$$= \int_0^{2\pi} \int_0^{\sqrt{2}} \int_r^{\sqrt{4-r^2}} r \, dz \, dr \, d\theta$$

$$= \int_0^{2\pi} \int_0^{\sqrt{2}} \left(\sqrt{4-r^2} - r\right) r \, dr \, d\theta$$

$$= 2\pi \int_0^{\sqrt{2}} \left(r\sqrt{4-r^2} - r^2\right) dr$$

$$= 2\pi \left[\frac{-(4-r^2)^{3/2}}{3} - \frac{r^3}{3}\right]_0^{\sqrt{2}}$$

$$= \frac{8\pi}{3}(2 - \sqrt{2})$$

23.
$$V = \iiint_Q dV$$

$$= \int_0^{2\pi} \int_0^1 \int_0^{6-r^2} r \, dz \, dr \, d\theta$$

$$= \int_0^{2\pi} \int_0^1 (6 - r^2) r \, dr \, d\theta$$

$$= \int_0^{2\pi} \frac{11}{4} \, d\theta = \frac{11\pi}{2}$$

25.
$$\int_0^2 \int_0^{x^2} f(x, y) \, dy \, dx \quad \int_0^4 \int_{\sqrt{y}}^2 f(x, y) \, dx \, dy$$

27.
$$\int_0^2 \int_{-\sqrt{4-x^2}}^{\sqrt{4-x^2}} 2x \, dy \, dx$$

$$= \int_{-\pi/2}^{\pi/2} \int_0^2 2r^2 \cos\theta \, dr \, d\theta$$

$$= \int_{-\pi/2}^{\pi/2} \frac{16}{3} \cos\theta \, d\theta = \frac{32}{3}$$

29.
$$m = \iint_R \rho(x, y) \, dA$$

$$= \int_0^2 \int_x^{2x} 2x \, dy \, dx$$

$$= \int_0^2 2x^2 \, dx = \frac{16}{3}$$

$$M_y = \iint_R x\rho(x, y)\, dA$$

$$= \int_0^2 \int_x^{2x} 2x^2\, dy\, dx$$

$$= \int_0^2 2x^3\, dx = 8$$

$$\bar{x} = \frac{M_y}{m} = \frac{3}{2}$$

$$M_x = \iint_R y\rho(x, y)\, dA$$

$$= \int_0^2 \int_x^{2x} 2xy\, dy\, dx$$

$$= \int_0^2 3x^3\, dx = 12$$

$$\bar{y} = \frac{M_x}{m} = \frac{9}{4}$$

31. $$m = \iiint_Q \rho(x, y, v)\, dV$$

$$= \int_{-1}^1 \int_0^{1-x^2} \int_0^{2-z} 2\, dy\, dz\, dx$$

$$= \int_{-1}^1 \int_0^{1-x^2} 2(2 - z)\, dz\, dx$$

$$= \int_{-1}^1 \left[4(1 - x^2) - (1 - x^2)^2\right] dx$$

$$= \int_{-1}^1 \left(3 - 2x^2 - x^4\right) dx = \frac{64}{15}$$

$$M_{yz} = \iiint_Q x\rho(x, y, v)\, dV$$

$$= \int_{-1}^1 \int_0^{1-x^2} \int_0^{2-z} 2x\, dy\, dz\, dx$$

$$= \int_{-1}^1 \int_0^{1-x^2} 2x(2 - z)\, dz\, dx$$

$$= \int_{-1}^1 \left(3x - 2x^3 - x^5\right) dx = 0$$

$$\bar{x} = \frac{M_{yz}}{m} = 0$$

$$M_{xz} = \iiint_Q y\rho(x, y, v)\, dV$$

$$= \int_{-1}^1 \int_0^{1-x^2} \int_0^{2-z} 2y\, dy\, dz\, dx$$

$$= \int_{-1}^1 \int_0^{1-x^2} (2 - z)^2\, dz\, dx$$

$$= \frac{1}{3} \int_{-1}^1 \left[8 - (1 + x^2)^3\right] dx$$

$$= \frac{1}{3} \int_{-1}^1 \left(7 - x^6 - 3x^4 - 3x^2\right) dx = \frac{368}{105}$$

$$\bar{y} = \frac{M_{xz}}{m} = \frac{23}{28}$$

$$M_{xy} = \iiint_Q z\rho(x, y, v)\, dV$$

$$= \int_{-1}^1 \int_0^{1-x^2} \int_0^{2-z} 2z\, dy\, dz\, dx$$

$$= \int_{-1}^1 \int_0^{1-x^2} 2z(2 - z)\, dz\, dx$$

$$= 2 \int_{-1}^1 \left(\frac{2}{3} - x^2 + \frac{1}{3}x^6\right) dx = \frac{32}{21}$$

$$\bar{z} = \frac{M_{zy}}{m} = \frac{5}{14}$$

33. The region of integration is shown below.

Region of Integration

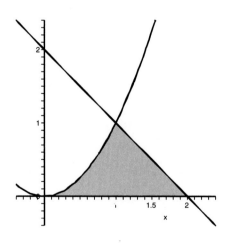

$$A = \iint_R dA$$

$$= \int_0^1 \int_{\sqrt{y}}^{2-y} dx\, dy$$

$$= \int_0^1 (2 - y - \sqrt{y})\, dy = \frac{5}{6}$$

It would work equally well to integrate with respect to y first. This would require splitting up the integral over two separate regions.

35.
$$a = \iint_R dA = \int_0^1 \int_x^{2x} dy\, dx$$

$$= \int_0^1 x\, dx = \frac{1}{2}$$

$$\iint_R x^2\, dA = \int_0^1 \int_x^{2x} x^2\, dy\, dx$$

$$= \int_0^1 x^3\, dy\, dx = \frac{1}{4}$$

Ave Val $= \dfrac{1}{a} \displaystyle\iint_R f(x, y)\, dA = \dfrac{1}{2}$

37. This is a plane over a triangle, T, in the xy-plane. The triangle T in the xy-plane has vertices $(0, 0)$, $(0, 2)$ and $(2, 2)$ and therefore has area of 2.

The surface area of the plane is

$$S = \iint_T \sqrt{(2)^2 + (4)^2 + 1}\, dA$$

$$= \iint_T \sqrt{21}\, dA$$

$$= \sqrt{21}(\text{Area of } T) = 2\sqrt{21}.$$

39.
$$S = \iint_R \sqrt{y^2 + x^2 + 1}\, dA$$

$$= \int_0^{\pi/2} \int_0^{\sqrt{8}} r\sqrt{r^2 + 1}\, dr\, d\theta$$

$$= \int_0^{\pi/2} \frac{26}{3}\, d\theta = \frac{13\pi}{3}$$

41. We will change to polar coordinates. Notice that since $z = r$, we have

$$z^2 = r^2 = x^2 + y^2$$

$$2z\frac{\partial z}{\partial x} = 2x$$

$$\frac{\partial z}{\partial x} = \frac{x}{z} = \frac{r\cos\theta}{r} = \cos\theta$$

Similarly, $\frac{\partial z}{\partial y} = \sin\theta$.

The region of integration is a disk of radius 4 in the xy-plane.

$$S = \iint_R \sqrt{(\cos\theta)^2 + (\sin\theta)^2 + 1}\, dA$$

$$= \iint_R \sqrt{2}\, dA$$

$$= \sqrt{2}\,(\text{Area of } R) = \sqrt{2}(16\pi)$$

43.
$$\iiint_Q z(x + y)\, dV$$

$$= \int_0^2 \int_{-1}^1 \int_{-1}^1 z(x + y)\, dz\, dy\, dx$$

$$= \int_0^2 \int_{-1}^1 \left[\frac{z^2}{2}(x + y)\right]_{-1}^1 dy\, dx$$

$$= 0$$

45. We use spherical coordinates.

$$\iiint_Q \sqrt{x^2 + y^2 + z^2}\, dV$$

$$= \int_0^{2\pi} \int_0^{\pi/4} \int_0^2 \rho^3 \sin\phi\, d\rho\, d\phi\, d\theta$$

$$= \int_0^{2\pi} \int_0^{\pi/4} 4\sin\phi\, d\phi\, d\theta$$

$$= \int_0^{2\pi} 4\left(1 - \frac{1}{\sqrt{2}}\right) d\theta$$

$$= 8\pi\left(1 - \frac{1}{\sqrt{2}}\right) = 4\pi(2 - \sqrt{2})$$

47.
$$\iiint_Q f(x, y, z)\, dV$$

$$= \int_0^2 \int_x^2 \int_0^{6-x-y} f(x, y, z)\, dz\, dy\, dz$$

49.
$$\iiint_Q f(x, y, z)\, dV$$

$$= \int_0^{2\pi} \int_0^{\pi/2} \int_0^2 \rho^2 \sin\phi$$

$$\cdot f(\rho \sin\phi \cos\theta, \rho \sin\phi \sin\theta, \rho \cos\phi)$$

$$d\rho\, d\phi\, d\theta$$

51.
$$\int_0^1 \int_x^{\sqrt{2-x^2}} \int_0^{\sqrt{x^2+y^2}} e^z\, dz\, dy\, dx$$

$$= \int_{\pi/4}^{\pi/2} \int_0^{\sqrt{2}} \int_0^r r e^z\, dz\, dr\, d\theta$$

$$= \int_{\pi/4}^{\pi/2} \int_0^{\sqrt{2}} r(e^r - 1)\, dr\, d\theta$$

$$= \int_{\pi/4}^{\pi/2} e^{\sqrt{2}}(\sqrt{2} - 1)\, d\theta$$

$$= \frac{\pi}{4} e^{\sqrt{2}}(\sqrt{2} - 1)$$

53.
$$\int_{-1}^1 \int_0^{\sqrt{1-x^2}} \int_{\sqrt{x^2+y^2}}^{\sqrt{2-x^2-y^2}} \sqrt{x^2+y^2+z^2}\, dz\, dy\, dx$$

$$= \int_0^{\pi} \int_0^{\pi/4} \int_0^{\sqrt{2}} \rho^3 \sin\phi\, d\rho\, d\phi\, d\theta$$

$$= \int_0^{\pi} \int_0^{\pi/4} \sin\phi\, d\phi\, d\theta$$

$$= \int_0^{\pi} \left(1 - \frac{1}{\sqrt{2}}\right) d\theta$$

$$= \pi \left(1 - \frac{1}{\sqrt{2}}\right) = \frac{\pi}{2}(2 - \sqrt{2})$$

55. (a) $y = 3$

$r \sin\theta = 3$

$r = 3\csc\theta$

(b) $y = 3$

$\rho \sin\theta \sin\phi = 3$

$\rho = 3\csc\theta\csc\phi$

57. (a) $x^2 + y^2 + z^2 = 4$

$r^2 + z^2 = 4$

(b) $x^2 + y^2 + z^2 = 4$

$\rho^2 = 4$

$\rho = 2$

59. (a) $z = \sqrt{x^2 + y^2}$

$z = r \quad r \geq 0$

(b) $z = \sqrt{x^2 + y^2}$

$\rho \cos\phi = \rho \sin\phi$

$\tan\phi = 1$

$\phi = \frac{\pi}{4}$

61.

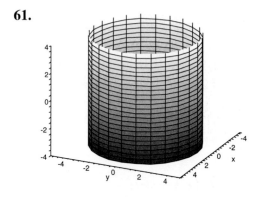

63.

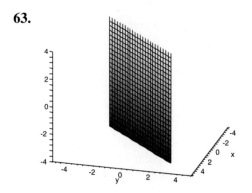

65.

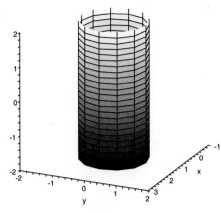

67.
$$y = 2x - 1 \quad \rightarrow \quad y - 2x = -1$$
$$y = 2x + 1 \quad \rightarrow \quad y - 2x = 1$$
$$\underbrace{}_{u = y - 2x}$$

$$\left.\begin{array}{ll} y = 2 - 2x & \rightarrow \quad y + 2x = 2 \\ y = 4 - 2x & \rightarrow \quad y + 2x = 4 \end{array}\right\} v = y + 2x$$

$$\left.\begin{array}{l} u = y - 2x \\ v = y + 2x \end{array}\right\} \rightarrow \left\{\begin{array}{l} x = \frac{-u+v}{4} \\ y = \frac{u+v}{2} \end{array}\right.$$

$$-1 \le u \le 1$$
$$2 \le v \le 4$$

69.
$$\frac{\partial(x, y)}{\partial(u, v)} = \begin{vmatrix} \frac{\partial x}{\partial u} & \frac{\partial x}{\partial v} \\ \frac{\partial y}{\partial u} & \frac{\partial y}{\partial v} \end{vmatrix}$$

$$= \begin{vmatrix} -\frac{1}{4} & \frac{1}{4} \\ \frac{1}{2} & \frac{1}{2} \end{vmatrix} = -\frac{1}{4}$$

$$\iint_R e^{y-2x} \, dA = \int_2^4 \int_{-1}^1 e^u \left| -\frac{1}{4} \right| \, du \, dv$$

$$= \frac{1}{4} \int_2^4 (e - e^{-1}) \, dv$$

$$= \frac{1}{2}(e - e^{-1})$$

71.
$$\frac{\partial(x, y)}{\partial(u, v)} = \begin{vmatrix} \frac{\partial x}{\partial u} & \frac{\partial x}{\partial v} \\ \frac{\partial y}{\partial u} & \frac{\partial y}{\partial v} \end{vmatrix}$$

$$= \begin{vmatrix} 2uv & u^2 \\ 4 & 2v \end{vmatrix} = 4uv^2 - 4u^2$$

Vector Calculus

15.1 VECTOR FIELDS

1.

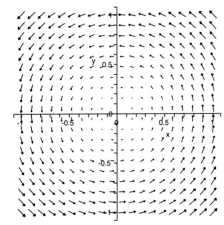

5.

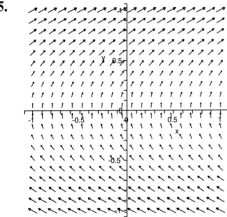

3.

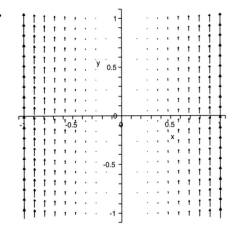

7.

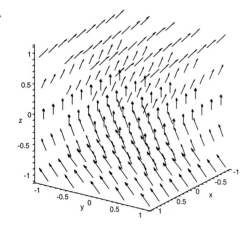

9.

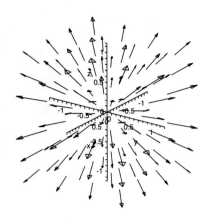

11. $\mathbf{F}_1 \leftrightarrow$ Graph D: Because all vectors point away from the origin and have the same length.

$\mathbf{F}_2 \leftrightarrow$ Graph B: Because all vectors point away from the origin and the lengths are proportional to the distances from the origin.

$\mathbf{F}_3 \leftrightarrow$ Graph A: Because the vectors point upward when $x > 0$ and downward for $x < 0$.

$\mathbf{F}_4 \leftrightarrow$ Graph C: Because the vectors depend only on y.

13. $\nabla f = \langle 2x, 2y \rangle$

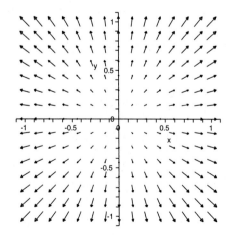

15. $\nabla f = \left\langle \dfrac{x}{\sqrt{x^2 + y^2}}, \dfrac{y}{\sqrt{x^2 + y^2}} \right\rangle$

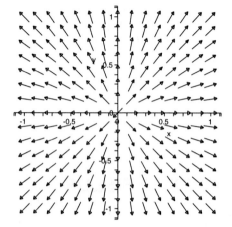

17. $\nabla f = \left\langle e^{-y}, -xe^{-y} \right\rangle$

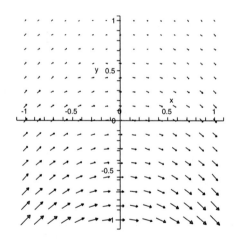

19. $\nabla f = \left\langle \dfrac{x}{\sqrt{x^2 + y^2 + z^2}}, \dfrac{y}{\sqrt{x^2 + y^2 + z^2}}, \dfrac{z}{\sqrt{x^2 + y^2 + z^2}} \right\rangle$

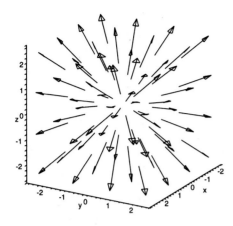

21. $\nabla f = \langle 2xy, x^2 + z, y \rangle$

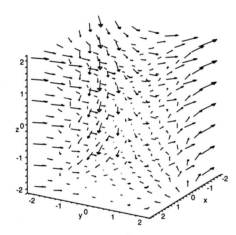

23. If $\nabla f(x, y) = \langle y, x \rangle$, then

$$\frac{\partial f}{\partial x} = y \text{ and } \frac{\partial f}{\partial y} = x.$$

$$f(x, y) = \int y \, dx = xy + g(y)$$

$$\frac{\partial f}{\partial y} = x + g'(y) = x$$

$$g'(y) = 0$$
$$g(y) = c$$
$$f(x, y) = xy + c$$

The vector field is conservative.

25. If $\nabla f(x, y) = \langle y, -x \rangle$, then
$\frac{\partial f}{\partial x} = y$ and $\frac{\partial f}{\partial y} = -x.$

$$f(x, y) = \int y \, dx = xy + g(y)$$

$$\frac{\partial f}{\partial y} = x + g'(y) = -x$$

$$g'(y) = -2x$$

Since this is not possible (g was to be a function of only y), the vector field is not conservative.

27. If $\nabla f(x, y) = \langle x - 2xy, y^2 - x^2 \rangle$, then
$\frac{\partial f}{\partial x} = x - 2xy$ and $\frac{\partial f}{\partial y} = y^2 - x^2.$

$$f(x, y) = \int (x - 2xy) \, dx$$

$$= \frac{x^2}{2} - x^2 y + g(y)$$

$$\frac{\partial f}{\partial y} = -x^2 + g'(y) = y^2 - x^2$$

$$g'(y) = y^2$$

$$g(y) = \frac{y^3}{3} + c$$

$$f(x, y) = \frac{x^2}{2} - x^2 y + \frac{y^3}{3} + c$$

The vector field is conservative.

29. If $\nabla f(x, y) = \langle y \sin xy, x \sin xy \rangle$, then
$\frac{\partial f}{\partial x} = y \sin xy$ and $\frac{\partial f}{\partial y} = x \sin xy.$

$$f(x, y) = \int y \sin xy \, dx = -\cos xy + g(y)$$

$$\frac{\partial f}{\partial y} = x \sin xy + g'(y) = x \sin xy$$

$$g(y) = c$$
$$f(x, y) = -\cos xy + c$$

The vector field is conservative.

31. If $\nabla f(x, y, z) = \langle 4x - z, 3y + z, y - x \rangle$,
then
$\frac{\partial f}{\partial x} = 4x - z, \frac{\partial f}{\partial y} = 3y + z, \frac{\partial f}{\partial z} = y - z.$

$$f(x, y, z) = \int (4x - z)\, dx$$

$$= 2x^2 - xz + g(y, z)$$

$$\frac{\partial f}{\partial y} = \frac{\partial g}{\partial y} = 3y + z$$

$$\frac{\partial g}{\partial y} = 3y + z$$

$$g(y, z) = \int (3y + z)\, dy$$

$$= \frac{3}{2}y^2 + yz + h(z)$$

$$f(x, y, z) = 2x^2 - xz + \frac{3}{2}y^2 + yz + h(z)$$

$$\frac{\partial f}{\partial z} = -x + y + h'(z) = y - x$$

$$h'(z) = 0$$

$$h(z) = c$$

$$f(x, y, z) = 2x^2 - xz + \frac{3}{2}y^2 + yz + c$$

The vector field is conservative.

33. If $\nabla f(x, y, z) = \left\langle y^2 z^2 - 1,\ 2xyz^2,\ 4z^3 \right\rangle$, then $\frac{\partial f}{\partial x} = y^2 z^2 - 1$, $\frac{\partial f}{\partial y} = 2xyz^2$, $\frac{\partial f}{\partial z} = 4z^3$.

$$f(x, y, z) = \int 4z^3\, dz$$

$$= z^4 + h(x, y)$$

$$\frac{\partial f}{\partial y} = \frac{\partial h}{\partial y} = 2xyz^2$$

Since h is supposed to be independent of z, this is impossible and the vector field is not conservative.

35. $$\frac{dy}{dx} = \frac{\cos x}{2}$$

$$\int dy = \frac{1}{2} \int \cos x\, dx$$

$$y = \frac{1}{2}\sin x + c$$

37. $$\frac{dy}{dx} = \frac{3x^2}{2y}$$

$$\int 2y\, dy = \int 3x^2\, dx$$

$$y^2 = x^3 + c$$

39. $$\frac{dy}{dx} = \frac{xe^y}{y}$$

$$\int ye^{-y}\, dy = \int x\, dx$$

$$-ye^{-y} - e^{-y} = \frac{x^2}{2} + c$$

$$(y + 1)e^{-y} = -\frac{x^2}{2} + k \quad (k = -c)$$

41. $$\frac{dy}{dx} = \frac{y^2 + 1}{y}$$

$$\int \frac{y}{y^2 + 1}\, dy = \int dx$$

$$\frac{1}{2}\ln(y^2 + 1) = x + c$$

$$\ln(y^2 + 1) = 2x + 2c$$

$$y^2 + 1 = e^{2x + 2c}$$

$$y^2 = Ae^{2x} - 1 \quad (A = e^{2c})$$

43. If we let $\phi_1(x) = \int_0^x f(u)\, du$ then $\phi_1'(x) = f(x)$.

Similarly, if we define $\phi_2(y) = \int_0^y g(u)\, du$ and $\phi_3(z) = \int_0^z h(u)\, du$ then $\phi_2'(y) = f(y)$ and $\phi_3'(z) = f(z)$.

So we can define
$$\phi(x, y, z) = \phi_1(x) + \phi_2(y) + \phi_3(z)$$
which gives us

$$\nabla\phi = \left\langle \phi_1'(x),\ \phi_2'(y),\ \phi_3'(z) \right\rangle$$

$$= \left\langle f(x),\ g(y),\ h(z) \right\rangle.$$

45. This is essentially exercise 15.

$$\nabla(r) = \frac{\langle x, y \rangle}{\sqrt{x^2 + y^2}} = \frac{\mathbf{r}}{r}$$

47. Here we use the chain rule for the functions
$f(u) = u^3$ and
$g(x, y, z) = (x^2 + y^2 + z^2)^{1/2} = r$. The chain
rule gives us, using exercise 45:

$$\nabla(r^3) = \nabla(f \circ g) = f'(g(x, y, z))\nabla g$$

$$= 3(g(x, y, z))^2 \left(\frac{\mathbf{r}}{r}\right)$$

$$= 3(r)^2 \left(\frac{\mathbf{r}}{r}\right) = 3r\mathbf{r}$$

49. Suppose this field is conservative, then there
is a function $f(x, y)$ such that
$f_x = \frac{1}{r} = \frac{1}{\sqrt{x^2 + y^2}}$ and

$f_y = \frac{1}{r} = \frac{1}{\sqrt{x^2 + y^2}}$.

Integrating these functions is a bit nasty, but
the tables can be used.

$$f(x, y) = \int \frac{1}{\sqrt{x^2 + y^2}} \, dx$$

$$= \ln\left(x + \sqrt{x^2 + y^2}\right) + g(y)$$

$$\frac{\partial f}{\partial y} = \left(\frac{1}{x + \sqrt{x^2 + y^2}}\right)\left(\frac{y}{\sqrt{x^2 + y^2}}\right)$$

$$+ g'(y)$$

$$= \frac{1}{\sqrt{x^2 + y^2}}$$

$$g'(y) = \frac{1}{\sqrt{x^2 + y^2}}$$

$$- \frac{y}{\left(x + \sqrt{x^2 + y^2}\right)\sqrt{x^2 + y^2}}$$

$$= \left(\frac{1}{\sqrt{x^2 + y^2}}\right)\left(1 - \frac{y}{x + \sqrt{x^2 + y^2}}\right)$$

But, g is supposed to depend only on y and be
independent of x. The vector field is therefore
not conservative.

51. This is the picture in the xy-plane. The wire is
represented by the z-axis, at the origin of the
graph below.

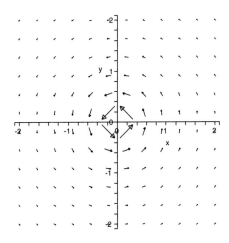

53. Since the field acts radially away from the
origin, we have
$$\mathbf{F} = c\langle x, y \rangle$$
for some c.

We are also told that $\|\mathbf{F}\| = 3$. This means that
$3 = c\sqrt{x^2 + y^2}$ and $c = \frac{3}{\sqrt{x^2 + y^2}}$.

$$\mathbf{F} = \frac{3\langle x, y \rangle}{\sqrt{x^2 + y^2}}$$

55. Since the field acts radially toward the origin,
we have
$$\mathbf{F} = c\langle -x, -y, -z \rangle$$
for some c.

We are also told that $\|\mathbf{F}\| = x^2 + y^2 + z^2$. This
means that
$c\sqrt{x^2 + y^2 + z^2} = x^2 + y^2 + z^2$ and
$c = \sqrt{x^2 + y^2 + z^2}$.
$$\mathbf{F} = \sqrt{x^2 + y^2 + z^2}\langle -x, -y, -z \rangle$$

57. This is very similar to Example 1.10. The
difference is the magnitude of the charges and
the charge at the origin.

$$\mathbf{E} = q\frac{\langle x - 1, y \rangle}{[(x - 1)^2 + y^2]^{3/2}}$$

$$+ q\frac{\langle x + 1, y \rangle}{[(x + 1)^2 + y^2]^{3/2}}$$

$$- q\frac{\langle x, y \rangle}{[x^2 + y^2]^{3/2}}$$

59. In a contour map of the isotherms (the level-
surfaces for T), the gradient is both orthogonal

to the isotherms and points in the direction of higher temperatures (the direction of greatest increase for T). The differential equation (with respect to time) for the flow lines is

$$\left\langle \frac{dx}{dt}, \frac{dy}{dt}, \frac{dz}{dt} \right\rangle = \mathbf{F}(x, y, z)$$

$$= -k\nabla T(x, y, z).$$

With $k > 0$, $-k\nabla T$ is in the direction from hot to cold and therefore heat must flow from hot to cold.

15.2 LINE INTEGRALS

1. $x = 1 + 2t, \quad y = 2 + 3t, \quad 0 \leq t \leq 1$

$$ds = \sqrt{[x'(t)]^2 + [y'(t)]^2}\, dt$$

$$= \sqrt{2^2 + 3^2}\, dt = \sqrt{13}\, dt$$

$$\int_C 2x\, ds = \int_0^1 2(1 + 2t)\sqrt{13}\, dt$$

$$= 4\sqrt{13}$$

3. $x = 5 - 4t, \quad y = 2 - t, \quad 0 \leq t \leq 1$

$$ds = \sqrt{[x'(t)]^2 + [y'(t)]^2}\, dt$$

$$= \sqrt{(-4)^2 + (-1)^2}\, dt = \sqrt{17}\, dt$$

$$\int_C (3x + y)\, ds$$

$$= \int_0^1 [3(5 - 4t) + (2 - t)]\sqrt{17}\, dt$$

$$= \sqrt{17} \int_0^1 (17 - 13t)\, dt = \frac{21}{2}\sqrt{17}$$

5. $x = 2t, \quad y = 2 + 4t, \quad 0 \leq t \leq 1$
$dx = 2\, dt$

$$\int_C 2x\, dx = \int_0^1 2(2t)(2)\, dt$$

$$= \int_0^1 8t\, dt = 4$$

7. $x = 2\cos t, \quad y = 2\sin t, \quad 0 \leq t \leq \frac{\pi}{2}$

$$ds = \sqrt{[x'(t)]^2 + [y'(t)]^2}\, dt$$

$$= \sqrt{(-2\sin t)^2 + (2\cos t)^2}\, dt = 2\, dt$$

$$\int_C 3x\, ds = \int_0^{\pi/2} 3(2\cos t)(2)\, dt$$

$$= 12 \int_0^{\pi/2} \cos t\, dt = 12$$

9. $x = 2\cos t, \quad y = 2\sin t, \quad 0 \leq t \leq \frac{\pi}{2}$
$dx = -2\sin t\, dt$

$$\int_C 2x\, dx = \int_0^{\pi/2} 2(2\cos t)(-2\sin t)\, dt$$

$$= -8 \int_0^{\pi/2} \cos t \sin t\, dt = -4$$

11. $x = 2\sin t, \quad y = \cos t, \quad 0 \leq t \leq \pi$
$dx = 2\cos t\, dt$

$$\int_C 3y\, dx = \int_0^\pi 3(\cos t)(2\cos t)\, dt$$

$$= 6 \int_0^\pi \cos^2 t\, dt = 3\pi$$

The student may be advised to review the list of trigonometric integrals and how they are integrated.

13. $x = t, \quad y = t^2, \quad 0 \leq t \leq 2$

$$ds = \sqrt{[x'(t)]^2 + [y'(t)]^2}\, dt$$

$$= \sqrt{(1)^2 + (2t)^2}\, dt = \sqrt{1 + 4t^2}\, dt$$

This integral can be computed using the tables or by using a trigonometric substitution.

$$\int_C 3y \, ds = \int_0^2 3t^2 \sqrt{1 + 4t^2} \, dt$$

$$= 6 \int_0^2 t^2 \sqrt{\frac{1}{4} + t^2} \, dt$$

$$= 6 \left[\frac{2\left(\frac{1}{4} + 8\right)\sqrt{\frac{1}{4} + 4}}{8} \right.$$

$$- \frac{\frac{1}{16}\ln\left|2 + \sqrt{\frac{1}{4} + 4}\right|}{8}$$

$$\left. + \frac{\frac{1}{16}\ln\left(\frac{1}{2}\right)}{8} \right]$$

$$= \frac{3}{64}\left[132\sqrt{17} - \ln(4 + \sqrt{17})\right]$$

$$\approx 25.4135$$

15. Using x as the parameter:

$$\int_C 2x \, dx = \int_2^0 2x \, dx = -4$$

17. Using x as the parameter:

$$\int_C 3y \, dx = \int_1^4 3\sqrt{x} \, dt = 14$$

19. $C_1 : x = t, \quad y = 0, \quad 0 \le t \le 1$

$ds = dt$

$$\int_{C_1} 3x \, ds = \int_0^1 3t \, dt = \frac{3}{2}$$

$C_2 : x = \cos t, \quad y = \sin t, \quad 0 \le t \le \frac{\pi}{2}$

$ds = dt$

$$\int_{C_2} 3x \, ds = \int_0^{\pi/2} 3\cos t \, dt = 3$$

$$\int_C 3x \, ds = \int_{C_1} 3x \, ds + \int_{C_2} 3x \, ds$$

$$= \frac{3}{2} + 3 = \frac{9}{2}$$

21. $x = 1 + t, \quad y = -2t, \quad z = 1 + t$
$0 \le t \le 1$

$$ds = \sqrt{[x'(t)]^2 + [y'(t)]^2 + [z'(t)]^2} \, dt$$

$$= \sqrt{(1)^2 + (-2)^2 + (1)^2} \, dt = \sqrt{6} \, dt$$

$$\int_C 4z \, ds = \int_0^1 4(1 + t)\sqrt{6} \, dt = 6\sqrt{6}$$

23. $x = t, \quad y = t^2, \quad z = 2$
$1 \le t \le 2$

$dx = dt$

$$\int_C 4(x - z)z \, dx = \int_1^2 4(t - 2)(2) \, dt$$

$$= 8 \int_1^2 (t - 2) \, dt = -4$$

25. $x = 3 + 2t, \quad y = 1 + 3t, \quad 0 \le t \le 1$

$$\int_C \mathbf{F} \cdot d\mathbf{r} = \int_C 2x \, dx + 2y \, dy$$

$$= \int_0^1 [2(3 + 2t)(2) + 2(1 + 3t)(3)] \, dt$$

$$= \int_0^1 (26t + 18) \, dt = 31$$

27. $x = 4\cos t, \quad y = 4\sin t, \quad 0 \le t \le \frac{\pi}{2}$

$$\int_C \mathbf{F} \cdot d\mathbf{r} = \int_C 2x \, dx + 2y \, dy$$

$$= \int_0^{\pi/2} [2(4\cos t)(-4\sin t)$$

$$+ 2(4\sin t)(4\cos t)] \, dt$$

$$= \int_0^{\pi/2} 0 \, dt = 0$$

29. $x = t, \quad y = t^2, \quad 0 \le t \le 1$

$$\int_C \mathbf{F} \cdot d\mathbf{r} = \int_C 2 \, dx + x \, dy$$

$$= \int_0^1 [2(1) + (t)(2t)] \, dt$$

$$= \int_0^1 (2 + 2t^2) \, dt = \frac{8}{3}$$

31. $C_1 : (0, 0)$ to $(0, 1) : \mathbf{F} = \langle 0, 2 \rangle, \; dx = 0$

$$\int_{C_1} \mathbf{F} \cdot d\mathbf{r} = \int_{C_1} 3x \, dx + 2 \, dy$$

$$= \int_{C_1} 2 \, dy = \int_0^1 2 \, dy = 2$$

$C_2 : (0, 1)$ to $(4, 1) : \mathbf{F} = \langle 3x, 2 \rangle, \; dy = 0$

$$\int_{C_2} \mathbf{F} \cdot d\mathbf{r} = \int_{C_2} 3x \, dx + 2 \, dy$$

$$= \int_{C_2} 3x \, dx = \int_0^4 3x \, dx = 24$$

$$\int_C \mathbf{F} \cdot d\mathbf{r} = \int_{C_1} \mathbf{F} \cdot d\mathbf{r} + \int_{C_2} \mathbf{F} \cdot d\mathbf{r}$$

$$= 2 + 24 = 26$$

33. $C_1 : (0, 0, 0)$ to $(2, 1, 2)$

$x = 2t, \quad y = t, \quad z = 2t$

$0 \le t \le 1$

$$\int_{C_1} \mathbf{F} \cdot d\mathbf{r} = \int_{C_1} y \, dx + 0 \, dy + z \, dz$$

$$= \int_0^1 [(t)(2) + 0 + 2t(2)] \, dt$$

$$= \int_0^1 6t \, dt = 3$$

$C_2 : (2, 1, 2)$ to $(2, 1, 0)$

$x = 2, \quad y = 1, \quad z = 2 - 2t$

$0 \le t \le 1$

$$\int_{C_2} \mathbf{F} \cdot d\mathbf{r} = \int_{C_2} y \, dx + 0 \, dy + z \, dz$$

$$= \int_0^1 [1(0) + 0(0) + (2 - 2t)(-2)] \, dt$$

$$= \int_0^1 4(t - 1) \, dt = -2$$

$C_3 : (2, 1, 0)$ to $(0, 0, 0)$

$x = 2 - 2t, \quad y = 1 - t, \quad z = 0$

$0 \le t \le 1$

$$\int_{C_3} \mathbf{F} \cdot d\mathbf{r} = \int_{C_3} y \, dx + 0 \, dy + z \, dz$$

$$= \int_0^1 [(1 - t)(-2) + (0)(-1) + 0(0)] \, dt$$

$$= \int_0^1 -2(1 - t) \, dt = -1$$

$$\int_C \mathbf{F} \cdot d\mathbf{r} = \int_{C_1} \mathbf{F} \cdot d\mathbf{r} + \int_{C_2} \mathbf{F} \cdot d\mathbf{r}$$

$$+ \int_{C_3} \mathbf{F} \cdot d\mathbf{r}$$

$$= 3 - 2 - 1 = 0$$

35. $x = \cos t, \quad y = \sin t, \quad z = 2t$

$0 \le t \le \pi/2$

$$\int_C \mathbf{F} \cdot d\mathbf{r} = \int_C xy \, dx + 3z \, dy + 1 \, dz$$

$$= \int_0^{\pi/2} [(\cos t \sin t)(- \sin t)$$

$$+ 3(2t)(\cos t) + 1(2)] \, dt$$

$$= \int_0^{\pi/2} \left[- \cos t \sin^2 t + 6t \cos t + 2 \right] dt$$

$$= -\frac{1}{3} + (3\pi - 6) + \pi = 4\pi - \frac{19}{3}$$

37. The motion is in the same direction as the force field, so the work done by the force field is positive.

39. The motion is perpendicular to the force field, so the work done by the force field must be zero.

41. Most of the motion is against the force field, so the work done by the force field must be negative.

43. $x = t, \quad y = t^2, \quad 0 \le t \le 3$

$$ds = \sqrt{[x'(t)]^2 + [y'(t)]^2} \, dt$$

$$= \sqrt{(1)^2 + (2t)^2} \, dt = \sqrt{1 + 4t^2} \, dt$$

$$m = \int_C \rho \, ds = \int_C x \, ds$$

$$= \int_0^3 t\sqrt{1 + 4t^2} \, dt$$

$$= \frac{37\sqrt{37} - 1}{12} \approx 18.672$$

45. The setup is as in exercise 43. The integrals here can be done using the table of integrals, a CAS, or by approximating using Simpson's Rule.

$$m\bar{x} = \int_C x\rho \, ds$$

$$= \int_0^3 t^2\sqrt{1 + 4t^2} \, dt$$

$$= \frac{1}{64}\left[438\sqrt{37} - \ln(6 + \sqrt{37})\right]$$

$$\approx 41.590$$

$$\bar{x} \approx 2.227$$

$$m\bar{y} = \int_C y\rho \, ds$$

$$= \int_0^3 t^3\sqrt{1 + 4t^2} \, dt$$

$$= \frac{1}{120}\left(1961\sqrt{37} + 1\right) \approx 99.411$$

$$\bar{y} \approx 5.324$$

47. The setup is as in exercise 43. The integral here can be done using the table of integrals, a CAS, or by approximating using Simpson's Rule.

$$I = \int_C x^2\rho \, ds$$

$$= \int_0^3 t^3\sqrt{1 + 4t^2} \, dt$$

$$= \frac{1}{120}\left(1961\sqrt{37} + 1\right) \approx 99.411$$

49. The setup is as in exercise 43. The integral here can be done using the table of integrals, a CAS, or by approximating using Simpson's

Rule.

$$I = \int_C w^2\rho \, ds$$

$$= \int_C (9 - y)^2 x \, ds$$

$$= \int_0^3 (9 - t^2)^2 t\sqrt{1 + 4t^2} \, dt$$

$$= \frac{50653\sqrt{37} - 5797}{840} \approx 359.897$$

51. $x = \cos 2t, \quad y = \sin 2t, \quad z = t$
$0 \le t \le \pi$

$$ds = \sqrt{[x'(t)]^2 + [y'(t)]^2 + [z'(t)]^2} \, dt$$

$$= \sqrt{(-2\sin 2t)^2 + (2\cos 2t)^2 + (1)^2} \, dt$$

$$= \sqrt{5} \, dt$$

$$m = \int_C \rho \, ds = \int_C z^2 \, ds$$

$$= \int_0^\pi t^2\sqrt{5} \, dt = \frac{\pi^3\sqrt{5}}{3}$$

53. The center of mass of the rod in exercise 45 is $(2.227, 5.324)$. Since $5.324 \ne (2.227)^2$, the point is not on the rod.

The center of mass of the rod in exercise 46 is $(0.9478, 2.8553)$. Since $2.8553 \ne 4 - (0.9478)^2$, the point is not on the rod.

Trying to balance a rod by a point off the rod is pretty much out of the question, but these points will be the points about which the rods will try to pivot.

55. This surface area does not require an integral, but here is one that works.
$x = 2\cos t, \quad y = 2\sin t, 0 \le t \le \pi/2$

$$ds = \sqrt{[x'(t)]^2 + [y'(t)]^2} \, dt$$

$$= \sqrt{(-2\sin t)^2 + (2\cos t)^2} \, dt = 2 \, dt$$

$$S = \int_C (\text{height})\, ds$$

$$= \int_C (x^2 + y^2)\, ds$$

$$= \int_0^{\pi/2} (4\cos^2 t + 4\sin^2 t)(2)\, dt$$

$$= 8\left(\frac{\pi}{2}\right) = 4\pi$$

57. $x = 2 - 4t, \quad y = 0,\ 0 \le t \le 1$

$$ds = \sqrt{[x'(t)]^2 + [y'(t)]^2}\, dt$$

$$= \sqrt{(-4)^2 + (0)^2}\, dt = 4\, dt$$

$$S = \int_C (\text{height})\, ds$$

$$= \int_C (4 - x^2 - y^2)\, ds$$

$$= \int_0^1 [4 - (2 - 4t)^2] 4\, dt$$

$$= 64 \int_0^1 (t - t^2)\, dt$$

$$= 64\left(\frac{1}{6}\right) = \frac{32}{3}$$

59. Again, no calculus is needed here—the surfaces in question are trapezoids. But, we will do this with calculus and let you check the work finding areas of these trapezoids.

The unit square is made of four sides, we will compute the line integral along each of these.

$C_1 : (0,0)$ to $(1,0)$, $ds = dx$

$$\int_{C_1} (\text{height})\, ds = \int_0^1 (4 - x - 0)\, dx = \frac{7}{2}$$

$C_2 : (1,0)$ to $(1,1)$, $ds = dy$

$$\int_{C_2} (\text{height})\, ds = \int_0^1 (4 - 1 - y)\, dx = \frac{5}{2}$$

$C_3 : (1,1)$ to $(0,1)$, $ds = -dx$

$$\int_{C_3} (\text{height})\, ds = \int_1^0 (4 - x - 1)(-1)\, dx = \frac{5}{2}$$

$C_4 : (0,1)$ to $(0,0)$, $ds = -dy$

$$\int_{C_4} (\text{height})\, ds = \int_1^0 (4 - 0 - y)(-1)\, dy = \frac{7}{2}$$

$$\int_C (\text{height})\, ds$$

$$= \int_{C_1} (\text{height})\, ds + \int_{C_2} (\text{height})\, ds$$

$$+ \int_{C_3} (\text{height})\, ds + \int_{C_4} (\text{height})\, ds$$

$$= \frac{7}{2} + \frac{5}{2} + \frac{5}{2} + \frac{7}{2} = 12$$

61. To make the estimate, we will essentially assume that the curve between points is a line segments. This will slightly underestimate the line integral.

For each line segment, let f^* denote the average of the function on the two endpoints.

	Δx	Δy	Δs	f^*
$(0,0) \to (1,0)$	1	0	1	2.5
$(1,0) \to (1,1)$	0	1	1	3.3
$(1,1) \to (1.5,1.5)$	0.5	0.5	$\frac{1}{\sqrt{2}}$	4
$(1.5,1.5) \to (2,2)$	0.5	0.5	$\frac{1}{\sqrt{2}}$	4.7
$(2,2) \to (3,2)$	1	0	1	4.5
$(3,2) \to (4,1)$	1	-1	$\sqrt{2}$	4

(a)
$$\int_C f\, ds \approx \sum_{i=1}^{6} f_i^* \Delta s_i$$

$$= (2.5)(1) + (3.3)(1)$$
$$+ (4.0)(0.7071) + (4.7)(0.7071)$$
$$+ (4.5)(1) + (4.0)(1.414)$$
$$\approx 22.11$$

(b)
$$\int_C f\, dx \approx \sum_{i=1}^{6} f_i^* \Delta x_i$$

$$= (2.5)(1) + (3.3)(0)$$
$$+ (4.0)(0.5) + (4.7)(0.5)$$
$$+ (4.5)(1) + (4.0)(1)$$
$$\approx 15.35$$

(c)
$$\int_C f\, dy \approx \sum_{i=1}^{6} f_i^* \Delta y_i$$
$$= (2.5)(0) + (3.3)(1)$$
$$+ (4.0)(0.5) + (4.7)(0.5)$$
$$+ (4.5)(0) + (4.0)(-1)$$
$$\approx 3.65$$

63. We follow the proof for Theorem 2.1, but "upgrade" it to curves in three-space.

We start with a curve in three dimensions. From Definition 2.1, we have

$$\int_C f(x,y,z)\, ds = \lim_{\|P\|\to 0} \sum_{i=1}^{n} f(x_i^*, y_i^*, z_i^*)\Delta s_i$$

where Δs_i represents the arc length of the section of the curve C between $(x_{i-1}, y_{i-1}, z_{i-1})$ and (x_i, y_i, z_i). Choose $t_0, t_1, \ldots, t_n$ so that $x(t_i) = x_i$, $y(t_i) = y_i$, and $z(t_i) = z_i$, for $i = 0, 1, \ldots, n$. We approximate the arc length of a section of the curve by the straight-line distance:

$$\Delta s_i \approx \sqrt{(x_i - x_{i-1})^2 + (y_i - y_{i-1})^2 + (z_i - z_{i-1})^2}$$

Since $x(t)$, $y(t)$ and $z(t)$ have continuous first derivatives, we have by the Mean Value Theorem:

$$\Delta s_i \approx \sqrt{(x_i - x_{i-1})^2 + (y_i - y_{i-1})^2 + (z_i - z_{i-1})^2}$$
$$\approx \sqrt{[x'(t_i^*)]^2 + [y'(t_i^*)]^2 + [z'(t_i^*)]^2}\Delta t_i$$

for some $t_i^* \in (t_{i-1}, t_i)$. Putting this together with Definition 2.1 gives

$$\int_C f(x,y,z)\, ds = \lim_{\|P\|\to 0} \sum_{i=1}^{n} f(x_i^*, y_i^*, z_i^*)\Delta s_i$$
$$= \lim_{\|P\|\to 0} \sum_{i=1}^{n} f(x_i^*, y_i^*, z_i^*)$$

$$\sqrt{[x'(t_i^*)]^2 + [y'(t_i^*)]^2 + [z'(t_i^*)]^2}\Delta t_i$$
$$= \int_a^b f(x,y,z)$$
$$\sqrt{[x'(t)]^2 + [y'(t)]^2 + [z'(t)]^2}\, dt$$

where the last equality is because we recognize the sum as a Riemann sum.

65. This is essentially done when arc length is described in Section 5.4. The main point is recognizing ds as the arc length element:

$$s = \lim_{\|P\|\to 0} \sum_{i=1}^{n} \sqrt{[x'(t_i)]^2 + [y'(t_i)]^2}\Delta t_i$$
$$= \int_a^b ds.$$

67. For **(i)**, note that the definition of the line integral is

$$\int_C f(x,y,z)\, dx$$
$$= \lim_{\|P\|\to 0} \sum_{i=1}^{n} f(x_i^*, y_i^*, z_i^*)\Delta x_i.$$

Of course, a partition for C gives a partition for $-C$. The only difference will be the sign of the Δx_i, which will be positive for C and negative for the curve $-C$, giving the result.

For **(ii)**, note that if we are given a partition of C, we can add the endpoints of each C_i to a finer partition of C. Then, this partition for C will give partitions for each of the curves C_i. Thus, when written as a Riemann sum, we will have

$$\int_C f(x,y,z)\, dx$$
$$= \lim_{\|P\|\to 0} \sum_{i=1}^{m} f(x_i^*, y_i^*, z_i^*)\Delta x_i$$
$$= \lim_{\|P_1\|\to 0} \sum_{i=1}^{m_1} f(x_i^*, y_i^*, z_i^*)\Delta x_i$$

$$+ \lim_{\|P_2\|\to 0} \sum_{i=1}^{m_2} f(x_i^*, y_i^*, z_i^*)\Delta x_i$$

$$+ \cdots +$$

$$+ \lim_{\|P_n\|\to 0} \sum_{i=1}^{m_n} f(x_i^*, y_i^*, z_i^*)\Delta x_i$$

$$= \int_{C_1} f(x,y,z)\,dx + \int_{C_2} f(x,y,z)\,dx$$

$$+ \cdots + \int_{C_n} f(x,y,z)\,dx$$

69. Recall that **n** is a unit vector.

There is a small issue here. The direction of the unit tangent vector is determined by the direction the curve is traversed.

$$\mathbf{T}(t) = \frac{\langle x'(t), y'(t)\rangle}{\sqrt{x'(t)^2 + y'(t)^2}}$$

Given a unit tangent vector in two dimensional space, there are two choices for the unit normal vector. We will choose to obtain **n** by rotating the unit tangent vector by $-\frac{\pi}{2}$ (so the vector rotates clockwise).

$$\mathbf{T}(t) = \frac{\langle y'(t), -x'(t)\rangle}{\sqrt{x'(t)^2 + y'(t)^2}}$$

Then, $ds = \sqrt{x'(t)^2 + y'(t)^2}\,dt$ and

$$\int_C \mathbf{F}\cdot\mathbf{T}\,ds = \int_C F_1\,dx + F_2\,dy$$

$$\int_C \mathbf{F}\cdot\mathbf{n}\,ds = \int_C F_1\,dy - F_2\,dx$$

15.3 INDEPENDENCE OF PATH AND CONSERVATIVE VECTOR FIELDS

1. $M_y = 2x$

$N_x = 2x$

Since $M_y = N_x$, the vector field is conservative.

If $\nabla f(x,y) = \mathbf{F}$, then
$\frac{\partial f}{\partial x} = 2xy - 1$ and $\frac{\partial f}{\partial y} = x^2$.

$$f(x,y) = \int (2xy - 1)\,dx = x^2 y - x + g(y)$$

$$\frac{\partial f}{\partial y} = x^2 + g'(y) = x^2$$

$$g'(y) = 0$$
$$g(y) = c$$
$$f(x,y) = x^2 y - x + c$$

3. $M_y = -\frac{1}{y^2}$

$N_x = -\frac{1}{y^2}$

Since $M_y = N_x$, the vector field is conservative.

If $\nabla f(x,y) = \mathbf{F}$, then $\frac{\partial f}{\partial x} = \frac{1}{y} - 2x$ and $\frac{\partial f}{\partial y} = y - \frac{x}{y^2}$

$$f(x,y) = \int\left(\frac{1}{y} - 2x\right)dx = \frac{x}{y} - x^2 + g(y)$$

$$\frac{\partial f}{\partial y} = -\frac{x}{y^2} + g'(y) = y - \frac{x}{y^2}$$

$$g'(y) = y$$
$$g(y) = \frac{y^2}{2} + c$$
$$f(x,y) = \frac{x}{y} - x^2 + \frac{y^2}{2} + c$$

5. $M_y = xe^{xy}$

$N_x = xye^{xy} + e^{xy}$

Since $M_y \neq N_x$, the vector field is not conservative.

7. $M_y = xye^{xy} + e^{xy}$

$N_x = xye^{xy} + e^{xy}$

Since $M_y = N_x$, the vector field is conservative.

If $\nabla f(x,y) = \mathbf{F}$, then $\frac{\partial f}{\partial x} = ye^{xy}$ and $\frac{\partial f}{\partial y} = xe^{xy} + \cos y$.

$$f(x, y) = \int ye^{xy}\,dx = e^{xy} + g(y)$$

$$\frac{\partial f}{\partial y} = xe^{xy} + g'(y) = xe^{xy} + \cos y$$

$$g'(y) = \cos y$$

$$g(y) = \sin y + c$$

$$f(x, y) = e^{xy} + \sin y + c$$

9. If $\nabla f(x, y) = \mathbf{F}$, then
$$\frac{\partial f}{\partial x} = z^2 + 2xy, \ \frac{\partial f}{\partial y} = x^2 + 1, \ \frac{\partial f}{\partial z} = 2xz - 3.$$

$$f(x, y, z) = \int (z^2 + 2xy)\,dx$$

$$= xz^2 + x^2y + g(y, z)$$

$$\frac{\partial f}{\partial y} = x^2 + \frac{\partial g}{\partial y} = x^2 + 1$$

$$\frac{\partial g}{\partial y} = 1$$

$$g(y, z) = \int 1\,dy = y + h(z)$$

$$f(x, y, z) = xz^2 + x^2y + y + h(z)$$

$$\frac{\partial f}{\partial z} = 2xz + h'(z) = 2xz - 3$$

$$h'(z) = -3$$

$$h(z) = -3z + c$$

$$f(x, y, z) = xz^2 + x^2y + y - 3z + c$$

Since f is a potential function for $\mathbf{F}$, $\mathbf{F}$ is conservative.

11. If $\nabla f(x, y) = \mathbf{F}$, then
$$\frac{\partial f}{\partial x} = y^2z^2 + x, \ \frac{\partial f}{\partial y} = y + 2xyz^2,$$
$$\frac{\partial f}{\partial z} = 2xy^2z.$$

$$f(x, y, z) = \int (y^2z^2 + x)\,dx$$

$$= xy^2z^2 + \frac{x^2}{2} + g(y, z)$$

$$\frac{\partial f}{\partial y} = 2xyz^2 + \frac{\partial g}{\partial y} = y + 2xyz^2$$

$$\frac{\partial g}{\partial y} = y$$

$$g(y, z) = \int y\,dy = \frac{y^2}{2} + h(z)$$

$$f(x, y, z) = xy^2z^2 + \frac{x^2}{2} + \frac{y^2}{2} + h(z)$$

$$\frac{\partial f}{\partial z} = 2xy^2z + h'(z) = 2xy^2z$$

$$h'(z) = 0$$

$$h(z) = c$$

$$f(x, y, z) = xy^2z^2 + \frac{x^2}{2} + \frac{y^2}{2} + c$$

Since f is a potential function for $\mathbf{F}$, $\mathbf{F}$ is conservative.

13. $M_y = 2x$

$N_x = 2x$

Since $M_y = N_x$, the vector field is conservative and the integral is independent of path.

The integral is $\displaystyle\int_C \mathbf{F} \cdot d\mathbf{r}$

where $\mathbf{F} = \langle 2xy, x^2 - 1 \rangle$.

A potential function is

$$f(x, y) = x^2y - y,$$

so the line integral is independent of path.

$$\int_C \mathbf{F} \cdot d\mathbf{r} = \left[x^2y - y \right]_{(1,0)}^{(3,1)}$$

$$= (8) - (0) = 8$$

15. $M_y = (1 + xy)e^{xy}$

$N_x = (1 + xy)e^{xy}$

Since $M_y = N_x$, the vector field is conservative and the integral is independent of path.

The integral is $\displaystyle\int_C \mathbf{F} \cdot d\mathbf{r}$ where

$\mathbf{F} = \langle ye^{xy}, xe^{xy} - 2y \rangle$.

A potential function is

$$f(x, y) = e^{xy} - y^2,$$

so the line integral is independent of path.

$$\int_C \mathbf{F} \cdot d\mathbf{r} = \left[e^{xy} - y^2 \right]_{(1,0)}^{(0,4)}$$
$$= (-15) - (1) = -16$$

17. The integral is $\int_C \mathbf{F} \cdot d\mathbf{r}$ where
$$\mathbf{F} = \left\langle z^2 + 2xy, \, x^2, \, 2xz \right\rangle.$$
This vector field is very similar to the field in exercise 9.

A potential function is
$f(x, y) = xz^2 + x^2y,$
so the line integral is independent of path.
$$\int_C \mathbf{F} \cdot d\mathbf{r} = \left[xz^2 + x^2y \right]_{(2,1,3)}^{(4,-1,0)}$$
$$= (-16) - (22) = -38$$

19. A potential function is
$f(x, y) = \frac{x^3}{3} + x + \frac{y^4}{4} - \frac{3}{2}y^2 + 2y,$
so the line integral is independent of path.
$$\int_C \mathbf{F} \cdot d\mathbf{r} = \left[\frac{x^3}{3} + x + \frac{y^4}{4} - \frac{3}{2}y^2 + 2y \right]_{(-4,0)}^{(4,0)}$$
$$= \left(\frac{76}{3} \right) - \left(-\frac{76}{3} \right) = \frac{152}{3}$$

21. A potential function is
$f(x, y) = \frac{1}{3}(x^3 + y^3 + z^3),$
so the line integral is independent of path.
$$\int_C \mathbf{F} \cdot d\mathbf{r} = \left[\frac{1}{3}(x^3 + y^3 + z^3) \right]_{(1,4,-3)}^{(1,4,3)}$$
$$= \frac{92}{3} - \frac{38}{3} = 18$$

23. A potential function is
$f(x, y) = r = \sqrt{x^2 + y^2 + z^2},$
so the line integral is independent of path.
$$\int_C \mathbf{F} \cdot d\mathbf{r} = \left[\sqrt{x^2 + y^2 + z^2} \right]_{(1,3,2)}^{(2,1,5)}$$
$$= \sqrt{30} - \sqrt{14}$$

25. $M_y = 3x^2$ and $N_x = 3y^2$, so the vector field is not conservative and the line integral is not independent of path.

$C : x = \cos t, \, y = -\sin t, \, 0 \le t \le \pi$
$$\int_C \mathbf{F} \cdot d\mathbf{r} = \int_C (3x^2y + 1)\, dx + 3xy^2\, dy$$
$$= \int_0^\pi \Big[3(\cos^2 t)(-\sin t)(-\sin t) + 1(-\sin t)$$
$$+ 3(\cos t)(\sin^2 t)(-\cos t) \Big]\, dt$$
$$= \int_0^\pi -\sin t\, dt = -2$$

27. A potential function is
$f(x, y) = e^{xy^2} - xy - y,$
so the line integral is independent of path.
$$\int_C \mathbf{F} \cdot d\mathbf{r} = \left[e^{xy^2} - xy - y \right]_{(2,3)}^{(3,0)}$$
$$= 1 - (e^{18} - 9) = 10 - e^{18}$$

29. A potential function is
$f(x, y) = \frac{x}{y} - \frac{1}{2}e^{2x} + y^2,$
so the line integral is independent of path.
Since C is a closed curve,
$\int_C \mathbf{F} \cdot d\mathbf{r} = 0.$

31. Since $\mathbf{F}$ is constant, $M_y = N_x = 0$. The vector field is conservative.

33. Consider $\int_C \mathbf{F} \cdot d\mathbf{r} = 0$ where C is a circle centered at the origin. This is a closed curve but the line integral is clearly not zero. The vector field is not conservative.

35. Taking a rectangular path counterclockwise around the entire visible part of the picture, it looks as though the top and bottom halves will cancel. But, the left edge (downward) and the right edge (upward) will support each other rather than cancel each other. Therefore, the field is not conservative.

37. $C_1 : x = t, \, y = 0, \, -2 \le t \le 2$
$$\int_{C_1} y\, dx - x\, dy = \int_{-2}^2 [0(1) - t(0)]\, dt$$
$$= \int_{-2}^2 (0)\, dt = 0$$

$C_2 : 2\cos t,\ y = 2\sin t,\ \pi \le t \le 2\pi$

$\int_{C_2} y\,dx - x\,dy$

$= \int_{\pi}^{2\pi} [2\sin t(-2\sin t) - (2\cos t)(2\cos t)]\,dt$

$= \int_{\pi}^{2\pi} -4\,dt = -4\pi$

The two paths from $(-2, 0)$ to $(2, 0)$ give different values, so the line integral is not independent of path.

39. $C_1 : x = -2 + 2t,\ y = 2 - 2t,\ 0 \le t \le 1$

$\int_{C_1} y\,dx - 3\,dy$

$= \int_0^1 [(2-2t)(2) - 3(-2)]\,dt$

$= \int_0^1 (10 - 4t)\,dt = 8$

$C_{2a} : x = t,\ y = 2,\ -2 \le t \le 0$
$C_{2b} : x = 0,\ y = 2 - t,\ 0 \le t \le 2$

$\int_{C_2} y\,dx - 3\,dy$

$= \left(\int_{C_{2a}} y\,dx - 3\,dy \right)$

$\quad + \left(\int_{C_{2b}} y\,dx - 3\,dy \right)$

$= \left(\int_{C_{2a}} [(2)(1) - 3(0)]\,dt \right)$

$\quad + \left(\int_{C_{2b}} [(2-t)(0) - 3(-1)]\,dt \right)$

$= \int_{-2}^0 2\,dt + \int_0^2 3\,dt$

$= 4 + 6 = 10$

The two paths from $(-2, 2)$ to $(0, 0)$ give different values, so the line integral is not independent of path.

41. False. This is only true if C is a closed curve.

43. True. See the boxed list titled "Conservative Vector Fields."

45. If $\nabla f(x, y) = \mathbf{F}$, then
$\frac{\partial f}{\partial x} = -\frac{y}{x^2+y^2},\ \frac{\partial f}{\partial y} = \frac{x}{x^2+y^2}.$

$f(x, y) = \int \frac{x}{x^2 + y^2}\,dy$

$= \tan^{-1}\frac{y}{x} + g(x)$

$\frac{\partial f}{\partial x} = -\frac{y}{x^2 + y^2} + \frac{\partial g}{\partial x}$

$= -\frac{y}{x^2 + y^2}$

$\frac{\partial g}{\partial x} = 0$

$f(x, y) = \tan^{-1}\frac{y}{x} + c$

This potential function is valid for $x > 0$.

$f(x, y) = \tan^{-1}\frac{y}{x}$ is often called the *polar angle*, θ. To the extent that the polar angle θ is ambiguously defined, $f = \theta$. The matter of identifying domains in which θ can be unambiguously and continuously defined is a delicate matter. Such a domain must be *simply connected* and must exclude the origin.

For the integral along the unit circle, let $x = \cos t,\ y = \sin t,\ 0 \le t \le 2\pi$, then

$\mathbf{F}(x(t), y(t)) = \langle -\sin t, \cos t \rangle$
and

$\int_C M\,dx + N\,dy$

$= \int_0^{2\pi} [(-\sin t)(-\sin t) + (\cos t)(\cos t)]\,dt$

$= \int_0^{2\pi} dt = 2\pi.$

The Fundamental Theorem for Line Integrals fails because the vector field is not continuous inside the circle.

Regarding the circle $K : (x - 2)^2 + (y - 3)^2 = 1$, which does lie entirely in the right half plane, the potential function is valid, and $\int_K M\,dx + N\,dy = 0$.

47. **(a)** A disk, simply connected.

(b) A ring (annulus), not simply connected.

49. $\mathbf{F} = \frac{kq}{r^3}\mathbf{r}$

Since $\nabla(r^n) = nr^{n-2}\mathbf{r}$,

$$\nabla\left(\frac{1}{r}\right) = -\frac{1}{r^3}\mathbf{r}.$$

(See, for example, exercise 48 of Section 15.1). This means that a potential function for $\mathbf{F}$ is $f(x, y, z) = -\frac{kq}{r}$ and

$$\int_{P_1}^{P_2} \mathbf{F} \cdot d\mathbf{r} = \left[-\frac{kq}{r}\right]_{P_1}^{P_2} = \frac{kq}{r_1} - \frac{kq}{r_2}.$$

51. $\displaystyle\int_{C_1} \frac{RT}{P}\,dP - R\,dT$

$$= \int_{T_1}^{T_2} -R\,dT + \int_{P_1}^{P_2} \frac{RT_2}{P}\,dP$$

$$= R(T_1 - T_2) + RT_2 \ln\left(\frac{P_2}{P_1}\right)$$

$\displaystyle\int_{C_2} \frac{RT}{P}\,dP - R\,dT$

$$= \int_{P_1}^{P_2} \frac{RT_1}{P}\,dP + \int_{T_1}^{T_2} -R\,dT +$$

$$= R(T_1 - T_2) + RT_1 \ln\left(\frac{P_2}{P_1}\right)$$

These are not generally equal unless $T_1 = T_2$, which means that the field

$$\left\langle \frac{RT}{P}, -R\right\rangle$$

is not conservative.

53. All the integrals in question are in fact zero. In fact, as shown in Theorem 3.1, $\int_{C_1} \mathbf{F}\cdot d\mathbf{r} = 2$ can not occur (it must be equal to 0).

This exercise can be compared to the polar angle (see exercise 45 this section and exercise 38 of Section 15.4). The difference is that

the polar angle can not be continuously and unambiguously defined in the entire region.

15.4 GREEN'S THEOREM

1. **(a)** $x = \cos t,\ y = \sin t,\ 0 \le t \le 2\pi$

$$\oint_C (x^2 - y)\,dx + y^2\,dy$$

$$= \int_0^{2\pi} [(\cos^2 t - \sin t)(-\sin t)$$

$$+ (\sin^2 t)(\cos t)]\,dt$$

$$= \int_0^{2\pi} (-\cos^2 t \sin t + \sin^2 t$$

$$+ \cos t \sin^2 t)\,dt = \pi$$

(b) $M = x^2 - y,\ N = y^2$

$$\oint_C M\,dx + N\,dy$$

$$= \iint_R \left(\frac{\partial N}{\partial x} - \frac{\partial M}{\partial y}\right)dA$$

$$= \iint_R [0 - (-1)]\,dA$$

$$= \iint_R dA$$

$$= (\text{Area of circle}) = \pi$$

3. Note that the curve is negatively oriented.

(a) Bottom: $(0, 0)$ to $(0, 2)$.
$x = 0,\ y = t,\ 0 \le t \le 2$

$$\oint_C x^2\,dx - x^3\,dy = \int_0^2 0\,dt = 0$$

Right: $(0, 2)$ to $(2, 2)$.
$x = t,\ y = 2,\ 0 \le t \le 2$

$$\oint_C x^2\,dx - x^3\,dy = \int_0^2 t^2\,dt = \frac{8}{3}$$

Top: $(2, 2)$ to $(2, 0)$.
$x = 2,\ y = 2 - t,\ 0 \le t \le 2$

$$\oint_C x^2\,dx - x^3\,dy$$

$$= \int_0^2 -8(-1)\,dt = 16$$

Left: $(2, 0)$ to $(0, 0)$.
$x = 2 - t,\ y = 0,\ 0 \le t \le 2$

$$\oint_C x^2\,dx - x^3\,dy$$

$$= \int_0^2 (2 - t)^2(-1)\,dt = -\frac{8}{3}$$

Summing up the integrals over the four subpaths, the line integral is 16.

(b) $M = x^2,\ N = -x^3$
Note that the curve is oriented negatively.

$$\oint_C M\,dx + N\,dy$$

$$= -\iint_R \left(\frac{\partial N}{\partial x} - \frac{\partial M}{\partial y}\right) dA$$

$$= -\iint_R [-3x^2 - 0]\,dA$$

$$= \int_0^2 \int_0^2 3x^2\,dy\,dx$$

$$= \int_0^2 6x^2\,dx = 16$$

5. $M = xe^{2x},\ N = -3x^2y$
Note that the curve has positive orientation.

$$\oint_C M\,dx + Ny$$

$$= \iint_R \left(\frac{\partial N}{\partial x} - \frac{\partial M}{\partial y}\right) dA$$

$$= \int_0^2 \int_0^3 (-6xy - 0)\,dx\,dy$$

$$= \int_0^2 -27y\,dy = -54$$

7. $M = \frac{x}{x^2+1} - y,\ N = 3x - 4\tan\frac{y}{2}$
Note that the curve has positive orientation.

$$\oint_C M\,dx + N\,dy$$

$$= \iint_R \left(\frac{\partial N}{\partial x} - \frac{\partial M}{\partial y}\right) dA$$

$$= \int_{-1}^1 \int_{x^2}^{2-x^2} (3 - (-1))\,dy\,dx$$

$$= \int_{-1}^1 4(2 - 2x^2)\,dx = \frac{32}{3}$$

9. $M = \tan x - y^3,\ N = x^3 - \sin y$
We will assume that the curve has positive orientation.

$$\oint_C M\,dx + N\,dy$$

$$= \iint_R \left(\frac{\partial N}{\partial x} - \frac{\partial M}{\partial y}\right) dA$$

$$= \iint_R (3x^2 + 3y^2)\,dA$$

$$= \int_0^{2\pi} \int_0^{\sqrt{2}} 3r^3\,dr\,d\theta$$

$$= \int_0^{2\pi} 3\,d\theta = 6\pi$$

11. $M = x^3 - y,\ N = x + y^3$
We will assume that the curve has positive orientation.

$$\oint_C M\,dx + N\,dy$$

$$= \iint_R \left(\frac{\partial N}{\partial x} - \frac{\partial M}{\partial y}\right) dA$$

$$= \int_0^1 \int_{x^2}^x (1 - (-1))\,dy\,dx$$

$$= \int_0^1 2(x - x^2)\,dx = \frac{1}{3}$$

13. $M = e^{x^2} - y,\ N = e^{2x} + y$
We will assume that the curve has positive orientation.

$$\oint_C M\,dx + N\,dy$$

$$= \iint_R \left(\frac{\partial N}{\partial x} - \frac{\partial M}{\partial y}\right) dA$$

$$= \int_{-1}^{1} \int_{0}^{1-x^2} (2e^{2x} - (-1))\,dy\,dx$$

$$= \int_{-1}^{1} (1 - x^2)(2e^{2x} + 1)\,dx$$

$$= \int_{-1}^{1} (2e^{2x} - 2x^2 e^{2x} + 1 - x^2)\,dx$$

$$= \frac{1}{2}e^2 + \frac{3}{2}e^{-2} + \frac{4}{3}$$

15. $M = y^3 - \ln(x+1)$, $N = \sqrt{y^2 + 1} + 3x$
We will assume that the curve has positive orientation.

$$\oint_C M\,dx + N\,dy$$

$$= \iint_R \left(\frac{\partial N}{\partial x} - \frac{\partial M}{\partial y}\right) dA$$

$$= \int_{-2}^{2} \int_{y^2}^{4} (3 - 3y^2)\,dx\,dy$$

$$= 3\int_{-2}^{2} (4 - y^2)(1 - y^2)\,dy$$

$$= 3\int_{-2}^{2} (y^4 - 5y^2 + 4)\,dy = \frac{32}{5}$$

17. Notice that in this case, the curve is in the plane $z = 2$ and $dz = 0$. This means that our integral becomes:

$$\oint_C x^2\,dx + 2x\,dy + (z - 2)\,dz$$

$$= \oint_C x^2\,dx + 2x\,dy$$

and the integral is actually independent of z and we can proceed as if the curve was in the xy-plane.
$M = x^2$, $N = 2x$
Note that the curve has positive orientation.

$$\oint_C M\,dx + N\,dy$$

$$= \iint_R \left(\frac{\partial N}{\partial x} - \frac{\partial M}{\partial y}\right) dA$$

$$= \iint_R (2)\,dA$$

$$= 2(\text{Area of Triangle}) = 2(2) = 4$$

19. Since the curve is in the plane $y = 0$, we can use
$$\mathbf{F} = \left\langle x^3,\ e^{x^2+z^2},\ x^2 \right\rangle.$$

Since $dy = 0$, the integral reduces to an integral in two dimensions (x and z):

$$\oint_C (x^3)\,dx + x^2\,dz$$

$$M = x^3,\ N = x^2$$

We will assume that the curve has positive orientation.

$$\oint_C M\,dx + N\,dy$$

$$= \iint_R \left(\frac{\partial N}{\partial x} - \frac{\partial M}{\partial z}\right) dA$$

$$= \iint_R 2x\,dA$$

$$= \int_0^{2\pi} \int_0^1 2r^2 \cos\theta\,dr\,d\theta$$

$$= \int_0^{2\pi} \frac{2}{3} \cos\theta\,d\theta = 0$$

21. $C: x = 2\cos t,\ y = 4\sin t,\ 0 \le t \le 2\pi$

$$A = \frac{1}{2}\oint_C x\,dy - y\,dx$$

$$= \frac{1}{2}\int_0^{2\pi} (8\cos^2 t + 8\sin^2 t)\,dt$$

$$= 4\int_0^{2\pi} 1\,dt = 8\pi$$

23. $C : x = \cos^3 t, \; y = \sin^3 t, \; 0 \le t \le 2\pi$

$$A = \frac{1}{2} \oint_C x \, dy - y \, dx$$

$$= \frac{1}{2} \int_0^{2\pi} [(\cos^3 t)(3 \sin^2 t \cos t)$$

$$+ (\sin^3 t)(3 \cos^2 t \sin t)] \, dt$$

$$= \frac{3}{2} \int_0^{2\pi} \sin^2 t \cos^2 t (\cos^2 t + \sin^2 t) \, dt$$

$$= \frac{3}{2} \int_0^{2\pi} \sin^2 t \cos^2 t \, dt = \frac{3}{2} \left(\frac{\pi}{4} \right) = \frac{3\pi}{8}$$

25. $C_1 : x = t, \; y = t^2, \; -2 \le t \le 2$
$C_2 : x = -t, \; y = 4, \; -2 \le t \le 2$

$$A = \frac{1}{2} \oint_{C_1} x \, dy - y \, dx + \frac{1}{2} \oint_{C_2} x \, dy - y \, dx$$

$$= \frac{1}{2} \int_{-2}^2 [t(2t) - t^2(1)] \, dt$$

$$+ \frac{1}{2} \int_{-2}^2 [(-t)(0) - (4)(-1)] \, dt$$

$$= \frac{1}{2} \int_{-2}^2 t^2 \, dt + \frac{1}{2} \int_{-2}^2 4 \, dt$$

$$= \frac{8}{3} + 8 = \frac{32}{3}$$

27. We apply Green's Theorem to the integrals in the problem:

$$\frac{1}{2A} \oint_C x^2 \, dy = \frac{1}{2A} \iint_R (2x - 0) \, dA$$

$$= \frac{1}{A} \iint_R x \, dA = \bar{x}$$

$$\frac{1}{2A} \oint_C -y^2 \, dx = \frac{1}{2A} \iint_R [0 - (-2y)] \, dA$$

$$= \frac{1}{A} \iint_R y \, dA = \bar{y}$$

29. Although the text gave the curve and the bounds for t, we don't know if the curve is negatively or positively oriented. You can check that there is symmetry ($x(t) = x(-t)$, $y(t) = y(-t)$) and no other self intersections other than $t = -1$ and $t = 1$.

This means that the integral below is either A or $-A$, depending on the orientation of the curve.

$$\pm A = \frac{1}{2} \oint_C x \, dy - y \, dx$$

$$= \frac{1}{2} \int_0^1 [(t^3 - t)(-2t) - (1 - t^2)(3t^2 - 1)] \, dt$$

$$= \frac{1}{2} \int_{-1}^1 (t^4 - 2t^2 + 1) \, dt = \frac{8}{15}$$

Therefore the curve is positively oriented and $A = \frac{8}{15}$.

$$\bar{x} = \frac{1}{2A} \oint_C x^2 \, dy = \frac{15}{16} \oint_{C_1} x^2 \, dy$$

$$= \frac{15}{16} \int_{-1}^1 (t^3 - t)^2 (-2t) \, dt$$

$$= \frac{15}{16} \int_{-1}^1 (-2t^7 + 4t^5 - 2t^3) \, dt = 0$$

$$\bar{y} = -\frac{1}{2A} \oint_C y^2 \, dx = -\frac{15}{16} \oint_{C_1} y^2 \, dx$$

$$= -\frac{15}{16} \int_{-1}^1 (1 - t^2)^2 (3t^2 - 1) \, dt$$

$$= -\frac{15}{16} \left(-\frac{64}{105} \right) = \frac{4}{7}$$

31. We apply Green's Theorem twice: first in the xy-plane and later in the uv-plane.

There will be a sign ambiguity because even if the curve C is positively oriented in the xy-plane the corresponding curve C^* in the uv-plane could be either positively or negatively oriented. Let R be the region in the xy-plane enclosed by C and let S be the region enclosed by C^* in the uv-plane.

$$\text{Area}(R) = \iint_R dA_{xy} = \oint_C x \, dy$$

$$= \oint_{C^*} x \left(\frac{\partial y}{\partial u} \, du + \frac{\partial y}{\partial v} \, dv \right)$$

$$= \pm \iint_S \left[\frac{\partial}{\partial v} \left(x \frac{\partial y}{\partial u} \right) - \frac{\partial}{\partial u} \left(x \frac{\partial y}{\partial v} \right) \right] dA_{uv}$$

$$= \pm \iint_S \left[\left(\frac{\partial x}{\partial v} \frac{\partial y}{\partial u} + x \frac{\partial^2 y}{\partial u \partial v} \right) \right.$$

$$\left. - \left(\frac{\partial x}{\partial u} \frac{\partial y}{\partial v} + x \frac{\partial^2 y}{\partial v \partial u} \right) \right] dA_{uv}$$

$$= \pm \iint_S \left[\frac{\partial x}{\partial v} \frac{\partial y}{\partial u} - \frac{\partial x}{\partial u} \frac{\partial y}{\partial v} \right] dA_{uv}$$

$$= \iint_S \left[\pm \frac{\partial(x, y)}{\partial(u, v)} \right] dA_{uv}$$

Since the starting sign is positive (it is an area), the choice of sign which works is the one which makes the final integrand positive. In any case, the final integrand will be the absolute value:

$$\iint_R dA_{xy} = \iint_S \left| \frac{\partial(x, y)}{\partial(u, v)} \right| dA_{uv}.$$

But, note that we have actually proved more. If the Jacobian is negative then orientation of the curve must have also changed—the transformation is orientation reversing if and only if the Jacobian is negative.

33. Let $M = \dfrac{x}{x^2 + y^2}$ and $N = \dfrac{y}{x^2 + y^2}$.

$$\frac{\partial N}{\partial x} = -\frac{2xy}{(x^2 + y^2)^2}$$

$$\frac{\partial M}{\partial y} = -\frac{2xy}{(x^2 + y^2)^2} = \frac{\partial N}{\partial x}$$

Define C, C_1, and R as in Example 4.5. Then,

$$\oint_C \mathbf{F} \cdot d\mathbf{r} - \oint_{C_1} \mathbf{F} \cdot d\mathbf{r}$$

$$= \iint_R \left(\frac{\partial N}{\partial x} - \frac{\partial M}{\partial y} \right) dA$$

$$= \iint_R 0 \, da = 0.$$

Therefore, the line integral has the same value on all positively oriented simple closed curves containing the origin. So, we may assume that C is $x = \cos t$, $y = \sin t$, $0 \le t \le 2\pi$.

$$\oint_C \mathbf{F} \cdot d\mathbf{r}$$

$$= \oint_C \frac{x}{x^2 + y^2} \, dx + \frac{y}{x^2 + y^2} \, dy$$

$$= \int_0^{2\pi} [(\cos t)(-\sin t) + (\sin t)(\cos t) \, dt$$

$$= \int_0^{2\pi} 0 \, dt = 0$$

But, there is another reason that this integral is zero: this field has a potential function $\phi = \dfrac{1}{2} \ln(x^2 + y^2)$.

Note that this is well defined for all $(x, y) \ne (0, 0)$. This is related to exercise 53 of Section 15.3.

35. Let $M = \dfrac{x^3}{x^4 + y^4}$ and $N = \dfrac{y^3}{x^4 + y^4}$.

$$\frac{\partial N}{\partial x} = -\frac{4x^3 y^3}{(x^4 + y^4)^2}$$

$$\frac{\partial M}{\partial y} = -\frac{4x^3 y^3}{(x^4 + y^4)^2}$$

$$= \frac{\partial N}{\partial x}$$

Define C, C_1 and R as in Example 4.5. Then,

$$\oint_C \mathbf{F} \cdot d\mathbf{r} - \oint_{C_1} \mathbf{F} \cdot d\mathbf{r}$$

$$= \iint_R \left(\frac{\partial N}{\partial x} - \frac{\partial M}{\partial y} \right) dA$$

$$= \iint_R 0 \, da = 0.$$

Therefore, the line integral has the same value on all positively oriented simple closed curves containing the origin. So, we may assume that C is $x = \cos t$, $y = \sin t$, $0 \le t \le 2\pi$.

$$\oint_C \mathbf{F} \cdot d\mathbf{r}$$

$$= \oint_C \frac{x^3}{x^4 + y^4}\, dx + \frac{y^3}{x^4 + y^4}\, dy$$

$$= \int_0^{2\pi} \frac{(\cos^3 t)(-\sin t) + (\sin^3 t)(\cos t)}{\cos^4 t + \sin^4 t}\, dt$$

$$= \int_0^{2\pi} \frac{(\sin^2 t - \cos^2 t)(\cos t \sin t)}{1 - 2\cos^2 t \sin^2 t}\, dt$$

$$= \int_0^{2\pi} \frac{-\cos 2t \sin 2t}{2\left(1 - \frac{\sin^2 2t}{2}\right)}\, dt$$

$$= -\frac{1}{2}\int_0^{4\pi} \frac{\cos u \sin u}{2 - \sin^2 u}\, du \qquad (u = 2t)$$

$$= \frac{1}{4}\int_2^2 \frac{dv}{v}\, dv = 0 \qquad (v = 2 - \sin^2 u)$$

37. The vector field in question is defined and differentiable everywhere except at the origin. Green's Theorem says nothing about the integral $\oint_C \mathbf{F} \cdot d\mathbf{r}$ where C encloses the origin. But, this vector field is double the vector field in exercise 33 and exercise 33 shows what we can say.

39. Yes, Green's Theorem will apply here and the integral will be equal to zero.

15.5 CURL AND DIVERGENCE

1. $\operatorname{curl} \mathbf{F} = \nabla \times \mathbf{F}$

$$= \begin{vmatrix} \mathbf{i} & \mathbf{j} & \mathbf{k} \\ \frac{\partial}{\partial x} & \frac{\partial}{\partial y} & \frac{\partial}{\partial z} \\ x^2 & -3xy & 0 \end{vmatrix}$$

$$= (0 - 0)\mathbf{i} - (0 - 0)\mathbf{j} + (3y - 0)\mathbf{k}$$

$$= \langle 0, 0, -3y \rangle$$

$\operatorname{div} \mathbf{F} = \nabla \cdot \mathbf{F}$

$$= \left\langle \frac{\partial}{\partial x}, \frac{\partial}{\partial y}, \frac{\partial}{\partial z}\right\rangle \cdot \left\langle x^2, -3xy, 0\right\rangle$$

$$= 2x - 3x = -x$$

3. $\operatorname{curl} \mathbf{F} = \nabla \times \mathbf{F}$

$$= \begin{vmatrix} \mathbf{i} & \mathbf{j} & \mathbf{k} \\ \frac{\partial}{\partial x} & \frac{\partial}{\partial y} & \frac{\partial}{\partial z} \\ 2xz & 0 & -3y \end{vmatrix}$$

$$= (-3 - 0)\mathbf{i} - (0 - 2x)\mathbf{j} + (0 - 0)\mathbf{k}$$

$$= \langle -3, 2x, 0 \rangle$$

$\operatorname{div} \mathbf{F} = \nabla \cdot \mathbf{F}$

$$= \left\langle \frac{\partial}{\partial x}, \frac{\partial}{\partial y}, \frac{\partial}{\partial z}\right\rangle \cdot \langle 2xz, 0, -3y \rangle$$

$$= 2z + 0 + 0 = 2z$$

5. $\operatorname{curl} \mathbf{F} = \nabla \times \mathbf{F}$

$$= \begin{vmatrix} \mathbf{i} & \mathbf{j} & \mathbf{k} \\ \frac{\partial}{\partial x} & \frac{\partial}{\partial y} & \frac{\partial}{\partial z} \\ xy & yz & x^2 \end{vmatrix}$$

$$= (0 - y)\mathbf{i} - (2x - 0)\mathbf{j} + (0 - x)\mathbf{k}$$

$$= \langle -y, -2x, -x \rangle$$

$\operatorname{div} \mathbf{F} = \nabla \cdot \mathbf{F}$

$$= \left\langle \frac{\partial}{\partial x}, \frac{\partial}{\partial y}, \frac{\partial}{\partial z}\right\rangle \cdot \left\langle xy, yz, x^2\right\rangle$$

$$= y + z + 0 = y + z$$

7. $\operatorname{curl} \mathbf{F} = \nabla \times \mathbf{F}$

$$= \begin{vmatrix} \mathbf{i} & \mathbf{j} & \mathbf{k} \\ \frac{\partial}{\partial x} & \frac{\partial}{\partial y} & \frac{\partial}{\partial z} \\ x^2 & y - z & xe^y \end{vmatrix}$$

$$= (xe^y + 1)\mathbf{i} - (e^y - 0)\mathbf{j} + (0 - 0)\mathbf{k}$$

$$= \langle xe^y + 1, -e^y, 0 \rangle$$

$\operatorname{div} \mathbf{F} = \nabla \cdot \mathbf{F}$

$$= \left\langle \frac{\partial}{\partial x}, \frac{\partial}{\partial y}, \frac{\partial}{\partial z}\right\rangle \cdot \left\langle x^2, y - z, xe^y\right\rangle$$

$$= 2x + 1 + 0 = 2x + 1$$

9. $\operatorname{curl} \mathbf{F} = \nabla \times \mathbf{F}$

$$= \begin{vmatrix} \mathbf{i} & \mathbf{j} & \mathbf{k} \\ \frac{\partial}{\partial x} & \frac{\partial}{\partial y} & \frac{\partial}{\partial z} \\ 3yz & x^2 & x\cos y \end{vmatrix}$$

$$= (-x\sin y)\mathbf{i} - (\cos y - 3y)\mathbf{j} + (2x - 3z)\mathbf{k}$$

$$= \langle -x\sin y, 3y - \cos y, 2x - 3z \rangle$$

$$\text{div } \mathbf{F} = \nabla \cdot \mathbf{F}$$

$$= \left\langle \frac{\partial}{\partial x}, \frac{\partial}{\partial y}, \frac{\partial}{\partial z} \right\rangle \cdot \left\langle 3yz, x^2, x \cos y \right\rangle$$

$$= 0 + 0 + 0 = 0$$

11. $\text{curl } \mathbf{F} = \nabla \times \mathbf{F}$

$$= \begin{vmatrix} \mathbf{i} & \mathbf{j} & \mathbf{k} \\ \frac{\partial}{\partial x} & \frac{\partial}{\partial y} & \frac{\partial}{\partial z} \\ 2xz & y + z^2 & zy^2 \end{vmatrix}$$

$$= (2yz - 2z)\,\mathbf{i} - (0 - 2x)\,\mathbf{j} + (0 - 0)\,\mathbf{k}$$

$$= \langle 2yz - 2z, 2x, 0 \rangle$$

$$\text{div } \mathbf{F} = \nabla \cdot \mathbf{F}$$

$$= \left\langle \frac{\partial}{\partial x}, \frac{\partial}{\partial y}, \frac{\partial}{\partial z} \right\rangle \cdot \left\langle 2xz, y + z^2, zy^2 \right\rangle$$

$$= 2z + 1 + y^2$$

13. $\text{curl } \mathbf{F} = \nabla \times \mathbf{F}$

$$= \begin{vmatrix} \mathbf{i} & \mathbf{j} & \mathbf{k} \\ \frac{\partial}{\partial x} & \frac{\partial}{\partial y} & \frac{\partial}{\partial z} \\ 2x & 2yz^2 & 2y^2z \end{vmatrix}$$

$$= (4yz - 4yz)\,\mathbf{i} - (0)\,\mathbf{j} + (0)\,\mathbf{k}$$

$$= \langle 0, 0, 0 \rangle$$

Since the components of **F** have continuous partial derivatives throughout $\mathbb{R}^3$ and curl $\mathbf{F} = \mathbf{0}$, **F** is conservative.

$$\text{div } \mathbf{F} = \nabla \cdot \mathbf{F}$$

$$= \left\langle \frac{\partial}{\partial x}, \frac{\partial}{\partial y}, \frac{\partial}{\partial z} \right\rangle \cdot \left\langle 2x, 2yz^2, 2y^2z \right\rangle$$

$$= 2 + 2z^2 + 2y^2$$

Since div $\mathbf{F} \neq 0$, **F** is not incompressible.

15. $\text{curl } \mathbf{F} = \nabla \times \mathbf{F}$

$$= \begin{vmatrix} \mathbf{i} & \mathbf{j} & \mathbf{k} \\ \frac{\partial}{\partial x} & \frac{\partial}{\partial y} & \frac{\partial}{\partial z} \\ 3yz & x^2 & x \cos y \end{vmatrix}$$

$$= (-x \sin y - 0)\,\mathbf{i} - (\cos y - 3y)\,\mathbf{j}$$
$$+ (2x - 3z)\,\mathbf{k}$$

$$= \langle -x \sin y, 3y - \cos y, 2x - 3z \rangle$$

Since curl $\mathbf{F} \neq \mathbf{0}$, **F** is not conservative.

$$\text{div } \mathbf{F} = \nabla \cdot \mathbf{F}$$

$$= \left\langle \frac{\partial}{\partial x}, \frac{\partial}{\partial y}, \frac{\partial}{\partial z} \right\rangle \cdot \left\langle 3yz, x^2, x \cos y \right\rangle$$

$$= 0 + 0 + 0$$

Since div $\mathbf{F} = 0$, **F** is incompressible.

17. $\text{curl } \mathbf{F} = \nabla \times \mathbf{F}$

$$= \begin{vmatrix} \mathbf{i} & \mathbf{j} & \mathbf{k} \\ \frac{\partial}{\partial x} & \frac{\partial}{\partial y} & \frac{\partial}{\partial z} \\ \sin z & e^{yz^2}z^2 & 2e^{yz^2}yz + x \cos z \end{vmatrix}$$

$$= \left(2ze^{yz^2} + 2yz^3e^{yz^2} - 2ze^{yz^2} \right.$$
$$\left. -2yz^3e^{yz^2} \right) \mathbf{i}$$

$$- (\cos z - \cos z)\,\mathbf{j} + (0 - 0)\,\mathbf{k}$$

$$= \langle 0, 0, 0 \rangle$$

Since the components of **F** have continuous partial derivatives throughout $\mathbb{R}^3$ and curl $\mathbf{F} = \mathbf{0}$, **F** is conservative.

$$\text{div } \mathbf{F} = \nabla \cdot \mathbf{F}$$

$$= \left\langle \frac{\partial}{\partial x}, \frac{\partial}{\partial y}, \frac{\partial}{\partial z} \right\rangle \cdot$$

$$\left\langle \sin z, e^{yz^2}z^2, 2e^{yz^2}yz + x \cos z \right\rangle$$

$$= 0 + z^4e^{yz^2} + 2(ye^{yz^2} + 2y^2z^2e^{yz^2})$$
$$- x \sin z$$

$$= z^4e^{yz^2} + 2ye^{yz^2} + 4y^2z^2e^{yz^2} - x \sin z$$

Since div $\mathbf{F} \neq 0$, **F** is not incompressible.

19. $\text{curl } \mathbf{F} = \nabla \times \mathbf{F}$

$$= \begin{vmatrix} \mathbf{i} & \mathbf{j} & \mathbf{k} \\ \frac{\partial}{\partial x} & \frac{\partial}{\partial y} & \frac{\partial}{\partial z} \\ z^2 - 3e^{3x}y & z^2 - e^{3x} & 2z\sqrt{xy} \end{vmatrix}$$

$$= \left(\frac{xz}{\sqrt{xy}} - 2z \right) \mathbf{i}$$

$$- \left(\frac{yz}{\sqrt{xy}} - 2z \right) \mathbf{j}$$

$$+ \left(-3e^{3x} + 3e^{3x} \right) \mathbf{k}$$

$$= \left\langle -2z + \frac{xz}{\sqrt{xy}}, \, 2z - \frac{yz}{\sqrt{xy}}, \, 0 \right\rangle$$

Since $\text{curl } \mathbf{F} \neq \mathbf{0}$, $\mathbf{F}$ is not conservative.

$\text{div } \mathbf{F} = \nabla \cdot \mathbf{F}$

$$= \left\langle \frac{\partial}{\partial x}, \frac{\partial}{\partial y}, \frac{\partial}{\partial z} \right\rangle \cdot$$

$$\left\langle z^2 - 3e^{3x}y, \, z^2 - e^{3x}, \, 2z\sqrt{xy} \right\rangle$$

$$= -9ye^{3x} + 0 + 2\sqrt{xy}$$

$$= -9ye^{3x} + 2\sqrt{xy}$$

Since $\text{div } \mathbf{F} \neq 0$, $\mathbf{F}$ is not incompressible.

21. $\text{curl } \mathbf{F} = \nabla \times \mathbf{F}$

$$= \begin{vmatrix} \mathbf{i} & \mathbf{j} & \mathbf{k} \\ \frac{\partial}{\partial x} & \frac{\partial}{\partial y} & \frac{\partial}{\partial z} \\ xy^2 & 3xz & 4 - y^2z \end{vmatrix}$$

$$= (-2yz - 3x)\mathbf{i} - (0)\mathbf{j}$$

$$+ (3z - 2xy)\mathbf{k}$$

$$= \langle -3x - 2yz, \, 0, \, 3z - 2xy \rangle$$

Since $\text{curl } \mathbf{F} \neq \mathbf{0}$, $\mathbf{F}$ is not conservative.

$\text{div } \mathbf{F} = \nabla \cdot \mathbf{F}$

$$= \left\langle \frac{\partial}{\partial x}, \frac{\partial}{\partial y}, \frac{\partial}{\partial z} \right\rangle \cdot$$

$$\left\langle xy^2, \, 3xz, \, 4 - y^2z \right\rangle$$

$$= y^2 + 0 - y^2 = 0$$

Since $\text{div } \mathbf{F} = 0$, $\mathbf{F}$ is incompressible.

23. $\text{curl } \mathbf{F} = \nabla \times \mathbf{F}$

$$= \begin{vmatrix} \mathbf{i} & \mathbf{j} & \mathbf{k} \\ \frac{\partial}{\partial x} & \frac{\partial}{\partial y} & \frac{\partial}{\partial z} \\ 4x & 3y^3 & e^z \end{vmatrix}$$

$$= (0 - 0)\mathbf{i} - (0 - 0)\mathbf{j} + (0 - 0)\mathbf{k}$$

$$= \langle 0, 0, 0 \rangle$$

Since the components of $\mathbf{F}$ have continuous partial derivatives throughout $\mathbb{R}^3$ and $\text{curl } \mathbf{F} = \mathbf{0}$, $\mathbf{F}$ is conservative.

$\text{div } \mathbf{F} = \nabla \cdot \mathbf{F}$

$$= \left\langle \frac{\partial}{\partial x}, \frac{\partial}{\partial y}, \frac{\partial}{\partial z} \right\rangle \cdot \left\langle 4x, 3y^3, e^z \right\rangle$$

$$= 4 + 9y^2 + e^z$$

Since $\text{div } \mathbf{F} \neq 0$, $\mathbf{F}$ is not incompressible.

25. $\text{curl } \mathbf{F} = \nabla \times \mathbf{F}$

$$= \begin{vmatrix} \mathbf{i} & \mathbf{j} & \mathbf{k} \\ \frac{\partial}{\partial x} & \frac{\partial}{\partial y} & \frac{\partial}{\partial z} \\ -2xy & -x^2 & 2yz \cos yz^2 \\ & +z^2 \cos yz^2 & \end{vmatrix}$$

$$= \left(2z \cos yz^2 - 2yz^3 \sin yz^2 \right.$$

$$\left. -2z \cos yz^2 + 2yz^3 \sin yz^2 \right) \mathbf{i}$$

$$- (0 - 0)\mathbf{j} + (-2x + 2x)\mathbf{k}$$

$$= \langle 0, 0, 0 \rangle$$

Since the components of $\mathbf{F}$ have continuous partial derivatives throughout $\mathbb{R}^3$ and $\text{curl } \mathbf{F} = \mathbf{0}$, $\mathbf{F}$ is conservative.

$\text{div } \mathbf{F} = \nabla \cdot \mathbf{F}$

$$= \left\langle \frac{\partial}{\partial x}, \frac{\partial}{\partial y}, \frac{\partial}{\partial z} \right\rangle \cdot$$

$$\left\langle -2xy, \, -x^2 + z^2 \cos yz^2, \right.$$

$$\left. 2yz \cos yz^2 \right\rangle$$

$$= -2y - z^4 \sin yz^2 + 2y \cos yz^2$$

$$- 4y^2z^2 \sin yz^2$$

Since $\text{div } \mathbf{F} \neq 0$, $\mathbf{F}$ is not incompressible.

27. (a) $\nabla \cdot (\nabla f) = \nabla \cdot (\nabla \text{scalar})$

$$= \nabla \cdot \text{vector}$$

$$= \text{scalar}$$

(b) $\nabla \times (\nabla \cdot \mathbf{F}) = \nabla \times (\nabla \cdot \text{vector})$

$$= \nabla \times \text{scalar}$$

$$= \text{undefined}$$

(c) $\nabla(\nabla \times \mathbf{F}) = \nabla(\nabla \times \text{vector})$

$$= \nabla \text{vector}$$

$$= \text{undefined}$$

(d) $\nabla(\nabla \cdot \mathbf{F}) = \nabla(\nabla \cdot \text{vector})$

$$= \nabla \text{scalar}$$

$$= \text{vector}$$

(e) $\nabla \times (\nabla f) = \nabla \times (\nabla \text{scalar})$

$$= \nabla \times \text{vector}$$

$$= \text{vector}$$

29. $\mathbf{r} = \langle x, y, z \rangle$

$\text{curl } \mathbf{r} = \nabla \times \mathbf{r}$

$$= \begin{vmatrix} \mathbf{i} & \mathbf{j} & \mathbf{k} \\ \frac{\partial}{\partial x} & \frac{\partial}{\partial y} & \frac{\partial}{\partial z} \\ x & y & z \end{vmatrix}$$

$$= (0 - 0)\mathbf{i} - (0 - 0)\mathbf{j} + (0 - 0)\mathbf{k}$$

$$= \langle 0, 0, 0 \rangle$$

$\text{div } \mathbf{r} = \nabla \cdot \mathbf{r}$

$$= \left\langle \frac{\partial}{\partial x}, \frac{\partial}{\partial y}, \frac{\partial}{\partial z} \right\rangle$$

$$\cdot \langle x, y, z \rangle$$

$$= 1 + 1 + 1 = 3$$

31. The divergence is positive because if we draw a box around P, the outflow (top and right sides) is greater than the inflow (bottom and left sides).

33. The divergence is negative because if we draw a box around P, the outflow (top and right sides) is less than the inflow (bottom and left sides).

35. The divergence is negative because if we draw a box around P, the outflow (mostly on the right side) is less than the inflow (mostly on the left side).

37. Let $\mathbf{F} = \langle F_1, F_2, F_3 \rangle$ and $\mathbf{G} = \langle G_1, G_2, G_3 \rangle$.

Then

$\nabla \cdot (\mathbf{F} \times \mathbf{G})$

$$= \left\langle \frac{\partial}{\partial x}, \frac{\partial}{\partial y}, \frac{\partial}{\partial z} \right\rangle$$

$$\cdot \langle F_2 G_3 - F_3 G_2,$$

$$F_3 G_1 - F_1 G_3,$$

$$F_1 G_2 - F_2 G_1 \rangle$$

$$= \frac{\partial F_2}{\partial x} G_3 + F_2 \frac{\partial G_3}{\partial x} - \frac{\partial F_3}{\partial x} G_2 - F_3 \frac{\partial G_2}{\partial x}$$

$$+ \frac{\partial F_3}{\partial y} G_1 + F_3 \frac{\partial G_1}{\partial y} - \frac{\partial F_1}{\partial y} G_3 - F_1 \frac{\partial G_3}{\partial y}$$

$$+ \frac{\partial F_1}{\partial z} G_2 + F_1 \frac{\partial G_2}{\partial z} - \frac{\partial F_2}{\partial z} G_1 - F_2 \frac{\partial G_1}{\partial z}.$$

And,

$\mathbf{G} \cdot (\nabla \times \mathbf{F})$

$$= \langle G_1, G_2, G_3 \rangle$$

$$\cdot \left\langle \frac{\partial F_3}{\partial y} - \frac{\partial F_2}{\partial z}, \frac{\partial F_1}{\partial z} - \frac{\partial F_3}{\partial x}, \frac{\partial F_2}{\partial x} - \frac{\partial F_1}{\partial y} \right\rangle$$

$$= G_1 \frac{\partial F_3}{\partial y} - G_1 \frac{\partial F_2}{\partial z} + G_2 \frac{\partial F_1}{\partial z} - G_2 \frac{\partial F_3}{\partial x}$$

$$+ G_3 \frac{\partial F_2}{\partial x} - G_3 \frac{\partial F_1}{\partial y}.$$

Similarly, we can find $\mathbf{F} \cdot (\nabla \times \mathbf{G})$. Putting everything together gives the result.

39. $\nabla \times (\nabla \times \mathbf{F}) = \nabla(\nabla \cdot \mathbf{F}) - \nabla^2 \mathbf{F}$

Let $\mathbf{F} = \langle A, B, C \rangle$. Then,

$\nabla \times (\nabla \times \mathbf{F})$

$$= \left\langle \frac{\partial}{\partial x}, \frac{\partial}{\partial y}, \frac{\partial}{\partial z} \right\rangle \times$$

$$\langle C_y - B_z, A_z - C_x, B_x - A_y \rangle$$

$$= \langle (B_x - A_y)_y - (A_z - C_x)_z,$$
$$(C_y - B_z)_z - (B_x - A_y)_x,$$
$$(A_z - C_x)_x - (C_y - B_z)_y \rangle$$

$$= \langle B_{xy} - A_{yy} - A_{zz} + C_{xz},$$
$$C_{yz} - B_{zz} - B_{xx} + A_{yx},$$
$$A_{zx} - C_{xx} - C_{yy} + B_{zy} \rangle$$

$$= \langle A_{xx} + B_{xy} + C_{xz},$$
$$A_{yx} + B_{yy} + C_{yz},$$
$$A_{zx} + B_{zy} + C_{zz} \rangle$$

$$- \langle A_{xx} + A_{yy} + A_{zz},$$
$$B_{xx} + B_{yy} + B_{zz},$$
$$C_{xx} + C_{yy} + C_{zz} \rangle$$

$$= \nabla (\nabla \cdot \mathbf{F}) - \nabla^2 \mathbf{F}.$$

41. Take a circle C in the xz-plane. Give C a positive orientation in that plane. On C, y is constant ($y = 0$) and therefore $dy = 0$. Therefore

$$\oint_C \mathbf{F} \cdot d\mathbf{r}$$

$$= \oint_C F_1(x, 0, z)\, dx + F_3(x, 0, z)\, dz$$

$$= \iint_R \left(\frac{\partial F_3}{\partial x} - \frac{\partial F_1}{\partial z} \right) dA_{xz} < 0.$$

43. Because $\mathbf{n} = \langle dy, -dx \rangle$,

$$\oint_C f(\nabla g) \cdot \mathbf{n}\, ds = \oint f \left(\frac{\partial g}{\partial x}\, dy - \frac{\partial g}{\partial y}\, dx \right)$$

$$= \oint M\, dx + N\, dy$$

where $M = -f \frac{\partial g}{\partial y}$. and $N = f \frac{\partial g}{\partial x}$
Therefore

$$\frac{\partial M}{\partial y} = -\frac{\partial f}{\partial y}\frac{\partial g}{\partial y} - f\frac{\partial^2 g}{\partial y^2}$$

$$\frac{\partial N}{\partial x} = \frac{\partial f}{\partial x}\frac{\partial g}{\partial x} + f\frac{\partial^2 g}{\partial x^2}$$

$$\frac{\partial N}{\partial x} - \frac{\partial M}{\partial y} = \nabla f \cdot \nabla g + f \nabla^2 g$$

Therefore, by Green's Theorem:

$$\oint_C f(\nabla g) \cdot \mathbf{n}\, ds$$

$$= \iint_R \left(\nabla f \cdot \nabla g + f \nabla^2 g \right) dA.$$

45. This is the vector form of Green's Theorem and exercise 69 in Section 15.2.

47. Let $\mathbf{F} = \langle A, B, C \rangle$. Then

$$\nabla \cdot (f\mathbf{F})$$
$$= \nabla \cdot \langle fA, fB, fC \rangle$$
$$= (fA)_x + (fB)_y + (fC)_z$$
$$= (f_x A + f A_x) + (f_y B + f B_y)$$
$$\qquad + (f_z C + f C_z)$$
$$= (f_x A + f_y B + f_z C) + f(A_x + B_y + C_z)$$
$$= \nabla f \cdot \mathbf{F} + f \nabla \cdot \mathbf{F}.$$

49. If $\nabla^2 f = 0$, then this means $\nabla \cdot \nabla f = 0$, or that ∇f is solenoidal.

Example 5.5 shows that $\nabla \times \nabla f = 0$, so ∇f is irrotational.

51.
$$\nabla f(r) = \nabla f \left(\sqrt{x^2 + y^2} \right)$$

$$= \left\langle \frac{\partial}{\partial x} \left[f \left(\sqrt{x^2 + y^2} \right) \right], \right.$$
$$\left. \frac{\partial}{\partial y} \left[f \left(\sqrt{x^2 + y^2} \right) \right] \right\rangle$$

$$= \left\langle \frac{x}{\sqrt{x^2 + y^2}} f' \left(\sqrt{x^2 + y^2} \right), \right.$$
$$\left. \frac{y}{\sqrt{x^2 + y^2}} f' \left(\sqrt{x^2 + y^2} \right) \right\rangle$$

$$= \left\langle \frac{x}{r} f'(r), \frac{y}{r} f'(r) \right\rangle = f'(r)\frac{\mathbf{r}}{r}$$

53. The problem should ask for $\nabla^2 f$ instead of ∇f.

From $r^2 = x^2 + y^2 + z^2$ we obtain

$$2r \frac{\partial r}{\partial x} = 2x \text{ or } r \frac{\partial r}{\partial x} = x.$$

Taking the second derivative of this gives

$$\frac{\partial}{\partial x}\left(r \frac{\partial r}{\partial x}\right) = \left(\frac{\partial r}{\partial x}\right)^2 + r \frac{\partial^2 r}{\partial x^2}$$

$$= \frac{\partial}{\partial x}(x) = 1.$$

Therefore

$$\frac{\partial^2 r}{\partial x^2} = \frac{1 - \left(\frac{\partial r}{\partial x}\right)^2}{r} = \frac{1 - \left(\frac{x}{r}\right)^2}{r}$$

$$= \frac{r^2 - x^2}{r^3}.$$

Similarly,

$$\frac{\partial^2 r}{\partial y^2} = \frac{r^2 - y^2}{r^3}$$

$$\frac{\partial^2 r}{\partial z^2} = \frac{r^2 - z^2}{r^3}$$

Therefore

$$\nabla^2 f = \nabla^2 r = \frac{\partial^2 r}{\partial x^2} + \frac{\partial^2 r}{\partial y^2} + \frac{\partial^2 r}{\partial z^2}$$

$$= \frac{r^2 - x^2}{r^3} + \frac{r^2 - y^2}{r^3} + \frac{r^2 - z^2}{r^3}$$

$$= \frac{2}{r} = \frac{2}{\sqrt{x^2 + y^2 + z^2}}.$$

55. For $\mathbf{F}(x, y) = \langle x^2, y^2 - 4x \rangle$, div $F = 2x + 2y$.

(a) div $\mathbf{F}(0, 0) = 0$ so the net flow will be 0 (flow in will equal flow out).

(b) div $\mathbf{F}(1, 0) = 2$ so the net flow will be positive (the flow in is less than the flow out).

57. Anything of the form
$\mathbf{F} = \langle xh(y), g(y), zk(y) \rangle$
will have

$\nabla \cdot \mathbf{F} = h(y) + g'(y) + k(y)$, which is a function of y. Now just choose h, g and k so that

their sum is positive. Here is one simple possibility:

$$\mathbf{F} = \langle xy^2, y, 0 \rangle$$

59. If $\frac{\rho}{\epsilon_0} = \nabla \cdot \mathbf{E}$ and $\mathbf{E} = -\nabla \phi$ then

$$\nabla^2 \phi = \nabla \cdot \nabla \phi = \nabla \cdot (-\mathbf{E})$$

$$= -\nabla \cdot \mathbf{E} = -\frac{\rho}{\epsilon_0}$$

61. The stream function for a two-dimensional vector field $\mathbf{v} = \langle v_1, v_2 \rangle$, if there is one, is a potential for the orthogonal field

$$\mathbf{v}^\perp = \langle -v_2, v_1 \rangle \qquad \text{(in general)}$$

$$= \langle y^2 - x, 2xy \rangle \qquad \text{(this case)}$$

A routine check shows that this field is conservative.

If $\nabla g(x, y) = \mathbf{v}^\perp$, then $\frac{\partial g}{\partial x} = y^2 - x$ and $\frac{\partial g}{\partial y} = 2xy$.

$$g(x, y) = \int (y^2 - x)\, dx$$

$$= y^2 x - \frac{1}{2} x^2 + h(y)$$

$$\frac{\partial g}{\partial y} = 2xy + h'(y) = 2xy$$

$$h'(y) = 0$$

$$h(y) = c$$

$$g(x, y) = y^2 x - \frac{1}{2} x^2 + c$$

63.

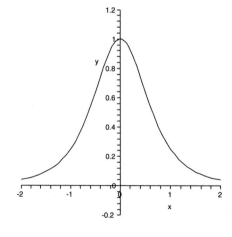

y=1/(1+x^2)^2

The vector field is perhaps easiest to visualize if it is graphed in the xy-plane:

F'in 'the 'xy-plane

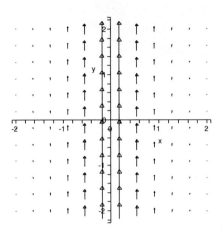

As can be seen from the graph of the vector field, if a paddle wheel is placed near the origin (but $x \neq 0$), then one side of the wheel will be pushed harder than the other side causing the wheel to turn. This is verified with the curl of **F**:

$$\text{curl } \mathbf{F} = \nabla \times \mathbf{F} = \nabla \times \left\langle 0, \frac{1}{1+x^2}, 0 \right\rangle$$

$$= \begin{vmatrix} \mathbf{i} & \mathbf{j} & \mathbf{k} \\ \frac{\partial}{\partial x} & \frac{\partial}{\partial y} & \frac{\partial}{\partial z} \\ 0 & \frac{1}{1+x^2} & 0 \end{vmatrix}$$

$$= \left\langle 0, 0, -\frac{2x}{(1+x^2)^2} \right\rangle.$$

65. In the special case that $\mathbf{H} = \langle h_1(x, y, z), 0, 0 \rangle$, then

$$\text{curl } \mathbf{H} = \nabla \times \mathbf{H} = \nabla \times \langle h_1, 0, 0 \rangle$$

$$= \begin{vmatrix} \mathbf{i} & \mathbf{j} & \mathbf{k} \\ \frac{\partial}{\partial x} & \frac{\partial}{\partial y} & \frac{\partial}{\partial z} \\ h_1 & 0 & 0 \end{vmatrix}$$

$$= \left\langle 0, \frac{\partial h_1}{\partial z}, -\frac{\partial h_1}{\partial y} \right\rangle$$

and

$$\text{div curl } \mathbf{H} = \nabla \cdot \nabla \times \mathbf{H}$$

$$= \nabla \cdot \left\langle 0, \frac{\partial h_1}{\partial z}, -\frac{\partial h_1}{\partial y} \right\rangle$$

$$= \frac{\partial^2 h_1}{\partial y \partial z} - \frac{\partial^2 h_1}{\partial z \partial y} = 0.$$

Similarly, this holds for vector fields of the form $\langle 0, h_2(x, y, z), 0 \rangle$ and $\langle 0, 0, h_3(x, y, z) \rangle$.

So,

$$\text{div curl } \left(h_1 \mathbf{i} + h_2 \mathbf{j} + h_3 \mathbf{k} \right)$$

$$= \text{div curl } \left(h_1 \mathbf{i} \right) + \text{div curl } \left(h_2 \mathbf{j} \right) + \text{div curl } \left(h_3 \mathbf{k} \right)$$

$$= 0 + 0 + 0 = 0.$$

15.6 SURFACE INTEGRALS

1. $x = x$
$y = y$
$z = 3x + 4y \quad -\infty < x < \infty, \quad -\infty < y < \infty$

3. $x = \cos u \cosh v$
$y = \sin u \cosh v$
$z = \sinh v$
$0 \leq u \leq 2\pi, \quad -\infty < v < \infty$

5. $x = 2 \cos \theta$
$y = 2 \sin \theta$
$z = z$
$0 \leq \theta \leq 2\pi, \quad 0 \leq z \leq 2$

7. $x = r \cos \theta$
$y = r \sin \theta$
$z = 4 - r^2$
$0 \leq \theta \leq 2\pi, \quad 0 \leq r \leq 2$

9.

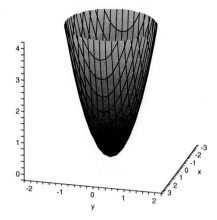

11.

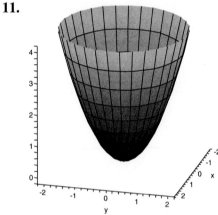

13.

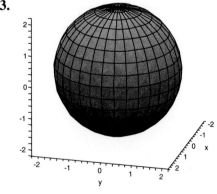

15.

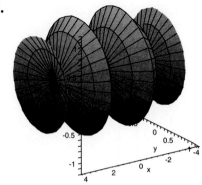

17. In **(a)**, (u, v) are polar coordinates in the xy-plane (v is the angle), and with $z = v^2$. This must be surface A.

In **(b)**, (u, v) are polar coordinates in the yz-plane (v is the angle), and with $x = v$. This must be surface C.

In **(c)**, (u, v) are polar coordinates in the yz-plane (v is the angle), and with $x = u$. This means that $x^2 = y^2 + z^2$, which must be surface B.

19. There are other possible ways to solve this, but we can parametrize this surface by polar coordinates in the xy-plane:

$(x, y, z) = \mathbf{r} = (r \cos \theta, r \sin \theta, r)$
$0 \le \theta \le 2\pi, \quad 0 \le r \le 4$

$\mathbf{r}_\theta = \langle -r \sin \theta, r \cos \theta, 0 \rangle$
$\mathbf{r}_r = \langle \cos \theta, \sin \theta, 1 \rangle$
$\mathbf{r}_\theta \times \mathbf{r}_r = \langle r \cos \theta, r \sin \theta, -r \rangle$
$\| \mathbf{r}_\theta \times \mathbf{r}_r \| = r \sqrt{2}$

$$\text{Area} = \iint_R \| \mathbf{r}_\theta \times \mathbf{r}_r \| \, dA_{\theta r}$$
$$= \iint_R r \sqrt{2} \, dA_{\theta r}$$
$$= \int_0^{2\pi} \int_0^4 r \sqrt{2} \, dr \, d\theta$$
$$= \int_0^{2\pi} 8 \sqrt{2} \, d\theta = 16\pi \sqrt{2}$$

21.
$$z = f(x, y) = \frac{6 - 3x - y}{2}$$

$$\mathbf{n} = \langle f_x, f_y, -1 \rangle = \left\langle -\frac{3}{2}, -\frac{1}{2}, -1 \right\rangle$$

$$\|\mathbf{n}\| = \frac{\sqrt{14}}{2}$$

$$\iint_S dS = \iint_R \|\mathbf{n}\| \, dA$$

$$= \iint_R \frac{\sqrt{14}}{2} \, dA = \frac{\sqrt{14}}{2} (\text{Area of Circle})$$

$$= \frac{\sqrt{14}}{2} (4\pi) = 2\pi\sqrt{14}$$

23.
$$z = f(x, y) = \sqrt{x^2 + y^2} = r$$

$$\mathbf{n} = \langle f_x, f_y, -1 \rangle$$

$$= \left\langle \frac{x}{\sqrt{x^2 + y^2}}, \frac{y}{\sqrt{x^2 + y^2}}, -1 \right\rangle$$

$$\|\mathbf{n}\| = \sqrt{2}$$

$$\iint_S dS = \iint_R \|\mathbf{n}\| \, dA = \iint_R \sqrt{2} \, dA$$

$$= \sqrt{2}(\text{Area of Triangle}) = \frac{\sqrt{2}}{2}$$

25.
$$z = f(x, y) = \sqrt{4 - x^2 - y^2}$$

$$\mathbf{n} = \langle f_x, f_y, -1 \rangle$$

$$= \left\langle -\frac{x}{\sqrt{4 - x^2 - y^2}}, \right.$$

$$\left. -\frac{y}{\sqrt{4 - x^2 - y^2}}, -1 \right\rangle$$

$$\|\mathbf{n}\| = \frac{2}{\sqrt{4 - x^2 - y^2}} = \frac{2}{\sqrt{4 - r^2}}$$

$$\iint_S dS = \iint_R \|\mathbf{n}\| \, dA$$

$$= \iint_R \frac{2}{\sqrt{4 - x^2 - y^2}} \, dA$$

$$= \int_0^{2\pi} \int_0^{\sqrt{3}} \frac{2r}{\sqrt{4 - r^2}} \, dr \, d\theta$$

$$= \int_0^{2\pi} 2 \, d\theta = 4\pi$$

27.
$$z = f(x, y) = 2x + 3y$$

$$\mathbf{n} = \langle f_x, f_y, -1 \rangle = \langle 2, 3, -1 \rangle$$

$$\|\mathbf{n}\| = \sqrt{14}$$

$$\iint_S xz \, dS = \iint_R x(2x + 3y) \|\mathbf{n}\| \, dA$$

$$= \iint_R x(2x + 3y)\sqrt{14} \, dA$$

$$= \int_1^2 \int_1^3 (2x^2 + 3xy)\sqrt{14} \, dy \, dx$$

$$= \sqrt{14} \int_1^2 \left(12x + 4x^2 \right) \, dx$$

$$= \frac{82\sqrt{14}}{3}$$

29.
$$z = f(x, y) = -\sqrt{9 - x^2 - y^2}$$

$$\mathbf{n} = \langle f_x, f_y, -1 \rangle$$

$$= \left\langle \frac{x}{\sqrt{9 - x^2 - y^2}}, \right.$$

$$\left. \frac{y}{\sqrt{9 - x^2 - y^2}}, -1 \right\rangle$$

$$\|\mathbf{n}\| = \frac{3}{\sqrt{9 - x^2 - y^2}} = \frac{3}{\sqrt{9 - r^2}}$$

$$\iint_S (x^2 + y^2 + z^2)^{3/2}\, dS = \iint_R 27\|\mathbf{n}\|\, dA$$

$$= 27\iint_R \frac{3}{\sqrt{9 - x^2 - y^2}}\, dA$$

$$= 81\int_0^{2\pi}\int_0^3 \frac{r}{\sqrt{9 - r^2}}\, dr\, d\theta$$

$$= 81\int_0^{2\pi} 3\, d\theta = 486\pi$$

$$\iint_S z^2\, dS = 2\iint_R z^2\|\mathbf{n}\|\, dA$$

$$= 2\iint_R r^2\sqrt{2}\, dA$$

$$= 2\sqrt{2}\int_0^4\int_0^{2\pi} r^3\, d\theta\, dr$$

$$= 4\pi\sqrt{2}\int_0^4 r^3\, dr = 256\pi\sqrt{2}$$

31.

$$z = f(x, y) = 4 - x^2 - y^2$$

$$\mathbf{n} = \langle f_x, f_y, -1\rangle = \langle -2x, -2y, -1\rangle$$

$$\|\mathbf{n}\| = \sqrt{4x^2 + 4y^2 + 1}$$

$$\iint_S (x^2 + y^2 - z)\, dS$$

$$= \iint_R (2x^2 + 2y^2 - 4)\|\mathbf{n}\|\, dA$$

$$= \iint_R (2x^2 + 2y^2 - 4)\sqrt{4x^2 + 4y^2 + 1}\, dA$$

$$= \int_{\sqrt{2}}^{\sqrt{3}}\int_0^{2\pi} (2r^2 - 4)r\sqrt{4r^2 + 1}\, d\theta\, dr$$

$$= 4\pi\int_{\sqrt{2}}^{\sqrt{3}} (r^2 - 2)r\sqrt{4r^2 + 1}\, dr$$

$$= 4\pi\left[\frac{2r^2 - 7}{40}\left(4r^2 + 1\right)^{3/2}\right]_{\sqrt{2}}^{\sqrt{3}}$$

$$= \frac{\pi(81 - 13\sqrt{13})}{10} \approx 10.7216$$

33.

$$z = f(x, y) = \sqrt{x^2 + y^2} = r$$

$$\mathbf{n} = \langle f_x, f_y, -1\rangle$$

$$= \left\langle \frac{x}{\sqrt{x^2 + y^2}}, \frac{y}{\sqrt{x^2 + y^2}}, -1\right\rangle$$

$$\|\mathbf{n}\| = \sqrt{2}$$

Note that due to symmetry the integral over the lower half is equal to that of the upper half. So, we integrate over the upper half of the cone and multiply by 2.

35.

$$z = f(x, y) = \sqrt{x^2 + y^2 - 1}$$

$$\mathbf{n} = \langle f_x, f_y, -1\rangle$$

$$= \left\langle \frac{x}{\sqrt{x^2 + y^2 - 1}}, \frac{y}{\sqrt{x^2 + y^2 - 1}}, -1\right\rangle$$

$$\|\mathbf{n}\| = \sqrt{\frac{2x^2 + 2y^2 - 1}{x^2 + y^2 - 1}} = \sqrt{\frac{2r^2 - 1}{r^2 - 1}}$$

$$\iint_S x\, dS = \iint_R x\|\mathbf{n}\|\, dA$$

$$= \iint_R x\sqrt{\frac{2r^2 - 1}{r^2 - 1}}\, dA$$

$$= \int_0^1\int_0^{2\pi} r^2\cos\theta\sqrt{\frac{2r^2 - 1}{r^2 - 1}}\, d\theta\, dr$$

$$= \int_0^1 (0)r^2\sqrt{\frac{2r^2 - 1}{r^2 - 1}}\, dr = 0$$

37.

$$\mathbf{n} = \frac{\langle 2x, 2y, 1\rangle}{\sqrt{4x^2 + 4y^2 + 1}}$$

$$dS = \sqrt{4x^2 + 4y^2 + 1}\, dA$$

$$\mathbf{F}\cdot\mathbf{n}\, dS$$

$$= \left(2x^2 + 2y^2 + z\right)\, dA$$

$$= \left(x^2 + y^2 + 4\right)\, dA$$

$$\iint_S \mathbf{F} \cdot \mathbf{n} \, dS$$

$$= \iint_R \left(x^2 + y^2 + 4 \right) \, dA$$

$$= \int_0^2 \int_0^{2\pi} \left(r^2 + 4 \right) r \, d\theta \, dr$$

$$= 2\pi \int_0^2 \left(r^2 + 4 \right) r \, dr = 24\pi$$

39.
$$\mathbf{n} = \frac{1}{\sqrt{2}} \left\langle \frac{x}{\sqrt{x^2 + y^2}}, \frac{y}{\sqrt{x^2 + y^2}}, -1 \right\rangle$$

$$dS = \sqrt{2} \, dA$$

$$\mathbf{F} \cdot \mathbf{n} \, dS$$

$$= \left(-\frac{z}{\sqrt{2}} \right) \sqrt{2} \, dA$$

$$= -z \, dA$$

$$\iint_S \mathbf{F} \cdot \mathbf{n} \, dS$$

$$= \iint_R -z \, dA$$

$$= \int_0^3 \int_0^{2\pi} -r^2 \, d\theta \, dr$$

$$= -2\pi \int_0^3 r^2 \, dr = -18\pi$$

41. Back of the box ($x = 0$):

$$\mathbf{n} = \langle -1, 0, 0 \rangle$$
$$\mathbf{F} \cdot \mathbf{n} = -xy = 0$$

Left of the box ($y = 0$):

$$\mathbf{n} = \langle 0, -1, 0 \rangle$$
$$\mathbf{F} \cdot \mathbf{n} = -y^2 = 0$$

Bottom of the box ($z = 0$):

$$\mathbf{n} = \langle 0, 0, -1 \rangle$$
$$\mathbf{F} \cdot \mathbf{n} = -z = 0$$

Front of the box, F ($x = 1$):

$$\mathbf{n} = \langle 1, 0, 0 \rangle$$
$$\mathbf{F} \cdot \mathbf{n} = y$$

$$\iint_F \mathbf{F} \cdot \mathbf{n} \, dS = \int_0^1 \int_0^1 y \, dy \, dz = \frac{1}{2}$$

Right of the box, R ($y = 1$):

$$\mathbf{n} = \langle 0, 1, 0 \rangle$$
$$\mathbf{F} \cdot \mathbf{n} = y^2 = 1$$

$$\iint_R \mathbf{F} \cdot \mathbf{n} \, dS = \int_R 1 \, dA = \text{Area}(R) = 1$$

Top of the box, T ($z = 1$):

$$\mathbf{n} = \langle 0, 0, 1 \rangle$$
$$\mathbf{F} \cdot \mathbf{n} = z = 1$$

$$\iint_T \mathbf{F} \cdot \mathbf{n} \, dS = \int_T 1 \, dA = \text{Area}(T) = 1$$

Therefore

$$\iint_S \mathbf{F} \cdot \mathbf{n} \, dS = \frac{1}{2} + 1 + 1 = \frac{5}{2}.$$

43. For the bottom, B ($z = 1$):

$$\mathbf{n} = \langle 0, 0, -1 \rangle$$
$$\mathbf{F} \cdot \mathbf{n} = -z = -1$$

$$\iint_B \mathbf{F} \cdot \mathbf{n} \, dS = \int_B -z \, dA$$
$$= -\text{Area}(B) = -3\pi$$

For the top, T ($z = 4 - x^2 - y^2$):

$$\mathbf{n} = \frac{\langle 2x, 2y, 1 \rangle}{\sqrt{4x^2 + 4y^2 + 1}}$$

$$dS = \sqrt{4x^2 + 4y^2 + 1} \, dA$$

$$\mathbf{F} \cdot \mathbf{n} \, dS$$
$$= (2x + z) \, dA$$
$$= \left(2x + 4 - x^2 - y^2 \right) \, dA$$

$$\iint_T \mathbf{F} \cdot \mathbf{n}\, dS$$

$$= \iint_R \left(2x + 4 - x^2 - y^2\right) dA$$

$$= \int_0^{\sqrt{3}} \int_0^{2\pi} \left(2r\cos\theta + 4 - r^2\right) r\, d\theta\, dr$$

$$= 2\pi \int_0^{\sqrt{3}} \left(4 - r^2\right) r\, dr = \frac{15\pi}{2}$$

Therefore

$$\iint_S \mathbf{F} \cdot \mathbf{n}\, dS$$

$$= -3\pi + \frac{15\pi}{2} = \frac{9\pi}{2}.$$

45.
$$\mathbf{n} = \frac{1}{\sqrt{3}} \langle 1, 1, 1 \rangle$$

$$dS = \sqrt{3}\, dA$$

$$\mathbf{F} \cdot \mathbf{n}\, dS$$
$$= (yx + 1 + x)\, dA$$

$$\iint_S \mathbf{F} \cdot \mathbf{n}\, dS$$

$$= \iint_R (yx + 1 + x)\, dA$$

$$= \int_0^1 \int_0^1 (yx + 1 + x)\, dy\, dx$$

$$= \int_0^1 \left(\frac{3}{2}x + 1\right) dx = \frac{7}{4}$$

47. This is part of a sphere (call it U) sitting on top of a cone (call it C). For the sphere, U:

$$\mathbf{n} = \frac{1}{\sqrt{8}} \langle x, y, \sqrt{8 - x^2 - y^2} \rangle$$

$$= \frac{1}{\sqrt{8}} \langle x, y, \sqrt{8 - r^2} \rangle$$

$$dS = \sqrt{\frac{8}{8 - x^2 - y^2}}\, dA$$

$$= \frac{\sqrt{8}}{\sqrt{8 - r^2}}\, dA$$

$$\mathbf{F} \cdot \mathbf{n}\, dS$$
$$= \left(\frac{xy}{\sqrt{8 - r^2}} + 2\right) dA$$

$$\iint_U \mathbf{F} \cdot \mathbf{n}\, dS$$

$$= \iint_R \left(\frac{xy}{\sqrt{8 - r^2}} + 2\right) dA$$

$$= \int_0^2 \int_0^{2\pi} \left(\frac{r^2 \cos\theta \sin\theta}{\sqrt{8 - r^2}} + 2\right) r\, d\theta\, dr$$

$$= \int_0^2 \left[\frac{r^3 \sin^2\theta}{2\sqrt{8 - r^2}} + 2r\theta\right]_{\theta=0}^{\theta=2\pi} dr$$

$$= \int_0^2 4\pi r\, dr = 8\pi$$

For the cone, C:

$$\mathbf{n} = \frac{1}{\sqrt{2}} \left\langle \frac{x}{\sqrt{x^2 + y^2}}, \frac{y}{\sqrt{x^2 + y^2}}, -1 \right\rangle$$

$$= \frac{1}{\sqrt{2}} \left\langle \frac{x}{r}, \frac{y}{r}, -1 \right\rangle$$

$$dS = \sqrt{2}\, dA$$

$$\mathbf{F} \cdot \mathbf{n}\, dS$$
$$= \left(\frac{xy}{r} - 2\right) dA$$

$$\iint_C \mathbf{F} \cdot \mathbf{n}\, dS$$

$$= \iint_R \left(\frac{xy}{r} - 2\right) dA$$

$$= \int_0^2 \int_0^{2\pi} \left(\frac{r^2 \cos\theta \sin\theta}{r} - 2\right) r\, d\theta\, dr$$

$$= \int_0^2 \int_0^{2\pi} \left(r^2 \cos\theta \sin\theta - 2r\right) d\theta\, dr$$

$$= -\int_0^2 4\pi r\, dr = -8\pi$$

Therefore,

$$\iint_S \mathbf{F} \cdot \mathbf{n}\, dS = -8\pi + 8\pi = 0.$$

$$= \frac{1}{8\pi} \int_0^2 \int_0^{2\pi} (6 - 3r\cos\theta - 2r\sin\theta)$$

$$(r^2\cos^2\theta + 1)r\, d\theta\, dr$$

$$= \frac{1}{8} \int_0^2 \left(6r^3 + 12r\right)\, dr = 6$$

49. $z = 6 - 3x - 2y$

$\mathbf{n} = \langle -3, -2, -1 \rangle$

$dS = \sqrt{14}\, dA$

$$m = \iint_S \rho\, dS = \iint_R (x^2 + 1)\sqrt{14}\, dA$$

$$= \sqrt{14} \int_0^{2\pi} \int_0^2 (r^2\cos^2\theta + 1)r\, dr\, d\theta$$

$$= \sqrt{14} \int_0^{2\pi} \left(4\cos^2\theta + 2\right)\, d\theta = 8\pi\sqrt{14}$$

$$\bar{x} = \frac{1}{m} \iint_S x\rho\, dS$$

$$= \frac{1}{8\pi\sqrt{14}} \iint_R x(x^2 + 1)\sqrt{14}\, dA$$

$$= \frac{1}{8\pi} \int_0^2 \int_0^{2\pi} (r^2\cos^2\theta + 1)r^2\cos\theta\, d\theta\, dr$$

$$= \frac{1}{8\pi} \int_0^2 (0)\, dr = 0$$

$$\bar{y} = \frac{1}{m} \iint_S y\rho\, dS$$

$$= \frac{1}{8\pi\sqrt{14}} \iint_R y(x^2 + 1)\sqrt{14}\, dA$$

$$= \frac{1}{8\pi} \int_0^2 \int_0^{2\pi} (r^2\cos^2\theta + 1)r^2\sin\theta\, d\theta\, dr$$

$$= \frac{1}{8\pi} \int_0^2 (0)\, dr = 0$$

$$\bar{z} = \frac{1}{m} \iint_S z\rho\, dS$$

$$= \frac{1}{8\pi\sqrt{14}} \iint_R z(x^2 + 1)\sqrt{14}\, dA$$

$$= \frac{1}{8\pi} \iint_R (6 - 3x - 2y)(x^2 + 1)\, dA$$

51. This is a good problem for spherical coordinates: Use spherical coordinates.

$x = \sin\phi\cos\theta$

$y = \sin\phi\sin\theta$

$z = \cos\phi$

$dS = \sin\phi\, d\phi\, d\theta$

$$m = \iint_S \rho\, dS = \iint_S (1 + x)\, dS$$

$$= \int_0^{2\pi} \int_0^{\pi/2} \sin\phi(1 + \sin\phi\cos\theta)\, d\phi\, d\theta$$

$$= \int_0^{2\pi} \left(\pi\frac{1}{4}\cos\theta + 1\right)\, d\theta = 0 + 2\pi = 2\pi$$

$$\bar{x} = \frac{1}{m} \iint_S x\rho\, dS = \frac{1}{2\pi} \iint_S (1 + x)x\, dS$$

$$= \frac{1}{2\pi} \int_0^{\pi/2} \int_0^{2\pi} \sin\phi(1 + \sin\phi\cos\theta)$$

$$(\sin\phi\cos\theta)\, d\theta\, d\phi$$

$$= \frac{1}{2} \int_0^{\pi/2} \sin^3\phi\, d\phi = \frac{1}{3}$$

$$\bar{y} = \frac{1}{m} \iint_S y\rho\, dS = \frac{1}{2\pi} \iint_S (1 + x)y\, dS$$

$$= \frac{1}{2\pi} \int_0^{\pi/2} \int_0^{2\pi} \sin\phi(1 + \sin\phi\cos\theta)$$

$$(\sin\phi\sin\theta)\, d\theta\, d\phi$$

$$= \frac{1}{2\pi} \int_0^{\pi/2} (0)\, d\phi = 0$$

$$\bar{z} = \frac{1}{m} \iint_S z\rho \, dS = \frac{1}{2\pi} \iint_S (1+x)z \, dS$$

$$= \frac{1}{2\pi} \int_0^{\pi/2} \int_0^{2\pi} \sin \phi (1 + \sin \phi \cos \theta)$$

$$(\cos \phi) \, d\theta \, d\phi$$

$$= \int_0^{\pi/2} \sin \phi \cos \phi \, d\phi = \frac{1}{2}$$

53. In this case, there is a function $x = f(y, z)$, $(y, z) \in R$. Then, surface area is given by

$$\iint_S g(x, y, z) \, dS$$

$$= \iint_R g(f(y, z), y, z)$$

$$\sqrt{1 + \left(f_y\right)^2 + \left(f_z\right)^2} \, dA_{yz}.$$

55. This is a segment of a half of a vertical circular cylinder.

$$x = f(y, z) = \sqrt{1 - y^2}$$

$$\|\mathbf{n}\| = \sqrt{(f_y)^2 + (f_z)^2 + 1}$$

$$= \sqrt{\left(\frac{-y}{\sqrt{1-y^2}}\right)^2 + (0)^2 + 1}$$

$$= \frac{1}{\sqrt{1-y^2}}$$

$$\iint_S z \, dS = \iint_R z \left(\frac{1}{\sqrt{1-y^2}}\right) dA$$

$$= \int_{-1}^1 \int_1^2 \frac{z}{\sqrt{1-y^2}} \, dz \, dy$$

$$= \frac{3}{2} \int_{-1}^1 \frac{1}{\sqrt{1-y^2}} \, dy$$

$$= \frac{3}{2} \left[\sin^{-1} y \right]_{-1}^1 = \frac{3}{2}\pi$$

57. $x = f(y, z) = 9 - y^2 - z^2$

$$\|\mathbf{n}\| = \sqrt{(f_y)^2 + (f_z)^2 + 1}$$

$$= \sqrt{(-2y)^2 + (-2z)^2 + 1}$$

$$= \sqrt{4y^2 + 4z^2 + 1}$$

$$\iint_S (y^2 + z^2) \, dS$$

$$= \iint_R (y^2 + z^2)\sqrt{4y^2 + 4z^2 + 1} \, dA$$

$$= \int_0^{2\pi} \int_0^3 r^3 \sqrt{4r^2 + 1} \, dr \, d\theta$$

$$= 2\pi \int_0^3 r^3 \sqrt{4r^2 + 1} \, dr$$

$$= 2\pi \int_1^{37} \left(\frac{u-1}{4}\right) \frac{\sqrt{u}}{8} \, du \qquad (u = 1 + 4r^2)$$

$$= \frac{\pi}{60} \left(1961\sqrt{37} + 1\right)$$

59. $y = f(x, z) = x^2 + z^2$

$$\|\mathbf{n}\| = \sqrt{(f_x)^2 + (f_z)^2 + 1}$$

$$= \sqrt{(2x)^2 + (2z)^2 + 1}$$

$$= \sqrt{4x^2 + 4z^2 + 1}$$

$$\iint_S x^2 \, dS = \iint_R x^2 \sqrt{4x^2 + 4z^2 + 1} \, dA$$

$$= \int_0^1 \int_0^{2\pi} r^3 \cos^2 \theta \sqrt{4r^2 + 1} \, d\theta \, dr$$

$$= \pi \int_0^1 r^3 \sqrt{4r^2 + 1} \, dr$$

$$= \pi \int_1^5 \left(\frac{u-1}{4}\right) \frac{\sqrt{u}}{8} \, du \qquad (u = 1 + 4r^2)$$

$$= \frac{\pi}{120} \left(25\sqrt{5} + 1\right)$$

61. $y = f(x, z) = 1 - x^2$

$\|\mathbf{n}\| = \sqrt{(f_x)^2 + (f_z)^2 + 1}$

$= \sqrt{(-2x)^2 + (0)^2 + 1}$

$= \sqrt{4x^2 + 1}$

We will need the table of integrals:

$\iint_S 4\, dS = \iint_R 4\sqrt{4x^2 + 1}\, dA$

$= \int_{-1}^{1} \int_0^2 4\sqrt{4x^2 + 1}\, dz\, dx$

$= \int_{-1}^{1} 8\sqrt{4x^2 + 1}\, dx$

$= 16 \int_{-1}^{1} \sqrt{x^2 + \frac{1}{4}}\, dx$

$= 8 \left[x\sqrt{x^2 + \frac{1}{4}} + \frac{1}{4}\ln\left(x + \sqrt{x^2 + \frac{1}{4}}\right) \right]_{-1}^{1}$

$= 8\sqrt{5} + 2\ln(9 + 4\sqrt{5}).$

63. The flux integral over any portion of the cone is zero because **F** is always perpendicular to the normal vector to the cone.

As in exercise 39,

$\mathbf{m} = \left\langle \frac{x}{z}, \frac{y}{z}, -1 \right\rangle$

$\mathbf{F} \cdot \mathbf{m} = \langle x, y, z \rangle \cdot \left\langle \frac{x}{z}, \frac{y}{z}, -1 \right\rangle$

$= \frac{x^2 + y^2}{z} - z = 0$

65. In cylindrical coordinates, the equation is $z = cr$. Converting to spherical coordinates gives
$c\rho \sin\phi = cr = z = \rho \cos\phi$
or $\tan\phi = \frac{1}{c}$.

Look at a right triangle with ϕ as one of its acute angles and with the length of the opposite side 1. $\tan\phi = \frac{1}{c}$ means that the length of the adjacent side is c and the length of the hypotenuse is $\sqrt{1 + c^2}$. Therefore

$\sin\phi = \frac{1}{\sqrt{1+c^2}}$ and $\cos\phi = \frac{c}{\sqrt{1+c^2}}$. This immediately gives the equations in the problem where $u = \rho$ and $v = \theta$ in spherical coordinates.

67. This is similar to exercise 63. The flux integral over any portion of the cone is zero because **F** is always perpendicular to the normal vector to the cone.

$\mathbf{m} = \left\langle \frac{c^2 x}{z}, \frac{c^2 y}{z}, -1 \right\rangle$

$\mathbf{F} \cdot \mathbf{m} = \langle x, y, z \rangle \cdot \left\langle \frac{c^2 x}{z}, \frac{c^2 y}{z}, -1 \right\rangle$

$= \frac{c^2(x^2 + y^2)}{z} - z = 0$

69. $x = \frac{u \cos v}{\sqrt{1+c^2}}$

$y = \frac{u \sin v}{\sqrt{1+c^2}}$

$z = \frac{cu}{\sqrt{1+c^2}}$

$0 \le v \le 2\pi \qquad 0 \le u \le \frac{\sqrt{c^2 + 1}}{c}$

$\mathbf{r}_v = \left\langle -\frac{u \sin v}{\sqrt{1+c^2}}, \frac{u \cos v}{\sqrt{1+c^2}}, 0 \right\rangle$

$\mathbf{r}_u = \left\langle \frac{\cos v}{\sqrt{1+c^2}}, \frac{\sin v}{\sqrt{1+c^2}}, \frac{c}{\sqrt{1+c^2}} \right\rangle$

$\mathbf{r}_v \times \mathbf{r}_u = \left\langle \frac{cu \cos v}{1+c^2}, \frac{cu \sin v}{1+c^2}, -\frac{u}{1+c^2} \right\rangle$

$\|\mathbf{r}_v \times \mathbf{r}_u\| = \frac{u}{\sqrt{c^2 + 1}}$

$\mathbf{F} \cdot (\mathbf{r}_v \times \mathbf{r}_u)$

$= \langle x, y, 0 \rangle \cdot \left\langle \frac{cu \cos v}{1+c^2}, \frac{cu \sin v}{1+c^2}, -\frac{u}{1+c^2} \right\rangle$

$= \frac{cu^2}{(c^2 + 1)^{3/2}}$

We are now ready to compute the flux:

$$\iint_S \mathbf{F} \cdot \mathbf{n} \, dS$$

$$= \int_0^{2\pi} \int_0^{\sqrt{c^2+1}/c} \frac{cu^2}{(c^2+1)^{3/2}} \, du \, dv$$

$$= \frac{2c\pi}{(c^2+1)^{3/2}} \int_0^{\sqrt{c^2+1}/c} u^2 \, du = \frac{2\pi}{3c^2}$$

15.7 THE DIVERGENCE THEOREM

1.   Back of cube ($x = 0$):

$$\mathbf{F} \cdot \mathbf{n} = \left\langle 2xz, \, y^2, \, -xz \right\rangle \cdot \langle -1, 0, 0 \rangle$$

$$= -2xz = 0$$

Left of cube ($y = 0$):

$$\mathbf{F} \cdot \mathbf{n} = \left\langle 2xz, \, y^2, \, -xz \right\rangle \cdot \langle 0, -1, 0 \rangle$$

$$= -y^2 = 0$$

Bottom of cube ($z = 0$):

$$\mathbf{F} \cdot \mathbf{n} = \left\langle 2xz, \, y^2, \, -xz \right\rangle \cdot \langle 0, 0, -1 \rangle$$

$$= xz = 0$$

Front of cube, F, ($x = 1$):

$$\mathbf{F} \cdot \mathbf{n} = \left\langle 2xz, \, y^2, \, -xz \right\rangle \cdot \langle 1, 0, 0 \rangle$$

$$= 2xz = 2z$$

$$\iint_F \mathbf{F} \cdot \mathbf{n} \, dS = \iint_F 2z \, dS$$

$$= \int_0^1 \int_0^1 2z \, dy \, dz = 1$$

Right of cube, R, ($y = 1$):

$$\mathbf{F} \cdot \mathbf{n} = \left\langle 2xz, \, y^2, \, -xz \right\rangle \cdot \langle 0, 1, 0 \rangle$$

$$= y^2 = 1$$

$$\iint_R \mathbf{F} \cdot \mathbf{n} \, dS = \iint_R 1 \, dS$$

$$= \int_0^1 \int_0^1 1 \, dz \, dx = 1$$

Top of cube, T, ($z = 1$):

$$\mathbf{F} \cdot \mathbf{n} = \left\langle 2xz, \, y^2, \, -xz \right\rangle \cdot \langle 0, 0, 1 \rangle$$

$$= -xz = -x$$

$$\iint_T \mathbf{F} \cdot \mathbf{n} \, dS = \iint_T -x \, dS$$

$$= \int_0^1 \int_0^1 -x \, dx \, dy = -\frac{1}{2}$$

Summing the above gives

$$\iint_{\partial Q} \mathbf{F} \cdot \mathbf{n} \, dS = 1 + 1 - \frac{1}{2} = \frac{3}{2}$$

On the other hand, div $\mathbf{F} = 2z + 2y - x$.

$$\iiint_Q \text{div}(\mathbf{F}) \, dV$$

$$= \iiint_Q (2x + 2y - x) \, dV$$

$$= \int_0^1 \int_0^1 \int_0^1 (2x + 2y - x) \, dz \, dy \, dz$$

$$= \frac{3}{2}$$

3.   The bottom of the surface ($z = 0$):

$$\mathbf{F} \cdot \mathbf{n} = \left\langle xz, \, zy, \, 2z^2 \right\rangle \cdot \langle 0, 0, -1 \rangle$$

$$= -2z^2 = 0$$

Therefore we only need to consider the top surface.

Top surface ($z = 1 - x^2 - y^2 = 1 - r^2$). Remember our unit normal must be unit and outward pointing.

$$\mathbf{m} = \langle -2x, -2y, -1 \rangle$$

$$\mathbf{n} = -\frac{1}{\|\mathbf{m}\|} \mathbf{m}$$

$$\mathbf{F} \cdot \mathbf{n} \, dS = -\mathbf{F} \cdot \mathbf{m} \, dA$$

$$= (2x^2 z + 2y^2 z + 2z^2) \, dA$$

$$= 2z(r^2 + z) \, dA = 2(1 - r^2) \, dA$$

$$\iint_{\partial Q} \mathbf{F} \cdot \mathbf{n} \, dS = \iint_{\partial Q} 2(1 - r^2) \, dS$$

$$= \int_0^{2\pi} \int_0^1 2(1 - r^2) r \, dr \, d\theta$$

$$= 2\pi \int_0^1 2(r - r^3) \, dr = \pi$$

On the other hand, $\operatorname{div}(\mathbf{F}) = z + z + 4z = 6z$.

$$\iiint_Q \nabla \cdot F \, dV = \iiint_Q 6z \, dV$$

$$= 6 \int_0^{2\pi} \int_0^1 \int_0^{1-r^2} zr \, dz \, dr \, d\theta$$

$$= 3 \int_0^{2\pi} \int_0^1 r(1 - r^2)^2 \, dr \, d\theta$$

$$= 6\pi \int_0^1 r(1 - r^2)^2 \, dr = \pi$$

5. $\nabla \cdot \mathbf{F} = 2 - 2 + 0 = 0$

$$\iint_{\partial Q} \mathbf{F} \cdot \mathbf{n} \, dS = \iiint_Q \nabla \cdot F \, dV = 0$$

7. $\nabla \cdot \mathbf{F} = 0 + 0 + 3z^2 = 3z^2$

$$\iint_{\partial Q} \mathbf{F} \cdot \mathbf{n} \, dS = \iiint_Q \nabla \cdot F \, dV$$

$$= \int_{-1}^1 \int_{-1}^1 \int_{-1}^1 3z^2 \, dz \, dy \, dz$$

$$= \int_{-1}^1 \int_{-1}^1 2 \, dy \, dz = 8$$

9. $\nabla \cdot \mathbf{F} = 3x^2 + 3y^2 + 0$

$$= 3x^2 + 3y^2 = 3r^2$$

$$\iint_{\partial Q} \mathbf{F} \cdot \mathbf{n} \, dS = \iiint_Q \nabla \cdot \mathbf{F} \, dV$$

$$= \iiint_Q 3r^2 \, dV$$

$$= \int_0^{2\pi} \int_0^2 \int_{r^2}^4 3r^3 \, dz \, dr \, d\theta$$

$$= \int_0^{2\pi} \int_0^2 3r^3(4 - r^2) \, dr \, d\theta$$

$$= 6\pi \int_0^2 r^3(4 - r^2) \, dr = 32\pi$$

11. $\nabla \cdot \mathbf{F} = 0 + x^2 + y^2 = r^2$

$$\iint_{\partial Q} \mathbf{F} \cdot \mathbf{n} \, dS = \iiint_Q \nabla \cdot \mathbf{F} \, dV$$

$$= \iiint_Q r^2 \, dV$$

$$= \int_0^{2\pi} \int_0^{\sqrt{3}} \int_0^1 r^3 \, dz \, dr \, d\theta$$

$$+ \int_0^{2\pi} \int_{\sqrt{3}}^2 \int_0^{4-r^2} r^3 \, dz \, dr \, d\theta$$

$$= \int_0^{2\pi} \int_0^{\sqrt{3}} r^3 \, dr \, d\theta$$

$$+ \int_0^{2\pi} \int_{\sqrt{3}}^2 r^3(4 - r^2) \, dr \, d\theta$$

$$= \int_0^{2\pi} \frac{9}{4} \, d\theta$$

$$+ \int_0^{2\pi} \frac{5}{6} \, d\theta$$

$$= \frac{9\pi}{2} + \frac{5\pi}{3} = \frac{37\pi}{6}$$

13. $\nabla \cdot \mathbf{F} = 1 + 0 + 3 = 4$

$$\iint_{\partial Q} \mathbf{F} \cdot \mathbf{n} \, dS = \iiint_Q \nabla \cdot \mathbf{F} \, dV$$

$$= \iiint_Q 4 \, dV$$

$$= 4 \, (\text{Volume of Cylinder}) = 4\pi$$

15. $\nabla \cdot \mathbf{F} = 3x^2 + 3y^2 + 3z^2 = 3\rho^2$

$$\iint_{\partial Q} \mathbf{F} \cdot \mathbf{n} \, dS = \iiint_Q \nabla \cdot \mathbf{F} \, dV$$

$$= \iiint_Q 3\rho^2 \, dV$$

$$= \int_0^{2\pi} \int_0^{\pi/2} \int_0^1 3\rho^4 \sin\phi \, d\rho \, d\phi \, d\theta$$

$$= \int_0^{2\pi} \int_0^{\pi/2} \frac{3}{5} \sin\phi \, d\phi \, d\theta$$

$$= \frac{6\pi}{5} \int_0^{\pi/2} \sin\phi \, d\phi = \frac{6\pi}{5}$$

17. $\nabla \cdot \mathbf{F} = 2x + 0 + 0 = 2x$

$$\iint_{\partial Q} \mathbf{F} \cdot \mathbf{n} \, dS = \iiint_Q \nabla \cdot \mathbf{F} \, dV$$

$$= \iiint_Q 2x \, dV$$

$$= \int_0^{2\pi} \int_0^1 \int_r^{\sqrt{2-r^2}} 2\rho^3 \sin^2\phi \cos\theta \, dz \, dr \, d\theta$$

$$= \left(\int_0^{2\pi} \cos\theta \, d\theta \right)$$

$$\cdot \left(\int_0^1 \int_r^{\sqrt{2-r^2}} 2\rho^3 \sin^2\phi \, dz \, dr \right)$$

$$= (0) \left(\int_0^1 \int_r^{\sqrt{2-r^2}} 2\rho^3 \sin^2\phi \, dz \, dr \right)$$

$$= 0$$

19. $\nabla \cdot \mathbf{F} = 0 + 0 + 2z = 2z$

$$\iint_{\partial Q} \mathbf{F} \cdot \mathbf{n} \, dS = \iiint_Q \nabla \cdot \mathbf{F} \, dV$$

$$= \iiint_Q 2z \, dV$$

$$= \int_0^{2\pi} \int_0^1 \int_0^r 2zr \, dz \, dr \, d\theta$$

$$= \int_0^{2\pi} \int_0^1 r^3 \, dr \, d\theta$$

$$= \int_0^{2\pi} \frac{1}{4} \, d\theta = \frac{\pi}{2}$$

21. $\nabla \cdot \mathbf{F} = 0 + 2 - 1 = 1$

$$\iint_{\partial Q} \mathbf{F} \cdot \mathbf{n} \, dS = \iiint_Q \nabla \cdot \mathbf{F} \, dV$$

$$= \iiint_Q 1 \, dV$$

$$= \text{Volume of Cylinder} = \pi$$

23. $\nabla \cdot \mathbf{F} = 3x^2 + 3y^2 + 3z^2$ Use polar coordinates: $r^2 = y^2 + z^2$
$dV = r \, dx \, dr \, d\theta$

$$\iint_{\partial Q} \mathbf{F} \cdot \mathbf{n} \, dS = \iiint_Q \nabla \cdot \mathbf{F} \, dV$$

$$= \iiint_Q 3(x^2 + y^2 + z^2) \, dV$$

$$= \int_0^{2\pi} \int_0^2 \int_{r^2}^4 3(x^2 + r^2) r \, dx \, dr \, d\theta$$

$$= \int_0^{2\pi} \int_0^2 (64 + 12r^2 - r^6 - 3r^4) r \, dr \, d\theta$$

$$= 2\pi \int_0^2 (64 + 12r^2 - r^6 - 3r^4) r \, dr$$

$$= 224\pi$$

25. $\nabla \cdot \mathbf{F} = y^2 + 0 + 0 = y^2$

$$\iint_{\partial Q} \mathbf{F} \cdot \mathbf{n} \, dS = \iiint_Q \nabla \cdot \mathbf{F} \, dV$$

$$= \int_0^2 \int_0^{6-3x} \int_0^{(6-3x-z)/2} y^2 \, dy \, dz \, dx$$

$$= \frac{1}{24} \int_0^2 \int_0^{6-3x} (6 - 3x - z)^3 \, dz \, dx$$

$$= \frac{1}{96} \int_0^2 (6 - 3x)^4 \, dx = \frac{27}{5}$$

27. $\nabla \cdot \mathbf{F} = 2x + 3y^2 + 0 = 2x + 2y^2$

$$\iint_{\partial Q} \mathbf{F} \cdot \mathbf{n} \, dS = \iiint_Q \nabla \cdot \mathbf{F} \, dV$$

$$= \int_{-2}^{2} \int_{-2}^{2} \int_{-3}^{1-x^2} (2x + 3y^2) \, dz \, dy \, dx$$

$$= \int_{-2}^{2} \int_{-2}^{2} (4 - x^2)(2x + 3y^2) \, dy \, dx$$

$$= \int_{-2}^{2} \left(64 - 16x^2 + 32x - 8x^3 \right) \, dx$$

$$= \frac{512}{3}$$

29. With a little bit of work, one can show that

$$\operatorname{div}\left(\frac{\mathbf{r}}{r^3} \right) = 0$$

But, this does not mean that the flux of $\mathbf{E}$ out of the sphere of radius a is zero! This is because the field is not defined at the origin. But, this does tell us that the flux is the same out of any sphere containing the origin (so the flux is independent of a).

This can also be seen by computing the flux. In this case, the unit normal is $\mathbf{n} = \frac{1}{r}\mathbf{r}$.

$$\text{Flux} = \iint_{\partial Q} \mathbf{E} \cdot \mathbf{n} \, dS = \iint_S \frac{q\mathbf{r}}{r^3} \cdot \frac{1}{r}\mathbf{r} \, dS$$

$$= q \iint_S \frac{\mathbf{r} \cdot \mathbf{r}}{r^4} \, dS = q \iint_S \frac{r^2}{r^4} \, dS$$

$$= q \iint_S \frac{1}{r^2} \, dS = q \iint_S \frac{1}{a^2} \, dS$$

$$= \frac{q}{a^2} (\text{Area of Sphere}) = \frac{q}{a^2}(4\pi a^2) = 4\pi q$$

31. There isn't much difference between this case and the two-dimensional case. The key to this is the fact that
$$\operatorname{div}(f\nabla g) = f(\nabla^2 g) + \nabla f \cdot \nabla g,$$
which we prove here:

$$\operatorname{div}(f\nabla g) = \operatorname{div}\langle fg_x, fg_y, fg_z \rangle$$

$$= \frac{\partial}{\partial x}(fg_x) + \frac{\partial}{\partial y}(fg_y) + \frac{\partial}{\partial z}(fg_z)$$

$$= (f_x g_x + fg_{xx}) + (f_y g_y + fg_{yy})$$

$$+ (f_z g_z + fg_{zz})$$

$$= f(\nabla^2 g) + \nabla f \cdot \nabla g$$

Now that we have this, the Divergence Theorem gives us

$$\iint_{\partial Q} f\nabla g \cdot \mathbf{n} \, dS = \iiint_Q \operatorname{div}(f\nabla g) \, dV$$

$$= \iiint_Q \left(f\nabla^2 g + \nabla f \cdot \nabla g \right) dV$$

$$= \iiint_Q f(\nabla^2 g) \, dV + \iiint_Q \nabla f \cdot \nabla g \, dV$$

and the result is now immediate (subtract one of the integrals to the other side of the equation).

33. Assuming that the electric field $\mathbf{E}$ is suitable for the Divergence Theorem and the function ρ is a charge-density with respect to volume, one could say that the total charge in a region Q is given by

$$q = \iiint_Q \rho \, dV = \iiint_Q \epsilon_0 \nabla \cdot \mathbf{E} \, dV$$

$$= \epsilon_0 \iiint_Q \operatorname{div}(\mathbf{E}) \, dV = \epsilon_0 \iint_S \mathbf{E} \cdot \mathbf{n} \, dS$$

35. In this problem we assume that "infinite plane of constant charge density ρ" means that ρ is a constant with respect to area in the xy-plane.

$$\mathbf{E} = \left\langle 0, 0, \frac{cz}{|z|} \right\rangle$$

$$= \begin{cases} c\mathbf{k} & \text{if } z > 0 \\ -c\mathbf{k} & \text{if } z < 0 \end{cases}$$

To continue, let $S = S(h, a)$ be the cylinder: $0 \le x^2 + y^2 \le a$, $-h \le z \le h$.

Then the total charge in S is the area-density (ρ) times the horizontal cross-sectional area

(πa^2):

$q = \pi \rho a^2$

The electric field, being only vertical, has only flux through the top (T) and the bottom (B) of the cylinder. Therefore

$$\text{Flux} = \iint_{\partial S} \mathbf{E} \cdot \mathbf{n}\, dS$$

$$= \iint_T \mathbf{E} \cdot \mathbf{n}\, dS + \iint_B \mathbf{E} \cdot \mathbf{n}\, dS$$

$$= \iint_T c\mathbf{k} \cdot \mathbf{k}\, dS + \iint_T -c\mathbf{k} \cdot (-\mathbf{k})\, dS$$

$$= c(\text{Area of } T) + c(\text{Area of } B) = 2c\pi a^2.$$

Finally, putting everything together gives

$$q = \pi \rho a^2 = \epsilon_0 \cdot \text{flux} = \epsilon_0 2c\pi a^2$$

$$c = \frac{\rho}{2\epsilon_0}$$

15.8 STOKES' THEOREM

1. ∂S is a circle in the plane $z = 0$ and therefore $dz = 0$.
    $x = 2\cos t, \ y = 2\sin t, \ z = 0$
    $0 \leq t \leq 2\pi$

$$\int_{\partial S} \mathbf{F} \cdot d\mathbf{r} = \int_{\partial S} xz\, dx + 2y\, dy + z^3\, dz$$

$$= \int_{\partial S} 0\, dx + 2y\, dy + z^3(0)$$

$$= \int_{\partial S} 2y\, dy$$

$$= \int_0^{2\pi} 2(2\sin t)(2\cos t)\, dt = 0$$

$$\nabla \times \mathbf{F} = \langle 0, x, 0 \rangle$$

$$\mathbf{n} = \frac{\langle 2x, 2y, 1 \rangle}{\sqrt{4x^2 + 4y^2 + 1}}$$

$$dS = \sqrt{4x^2 + 4y^2 + 1}\, dA$$

$$\iint_S (\nabla \times \mathbf{F}) \cdot \mathbf{n}\, dS$$

$$= \iint_S 2xy\, dA$$

$$= \int_0^{2\pi} \int_0^2 2(r\cos\theta)(r\sin\theta) r\, dr\, d\theta$$

$$= \int_0^{2\pi} \int_0^2 2r^3 \cos\theta \sin\theta\, dr\, d\theta$$

$$= \int_0^{2\pi} 8\cos\theta \sin\theta\, d\theta = 0$$

3. For ∂S: $x = 2\cos t, \ y = 2\sin t, \ z = 0$
    $0 \leq t \leq 2\pi$

$$\int_{\partial S} \mathbf{F} \cdot d\mathbf{r} = \int_{\partial S} (2x - y)\, dx + yz^2\, dy + y^2 z\, dz$$

$$= \int_{\partial S} (2x - y)\, dx$$

$$= \int_0^{2\pi} [2(2\cos t) - 2\sin t](-2\sin t)\, dt$$

$$= \int_0^{2\pi} [-8\cos t \sin t + 4\sin^2 t]\, dt$$

$$= 0 + 4\pi = 4\pi$$

$$\nabla \times \mathbf{F} = \langle 0, 0, 1 \rangle$$

$$\mathbf{n} = \frac{1}{2}\langle x, y, z \rangle$$

Use spherical coordinates:
$x = 2\sin\phi \cos\theta, \ y = 2\sin\phi \sin\theta,$
$z = 2\cos\phi, \ dS = 4\sin\phi\, d\phi\, d\theta.$

$$\iint_S (\nabla \times \mathbf{F}) \cdot \mathbf{n}\, dS$$

$$= \iint_S \frac{z}{2}\, dS$$

$$= \int_0^{2\pi} \int_0^{\pi/2} \frac{2\cos\phi}{2} 4\sin\phi\, d\phi\, d\theta$$

$$= 4 \int_0^{2\pi} \int_0^{\pi/2} \cos\phi \sin\phi\, d\phi\, d\theta$$

$$= 8\pi \int_0^{\pi/2} \cos\phi \sin\phi\, d\phi = 4\pi$$

5. ∂S is the triangle with vertices at $(0, 0, 0)$, $(2, 0, 0)$, and $(0, 2, 0)$ and is therefore made up of three line segments.
C_1: from $(0, 0, 0)$ to $(2, 0, 0)$.
$x = t, y = 0, z = 0, 0 \leq t \leq 2$.

$$\int_{C_1} \mathbf{F} \cdot d\mathbf{r}$$

$$= \int_{C_1} (zy^4 - y^2)\, dx + (y - x^3)\, dy + z^2\, dz$$

$$= \int_0^2 [0 + 0 + 0]\, dt = 0$$

C_2: from $(2, 0, 0)$ to $(0, 2, 0)$.
$x = 2 - t, y = t, z = 0, 0 \leq t \leq 2$.

$$\int_{C_2} \mathbf{F} \cdot d\mathbf{r}$$

$$= \int_{C_2} (zy^4 - y^2)\, dx + (y - x^3)\, dy + z^2\, dz$$

$$= \int_0^2 [-t^2(-1) + (t - (2 - t)^3)(1) + 0]\, dt$$

$$= \int_0^2 (t^3 - 5t^2 + 13t - 8)\, dt = \frac{2}{3}$$

C_3: from $(0, 2, 0)$ to $(0, 0, 0)$.
$x = 0, y = 2 - t, z = 0, 0 \leq t \leq 2$.

$$\int_{C_3} \mathbf{F} \cdot d\mathbf{r}$$

$$= \int_{C_3} (zy^4 - y^2)\, dx + (y - x^3)\, dy + z^2\, dz$$

$$= \int_0^2 [0 + (2 - t)(-1) + 0]\, dt$$

$$= \int_0^2 (t - 2)\, dt = -2$$

Summing these up:

$$\iint_S (\nabla \times \mathbf{F}) \cdot \mathbf{n}\, dS$$

$$= \frac{2}{3} - 2 = -\frac{4}{3},$$

7. ∂S is the circle:
$x = \cos t, y = \sin t, z = 0, 0 \leq t \leq 2\pi$.

$$\iint_S (\nabla \times \mathbf{F}) \cdot \mathbf{n}\, dS = \int_{\partial S} \mathbf{F} \cdot d\mathbf{r}$$

$$= \int_{\partial S} xz^2\, dx + (ze^{xy^2} - x)\, dy + x \ln y^2\, dz$$

$$= \int_0^{2\pi} [0 + -(\cos t)(\cos t) + 0]\, dt$$

$$= \int_0^{2\pi} -\cos^2 t\, dt = -\pi$$

9. ∂S is the triangle with vertices at $(0, 0, 0)$, $(0, 0, 1)$, and $(2, 0, 0)$ and is therefore made up of three line segments.
C_1: from $(0, 0, 0)$ to $(0, 0, 1)$.
$x = 0, y = 0, z = t, 0 \leq t \leq 1$.

$$\int_{C_1} \mathbf{F} \cdot d\mathbf{r}$$

$$= \int_{C_1} (zy^4 - y^2)\, dx + (y - x^3)\, dy + z^2\, dz$$

$$= \int_0^1 [0 + 0 + t^2(1)]\, dt = \frac{1}{3}$$

C_2: from $(0, 0, 1)$ to $(2, 0, 0)$.
$x = 2t, y = 0, z = 1 - t, 0 \leq t \leq 1$.

$$\int_{C_2} \mathbf{F} \cdot d\mathbf{r}$$

$$= \int_{C_2} (zy^4 - y^2)\, dx + (y - x^3)\, dy + z^2\, dz$$

$$= \int_0^1 [0 + 0 + (1 - t)^2(-1)]\, dt$$

$$= -\int_0^1 (1 - t)^2\, dt = -\frac{1}{3}$$

C_3: from $(2, 0, 0)$ to $(0, 0, 0)$.
$x = 2 - t, y = 0, z = 0, 0 \leq t \leq 2$.

$$\int_{C_3} \mathbf{F} \cdot d\mathbf{r}$$

$$= \int_{C_3} (zy^4 - y^2)\, dx + (y - x^3)\, dy + z^2\, dz$$

$$= \int_0^2 [0 + 0 + 0]\, dt = 0$$

Summing these up:

$$\iint_S (\nabla \times \mathbf{F}) \cdot \mathbf{n} \, dS = \frac{1}{3} - \frac{1}{3} = 0.$$

11. ∂S is the unit square with vertices at $(0, 0, 1)$, $(0, 1, 1)$, $(1, 1, 1)$, $(1, 0, 1)$.

If $\mathbf{n}$ is upward, then this means that we must traverse the unit square in counterclockwise direction.

C_1: $(0, 0, 1)$ to $(1, 0, 1)$.
$x = t, y = 0, z = 1, 0 \le t \le 1.$

$$\int_{C_1} \mathbf{F} \cdot d\mathbf{r}$$

$$= \int_{C_1} xyz \, dx + (4x^2y^3 - z) \, dy$$

$$+ 8\cos(xz^2) \, dz$$

$$= \int_0^1 (0 + 0 + 0) \, dt = 0$$

C_2: $(1, 0, 1)$ to $(1, 1, 1)$.
$x = 1, y = t, z = 1, 0 \le t \le 1.$

$$\int_{C_2} \mathbf{F} \cdot d\mathbf{r}$$

$$= \int_{C_2} xyz \, dx + (4x^2y^3 - z) \, dy$$

$$+ 8\cos(xz^2) \, dz$$

$$= \int_0^1 [0 + (4(1)(t)^3 - 1) + 0] \, dt$$

$$= \int_0^1 (4t^3 - 1) \, dt = 0$$

C_3: $(1, 1, 1)$ to $(0, 1, 1)$.
$x = 1 - t, y = 1, z = 1, 0 \le t \le 1.$

$$\int_{C_2} \mathbf{F} \cdot d\mathbf{r}$$

$$= \int_{C_2} xyz \, dx + (4x^2y^3 - z) \, dy$$

$$+ 8\cos(xz^2) \, dz$$

$$= \int_0^1 [(1 - t)(-1) + 0 + 0] \, dt$$

$$= \int_0^1 (t - 1) \, dt = -\frac{1}{2}$$

C_4: $(0, 1, 1)$ to $(0, 0, 1)$.
$x = 0, y = 1 - t, z = 1, 0 \le t \le 1.$

$$\int_{C_1} \mathbf{F} \cdot d\mathbf{r}$$

$$= \int_{C_1} xyz \, dx + (4x^2y^3 - z) \, dy$$

$$+ 8\cos(xz^2) \, dz$$

$$= \int_0^1 [0 + (-1)(-1) + 0] \, dt$$

$$= \int_0^1 1 \, dt = 1$$

$$\iint_S (\nabla \times \mathbf{F}) \cdot \mathbf{n} \, dS$$

$$= \int_{C_1} \mathbf{F} \cdot d\mathbf{r} + \int_{C_2} \mathbf{F} \cdot d\mathbf{r}$$

$$+ \int_{C_3} \mathbf{F} \cdot d\mathbf{r} + \int_{C_4} \mathbf{F} \cdot d\mathbf{r}$$

$$= 0 + 0 - \frac{1}{2} + 1 = \frac{1}{2}$$

13. ∂S is the circle (note the orientation):
$x = \cos t, y = -\sin t, z = 1,$
$0 \le t \le 2\pi.$

$$\iint_S (\nabla \times \mathbf{F}) \cdot \mathbf{n} \, dS = \int_{\partial S} \mathbf{F} \cdot d\mathbf{r}$$

$$= \int_{\partial S} (x^2 + y^2) \, dx + ze^{x^2+y^2} \, dy + e^{x^2+z^2} \, dz$$

$$= \int_0^{2\pi} [(1)(-\sin t) + e^1(-\cos t) + 0] \, dt$$

$$= 0$$

15. C is the circle $x^2 + z^2 = 4$ in the xz-plane. It is easier to just let S be the disk $x^2 + z^2 \le 4$ in the plane $y = 0$ (we can do this because the disk has the same boundary as the portion of the paraboloid and therefore the integrals will be equal).

$$\nabla \times \mathbf{F} = \langle 0, x^2 \rangle$$
$$\mathbf{n} = \langle 0, 1, 0 \rangle$$
$$dS = dA$$

$$\int_C \mathbf{F} \cdot d\mathbf{r} = \iint_S (\nabla \times \mathbf{F}) \cdot \mathbf{n} \, dS$$
$$= \iint_S x^2 \, dA$$
$$= \int_0^{2\pi} \int_0^2 r^3 \cos^2 \theta \, dr \, d\theta$$
$$= \int_0^{2\pi} 4 \cos^2 \theta \, d\theta$$
$$= 4\pi$$

17. C is the circle $x^2 + y^2 = 4$ in the plane $z = 0$. It is easier to just let S be the disk $x^2 + y^2 \leq 4$ in the plane $z = 0$ (we can do this because the disk has the same boundary as the portion of the paraboloid and therefore the integrals will be equal).

$$\nabla \times \mathbf{F} = \langle 0, 0, 1 \rangle$$
$$\mathbf{n} = \langle 0, 0, 1 \rangle$$
$$dS = dA$$
$$\int_C \mathbf{F} \cdot d\mathbf{r} = \iint_S (\nabla \times \mathbf{F}) \cdot \mathbf{n} \, dS$$
$$= \iint_S dA$$
$$= \text{Area of } S = 4\pi$$

19. $\nabla \times \mathbf{F} = \langle 0, 0, 0 \rangle$ Therefore,

$$\int_C \mathbf{F} \cdot d\mathbf{r} = \iint_S (\nabla \times \mathbf{F}) \cdot \mathbf{n} \, dS$$
$$= \iint_S (0) \, dA = 0$$

21.
$$\nabla \times \mathbf{F} = \langle 0, 0, 0 \rangle$$
$$\int_C \mathbf{F} \cdot d\mathbf{r} = \iint_S (\nabla \times \mathbf{F}) \cdot \mathbf{n} \, dS$$
$$= \iint_S 0 \, dA = 0$$

23. If we let S be the portion of the paraboloid $z = 4 - x^2 - y^2$ inside the cylinder $x^2 + z^2 = 1$, with $\mathbf{n}$ pointing to the left, then the surface is described by the function $f(x, z) = (4 - x^2 - z)^{1/2}$.

$$\nabla \times \mathbf{F} = \langle 0, -3, 0 \rangle$$
$$\mathbf{n} = \frac{\langle f_x, -1, f_z \rangle}{\sqrt{1 + f_x^2 + f_z^2}}$$
$$dS = \sqrt{1 + f_x^2 + f_z^2} \, dA$$

$$\int_C \mathbf{F} \cdot d\mathbf{r} = \iint_S (\nabla \times \mathbf{F}) \cdot \mathbf{n} \, dS$$
$$= \iint_R \langle 0, -3, 0 \rangle \cdot \langle f_x, -1, f_z \rangle \, dA$$
$$= \iint_R 3 \, dA$$
$$= 3 \, (\text{Area of Circle}) = 3\pi$$

25. The chain rule tells us that $\nabla(f^2) = 2f \nabla f$.

Therefore, the vector field $f \nabla f$ has a potential function $\frac{1}{2} f^2$. Therefore the curl of $f \nabla f$ is zero and Stokes' Theorem tells us that

$$\int_C \mathbf{F} \cdot d\mathbf{r} = \iint_S (\nabla \times \mathbf{F}) \cdot \mathbf{n} \, dS = 0.$$

27. The vector form of Green's Theorem is

$$\int_{\partial D} \mathbf{F} \cdot \mathbf{n} \, ds = \iint_D \nabla \cdot \mathbf{F} \, dA.$$

Using this, it is easy to see that if $\nabla \cdot \mathbf{F} = 0$ then

$$\int_{\partial D} \mathbf{F} \cdot \mathbf{n} \, ds = \iint_D \nabla \cdot \mathbf{F} \, dA = 0.$$

For the converse, the issue is whether a continuous integrand f (in our case, div $\mathbf{F}$) which integrates to zero over every set D must be identically zero. Suppose there is a single point P such that $f(P) > 0$. Then, there is a small disc D containing P such that $f(x, y) > \frac{1}{2} f(P)$ (such a disc exists because f is assumed to be continuous). Then,

$$\iint_D f\, dA \ge \frac{1}{2} f(P) \cdot \text{Area}(D) > 0$$

which is a contradiction.

If there were a point $f(P) < 0$, then the argument would still work but the resulting integral would be negative (but the contradiction is that the integral is not zero).

29. Given a vector field $\mathbf{F} = \langle M, N \rangle$, with div $\mathbf{F} = 0$, consider the perpendicular field $\mathbf{F}^\perp = \langle P, Q \rangle = \langle -N, M \rangle$. Then,

$$0 = \text{div } \mathbf{F} = M_x + N_y = Q_x - P_y$$

which is the condition for independence of path, which means that $\mathbf{F}^\perp$ has a potential function, $\nabla g = \mathbf{F}^\perp$. Therefore $g_x = P = -N$ and $g_y = Q = M$ which means that g is also a stream function for $\mathbf{F}$.

31. The only assumption is that the selection for normals for S_1 and S_2 induce the same orientation on the common boundary. Therefore the absolute value of integrals will be equal.

33. Exercise 48 of Section 15.5 gives the identity $\nabla \times (f\mathbf{F}) = \nabla f \times \mathbf{F} + f(\nabla \times \mathbf{F})$.

We apply this to the case $\mathbf{F} = \nabla g$, which gives

$$\nabla \times (f\nabla g) = \nabla f \times \nabla g + f(\nabla \times \nabla g)$$
$$= \nabla f \times \nabla g + f\mathbf{0}$$
$$= \nabla f \times \nabla g.$$

Thus, using Stokes' Theorem,

$$\int_C (f\nabla g) \cdot d\mathbf{r} = \iint_S \left[\nabla \times (f\nabla g)\right] \cdot \mathbf{n}\, dS$$
$$= \iint_S \left[\nabla f \times \nabla g\right] \cdot \mathbf{n}\, dS.$$

15.9   APPLICATIONS OF VECTOR CALCULUS

1. The surface S in Example 9.2 shares its boundary with the unit square, S_0, in the xy-plane ($z = 0$), and if we equip S_0 with the upward unit normal, $\langle 0, 0, 1 \rangle$, then the orientation of

$C = \partial S_0$ will be counterclockwise, and the same as that induced by an "outer" normal to S. Therefore

$$\iint_S \nabla \times \mathbf{F} \cdot \mathbf{n}\, dS = \oint_{\partial S} \mathbf{F} \cdot d\mathbf{r}$$
$$= \oint_{\partial S_0} \mathbf{F} \cdot d\mathbf{r}$$
$$= \oint_{\partial S_0} (e^{x^2} - 2xy)\, dx + \sin y^2\, dy$$
$$+ (3yz - 2x)\, dz$$
$$= \oint_{\partial S_0} (e^{x^2} - 2xy)\, dx + \sin y^2\, dy$$

We split this integral into the four parts of the ∂S.

C_1: from $(0, 0)$ to $(1, 0)$.
$y = 0$ and

$$\oint_{\partial C_1} (e^{x^2} - 2xy)\, dx + \sin y^2\, dy$$
$$= \int_0^1 e^{x^2}\, dx$$

C_2: from $(1, 0)$ to $(1, 1)$.
$x = 1$ and

$$\oint_{\partial C_2} (e^{x^2} - 2xy)\, dx + \sin y^2\, dy$$
$$= \int_0^1 \sin y^2\, dy$$

C_3: from $(1, 1)$ to $(0, 1)$.
$y = 1$ and

$$\oint_{\partial C_3} (e^{x^2} - 2xy)\, dx + \sin y^2\, dy$$
$$= \int_1^0 (e^{x^2} - 2x)\, dx$$
$$= \left(-\int_0^1 e^{x^2}\, dx\right) + 1$$

C_4: from $(0, 1)$ to $(0, 0)$.
$x = 0$ and

$$\oint_{\partial C_4} (e^{x^2} - 2xy)\, dx + \sin y^2\, dy$$

$$= \int_1^0 \sin y^2\, dy = -\int_0^1 \sin y^2\, dy$$

Adding these all up gives

$$\oint_{\partial S} \mathbf{F} \cdot d\mathbf{r}$$

$$= \int_0^1 e^{x^2}\, dx + \int_0^1 \sin y^2\, dy$$

$$- \int_0^1 e^{x^2}\, dx + 1 - \int_0^1 \sin y^2\, dy$$

$$= 1$$

3. Since div $\mathbf{E} = 0$, there is no charge on the sphere.

5. $\nabla \cdot \mathbf{E} = 6$ and therefore

$$q = \iiint_H \epsilon_0 \nabla \cdot \mathbf{E}\, dV = \iiint_H 6\epsilon_0\, dV$$

$$= 6\epsilon_0 (\text{ Volume of } H)$$

$$= 6\epsilon_0 \left(\frac{2}{3}\pi R^3\right) = 4\epsilon_0 \pi R^3.$$

7. The solid is unbounded and the divergence in nonzero, $\nabla \cdot \mathbf{E} = 4y$, therefore the charge must be infinite.

9. Note that $\mathbf{B}$ is a time-dependent field. If we let

$$\phi_S(t) = \iint_S \mathbf{B} \cdot \mathbf{n}\, dS$$

then

$$\iint_S \nabla \times \mathbf{E} \cdot \mathbf{n}\, dS = \oint_C \mathbf{E} \cdot d\mathbf{r}$$

$$= -\phi_S'(t) = -\iint_S \frac{d\mathbf{B}}{dt} \cdot \mathbf{n}\, dS.$$

Since this is true for all t and for all S, we must have $\nabla \times \mathbf{E} = -\frac{d\mathbf{B}}{dt}$.

Sufficient conditions would be that B and $\frac{d\mathbf{B}}{dt}$ be continuous in all variables (both time and space) and that $\mathbf{E}$ have continuous derivatives in all variables.

11. From Ampere's Law we have

$$\mathbf{J} = \epsilon_0 c^2 \nabla \times \mathbf{B} - \epsilon_0 \frac{d\mathbf{E}}{dt}.$$

Therefore

$$\nabla \cdot \mathbf{J} = \epsilon_0 c^2 \nabla \cdot (\nabla \times \mathbf{B}) - \epsilon_0 \nabla \cdot \frac{d\mathbf{E}}{dt}$$

$$= \epsilon_0 c^2 (0) - \epsilon_0 \frac{d}{dt} (\nabla \cdot \mathbf{E})$$

$$= -\epsilon_0 \frac{d}{dt} \left(\frac{\rho}{\epsilon_0}\right) = -\frac{d\rho}{dt}$$

The argument depends on the equality of mixed partial derivative (in two places). We also need to require that $\mathbf{E}$ and $\mathbf{B}$ have continuous second derivatives.

13. We are given

$$\iint_S \mathbf{J} \cdot \mathbf{n}\, dS = I = \oint_C \mathbf{B} \cdot d\mathbf{r}.$$

Applying Stokes' Theorem to this last integral gives the equality

$$\iint_S \mathbf{J} \cdot \mathbf{n}\, dS = \iint_S \nabla \times B \cdot \mathbf{n}\, dS.$$

Since this is true for all S, we must have equal integrands, $\mathbf{J} = \nabla \times B$.

15. With a little bit of work, one can show that $\nabla \cdot \mathbf{E} = 0$ and therefore the integral is zero.

Here's the work to show this. We need to show that $\nabla \cdot \left(\frac{\mathbf{r}}{r^3}\right) = 0$.

$$r^2 = x^2 + y^2 + z^2$$

$$2r\frac{\partial r}{\partial x} = 2x$$

$$\frac{\partial r}{\partial x} = \frac{x}{r}$$

$$\frac{\partial}{\partial x}\left(xr^{-3}\right) = r^{-3} + x(-3)r^{-4}\frac{\partial r}{\partial x}$$

$$= \frac{r^2 - 3x^2}{r^5}$$

Similarly, we can compute

$$\frac{\partial}{\partial x}\left(yr^{-3}\right) = \frac{r^2 - 3y^2}{r^5}$$

$$\frac{\partial}{\partial x}\left(zr^{-3}\right) = \frac{r^2 - 3z^2}{r^5}$$

Therefore,

$$\nabla \cdot \left(\frac{\mathbf{r}}{r^3}\right) = \frac{r^2 - 3x^2}{r^5} + \frac{r^2 - 3y^2}{r^5} + \frac{r^2 - 3z^2}{r^5}$$

$$= \frac{3r^2 - 3(x^2 + y^2 + z^2)}{r^5} = 0.$$

17. This problem is nearly identical to exercise 29 of Section 15.7.

In this case, the unit normal is $\mathbf{n} = \frac{1}{r}\mathbf{r}$. Also notice that for all points on the sphere, $r = R$.

$$\text{Flux} = \iint_{\partial Q} \mathbf{E} \cdot \mathbf{n}\, dS$$

$$= \iint_{S} \left(\frac{q}{4\pi\epsilon_0 r^3}\mathbf{r}\right) \cdot \left(\frac{1}{r}\mathbf{r}\right) dS$$

$$= \frac{q}{4\pi\epsilon_0} \iint_{S} \frac{\mathbf{r} \cdot \mathbf{r}}{r^4}\, dS = \frac{q}{4\pi\epsilon_0} \iint_{S} \frac{r^2}{r^4}\, dS$$

$$= \frac{q}{4\pi\epsilon_0} \iint_{S} \frac{1}{r^2}\, dS = \frac{q}{4\pi\epsilon_0 R^2} \iint_{S} dS$$

$$= \frac{q}{4\pi\epsilon_0 R^2}\,(\text{Area of Sphere})$$

$$= \frac{q}{4\pi\epsilon_0 R^2}\,(4\pi R^2) = \frac{q}{\epsilon_0}$$

19. Gauss' Law is introduced in Section 15.7 (exercises 33–36) (where ρ is used for charge density instead of Q as in this problem). Gauss's Law says that

$$Q = \epsilon_0 \nabla \cdot \mathbf{E} = \nabla \cdot (\epsilon_0 \mathbf{E}) = \nabla \cdot \mathbf{D}$$

21. We apply the Divergence Theorem:

$$\iint_{S} \frac{\partial u}{\partial n}\, dS = \iint_{S} \nabla u \cdot \mathbf{n}\, dS$$

$$= \iiint_{Q} \nabla \cdot (\nabla u)\, dV$$

$$= \iiint_{Q} \nabla^2 u\, dV.$$

23. We can pick up from Example 9.3 at the point:

$$\iiint_{Q} \nabla \cdot (k\nabla T)\, dV = \iiint_{Q} \rho\sigma\frac{\partial T}{\partial t}\, dV$$

From exercise 47 of Section 15.5, with $f = k$ and $\mathbf{F} = \nabla T$, $\nabla \cdot (k\nabla T) = \nabla k \cdot \nabla T + k\nabla^2 T$. Therefore,

$$\iiint_{Q} \left(k\nabla^2 T + \nabla k \cdot \nabla T - \rho\sigma\frac{\partial T}{\partial t}\right) dV$$

is zero for all Q. Therefore we conclude that the integrand is identically zero, so the heat equation takes the form:

$$\rho\sigma\frac{\partial T}{\partial t} = k\nabla^2 T + \nabla k \cdot \nabla T.$$

25. We apply exercise 24 (where we let the $h = f - g$). Remember that $h = 0$ on the surface $S = \partial Q$.

$$0 = \iint_{\partial Q} (0\nabla h) \cdot \mathbf{n}\, dS$$

$$= \iint_{\partial Q} (h\nabla h) \cdot \mathbf{n}\, dS$$

$$= \iiint_{Q} \left(h\nabla^2 h + \nabla h \cdot \nabla h\right) dS$$

$$= \iiint_{Q} \left[h\,(0 - 0) + \|\nabla h\|^2\right] dS$$

$$= \iiint_{Q} \|\nabla h\|^2\, dS$$

But, the last integrand ($\|\nabla h\|^2$) is nonnegative. Therefore, since the integral is zero we conclude that $\|\nabla h\| = 0$ and therefore $h = f - g$ is a constant. Since we know that $f - g = 0$ on the boundary, the constant must always be equal to zero. Thus, we have $f = g$.

Chapter 15 Review exercises

1.

3. $\mathbf{F}_4 = \langle 3, x^2 \rangle$: This is Graph A because all vectors have the same (positive) horizontal component.

$\mathbf{F}_1$ and $\mathbf{F}_2$ have the symmetry $\mathbf{F}_1(-x, -y) = -\mathbf{F}_1(x, y)$ and $\mathbf{F}_2(-x, -y) = -\mathbf{F}_2(x, y)$ which is visible in Graphs C and D. To sort these out, one telling feature of Graph C is that fourth quadrant vectors ($x > 0$, $y < 0$) point toward the origin whereas in Graph D they point away.

Therefore, $\mathbf{F}_1 = \langle \sin x, y \rangle$ is Graph D. and, $\mathbf{F}_2 = \langle \sin y, x \rangle$ is Graph C.

Finally, $\mathbf{F}_3 = \langle y^2, 2x \rangle$ is Graph B—all horizontal components are positive and all vertical components are positive for $x > 0$.

5. If
$\nabla f(x, y) = \langle y - 2xy^2, x - 2yx^2 + 1 \rangle$,
then
$\frac{\partial f}{\partial x} = y - 2xy^2$ and $\frac{\partial f}{\partial y} = x - 2yx^2 + 1$.

$f(x, y) = \int (y - 2xy^2) \, dx = xy - x^2y^2 + g(y)$

$\frac{\partial f}{\partial y} = x - 2yx^2 + g'(y) = x - 2yx^2 + 1$

$g'(y) = 1$
$g(y) = y + c$
$f(x, y) = xy - x^2y^2 + y + c$

The vector field is conservative.

7. If $\nabla f = \langle 2xy - 1, x^2 + 2xy \rangle$, then

$f(x, y) = \int (2xy - 1) \, dx$

$= x^2 y - x + g(y)$

$\frac{\partial f}{\partial y} = x^2 + g'(y) = x^2 + 2xy$

$g'(y) = 2xy$

But this is impossible since $g(y)$ is to be a function of y. Therefore the field is not conservative.

9. $\frac{dy}{dx} = \frac{2x/y}{y} = \frac{2x}{y^2}$

$\int y^2 \, dy = \int 2x \, dx$

$\frac{y^3}{3} = x^2 + c$

$y^3 = 3x^2 + k \quad (k = 3c)$

11. From $r^2 = x^2 + y^2$ we get $\frac{\partial r}{\partial x} = \frac{x}{r}$. Therefore

$\frac{\partial}{\partial x} \ln r = \frac{1}{r} \frac{\partial r}{\partial x} = \frac{x}{r^2}$.

Similarly, $\frac{\partial}{\partial y} \ln r = \frac{y}{r^2}$. Therefore,

$\nabla \ln r = \left\langle \frac{x}{r^2}, \frac{y}{r^2} \right\rangle = \frac{1}{r^2} \langle x, y \rangle = \frac{\mathbf{r}}{r^2}$.

13. Parameterize the segment by x, running from 2 to 4 ($y = 3$):

$\int_C 3y \, dx = \int_2^4 3(3) \, dx = 18$

15. In this case, the integrand is constant on the curve:

$\int_C \sqrt{x^2 + y^2} \, ds = \int_C 3 \, ds$

$= 3(\text{Length of } C)$
$= 3(6\pi) = 18\pi.$

17. C is a closed curve and the integral is of the form $\int_C f(x) \, dx$. All integrals of this form are equal to zero, basically being equal to a definite $\int_a^a f(x) \, dx$.

19. $x = 2\cos t, \ y = 2\sin t, \ 0 \le t \le 2\pi$

$$\oint_C \mathbf{F} \cdot d\mathbf{r} = \oint_C x \, dx - y \, dy$$

$$= \int_0^{2\pi} [(2\cos t)(-2\sin t) - (2\sin t)(2\cos t)] \, dt$$

$$= \int_0^{2\pi} -8(\sin t \cos t) \, dt = 0$$

21. We first parameterize the quarter-circle, C_1, by $x = 2\cos t, \ y = 2\sin t, \ 0 \le t \le \frac{\pi}{2}$.

$$\int_{C_1} \mathbf{F} \cdot d\mathbf{r} = \int_{C_1} 2 \, dx + 3x \, dy$$

$$= \int_0^{\pi/2} [2(-2\sin t) + 3(2\cos t)(2\cos t)] \, dt$$

$$= \int_0^{\pi/2} [-4\sin t + 12\cos^2 t] \, dt$$

$$= -4 + 3\pi$$

23. This one is a bit difficult to say. The motion is somewhat against the force in the early part of the trajectory and with the force in the later part of the trajectory. It appears that the later part will dominate; the work will be small but positive.

25. $x = \cos 3t, \ y = \sin 3t, \ z = 4t, \ 0 \le t \le 2\pi$

$$ds = \sqrt{[x'(t)]^2 + [y'(t)]^2 + [z'(t)]^2} + dt$$

$$= \sqrt{(-3\sin 3t)^2 + (3\cos 3t)^2 + 4^2} \, dt$$

$$= \sqrt{3^2 + 4^2} \, dt = 5 \, dt$$

$$m = \int_C \rho \, dS = \int_0^{2\pi} 5(4) \, dt = 40\pi$$

27. For $\mathbf{F} = \langle 3x^2 y - x, \ x^3 \rangle$, a potential function is $f(x, y) = x^3 y - \frac{1}{2}x^2$. Therefore,

$$\int_C (3x^2 y) \, dx - x^3 \, dy$$

$$= \left[x^3 y - \frac{1}{2}x^2 \right]_{(2,-1)}^{(4,1)}$$

$$= 56 - (-10) = 66.$$

29. A quick check shows that $\frac{\partial N}{\partial x} = \frac{\partial M}{\partial y}$ and therefore the field has a potential function. A potential function for $\mathbf{F}$ is

$f(x, y) = x^2 y - y\cos x + e^{x+y}$. Therefore,

$$\int_C \mathbf{F} \cdot d\mathbf{r} = \left[f(x, y) \right]_{(0,3)}^{(3,0)}$$

$$= f(3, 0) - f(0, 3)$$

$$= e^3 - (-3 + e^3) = 3$$

31. A bit of work shows that $\mathrm{curl}(\mathbf{F}) = \mathbf{0}$ and therefore the integral is independent of path. A potential function is

$f(x, y, z) = x^2 y - \frac{1}{2}y^2 + z^2$. Therefore,

$$\int_C \mathbf{F} \cdot d\mathbf{r} = \left[x^2 y - \frac{1}{2}y^2 + z^2 \right]_{(1,3,2)}^{(2,1,-3)}$$

$$= \frac{25}{2} - \frac{5}{2} = 10.$$

33. In inspecting the graph, we are looking for a closed path over which there is non-zero work. No such path is obvious.

Note also that the field appears to be independent of y and the vertical component is independent of x. This means that $M_y = 0 = N_y$ and that $N_x = 0$ from which we conclude that the field is conservative ($N_x = M_y$).

35. $M = x^3 - y$ and $N = x + y^3$. By Green's Theorem:

$$\oint M \, dx + N \, dy = \iint_R (N_x - M_y) \, dA$$

$$= \iint_R [1 - (-1)] \, dA = \iint_R 2 \, dA$$

$$= 2 \int_0^1 \int_{x^2}^x dy \, dx$$

$$= 2 \int_0^1 (x - x^2) \, dx = \frac{1}{3}.$$

37. Note that this curve is oriented negatively. $M = \tan(x^2)$ and $N = x^2$ By Green's Theorem:

$$\oint M\,dx + N\,dy = -\iint_R (N_x - M_y)\,dA$$

$$= -\iint_R (2x - 0)\,dA = -2\iint_R x\,dA$$

$$= -2\int_0^1 \int_y^{2-y} x\,dx\,dy$$

$$= -\int_0^1 [(2-y)^2 - y^2]\,dy$$

$$= -\int_0^1 (4 - 4y)\,dy = -2.$$

39. In this case, we have $x = 0$ and the integral becomes:

$$\oint_C \mathbf{F} \cdot d\mathbf{r}$$

$$= \oint_C 3x^2\,dx + (4y^3 - z)\,dy + z^2\,dz$$

$$= \oint_C (4y^3 - z)\,dy + z^2\,dz$$

$$= \oint_C M\,dy + N\,dz$$

$$= \iint_R \left(\frac{\partial N}{\partial y} - \frac{\partial M}{\partial z}\right) dA$$

$$= \int_{-2}^2 \int_{y^2}^4 [0 - (-1)]\,dz\,dy$$

$$= \int_{-2}^2 (4 - y^2)\,dy = \frac{32}{3}.$$

41. Parameterize the ellipse:
$x = 3\cos t,\ y = 2\sin t,\ 0 \le t \le 2\pi$

$$A = \frac{1}{2}\oint_C x\,dy - y\,dx$$

$$= \frac{1}{2}\int_0^{2\pi} [(3\cos t)(2\cos t)$$

$$\qquad - (2\sin t)(-3\sin t)]\,dt$$

$$= 3\int_0^{2\pi} (\cos^2 t + \sin^2 t)\,dt = 6\pi$$

43. $\operatorname{curl}\mathbf{F} = \nabla \times \mathbf{F}$

$$= \begin{vmatrix} \mathbf{i} & \mathbf{j} & \mathbf{k} \\ \frac{\partial}{\partial x} & \frac{\partial}{\partial y} & \frac{\partial}{\partial z} \\ x^3 & -y^3 & 0 \end{vmatrix}$$

$$= (0)\,\mathbf{i} - (0)\,\mathbf{j} + (0)\,\mathbf{k}$$

$$= \langle 0, 0, 0 \rangle$$

$\operatorname{div}\mathbf{F} = \nabla \cdot \mathbf{F}$

$$= \left\langle \frac{\partial}{\partial x}, \frac{\partial}{\partial y}, \frac{\partial}{\partial z} \right\rangle \cdot \left\langle x^3, -y^3, 0 \right\rangle$$

$$= 3x^2 - 3y^2 + 0 = 3x^2 - 3y^2$$

45. $\operatorname{curl}\mathbf{F} = \nabla \times \mathbf{F}$

$$= \begin{vmatrix} \mathbf{i} & \mathbf{j} & \mathbf{k} \\ \frac{\partial}{\partial x} & \frac{\partial}{\partial y} & \frac{\partial}{\partial z} \\ 2x & 2yz^2 & 2y^2z \end{vmatrix}$$

$$= (4yz - 4yz)\,\mathbf{i} - (0 - 0)\,\mathbf{j} + (0 - 0)\,\mathbf{k}$$

$$= \langle 0, 0, 0 \rangle$$

$\operatorname{div}\mathbf{F} = \nabla \cdot \mathbf{F}$

$$= \left\langle \frac{\partial}{\partial x}, \frac{\partial}{\partial y}, \frac{\partial}{\partial z} \right\rangle \cdot \left\langle 2x, 2yz^2, 2y^2z \right\rangle$$

$$= 2 + 2z^2 + 2y^2$$

47. $\operatorname{curl}\mathbf{F} = \nabla \times \mathbf{F}$

$$= \begin{vmatrix} \mathbf{i} & \mathbf{j} & \mathbf{k} \\ \frac{\partial}{\partial x} & \frac{\partial}{\partial y} & \frac{\partial}{\partial z} \\ 2x - y^2 & z^2 - 2xy & xy^2 \end{vmatrix}$$

$$= (2xy - 2z)\,\mathbf{i} - \left(y^2 - 0\right)\mathbf{j}$$

$$\qquad + (-2y - (-2y))\,\mathbf{k}$$

$$= \left\langle 2xy - 2z, y^2, 0 \right\rangle \ne \mathbf{0}$$

This vector field is not conservative.

$\operatorname{div}\mathbf{F} = \nabla \cdot \mathbf{F}$

$$= \left\langle \frac{\partial}{\partial x}, \frac{\partial}{\partial y}, \frac{\partial}{\partial z} \right\rangle \cdot$$

$$\qquad \left\langle 2x - y^2, z^2 - 2xy, xy^2 \right\rangle$$

$$= 2 - 2x + 0 = 2 - 2x$$

This vector field is not incompressible.

49. curl $\mathbf{F} = \nabla \times \mathbf{F}$

$$= \begin{vmatrix} \mathbf{i} & \mathbf{j} & \mathbf{k} \\ \frac{\partial}{\partial x} & \frac{\partial}{\partial y} & \frac{\partial}{\partial z} \\ 4x - y & 3 - x & 2 - 4z \end{vmatrix}$$

$$= (0 - 0)\,\mathbf{i} - (0 - 0)\,\mathbf{j}$$
$$+ (-1 + 1)\,\mathbf{k}$$
$$= \langle 0, 0, 0 \rangle$$

This vector field is conservative.

div $\mathbf{F} = \nabla \cdot \mathbf{F}$

$$= \left\langle \frac{\partial}{\partial x}, \frac{\partial}{\partial y}, \frac{\partial}{\partial z} \right\rangle \cdot$$
$$\langle 4x - y, 3 - x, 2 - 4z \rangle$$
$$= 4 + 0 - 4 = 0$$

This vector field is incompressible.

51. The divergence at P is positive because if we draw a small box around P, the flow in on the left exactly matches the flow our on the right whereas the flow out through the top is much greater than the flow in at the bottom.

53. This is a bit difficult to visualize. Remember to draw level curves and traces. Also notice that $x \geq 0$ and $y \geq 0$.

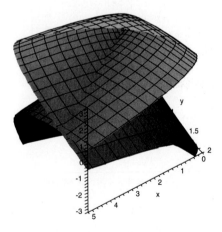

55. Surfaces B and C appear to have $x \geq 0$, which is true of **a** and **b**. By elimination **c** must be surface A, and this fits because A also has $z \geq 0$. Surface B also has $z \geq 0$, and therefore must correspond to **a**. This leaves **b** for surface C, which also looks reasonable.

(a) B

(b) C

(c) A

57. $f(x, y) = x^2 + y^2$

$$\mathbf{n} = \langle 2x, 2y, -1 \rangle$$
$$\|\mathbf{n}\| = \sqrt{4x^2 + 4y^2 + 1}$$
$$S = \iint_S dS$$
$$= \iint_R \|\mathbf{n}\|\, dA$$
$$= \iint_R \sqrt{4x^2 + 4y^2 + 1}\, dA$$
$$= \int_0^{2\pi} \int_1^2 r\sqrt{4r^2 + 1}\, dr\, d\theta$$
$$= \frac{1}{12} \int_0^{2\pi} \left(17^{3/2} - 5^{3/2} \right)\, d\theta$$
$$= \frac{\pi}{6} \left(17^{3/2} - 5^{3/2} \right)$$

59. $y = f(x, z) = 12 - 3x - 2y$

$$\mathbf{n} = \langle -3, -2, -1 \rangle$$
$$\|\mathbf{n}\| = \sqrt{9 + 4 + 1} = \sqrt{14}$$

The surface projects onto a triangle T in the xy-plane with vertices

$(0, 0)$, $(4, 0)$ and $(0, 6)$.

$$\iint_S (x - y)\, dS$$
$$= \iint_T (x - y)\sqrt{14}\, dA$$
$$= \sqrt{14} \int_0^4 \int_0^{6 - 3x/2} (x - y)\, dy\, dx$$
$$= \sqrt{14} \int_0^4 \left(-\frac{21}{8}x^2 + 15x - 18 \right)\, dx$$
$$= -8\sqrt{14}$$

61.
$$z = f(x, y) = 4 - \frac{4}{3}x - \frac{1}{3}y$$

$$\mathbf{n} = \left\langle -\frac{4}{3}, -\frac{1}{3}, -1 \right\rangle$$

$$\|\mathbf{n}\| = \sqrt{\frac{16}{9} + \frac{1}{9} + 1} = \frac{\sqrt{26}}{3}$$

The surface projects onto the unit disk D.

$$\iint_S (4x + y + 3z)\, dS$$

$$= \iint_D (12)\frac{\sqrt{26}}{3}\, dA = 4\sqrt{26}\iint_D dA$$

$$= 4\sqrt{26}(\text{Area of } D) = 4\pi\sqrt{26}$$

63.
$$y = f(x, z) = \sqrt{x^2 + z^2}$$

$$\mathbf{n} = \left\langle \frac{x}{\sqrt{x^2 + z^2}}, -1, \frac{z}{\sqrt{x^2 + z^2}} \right\rangle$$

$$\|\mathbf{n}\| = \sqrt{2}$$

We integrate over a disk and convert to polar coordinates.

$$\iint_S yz\, dS$$

$$= \iint_R z\sqrt{x^2 + z^2}\sqrt{2}\, dA$$

$$= \sqrt{2}\int_0^2 \int_0^{2\pi} (r\sin\theta)(r)r\, d\theta\, dr$$

$$= \sqrt{2}\int_0^2 r^3(0)\, dr = 0$$

65. $f(x, y) = x^2 + y^2$

$$\mathbf{n} = \langle 2x, 2y, -1 \rangle$$

$$\|\mathbf{n}\| = \sqrt{4x^2 + 4y^2 + 1} = \sqrt{4r^2 + 1}$$

$$dS = \sqrt{4r^2 + 1}\, dA$$

$$m = \iint_S \rho\, dS$$

$$= \iint_R 2\sqrt{4r^2 + 1}\, dA$$

$$= \int_0^{2\pi} \int_0^2 2r\sqrt{4r^2 + 1}\, dr\, d\theta$$

$$= 4\pi \int_0^2 r\sqrt{4r^2 + 1}\, dr$$

$$= 4\pi \left[\frac{1}{12}(1 + 4r^2)^{3/2} \right]_0^2$$

$$= \frac{\pi(17^{3/2} - 1)}{3}$$

Note that by symmetry, we must have $\bar{x} = \bar{y} = 0$.

$$\bar{z} = \frac{1}{m} \iint_S \rho z\, dS$$

$$= \frac{1}{m} \iint_R 2r^2\sqrt{4r^2 + 1}\, dA$$

$$= \frac{2}{m} \int_0^{2\pi} \int_0^2 r^3\sqrt{4r^2 + 1}\, dr\, d\theta$$

$$= \frac{4\pi}{m} \int_0^2 r^3\sqrt{4r^2 + 1}\, dr$$

$$= \frac{\pi}{30m}(391\sqrt{17} + 1)$$

$$= \frac{391\sqrt{17} + 1}{10(17^{3/2} - 1)}$$

67. The integrals in exercises 67–70 are integrated over ∂Q and not Q, as stated in the text.

$$\nabla \cdot \mathbf{F} = 0 + 2y + 0 = 2y$$

$$\iint_{\partial Q} \mathbf{F} \cdot \mathbf{n} \, dS = \iiint_Q \nabla \cdot \mathbf{F} \, dV$$

$$= \iiint_Q 2y \, dV$$

$$= \int_0^2 \int_0^{4-2y} \int_0^{4-2y-x} 2y \, dz \, dx \, dy$$

$$= \int_0^2 \int_0^{4-2y} (4 - 2y - x) 2y \, dx \, dy$$

$$= \int_0^2 \left(4y^3 - 16y^2 + 16y \right) dy = \frac{16}{3}$$

69. The integrals in exercises 67–70 are integrated over ∂Q and not Q, as stated in the text.

$$\nabla \cdot \mathbf{F} = 2y + 7x + 0 = 2y + 7x$$

The region Q is part of a cylinder set in the x-direction, whose cross section in the yz-plane is $\{(y, z) : 0 \le z \le 1 - y^2, -1 \le y \le 1\}$.

$$\iint_{\partial Q} \mathbf{F} \cdot \mathbf{n} \, dS = \iiint_Q \nabla \cdot \mathbf{F} \, dV$$

$$= \iiint_Q (2y + 7x) \, dV$$

$$= \int_{-1}^1 \int_0^{1-y^2} \int_0^{4-z} (2y + 7x) \, dx \, dz \, dy$$

$$= \int_{-1}^1 \int_0^{1-y^2} (8y - 2yz + 56$$

$$- 28z + \frac{7}{2}z^2 \Bigg) \, dz \, dy$$

$$= \int_{-1}^1 \left(\frac{259}{6} + 7y - \frac{63}{2}y^2 - 6y^3 \right.$$

$$\left. - \frac{21}{2}y^4 - y^5 - \frac{7}{6}y^6 \right) dy$$

$$= \int_{-1}^1 \left(\frac{259}{6} - \frac{63}{2}y^2 - \frac{21}{2}y^4 - \frac{7}{6}y^6 \right) dy$$

$$= \frac{304}{5}$$

71. We will use cylindrical coordinates:
$$Q = \{(r, \theta, z) : 0 \le z \le r, 0 \le r \le 2\}.$$

$$\nabla \cdot \mathbf{F} = z + z - 1 = 2z - 1$$

$$\iint_{\partial Q} \mathbf{F} \cdot \mathbf{n} \, dS = \iiint_Q \nabla \cdot \mathbf{F} \, dV$$

$$= \iiint_Q (2z - 1) \, dV$$

$$= \int_0^{2\pi} \int_0^2 \int_0^r (2z - 1) r \, dz \, dr \, d\theta$$

$$= \int_0^{2\pi} \int_0^2 (r^3 - r^2) \, dr \, d\theta$$

$$= 2\pi \int_0^2 (r^3 - r^2) \, dr = \frac{8\pi}{3}$$

73. ∂S is the triangle T in the yz-plane with vertices at $(0, 0, 0)$, $(0, 2, 0)$ and $(0, 0, 1)$.

$$\iint_S (\nabla \times \mathbf{F}) \cdot \mathbf{n} \, dS = \int_{\partial S} \mathbf{F} \cdot d\mathbf{r}$$

$$= \int_{\partial S} (zy^4 - y^2) \, dx + (y - x^3) \, dy + z^2 \, dz$$

We will assume that $\mathbf{n}$ is pointed in the direction of the positive x-axis.

For the segment C_1 from $(0, 0, 0)$ to $(0, 2, 0)$: $x = 0, y = t, z = 0, 0 \le t \le 2$.

$$\int_{C_1} \mathbf{F} \cdot d\mathbf{r} = \int_0^2 t \, dt = 2$$

For the segment C_2 from $(0, 2, 0)$ to $(0, 0, 1)$: $x = 0, y = 2 - 2t, z = t, 0 \le t \le 1$.

$$\int_{C_2} \mathbf{F} \cdot d\mathbf{r} = \int_0^1 [(2 - 2t)(-2) + t^2] \, dt$$

$$= \int_0^1 (t^2 + 4t - 4) \, dt = -\frac{5}{3}$$

For the segment C_3 from $(0, 0, 1)$ to $(0, 0, 0)$: $x = 0, y = 0, z = -t, -1 \le t \le 0$.

$$\int_{C_1} \mathbf{F} \cdot d\mathbf{r} = \int_{-1}^0 (-t)^2 (-1) \, dt = -\frac{1}{3}$$

Therefore,

$$\iint_S (\nabla \times \mathbf{F}) \cdot \mathbf{n} \, dS = \int_{\partial S} \mathbf{F} \cdot d\mathbf{r}$$

$$= 2 - \frac{5}{3} - \frac{1}{3} = 0.$$

75. $\operatorname{curl} \mathbf{F} = \nabla \times \mathbf{F}$

$$= \begin{vmatrix} \mathbf{i} & \mathbf{j} & \mathbf{k} \\ \frac{\partial}{\partial x} & \frac{\partial}{\partial y} & \frac{\partial}{\partial z} \\ 4x^2 & 2ye^y & \sqrt{z^2+1} \end{vmatrix}$$

$$= (0)\,\mathbf{i} - (0)\,\mathbf{j} + (0)\,\mathbf{k}$$

$$= \langle 0,\, 0,\, 0 \rangle$$

$$\iint_S (\nabla \times \mathbf{F}) \cdot \mathbf{n}\, dS = \iint_S (0) \cdot \mathbf{n}\, dS = 0$$

77. This field has zero curl, $\nabla \times \mathbf{F} = \mathbf{0}$. This can be seen by direct calculation (which is quite a bit of work) or by finding a potential function (which is also some work). A potential function for $\mathbf{F}$ is
$\phi = \frac{y^3}{3} + \frac{z^2}{2} + yx^2 \cos z$.

$$\iint_S (\nabla \times \mathbf{F}) \cdot \mathbf{n}\, dS = \iint_S (0) \cdot \mathbf{n}\, dS = 0$$

Second-Order Differential Equations

16.1 SECOND-ORDER EQUATIONS WITH CONSTANT COEFFICIENTS

1. The characteristic equation is
$$r^2 - 2r - 8 = 0$$
which has solutions
$r = 4$ and $r = -2$.

Thus, we are in "Case 1" and the general solution is
$$y = c_1 e^{4t} + c_2 e^{-2t}.$$

3. The characteristic equation is
$$r^2 - 4r + 4 = 0$$
which has solutions
$r = 2$ and $r = 2$.

Thus, we are in "Case 2" and the general solution is
$$y = c_1 e^{2t} + c_2 t e^{2t}.$$

5. The characteristic equation is
$$r^2 - 2r + 5 = 0$$
which has solutions
$r = 1 + 2i$ and $r = 1 - 2i$.

Thus, we are in "Case 3" and the general solution is
$$y = c_1 e^t \cos(2t) + c_2 e^t \sin(2t).$$

7. The characteristic equation is
$$r^2 - 2r = 0$$
which has solutions
$r = 0$ and $r = 2$.

Thus, we are in "Case 1" and the general solution is
$$y = c_1 + c_2 e^{2t}.$$

9. The characteristic equation is
$$r^2 - 2r - 6 = 0$$
which has solutions
$r = 1 + \sqrt{7}$ and $r = 1 - \sqrt{7}$.

Thus, we are in "Case 1" and the general solution is
$$y = c_1 e^{(1+\sqrt{7})t} + c_2 e^{(1-\sqrt{7})t}.$$

11. The characteristic equation is
$$r^2 - \sqrt{5}r + 1 = 0$$
which has solutions
$r = \frac{\sqrt{5}+1}{2}$ and $r = \frac{\sqrt{5}-1}{2}$.

Thus, we are in "Case 1" and the general solution is
$$y = c_1 e^{(\sqrt{5}+1)t/2} + c_2 e^{(\sqrt{5}-1)t/2}.$$

13. The characteristic equation is
$$r^2 + 4 = 0$$
which has solutions
$r = 2i$ and $r = -2i$.

Thus, we are in "Case 3" and the general solution is

$y = c_1 \cos(2t) + c_2 \sin(2t)$.
To solve the initial value problem, we use the initial conditions.
$2 = y(0) = c_1 - 3 = y'(0) = 2c_2$.

Solving gives $c_1 = 2$ and $c_2 = -\frac{3}{2}$ and therefore the solution is
$y = 2\cos(2t) - \dfrac{3}{2}\sin(2t)$.

15. The characteristic equation is
$r^2 - 3r + 2 = 0$
which has solutions
$r = 1$ and $r = 2$.

Thus, we are in "Case 1" and the general solution is
$y = c_1 e^t + c_2 e^{2t}$.

To solve the initial value problem, we use the initial conditions.
$0 = y(0) = c_1 + c_2$
$1 = y'(0) = c_1 + 2c_2$

Solving gives $c_1 = -1$ and $c_2 = 1$ and therefore the solution is
$y = -e^t + e^{2t}$.

17. The characteristic equation is
$r^2 - 2r + 5 = 0$
which has solutions
$r = 1 + 2i$ and $r = 1 - 2i$.

Thus, we are in "Case 3" and the general solution is
$y = c_1 e^t \cos(2t) + c_2 e^t \sin(2t)$.

To solve the initial value problem, we use the initial conditions.
$2 = y(0) = c_1$
$0 = y'(0) = c_1 + 2c_2$

Solving gives $c_1 = 2$ and $c_2 = -1$ and therefore the solution is
$y = 2e^t \cos(2t) - e^t \sin(2t)$.

19. The characteristic equation is
$r^2 - 2r + 1 = 0$
which has solutions
$r = 1$ and $r = 1$ (repeated).

Thus, we are in "Case 2" and the general solution is
$y = c_1 e^t + c_2 t e^t$.

To solve the initial value problem, we use the initial conditions.
$-1 = y(0) = c_1$
$2 = y'(0) = c_1 + c_2$.

Solving gives $c_1 = -1$ and $c_2 = 3$ and therefore the solution is
$y = -e^t + 3t e^t$.

21. Trigonometric identities give:

$A \sin(kt + \delta) = A\,(\sin kt \cos \delta + \cos kt \sin \delta)$.

Therefore $A \sin(kt + \delta) = c_1 \cos kt + c_2 \sin kt$ where $c_1 = A \sin \delta$ and $c_2 = A \cos \delta$. Then,

$c_1^2 + c_2^2 = A^2 \sin^2 \delta + A^2 \cos^2 \delta = A^2$.

and

$\tan \delta = \frac{\sin \delta}{\cos \delta} = \frac{A \sin \delta}{A \cos \delta} = \frac{c_1}{c_2}$.

Note that it is important to look at the signs of c_1 and c_2 to determine the correct quadrant for δ.

For the differential equation $y'' + 9y = 0$, the characteristic equation is
$r^2 + 9 = 0$
which has solutions
$r = 3i$ and $r = -3i$.

The general solution is
$y = c_1 \cos(3t) + c_2 \sin(3t)$.

To solve the initial value problem, we use the initial conditions.
$3 = y(0) = c_1$
$-6 = y'(0) = 3c_2$

Solving gives $c_1 = 3$ and $c_2 = -2$ and therefore the solution is
$y = 3\cos(3t) - 2\sin(3t) = \sqrt{13}\sin(2t + \delta)$
where $\delta = \pi + \tan^{-1}(-3/2) \approx 2.16$(rad) (second quadrant).

So, the amplitude is $\sqrt{13}$ and the phase shift is $\pi + \tan^{-1}(-3/2)$.

23. The characteristic equation is
$r^2 + 20 = 0$

which has solutions
$r = 2\sqrt{5}i$ and $r = -2\sqrt{5}i$.

The general solution is
$y = c_1 \cos(2\sqrt{5}t) + c_2 \sin(2\sqrt{5}t)$.

To solve the initial value problem, we use the initial conditions.
$-2 = y(0) = c_1$
$2 = y'(0) = 2\sqrt{5}c_2$

Solving gives $c_1 = -2$ and $c_2 = \frac{1}{\sqrt{5}}$ and therefore the solution is

$$y = -2\cos(2\sqrt{5}t) + \frac{1}{\sqrt{5}}\sin(2\sqrt{5}t)$$

$$= \sqrt{\frac{21}{5}}\sin(2\sqrt{5}t + \delta)$$

where $\tan\delta = 2\sqrt{5}$.

So, the amplitude is $\sqrt{\frac{21}{5}}$ and the phase shift is $\tan^{-1}(-2\sqrt{5}) \approx -1.351$ (fourth quadrant).

25. The mass is $m = \frac{12}{32} = \frac{3}{8}$. We now find k using $F = kx$:
$12 = k(1/2)$, so $k = 24$.

This leads to the differential equation
$$\frac{3}{8}u'' + 24u = 0, \quad u(0) = \frac{2}{3}, \quad u'(0) = 0.$$

The characteristic equation is $\frac{3}{8}r^2 + 24 = 0$, which has solutions $r = \pm 8i$. This gives the general solution
$u = c_1 \cos(8t) + c_2 \sin(8t)$.

To solve the initial value problem, we use the initial conditions.
$\frac{2}{3} = u(0) = c_1$
$0 = u'(0) = 8c_2$

Solving gives $c_1 = \frac{2}{3}$ and $c_2 = 0$ and therefore the solution is
$$u = \frac{2}{3}\cos(8t).$$

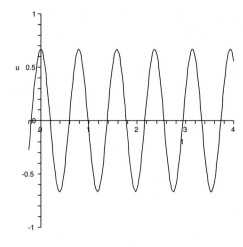

27. We first find k using $F = kx$:
$4(9.8) = k(0.1)$, so $k = 392$.

This leads to the differential equation
$4u'' + 392u = 0, \quad u(0) = 0.2,$
$u'(0) = -4$.

The characteristic equation is $4r^2 + 392 = 0$, which has solutions $r = \pm 7\sqrt{2}i$. This gives the general solution
$u = c_1 \cos(7\sqrt{2}t) + c_2 \sin(7\sqrt{2}t)$.

To solve the initial value problem, we use the initial conditions.
$0.2 = u(0) = c_1$
$-4 = u'(0) = 7\sqrt{2}c_2$

Solving gives $c_1 = \frac{1}{5}$ and $c_2 = -\frac{2\sqrt{2}}{7}$ and therefore the solution is
$$u = \frac{1}{5}\cos(7\sqrt{2}t) - \frac{2\sqrt{2}}{7}\sin(7\sqrt{2}t).$$

The amplitude is

$$A = \sqrt{\frac{1}{25} + \frac{8}{49}} = \frac{\sqrt{249}}{35} \approx 45 \text{ cm}.$$

The phase shift is

$$\delta = \pi + \tan^{-1}\left(\frac{1/5}{-2\sqrt{2}/7}\right)$$

$$= \pi + \tan^{-1}\left(-\frac{7\sqrt{2}}{20}\right)$$

$$\approx 2.682 \text{ (second quadrant)}.$$

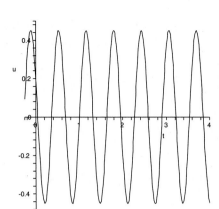

29. The mass is $m = \dfrac{16}{32} = \dfrac{1}{2}$. We now find k using $F = kx$:
$16 = k(1/3)$, so $k = 48$.

This leads to the differential equation
$\frac{1}{2}u'' + 10u' + 48u = 0$, $\quad u(0) = -\frac{1}{2}$,
$u'(0) = 0$.

The characteristic equation is
$\frac{1}{2}r^2 + 10r + 48 = 0$
which has solutions
$r = -12, -8$.

This gives the general solution
$u = c_1 e^{-12t} + c_2 e^{-8t}$.

To solve the initial value problem, we use the initial conditions.
$-\dfrac{1}{2} = u(0) = c_1 + c_2$
$0 = u'(0) = -12c_2 - 8c_2$

Solving gives $c_1 = 1$ and $c_2 = -\frac{3}{2}$ and therefore the solution is
$u = e^{-12t} - \dfrac{3}{2}e^{-8t}$.

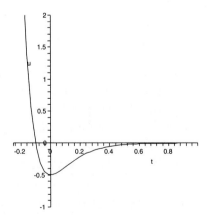

31. We first find k using $F = kx$:
$4(9.8) = k(0.25)$, so $k = 156.8$.

This leads to the differential equation
$4u'' + 2u + 156.8u = 0$,
$u(0) = -\frac{1}{2}$, $\quad u'(0) = 0$.

The characteristic equation is
$4r^2 + 2r + 156.8 = 0$
which has solutions
$r = -\dfrac{1}{4} \pm \dfrac{i}{20}\sqrt{15,655}$

This gives the general solution

$u = c_1 e^{-t/4} \cos\left(\dfrac{t\sqrt{15,655}}{20}\right)$

$\qquad + c_2 e^{-t/4} \sin\left(\dfrac{t\sqrt{15,655}}{20}\right)$.

To solve the initial value problem, we use the initial conditions.
$-\dfrac{1}{2} = u(0) = c_1$

$0 = u'(0) = \dfrac{\sqrt{15,655}}{20}c_2 - \dfrac{c_1}{4}$

Solving gives
$c_1 = -\frac{1}{2}$ and $c_2 = -\dfrac{\sqrt{15,655}}{6262}$
and therefore the solution is

$u = -\dfrac{1}{2}e^{-t/4} \cos\left(\dfrac{t\sqrt{15,655}}{20}\right)$

$\qquad - \dfrac{\sqrt{15,655}}{6262}e^{-t/4} \sin\left(\dfrac{t\sqrt{15,655}}{20}\right)$.

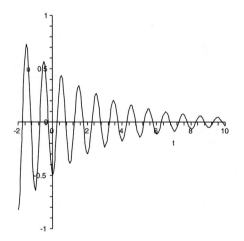

33. If the differential equation is
$ay'' + by' + cy = 0$, and if r_1 is a repeated root
to the characteristic polynomial, $P(r)$, then
$P(r) = ar^2 + br + c = a(r - r_1)^2$
and therefore
$0 = P(r_1) = ar_1^2 + br_1 + c$ and
$0 = P'(r_1) = 2ar_1 + b$.

We know that $y_1 = e^{r_1 t}$ is a solution. We now
test $y_2 = te^{r_1 t}$ in the differential equation.

$$y_2 = te^{r_1 t}$$

$$y_2' = r_1 te^{r_1 t} + e^{r_1 t}$$

$$y_2'' = r_1^2 te^{r_1 t} + 2r_1 e^{r_1 t}$$

and

$$ay_2'' + by_2' + cy_2$$

$$= a\left(r_1^2 te^{r_1 t} + 2r_1 e^{r_1 t}\right)$$

$$\quad + b\left(r_1 te^{r_1 t} + e^{r_1 t}\right) + c\left(te^{r_1 t}\right)$$

$$= \left(ar_1^2 + br_1 + c\right) te^{r_1 t} + \left(2ar_1 + b\right) e^{r_1 t}$$

$$= 0te^{r_1 t} + 0e^{r_1 t} = 0$$

35. The differential equation $u'' + cu' + 16u = 0$
has characteristic equation
$r^2 + cr + 16 = 0$, which has solutions
$r = \frac{-c \pm \sqrt{c^2 - 64}}{2}$.

If $0 < c < 8$ the roots are complex with negative real part. In this case, the solution will
show oscillatory behavior (similar to the $c = 0$
case) but the oscillations will slowly die off.
"Underdamped" suggests that the damping is
insufficient to completely eliminate oscillations.

If $c > 8$ then the roots are real, distinct, and
both negative. In this case there will be no
oscillatory behavior. "Overdamped" suggests
that the damping overrides the tendency to
oscillate, causing the solution to quickly go to
zero.

If $c = 8$ then there is a repeated root ($r = -4$).
This case separates the underdamped and
overdamped case, but in reality it is very

similar to the overdamped case—the solution
goes to zero with no tendency to oscillate.

37. Critical damping occurs when $c = 2\sqrt{mk}$. In
this case, we have
$m = \frac{16}{32} = \frac{1}{2}$
and
$k = \frac{16}{1/4} = 64$.

So, critical damping is when
$c = 2\sqrt{\frac{1}{2}(64)} = 8\sqrt{2}$.

39. Underdamping would mean that the system
would oscillate, probably not good. Over-
damping would mean that the door would take
a long time to shut, probably not so terrible.

41. It is routine to check that $\sinh at$ and $\cosh at$
are solutions to the differential equation. If we
start with the solution
$y = c_1 \sinh at + c_2 \cosh at$ and if we have
initial conditions $y(0) = y_0$ and $y'(0) = y_1$,
then solving for c_1 and c_2 amounts to solving
the equations

$$c_2 = y_0$$

$$ac_1 = y_1 \Rightarrow c_1 = \frac{y_1}{a}$$

Therefore, any solution is of the form $y =
c_1 \sinh at + c_2 \cosh at$ and this is therefore a
general solution.

The general solution of $y'' + a^2 y = 0$ is $y =
c_1 \sin at + c_2 \cos at$.

43. The roots of the characteristic equation are
$$r_1 = \frac{-b - \sqrt{b^2 - 4ac}}{2a}$$
$$r_2 = \frac{-b + \sqrt{b^2 - 4ac}}{2a}$$
If $a > 0$, we have $r_1 < r_2$ (if $a < 0$ then we
have $r_2 < r_1$ and the rest of the analysis still
holds with only slight modification). Since
$r_1 < r_2$, we only have to show that $r_2 < 0$.
Since $ac > 0$, we have $b^2 - 4ac < b^2$ and
therefore $\sqrt{b^2 - 4ac} < b$ (remember $b > 0$).
This means that $-b + \sqrt{b^2 - 4ac} < 0$ and
therefore $r_2 < 0$ (since we assumed $a > 0$).

Therefore the general solution is
$y = c_1 e^{r_1 t} + c_2 e^{r_2 t}$.

Since $r_1 < r_2 < 0$ we have $y \to 0$ as $t \to \infty$.

45. If there are complex roots, then we are in the situation of exercise 42. If there are distinct real roots, then we are in the situation of exercise 43. If there is a repeated root, then we are in the situation of exercise 44. In any case, we always have $y \to 0$ as $t \to \infty$. (Notice that each of these cases needed the fact that a, b and c are positive.)

16.2 NONHOMOGENEOUS EQUATIONS: UNDETERMINED COEFFICIENTS

1. The characteristic equation is
$r^2 + 2r + 5 = 0$
which has solutions $r = -1 \pm 2i$. Therefore, the general solution to the homogeneous equation is
$u = e^{-t} c_1 \cos 2t + e^{-t} c_2 \sin 2t$.
Therefore the general solution to the given equation is
$u = e^{-t} c_1 \cos 2t + e^{-t} c_2 \sin 2t + 3e^{-2t}$.

3. The characteristic equation is
$r^2 + 4r + 4 = 0$
which has solutions $r = -2$ (repeated root). Therefore, the general solution to the homogeneous equation is
$u = c_1 e^{-2t} + c_2 t e^{-2t}$.
Therefore the general solution to the given equation is
$u = c_1 e^{-2t} + c_2 t e^{-2t} + t^2 - 2t + \frac{3}{2}$.

5. The characteristic equation is
$r^2 + 2r + 10 = 0$
which has solutions $r = -1 \pm 3i$. Therefore, the general solution to the homogeneous equation is
$u = c_1 e^{-t} \cos 3t + c_2 e^{-t} \sin 3t$.

To find a particular solution, we try the guess
$u_p = A e^{-3t}$.

$26 e^{-3t} = u_p'' + 2u_p' + 10 u_p$
$\qquad = 9A e^{-3t} - 6A e^{-3t} + 10A e^{-3t}$
$\qquad = 13A e^{-3t}$

Solving gives $A = 2$. Therefore the general solution to the given equation is
$u = c_1 e^{-t} \cos 3t + c_2 e^{-t} \sin 3t + 2e^{-3t}$.

7. The characteristic equation is
$r^2 + 2r + 1 = 0$
which has solutions $r = -1$ (repeated root). Therefore, the general solution to the homogeneous equation is
$u = c_1 e^{-t} + c_2 t e^{-t}$.

To find a particular solution, we try the guess
$u_p = A \cos t + B \sin t$.

$25 \sin t = u_p'' + 2u_p' + u_p$
$\qquad = (-A \cos t - B \sin t)$
$\qquad\qquad + 2(-A \sin t + B \cos t)$
$\qquad\qquad + (A \cos t + B \sin t)$
$\qquad = 2B \cos t - 2A \sin t$

Solving gives $A = -\frac{25}{2}$ and $B = 0$. Therefore the general solution to the given equation is
$u = c_1 e^{-t} + c_2 t e^{-t} - \frac{25}{2} \cos t$.

9. The characteristic equation is
$r^2 - 4 = 0$
which has solutions $r = \pm 2$. Therefore, the general solution to the homogeneous equation is
$u = c_1 e^{-2t} + c_2 e^{2t}$.

To find a particular solution, we try the guess
$u_p = At^3 + Bt^2 + Ct + D$.

$2t^3 = u_p'' - 4u_p$
$\qquad = (6At + 2B) - 4(At^3 + Bt^2 + Ct + D)$
$\qquad = -4At^3 - 4Bt^2 + (6A - 4C)t + 2B - 4D$

Therefore we must have $-4A = 2$, $-4B = 0$, $6A - 4C = 0$ and $2B - 4D = 0$.

Solving gives $A = -\frac{1}{2}$, $B = 0$, $C = -\frac{3}{4}$ and $D = 0$. Therefore the general solution to the given equation is

$$u = c_1 e^{-2t} + c_2 e^{2t} - \frac{1}{2} t^3 - \frac{3}{4} t.$$

11. The characteristic equation is
$r^2 + 2r + 10 = 0$
which has solutions $r = -1 \pm 3i$. Therefore, the general solution to the homogeneous equation is
$$u = c_1 e^{-t} \cos 3t + c_2 e^{-t} \sin 3t.$$

The initial guess for the particular solution is

$$u_p = Ae^{-t} + e^{-t}(B \cos 3t + C \sin 3t)$$
$$+ D \cos 3t + E \sin 3t.$$

Unfortunately part of this solution is a solution to the homogeneous equation, so our modified guess is

$$u_p = Ae^{-t} + te^{-t}(B \cos 3t + C \sin 3t)$$
$$+ D \cos 3t + E \sin 3t.$$

13. The characteristic equation is
$r^2 + 2r = 0$
which has solutions $r = 0, -2$. Therefore, the general solution to the homogeneous equation is
$$u = c_1 + c_2 e^{-2t}.$$

The initial guess for the particular solution is

$$u_p = At^3 + Bt^2 + Ct + D + Ee^{2t}.$$

Unfortunately part of this solution is a solution to the homogeneous equation, so our modified guess is

$$u_p = t(At^3 + Bt^2 + Ct + D) + Ee^{2t}.$$

15. The characteristic equation is
$r^2 + 9 = 0$
which has solutions $r = \pm 3i$. Therefore, the general solution to the homogeneous equation is
$$u = c_1 \cos 3t + c_2 \sin 3t.$$

The initial guess for the particular solution is

$$u_p = e^t(A \cos 3t + B \sin 3t)$$
$$+ (Ct + D) \cos 3t + (Et + F) \sin 3t.$$

Unfortunately, part of this solution is a solution to the homogeneous equation, so our modified guess is

$$u_p = e^t(A \cos 3t + B \sin 3t)$$
$$+ t(Ct + D) \cos 3t$$
$$+ t(Et + F) \sin 3t.$$

17. The characteristic equation is
$r^2 + 4r + 4 = 0$
which has solutions $r = -2$ (repeated root). Therefore, the general solution to the homogeneous equation is
$$u = c_1 e^{-2t} + c_2 t e^{-2t}.$$

The initial guess for the particular solution is

$$u_p = (At^2 + Bt + C)e^{-2t}$$
$$+ (Dt + E)e^{-2t} \cos t$$
$$+ (Ft + G)e^{-2t} \sin t.$$

Unfortunately part of this solution is a solution to the homogeneous equation, so our modified guess is

$$u_p = t^2(At^2 + Bt + C)e^{-2t}$$
$$+ (Dt + E)e^{-2t} \cos t$$
$$+ (Ft + G)e^{-2t} \sin t.$$

19. We find k:
$0.1(9.8) = k(0.002)$ gives $k = 490$.

The differential equation is given by
$0.1u'' + 0.2u' + 490u = 0.1 \cos 4t$,
$u'' + 2u' + 4900u = \cos 4t$
$u'(0) = 0, \quad u(0) = 0.$

The characteristic equation is
$r^2 + 2r + 4900 = 0$
which has complex solutions
$r = -1 \pm i\sqrt{4899}.$

Therefore the general solution is

$$u = c_1 e^{-t} \cos(t\sqrt{4899}) + c_2 e^{-t} \sin(t\sqrt{4899})$$
$$+ A \cos 4t + B \sin 4t.$$

After some work, one finds that
$$A = \frac{1221}{5,963,380} \qquad B = \frac{1}{2,981,690}$$

Using the initial conditions we find that

$$c_1 = \frac{-1221}{5{,}963{,}380} \qquad c_2 = -\frac{1229}{5{,}963{,}380\sqrt{4899}}$$

21. The mass is $\dfrac{0.4}{32} = \dfrac{1}{80}$. We find k:
$0.4 = k(1/4)$ gives $k = 1.6$.

The differential equation is given by
$\frac{0.4}{32}u'' + 0.4u' + 1.6u = 0.2e^{-t/2}$
$u'' + 32u' + 128u = 16e^{-t/2}$
$u(0) = 0, \quad u'(0) = 1$

The characteristic equation is
$r^2 + 32r + 128 = 0$
which has complex solutions
$r = -16 \pm 8\sqrt{2}$.

Therefore the general solution is

$$u = c_1 e^{(-16+8\sqrt{2})t} + c_2 e^{(-16-8\sqrt{2})t} + Ae^{-t/2}.$$

After some work, one finds that
$A = \dfrac{64}{449}$.

Using the initial conditions we find that

$$c_1 = -\frac{32}{449} - \frac{543\sqrt{2}}{14{,}368} \qquad c_2 = -\frac{32}{449} + \frac{543\sqrt{2}}{14{,}368}$$

23. The characteristic equation is
$r^2 + 2r + 6 = 0$
which has solutions $r = -1 \pm i\sqrt{5}$. Therefore the homogeneous equation has general solution
$u = c_1 e^{-t}\cos(t\sqrt{5}) + c_2 e^{-t}\sin(t\sqrt{5})$
which goes to 0 as $t \to \infty$.

The particular solution is of the form
$u_p = A\cos 3t + B\sin 3t$.

$15\cos 3t = u_p'' + 2u_p' + 6u_p$
$\quad = (-9A\cos 3t - 9B\sin 3t)$
$\qquad + 2(-3A\sin 3t + 3B\cos 3t)$
$\qquad + 6(A\cos 3t + B\sin 3t)$
$\quad = (-3A + 6B)\cos 3t + (-6A - 3B)\sin 3t$

So we must solve $-3A + 6B = 15$ and $-6A - 3B = 0$. Solving gives $A = -1$ and $B = 2$. Therefore the steady-state solution to the given equation is

$u_p = -\cos 3t + 2\sin 3t$
$\quad = \sqrt{5}\sin(t + \delta)$

where $\delta = \tan^{-1}\left(-\frac{1}{2}\right) \approx -0.4636$ radians (fourth quadrant).

25. The characteristic equation is
$r^2 + 4r + 8 = 0$
which has solutions $r = -2 \pm 2i$. Therefore the homogeneous equation has the general solution
$u = c_1 e^{-2t}\cos 2t + c_2 e^{-2t}\sin 2t$
which goes to 0 as $t \to \infty$.

The particular solution is of the form
$u_p = A\cos t + B\sin t$.

$15\cos t + 10\sin t = u_p'' + 4u_p' + 8u_p$
$\quad = (-A\cos t - B\sin t)$
$\qquad + 4(-A\sin t + B\cos t)$
$\qquad + 8(A\cos t + B\sin t)$
$\quad = (7A + 4B)\cos t + (-4A + 7B)\sin t$

and therefore
$7A + 4B = 15$ and $-4A + 7B = 10$.

Solving gives $A = 1$ and $B = 2$. Therefore the steady-state solution to the given equation is

$u_p = \cos t + 2\sin t = \sqrt{5}\sin(t + \delta)$

where $\delta = \tan^{-1}(1/2) \approx 0.4636$ radians (first quadrant).

27. The mass is $\dfrac{2}{32} = \dfrac{1}{16}$. We now find k:
$k = \frac{2}{1/2} = 4$.

Therefore the differential equation is
$\frac{1}{16}u'' + 0.4u' + 4u = 2\sin 2t$
$u'' + 6.4u' + 64u = 32\sin 2t$

The characteristic equation is
$r^2 + 6.4r + 64 = 0$
which has solutions $r = -\frac{16}{5} \pm \frac{8i\sqrt{21}}{5}$. Therefore the solution to the homogeneous equation goes to 0 as $t \to \infty$.

The particular solution is of the form
$u_p = A\cos 2t + B\sin 2t$.

$32 \sin 2t = u_p'' + 6.4u_p' + 64u_p$

$= (-4A \cos 2t - 4B \sin 2t)$

$\quad + 6.4(-2A \sin 2t + 2B \cos 2t)$

$\quad + 64(A \cos 2t + B \sin 2t)$

$= (-4A + 12.8B + 64A) \cos 2t$

$\quad + (-4B - 12.8A + 64B) \sin 2t$

So we must solve the equations

$60A + 12.8B = 0$

$-12.8A + 60B = 32$

Solving gives $A = -\frac{640}{5881}$ and $B = \frac{3000}{5881}$. Therefore the steady-state solution to the given equation is

$$u_p = -\frac{640}{5881} \cos 2t + \frac{3000}{5881} \sin 2t$$

$$= \frac{40}{\sqrt{5881}} \sin (t + \delta)$$

where $\delta = \tan^{-1}\left(-\frac{64}{300}\right) \approx -0.2102$ (fourth quadrant).

29. For $u'' + 3u = 4 \sin \omega t$, the natural frequency of the system is $\sqrt{3}$. Resonance occurs if $\omega = \sqrt{3}$. Beats will occur if ω is close to $\sqrt{3}$, for example $\omega = 1.8$ will produce beats.

31. In this spring problem we have $m = \frac{0.4}{32}$ and $k = \frac{0.4}{1/4} = 1.6$. Therefore the differential equation is

$\frac{0.4}{32}u'' + 1.6u = 2 \sin \omega t$

$u'' + 128u$

$= 160 \sin \omega t$

The natural frequency is $\omega = \sqrt{\frac{k}{m}} = \sqrt{128}$.

Therefore resonance occurs if $\omega = \sqrt{128}$ and beats occur if ω is near $\sqrt{128}$. So, beats will occur, for example, if $\omega = 11$.

33. The general solution to the differential equation

$y'' + 9y = 12 \cos 3t$

is

$y_1 = c_1 \cos 3t + c_2 \sin 3t + 2t \sin 3t$.

Using our initial conditions and solving for c_1 and c_2 gives $c_1 = 1$ and $c_2 = 0$. Therefore, our

solution is

$y_1 = \cos 3t + 2t \sin 3t$.

In the case of damping, the general solution to the differential equation

$y'' + 0.1y' + 9y = 12 \cos 3t$

is

$$y_2 = c_1 e^{-t/20} \cos\left(\frac{t\sqrt{3599}}{20}\right)$$

$$\quad + c_2 e^{-t/20} \sin\left(\frac{t\sqrt{3599}}{20}\right)$$

$$\quad + 40 \sin 3t.$$

Using our initial conditions and solving for c_1 and c_2 gives $c_1 = 1$ and $c_2 = -\frac{2399}{\sqrt{3599}}$. Therefore, our solution is

$$y_2 = e^{-t/20} \cos\left(\frac{t\sqrt{3599}}{20}\right)$$

$$\quad - \frac{2399}{\sqrt{3599}} e^{-t/20} \sin\left(\frac{t\sqrt{3599}}{20}\right)$$

$$\quad + 40 \sin 3t.$$

If these two curves are plotted on the same axes, the remain very close for a long time, but the undamped solution eventually becomes unbounded. In the plot, the darker curve represents the damped solution.

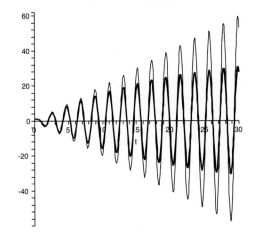

35. The form of the trial solution is

$u_p = A \cos \omega t + B \sin \omega t$.

Finding A and B:

$$\sin \omega t = u_p'' + 4u_p$$

$$= (-A\omega^2 \cos \omega t - B\omega^2 \sin \omega t)$$
$$+ 4(A \cos \omega t + B \sin \omega t)$$
$$= (-A\omega^2 + 4A) \cos \omega t$$
$$+ (-B\omega^2 + 4B) \sin \omega t$$

which requires $A(4 - \omega^2) = 0$, so $A = 0$ (unless $\omega = \pm 2$).

Of course, this happens only because there is no u' term—no damping. If there is damping, this will fail.

37. As explained in exercise 35, the form of the particular solution is $y_p = A \sin(2.1t)$. Solving for A gives the particular solution
$$y_p = -\frac{200}{41} \sin(2.1t)$$
The general solution is
$$y = c_1 \cos 2t + c_2 \sin 2t - \frac{200}{41} \sin(2.1t).$$
The initial conditions give equations:

$$0 = c_1$$

$$0 = 2c_2 - \frac{200(2.1)}{41}$$

So, $c_2 = \frac{210}{41}$ and the solution is
$$y = \frac{210}{41} \sin 2t - \frac{200}{41} \sin(2.1t).$$

39. If $\omega = 2$ then there is resonance.

If $\omega \neq 2$ then the general solution is
$$u = c_1 \cos 2t + c_2 \sin 2t + \frac{\sin \omega t}{4 - \omega^2}.$$

Using the initial conditions $u(0) = u'(0) = 0$ gives
$$0 = c_1 0 = 2c_2 + \frac{\omega}{4 - \omega^2}$$
and therefore $c_2 = -\frac{\omega}{2(4 - \omega^2)}$.

$$u = -\frac{\omega \sin 2t}{2(4 - \omega^2)} + \frac{\sin \omega t}{4 - \omega^2}$$

$\omega = 0.5$

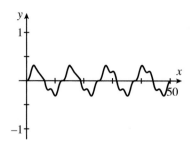

$\omega = 0.9$

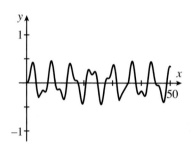

$\omega = 1$

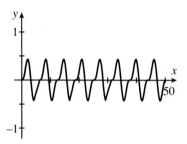

41. The characteristic equation is
$$r^2 + \frac{1}{10}r + 4 = 0$$
which has solutions $r = -\frac{1}{20} \pm \frac{i\sqrt{1599}}{20}$. Therefore the homogeneous equation goes to 0 as $t \to \infty$.

The particular solution is of the form
$$u_p = A \cos \omega t + B \sin \omega t.$$

$$\sin \omega t = u_p'' + \frac{1}{10}u_p' + 4u_p$$

$$= \left(-A\omega^2 + \frac{B\omega}{10} + 4A\right) \cos \omega t$$

$$+ \left(-B\omega^2 - \frac{A\omega}{10} + 4B\right) \sin \omega t$$

So we must solve

$$(4 - \omega^2)A + \frac{B\omega}{10} = 0$$

$$-\frac{A\omega}{10} + (4 - \omega^2)B = 1$$

Solving gives

$$A = -\frac{10\omega}{1600 - 799\omega^2 + 100\omega^4}$$

$$B = \frac{100(4 - \omega^2)}{1600 - 799\omega^2 + 100\omega^4}$$

The amplitude is therefore

$$\text{Amp} = \left[\left(-\frac{10\omega}{1600 - 799\omega^2 + 100\omega^4}\right)^2\right.$$

$$\left. + \left(\frac{100(4 - \omega^2)}{1600 - 799\omega^2 + 100\omega^4}\right)^2\right]^{1/2}$$

$$= \frac{1}{\sqrt{\omega^4 - 7.99\omega^2 + 16}}.$$

16.3 APPLICATIONS OF SECOND-ORDER EQUATIONS

1. We have
$R = 200$, $C = 10^{-4}$ and $L = 0.4$
which gives the differential equation
$0.4Q''(t) + 200Q'(t) + 10^4 Q(t) = 0$,
$Q(0) = 10^{-5}$, $Q'(0) = 0$
The characteristic equation is
$0.4r^2 + 200r + 10^4 = 0$
which has solutions
$r = -250 \pm 50\sqrt{15}$.
This gives the general solution

$$Q(t) = c_1 e^{-t(250 - 50\sqrt{15})} + c_2 e^{-t(250 + 50\sqrt{15})}.$$

Using the initial conditions

$10^{-5} = Q(0) = c_1 + c_2$

$0 = Q'(0) = c_1\left(-250 + 50\sqrt{15}\right)$

$\qquad\qquad + c_2\left(-250 - 50\sqrt{15}\right)$

Solving gives $c_1 = \left(\frac{3 + \sqrt{15}}{6}\right)10^{-5}$
and $c_2 = \left(\frac{3 - \sqrt{15}}{6}\right)10^{-5}$
and therefore the solution is

$$Q(t) = \left(\frac{3 + \sqrt{15}}{6}\right)10^{-5}e^{-t(250 - 50\sqrt{15})}$$

$$+ \left(\frac{3 - \sqrt{15}}{6}\right)10^{-5}e^{-t(250 + 50\sqrt{15})}$$

$$\approx 10^{-5}\left(1.1455e^{-56.35t} - 0.1455e^{-443.65t}\right)$$

$I(t) = Q'(t)$

$$\approx 6.45497 \times 10^{-4}4\left(+e^{-443.65t} - e^{-56.35t}\right)$$

3. We have
$R = 0$, $C = 10^{-5}$ and $L = 0.2$
which gives the differential equation
$0.2Q''(t) + 10^5 Q(t) = 0$,
$Q(0) = 10^{-6}$, $Q'(0) = 0$
The characteristic equation is
$0.2r^2 + 10^5 = 0$
which has solutions $r = \pm 500i\sqrt{2}$.
This gives the general solution
$Q(t) = c_1 \cos(500t\sqrt{2}) + c_2 \sin(500t\sqrt{2})$.
Using the initial conditions
$10^{-6} = Q(0) = c_1 0 = Q'(0) = 500c_2\sqrt{2}$.
Solving gives $c_1 = 10^{-6}$ and $c_2 = 0$ and therefore the solution is

$$Q(t) = 10^{-6}\cos(500t\sqrt{2})$$

$$= 10^{-6}\sin(500t\sqrt{2} + \pi/2).$$

So the amplitude is 10^{-6} and the phase shift is $\pi/2$.

5. We set up the differential equation
$0.5Q'' + 2Q' + 20Q = 3\cos 2t$

$Q'' + 4Q' + 40Q = 6 \cos 2t$
$Q(0) = 0 \quad Q'(0) = 1$

The characteristic equation is
$r^2 + 4r + 40 = 0$
which has solutions
$r = -2 \pm 6i$.

Therefore the general solution is
$Q = c_1 e^{-2t} \cos 6t + c_2 e^{-2t} \sin 6t \, A \cos 2t$
$\quad + B \sin 2t$.

After some work, one finds that
$A = \dfrac{27}{170}$ and $B = \dfrac{3}{85}$.

Using the initial conditions, one finds that
$c_1 = -\dfrac{27}{170}$ and $c_2 = \dfrac{26}{255}$.

Thus,

$$Q(t) = e^{-2t}\left(-\frac{27}{170}\cos 6t + \frac{26}{255}\sin 6t\right)$$
$$+ \frac{27}{170}\cos 2t + \frac{3}{85}\sin 2t$$

and

$$I(t) = Q'(t) = e^{-2t}\left(\frac{79}{85}\cos 6t + \frac{191}{255}\sin 6t\right)$$
$$+ \frac{6}{85}\cos 2t - \frac{27}{85}\sin 2t.$$

7. We set up the differential equation
$Q'' + 10Q' + 2Q = 0.1 \cos 2t$.

The characteristic equation is given by
$r^2 + 10r + 2 = 0$
which has solutions
$r = -5 \pm \sqrt{23}$.

Both of these are real and negative, therefore the solution to the homogeneous equation goes to 0 as $t \to \infty$.

The form of the particular solution (and the steady-state solution) is

$Q_p = A \cos 2t + B \sin 2t$.

After some work, one finds that
$A = -\dfrac{1}{2020} \quad B = \dfrac{1}{202}$
Therefore, the steady state solution is

$$Q_p = -\frac{1}{2020}\cos 2t + \frac{1}{202}\sin 2t$$
$$= \frac{\sqrt{101}}{2020}\sin(2t + \delta)$$

where $\delta = \tan^{-1}(-0.1)$ (fourth quadrant).

The amplitude is ≈ 0.00498 and the phase shift is ≈ 0.00498.

9. The characteristic equation is
$r^2 + 2r + 5 = 0$
and has solutions
$r = -1 \pm 2i$ so the solution to the homogeneous equation goes to 0 as $t \to \infty$. The steady-state solution has form
$x = A \cos \omega t + B \sin \omega t$,

Finding A and B:

$A_1 \sin \omega t = x'' + 2x' + 5x$
$\quad = (-A\omega^2 \cos \omega t - B\omega^2 \sin \omega t)$
$\quad\quad + 2(-A\omega \sin \omega t + B\omega \cos \omega t)$
$\quad\quad + 5(A \cos \omega t + B \sin \omega t)$.

Therefore we must have
$-A\omega^2 + 2B\omega + 5A = 0$
$-B\omega^2 - 2A\omega + 5B = A_1$
Solving gives

$$B = \frac{A_1(5 - \omega^2)}{(5 - \omega^2)^2 + 4\omega^2}$$

$$A = \frac{2B\omega}{\omega^2 - 5} = -\frac{2\omega A_1}{(5 - \omega^2)^2 + 4\omega^2}$$

If we choose to write the solution in the form
$x_p = A_2 \sin(\omega t + \delta)$
then we must have

$$A_2^2 = A^2 + B^2$$
$$= \frac{A_1^2\left[(-2\omega)^2 + (5 - \omega^2)^2\right]}{[(5 - \omega^2)^2 + 4\omega^2]^2}$$
$$= \frac{A_1^2}{(5 - \omega^2)^2 + 4\omega^2}$$
$$A_2 = \frac{A_1}{\sqrt{(5 - \omega^2)^2 + 4\omega^2}}$$

$$\frac{A_2}{A_1} = \frac{1}{\sqrt{(5 - \omega^2)^2 + 4\omega^2}}$$

and

$$\tan \delta = \frac{A}{B} = \frac{2\omega}{\omega^2 - 5}.$$

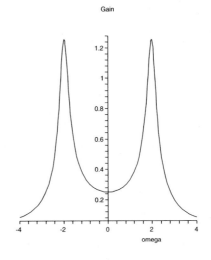

Gain

11. Looking back to exercise 9, the role of the 5 and the 2 can (and should) be generalized. If the differential equation is

$x'' + \beta x' + \gamma x = A_1 \sin \omega t$

and the steady-state solution is in the form

$x_p = A_2 \sin(\omega t + \delta)$

then the gain is the ratio $\frac{A_2}{A_1}$ and can be calculated as

$$g = \frac{A_2}{A_1} = \frac{1}{\sqrt{(\gamma - \omega^2)^2 + \beta^2 \omega^2}}$$

To maximize the gain, we need to minimize the radicand, which is equal to

$$(\gamma - \omega^2)^2 + \beta^2 \omega^2 = \omega^4 + (\beta^2 - 2\gamma)\omega^2 + \gamma^2.$$

This is a quadratic in ω^2 and has a minimum (the resonant frequency)

$$\omega_r^2 = \frac{2\gamma - \beta^2}{2}$$

$$\omega_r = \sqrt{\gamma - \frac{\beta^2}{2}}$$

for which

$$g_{\max} = \frac{2}{\beta\sqrt{4\gamma - \beta^2}}.$$

For the problem at hand, we have $\gamma = 4$ and $\beta = 0.4$. This gives us

$$g = \frac{1}{\sqrt{\omega^4 - 7.84\omega^2 + 16}}$$

$$\omega_r = \sqrt{4 - 0.08} = \sqrt{3.92}$$

$$g_{\max} = \frac{2}{0.4\sqrt{16 - 0.16}} \approx 1.2563$$

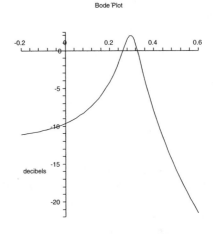

Bode Plot

13. We use the principle described in the solution to exercise 11, with $\gamma = 4$ and $\beta = 0.2$. This gives

$$g = \frac{1}{\sqrt{\omega^4 - 7.96\omega^2 + 16}}$$

$$\omega_r^2 = 4 - 0.02 = 3.98$$

$$g_{\max} = \frac{2}{0.2\sqrt{16 - 0.04}} \approx 2.503$$

Bode Plot

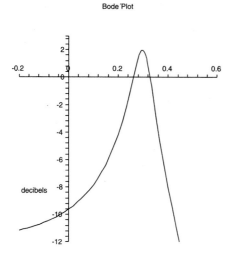

15. The computation of the gain function is the same whether the forcing function is a sine or a cosine. The presence of the ω^2 is also irrelevant, as it plays the role of A_1. Therefore, following the formulas of exercise 11, we use $\beta = 1$ and $\gamma = 4$. This gives

$$\omega_r = \sqrt{4 - \frac{1}{2}} \approx 1.8708$$

$$g_{max} = \frac{2}{\sqrt{16 - 1}} \approx 0.5164$$

From a seismological perspective, however, one must be more interested in the actual amplitude of the steady-state solution, which would be

$$A = \omega^2 g = \frac{\omega^2}{\sqrt{(\omega^2 - c)^2 + b^2 \omega^2}}.$$

In this case, we will have

$$A = \frac{1}{\sqrt{1 - \frac{7}{\omega^2} + \frac{16}{\omega^4}}}$$

which will have a maximum of
$A_{max} = \frac{8}{\sqrt{15}} \approx 2.656$
when $\omega = 4\sqrt{\frac{2}{7}} \approx 2.138$.

17. Both the functions $\sin^{-1}$ and $\tan^{-1}$ produce, as a matter of definition, numerical values in the interval $\left[-\frac{\pi}{2}, \frac{\pi}{2}\right]$. Geometrically, this is Quadrant I or IV. Thus no angle ϕ in Quadrant

III can obey
$$\phi = \sin^{-1}(\sin(\phi))$$
or
$$\phi = \tan^{-1}(\tan(\phi)).$$

The function $\cos^{-1}$, on the other hand, produces values in the interval $[0, \pi]$, and for angles ϕ in that interval, it is true that

$$\phi = \cos^{-1}(\cos(\phi)).$$

Thus, if indeed an angle θ is in Quadrant III or IV (recognized by the condition $\sin(\theta) < 0$), we may assume that, numerically, $-\pi \le \theta \le 0$. Under those circumstances, $-\theta$ lies in $[0, \pi]$ and because the cosine is an even function,

$$-\theta = \cos^{-1}(\cos(-\theta)) = \cos^{-1}(\cos(\theta))$$

which is another way of saying that

$$\theta = -\cos^{-1}(\cos(\theta)) \quad \text{if } -\pi \le \theta \le 0.$$

Without the proviso, this formula is correct up to added integral multiples of 2π whenever $\sin \theta < 0$.

In the case at hand, we have

$$\theta = -\cos^{-1}\left(\frac{5 - \omega^2}{\sqrt{(5 - \omega^2)^2 + (2\omega)^2}}\right).$$

Phase Shift

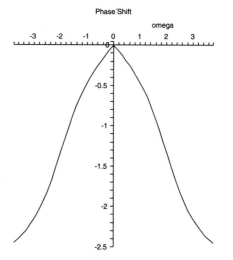

19. This is exercise 45 of Section 15.1.

21. The gain in the circuit described by
$$ax'' + bx' + cx = A\sin\omega t$$

will be the ratio $\frac{A_1}{A}$ where the steady-state solution is $A_1 \sin(\omega t + \delta)$.

We use the normalized equation:
$$x'' + \frac{b}{a}x' + \frac{c}{a}x = \frac{A}{a}\sin \omega t$$
and the results of exercise 11, with $\beta = \frac{b}{a}$ and $\gamma = \frac{c}{a}$, which gives the gain of the normalized equation as

$$\frac{A_1}{A/a} = \frac{1}{\sqrt{(\gamma - \omega^2)^2 + \beta^2 \omega^2}}$$

$$= \frac{1}{\sqrt{\left(\frac{c}{a} - \omega^2\right)^2 + \frac{b^2}{a^2}\omega^2}}$$

$$= \frac{a}{\sqrt{(c - a\omega^2)^2 + b^2\omega^2}}.$$

Therefore, canceling the a terms gives

$$g = \frac{A_1}{A} = \frac{a}{\sqrt{(c - a\omega^2)^2 + b^2\omega^2}}.$$

23. Substituting into Equation (3.7):

$$\theta'' + \frac{9.8}{0.10}\theta = 0 \qquad \theta(0) = 0.2, \ \theta'(0) = 0$$

The characteristic equation is
$$r^2 + 98 = 0$$
and has solutions
$r = \pm 7i\sqrt{2}$. Therefore the general solution is
$\theta = c_1 \cos(7t\sqrt{2}) + c_2 \sin(7t\sqrt{2})$.

Using the initial conditions to find c_1 and c_2:

$0.2 = \theta(0) = c_1$

$0 = \theta'(0) = c_2 7\sqrt{2} \Rightarrow c_2 = 0$

So the solution is
$\theta = 0.2 \cos(7t\sqrt{2})$.

The amplitude of the motion is 0.2 and the period of the motion is $\frac{2\pi}{7\sqrt{2}}$.

The length of the pendulum affects the period of the motion.

25. See the solution to exercise 23. This change will change the initial conditions to $\theta(0) = 0$ and $\theta'(0) = 0.1$.

The general solution remains as in exercise 23:
$\theta = c_1 \cos(7t\sqrt{2}) + c_2 \sin(7t\sqrt{2})$.

Using the initial conditions to find c_1 and c_2:

$0 = \theta(0) = c_1$

$0.1 = \theta'(0) = c_2 7\sqrt{2} \Rightarrow c_2 = \frac{1}{70\sqrt{2}}$

So the solution is
$\theta = \frac{1}{70\sqrt{2}}\sin(7t\sqrt{2})$

The amplitude of the motion is $\frac{1}{70\sqrt{2}} \approx 0.01$ and the period of the motion remains as in exercise 23, $\frac{2\pi}{7\sqrt{2}}$.

27. Substituting into Equation (3.8):
$$\theta'' + 0.2\theta' + \frac{32}{2/3}\theta = \frac{32}{6}\cos 3t.$$

The characteristic equation is
$$r^2 + 0.2r + 48 = 0$$
and has solutions
$r = -\frac{1}{10} \pm \frac{i}{10}\sqrt{4799}$. Therefore the steady-state solution is the particular solution:
$\theta = A \cos 3t + B \sin 3t$.

Finding A and B:

$$\frac{16}{3}\cos 3t = \theta'' + \frac{1}{5}\theta' + 48\theta$$

$$= (-9A\cos 3t - 9B\sin 3t)$$

$$+ \frac{1}{5}(-3A\sin 3t + 3B\cos 3t)$$

$$+ 48(A\cos 3t + B\sin 3t)$$

$$= \left(39A + \frac{3}{5}B\right)\cos 3t + \left(39B - \frac{3}{5}A\right)\sin 3t.$$

Solving the equations
$39A + \frac{3}{5}B = \frac{16}{3}$ $39B - \frac{3}{5}A = 0$
gives $A = \frac{2600}{19,017}$ and $B = \frac{40}{19,017}$.

Therefore the steady state solution is
$\theta = \frac{2600}{19,017}\cos 3t + \frac{40}{19,017}\sin 3t.$

The amplitude of the motion is
$\sqrt{A^2 + B^2} = \frac{40\sqrt{4226}}{19,017} \approx 0.1367$
and the period of the motion is $\frac{2\pi}{3}$.

29. The differential equation in Example 3.3 is
$u'' + 8u' + 2532u = \sin \omega t$.

The characteristic equation is
$r^2 + 8r + 2532 = 0$

and has solutions
$r = -4 \pm 2i\sqrt{629}$.

Therefore the homogeneous solution is

$$u = c_1 e^{-4t} \cos(2t\sqrt{629}) + c_2 e^{-4t} \sin(2t\sqrt{629})$$

which clearly goes to 0 as $t \to \infty$.

31. We know that the Taylor series for $\sin\theta$ is

$$\sin\theta = \theta - \frac{\theta^3}{3!} + \frac{\theta^5}{5!} - \cdots$$

Taylor's Theorem says that the remainder satisfies

$$|R_n(\theta)| = \left| \frac{f^{(n+1)}(z)}{(n+1)!} \theta^{n+1} \right|$$

where z is some point in the interval $[-\theta, \theta]$. We are interested in R_2 since the Taylor polynomial of degree 2 is $P_2 = \theta$.

Note that in all cases, $|f^{(n+1)}(z)| \le 1$ since $f(\theta) = \sin\theta$.

Therefore

$$|R_2(\theta)| \le \left| \frac{f^{(3)}(z)}{(3)!} \theta^3 \right| = \frac{|\theta|^3}{6}.$$

33. $\theta'' + \frac{g}{L}\theta = 0$
has characteristic equation is
$r^2 + \frac{g}{L} = 0$
and has solutions
$r = \pm i\sqrt{\frac{g}{L}}$. Therefore the general solution is

$$\theta = c_1 \cos\left(t\sqrt{\frac{g}{L}}\right) + c_2 \sin\left(t\sqrt{\frac{g}{L}}\right)$$

which has period
$T = 2\pi\sqrt{\frac{L}{g}}$.

This equation is equivalent to
$T^2 = \frac{4\pi^2}{g} L$
which says that the square of the period varies directly with the length.

35. In this model, Galileo was correct, the amplitude does not affect the period. This is demonstrated in exercise 24 but can also be seen in exercise 33—the amplitude is deter-

mined by the initial conditions, which do not affect the period of the solution.

37. $y'' + 2\alpha y' + \alpha^2 y = 0$
has characteristic equation is
$r^2 + 2\alpha r + \alpha^2 = 0$
which has solution $r = -\alpha$ (repeated root).

Therefore the general solution is

$y = c_1 e^{-\alpha t} + c_2 t e^{-\alpha t}$.

Using the initial condition to find c_1 and c_2:

$$0 = y(0) = c_1$$
$$100 = y'(0) = -\alpha c_1 + c_2$$

So, $c_1 = 0$ and $c_2 = 100$ and the solution is

$y = 100t e^{-\alpha t}$.

Therefore,

$y^2 + (y')^2 = 10{,}000 e^{-2\alpha t}$
$[(\alpha^2 + 1)t^2 - 2\alpha t + 1]$.

Evaluated at $t = 1$, this is equal to
$10{,}000 e^{-2\alpha}[(\alpha^2 + 1) - 2\alpha + 1]$.

Graphing, as seen below, shows that if $\alpha > 9$ then this quantity is less than 0.01.

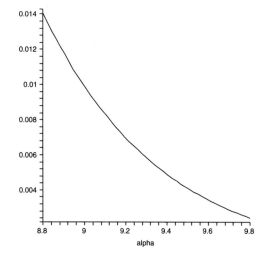

39. If $0 < \alpha < \omega$, then the differential equation is
$g'' + 2\alpha g' + \omega^2 g = 0$
and the characteristic equation is
$r^2 + 2\alpha r + \omega^2 = 0$, which has solutions

$$r = -\alpha \pm i\sqrt{\omega^2 - \alpha^2}.$$

Therefore the general solution to the differential equation will be

$$g(t) = e^{-\alpha t} \left[c_1 \cos \left(t\sqrt{\omega^2 - \alpha^2} \right) \right.$$
$$\left. + c_2 \sin \left(t\sqrt{\omega^2 - \alpha^2} \right) \right].$$

We can convert this solution to

$$g(t) = Ae^{-\alpha t} \sin \left(t\sqrt{\omega^2 - \alpha^2} + \delta \right)$$

where $A = \sqrt{c_1^2 + c_2^2}$ and $\tan \delta = \frac{c_1}{c_2}$.

Therefore there will be zeros only where

$$\sin \left(t\sqrt{\omega^2 - \alpha^2} + \delta \right) = 0$$

and the distance between zeros will be

$$\frac{\pi}{\sqrt{\omega^2 - \alpha^2}} > \frac{\pi}{\omega}.$$

The data given can be put in a table:

t	$G(t)$	$g(t) = G(t) - G_0$
1	0.9	0.15
2	0.7	-0.15
3	0.78	0.03

Therefore there are at least two zeros, less than 2 hours apart. It appears that the data is a damped exponential. Putting this all together gives

$$\frac{\pi}{\omega} < \text{(distance between zeros)} < 2$$

$$\frac{\pi}{\omega} < 2 \quad \Rightarrow \quad \frac{2\pi}{\omega} < 4$$

and the patient is not likely to be diabetic.

41. $Q'' + \frac{R}{L}Q' + \frac{1}{LC}Q = 0$
has characteristic equation
$r^2 + \frac{R}{L}r + \frac{1}{LC} = 0$, which has solutions

$$r = -\frac{R}{2L} \pm \frac{\sqrt{R^2 - \frac{4L}{C}}}{2L}.$$

If $R^2 < \frac{4L}{C}$ then the solutions are complex and the general solution to the differential equation will be

$$Q(t) = e^{-Rt/(2L)} \left[c_1 \cos \left(\frac{t\sqrt{4L/C - R^2}}{2L} \right) \right.$$
$$\left. + c_2 \sin \left(\frac{t\sqrt{4L/C - R^2}}{2L} \right) \right]$$
$$= e^{-Rt/(2L)} \left[Q_0 \cos (\omega t) + c_2 \sin (\omega t) \right].$$

It is clear that $c_1 = Q_0 = Q(0)$, as stated in the textbook.

Now, to find the inductance quality factor.

$$u(t) = \frac{Q(t)^2}{2C}$$

$$u(0) = \frac{Q(0)^2}{2C} = \frac{Q_0^2}{2C}$$

$$u(2\pi/\omega) = \frac{[Q(2\pi/\omega)]^2}{2C}$$
$$= \frac{Q_0^2 e^{-2R\pi/(L\omega)}}{2C}$$

$$\text{IQF} = \frac{2\pi}{U_{\text{loss}}}$$
$$= \frac{2\pi u(0)}{u(2\pi/\omega) - u(0)}$$
$$= \frac{\frac{2\pi Q_0^2}{2C}}{\frac{Q_0^2 e^{-2R\pi/(L\omega)}}{2C} - \frac{Q_0^2}{2C}}$$
$$= \frac{2\pi}{e^{-2R\pi/(L\omega)} - 1}$$
$$= \frac{2\pi}{-\frac{2R\pi}{L\omega} + \frac{1}{2!}\left(-\frac{2R\pi}{L\omega} \right)^2 + \cdots}$$
$$\approx \frac{2\pi}{-\frac{2R\pi}{L\omega}} = -\frac{L\omega}{R}$$

This is negative because energy is lost—$u(t)$ is decreasing and therefore U_{loss} is negative.

16.4 POWER SERIES SOLUTIONS OF DIFFERENTIAL EQUATIONS

1. Assume that $y = \sum\limits_{n=0}^{\infty} a_n x^n$.

$$0 = y'' + 2xy' + 4y$$

$$= \sum_{n=2}^{\infty} n(n-1)a_n x^{n-2}$$

$$+ 2x \sum_{n=1}^{\infty} na_n x^{n-1} + 4 \sum_{n=0}^{\infty} a_n x^n$$

$$= \sum_{n=0}^{\infty} (n+2)(n+1)a_{n+2} x^n$$

$$+ 2 \sum_{n=0}^{\infty} na_n x^n + 4 \sum_{n=0}^{\infty} a_n x^n$$

$$= \sum_{n=0}^{\infty} [(n+2)(n+1)a_{n+2}$$

$$+ 2na_n + 4a_n]x^n$$

So, $(n+2)(n+1)a_{n+2} + 2(n+2)a_n = 0$, or

$$a_{n+2} = -\frac{2(n+2)}{(n+2)(n+1)}a_n = -\frac{2}{n+1}a_n$$

so

$$a_{2n} = \frac{(-1)^n 2^{2n} n!}{(2n)!}a_0$$

$$a_{2n+1} = \frac{(-1)^n}{n!}a_1$$

The general solution is
$y = a_0 y_1 + a_1 y_2$
where

$$y_1(x) = \sum_{n=0}^{\infty} \frac{(-1)^n 2^{2n} n!}{(2n)!}x^{2n}$$

$$y_2(x) = \sum_{n=0}^{\infty} \frac{(-1)^n}{n!}x^{2n+1}$$

3. Assume that $y = \sum\limits_{n=0}^{\infty} a_n x^n$.

$$0 = y'' - xy' - y$$

$$= \sum_{n=2}^{\infty} n(n-1)a_n x^{n-2}$$

$$- x \sum_{n=1}^{\infty} na_n x^{n-1} - \sum_{n=0}^{\infty} a_n x^n$$

$$= \sum_{n=0}^{\infty} (n+2)(n+1)a_{n+2} x^n$$

$$- \sum_{n=0}^{\infty} na_n x^n - \sum_{n=0}^{\infty} a_n x^n$$

$$= \sum_{n=0}^{\infty} [(n+2)(n+1)a_{n+2} - (n+1)a_n]x^n$$

So, $(n+2)(n+1)a_{n+2} - (n+1)a_n = 0$, or

$$a_{n+2} = \frac{n+1}{(n+2)(n+1)}a_n = \frac{1}{n+2}a_n$$

so

$$a_{2n} = \frac{1}{2^n n!}a_0$$

$$a_{2n+1} = \frac{2^n n!}{(2n+1)!}a_1$$

The general solution is
$y = a_0 y_1 + a_1 y_2$
where

$$y_1(x) = \sum_{n=0}^{\infty} \frac{1}{2^n n!}x^{2n}$$

$$y_2(x) = \sum_{n=0}^{\infty} \frac{2^n n!}{(2n+1)!}x^{2n+1}$$

5. Assume that $y = \sum\limits_{n=0}^{\infty} a_n x^n$.

$0 = y'' - xy'$

$$= \sum_{n=2}^{\infty} n(n-1)a_n x^{n-2} - x\sum_{n=1}^{\infty} na_n x^{n-1}$$

$$= \sum_{n=0}^{\infty} (n+2)(n+1)a_{n+2}x^n - \sum_{n=0}^{\infty} na_n x^n$$

$$= \sum_{n=0}^{\infty}[(n+2)(n+1)a_{n+2} - na_n]x^n$$

So, $(n+2)(n+1)a_{n+2} - na_n = 0$, or

$$a_{n+2} = \frac{n}{(n+2)(n+1)}a_n$$

so

$$a_{2n} = 0 \quad \text{for } n > 0$$

$$a_{2n+1} = \frac{1}{(2n+1)n!2^n}a_1$$

The general solution is
$y = a_0 y_1 + a_1 y_2$
where

$y_1(x) = 1$

$$y_2(x) = \sum_{n=0}^{\infty}\frac{1}{(2n+1)n!2^n}x^{2n+1}$$

7. Assume that $y = \sum_{n=0}^{\infty} a_n x^n$.

$0 = y'' - x^2 y'$

$$= \sum_{n=2}^{\infty} n(n-1)a_n x^{n-2} - x^2\sum_{n=1}^{\infty} na_n x^{n-1}$$

$$= \sum_{n=2}^{\infty} n(n-1)a_n x^{n-2} - \sum_{n=1}^{\infty} na_n x^{n+1}$$

$$= 2a_2 + \sum_{n=3}^{\infty} n(n-1)a_n x^{n-2} - \sum_{n=1}^{\infty} na_n x^{n+1}$$

$$= 2a_2 + \sum_{n=0}^{\infty}(n+3)(n+2))a_{n+3}x^{n+1}$$

$$- \sum_{n=0}^{\infty} na_n x^{n+1}$$

$$= 2a_2 + \sum_{n=0}^{\infty}[(n+3)(n+2)a_{n+3} - na_n]x^{n+1}$$

So, $(n+3)(n+2)a_{n+3} - na_n = 0$, or

$$a_{n+3} = \frac{n}{(n+3)(n+2)}a_n.$$

Then $a_2 = 0$ and

$$0 = a_2 = a_5 = a_8 = \cdots$$

$$0 = a_3 = a_6 = a_9 = \cdots$$

$$a_{3n+1} = \frac{1}{(3n+1)n!3^n}a_1$$

The general solution is
$y = a_0 y_1 + a_1 y_2$
where

$y_1(x) = 1$

$$y_2(x) = \sum_{n=0}^{\infty}\frac{1}{(3n+1)n!3^n}x^{3n+1}$$

9. $0 = y'' - (x-1)y' - y$

$$= \sum_{n=2}^{\infty} n(n-1)a_n(x-1)^{n-2}$$

$$- (x-1)\sum_{n=1}^{\infty} na_n(x-1)^{n-1}$$

$$- \sum_{n=0}^{\infty} a_n(x-1)^n$$

$$= \sum_{n=0}^{\infty}(n+2)(n+1)a_{n+2}(x-1)^n$$

$$- \sum_{n=0}^{\infty} na_n(x-1)^n - \sum_{n=0}^{\infty} a_n(x-1)^n$$

$$= \sum_{n=0}^{\infty}[(n+2)(n+1)a_{n+2}$$

$$- (n+1)a_n](x-1)^n$$

So, $(n+2)(n+1)a_{n+2} - (n+1)a_n = 0$, or

$$a_{n+2} = \frac{n+1}{(n+2)(n+1)}a_n = \frac{1}{n+2}a_n$$

so

$$a_{2n} = \frac{1}{2^n n!}a_0$$

$$a_{2n+1} = \frac{2^n n!}{(2n+1)!}a_1$$

The general solution is
$y = a_0 y_1 + a_1 y_2$
where

$$y_1(x) = \sum_{n=0}^{\infty} \frac{1}{2^n n!}(x-1)^{2n}$$

$$y_2(x) = \sum_{n=0}^{\infty} \frac{2^n n!}{(2n+1)!}(x-1)^{2n+1}$$

11. $0 = y'' - (x-1)y - y$

$$= \sum_{n=2}^{\infty} n(n-1)a_n(x-1)^{n-2}$$

$$- (x-1)\sum_{n=0}^{\infty} a_n(x-1)^n$$

$$- \sum_{n=0}^{\infty} a_n(x-1)^n$$

$$= \sum_{n=0}^{\infty} (n+2)(n+1)a_{n+2}(x-1)^n$$

$$- \sum_{n=0}^{\infty} a_n(x-1)^{n+1}$$

$$- \sum_{n=0}^{\infty} a_n(x-1)^n$$

$$= 2a_2 - a_0$$

$$+ \sum_{n=1}^{\infty} (n+2)(n+1)a_{n+2}(x-1)^n$$

$$- \sum_{n=1}^{\infty} a_{n-1}(x-1)^n$$

$$- \sum_{n=1}^{\infty} a_n(x-1)^n$$

$$= 2a_2 - a_0 + \sum_{n=1}^{\infty}[(n+2)(n+1)a_{n+2}$$

$$- a_{n-1} - a_n](x-1)^n$$

So, a_0 and a_1 are arbitrary, $a_2 = \frac{1}{2}a_0$ and $(n+2)(n+1)a_{n+2} - a_{n-1} - a_n = 0$ or

$$a_{n+2} = \frac{a_n + a_{n-1}}{(n+2)(n+1)} \text{ for } n \geq 1.$$

This is the recursion formula, which gives the solution

$$y = a_0\left(1 + \frac{1}{2}(x-1)^2 + \frac{1}{12}(x-1)^3\right.$$

$$\left. + \frac{7}{144}(x-1)^4 + \frac{19}{2880}(x-1)^5 + \cdots\right)$$

$$+ a_1\left((x-1) + \frac{1}{6}(x-1)^3 + \frac{1}{72}(x-1)^4\right.$$

$$\left. + \frac{13}{1440}(x-1)^5 + \cdots\right)$$

13. We use the solution found in exercise 1 and substitute $y(0) = a_0 = 5$ and $y'(0) = a_1 = -7$:

$$y(x) = 5\sum_{n=0}^{\infty} \frac{(-1)^n 2^{2n} n!}{(2n)!}x^{2n}$$

$$- 7\sum_{n=0}^{\infty} \frac{(-1)^n}{n!}x^{2n+1}$$

15. We use the solution found in exercise 9 and substitute $y(1) = a_0 = -3$ and $y'(1) = a_1 = 12$:

$$y(x) = -3\sum_{n=0}^{\infty} \frac{1}{2^n n!}(x-1)^{2n}$$

$$+ 12\sum_{n=0}^{\infty} \frac{2^n n!}{(2n+1)!}(x-1)^{2n+1}$$

17. Using the ratio test, we can look separately at the two series in the solution of exercise 3:

$$\sum_{n=0}^{\infty} \frac{1}{2^n n!}x^{2n} \quad \text{and} \quad \sum_{n=0}^{\infty} \frac{2^n n!}{(2n+1)!}x^{2n+1}.$$

Letting $a_n = \frac{x^{2n}}{2^n n!}$, then

$$\lim_{n \to \infty} \left| \frac{a_{n+1}}{a_n} \right| = \lim_{n \to \infty} \left| \frac{x^{2n+2} 2^n n!}{2^{n+1} (n+1)! x^{2n}} \right|$$

$$= \lim_{n \to \infty} \left| \frac{x^2}{2(n+1)} \right| = 0.$$

So this series converges for all x.

Now letting $a_n = \frac{2^n n! x^{2n+1}}{(2n+1)!}$, then

$$\lim_{n \to \infty} \left| \frac{a_{n+1}}{a_n} \right|$$

$$= \lim_{n \to \infty} \left| \frac{2^{n+1} (n+1)! x^{2n+3} (2n+1)!}{(2n+3)! 2^n n! x^{2n+1}} \right|$$

$$= \lim_{n \to \infty} \left| \frac{x^2}{2n+3} \right| = 0.$$

So this series converges for all x.

Hence the sum of these two series converges for all x and the radius of convergence of the power series is $r = \infty$.

19. Using the Ratio Test, we can look separately at the two series in the solution of exercise 9:

$$\sum_{n=0}^{\infty} \frac{1}{2^n n!} (x-1)^{2n}$$

$$\sum_{n=0}^{\infty} \frac{2^n n!}{(2n+1)!} (x-1)^{2n+1}$$

These are the exact same series in exercise 17 except that x is replaced by $x - 1$. Therefore the radius of convergence of the power series is $r = \infty$.

21. $0 = x^2 y'' + xy' + x^2 y$

$$= x^2 \sum_{n=2}^{\infty} n(n-1) a_n x^{n-2}$$

$$+ x \sum_{n=1}^{\infty} n a_n x^{n-1} + x^2 \sum_{n=0}^{\infty} a_n x^n$$

$$= \sum_{n=2}^{\infty} n(n-1) a_n x^n$$

$$+ \sum_{n=1}^{\infty} n a_n x^n + \sum_{n=0}^{\infty} a_n x^{n+2}$$

$$= \sum_{n=0}^{\infty} (n+2)(n+1) a_{n+2} x^{n+2}$$

$$+ \sum_{n=-1}^{\infty} (n+2) a_{n+2} x^{n+2} + \sum_{n=0}^{\infty} a_n x^{n+2}$$

$$= a_1 x$$

$$+ \sum_{n=0}^{\infty} [(n+2)(n+1) a_{n+2}$$

$$+ (n+2) a_{n+2} + a_n] x^{n+2}$$

So, $a_1 = 0$ and

$$0 = (n+2)(n+1) a_{n+2} + (n+2) a_{n+2} + a_n$$

or

$$a_{n+2} = -\frac{a_n}{(n+2)^2}$$

so

$$0 = a_1 = a_3 = a_5 = \cdots$$

$$a_{2n} = \frac{(-1)^n}{2^{2n} (n!)^2} a_0$$

So a solution is

$$y = a_0 \sum_{n=0}^{\infty} \frac{(-1)^n}{2^{2n} (n!)^2} x^{2n}.$$

23. Let y_1 and y_2 be the two solutions described in Example 4.3. Then y_1 is of the form $\sum_{n=0}^{\infty} b_{3n} x^{3n}$, where the coefficients satisfy the recurrence relation $b_{n+3} = \frac{b_n}{(n+3)(n+2)}$. We now apply the Ratio Test to this series.

$$\lim_{n \to \infty} \left| \frac{b_{3n+3} x^{3n+3}}{b_{3n} x^{3n}} \right|$$

$$= \lim_{n \to \infty} \left| \frac{x^3}{(3n+3)(3n+2)} \right| = 0$$

So this series converges for all x. The same argument applies to the series for y_2, so the radius of convergence of the series solution is $r = \infty$.

25. We are given

$$y'' + 2xy' - xy = 0$$
$$y(0) = 2, \ y'(0) = -5$$

Substituting $x = 0$ we get

$$y''(0) + 2(0)y'(0) - 0y(0) = 0$$

so $y''(0) = 0$.

We now differentiate the differential equation:

$$y'' = -2xy' + xy$$
$$y''' = -2y' - 2xy'' + y + xy'$$
$$= -2xy'' + (x - 2)y' + y$$

Substituting $x = 0$,

$$y'''(0) = -2(0)y''(0) + (0 - 2)y'(0) + y(0)$$
$$= 0 - 2(-5) + 2 = 12$$

Differentiating again:

$$y''' = -2xy'' + (x - 2)y' + y$$
$$y^{(4)} = -2y'' - 2xy''' + y' + (x - 2)y'' + y'$$
$$= -2xy''' + (x - 4)y'' + 2y'$$

Substituting $x = 0$,

$$y^{(4)}(0) = -2(0)y'''(0) + (0 - 4)y''(0) + 2y'(0)$$
$$= 0 - 4(0) + 2(-5) = -10$$

Differentiating again:

$$y^{(4)} = -2xy''' + (x - 4)y'' + 2y'$$
$$y^{(5)} = -2y''' - 2xy^{(4)} + y'' + (x - 4)y''' + 2y''$$
$$= -2xy^{(4)} + (x - 6)y''' + 3y''$$

Substituting $x = 0$,

$$y^{(5)}(0) = -2(0)y^{(4)}(0) + (0 - 6)y'''(0) + 3y''(0)$$
$$= 0 - 6(12) + 3(0) = -72$$

Therefore,

$$P_5(x) = y(0) + y'(0)x + y''(0)\frac{x^2}{2} + y'''(0)\frac{x^3}{3!}$$
$$+ y^{(4)}(0)\frac{x^3}{4!} + y^{(5)}\frac{x^4}{5!}$$
$$= 2 - 5x - 2x^3 - \frac{5x^4}{12} - \frac{3x^5}{5}$$

27. Following the technique of exercise 25, we first substitute:

$$y'' = -e^x y' + (\sin x)y$$
$$y''(0) = -e^0 y'(0) + (\sin 0)y(0)$$
$$= -1(1) + 0 = -1$$

Differentiating and substituting,

$$y''' = -e^x y' - e^x y'' + (\cos x)y + (\sin x)y'$$
$$= -e^x y'' + (\sin x - e^x)y' + (\cos x)y$$
$$y'''(0) = -e^0 y''(0) + (\sin 0 - e^0)y'(0)$$
$$+ (\cos 0)y(0)$$
$$= -1(-1) + (-1)(1) + (1)(-2) = -2$$

Differentiating and substituting,

$$y^{(4)} = -e^x y'' - e^x y''' + (\cos x - e^x)y'$$
$$+ (\sin x - e^x)y'' + (-\sin x)y$$
$$+ (\cos x)y'$$
$$= -e^x y''' + (\sin x - 2e^x)y''$$
$$+ (2\cos x - e^x)y' - (\sin x)y$$
$$y^{(4)}(0) = -e^0 y'''(0) + (\sin 0 - 2e^0)y''(0)$$
$$+ (2\cos 0 - e^0)y'(0) - (\sin 0)y(0)$$
$$= (-1)(-2) + (-2)(-1) + (1)(1) - 0$$
$$= 5$$

Differentiating and substituting,

$$y^{(5)} = -e^x y''' - e^x y^{(4)} + (\cos x - 2e^x)y''$$
$$+ (\sin x - 2e^x)y'''$$
$$+ (-2\sin x - e^x)y' + (2\cos x - e^x)y''$$
$$- (\cos x)y - (\sin x)y'$$
$$= -e^x y^{(4)} + (\sin x - 3e^x)y'''$$
$$+ (3\cos x - 3e^x)y''$$
$$+ (-3\sin x - e^x)y' - (\cos x)y$$

$$y^{(5)}(0) = -e^0 y^{(4)}(0) + (\sin 0 - 3e^0) y'''(0)$$
$$+ (3\cos 0 - 3e^0) y''(0)$$
$$+ (-3\sin 0 - e^0) y'(0) - (\cos 0) y(0)$$
$$= (-1)(5) + (-3)(-2) + (0)(-1)$$
$$+ (-1)(1) - (1)(-2)$$
$$= 2$$

Therefore,

$$P_5(x) = -2 + x - \frac{x^2}{2} - \frac{x^3}{3} + \frac{5x^4}{24} + \frac{x^5}{60}.$$

29. Following the technique of exercise 25, we first substitute:

$$y'' = -xy' - (\sin x)y$$
$$y''(\pi) = -\pi y'(\pi) - (\sin \pi)y(\pi)$$
$$= -4\pi$$

Differentiating and substituting,

$$y''' = -y' - xy'' - (\cos x)y - (\sin x)y'$$
$$= -xy'' + (-1 - \sin x)y' - (\cos x)y$$
$$y'''(\pi) = -\pi y''(\pi) + (-1 - \sin \pi)y'(\pi)$$
$$- (\cos \pi)y(\pi)$$
$$= 4\pi^2 - 4$$

Differentiating and substituting,

$$y^{(4)} = -y'' - xy''' + (-\cos x)y'$$
$$+ (-1 - \sin x)y'' + (\sin x)y$$
$$- (\cos x)y'$$
$$= -xy''' + (-2 - \sin x)y''$$
$$+ (-2\cos x)y' + (\sin x)y$$
$$y^{(4)}(\pi) = -\pi y'''(\pi) + (-2 - \sin \pi)y''(\pi)$$
$$+ (-2\cos \pi)y'(\pi) + (\sin \pi)y(\pi)$$
$$= -\pi(4\pi^2 - 4) + (-2)(-4\pi)$$
$$+ 2(4) + 0$$
$$= -4\pi^3 + 12\pi + 8$$

Differentiating and substituting,

$$y^{(5)} = -y''' - xy^{(4)} + (-\cos x)y''$$
$$+ (-2 - \sin x)y''' + (2\sin x)y'$$
$$+ (-2\cos x)y'' + (\cos x)y$$
$$+ (\sin x)y'$$
$$= -xy^{(4)} + (-3 - \sin x)y'''$$
$$+ (-3\cos x)y'' + (3\sin x)y'$$
$$+ (\cos x)y$$
$$y^{(5)}(\pi) = -\pi y^{(4)}(\pi) + (-3 - \sin \pi)y'''(\pi)$$
$$+ (-3\cos \pi)y''(\pi) + (3\sin \pi)y'(\pi)$$
$$+ (\cos \pi)y(\pi)$$
$$= (-\pi)(-4\pi^3 + 12\pi + 8)$$
$$+ (-3)(4\pi^2 - 4) + (3)(-4\pi)$$
$$+ 0 + 0$$
$$= 4\pi^4 - 24\pi^2 - 20\pi + 12$$

Therefore,

$$P_5(x) = 0 + 4(x - \pi) + (-4\pi)\frac{(x - \pi)^2}{2}$$
$$+ (4\pi^2 - 4)\frac{(x - \pi)^3}{3!}$$
$$+ (-4\pi^3 + 12\pi + 8)\frac{(x - \pi)^4}{4!}$$
$$+ (4\pi^4 - 24\pi^2 - 20\pi + 12)\frac{(x - \pi)^5}{5!}.$$

Chapter 16 Review exercises

1. The characteristic equation is
$$r^2 + r - 12 = 0$$
which has solutions $r = -4, 3$. Thus the general solution is
$$y = c_1 e^{-4t} + c_2 e^{3t}.$$

3. The characteristic equation is
$$r^2 + r + 3 = 0$$
which has solutions $r = \frac{-1 \pm i\sqrt{11}}{2}$. Thus the general solution is

$$y = c_1 e^{-t/2} \cos\left(\frac{t\sqrt{11}}{2}\right) + c_2 e^{-t/2} \sin\left(\frac{t\sqrt{11}}{2}\right).$$

5. The characteristic equation is
$$r^2 - r - 6 = 0$$
which has solutions $r = -2, 3$. Thus the general solution to the homogeneous equation is
$$y = c_1 e^{-2t} + c_2 e^{3t}.$$

The guess for a particular solution is
$$y_p = Ate^{3t} + Bt^2 + Ct + D$$
(note that we had to add a t to the e^{3t} term since that is a solution to the homogeneous equation). We solve for the constants.

$$e^{3t} + t^2 + 1 = y_p'' - y_p' - 6y_p$$
$$= A(9t + 6)e^{3t} + 2B$$
$$\quad - \left[A(3t + 1)e^{3t} + 2Bt + C \right]$$
$$\quad - 6 \left[Ate^{3t} + Bt^2 + Ct + D \right]$$
$$= 5Ae^{3t} - 6Bt^2 + (-2B - 6C)t$$
$$\quad + 2B - C - 6D$$

For this to work we require
$$5A = 1 \Rightarrow A = \tfrac{1}{5}$$
$$-6B = 1 \Rightarrow B = -\tfrac{1}{6}$$
$$-2B - 6C = 0 \Rightarrow C = -\tfrac{B}{3} = \tfrac{1}{18}$$
$$2B - C - 6D = 1 \Rightarrow D = \tfrac{2B - C - 1}{6} = -\tfrac{25}{108}$$

Thus, the general solution is
$$y = c_1 e^{-2t} + c_2 e^{3t}$$
$$\quad + \tfrac{1}{5}te^{3t} - \tfrac{1}{6}t^2 + \tfrac{1}{18}t - \tfrac{25}{108}.$$

7. The characteristic equation is
$$r^2 + 2r - 8 = 0$$
which has solutions $r = -4, 2$. Thus the general solution is
$$y = c_1 e^{-4t} + c_2 e^{2t}.$$

We use the initial conditions to find c_1 and c_2:
$$5 = y(0) = c_1 + c_2$$
$$-2 = y'(0) = -4c_1 + 2c_2$$

Solving gives $c_1 = 2$ and $c_2 = 3$. Thus the solution is
$$y = 2e^{-4t} + 3e^{2t}.$$

9. The characteristic equation is
$$r^2 + 4 = 0$$

which has solutions $r = \pm 2i$. Thus the general solution to the homogeneous equation is
$$y = c_1 \cos 2t + c_2 \sin 2t.$$

The guess for a particular solution is
$$y_p = A \cos t + B \sin t.$$

We solve for the constants.

$$3 \cos t = y_p'' + 4y_p$$
$$= (-A \cos t - B \sin t)$$
$$\quad + 4(A \cos t + B \sin t)$$
$$= 3A \cos t + 3B \sin t$$

For this to work, we require $3A = 3$ and $3B = 0$. Therefore, $A = 1$ and $B = 0$. Thus, the general solution is
$$y = c_1 \cos 2t + c_2 \sin 2t + \cos t.$$

We use the initial conditions to find c_1 and c_2:
$$1 = y(0) = c_1 + 1$$
$$2 = y'(0) = 2c_2$$

Solving gives $c_1 = 0$ and $c_2 = 1$. Thus the solution is
$$y = \sin 2t + \cos t.$$

11. Solving for k gives
$$k = \tfrac{4}{1/3} = 12.$$

Thus we get the differential equation

$$\frac{4}{32}u'' + 12u = 0$$
$$u'' + 96u = 0$$

The characteristic equation is
$$r^2 + 96 = 0$$
which has solutions $r = \pm 4i\sqrt{6}$. Thus the general solution is
$$y = c_1 \cos\left(4t\sqrt{6}\right) + c_2 \sin\left(4t\sqrt{6}\right).$$

We use the initial conditions to find c_1 and c_2:
$$\frac{2}{12} = u(0) = c_1$$
$$0 = u'(0) = 4c_2\sqrt{6}$$

Therefore. $c_1 = \tfrac{1}{6}$ and $c_2 = 0$. Thus the solution is
$$u = \frac{1}{6}\cos\left(4t\sqrt{6}\right).$$

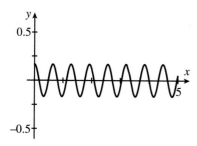

13. The model
$$LQ'' + RQ' + \tfrac{1}{C}Q = 0$$
becomes, after substitutions, the differential equation
$$0.2Q'' + 160Q' + 100Q = 0, \text{ or}$$
$$Q'' + 800Q' + 500Q = 0.$$

The characteristic equation is
$$r^2 + 800r + 500 = 0$$
which has solutions $r = -400 \pm 10\sqrt{1595}$. Thus the general solution is
$$Q = c_1 e^{(-400+10\sqrt{1595})t} + c_2 e^{(-400-10\sqrt{1595})t}.$$

We use the initial conditions to find c_1 and c_2:

$$10^{-4} = Q(0) = c_1 + c_2$$
$$\begin{aligned} 0 = Q'(0) &= (-400 + 10\sqrt{1595})c_1 \\ &\quad + (-400 - 10\sqrt{1595})c_2 \\ &= 10\sqrt{1595}(c_1 - c_2) - 400(c_1 + c_2). \end{aligned}$$

Solving (which is a bit messy) gives

$$c_1 = \frac{1}{20,000} + \frac{\sqrt{1595}}{797,500} \approx 0.000100$$

$$c_2 = \frac{1}{20,000} - \frac{\sqrt{1595}}{797,500} \approx -7.8 \times 10^{-8} \approx 0$$

This gives $Q(t)$. The current is

$$\begin{aligned} I(t) &= Q'(t) \\ &= c_1(-400 + 10\sqrt{1595})e^{(-400+10\sqrt{1595})t} \\ &\quad + c_2(-400 - 10\sqrt{1595})e^{(-400-10\sqrt{1595})t} \\ &= -\frac{\sqrt{1595}}{638,000}\left[e^{(-400+10\sqrt{1595})t} \right. \\ &\qquad\qquad\qquad \left. - e^{(-400-10\sqrt{1595})t} \right] \end{aligned}$$

15. The characteristic equation is
$$r^2 + 2r + 5 = 0$$
which has solutions $r = -1 \pm 2i$. The homogeneous equation has solution
$$u = c_1 e^{-t}\cos 2t + c_2 e^{-t}\sin 2t.$$

Our initial guess for a particular solution is

$$\begin{aligned} u_p &= A\cos 2t + B\sin 2t \\ &\quad + e^{-t}(C\cos 2t + D\sin 2t) \\ &\quad + Et^3 + Ft^2 + Gt + H. \end{aligned}$$

But, we have to modify this to

$$\begin{aligned} u_p &= A\cos 2t + B\sin 2t \\ &\quad + te^{-t}(C\cos 2t + D\sin 2t) \\ &\quad + Et^3 + Ft^2 + Gt + H. \end{aligned}$$

17. We first find k:
$$4 = \tfrac{1}{3}k \text{ which gives } k = 12.$$

The initial value problem is then
$$\frac{4}{32}u'' + 0.4u' + 12u = 2\sin 2t\, u'' + 3.2u' +$$
$$96u = 16\sin 2t$$
$$u(0) = \frac{1}{6}, \quad u'(0) = 2$$

The characteristic equation is
$$r^2 + 3.2r + 96 = 0$$
which has solutions
$$r = -1.6 \pm i\sqrt{93.44}.$$

Thus the general solution to the homogeneous equation is

$$\begin{aligned} u &= c_1 e^{-1.6t}\cos\left(t\sqrt{93.44}\right) \\ &\quad + c_2 e^{-1.6t}\sin\left(t\sqrt{93.44}\right). \end{aligned}$$

The particular solution is of the form
$$u_p = A\cos 2t + B\sin 2t.$$

Substituting this into the differential equation to solve for A and B gives

$16 \sin 2t = u''_p + 3.2u'_p + 96u_p$

$\quad = (-4A \cos 2t - 4B \sin 2t)$

$\qquad + 3.2(-2A \sin 2t + 2B \cos 2t)$

$\qquad + 96(A \cos 2t + B \sin 2t)$

$\quad = (92A + 6.4B) \cos 2t$

$\qquad + (92B - 6.4A) \sin 2t.$

This gives us the equations

$92A + 6.4B = 0$

$-6.4A + 92B = 16$

Solving gives

$$A = -\frac{160}{13,289} \quad B = \frac{2300}{13,289}$$

Thus, our general solution is:

$$u = c_1 e^{-1.6t} \cos\left(t\sqrt{93.44}\right)$$

$$+ c_2 e^{-1.6t} \sin\left(t\sqrt{93.44}\right)$$

$$- \frac{160}{13,289} \cos 2t + \frac{2300}{13,289} \sin 2t.$$

The steady-state solution is

$$-\frac{160}{13,289} \cos 2t + \frac{2300}{13,289} \sin 2t.$$

19. $y'' - 2xy' - 4y$

$$= \sum_{n=2}^{\infty} n(n-1)a_n x^{n-2}$$

$$- 2x \sum_{n=1}^{\infty} n a_n x^{n-1}$$

$$- 4 \sum_{n=0}^{\infty} a_n x^n$$

$$= \sum_{n=0}^{\infty} (n+2)(n+1)a_{n+2} x^n$$

$$- 2 \sum_{n=0}^{\infty} n a_n x^n$$

$$- 4 \sum_{n=0}^{\infty} a_n x^n$$

$$= \sum_{n=0}^{\infty} [(n+2)(n+1)a_{n+2} - 2(n+2)a_n] x^n$$

So, $(n+2)(n+1)a_{n+2} - 2(n+2)a_n = 0$, or $a_{n+2} = \frac{2(n+2)}{(n+2)(n+1)} a_n = \frac{2}{n+1} a_n.$

This gives us

$$a_{2n} = \frac{2^{2n} n!}{(2n)!} a_0$$

$$a_{2n+1} = \frac{1}{n!} a_1$$

Hence a_0 and a_1 are arbitrary, and the general solution is

$y = a_0 y_1 + a_1 y_2$

where

$$y_1(x) = \sum_{n=0}^{\infty} \frac{2^{2n} n!}{(2n)!} x^{2n}$$

$$y_2(x) = \sum_{n=0}^{\infty} \frac{1}{n!} x^{2n+1}$$

21. $y'' - 2xy' - 4y = y'' - 2(x-1)y' - 2y' - 4y$

$$= \sum_{n=2}^{\infty} n(n-1)a_n (x-1)^{n-2}$$

$$- 2(x-1) \sum_{n=0}^{\infty} n a_n (x-1)^{n-1}$$

$$- 2 \sum_{n=0}^{\infty} n a_n (x-1)^{n-1} - 4 \sum_{n=0}^{\infty} a_n (x-1)^n$$

$$= \sum_{n=0}^{\infty} (n+2)(n+1)a_{n+2}(x-1)^n$$

$$- 2\sum_{n=0}^{\infty} n a_n (x-1)^n$$

$$- 2\sum_{n=0}^{\infty} (n+1)a_{n+1}(x-1)^n$$

$$- 4\sum_{n=0}^{\infty} a_n (x-1)^n$$

$$= \sum_{n=0}^{\infty} \Big[(n+2)(n+1)a_{n+2} - 2na_n$$

$$- 2(n+1)a_{n+1} - 4a_n \Big] (x-1)^n$$

So,

$$0 = (n+2)(n+1)a_{n+2} - 2(n+1)a_{n+1}$$
$$- 2(n+2)a_n$$

or

$$a_{n+2} = \frac{2a_n}{n+1} + \frac{2a_{n+1}}{n+2}$$

which gives the general solution.

$$y(x) = \sum_{n=0}^{\infty} a_n (x-1)^n$$

$$= a_0 \left[1 + 2(x-1)^2 + \frac{4}{3}(x-1)^3 \right.$$

$$+ 2(x-1)^4 + \frac{22}{15}(x-1)^5$$

$$\left. + \frac{58}{45}(x-1)^6 + \cdots \right]$$

$$+ a_1 \left[(x-1) + (x-1)^2 \right.$$

$$+ \frac{5}{3}(x-1)^3 + \frac{3}{2}(x-1)^4$$

$$\left. + \frac{43}{30}(x-1)^5 + \frac{97}{90}(x-1)^6 \cdots \right]$$

23. We use the solution found in exercise 19 and substitute $y(0) = a_0 = 4$ and $y'(0) = a_1 = 2$:

$$y = 4\sum_{n=0}^{\infty} \frac{2^{2n}n!}{(2n)!} x^{2n} + 2\sum_{n=0}^{\infty} \frac{1}{n!} x^{2n+1}$$

Notes